Smart and Connected Health: AI, IoT, and Trustworthy Technologies

Farshad Firouzi · Bahar Farahani ·
Krishnendu Chakrabarty
Editors

Smart and Connected Health: AI, IoT, and Trustworthy Technologies

Springer

Editors
Farshad Firouzi
Arizona State University
Tempe, AZ, USA

Bahar Farahani
Shahid Beheshti University
Tehran, Iran

Krishnendu Chakrabarty
Arizona State University
Tempe, AZ, USA

ISBN 978-3-032-06285-7 ISBN 978-3-032-06286-4 (eBook)
https://doi.org/10.1007/978-3-032-06286-4

This Springer imprint is published by the registered company Springer Nature Switzerland AG
The registered company address is: Gewerbestrasse 11, 6330 Cham, Switzerland

To my father, Fereidoun Firouzi—always with me.
To my mother, Latifeh (Fatemeh), with deepest love.

—Farshad Firouzi

To my beloved father and mother, my first teachers, with gratitude and love.

—Bahar Farahani

In loving memory of my father, Ramendra Narayan Chakrabarty.

—Krishnendu Chakrabarty

Preface

The rapid convergence of Artificial Intelligence (AI), the Internet of Things (IoT), edge-fog-cloud computing, blockchain, and other emerging technologies is fundamentally transforming health care, driving the evolution of Smart and Connected Health (SCH). This paradigm shift represents a profound transition from traditional, hospital-centric, and reactive healthcare models to a patient-centric, proactive, and data-driven model, facilitating personalized, efficient, and accessible medical solutions through real-time monitoring, intelligent data processing, and advanced analytics. This two-volume book serves as a comprehensive resource, offering both foundational knowledge and practical applications of these cutting-edge innovations.

Volume 1, *Smart and Connected Health: AI, IoT, and Trustworthy Technologies*, is designed to serve as both a reference guide and a textbook for students, researchers, and professionals. Covering essential topics such as machine learning, deep learning, Large Language Models (LLMs), sensor technologies, cloud computing, software architecture, security, and blockchain, this volume provides a structured introduction to the principles shaping modern healthcare systems. With a strong focus on foundational concepts and methodologies, this resource is well-suited for use in academic courses and training programs, equipping readers with the theoretical and technical knowledge necessary to develop SHC solutions.

Volume 2, *Smart and Connected Health: Application and Real-World Use Cases*, presents real-world implementations and case studies. From AI-driven clinical decision support to smart toilets for health monitoring and AI in diabetes management, this volume showcases the transformative impact of emerging technologies in health care.

Whether you are a student seeking a structured learning path, a researcher exploring new frontiers, or a practitioner striving to implement and integrate emerging technologies, we hope this book provides valuable insights into the future of smart and connected health systems.

Tempe, USA Farshad Firouzi
Tehran, Iran Bahar Farahani
Tempe, USA Krishnendu Chakrabarty

Contents

Smart and Connected Health

Farshad Firouzi, Bahar Farahani, Negin Firouzi, and Mahmoud Daneshmand

"The greatest wealth is health."

– Virgil

1 Smart and Connected Health (SCH)

Smart and Connected Health (SCH) represents a transformative approach to healthcare, integrating advanced digital technologies to revolutionize healthcare delivery, management, and outcomes. SCH draws upon a broad spectrum of innovations, including but not limited to the Internet of Things (IoT), Artificial Intelligence (AI), Distributed Ledger Technology (DLT), and Edge-Fog-Cloud computing, among others (See Fig. 1). These technologies converge to create a dynamic, interconnected ecosystem that enhances patient care, streamlines operational efficiency, and enables data-driven decision-making. The SCH paradigm addresses all health journey touchpoints, including continuous monitoring for proactive care, predictive analytics to anticipate health outcomes, personalized medicine tailored to individual needs, advanced diagnostics for early and precise detection, intelligent interventions and treatments, enhanced patient engagement through tailored communication, and

F. Firouzi (✉)
Arizona State University, Tempe, USA
e-mail: farshad.firouzi@asu.edu

B. Farahani
Shahid Beheshti University, Tehran, Iran

N. Firouzi
Shiraz University of Medical Sciences, Shiraz, Iran

M. Daneshmand
Stevens Institute of Technology, Hoboken, USA

F. Firouzi et al. (eds.), *Smart and Connected Health: AI, IoT, and Trustworthy Technologies*, https://doi.org/10.1007/978-3-032-06286-4_1

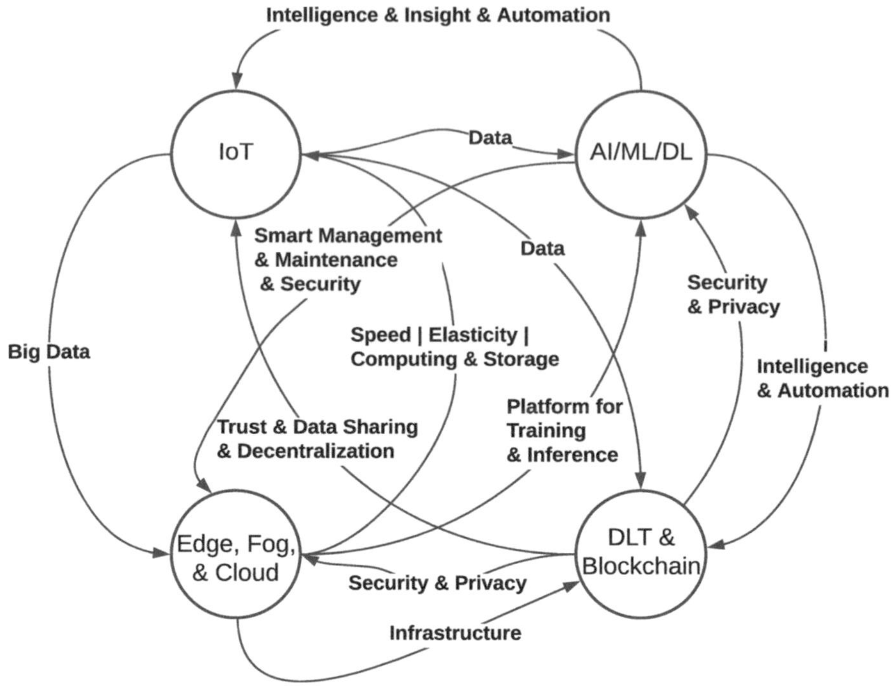

Fig. 1 Illustration of SCH integrating IoT, AI, DLT, and edge-fog-cloud computing [1]

seamless post-treatment follow-up to support recovery and long-term health management. By fostering a healthcare system that is more integrated, precise, and patient-centric, SCH sets the stage for a smarter, more efficient, and sustainable future in healthcare.

In this context, the convergence of IoT, AI, DLT, and Edge-Fog-Cloud computing serves as a transformative force in healthcare. IoT enables the interconnection of a vast network of medical devices and sensors, generating continuous streams of real-time data that are foundational for advanced healthcare applications. AI complements this ecosystem by processing and analyzing complex, multi-modal healthcare data to derive actionable insights, automate clinical workflows, and power predictive analytics. The integration of AI with IoT, often referred to as AIoT, adds an intelligent layer to IoT systems, enhancing their ability to transform raw data into meaningful information. This synergy supports real-time health monitoring, predictive diagnostics, and personalized care while enabling predictive maintenance of medical equipment. By leveraging IoT's data-generating capabilities alongside AI's analytical power, these technologies improve patient outcomes, optimize healthcare processes, and create a more patient-centric and responsive healthcare system.

Edge-Fog-Cloud computing further enhances the capabilities of Cloud IoT by addressing its inherent limitations in latency, connectivity, and bandwidth. While the integration of Cloud computing and IoT, referred to as Cloud IoT, has unlocked significant opportunities, it struggles to meet the stringent time-sensitive and context-aware requirements of diverse IoT applications. To overcome these challenges, distributed computing paradigms such as Mobile Edge Computing, Fog Computing (FC), Cloudlets, and Transparent Computing have emerged. Edge and Fog Computing specifically bridge the gap between centralized Cloud IoT systems and endpoint IoT devices by deploying computational power closer to the data source at the network edge. This proximity enables real-time performance and resilience, which are critical for latency-sensitive applications like remote surgery, emergency response, and continuous health monitoring. Rather than replacing Cloud IoT, Edge-Fog-Cloud computing extends and complements its functionality, creating a hierarchical, heterogeneous, and dynamic architecture. This paradigm processes large volumes of IoT-generated data locally at the edge, reducing reliance on network stability, minimizing latency, and improving scalability for real-time applications. The combination of Edge, Fog, and Cloud computing enhances resource utilization, supports distributed applications, and optimizes the quality of experience for IoT services, ensuring that SCH solutions can effectively address the real-time and context-aware challenges inherent in the modern healthcare domain.

Additionally, DLT, exemplified by blockchain, addresses longstanding security, scalability, and trust challenges in healthcare by enabling tamper-proof, decentralized data management. DLT ensures transparency, accountability, and integrity in data sharing while protecting patient privacy, fostering trust among stakeholders. By replacing traditional centralized systems with a peer-to-peer architecture, DLT alleviates vulnerabilities inherent in conventional healthcare solutions, such as security breaches and limited scalability. Its robust framework offers automation through smart contracts, cost-efficiency, and improved data integrity, providing an unprecedented level of reliability and transparency. The convergence of DLT with IoT represents a paradigm shift that empowers healthcare applications to overcome the limitations of traditional systems. DLT's ability to securely manage large-scale, multi-stakeholder interactions enables transformative innovations, including real-time diagnostics, secure patient data exchange, and scalable telemedicine solutions. Furthermore, DLT's tamper-resistant records and decentralized nature accelerate the adoption of data-driven healthcare marketplaces, ensuring interoperability, security, and enhanced patient trust. Together, these advancements position DLT as a cornerstone of the next wave of digital transformation in healthcare.

These four transformational technologies—IoT, AI, DLT, and Edge-Fog-Cloud computing—offer unparalleled opportunities to revolutionize healthcare individually but become exponentially more impactful when combined under the SCH paradigm. Together, they synergize their capabilities to create an integrated, intelligent, and patient-centric healthcare ecosystem, addressing critical needs such as real-time monitoring, predictive analytics, personalized treatments, secure data sharing, and operational efficiency. However, realizing their full potential in healthcare requires addressing key challenges related to design, deployment, scalability, security, and interoperability.

2 Internet of Things (IoT)

IoT represents a significant evolution of the Internet, extending its reach into the physical world through a network of connected smart devices. These "things" are embedded with sensors, actuators, and communication technologies, enabling them to gather and exchange data about themselves and their environment. IoT has been aptly described as the next logical stage of the Internet, providing a framework where devices interact with one another and with larger networks to generate insights, automate processes, and support informed decision-making. At its core, IoT relies on several foundational elements [2]:

- **Things or Devices**—Physical objects embedded with sensors, actuators, and communication modules, capable of sensing, processing, and transmitting data.
- **Connectivity**—The communication framework that enables devices to connect to networks and exchange data using protocols such as Wi-Fi, Bluetooth, or Zigbee.
- **Data**—The information collected by IoT devices, which serves as the foundation for intelligence and action. This includes raw sensor data, diagnostic reports, and commands sent to or from devices.
- **Intelligence and Action**—The ability to analyze data using technologies such as AI and Machine Learning (ML), enabling predictive analytics, operational optimization, and automated or stakeholder-driven actions.
- **Ecosystem and Interoperability**—The broader environment of IoT, encompassing devices, platforms, protocols, and stakeholders, which must function cohesively to achieve system goals. Interoperability ensures seamless communication between heterogeneous systems.
- **Dynamic Changes and Scalability**—The capacity of IoT systems to adapt to changing device states, contexts, and environmental conditions, while handling the enormous scale of connected devices and data.
- **Security, Privacy, and Energy Efficiency**—Ensuring the confidentiality and integrity of data, protecting user privacy, and addressing the energy constraints of IoT devices for sustainable and secure operation.

IoT ecosystem comprises four fundamental components: Things, Data, People, and Processes. Each component plays a critical role in realizing the potential of IoT across various domains, including healthcare (See Fig. 2) [2]:

1. **Things**—Physical devices that form the foundational elements of the IoT network. These devices, such as sensors, actuators, and smart systems, are designed to collect, transmit, and receive data while interacting with other devices or systems. For instance, in healthcare, wearable health monitors, connected infusion pumps, and diagnostic imaging systems exemplify IoT-enabled devices that gather patient data and transmit it for analysis or immediate action.
2. **Data**—The driving force of IoT, encompassing raw inputs, processed insights, and actionable intelligence. IoT devices generate vast amounts of data, which

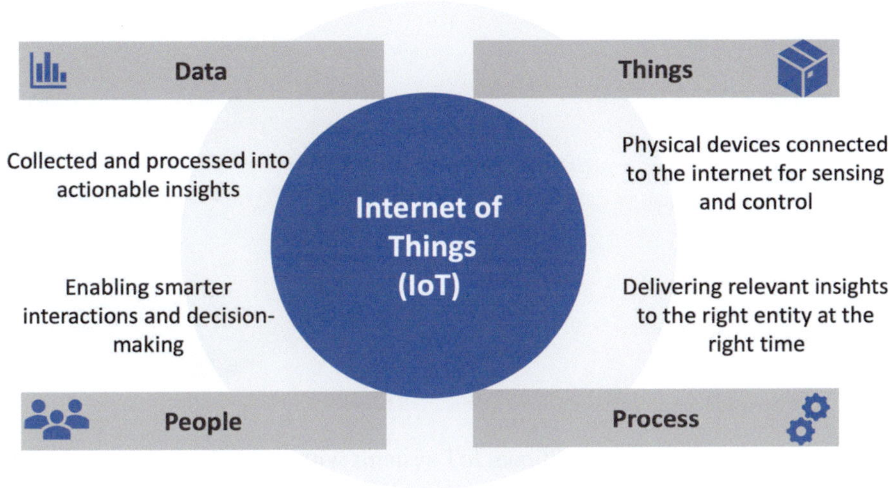

Fig. 2 A visual representation of the Internet of Things (IoT), defined by four foundational components: Things (connected devices), Data (processed for insights), Process (timely delivery of information and insights), and People (end-users and stakeholders) [2]

can be processed at the edge for time-sensitive applications or in centralized cloud systems for detailed analysis. In healthcare, this might include patient vitals from remote monitoring devices, diagnostic data from imaging equipment, or aggregated health metrics for predictive analytics.

3. **People**—The stakeholders who interact with and benefit from IoT systems. This includes the creators and maintainers of IoT infrastructure, as well as the end-users who leverage its capabilities. In healthcare, clinicians use IoT systems to monitor patient conditions, make informed decisions, and streamline workflows, while patients benefit from personalized care, real-time health insights, and improved health outcomes.
4. **Processes**—The workflows and systems that convert IoT data into meaningful outcomes. Processes ensure the right information is delivered to the right stakeholders at the right time, enabling actionable insights and system optimization. For example, in healthcare, IoT-enabled processes can automate alerts for critical patient conditions, streamline the allocation of medical resources, or enhance operational efficiency in hospitals through predictive maintenance of equipment.

2.1 Artificial Intelligence of Things (AIoT) and Generative IoT (GIoT)

Artificial Intelligence of Things (AIoT) refers to the integration of AI into the IoT ecosystem (See Fig. 3). AI adds a layer of intelligence to connected devices by enabling them to analyze data, learn from patterns, and make autonomous decisions.

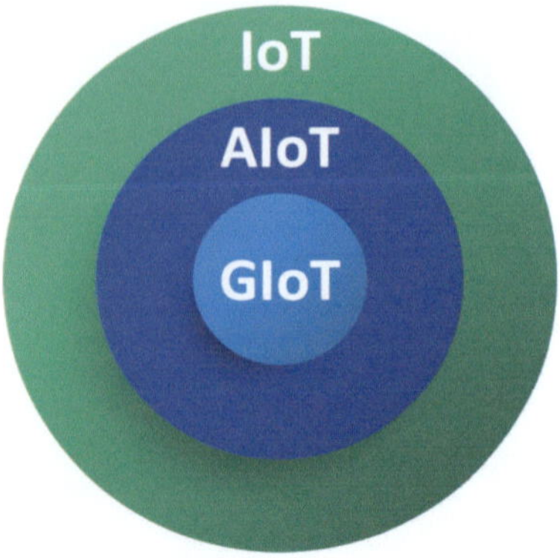

Fig. 3 Conceptual hierarchy of IoT paradigms. IoT represents the foundation of connected physical devices. Artificial Intelligence of Things (AIoT) builds on IoT by adding intelligent data processing and autonomous behavior. Generative IoT (GIoT) further evolves this by integrating generative AI capabilities, making it a specialized subset of AIoT

This convergence enhances traditional IoT systems with capabilities such as predictive analytics, anomaly detection, and adaptive behavior, allowing for more efficient, context-aware, and optimized operations across various domains.

Generative Internet of Things (GIoT) represents a significant advancement in the evolution of intelligent connected systems. Building upon the capabilities of AIoT, GIoT integrates Generative Artificial Intelligence (GenAI)—including Large Language Models (LLMs), Vision–Language Models (VLMs), and other foundation models—into IoT ecosystems to enable dynamic content creation, contextual reasoning, and real-time adaptation (See Fig. 4). By fusing generative intelligence with sensor-rich environments, GIoT empowers IoT systems to move beyond traditional automation and analytics. It enables devices to learn continuously, adapt autonomously, and generate personalized outputs based on complex, multimodal data. As a result, GIoT systems support real-time responsiveness, context-aware decision-making, intuitive human–machine interaction, and self-optimizing behaviors, ultimately delivering richer, more meaningful user experiences.

2.2 Internet of Medical Things (IoMT) and Healthcare IoT

The interaction between technology and healthcare has a long history. However, in recent years, the rapid growth of the IoT paradigm, the emergence of miniature wearable biosensors, and advancements in Big Data techniques have brought transformative changes to the field. These innovations enable the effective handling of large, multiscale, multimodal, distributed, and heterogeneous datasets, opening up new possibilities for customized healthcare services. Collectively known as Healthcare IoT or the Internet of Medical Things (IoMT), this paradigm shift offers significant advantages, including improved accessibility, enhanced availability of services, personalized care delivery, and cost-efficient solutions. IoMT encompasses a wide range of interconnected devices and systems, such as wearable health monitors, remote

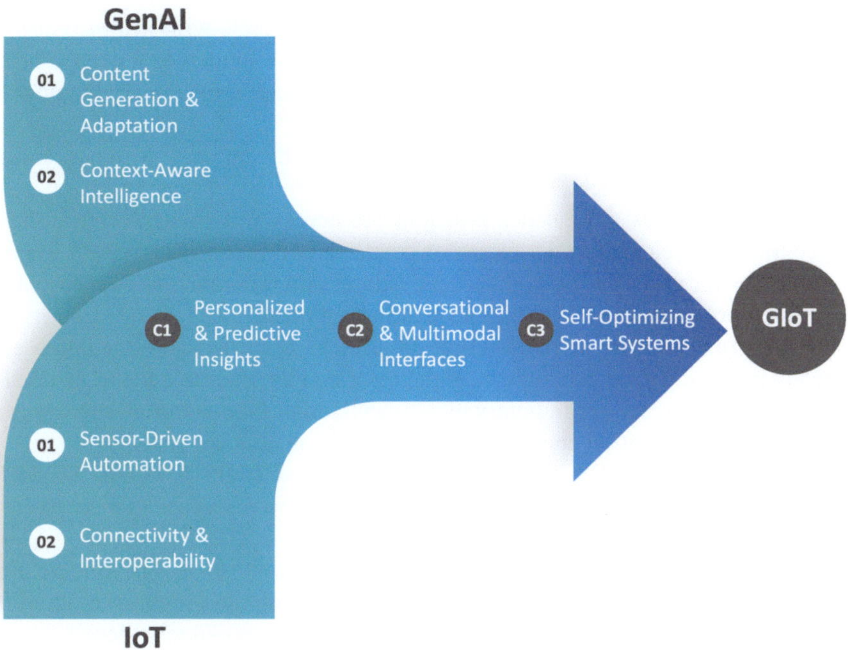

Fig. 4 Illustration of GIoT as the convergence of GenAI and IoT. GIoT transforms IoT devices into intelligent, self-learning systems with real-time adaptability, multimodal understanding, autonomous decision-making, seamless human-machine interaction, contextual awareness, and personalized user experiences

patient monitoring systems, and smart diagnostic tools, all working to deliver more efficient and tailored healthcare.

2.2.1 Modern Healthcare Challenges

Many regions worldwide are grappling with significant healthcare challenges, including managing an aging population, rising chronic health conditions, disease outbreaks, child mortality, inadequate living conditions, lack of sanitation, insufficient access to clean water, and escalating pollution levels. Despite the increasing demand for medical care, the traditional hospital-centric model—where patients visit physicians primarily when ill—remains prevalent. This reactive, physician-centered approach often excludes patients as active participants in their healthcare journey. Key obstacles in this traditional healthcare model include [3]:

- **Healthcare Workforce Shortages**: There is a growing shortage of healthcare professionals, including physicians, nurses, and caregivers, particularly in underserved urban and rural areas. Expanding telemedicine services is a potential solution to address this gap.

- **Time Constraints**: The growing number of individuals with health issues limits the time physicians can spend with each patient, hindering a comprehensive understanding of patients' daily lives, which is crucial for accurate diagnosis and effective treatment.
- **Escalating Medical Costs**: The rising cost of healthcare poses a significant challenge, placing a financial strain on patients and healthcare systems alike.
- **Health Inequities**: Disparities in healthcare access and quality, often driven by geographic, socioeconomic, racial, or gender-based inequalities, result in poorer outcomes for underserved populations. Addressing these inequities is essential to create more inclusive and equitable healthcare systems.
- **Monitoring Treatment Adherence**: Physicians often lack the means to effectively monitor patient compliance with prescribed treatments, such as medication, rehabilitative exercises, or dietary recommendations. Non-adherence increases the risks of hospitalization and contributes to higher healthcare costs.
- **Aging Population**: The global population of older adults is rapidly increasing, requiring additional medical facilities and healthcare resources to meet their specific needs.
- **Urbanization**: Rising urban populations necessitate substantial healthcare infrastructure to accommodate increasing demands. Urban areas are also more vulnerable to the rapid spread of contagious diseases due to population density.
- **Infectious Disease Resurgence and Emerging Threats**: The resurgence of infectious diseases, exacerbated by antimicrobial resistance (AMR), global travel, and climate change, poses significant challenges. Emerging pandemics, such as COVID-19, underscore the need for robust preparedness, surveillance, and rapid response systems.

2.2.2 Transitioning from Hospital-Centric to Patient-Centric

Patient-Centered Care (PCC) is an emerging healthcare model focused on addressing the individual medical needs of patients. Coined by the Picker/Commonwealth Program developed by the Picker Institute in 1988, PCC emphasizes personalized care and active patient participation. The Institute of Medicine defines PCC as: *"Healthcare that establishes a partnership among practitioners, patients, and their families (when appropriate) to ensure that decisions respect patients' wants, needs, and preferences and that patients have the education and support they need to make decisions and participate in their own care."*

While PCC has demonstrated success on smaller scales, its full potential remains unrealized due to conflicts with the traditional hospital-centric model. PCC does not eliminate the need for hospitals or clinics but integrates them into a shared care model. IoT, a convergence of telecommunication, sensors, actuators, cloud computing, and Big Data, enables this transformation by tailoring healthcare services to diverse environments. IoT's role in healthcare can be categorized into three main areas [3]:

- **Hospitals**: Hospitals are increasingly adopting IoT to enhance operational efficiency and patient care. Examples include:
  - *Smart Ambulances*: Equipped with IoT-enabled diagnostic sensors, these ambulances transmit critical patient data to hospitals en route, enabling staff to prepare operating rooms or equipment in advance.
  - *Operating Rooms*: IoT improves coordination among medical devices, physicians, and support staff, streamlining surgeries and other critical procedures.
  - *Specialized Units*: Intensive care units (ICUs) and primary care units (PCUs) benefit from real-time monitoring and data sharing across hospital systems, improving patient outcomes.
- **Clinics**: IoT optimizes primary care by reducing administrative burdens and enhancing patient interactions. Examples include:
  - Real-time access to virtual laboratory reports prior to patient visits, enabling faster and more informed diagnoses.
  - IoT-enabled systems for insurance verification, appointment scheduling, and pre-visit screenings, improving workflow efficiency.
  - Low-cost, IoT-based smart devices for at-home self-testing, empowering patients to manage their health proactively.
- **Non-Clinical Settings**: IoT extends healthcare beyond traditional settings, providing innovative solutions for remote and everyday environments, such as:
  - *Mobile Clinics*: IoT strengthens the capabilities of mobile clinics by enhancing their connectivity and infrastructure, enabling seamless collaboration with hospitals and specialists for remote diagnoses and treatment planning. This is especially impactful in rural and underserved areas, where access to traditional healthcare facilities is limited. Through IoT, mobile clinics can offer timely and effective care, improving health outcomes in these communities.
  - *Telemedicine*: IoT revolutionizes telemedicine by facilitating remote consultations, diagnostics, and tele-screening using smartphones, wearable devices, and other connected sensors. This technology expands healthcare access to remote or underserved populations, reduces the need for in-person visits, and provides real-time data to healthcare providers for timely interventions.
  - *Wellness and Preventive Care*: Wearable devices and fitness trackers promote wellness by monitoring health metrics and encouraging preventive care. Insurance cost reductions through wellness rewards can further incentivize healthy behaviors.
  - *Smart Homes*: IoT-enabled homes provide a seamless and supportive environment that enhances the quality of life for individuals, particularly the elderly, disabled, and chronically ill. By leveraging IoT technologies, these homes enable continuous health monitoring, proactive care, and responsive assistance tailored to individual needs. For example, smart toilets can perform routine health analyses, while adaptable home environments can optimize lighting or noise levels

for sensory-sensitive individuals. Such systems reduce dependency on frequent clinical visits, offering patients greater independence and comfort [3].
- *Smart Cities and Connected Communities*: IoT technologies play a pivotal role in integrating healthcare into the broader urban infrastructure, enabling a more connected and efficient community. By utilizing environmental sensors, cities can monitor air quality, water contamination, and allergen levels to provide timely health alerts to residents. For example, asthma patients can be advised to avoid high-pollution areas. Smart traffic systems further enhance healthcare delivery by prioritizing ambulances, ensuring faster response times during emergencies. These IoT-enabled frameworks contribute to healthier urban living and more sustainable city management.

IoT-driven healthcare represents a shift from reactive, hospital-centric care to a proactive, patient-centric approach. By creating a smart ecosystem that connects physicians, patients, hospitals, and communities, IoT paves the way for more efficient, accessible, and personalized healthcare services.

2.2.3 Transformative Benefits of IoT in Healthcare

Healthcare research spans a wide range of disciplines, including medicine, microbiology, biomedical engineering, computer science, and big data analytics. While these fields often collaborate on shared projects, significant gaps persist between them. These gaps present critical challenges to developing a unified, adaptive, and scalable healthcare framework. IoT-driven solutions have emerged as a transformative approach, bridging these divides and enabling advancements across all domains of healthcare.

IoT technologies are already revolutionizing healthcare through innovative diagnostics, monitoring, and treatment systems. Devices such as microfluidic biochips, wearable biosensors, and connected diagnostic tools are enhancing clinical capabilities across settings—from laboratories to hospitals and patient homes. IoT-enabled systems empower healthcare providers to perform real-time monitoring and point-of-care molecular testing for conditions such as breast, lung, and colorectal cancers. These tools are also enabling data-driven treatment plans, improving recovery rates and creating more personalized patient care experiences.

IoT systems go beyond individual applications by leveraging timestamped, location-tagged, and context-aware data, combined with personal health metrics (e.g., age, weight, gender). This comprehensive and dynamic data fabric not only reveals individual health trends but also uncovers broader population patterns, such as emerging epidemics or community health risks. IoT-driven healthcare integrates these components to assess, predict, and adapt in near real-time, offering unprecedented precision, responsiveness, and efficiency.

The following 12 key benefits underscore the transformative potential of IoT in healthcare [3]:

- **All-Encompassing Solutions**: IoT eHealth offers holistic solutions tailored to a wide range of needs, including health monitoring, wellness, exercise, and patient safety, ensuring comprehensive care for diverse populations.
- **Resiliency and Adaptability**: IoT-enabled systems are self-learning and adaptive, capable of evolving with changing demands, recovering from data inaccuracies, and optimizing healthcare delivery over time.
- **Big Data Analytics and Integration**: IoT systems seamlessly process and integrate multimodal datasets generated by connected devices. This enables timely and actionable insights that improve clinical decision-making and public health strategies.
- **Personalized Services and Preventive Care**: IoT empowers predictive care through big data analytics and machine learning. For instance, it can forecast health risks such as heart attacks or infections, allowing early intervention and improving patient outcomes.
- **Lifetime Monitoring**: Continuous tracking of historical, current, and predictive health data enables patients and healthcare providers to make informed decisions, supporting long-term health management.
- **Ease of Use and Patient Engagement**: Intuitive IoT devices, such as wearables and smartphone apps, simplify health tracking and actively involve patients in managing their well-being, fostering greater engagement and compliance.
- **Cost Reduction and Efficiency**: IoT eliminates redundancies by integrating diverse healthcare technologies into unified platforms. This reduces operational costs and allows patients to self-monitor, minimizing unnecessary clinical visits.
- **Optimized Physician Workflows**: IoT provides real-time data that streamlines diagnostic and monitoring processes, enabling physicians to manage more patients without compromising care quality.
- **Real-Time, Location-Independent Access**: IoT ensures that patients, caregivers, and healthcare professionals can access healthcare services and data anytime, anywhere, eliminating geographical barriers and enabling timely interventions.
- **Efficient Healthcare Resource Management**: IoT optimizes resource allocation by enabling remote monitoring and proactive care. This reduces system strain and ensures healthcare resources are used effectively.
- **Global Collaboration**: IoT systems connect healthcare professionals and institutions worldwide, offering patients access to international expertise and fostering global knowledge-sharing.
- **Scalability and Future-Readiness**: IoT systems are designed to scale across healthcare environments, from small rural clinics to large urban hospitals. This ensures that IoT remains a vital component as healthcare infrastructure grows and evolves to meet future demands.

2.2.4 Key IoT Applications in Healthcare

IoT has introduced transformative changes to healthcare through diverse applications. Below are some of the key IoT applications in healthcare [3]:

- **Remote Patient Monitoring (RPM)**: IoT devices such as wearable health monitors (e.g., Fitbit or continuous glucose monitors) enable real-time tracking of vital signs, reducing hospital visits and allowing early detection of abnormalities.
- **Telemedicine and mHealth**: Platforms like Teladoc Health leverage IoT-enabled devices to facilitate virtual consultations and diagnoses, ensuring patients in remote areas receive timely medical attention.
- **Medication Management**: Smart pill bottles such as AdhereTech notify patients when doses are missed, improving treatment adherence and preventing complications.
- **Early Warning Systems (EWS)**: IoT-based EWS are widely used in hospitals to monitor vital signs and predict conditions like sepsis up to 24 h before symptoms become critical, enabling timely interventions.
- **Population Health Management**: IoT facilitates large-scale data collection to track public health trends, identify outbreaks, and allocate healthcare resources efficiently. For example, big data analytics was used during the COVID-19 pandemic to monitor infection trends.
- **Personalized Healthcare**: IoT integrates data from wearable devices, Electronic Health Records (EHRs), and lifestyle tracking to enable tailored treatment plans. For instance, smartwatches like Apple Watch provide insights into heart health for personalized care.
- **Assistance for Disabilities**: IoT devices like smart gloves help individuals with hearing impairments communicate, while smart wheelchairs improve accessibility for those with mobility challenges in urban environments.
- **Smart Hospitals**: Real-time tracking of medical equipment, automated workflows, and patient movement monitoring optimize hospital operations. IoT-enabled RTLS (Real-Time Location Systems) help locate critical equipment in emergencies.
- **IoT in Rehabilitation and Physiotherapy**: Connected devices and wearables provide real-time feedback during rehabilitation, such as robotic exoskeletons guiding patients through recovery exercises.
- **IoT for Emergency Response**: Ambulance tracking systems integrated with hospital IoT networks provide real-time patient condition updates, allowing emergency rooms to prepare in advance for critical cases.

2.2.5 Key Challenges in IoT Healthcare

While IoT has immense potential to transform healthcare, several critical challenges need to be addressed for its widespread adoption and effective implementation. Below is a focused list of key challenges in IoT healthcare [2]:

- **Interoperability and Standardization:** The lack of universal standards across IoT devices, platforms, and communication protocols creates difficulties in integrating solutions. Additionally, regulatory compliance varies across regions and industries, further complicating adoption.

- **Heterogeneity in Systems and Devices:** IoT healthcare systems involve diverse devices, sensors, and data formats. This diversity hinders seamless data exchange and interoperability, especially when combining technologies from different manufacturers.
- **Data Management, Integration, and Governance:** The vast volume of multimodal, real-time data generated by IoT devices requires efficient processing, storage, and analysis. Challenges include inconsistent data formats, high storage demands, real-time processing capabilities, managing data variety, velocity, and veracity, as well as addressing ownership, access rights, and compliance with data protection regulations like GDPR or HIPAA.
- **Energy Efficiency and Sustainability:** Many IoT devices rely on limited battery life, particularly in remote monitoring systems or wearable devices. Frequent recharging or replacement can disrupt operations, reducing their effectiveness in long-term care applications. Developing energy-efficient devices and sustainable power solutions is vital for long-term scalability.
- **Security and Privacy Concerns:** Healthcare IoT systems are vulnerable to cyberattacks, data breaches, and unauthorized access, which can jeopardize sensitive patient information. Ensuring robust security measures and privacy safeguards is a critical priority.
- **Scalability of IoT Systems:** Scaling IoT systems from small implementations to larger networks, such as hospitals or entire cities, requires advanced architectures capable of managing increased data traffic and device connectivity.
- **User-Centric Design and Accessibility:** IoT devices must be intuitive and accessible, particularly for elderly patients or those with limited technical skills. Participatory design approaches that prioritize user feedback are essential for widespread adoption.
- **Intrusiveness and Patient Comfort:** IoT devices must balance functionality with patient comfort, especially for wearables or implants. Devices that are too invasive or interfere with daily life may lead to poor adherence and diminished user satisfaction.
- **Reliability and Real-Time Availability:** Reliable data delivery is critical for timely healthcare interventions. Device malfunctions, network failures, or latency issues can disrupt services and compromise patient outcomes.
- **Context-Aware Systems:** IoT healthcare systems need to interpret contextual data, such as environmental conditions and patient activity, to provide accurate insights. Developing reliable, context-aware applications remains a significant challenge.
- **Design and Deployment Complexity:** IoT healthcare systems require multidisciplinary expertise to integrate hardware, software, and networking components. Managing dynamic, large-scale, and distributed systems adds complexity to design and deployment.
- **Ethical and Legal Considerations:** IoT applications in healthcare must address ethical concerns such as data ownership, informed consent, and the potential misuse of personal health data. Clear legal frameworks are needed to govern these aspects.

- **Integration with Legacy Systems:** Many healthcare facilities rely on legacy systems that are not designed for IoT integration. Retrofitting these systems to accommodate IoT devices requires careful planning and significant investment.
- **Environmental Impact:** The rapid proliferation of IoT devices raises concerns about e-waste and environmental sustainability. Designing recyclable hardware and reducing the ecological footprint of IoT systems are important considerations.
- **Cost of Implementation:** While IoT promises long-term cost savings, initial setup costs for devices, infrastructure, and training can be prohibitive, especially in low-resource settings. Ensuring affordability while maintaining quality is a key challenge.

Addressing these challenges through technological innovation, regulatory support, and collaborative research will enable IoT healthcare systems to deliver secure, efficient, and scalable solutions for the future.

3 IoT System Architectures and Reference Models

The development and deployment of IoT systems have been guided by several reference architectures and models proposed by organizations, research institutions, and industry leaders. These frameworks aim to standardize IoT solutions, ensuring interoperability, scalability, and security across diverse applications.

3.1 IoT World Forum (IoTWF) Reference Model

The **IoT World Forum (IoTWF)**, led by Cisco, introduced a standardized IoT Reference Model in 2014. Developed in collaboration with industry leaders such as Cisco, IBM, and Rockwell Automation, this model provides a **seven-layer framework** to standardize the design, development, and deployment of IoT systems. It defines a **hierarchical flow of data**, starting from physical devices, progressing through networking layers, and culminating in actionable insights at the application and collaboration layers. The IoTWF model emphasizes modularity, security at each layer, and a clear separation of concerns, enabling various stakeholders to contribute interoperable components. The seven layers of the IoTWF Reference Model are described as follows (See Fig. 5) [2]:

1. **Physical Devices and Controllers (Things):** This layer consists of *physical devices, sensors, actuators, and controllers* that form the foundation of IoT. Devices collect real-world data (e.g., temperature, pressure, motion) and send it upstream for further processing. They can also receive control commands to execute actions.

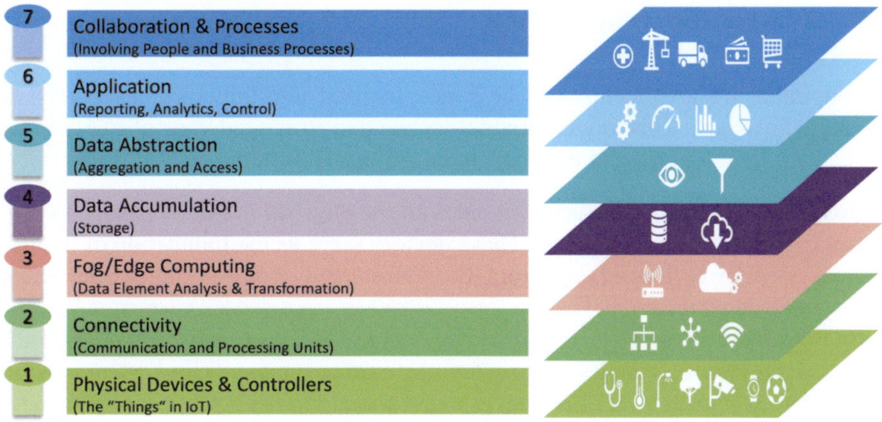

Fig. 5 IoT Reference model by IoT world forum

2. **Connectivity (Networking):** The connectivity layer ensures *reliable, secure, and timely data transmission* from physical devices to higher layers. It manages networking protocols, routing, and switching to support communication between devices and systems.
3. **Edge Data Analysis and Transformation:** This layer prepares data for storage and analysis as it progresses toward higher levels. It handles *data cleaning, aggregation, and transformation* to reduce noise and ensure meaningful data is transmitted. This intermediate step optimizes bandwidth usage and improves processing efficiency at the upper layers.
4. **Data Accumulation (Storage):** At this layer, data is *stored in appropriate formats* within centralized databases, data centers, or cloud environments. The emphasis is on ensuring persistence and enabling efficient data retrieval for downstream operations.
5. **Data Abstraction (Aggregation and Access):** The data abstraction layer ensures data is *consistent, validated, and accessible*. It combines data from multiple sources and allows unified querying across distributed datasets, providing a reliable foundation for analysis.
6. **Application (Analytics, Reporting, Control):** This layer hosts *software applications* that interact with the processed data. Applications perform functions such as *monitoring, reporting, analytics, control of devices, and visualization* to deliver actionable insights.
7. **Collaboration and Processes (Business Integration):** The final layer integrates *business processes and workflows* with IoT systems. Insights generated from the application layer are utilized to optimize operations, support decision-making, and enable innovation. This layer represents the interface where IoT systems deliver measurable value to organizations and stakeholders.

3.2 3-Layer Reference Model

A simplified IoT architecture can be conceptualized into three primary layers, as illustrated in Fig. 6. This model provides a clear and structured view of IoT components, their interactions, and their role in enabling IoT systems [2]:

1. **IoT Things Layer:** This layer includes all the *physical IoT devices* such as *sensors, actuators, and gateways*. These devices serve as the foundation of the IoT ecosystem by *collecting data* from the environment or performing actions based on received commands.
2. **IoT Network Layer:** The network layer comprises *connectivity components* such as IoT gateways, switches, and routers that ensure the *timely and reliable transmission of data* between IoT devices and the cloud. This layer also integrates *fog/edge computing nodes* to perform *data analysis and transformation* closer to the data source. By processing information locally, this reduces latency, minimizes bandwidth usage, and enables *faster responses* in real-time applications. For instance, in *IoT healthcare*, edge computing supports low-latency emergency responses such as monitoring patient vitals and triggering alerts.
3. **IoT Cloud and Application Layer:** This layer serves as the backbone for *managing, processing, and storing data* collected from the IoT Things and Network layers. It transforms raw data into meaningful insights and supports decision-making processes. The key functions of this layer include:

   - **Data Ingestion:** Seamlessly collects and aggregates data from IoT devices for further processing.

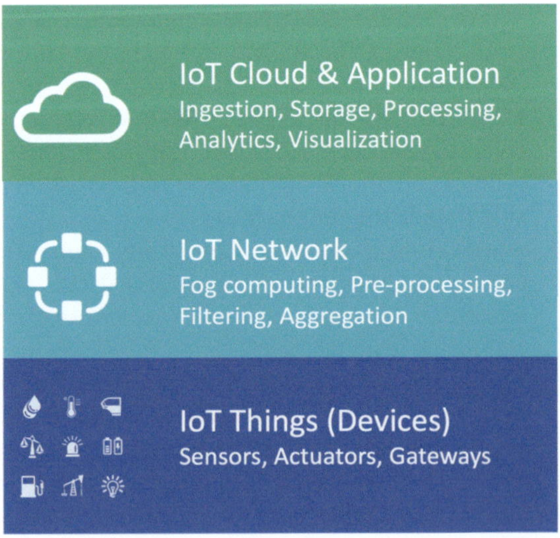

Fig. 6 3-Layer IoT Architecture

- **Data Interpretation:** Processes, analyzes, and transforms data using advanced software applications, enabling tasks such as *analytics*, *visualization*, and *report generation*.
- **Integration:** Connects with other platforms, systems, and business workflows to drive *informed decision-making* and deliver *actionable insights* that enhance operational efficiency and business outcomes.

4 Smart Things

4.1 Definition and Architecture

The **Things Layer** is the foundation of the IoT ecosystem, bridging the gap between the digital and physical worlds. These devices—commonly called *smart objects*, *smartdevices*, *IoT devices*, or *endpoints*—are embedded with electronics that enable them to *sense, process, and exchange data* over a network with minimal or no human intervention. These smart things are the enablers of IoT applications, empowering innovation in domains such as healthcare, smart cities, and industrial automation. A typical IoT device comprises the following essential components [2]:

1. **Processing Unit** The processing unit is the *intelligence hub* of an IoT device, managing all computational tasks. It is responsible for:

    - Collecting and processing data from sensors.
    - Analyzing data to produce actionable insights and enable decision-making.
    - Coordinating communication with external devices or networks.
    - Managing actuator signals and regulating energy consumption.

    Most IoT devices rely on *microcontrollers* as their processing units due to their *small size, low cost, energy efficiency, and programmability*. More advanced IoT devices may also integrate AI-enabled processing units for on-device intelligence.
2. **Transducers** Transducers are the *sensing and acting components* of IoT devices, enabling interaction with the physical environment. They convert one form of energy into another and are categorized into two primary types:

    - **Sensors:** Sensors detect physical, chemical, or environmental phenomena such as temperature, light, pressure, or motion and convert these into measurable electrical signals (analog or digital). Examples include:

        – *Temperature sensors* for climate monitoring.
        – *Motion detectors* for security systems.
        – *Chemical sensors* for air quality assessment.

    - **Actuators:** Actuators enable IoT devices to act upon their environment by converting electrical energy into motion or mechanical output. Examples of actuators include:

  - *Thermal Actuators:* Convert heat into motion, often used in thermostats.
  - *Mechanical Actuators:* Transform mechanical input into linear motion, such as in hydraulic lifts or screw jacks.
  - *Electrical Actuators:* The most common in IoT applications, these convert electrical energy into motion, such as in electric motors or servo systems.

3. **Communication Device** The communication device is the *connectivity enabler*, allowing IoT devices to interact with other devices, systems, and networks. It facilitates data exchange and supports the seamless integration of IoT systems. Connectivity can be achieved through:

   - **Wired technologies,** such as Ethernet, which offer high-speed, stable, and secure communication.
   - **Wireless technologies,** such as Wi-Fi, Bluetooth, Zigbee, and LoRa, which are favored for their ease of deployment, scalability, and suitability for remote and infrastructure-limited environments.

   The choice of communication technology depends on the specific requirements of the application, including range, bandwidth, latency, power consumption, and reliability. Wireless protocols dominate most IoT use cases due to their flexibility and cost-effectiveness.
4. **Power Source** The power source provides the energy needed to operate all the components of the IoT device. Power efficiency is a critical consideration, as many IoT devices are deployed in *remote, hard-to-access locations* with limited energy resources. Key considerations for power management include:

   - Using *low-power communication protocols and hardware* to minimize energy consumption.
   - Incorporating *sleep modes* and energy-saving mechanisms to reduce power usage during idle periods.
   - Leveraging *renewable energy solutions*, such as solar panels, for devices in remote areas.

   Batteries are the most common power source for IoT devices; however, advancements in energy harvesting and efficient power management are helping extend operational lifetimes and reduce maintenance.

By integrating these components—**processing units, transducers, communication devices, and power sources**—IoT devices form the backbone of the IoT ecosystem. They facilitate data collection, real-time decision-making, and environmental interaction, driving diverse applications across industries.

4.2 Sensors in IoMT

Ssensors play a crucial role in collecting data that is essential for monitoring, diagnosing, and managing patient health. Sensors can be classified based on **what they measure** (functional classification) or **where and how they are used** (contextual classification). Below is a detailed explanation of these classifications [2].

A. Functional Classification

This classification is based on the type of data the sensors capture. It helps categorize sensors by their primary function or purpose.

1. **Physiological Sensors:** These sensors measure biological signals and are commonly used in healthcare for real-time monitoring and diagnosis. Examples include:

   - **Heart Rate Sensors**, such as those using electrocardiography (ECG) or photoplethysmography (PPG) to measure heartbeats per minute (BPM).
   - **Blood Pressure Sensors**, such as oscillometric or piezoelectric devices that monitor systolic and diastolic pressure.
   - **Respiration Rate Sensors**, such as those using impedance pneumography or airflow sensors to measure the rate of breathing.
   - **Pulse Oximeters**, such as optical sensors that monitor oxygen saturation (SpO_2) in the blood.
   - **Body Temperature Sensors**, such as thermistors or infrared devices to record body temperature.
   - **Electroencephalogram (EEG) Sensors**, such as electrodes that detect and record brain activity.
   - **Electromyography (EMG) Sensors**, such as those that measure muscle activity and detect neuromuscular issues.
   - **Electrocardiography (ECG) Sensors**, such as devices that capture the electrical activity of the heart.

2. **Biochemical Sensors:** These sensors analyze chemical compositions in bodily fluids or tissues. Examples include:

   - **Glucose Sensors**, such as electrochemical devices used for diabetes management.
   - **Lactate Sensors**, such as those for metabolic monitoring or athletic performance analysis.
   - **pH Sensors**, such as devices used to measure acidity or alkalinity for wound management or metabolic analysis.
   - **Electrolyte Sensors**, such as those that detect concentrations of ions like sodium or potassium.

- **Hormone Sensors**, such as cortisol or insulin monitoring devices.
- **Cholesterol Sensors**, such as lipid profile analyzers used for cardiovascular health assessment.

3. **Imaging Sensors:** These sensors capture visual or diagnostic data. Examples include:

- **Infrared Sensors**, such as those used to monitor thermal data for body temperature or vascular imaging.
- **Optical Sensors**, such as cameras that capture visible or near-infrared light for diagnostic purposes.
- **Camera-Based Sensors**, such as those used in telemedicine for remote diagnosis or dermatological imaging.
- **Ultrasound Sensors**, such as devices that use high-frequency sound waves for non-invasive imaging.
- **X-Ray Detectors**, such as those used for skeletal, dental, or other diagnostic imaging.

4. **Environmental Sensors:** These sensors monitor the external environment that may affect patient health. Examples include:

- **Temperature Sensors**, such as thermistors that measure ambient temperature for comfort and equipment safety.
- **Humidity Sensors**, such as those that track air moisture levels to prevent dehydration or respiratory discomfort.
- **Air Quality Sensors**, such as devices that detect pollutants, allergens, or harmful gases.
- **Pressure Sensors**, such as barometers used to monitor atmospheric pressure in specialized treatments.

5. **Motion and Activity Sensors:** These sensors detect movement, posture, or activity levels. Examples include:

- **Accelerometers**, such as those used to measure linear motion for fall detection or physical activity tracking.
- **Gyroscopes**, such as devices that monitor angular velocity for balance or posture analysis.
- **Inertial Measurement Units (IMUs)**, such as sensors combining accelerometers and gyroscopes for 3D motion tracking.
- **Pressure Sensors**, such as those used in gait analysis or rehabilitation to monitor force distribution.

6. **Acoustic and Sound Sensors:** These sensors capture sound or vibrations for diagnostic purposes. Examples include:

- **Stethoscope Sensors**, such as those used to detect heart and lung sounds for remote auscultation.

- **Microphones**, such as those that analyze speech, breathing, or snoring patterns (e.g., sleep apnea monitoring).
- **Vibration Sensors**, such as devices used to measure mechanical vibrations for tremor analysis in conditions like Parkinson's disease.

7. **Radiation Sensors:** These sensors measure radiation exposure in specific medical contexts. Examples include:

- **Dosimeters**, such as devices that track radiation exposure for patients undergoing radiotherapy.
- **Gamma Ray Detectors**, such as those used to detect radioactive isotopes in cancer treatments or imaging.

8. **Chemical Sensors:** These sensors monitor specific chemical components. Examples include:

- **CO_2 Sensors**, such as those used to measure carbon dioxide levels in respiratory systems.
- **Oxygen Sensors**, such as devices used to track oxygen levels in ventilators or patient surroundings.

9. **Tactile or Pressure Sensors:** These sensors detect physical pressure or force. Examples include:

- **Force Sensors**, such as those used to measure applied pressure in prosthetics or rehabilitation devices.
- **Skin Pressure Sensors**, such as those that monitor pressure points to prevent bedsores or prosthetic discomfort.

10. **Proximity Sensors:** These sensors detect the presence of objects or body parts without physical contact. Examples include:

- **Infrared Proximity Sensors**, such as those used to monitor limb or hand movements in assistive or rehabilitation devices.

11. **Energy Sensors:** These sensors measure energy-related parameters. Examples include:

- **Bioimpedance Sensors:** These sensors measure parameters related to body composition, hydration, and fat/muscle analysis by detecting the resistance and reactance of biological tissues to an electrical current. These are commonly used in body composition analysis devices and wearable health trackers.
- **Thermal Energy Sensors:** These sensors are utilized to monitor heat transfer or thermal dynamics in the body. This can be useful in metabolic studies, thermoregulation monitoring, or even for devices like thermographic cameras in healthcare settings.

B. Contextual Classification

This classification is based on the usage context, form factor, or integration of the sensors in IoMT devices.

- **Wearable Sensors**: Integrated into devices worn on the body, such as fitness bands or smartwatches. These can measure physiological parameters (e.g., heart rate, SpO_2) or motion and activity (e.g., accelerometers for fall detection).
- **Implantable Sensors**: Placed inside the body for continuous monitoring, such as glucose sensors for diabetes management or pressure sensors for cardiovascular monitoring.
- **Ambient or Environmental Sensors**: Positioned in the patient's surroundings to track environmental conditions, such as air quality, temperature, or humidity sensors in patient rooms.
- **Portable/Handheld Sensors**: Used in devices like portable ECG monitors or glucometers for home diagnostics.

4.3 Actuators in IoMT

Actuators are devices that perform specific actions in response to data collected by sensors or commands issued by control systems. Actuators enable a range of critical functions, such as:

- **Delivery and Dispensing:** Controlled release or administration of substances, such as medications, fluids, or gases.
- **Stimulation and Regulation:** Application of controlled energy—electrical, thermal, mechanical, or magnetic—to influence biological or physiological functions.
- **Motion and Force Generation:** Execution of precise movements or application of force for medical assistance, intervention, or therapeutic support.
- **Feedback and Alerts:** Provision of sensory outputs, including visual, auditory, or tactile signals, to guide or inform patients and healthcare providers.
- **Environmental and Process Control:** Regulation of temperature, pressure, fluid flow, or chemical processes within medical systems.

Actuators can be classified based on two primary perspectives: Functionality-Based Classification and Mechanism-Based Classification.

A. Functionality-Based Classification:

This classification emphasizes **what the actuator achieves** in terms of its purpose or outcome in healthcare applications. Examples include:

1. **Drug and Fluid Delivery:** Actuators that deliver or dispense medications, fluids, or gases to patients for therapeutic or diagnostic purposes. Examples include:
   - Wearable insulin pumps, such as those that automatically deliver insulin doses based on continuous glucose monitoring.
   - Portable nebulizers, such as devices that aerosolize liquid medication for respiratory therapy.
   - IoT-enabled pill dispensers, such as automated systems that remind users to take medications and notify caregivers of missed doses.

2. **Therapeutic Stimulation:** Actuators that apply controlled stimuli to regulate biological functions or promote healing. Examples include:
   - Pacemakers, such as devices that regulate heart rhythms by delivering electrical impulses.
   - Wearable cryotherapy wraps, such as cooling patches for joint or muscle therapy.
   - Photodynamic therapy systems, such as light-activated devices for wound healing or skin treatments.

3. **Motion and Force Assistance:** Actuators that generate motion or force to assist mobility, rehabilitation, or surgical precision. Examples include:
   - Smart prosthetics, such as robotic limbs that adapt to user movements and transmit usage data to healthcare providers.
   - Powered exoskeletons, such as wearable systems that enable walking and transmit movement data to IoT platforms.
   - Rehabilitation robots, such as portable systems that provide guided exercises and feedback during home-based therapy.

4. **Feedback and Sensory Outputs:** Actuators that provide alerts or sensory feedback to guide or inform patients and caregivers. Examples include:
   - Haptic feedback wearables, such as devices that vibrate to alert users about medication schedules or abnormal health readings.
   - Smart auditory alarms, such as systems that notify users or caregivers of critical conditions via IoT-connected platforms.
   - LED-enabled wearable devices, such as visual indicators showing real-time health data like glucose levels or hydration status.

5. **Process and Environmental Control:** Actuators that regulate internal processes or external environmental conditions for therapeutic purposes. Examples include:
   - Wearable cryotherapy patches, such as devices that apply cooling therapy and track session data via apps.
   - pH-responsive drug delivery systems, such as systems that release medication in response to pH changes in the body.

- Portable ventilation systems, such as compact devices that adjust airflow based on patient vitals monitored through IoT platforms.

6. **Advanced Regenerative Applications:** Actuators designed for cutting-edge applications in regenerative medicine and tissue engineering. Examples include:

   - 3D bioprinting actuators, such as systems that construct tissues or organs layer by layer for transplantation or research.
   - Cell manipulation devices, such as systems that enable precise control of cell-based therapies.

B. Mechanism-Based Classification

This classification emphasizes **how the actuator works**, focusing on the energy source or operating principle driving the actuator. Examples include:

1. **Electrical Actuators:** Operate using electrical energy to perform actions. Examples include:

   - Pacemakers, such as devices that deliver electrical impulses to regulate heart rhythms.
   - Neuromodulation devices, such as systems that stimulate nerves or the spinal cord for therapeutic purposes.

2. **Thermal Actuators:** Use heating or cooling mechanisms for therapeutic purposes. Examples include:

   - Cryotherapy systems, such as wearable devices that apply localized cooling for pain relief.
   - Hyperthermia devices, such as systems that deliver heat for wound healing or cancer therapy.

3. **Magnetic Actuators:** Leverage magnetic fields to perform actions. Examples include:

   - Magnetic stimulators, such as non-invasive devices used for brain stimulation in depression therapy.
   - Implantable magnetic systems, such as devices that use external magnetic fields for valve control or targeted drug delivery.

4. **Light-Based Actuators:** Use light energy for therapeutic or surgical applications. Examples include:

   - Photodynamic therapy systems, such as light-activated devices for cancer treatment.

- Laser surgery tools, such as systems that perform minimally invasive procedures with high precision.

5. **Chemical Actuators:** Operate based on chemical reactions to trigger actions. Examples include:

   - pH-responsive drug delivery systems, such as those that release medication based on pH changes in the body.
   - Smart polymers, such as materials that change shape or behavior in response to chemical stimuli.

6. **Mechanical Actuators:** Generate motion or force through mechanical means. Examples include:

   - Prosthetic actuators, such as robotic limbs that mimic natural movement.
   - Robotic surgery tools, such as systems that translate surgeon commands into precise mechanical actions.

7. **Microfluidic Actuators:** Control small fluid volumes for precise therapeutic or diagnostic purposes. Examples include:

   - Lab-on-a-chip devices, such as systems that regulate reagent flow for rapid diagnostics.
   - Microfluidic drug delivery systems, such as systems that dispense precise doses of medication.

8. **Radiation-Based Actuators:** Use targeted radiation for medical interventions. Examples include:

   - Radiotherapy devices, such as systems that deliver precise doses of radiation to cancer cells.
   - Gamma knife systems, such as tools for non-invasive brain surgeries.

4.4 Key Communication Interfaces for Sensors and Actuators

The interaction between sensors, actuators, and processing units is crucial for the functioning of IoT devices. These interfaces and protocols enable the exchange of data and control signals, forming the communication backbone for IoT systems. Effective interfaces must meet key requirements such as speed and reliability while accommodating the specific needs of the application. By leveraging the right interfaces, IoT devices can ensure seamless data flow and precise control, enabling optimal performance across various use cases [2].

Parallel Versus Serial Interfaces

Communication between sensors, actuators, and processing units in IoT devices can be broadly categorized into **Parallel Interfaces** and **Serial Interfaces** [2]:

1. **Parallel Interfaces**: Parallel interfaces transmit multiple bits simultaneously over multiple data lines, typically requiring 8, 16, or 32 wires. These interfaces offer high-speed communication, making them ideal for applications requiring rapid data transfer. However, the increased wiring complexity, cost, and power consumption often limit their use in compact IoT systems.
2. **Serial Interfaces**: Serial interfaces stream data one bit at a time over fewer wires (usually 1 to 4), making them more efficient and cost-effective for IoT applications. Due to their simplicity and reduced hardware requirements, serial interfaces dominate IoT systems. They are further divided into:

    - **Synchronous Interfaces**: Use a shared clock signal to synchronize data transfer between devices (e.g., SPI, I2C).
    - **Asynchronous Interfaces**: Transmit data without a shared clock, reducing wiring needs but requiring additional mechanisms for synchronization (e.g., UART).

Serial Interfaces in IoT Systems

In IoT systems, **serial interfaces** are the preferred choice due to their flexibility, lower cost, and simplicity. Below are the key serial interfaces commonly used to connect sensors and actuators to processing units:

1. **Universal Asynchronous Receiver/Transmitter (UART)**

    - UART is a widely used asynchronous serial communication interface that supports reliable, full-duplex (two-way) communication.
    - It uses three signals: TX (transmit), RX (receive), and Ground. Data is transmitted in packets that include start bits, data bits, optional parity bits, and stop bits, eliminating the need for an external clock.
    - UART is ideal for simple, low-speed, point-to-point communication in IoT applications, thanks to its minimal hardware requirements.

2. **Serial Peripheral Interface (SPI)**

    - SPI is a synchronous serial communication protocol designed for high-speed data exchange between microcontrollers and peripherals like sensors and SD cards.
    - It uses four lines: CLK (clock), MOSI (master out, slave in), MISO (master in, slave out), and CS (chip select).

- SPI operates in a master-slave configuration, supporting multiple slaves through chip-select signals or daisy-chaining. It is commonly used in scenarios requiring high data rates and full-duplex communication.

3. **Inter-Integrated Circuit (I2C)**

    - I2C is a versatile synchronous protocol that uses only two bidirectional lines: SDA (data) and SCL (clock).
    - It supports multiple masters and slaves on a single bus, with each device identified by a unique address.
    - I2C is particularly advantageous in applications requiring many devices to communicate over short distances with minimal wiring.

4. **Universal Synchronous Asynchronous Receiver Transmitter (USART)**

    - USART combines the features of UART with synchronous operation, allowing for higher data rates when an external clock is used.
    - It supports both synchronous and asynchronous communication, making it a flexible option for IoT systems requiring diverse data transfer mechanisms.

5. **RS232 and RS422**

    - **RS232**: Extends UART functionality by using line drivers to boost signal strength and support longer distances (up to 10 m) with voltage levels ranging from ± 3 V to ± 15 V.
    - **RS422**: Builds on RS232 by employing differential signaling, providing improved noise immunity and supporting even longer distances. These features make RS422 particularly suitable for industrial IoT applications.

Key Considerations for Choosing Interfaces

When selecting an interface for IoT devices, the choice depends on the specific requirements of the application, including [2]:

- **Data Speed:** SPI supports high-speed communication, typically up to several Mbps, making it ideal for time-sensitive applications. In contrast, I^2C and UART are better suited for lower-speed applications, with I^2C generally operating up to a few hundred Kbps and UART speeds depending on configuration and hardware.
- **Complexity and Wiring:** UART and I^2C require minimal wiring (2–4 lines), while SPI demands additional lines for chip select (CS) to manage multiple devices, increasing complexity.
- **Device Count:** I^2C excels in multi-device setups thanks to its addressing system and shared two-wire bus, supporting multiple masters and slaves with minimal wiring.

- **Distance and Noise Resistance:** RS-422 provides excellent noise immunity and supports long-distance communication (up to 1,200 m at lower speeds), making it suitable for industrial and noisy environments.

5 Communications and Networks

The IoT Network plays a critical role in connecting billions of devices, sensors, and systems across diverse environments to enable seamless data exchange and communication. The growing scale, heterogeneity, and dynamic nature of IoT systems present significant challenges for network design and operation. Key requirements include low latency, high reliability, scalability, energy efficiency, and security to accommodate the unique demands of IoT applications, such as real-time monitoring in healthcare [2].

To meet these requirements, numerous network technologies, protocols, and architectures have been proposed, each designed to address specific needs, such as constrained devices, long-range communication, or high-speed data transfer. Based on the system's design objectives and application-specific requirements, the most suitable network solution must be selected to ensure seamless connectivity, ultimately enabling a robust and efficient IoT ecosystem.

5.1 Key Features of IoT Networks

IoT systems possess unique characteristics that distinguish them from traditional networks. These features emphasize their complexity, heterogeneity, and integration challenges [2]:

- **Dependence Between Application and Communication:** IoT applications are deeply interconnected with both application requirements and communication needs. The specific demands of IoT use cases vary significantly, ranging from low data rates in wearable health monitors to the high bandwidth requirements of medical imaging and smart home surveillance systems. As a result, highly tailored communication protocols are essential to meet these diverse needs. Energy efficiency becomes particularly critical for devices such as implantable sensors, battery-powered monitors, and remote nodes operating in constrained environments like hospitals or smart homes. These conditions call for careful design to ensure a balance between performance, energy efficiency, and reliability. To achieve this, communication protocols must address various challenges, such as managing trade-offs between performance and energy efficiency, especially for resource-constrained devices. Moreover, real-time communication is crucial in applications where timely data transmission is vital, such as monitoring vital signs in healthcare or transmitting high-resolution images for medical diagnostics. Surveillance systems in smart homes also require real-time communication, particularly in monitoring individuals with Alzheimer's disease or other cognitive impairments. Additionally, efficient data aggregation from distributed sensors—like environmental monitors in hospitals or air quality sensors in urban healthcare settings—is necessary to provide comprehensive insights and ensure system-level reliability. Finally, adaptive communication strategies are vital to account for dynamic conditions, adjusting transmission rates based on factors like battery levels, network congestion, or changes in the monitored individual's health or activity.

- **Scalability and Heterogeneity:** IoT systems must accommodate a wide range of devices, protocols, and data rates while scaling from small deployments to massive interconnected networks. This heterogeneity introduces challenges in device integration, protocol compatibility, and efficient resource allocation. Incorporating a variety of devices, from simple sensors to complex embedded systems, requires careful design to ensure compatibility. Integrating communication protocols like Bluetooth, Zigbee, or 5G within a unified framework is essential for ensuring seamless interaction between devices. Additionally, managing resource allocation to meet diverse device and application requirements is crucial for maintaining optimal performance across a heterogeneous network.
- **Environmental Integration:** IoT devices operate within physical environments that impose significant constraints on their placement, connectivity, and performance. The positioning of devices is often dictated by the need for data collection or action execution, such as with body-worn sensors or fixed actuators. Environmental factors, such as walls in hospital settings, also play a crucial role in signal propagation and influence the design of the communication network. Therefore, careful consideration must be given to ensure that device placement aligns with the specific data collection or action requirements. Additionally, selecting the appropriate radio frequency bands is essential to overcome environmental challenges, such as the presence of walls in healthcare environments that can obstruct signals. The overall communication architecture must be designed with these environmental factors in mind, taking into account the positioning of nodes and the density of the network to ensure efficient and reliable performance.
- **System-of-Systems Nature:** IoT applications often function as interconnected systems, where individual nodes operate autonomously but collaborate to achieve global objectives. For instance, in healthcare automation, a network of devices works together to monitor patient health while meeting system-wide goals. Determining the optimal number of nodes to fulfill system objectives is a critical consideration. Assigning tasks to nodes based on their capabilities and requirements ensures efficient system operation. Additionally, selecting appropriate communication protocols while accommodating legacy infrastructures is important for achieving compatibility and smooth operation across diverse systems.
- **Real-Time and Adaptive Operations:** Many IoT applications require real-time responsiveness and adaptability to changing conditions. For example, healthcare systems must prioritize life-critical alerts over routine data or adjust device operation based on available energy resources. Adapting device communication frequency based on environmental or operational conditions is essential for efficient

operation. Prioritizing mission-critical tasks in dynamic, real-time environments ensures that the most important actions are taken first. Additionally, balancing performance and energy efficiency in resource-constrained devices is crucial to maintain functionality while managing limited resources.

5.2 Fundamentals of Networking

Networking is a fundamental element of modern IoT systems, enabling devices to connect, share data, and collaborate efficiently. It involves a wide range of network types, each designed to meet the unique requirements of different IoT applications, scales, and communication technologies. At its core, a network is a collection of interconnected devices, known as nodes, that communicate to exchange data and resources. These nodes can include IoT devices such as sensors, smartphones, and servers. Networks play a crucial role in facilitating communication and ensuring the smooth operation of IoT systems, connecting devices across various environments and enabling real-time data sharing. This connectivity supports everything from personal applications to the functioning of large-scale IoT ecosystems, making networks essential for the seamless integration and management of IoT devices.

5.2.1 Types of Networks

Networks can be categorized based on their scale, purpose, and the technologies they employ to facilitate communication and connectivity [2]. The following classification provides an overview of various network types, ranging from small-scale personal networks to large-scale global communication infrastructures.

- **Personal Area Network (PAN):** Operates within a few meters, typically around an individual. Designed to connect personal devices such as smartphones, laptops, tablets, and wearable devices. Bluetooth is a widely used PAN technology that connects peripherals like wireless headphones and fitness trackers.
- **Body Area Network (BAN):** Functions over a very short range, typically within or in close proximity to the human body. Commonly used in healthcare and fitness applications, BANs connect sensors and actuators to a control device, such as a smartphone, enabling real-time physiological monitoring. Examples include heart rate monitors, smartwatches, and implanted medical devices.
- **Home Area Network (HAN):** Covers a small area such as a home or apartment, typically extending over a few tens of meters. Used for connecting smart home devices, including IoT-enabled appliances, thermostats, and entertainment systems. Common wireless technologies in HANs include ZigBee, Wi-Fi, and Bluetooth Low Energy (BLE).
- **Local Area Network (LAN):** Typically spans a single building or a campus, with coverage extending up to 100 m. LANs facilitate high-speed communication and resource sharing within localized environments such as offices, schools, and

enterprises. Wired LANs rely on Ethernet, while wireless LANs utilize Wi-Fi for connectivity.

- **Field Area Network (FAN):** Extends over a larger outdoor area, ranging from tens to hundreds of meters. Predominantly used in industrial, agricultural, and smart grid applications to connect distributed sensors and control systems. Technologies such as 6LoWPAN support low-power communication for IoT devices in these environments.
- **Metropolitan Area Network (MAN):** Spans a city or metropolitan region, typically covering a few kilometers. MANs provide connectivity for businesses, governments, and public services within urban areas. These networks are widely used in smart city infrastructures, including intelligent traffic systems and city-wide Wi-Fi networks.
- **Wide Area Network (WAN):** Covers extensive geographic regions, often spanning entire countries or continents. WANs enable long-distance communication and interconnect remote locations using wired and wireless technologies, including fiber optics, satellite communication, and mobile networks such as 4G and 5G. The Internet is the largest WAN, linking billions of devices worldwide.

5.2.2 Network Topologies

Network topology refers to the arrangement of nodes and how they are connected in a network. Different topologies suit different applications (See Fig. 7) [2]:

- **Star Topology:** A centralized topology where all devices are connected to a central hub or switch. This structure simplifies network management and is easy to scale.

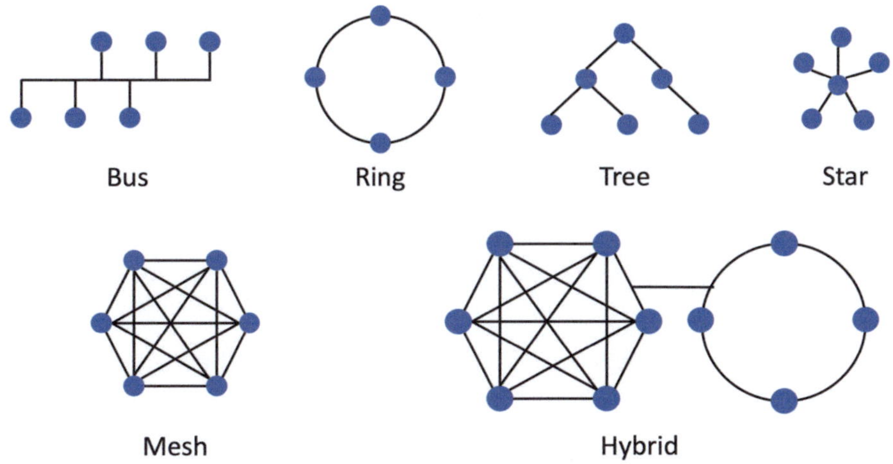

Fig. 7 Different types of network topologies illustrating various ways devices can be connected in a network

A key advantage is that failure of an individual device does not impact the rest of the network. However, the central hub represents a single point of failure, and if it malfunctions, the entire network is disrupted. A common example is a home Wi-Fi network, where a router acts as the central hub.

- **Tree Topology:** An extension of the star topology that organizes devices in a hierarchical structure. This topology is commonly used in large-scale network deployments that require structured management. It supports scalability by enabling hierarchical organization but introduces vulnerabilities, as a failure at a higher-level node can affect all dependent devices.
- **Mesh Topology:** A decentralized topology where each node is connected to multiple other nodes, ensuring multiple data paths for redundancy and reliability. This setup enhances fault tolerance and minimizes network downtime. However, managing and maintaining a mesh network is complex and resource-intensive. Mesh topologies are frequently used in industrial IoT applications, where reliability is crucial.
- **Bus Topology:** A linear topology where all devices share a single central communication cable. This structure is cost-effective and easy to implement in small-scale networks. However, network performance can degrade as more devices are added, leading to increased data collisions. Additionally, failure of the central cable can disrupt the entire network.
- **Ring Topology:** A circular topology where each device is connected to two neighboring devices, forming a closed loop. Data travels in a unidirectional or bidirectional manner, depending on the configuration. The advantage of this topology is that it provides structured data transmission with reduced collision risks. However, in a single-ring setup, failure of any single device or connection can disrupt the network. A dual-ring configuration can mitigate this issue but increases complexity and cost.
- **Hybrid Topology:** A combination of two or more different topologies, designed to meet specific network requirements. Hybrid networks offer flexibility and scalability by leveraging the strengths of multiple topologies. For instance, a corporate network might use a star topology within individual departments while employing a mesh topology for inter-department communication. Although hybrid topologies provide adaptability, they can be complex and expensive to implement due to the integration of multiple network structures.

5.3 OSI Model

The Open Systems Interconnection (OSI) Model, developed by the International Organization for Standardization (ISO), provides a standardized framework for understanding and implementing network communication. It comprises seven distinct layers, each with specific roles and responsibilities that work collaboratively to enable seamless data exchange. Since data transmission between multiple end nodes in a network is inherently complex, a common engineering approach is to break it

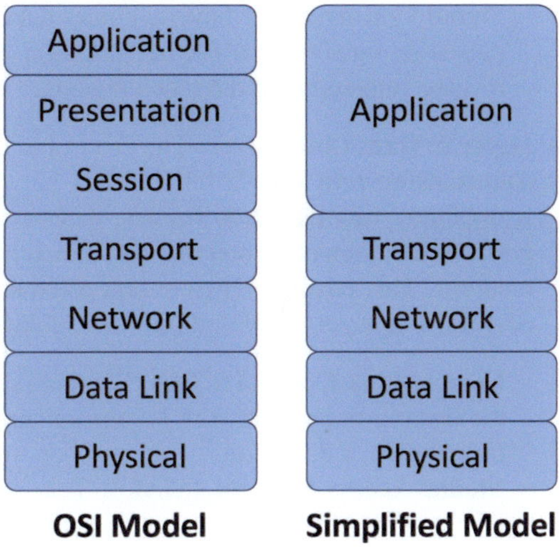

Fig. 8 The original OSI Reference Model consists of seven layers, whereas the commonly used Simplified OSI Model has five layers. The difference arises because the top three layers of the original model are combined into a single layer in the simplified version

down into smaller, more manageable tasks. While the original OSI model consists of seven layers, there is also an implied version with only five layers, where the upper three layers are merged into a single layer. This layered approach simplifies the design, development, and troubleshooting of networks, ensuring compatibility and interoperability across diverse systems and technologies (See Fig. 8) [2].

- **Layer 1: Physical Layer:** The physical layer governs the transmission of raw binary data between devices over a physical medium. It facilitates the conversion of bits into physical signals (electrical, optical, or radio) and defines the hardware specifications required for communication, including cable types, signal voltages, and modulation techniques.
  - **Bit Synchronization:** Ensures alignment of data transmission between the sender and receiver, typically achieved through a shared clock or encoding methods such as Manchester encoding. This synchronization enables accurate interpretation of transmitted bits. **Note:** Some physical layer implementations, including asynchronous communication protocols (e.g., certain Bluetooth modes), do not rely on continuous synchronization.
  - **Bit Rate Control:** Specifies the transmission rate, which defines the number of bits transmitted per second (e.g., Mbps, Gbps).
  - **Physical Topologies:** Describes the physical arrangement of devices in a network, such as bus, star, or mesh topologies, which influence scalability and performance.
  - **Transmission Modes:** Defines the mode of data flow between devices, supporting simplex (unidirectional), half-duplex (bidirectional but alternating), and full-duplex (simultaneous bidirectional) modes.

  - **Signal Conversion:** Converts binary data into physical signals (e.g., electrical pulses or light waves) for transmission through a medium and performs the reverse conversion at the receiving end.

- **Layer 2: Data Link Layer:** The Data Link Layer is responsible for node-to-node data transfer, providing a reliable link between two directly connected devices. It organizes raw data into frames, detects and optionally corrects transmission errors to maintain data integrity, and manages access to the shared communication medium. This ensures efficient and reliable communication between connected devices. The layer is divided into two sublayers:

  - **Logical Link Control (LLC):** This sublayer manages communication between the Data Link and Network Layers, ensuring frame synchronization, flow control, and error detection.
  - **Media Access Control (MAC):** This sublayer handles access to the physical transmission medium, defines rules for addressing devices using MAC addresses, and ensures proper data encapsulation for transmission.

  The primary functions of the Data Link Layer include:

  - **Framing:** Encapsulates network layer packets into frames, adding headers and trailers for efficient and error-detectable transmission over the medium.
  - **Error Control:** Detects errors introduced during transmission using mechanisms such as cyclic redundancy check (CRC). Depending on the protocol, it may request retransmission of corrupted frames.
  - **Flow Control:** Ensures a consistent transmission rate between sender and receiver, preventing data loss or buffer overflows caused by mismatched speeds.
  - **Access Control:** Regulates access to the shared communication medium, determining which device can transmit at any given time using mechanisms like CSMA/CD (Ethernet) or CSMA/CA (Wi-Fi).
  - **Physical Addressing:** Adds MAC addresses to frames to uniquely identify devices within a local network, ensuring proper delivery to the intended recipient.

- **Layer 3: Network Layer:** The Network Layer is responsible for enabling communication between devices on different networks by handling data transfer across interconnected networks. It uses logical addressing, such as IP addresses, to identify devices and employs routing protocols to determine the most efficient path for data packets to travel from source to destination. While IP (IPv4/IPv6) is the most widely used protocol for addressing and routing, alternative protocols like CLNP, IPX, and AppleTalk have also been used historically in specific networks. The key functions of the Network Layer include:

  - **Routing:** Determines the optimal path for data packets to travel through the network, using routing protocols like OSPF and BGP to make decisions based on factors such as network topology, link costs, and congestion.

  - **Logical Addressing:** Assigns unique addresses (e.g., IP addresses) to devices, enabling global identification and communication across different networks. This ensures that data is sent to the correct destination.
  - **Packet Forwarding:** Breaks down data into smaller units called packets and forwards them through intermediate routers and networks until they reach their final destination.

- **Layer 4: Transport Layer:** The Transport Layer is responsible for ensuring end-to-end communication between devices. Depending on the protocol, it may provide reliability, managing data segmentation, flow control, and error correction, or offer lightweight, best-effort delivery for time-sensitive applications. The Transport Layer interfaces directly with applications or processes running on devices, directing data to the appropriate destination. In protocols such as TCP, UDP, and SCTP, this is achieved through the use of port numbers, which uniquely identify the application or process. These port numbers allow the Transport Layer to distinguish between multiple simultaneous connections or services on the same device, ensuring accurate delivery of data. While most modern Transport Layer protocols use port numbers, some specialized or legacy protocols may employ alternative mechanisms. The key functions of the Transport Layer include:

  - **Segmentation and Reassembly:** Divides large messages into smaller segments for efficient transmission and reassembles them at the destination.
  - **Flow Control:** Regulates the sender's transmission rate based on the receiver's capacity to prevent buffer overflow. (Note: This is distinct from congestion control, which addresses network-wide congestion.)
  - **Error Correction:** Ensures reliable delivery of data by detecting and retransmitting lost or corrupted segments (This applies to reliable protocols like TCP, but not to UDP).
  - **Service Point Addressing:** Directs data to the correct application or process on the destination device using port numbers.

- **Layer 5: Session Layer:** The Session Layer is responsible for establishing, managing, and terminating communication sessions between two or more devices, which are referred to as"sessions." This layer creates the setup, controls the connections, and ends the teardown process, ensuring the orderly flow of data. While the Session Layer provides synchronization and dialog control for efficient data exchange, many of its functions are now handled by Transport Layer (e.g., TCP) or Application Layer protocols in modern networking environments. The key functions of the Session Layer include:

  - **Session Management:** Handles the initiation, maintenance, and orderly termination of communication sessions.
  - **Synchronization:** Inserts checkpoints in data streams to allow recovery from interruptions without restarting the session.

  - **Dialog Control:** Manages communication modes, such as half-duplex (one direction at a time) or full-duplex (both directions simultaneously), ensuring orderly data exchange.

- **Layer 6: Presentation Layer:** The Presentation Layer is responsible for translating data between application-layer formats to ensure compatibility between communicating systems. It also conceptually supports encryption for secure data transmission and compression to optimize data for faster transmission. In modern networking, many of these functions are handled by Application Layer protocols or specialized libraries, but the Presentation Layer remains a key conceptual part of the OSI model for handling data representation. The key functions of the Presentation Layer include:

  - **Data Translation:** Converts data formats, such as from ASCII to EBCDIC or between data serialization formats like XML, JSON, or protobuf.
  - **Encryption/Decryption:** Conceptually secures data by encoding and decoding it, although this function is often implemented at the Application or Transport Layer in modern systems.
  - **Compression:** Reduces the size of data to optimize transmission speed and bandwidth usage.

- **Layer 7: Application Layer:** The Application Layer is the topmost layer of the OSI model and provides network services that enable software applications to interact with the network. It facilitates communication between applications and the underlying network by providing protocols and services that support end-user tasks, such as email communication, web browsing, and file transfers. The key functions of the Application Layer include:

  - **Protocol Support:** Provides protocols that enable applications to access network resources, such as HTTP for web browsing, SMTP for email, and FTP for file transfers.
  - **Application Services:** Supports functionalities such as file sharing, email transmission, domain name resolution, and web browsing.

5.4 Wireless Communication Protocols

This section discusses protocols specifically designed for wireless communication environments. These protocols often span multiple layers of the OSI model to ensure reliable, secure, and efficient communication over wireless media. Examples include Bluetooth, Zigbee, Wi-Fi, and 5G.

- **Bluetooth Low Energy (BLE):** Bluetooth Low Energy (BLE) is a wireless communication technology optimized for short-range, low-power applications. It is widely used in devices such as fitness trackers, wearables, and medical monitoring systems, where minimal power consumption is essential. BLE operates within a range of 10 to 100 m for BLE 4.x versions, while BLE 5.x extends the range up to 400 m with reduced bandwidth. One of BLE's primary advantages is its extremely low power consumption, allowing devices to function for months or even years on small batteries. In terms of data transfer, BLE 4.x supports a bandwidth of 1 megabit per second, while BLE 5.x improves the rate to 2 megabits per second. Different versions of BLE offer various enhancements. BLE 4.0 and 4.1 provide basic low-power, low-bandwidth communication. BLE 4.2 introduced IPv6 support and improved security features, including LE Secure Connections. BLE 5.x further enhanced the technology by introducing long-range communication modes, higher data transfer rates, and improved advertising mechanisms, making it more suitable for IoT applications. Security in BLE is ensured through AES-128 encryption and multiple pairing options, such as Just Works and Passkey authentication, providing secure communication for connected devices. From an OSI model perspective, BLE defines the physical layer and the data link layer. The physical layer is based on the IEEE 802.15.1 standard, which manages radio communication and modulation. The data link layer is responsible for medium access control and link-layer encryption, ensuring reliable low-power communication. The network layer and higher layers are not explicitly defined by BLE and depend on external implementations. BLE devices commonly use the Generic Attribute Profile (GATT) and the Attribute Protocol (ATT) at higher layers to enable efficient data exchange in IoT applications. BLE is widely used in applications requiring short-range, energy-efficient wireless connectivity, including smart home automation, medical devices, and industrial monitoring systems.

- **Zigbee:** Zigbee is a wireless mesh networking protocol designed for low-power devices and short-range communication. It operates within a range of approximately 100 m, with the capability to extend coverage through mesh networking. The protocol is optimized for minimal power consumption, making it well-suited for battery-operated devices. With a data transfer rate of 250 kilobits per second, Zigbee provides efficient communication for applications that do not require high bandwidth. Security is ensured through AES-128 encryption, offering a secure data transfer mechanism. From an OSI model perspective, Zigbee defines the physical layer, the data link layer, and the network layer. The physical layer is based on the IEEE 802.15.4 standard, handling radio communication and modulation. The data link layer includes the medium access control (MAC) sublayer, which manages channel access, and the logical link control (LLC) sublayer, which ensures reliable frame delivery. The network layer in Zigbee is responsible for packet routing and addressing within the mesh network, enabling multi-hop communication to extend coverage. The transport layer and higher layers are not explicitly defined by Zigbee but are left flexible for application-specific implementations. Zigbee is widely used in smart home automation, including lighting and HVAC control, as well as in industrial automation for monitoring and control systems.
- **Thread:** Thread is a low-power mesh networking protocol designed for IoT applications, built on the IEEE 802.15.4 standard. It supports IPv6 addressing and routing, making it suitable for modern interconnected systems. Thread is compatible with Zigbee hardware but operates as a distinct protocol. It supports a range

of approximately 30 m per hop in indoor environments, with coverage extendable through mesh networking. Optimized for battery-powered devices, Thread ensures low power consumption and reliable communication, with a data transfer rate of 250 kilobits per second. Security is enforced through AES-128 encryption and secure commissioning, ensuring protected data transmission. From an OSI model perspective, Thread defines the physical layer, the data link layer, and the network layer. The physical layer is based on IEEE 802.15.4, managing radio communication and modulation. The data link layer provides medium access control and frame handling, ensuring efficient communication in low-power environments. The network layer in Thread is fully IPv6-based, allowing for seamless routing and addressing, distinguishing it from other low-power mesh protocols. The transport layer and higher layers are not defined by Thread but rely on standard networking protocols such as UDP, CoAP, or MQTT for data exchange, depending on the application.

- **Wi-Fi:** Wi-Fi is a high-speed wireless networking technology designed for devices requiring continuous connectivity and large bandwidth. It supports a typical indoor range of approximately 30 to 100 m, depending on environmental conditions and interference. Compared to low-power wireless protocols such as Bluetooth Low Energy or Zigbee, Wi-Fi consumes significantly more power, making it less suitable for battery-operated devices. Modern Wi-Fi standards offer high data transfer rates, reaching up to 1 gigabit per second with Wi-Fi 6. Security is ensured through WPA2 or WPA3 encryption protocols, providing robust protection against unauthorized access. From an OSI model perspective, Wi-Fi explicitly defines the physical layer and the data link layer. The physical layer is responsible for radio transmission, modulation, and channel access using technologies such as orthogonal frequency-division multiplexing (OFDM). The data link layer includes both the medium access control (MAC) sublayer, which manages channel contention and access control, and the logical link control (LLC) sublayer, which ensures reliable frame delivery. The network layer and higher layers are not defined by Wi-Fi itself but rely on standard networking protocols such as IP for data routing and transport layer protocols like TCP and UDP for communication. Wi-Fi is widely used in applications such as smart IoT hubs and home automation systems, where high-speed connectivity is essential for seamless communication and data streaming.
- **LoRaWAN:** LoRaWAN is a long-range, low-power wireless protocol specifically designed for IoT applications in both rural and urban environments. It provides extensive coverage, reaching distances of up to 10 to 15 km in rural areas and 2 to 5 km in urban settings, making it suitable for large-scale deployments. LoRaWAN is highly energy-efficient, allowing devices to operate for several years on small batteries, making it ideal for applications where frequent battery replacement is impractical. The protocol supports data rates ranging from 27 to 50 kilobits per second, balancing communication efficiency with low power consumption. Security is ensured through AES-128 encryption, protecting transmitted data against unauthorized access. From an OSI model perspective, LoRaWAN defines the physical layer and the data link layer. The physical layer is based on LoRa modulation, which

enables long-range communication with minimal power consumption. The data link layer manages medium access control, ensuring efficient use of available spectrum while supporting adaptive data rate mechanisms to optimize performance. The network layer and higher layers are not explicitly defined by LoRaWAN.

- **Sigfox:** Sigfox is a proprietary low-power wide-area network technology designed for ultra-narrowband communication with low data rates. It provides extensive coverage, reaching distances of 10 to 50 km in rural areas and 3 to 10 km in urban environments, making it suitable for large-scale IoT deployments. Sigfox is optimized for ultra-low power consumption, allowing devices to operate for extended periods with minimal energy usage. This makes it particularly ideal for applications that require the transmission of small, infrequent messages. The protocol supports data rates of approximately 100 bits per second, prioritizing energy efficiency over high-speed communication. Security is ensured through encrypted payloads, providing data protection during transmission. From an OSI model perspective, Sigfox primarily defines the physical layer and data link layer. The physical layer is responsible for the ultra-narrowband modulation used to transmit data over long distances with minimal power. The data link layer manages message transmission and reception while optimizing energy consumption. The network layer and higher layers are not explicitly defined by Sigfox and depend on external implementations, often relying on cloud-based architectures for data processing. Sigfox is commonly used in applications such as asset tracking, remote monitoring, and smart metering, where low-power, long-range connectivity is essential for operational efficiency.
- **NB-IoT (Narrowband IoT):** Narrowband IoT (NB-IoT) is a low-power wide-area network (LPWAN) protocol that belongs to the LTE family, specifically designed for IoT applications requiring low data rates and extended battery life. It operates within existing cellular infrastructure, providing wide-area coverage of up to several kilometers, making it highly suitable for large-scale deployments. NB-IoT is optimized for minimal power consumption, enabling devices to function for up to 10 years on small batteries, significantly reducing maintenance costs. The protocol supports data rates ranging from 20 to 250 Kbps, ensuring efficient communication for IoT applications that do not require high throughput. Security is based on cellular-grade encryption mechanisms inherited from LTE, ensuring robust protection against cyber threats. NB-IoT primarily defines the physical layer (Layer 1) and data link layer (Layer 2) within the OSI model. It leverages LTE's physical layer for radio transmission and a simplified version of LTE's MAC layer for efficient data transmission. Higher layers, such as the network layer (Layer 3), rely on standard IP-based or non-IP protocols for data transport. NB-IoT is widely used for static IoT sensor applications, such as smart parking meters, utility metering, and environmental monitoring, where long-range, low-power connectivity is essential.
- **LTE-M (Cat-M1):** LTE-M (Cat-M1) is a cellular IoT standard optimized for low-to-medium data rates and mobility, making it suitable for IoT devices requiring real-time or mobile connectivity. It operates within the coverage area of LTE networks, providing wide-area connectivity that extends up to several kilometers. LTE-M is designed for low power consumption, incorporating battery-saving fea-

tures that enable mobile IoT devices to operate efficiently over extended periods. The protocol supports data rates of up to 1 Mbps, offering a balance between energy efficiency and communication speed. Security in LTE-M is based on cellular-grade encryption, following LTE security standards to ensure secure data transmission. The LTE-M communication stack primarily defines the physical layer (Layer 1) and data link layer (Layer 2) of the OSI model, leveraging LTE's radio access technology for reliable communication. At the network layer (Layer 3), LTE-M supports both IP-based and non-IP data delivery, depending on the application. The transport layer (Layer 4) typically employs UDP or TCP for data transmission. The session, presentation, and application layers (Layers 5–7) are determined by the specific IoT application, utilizing protocols such as MQTT and CoAP for lightweight communication. LTE-M is widely used in applications such as fleet tracking, connected vehicles, wearable devices, and industrial IoT, where reliable mobile connectivity, extended coverage, and low power consumption are essential.

- **5G:** 5G is a next-generation wireless communication standard designed to enable high-speed data transfer, ultra-low latency, and massive connectivity for IoT networks. It introduces significant advancements over previous generations, including enhanced mobile broadband, ultra-reliable low-latency communication, and massive machine-type communication, making it suitable for applications such as autonomous vehicles, industrial automation, augmented reality, and smart cities. The range of 5G depends on the frequency band used. The sub-6 GHz spectrum provides coverage of approximately one kilometer with better penetration through obstacles, ensuring more reliable communication. In contrast, millimeter wave technology offers extremely high data speeds but is limited to a range of around 100 m, making it suitable for dense urban environments requiring high-bandwidth connectivity. Power consumption in 5G is higher than low-power wide-area network protocols like Narrowband IoT. However, it is optimized for massive IoT networks by incorporating energy-efficient communication techniques. 5G supports bandwidths of up to 10 gigabits per second in enhanced mobile broadband applications, enabling data-intensive use cases. Security in 5G is based on advanced encryption mechanisms, secure access protocols, and mutual authentication, ensuring robust protection against cyber threats. The introduction of network slicing in 5G enables customized security and performance optimizations based on application requirements. From an OSI model perspective, 5G explicitly defines the physical layer, the data link layer, and the network layer. The physical layer is responsible for radio communication, utilizing orthogonal frequency-division multiplexing-based waveforms for efficient spectrum utilization. The data link layer ensures reliable transmission, incorporating error correction mechanisms and traffic scheduling. The network layer manages both IPv4 and IPv6-based data transmission, supports advanced routing techniques, and enables network slicing for differentiated services. The transport layer and application layer are not defined by 5G. Instead, these layers depend on existing and application-specific protocols. For instance, the transport layer typically employs TCP or UDP, while applications leverage protocols such as HTTP/2, MQTT, and WebRTC for real-time services.

Selecting the most suitable communication solution requires a careful evaluation of application requirements, operational constraints, and environmental factors. Key considerations include the specific use case (e.g., data transfer, real-time monitoring, or control), environmental challenges (e.g., physical obstructions like walls, required range, or device density), and infrastructure limitations (e.g., accessibility for installing antennas or wiring). Additionally, the power characteristics of devices (e.g., battery-powered, energy-harvesting, or mains-powered) and installation feasibility must be taken into account.

5.5 Popular Protocols Across OSI Layers

This section highlights protocols designed for *specific layers of the OSI model*. These protocols provide specialized functionality, often addressing one key task within a single OSI layer.

- **Layers 5, 6, and 7 (Session, Presentation, and Application Layers):**
  - **HTTP:** Hypertext Transfer Protocol (HTTP) is the foundational communication protocol of the internet, facilitating data exchange between web browsers and web servers. As a stateless protocol, HTTP treats each request independently, simplifying communication but requiring additional mechanisms for session management when needed. It supports various request methods, including GET, POST, PUT, DELETE, PATCH, HEAD, and OPTIONS, enabling diverse interactions with web resources. The protocol is highly extensible, enabling the inclusion of metadata in headers for functionalities such as caching, authentication, and content negotiation. A secure variant, HTTPS, incorporates SSL/TLS encryption to ensure data confidentiality and integrity during transmission.
  - **FTP:** File Transfer Protocol (FTP) is a standard network protocol used for transferring files between a client and a server over a network. It supports both binary and text file transfers, making it suitable for various applications requiring efficient data exchange. FTP employs authentication mechanisms that typically require a username and password for access, though it also allows anonymous login for public file repositories. The protocol operates in two modes: active and passive. Active mode establishes a direct connection between the client and server, while passive mode improves compatibility with firewalls and network address translation (NAT) by allowing the client to initiate both control and data connections. Despite its widespread use, FTP lacks inherent encryption, making it susceptible to security risks unless secured variants such as FTPS (FTP Secure) or SFTP (SSH File Transfer Protocol) are used.
  - **AMQP:** The Advanced Message Queuing Protocol (AMQP) is a standardized messaging protocol designed to enable reliable, secure, and interoperable communication in distributed systems. It facilitates message-oriented middleware by supporting message queuing, ensuring that messages are delivered even

if the recipient is temporarily unavailable. AMQP promotes interoperability by allowing communication across different platforms and programming languages, making it a widely adopted solution in heterogeneous environments. A key feature of AMQP is its reliable delivery mechanism, which includes message acknowledgment, message persistence, and guaranteed delivery, ensuring robustness in message exchange. The protocol supports flexible messaging patterns, including publish-subscribe, point-to-point communication, and request-response, making it suitable for diverse use cases. AMQP operates over TCP, ensuring reliable transport, and it includes built-in security features such as authentication, encryption (via TLS), and access control mechanisms to protect data integrity and confidentiality. AMQP is commonly used in enterprise messaging, cloud-based services, financial transactions, IoT systems, and event-driven architectures, where efficient, scalable, and fault-tolerant message queuing is essential for real-time data exchange. Its ability to handle complex messaging workflows makes it a preferred choice for mission-critical applications.

– **MQTT:** The Message Queuing Telemetry Transport (MQTT) is a lightweight messaging protocol optimized for IoT applications and constrained environments. It is designed to operate efficiently in low-bandwidth, high-latency, or unreliable networks, making it well-suited for scenarios where minimal network overhead is required. MQTT follows a publish-subscribe communication model, which decouples message producers (publishers) from message consumers (subscribers), enabling scalable and flexible message distribution. The protocol incorporates three QoS levels to ensure different degrees of message delivery reliability: at most once, where a message is delivered with no guarantee of receipt; at least once, where a message is delivered but may be duplicated; and exactly once, ensuring that a message is delivered only once. This flexibility allows MQTT to balance reliability and efficiency based on application requirements. By default, MQTT operates over TCP, ensuring reliable and ordered message delivery between clients and brokers. Secure implementations, such as MQTT over TLS (MQTTS), provide encryption for enhanced security. A variation called MQTT-SN (MQTT for Sensor Networks) is designed for constrained devices and can use UDP instead of TCP, enabling lightweight communication in low-power environments. Security is not built into the core MQTT protocol, but it can be strengthened through Transport Layer Security (TLS) for encryption, as well as authentication and authorization mechanisms such as username-password authentication, token-based authentication, and access control lists (ACLs). Without these security measures, MQTT is vulnerable to threats such as man-in-the-middle attacks, unauthorized subscriptions, and denial-of-service attacks. MQTT is widely used in IoT applications, including smart home automation, wearable devices, and industrial automation, where low-power, reliable, and asynchronous communication is essential for seamless data exchange. Secure implementations, such as MQTT over TLS (MQTTS) and enterprise-grade brokers with robust access controls, enhance its security and reliability in critical applications.

- **CoAP:** The Constrained Application Protocol (CoAP) is a lightweight RESTful protocol designed for constrained devices and networks, particularly in IoT environments. It is optimized to operate over low-power and lossy networks (LLNs), making it suitable for resource-constrained devices that require efficient communication. CoAP follows a RESTful architecture similar to HTTP, supporting request methods such as GET, POST, PUT, and DELETE, enabling seamless integration with web-based applications. The protocol is designed to minimize overhead, featuring a compact binary header structure and low processing requirements, making it well-suited for embedded systems. By default, CoAP operates over UDP (User Datagram Protocol), reducing latency and communication overhead compared to TCP-based protocols. However, it can also be secured using DTLS (Datagram Transport Layer Security) to provide authentication and encryption. CoAP supports two types of messaging modes: Confirmable (CON) messages, which require an acknowledgment to ensure reliable delivery, and Non-Confirmable (NON) messages, which prioritize speed and efficiency without requiring acknowledgment. CoAP is widely used in smart city applications and IoT deployments, particularly in scenarios where constrained networks demand efficient, scalable, and interoperable communication. Its lightweight nature and support for asynchronous communication make it ideal for applications such as environmental monitoring, smart lighting, industrial automation, and smart agriculture.
- **RTSP:** The Real-Time Streaming Protocol (RTSP) is a network control protocol designed for managing media streaming sessions. It enables clients to control real-time media playback by providing commands such as PLAY, PAUSE, and TEARDOWN, allowing for an interactive streaming experience. RTSP operates at the application layer and is commonly used in conjunction with RTP (Real-time Transport Protocol) and RTCP (Real-time Control Protocol) for media transmission and stream synchronization. A key feature of RTSP is its session-based control over media streams, making it well-suited for low-latency streaming and live media applications. RTSP itself does not handle media transport but provides control mechanisms that facilitate functions such as seeking within media streams, adjusting playback, and managing multiple media streams in synchronization. RTSP typically runs over TCP, ensuring reliable session control. However, it can also work alongside RTP over UDP for media transport, allowing for lower-latency streaming at the cost of reliability. The protocol supports both unicast and multicast streaming, making it versatile for various media applications. RTSP is widely used in video streaming platforms, IP cameras, surveillance systems, and multimedia applications, where real-time control and efficient media delivery are critical. While RTSP allows interactive playback and session management, it primarily functions as a request-response protocol rather than a bidirectional media transport protocol. Its ability to support stream synchronization, interactive playback, and media session control makes it a fundamental protocol for on-demand and live streaming services.
- **HLS:** HTTP Live Streaming (HLS) is a media streaming protocol that delivers video content in small segments over HTTP, enabling adaptive bitrate stream-

ing. Developed by Apple, HLS is widely used for live and on-demand video distribution across various platforms, including web browsers, mobile devices, and smart TVs. The protocol segments video into small chunks, allowing efficient buffering and seamless playback. A key feature of HLS is its adaptive bitrate streaming capability, which dynamically adjusts video quality based on network conditions to ensure a smooth viewing experience. Since HLS operates over HTTP, it is compatible with existing network infrastructure, including firewalls and content delivery networks, ensuring broad accessibility. The protocol's cross-platform compatibility makes it a preferred choice for delivering high-quality video content across diverse devices. HLS is widely adopted in video-on-demand platforms and live streaming services, where efficient content delivery and network adaptability are essential for maintaining playback quality and reducing buffering delays.
- **MPEG-DASH:** Dynamic Adaptive Streaming over HTTP (MPEG-DASH) is an HTTP-based streaming protocol designed to provide adaptive bitrate media delivery across various devices and network conditions. Unlike proprietary solutions such as HLS, MPEG-DASH is an open standard, ensuring platform independence and broader adoption in multimedia streaming applications. The protocol enables efficient media transmission by segmenting video content into smaller chunks, allowing seamless playback while dynamically adjusting the bitrate based on network bandwidth and device capabilities. One of the key advantages of MPEG-DASH is its support for adaptive streaming, which optimizes video quality by selecting the most appropriate bitrate in real time. Additionally, it offers broad compatibility with various audio and video codecs, making it a flexible choice for content providers. As it operates over HTTP, MPEG-DASH is compatible with standard network infrastructures, facilitating global content delivery through content distribution networks (CDNs). MPEG-DASH is widely used in media streaming services and over-the-top (OTT) platforms, where uninterrupted playback, high-quality video delivery, and cross-platform accessibility are essential for user experience.
- **DNS:** The Domain Name System (DNS) is a hierarchical and distributed system that resolves human-readable domain names (e.g., www.google.com) into numerical IP addresses. Reverse DNS lookups (via PTR records) allow IP addresses to be mapped back to domain names when required. DNS enables seamless access to network resources by mapping domain names to corresponding IP addresses through a structured system involving root servers, top-level domains (TLDs), and authoritative name servers. Caching mechanisms improve efficiency by storing resolved queries, reducing response times and network load. To enhance security, DNS-over-HTTPS (DoH) and DNS-over-TLS (DoT) encrypt DNS queries to prevent eavesdropping and tampering. DoH sends DNS queries over HTTPS, while DoT transmits them over TLS. Standard DNS queries primarily use UDP due to its low latency, while TCP is employed for larger responses and zone transfers. DNS plays a crucial role in any service requiring name-to-IP translation.

  - **TLS/SSL:** Transport Layer Security (TLS) and its predecessor, Secure Sockets Layer (SSL), are cryptographic protocols designed to ensure secure communication between two endpoints by providing encryption, authentication, and data integrity. Operating primarily on top of transport layer protocols such as TCP, TLS/SSL secures data transmission for applications requiring confidentiality and protection against eavesdropping and tampering. While TLS is most commonly used with TCP, a variation called Datagram TLS (DTLS) allows it to function over UDP, enabling secure communication for real-time applications such as VoIP and video conferencing. TLS/SSL employs encryption to protect data exchanged between a client and server, preventing unauthorized access. Authentication is achieved through X.509 digital certificates, which verify the identity of the server and, optionally, the client. Additionally, integrity checks use mechanisms such as Hash-based Message Authentication Codes (HMAC) to detect and prevent unauthorized modifications to transmitted data. TLS/SSL is widely used with application-layer protocols, including HTTP as HTTPS, FTP as FTPS, and email protocols such as SMTP and IMAP as SMTPS and IMAPS.

- **Layer 4 (Transport Layer):**

  - **TCP:** The Transmission Control Protocol (TCP) is a connection-oriented transport layer protocol that ensures reliable and ordered communication between devices. It establishes a connection using a three-way handshake before data transmission and guarantees that all data is delivered accurately and in sequence. TCP provides reliable delivery by implementing acknowledgment mechanisms and retransmitting lost or corrupted packets. While error checking (via checksums) detects corrupted data, TCP relies on retransmission for error correction. Flow control mechanisms regulate data transmission to prevent overwhelming the receiver, while congestion control manages network traffic to prevent excessive data loss due to network congestion. These features make TCP essential for applications requiring accurate and complete data delivery.
  - **UDP:** User Datagram Protocol (UDP) is a lightweight, connectionless transport layer protocol that prioritizes speed and low latency over reliability. Unlike TCP, UDP does not establish a connection before data transmission, making it more efficient for time-sensitive applications. UDP operates without acknowledgment or retransmission mechanisms, reducing overhead and enabling fast data delivery. It supports multicast communication, allowing efficient data distribution to multiple recipients simultaneously. Due to its low-latency nature, UDP is widely used in applications such as video streaming, online gaming, Voice over IP (VoIP), and IoT networks, where real-time performance is prioritized over guaranteed delivery.

- **Layer 3 (Network Layer):**

  - **IP:** Internet Protocol (IP) is a network layer protocol responsible for routing and addressing data packets across interconnected networks. It serves as the foundation of internet communication, enabling data exchange between devices by assigning unique addresses and facilitating packet delivery through a structured routing system. IP exists in two primary versions: IPv4 and IPv6. IPv4 uses a 32-bit addressing scheme and remains widely deployed, although its limited address space necessitated the development of IPv6. IPv6 expands the address space with a 128-bit structure, accommodating a vastly larger number of connected devices. Additionally, IP supports packet fragmentation, allowing large data packets to be divided into smaller fragments for efficient transmission and reassembly at the destination. IP is universally used in internet communication, underpinning data exchange for all networked applications, including web browsing, streaming, and cloud computing.

  - **6LoWPAN:** 6LoWPAN (IPv6 over Low-Power Wireless Personal Area Networks) is a network protocol designed to enable IPv6 communication over resource-constrained wireless networks, particularly those based on IEEE 802.15.4. As a key **adaptation layer**, 6LoWPAN allows low-power, low-data-rate devices to communicate using standard IP, facilitating seamless integration with existing IP-based networks. It introduces header compression techniques to reduce the overhead of IPv6 packets, making them suitable for transmission over networks with limited bandwidth and energy constraints. From an OSI model perspective, 6LoWPAN operates at the adaptation layer between the network layer (Layer 3) and the data link layer (Layer 2). It bridges the gap between IPv6 and the underlying IEEE 802.15.4 MAC and physical layers, ensuring efficient packet fragmentation and reassembly while optimizing routing mechanisms for constrained environments. While originally designed for IEEE 802.15.4-based networks, 6LoWPAN can also be used on top of other low-power wireless technologies, such as BLE, to enable IPv6 communication across diverse IoT ecosystems. 6LoWPAN is widely used in IoT applications, including smart grids, industrial automation, and wireless sensor networks, where scalable connectivity is essential.

- **Layers 1 and 2 (Physical and Data Link Layers):**

  - **IEEE 802.15.4:** IEEE 802.15.4 is a wireless communication standard that defines the physical layer and the MAC) sublayer of the data link layer in the OSI model. It is designed for low-power, low-data-rate, and short-range communication, making it a fundamental technology fo WSNs and IoT applications. At the physical layer, IEEE 802.15.4 specifies radio frequency bands, modulation techniques, transmission power levels, and data rates, supporting multiple frequency bands such as 2.4 GHz, 868 MHz, and 915 MHz. It employs techniques like direct-sequence spread spectrum (DSSS) to enhance signal robustness in low-power environments. The MAC sublayer provides mechanisms for reliable communication, including carrier sense multiple access with collision avoidance (CSMA/CA), frame validation, and acknowledgment-based transmissions. It supports star, peer-to-peer, and mesh topologies, allowing flexibility in network design. However, higher-layer protocols such as Zigbee, Thread, and 6LoWPAN build upon IEEE 802.15.4 to provide network and application-layer functionalities. IEEE 802.15.4 is widely used in IoT, industrial automation, and smart home applications due to its energy-efficient operation and ability to support large-scale, low-power wireless networks.

- **IEEE 802.11:** IEEE 802.11 is a set of wireless networking standards that define the physical layer and the MAC sublayer of the data link layer, enabling WLAN communication. It operates across multiple frequency bands, such as 2.4 GHz, 5 GHz, and 6 GHz, supporting various modulation techniques such as orthogonal frequency-division multiplexing (OFDM) to enhance data transmission efficiency. IEEE 802.11 includes multiple amendments, such as 802.11n, 802.11ac, and 802.11ax (Wi-Fi 6), which introduce improvements in speed, capacity, and spectral efficiency. The MAC sublayer employs carrier sense multiple access with collision avoidance (CSMA/CA) for managing medium access and minimizing interference in shared wireless environments. Security mechanisms such as WPA2 and WPA3 enhance data confidentiality and integrity in wireless transmissions. IEEE 802.11 is the foundation of modern Wi-Fi technology and is widely used in residential, commercial, and industrial applications for high-speed, scalable, and flexible wireless connectivity.
- **Ethernet:** Ethernet, defined by the IEEE 802.3 standard, specifies both the physical layer and the MAC sublayer of the data link layer, enabling wired LAN communication. It supports various transmission media, including twisted-pair copper cables, fiber optics, and, historically, coaxial cables. In early Ethernet networks that used hubs, carrier sense multiple access with collision detection (CSMA/CD) was employed for network access control in half-duplex mode to manage collisions. However, modern Ethernet networks operate in full-duplex mode using network switches, eliminating collisions and making CSMA/CD obsolete. Full-duplex Ethernet allows simultaneous bidirectional communication, enhancing efficiency and performance, especially in high-speed networks. Ethernet is widely used in residential, commercial, and industrial environments, providing reliable, scalable, and high-speed connectivity.

6 Cloud Computing

Cloud computing has revolutionized the way organizations access and manage computing resources, offering on-demand scalability, cost-efficiency, and flexibility. In the context of SCH, cloud computing enables seamless data collection, storage, and analysis, facilitating real-time insights, improved patient care, and enhanced collaboration among healthcare providers. By leveraging cloud-based solutions, SCH integrates diverse technologies such as IoT, Big Data analytics, and AI to deliver smarter healthcare services [2].

6.1 Cloud Computing Essentials: Concepts, Models, and Deployment

Cloud computing has evolved over decades, starting from the **1950s**, when centralized **mainframe computers** allowed multiple users to access computing resources through *time-sharing systems*. This was followed by advancements in **grid computing**, where distributed systems collaborated to solve large-scale problems, and later **virtualization**, which enabled hardware abstraction and efficient resource sharing [2].

Today, cloud computing delivers scalable and flexible access to shared computing resources over the internet. The **National Institute of Standards and Technology (NIST)** provides a widely accepted definition [2]:

> A model for enabling ubiquitous, convenient, on-demand network access to a shared pool of configurable computing resources (e.g., networks, servers, storage, applications, and services) that can be rapidly provisioned and released with minimal management effort or service provider interaction.

6.1.1 Key Characteristics of Cloud Computing

Cloud computing, as defined by NIST, is characterized by **five essential characteristics**, delivered through **three primary service models**, and implemented using **three main deployment models**. The five essential characteristics form the foundation of cloud computing systems, enabling scalable, efficient, and accessible solutions for users:

- **On-Demand Self-Service**: Cloud computing allows users to unilaterally provision and manage computing resources, such as virtual machines, storage, or applications, as needed, without requiring human intervention from the cloud provider. This is enabled through automated systems, intuitive interfaces, and APIs. For example, a developer can instantly provision a virtual server using the *AWS Management Console*, reducing setup times and increasing agility.
- **Broad Network Access**: Cloud resources are available over the network and can be accessed through standard protocols and mechanisms that support heterogeneous platforms, such as laptops, mobile phones, tablets, workstations, and IoT systems. This characteristic enables accessibility across diverse devices, facilitating remote work and global collaboration.
- **Resource Pooling**: Cloud providers pool computing resources to serve multiple consumers using a multi-tenant model. Physical and virtual resources, such as processing power, storage, and network bandwidth, are dynamically allocated and reassigned based on demand. This approach ensures efficient resource utilization while providing location independence, as users generally do not have control or knowledge of the exact physical location of their resources. For example, *Amazon Web Services (AWS)* operates large-scale data centers to pool resources for thousands of customers, reallocating unused capacity efficiently to meet demand.

- **Rapid Elasticity**: Cloud platforms can scale resources up or down dynamically to accommodate changing workload demands. This elasticity ensures organizations can handle spikes in traffic or computing needs without over-provisioning during low-demand periods. For example, an e-commerce platform can scale server capacity during a *Black Friday* sale to manage increased traffic and scale back afterward to reduce costs.
- **Measured Service**: Cloud systems automatically monitor, meter, and report resource usage, allowing for transparency and cost optimization. Customers are billed based on actual consumption of compute power, storage, and bandwidth, following a *pay-as-you-go* model. For example, *Google Cloud Platform* tracks storage usage and bills customers based on the exact volume of data stored per month.

6.1.2 Cloud Computing Service Models

Cloud services are categorized into three primary models, each providing different levels of abstraction, management, and control over computing resources. These models allow users to select the appropriate solution based on their requirements [2].

- **Infrastructure as a Service (IaaS)**: IaaS provides fundamental computing resources, such as virtualized servers, storage, and networking, on a *pay-per-use* basis. Users have control over operating systems, applications, and configurations while the provider manages the underlying physical infrastructure. For example, *Amazon EC2* allows users to deploy virtual servers, and *Google Compute Engine* provides scalable compute power for enterprise workloads.
- **Platform as a Service (PaaS)**: PaaS offers a complete development and deployment platform, including tools, frameworks, and runtime environments, enabling developers to build, test, and deploy applications without managing hardware or software infrastructure. For instance, *Google App Engine* and *Microsoft Azure App Services* provide platforms for deploying web applications, reducing complexity and accelerating development.
- **Software as a Service (SaaS)**: SaaS delivers fully managed software applications over the internet. Users access these applications via web browsers or APIs, eliminating the need for installation, maintenance, or updates. Providers handle infrastructure, software management, and security. Examples include *Google Workspace* for productivity tools, *Salesforce* for customer relationship management, and *Microsoft Office 365* for cloud-based office applications.

Beyond these, recently, additional service models have gained popularity to address specialized needs. Particularly, **Function as a Service (FaaS)** enables serverless computing, allowing developers to execute discrete functions on-demand without the complexity of managing server infrastructure. Similarly, **Backend as a Service (BaaS)**, also known as **Mobile Backend as a Service (MBaaS)**, provides preconfigured backend functionalities such as cloud storage, authentication, and real-time databases, streamlining application development. These emerging platforms cater to diverse use cases, further enhancing the flexibility, scalability, and efficiency of cloud environments [2].

6.1.3 Cloud Deployment Models

Cloud deployment models describe how cloud services are implemented and accessed. NIST outlines three main deployment models to meet varying organizational needs [2].

- **Private Cloud**: A private cloud is dedicated to a single organization and can be hosted on-premises (within the organization's facilities) or externally by a third-party provider. Private clouds offer greater control, security, and compliance, making them ideal for organizations handling sensitive data. For example, a financial institution may deploy a private cloud to meet regulatory requirements for data security and privacy.
- **Public Cloud**: Public cloud services are delivered over the internet and shared among multiple customers. These services are hosted, managed, and maintained by third-party providers, offering high scalability, cost-efficiency, and minimal maintenance for users. For instance, *Amazon Web Services (AWS)*, *Microsoft Azure*, and *Google Cloud Platform (GCP)* provide public cloud solutions for enterprises, startups, and individuals.
- **Hybrid Cloud**: A hybrid cloud integrates private and public cloud environments, allowing data and applications to move seamlessly between them. Organizations can leverage the control of private clouds for sensitive workloads while benefiting from the scalability and cost-efficiency of public clouds for less critical tasks. For example, an enterprise might store sensitive financial data in a private cloud while using public cloud resources for analytics and processing during peak demand.

6.2 Architecture of Cloud Computing

6.2.1 Overview

Cloud computing architecture integrates four fundamental components, enabling the abstraction, pooling, and dynamic allocation of scalable resources across single or multi-cloud environments. These components are [2]:

- **Front-End:** The front-end is the client-facing layer of the cloud system, responsible for facilitating user interactions with cloud services. It includes graphical user interfaces (GUIs), dashboards, and access tools that allow users to request and manage resources on demand. This layer relies on client-side applications or browsers installed on devices like smartphones, laptops, or desktops to interact with the cloud.

- **Back-End:** The back-end, often referred to as the "server-side," forms the foundational architecture and programming that powers the core functionalities of a system. It handles all the behind-the-scenes technologies, including cloud servers, databases, and application programming interfaces (APIs), provided by the cloud service provider (CSP) to seamlessly support and enable front-end operations.
- **Network:** The network serves as the communication backbone that connects the front-end and back-end. It typically involves public internet connections, private intranets, or intercloud systems that enable seamless data transfer. High-performance networks are designed with low latency, high bandwidth, and redundancy to ensure consistent availability and performance. Advanced features such as load balancing optimize traffic distribution, while content delivery networks (CDNs) improve the speed of data delivery to end-users. Software-defined networking (SDN) and network virtualization allow for dynamic reconfiguration of network resources, ensuring agility and scalability to meet changing demands.
- **Cloud-Based Delivery Platform:** Cloud delivery platforms define the service models through which cloud resources are provisioned, including Infrastructure as a Service (IaaS), Platform as a Service (PaaS), and Software as a Service (SaaS).

6.2.2 Cloud Architecture Layers

Cloud architecture can also be conceptualized as a set of interconnected layers, complementing the perspective of the four core components. Each layer performs distinct functions that work together to collectively deliver cloud services, providing an alternative yet cohesive way to understand the architecture. The key layers of cloud architecture include [2]:

- **Hardware:** This foundational layer consists of physical infrastructure such as servers, storage devices, networking equipment, and other hardware components that provide the computational power and connectivity necessary for cloud operations.
- **Virtualization:** Serving as an abstraction layer, virtualization converts physical resources into virtual instances. This enables efficient resource sharing, allowing multiple applications and users to leverage the same underlying hardware seamlessly.
- **Application and Service:** Positioned at the top, this layer manages requests from the user interface and provides various functionalities based on the cloud service model. It includes resource provisioning, development tools, and web-based applications, ensuring a smooth and user-friendly cloud experience.

6.2.3 Back-End Components

The backend of cloud architecture comprises essential components that work in harmony to process client requests, manage resources, and provide secure, efficient services. These include:

- **Applications:** Backend applications are the software or platforms that handle client service requests initiated from the frontend. They coordinate workflows, process data, and deliver required services, serving as the operational core of cloud systems.
- **Cloud Computing Services:** The backend services deliver essential functionality in cloud architecture by managing the accessibility and utilization of cloud-based resources, including storage solutions, application development environments, web services, security systems, and other related capabilities.
- **Analytics:** Analytics services enable advanced data processing and insights generation. Tools for machine learning, data warehousing, and business intelligence empower organizations to derive actionable insights from their data, driving optimization and innovation.
- **Cloud Runtime:** The runtime environment provides the necessary resources (e.g., operating systems, memory, and hardware) to execute services. It mostly leverages virtualization to create multiple isolated environments on the same physical infrastructure, ensuring efficient and reliable task execution.
- **Cloud Storage:** Cloud storage refers to scalable and flexible solutions for storing and managing data required to operate applications. Common types include object storage, block storage, and file storage. These systems are optimized for high performance, availability, and security.
- **Database:** Databases are crucial for storing and managing both structured and unstructured data. Backend cloud databases, such as Amazon RDS, Microsoft Azure SQL Database, and Google Cloud SQL, provide scalable and efficient solutions for handling large-scale transactional and analytical workloads.
- **Infrastructure:** Infrastructure encompasses the physical and virtual resources that power cloud services. This includes hardware components such as CPUs, GPUs, servers, storage devices, and network devices like routers and switches. It also includes the software necessary for virtualization, orchestration, and management of resources.
- **Management Software (Middleware):** Middleware, also known as management software, plays a crucial role in coordinating communication between frontend and backend components in cloud architecture, ensuring seamless service delivery. It facilitates real-time management of resources to meet dynamic user requirements, handling critical tasks such as resource allocation, workload monitoring, application deployment, and disaster recovery. Additionally, middleware extends its capabilities to include usage tracking, data integration, and efficient management of backend operations, making it indispensable for maintaining operational efficiency and delivering responsive cloud services.
- **Security Tools:** Security tools safeguard backend systems from cyber threats, data breaches, and system failures. Key components include virtual firewalls, encryption mechanisms, identity and access management (IAM), regular backups, and disaster recovery solutions. These tools ensure data integrity, compliance, and continuous availability.

6.3 Virtualization and Containerization

Virtualization is a transformative technology that has reshaped the IT landscape and serves as the foundation for modern computing environments, including cloud computing. Its origins trace back to the 1960s, when IBM pioneered techniques to partition mainframe systems into isolated virtual machines, enabling multiple applications to run simultaneously. This groundbreaking innovation set the stage for virtualization's evolution over the decades. Today, virtualization involves creating virtual versions of physical resources—such as servers, storage devices, and networks—abstracting hardware to optimize resource utilization, enhance flexibility, and simplify management [2].

6.3.1 Benefits of Virtualization

Virtualization offers numerous advantages that make it a cornerstone of modern IT infrastructure. Key benefits include:

- **Cost Efficiency:** Virtualization reduces the reliance on dedicated physical servers by enabling a single machine to host multiple virtual environments. This improves hardware utilization, addressing the inefficiency where servers historically operated at just 10–15% capacity. By consolidating workloads onto fewer machines, organizations save significantly on hardware investments and associated maintenance costs.
- **Energy Savings:** With fewer physical servers required, virtualization dramatically lowers energy consumption. The need for powering and cooling underutilized servers is minimized, leading to reduced energy bills and a smaller carbon footprint. This makes virtualization a vital tool for adopting greener and more sustainable IT practices.
- **Streamlined IT Operations:** Virtualization accelerates and simplifies key IT processes. Tasks like deploying servers, creating backups, and recovering data, which once required weeks, can now be completed in a matter of hours or even minutes. This agility empowers IT teams to focus on innovation and strategic initiatives instead of routine maintenance.
- **Scalability and Flexibility:** Virtualization enables organizations to easily scale their IT resources up or down based on demand. New virtual machines can be provisioned quickly without needing additional physical hardware, allowing businesses to adapt to changing workloads with minimal delay.
- **Disaster Recovery and Business Continuity:** Virtualization enhances disaster recovery by enabling faster restoration of systems and data. Virtual machines can be quickly replicated or migrated to alternative servers, ensuring minimal downtime during hardware failures or disasters.
- **Enhanced Security:** Virtualization allows for better isolation of applications and workloads. Segmenting systems into separate virtual environments reduces the risk of cross-contamination and minimizes the impact of potential security breaches.

- **Optimized Development and Testing:** Developers and testers can quickly create and manage isolated environments for their applications, making it easier to test software across different operating systems and configurations without requiring multiple physical machines.

6.3.2 Types of Virtualization

Virtualization can be categorized into three primary types, each addressing specific IT resources [2]:

- **Network Virtualization:** Network virtualization integrates physical networking components—such as switches, routers, and cables—into a unified, software-defined system. This allows bandwidth to be divided into multiple logical channels, dynamically assigned to devices or applications as needed. By abstracting the underlying hardware, network virtualization improves scalability, enhances bandwidth allocation, and simplifies network management.
- **Storage Virtualization:** Storage virtualization consolidates multiple physical storage devices into a single logical storage system. This abstraction simplifies management, centralizes resource provisioning, and enhances scalability. Organizations benefit from streamlined backups, improved data redundancy, and better overall storage utilization. Common implementations include storage area networks (SANs) and virtual SANs.
- **Server Virtualization:** Server virtualization abstracts a physical server into multiple virtual servers, each capable of running its own operating system and applications. A hypervisor is used to manage these virtual environments, maximizing resource utilization, reducing hardware dependencies, and increasing operational flexibility. This is the most widely adopted form of virtualization, serving as a foundation for many IT infrastructures.

6.3.3 Server Virtualization Approaches

Server virtualization can be implemented through three main approaches [2]:

- **System Virtualization (Virtual Machine):** This approach partitions a physical server into multiple virtual machines (VMs), each with its own operating system and applications. A *hypervisor* (e.g., VMware ESXi) manages these VMs and provides strong isolation between them. This ensures that a failure or compromise in one VM does not impact others. For example, a physical server running Windows can host a Linux VM. Virtual machine-based virtualization is versatile, enabling organizations to run diverse workloads on a single physical server while maintaining high levels of security and stability. It is widely used in enterprise IT environments and forms the backbone of many private and public cloud platforms.

- **Containerization (Operating System Virtualization):** Containerization creates lightweight, isolated environments—called containers—within a shared operating system kernel. Unlike VMs, containers do not include a full OS stack, which makes them faster and more resource-efficient. Containers bundle applications with their dependencies, ensuring consistent behavior across development, testing, and production environments. Tools like Docker popularized containerization, while orchestration platforms such as Kubernetes allow efficient scaling and management of containerized workloads. Containers are ideal for modern development paradigms like microservices, where each service is deployed independently in its own container. However, containers provide less isolation than VMs, as they share the host OS kernel.
- **Serverless Computing (Function-as-a-Service):** Serverless computing, or *Function-as-a-Service (FaaS)*, is a cloud computing model that allows developers to execute code without provisioning or managing underlying server infrastructure. Instead of deploying entire applications, developers write small, single-purpose functions that are triggered by specific events. For instance, AWS Lambda enables developers to upload code that automatically runs when triggered by HTTP requests, database updates, or other predefined events. Serverless computing scales dynamically, charging only for execution time, making it highly cost-efficient. This approach is ideal for event-driven applications such as IoT data processing or real-time analytics. However, serverless architectures face challenges like cold start latency, execution time limits, and potential vendor lock-in.

6.4 Cloud Storage and Databases

Cloud storage and databases are foundational to cloud computing, offering flexible, scalable, and reliable solutions for handling diverse workloads. These technologies enable efficient storage and management of both structured and unstructured data, meeting the demands of modern applications and enterprises [2].

6.4.1 Cloud Storage

Storage is designed to hold raw data or files in a simple, unstructured format, serving as a repository for diverse data types like documents, images, videos, backups, and more. Depending on the nature of the data and specific use cases, storage is categorized into three main types: File Storage, Block Storage, and Object Storage. Each type has distinct features and applications (See Table 1) [2]:

- **File Storage:** File storage organizes data in a hierarchical structure using directories and subdirectories, similar to traditional file systems on personal computers. This approach is well-suited for scenarios where file-level data organization and

Table 1 Comparison of file storage, block storage, and object storage

Feature	File storage	Block storage	Object storage
Data organization	Hierarchical (files and folders)	Divided into fixed-sized blocks; no hierarchy	Managed as objects with metadata and unique identifiers
Access protocols	NFS, SMB, CIFS	Accessed via OS	HTTP/HTTPS
File system	Built-in (e.g., POSIX-compliant)	Added by the OS/application	No file system; uses unique identifiers and metadata
Scalability	Moderate scalability; limited by hierarchy	Moderate; depends on storage volume and system	Highly scalable; supports vast amounts of unstructured data
Performance	Suitable for shared access and concurrent workflows	Low latency and high IOPS; ideal for performance-sensitive workloads	High throughput; optimized for large-scale data retrieval
Use cases	Collaborative tools, content management, media editing	Transactional databases, virtual machines, HPC workloads	Multimedia storage, backups, data lakes, disaster recovery
Examples	Amazon Elastic File System (EFS), Google Filestore	Amazon Elastic Block Store (EBS), Google Persistent Disks	Amazon Simple Storage Service (S3), Google Cloud Storage
Key advantages	Easy file sharing; POSIX compliance; concurrent access	High speed and low latency; mounted and treated as local storage by OS	Extreme scalability; durability; integration with analytics and big data tools
Key limitations	Limited scalability; slower for high IOPS workloads	Typically attached to a single instance; requires file system configuration and management	Higher latency for small, frequent changes; not ideal for structured data
Best fit for	Legacy applications, file-level data organization	High-performance, low-latency applications	Unstructured data, long-term archiving, analytics workloads

shared access across multiple systems are necessary. For instance, cloud-based development environments, content management systems, and collaborative tools heavily rely on file storage due to its structured and user-friendly format. One notable example of file storage in the cloud is Amazon Elastic File System (EFS). EFS is designed to provide elastic scalability, automatically adjusting capacity to meet changing demands without manual intervention. It uses Network File System (NFS) protocols, ensuring compatibility with existing applications and workflows. Additionally, EFS supports POSIX-compliant interfaces, which maintain standards for compatibility across different operating systems, allowing seamless integration with legacy systems. EFS is particularly advantageous for use cases such as media processing workflows, where multiple users or systems require simultaneous access to shared files. Its ability to handle high levels of concurrency also makes it suitable for containerized applications and real-time collaboration tools, where users need access to shared resources without performance degradation.

- **Block Storage:** Block storage is a type of data storage that divides data into fixed-sized pieces called "blocks." Each block has a unique address and is managed independently. Unlike file storage, which organizes data within a predefined hierarchy of directories and files, block storage does not include an inherent file system. Instead, the operating system or application using the storage determines how the blocks are organized by applying a file system (e.g., NTFS, ext4, or XFS). This design allows block storage to remain flexible and adaptable to a wide range of use cases. The lack of a predefined file system enables block storage to provide low-latency, high-performance data access. This makes it particularly suitable for demanding workloads such as transactional databases, high-performance computing, and virtual machines, where rapid data retrieval and processing are critical. A prominent example of block storage is Amazon Elastic Block Store (EBS), which provides persistent, block-level storage for Amazon EC2 instances. EBS supports multiple storage options tailored to specific workload requirements, including SSD-backed storage for high-speed performance and HDD-backed storage for cost-effective, large-scale data retention. Additionally, EBS offers advanced features such as snapshots for point-in-time backups and encryption for secure data storage. This architecture enables applications to access storage as if it were a physical hard drive, providing low-latency, high-speed data access. The combination of raw storage independence and configurability makes block storage indispensable in enterprise-grade cloud architectures, where speed, reliability, and scalability are critical.
- **Object Storage:** Object storage takes a different approach by managing data as discrete objects, each containing three components: the data itself, metadata, and a unique identifier. This architecture provides immense scalability and flexibility, making object storage particularly effective for managing unstructured data such as multimedia files, backups, and large datasets. A widely recognized example of object storage is Amazon Simple Storage Service (S3). S3 uses HTTP-based access protocols, which make it easy to integrate with web and mobile applications. Its design ensures seamless scalability, accommodating vast amounts of data without performance bottlenecks. S3 includes features like lifecycle management, which automates tasks such as archiving or deleting data based on predefined policies. This functionality is essential for long-term cost management, especially for organizations handling archival storage, disaster recovery, or data lakes for analytics. The ability to retrieve data efficiently from S3 also supports machine learning workflows, where large datasets are a necessity. Object storage's scalabil-

ity and flexibility make it a natural fit for modern applications, especially those that require global accessibility or need to handle data with varying degrees of usage frequency. Its architecture allows organizations to focus on innovation without worrying about storage limitations.

6.4.2 Databases

Databases are essential for efficiently storing, retrieving, and processing data in cloud and traditional environments. They enable users to interact with data meaningfully using query languages like SQL. Databases are broadly categorized into (1) Relational Databases and (2) Non-Relational Databases (NoSQL), each catering to specific types of data and application requirements [2].

(1) Relational Databases:
Relational databases organize data into structured tables with predefined schemas and relationships between entities. This structured approach makes them highly suitable for applications requiring **ACID compliance** (Atomicity, Consistency, Isolation, Durability) to ensure data integrity and support for complex queries. These databases typically use **SQL (Structured Query Language)** as the standard language for data manipulation and querying. One prominent example is **Amazon Relational Database Service (RDS)**, a fully managed solution supporting popular database engines like MySQL, PostgreSQL, Oracle, and SQL Server. Amazon RDS automates critical administrative tasks such as backups, patch management, and scaling, allowing developers to focus on application logic. It also supports **multi-AZ (Availability Zone)** deployments for high availability and offers **read replicas** to enable horizontal scaling for read-intensive workloads. Relational databases are particularly well-suited for transactional systems, financial applications, and ERP systems where **data consistency and integrity** are critical [2]. RDBMS organizes data into structured components that provide the foundation for data storage and manipulation [2]:

- **Table:** A table is the primary component of an RDBMS, used to store collections of related data. Each table is structured according to a predefined schema, composed of rows (records) and columns (fields), where rows represent individual data entries and columns represent attributes of those entries.
- **Record (Row):** A record, also known as a row, is a horizontal entity in a table. Each record corresponds to a unique entry in the database, holding information about a specific object or entity, with values defined for every column based on the table's schema.
- **Field (Column):** A field, or column, is a vertical entity in a table that contains a specific type of data. Each field is defined with a data type and may include constraints (e.g., `NOT NULL`, `UNIQUE`), ensuring consistent and valid data for all records in the table.
- **Key:** Keys are essential for establishing relationships between tables and ensuring data integrity. A primary key uniquely identifies each row in a table, ensuring that

no two rows share the same identifier. Foreign keys link related tables, enabling relational integrity. Other types of keys, such as composite keys (combining multiple fields) and candidate keys, may also be used depending on the database design.

As an example, consider the following `Users` table in an RDBMS. In this example, the `Users` table stores related data about users. Each row (record) represents a unique user: `Bahar` and `Farshad`. Each column (field) represents an attribute: `ID`, `Name`, `Email`. The `ID` column serves as the **Primary Key**, uniquely identifying each user.

ID	Name	Email
1	Bahar	bahar@email.com
2	Farshad	farshad@email.com

Normalization: To optimize data storage and minimize redundancy, RDBMS employs a process known as normalization. This ensures that data is stored logically, eliminating unnecessary duplication and enforcing dependencies [2]. Normalization is the systematic organization of data in a database to reduce redundancy and enhance efficiency while ensuring data integrity. The main goals of normalization include:

1. *Remove Redundancy:* Avoid duplicate data across tables to save space and maintain consistency.
2. *Enforce Dependencies:* Ensure that all relationships between data elements are properly defined.

Normalization is a stepwise process designed to organize a database efficiently, minimizing redundancy and ensuring logical data storage. This process is structured into various stages, known as normal forms, each building upon the previous to progressively refine the database structure. Each normal form addresses specific aspects of database organization, enhancing clarity and consistency in data relationships [2]:

- **1st Normal Form (1NF):** This foundational stage ensures that tables have no repeating groups or arrays, enforcing atomicity by requiring each column to contain atomic (indivisible) values, and ensuring each row is unique.
- **2nd Normal Form (2NF):** Building on 1NF, this stage eliminates partial dependencies by ensuring that all non-key columns are fully functionally dependent on the entire primary key (for composite keys).
- **3rd Normal Form (3NF):** Extending 2NF, this stage removes transitive dependencies, ensuring that non-key attributes depend only on the primary key and not indirectly through other non-key attributes.

ACID: While normalization ensures static data integrity, ACID (Atomicity, Consistency, Isolation, Durability) compliance governs the dynamic reliability of transactions. A transaction is a set of operations performed as a single logical unit, and ACID properties ensure the accuracy and integrity of these operations. The key components of ACID are [2]:

- **Atomicity:** Guarantees that a transaction is executed fully or not at all. Partial execution is not tolerated, ensuring that all database operations succeed or the entire transaction is rolled back.
- **Consistency:** Maintains the database's stable state before and after a transaction, ensuring that operations do not violate data integrity.
- **Isolation:** Ensures that concurrent transactions do not interfere with each other, behaving as if each transaction were executed independently.
- **Durability:** Ensures that once a transaction is committed, the changes it made are permanent, even in the event of a system failure.

ACID compliance is crucial for applications requiring reliable and consistent data handling, such as financial systems, e-commerce platforms, and real-time monitoring.

(2) Non-Relational Databases (NoSQL):
The term **NoSQL** originally referred to "non-SQL" or "non-relational" databases, highlighting their departure from the tabular data model used in traditional relational databases. NoSQL databases are designed to handle semi-structured or unstructured data and are built to store and retrieve data in ways that differ from relational systems, utilizing alternative models such as key-value pairs, documents, graphs, or wide-column stores. These databases are particularly well-suited for managing data that does not conform to the rigid structure of relational databases. The term **NoSQL** has since evolved to also mean **"Not Only SQL"**, underscoring that many NoSQL systems can support SQL-like query languages or coexist with relational databases [2].

Unlike traditional relational databases, NoSQL systems excel in dynamic and high-throughput workloads, making them ideal for use cases such as **IoT**, **real-time analytics**, and **content management**. For example, **Amazon DynamoDB** is a fully managed NoSQL database that supports key-value and document-based data models. It delivers *single-digit millisecond latency* at scale and features a distributed architecture that ensures high availability. Additionally, DynamoDB Streams enable real-time event processing, making it particularly suitable for applications with demanding read/write operations, such as **mobile applications**, **gaming leaderboards**, and **IoT telemetry**. Non-relational databases are highly adaptable to modern workloads, especially in scenarios where traditional relational models may impose limitations on scalability or performance. There are four primary types of NoSQL databases, each tailored to specific data requirements [2]:

- **Graph Database:** Graph databases are designed to store and query data that is best represented as a graph, consisting of nodes (entities) and edges (relationships). These databases excel in scenarios where the data has complex and dynamic interconnections, such as social networks, recommendation systems, and fraud detection. They use graph theory to model and process relationships efficiently, making queries involving pathfinding, clustering, and ranking seamless. *Examples:* Neo4j, JanusGraph. **Example of Graph Data:** A social network in a graph database might represent users and their connections as follows:

```
Nodes:
    1. { "id": "123", "name": "Farshad", "age": 30 }
    2. { "id": "124", "name": "Bahar", "age": 25 }
    3. { "id": "125", "name": "Negin", "age": 28 }

Edges:
    1. { "from": "123", "to": "125", "type": "sister", "since": "1995-01-01" }
    2. { "from": "123", "to": "124", "type": "friend", "since": "2011-06-15" }
    3. { "from": "125", "to": "124", "type": "friend", "since": "2024-05-20" }
```


In this example, nodes represent entities, such as users like Farshad, Bahar, and Negin. Edges represent relationships between nodes. Farshad and Negin are siblings (`"type": "sister"`). Bahar and Farshad are friends (`"type": "friend"`). Negin and Bahar are friends (`"type": "friend"`). Additional metadata like `"since"` indicates when the relationships began. Queries might involve finding the type of relationship between two people or identifying all friends of Negin.

- **Key-Value Store:** Key-value stores are among the simplest and most efficient types of NoSQL databases. They store data as key-value pairs, where each key is unique and associated with a value that can be a string, object, or more complex data. This schema-less structure makes them highly flexible and ideal for caching, session management, and simple configuration storage. They are optimized for fast lookups and retrievals, ensuring high performance even at scale. *Examples:* DynamoDB, Redis, Azure Table, BerkeleyDB. **Example of Key-Value Data:**

Key	Value
User:123	"Name": "Bahar", "Age": 25
User:456	"Name": "Farshad", "Age": 30
Session:ABC123	"LoginTime": "2023-01-01T10:00:00"
Config:Theme	"DarkMode"

In this example, each **key** (e.g., `User:123`, `Session:ABC123`) is unique and associated with a **value**. The **value** can be simple (e.g., `"DarkMode"`) or complex (e.g., a JSON object like `"Name": "Bahar", "Age": 25`). Key-value pairs are highly flexible and require no predefined schema.

- **Column Store (Wide-Column Stores):** Column stores organize data into columns rather than rows, optimizing for analytical queries that involve filtering, aggregation, or accessing specific columns of large datasets. In contrast, row-oriented databases (e.g., MySQL, PostgreSQL, and Oracle Database) store records row by row, with all fields of a record stored together. This makes row-oriented databases more efficient for transactional operations like retrieving or updating an entire record. For example, in a table with columns `Name`, `Age`, and `City`, a row-oriented database stores data as `[Farshad, 30, Tempe], [Bahar, 25, LA]`, while a column-oriented database stores `[Farshad, Bahar], [30, 25], [Tempe, LA]`. This columnar organization reduces the overhead of reading irrelevant data, enabling faster filtering, aggregation, and analytical operations,

especially on large datasets. Analytics-oriented column stores (e.g., Amazon Redshift, Google BigQuery, and ClickHouse) are commonly used for data warehousing and business intelligence tasks.
While both column stores and wide-column stores organize data around columns, their implementation and purpose are distinct; column stores store data contiguously on disk for analytical queries, whereas wide-column stores use a sparse, distributed key-value storage model optimized for high-throughput transactional workloads.

- **Column Stores** are typically used for analytical workloads. They store columns contiguously on disk, allowing for high-performance filtering and aggregation over large datasets. These systems are optimized for queries requiring access to specific columns rather than entire rows.
- **Wide-Column Stores** are designed for distributed, high-throughput transactional workloads. Unlike column stores, wide-column stores use a sparse key-value storage model, where each row (or key) can contain a dynamic set of column-value pairs. This design is highly flexible and scalable, making it suitable for distributed systems handling massive datasets.

As an example, consider a table storing user activity logs:

- A **Column Store** (e.g., Amazon Redshift) would store all timestamps in one column, all user IDs in another, and all activities in a third, making it efficient for analytical queries like "count the number of login activities during a specific time range."
- A **Wide-Column Store** (e.g., Cassandra) would represent the same data as a sparse, distributed structure, where each row key is a user ID, and the columns are dynamically defined key-value pairs representing activities over time. Unlike column stores, which store each column contiguously on disk for analytical purposes, wide-column stores organize data as follows:

```
UserID: 123
{
    "2023-01-01 10:00:00": "login",
    "2023-01-01 10:15:00": "browse"
}

UserID: 456
{
    "2023-01-01 11:00:00": "purchase",
    "2023-01-01 11:30:00": "logout"
}
```

Here, each timestamp (e.g., '2023-01-01 10:00:00') is a column, and its associated activity (e.g., 'login') is the value stored in that column. This structure highlights that each row (identified by the row key, such as UserID) contains its own set of timestamped columns with corresponding values. This design allows for fast writes and efficient retrievals in distributed workloads but is less suited for complex analytical queries compared to column stores.

- **Document Database:** Document databases expand on the concept of key-value stores by organizing data into documents, often in formats like JSON, BSON, or XML. Each document contains a unique key and stores data in a semi-structured manner, which can include nested elements and arrays. This makes them ideal for applications requiring flexible schemas, such as content management systems, e-commerce catalogs, and real-time analytics. Document databases are highly adaptable and allow developers to store data in a way that closely matches their application's object model. *Examples:* MongoDB, CouchDB. **Example of Document Data:** A `Users` collection in a document database might store data like this:

```
{
    "_id": "123",
    "name": "Farshad",
    "age": 30,
    "contact": {
        "email": "farshad@example.com",
        "phone": "123-456-7890"
    },
    "hobbies": ["Reading", "Traveling"]
},
{
    "_id": "124",
    "name": "Bahar",
    "age": 25,
    "contact": {
        "email": "bahar@example.com",
        "phone": "098-765-4321"
    },
    "hobbies": ["Photography", "Traveling"]
}
```

In this example, each document represents a user, uniquely identified by the `"_id"` field (e.g., `"123"` for Farshad, `"124"` for Bahar). Fields like `"name"` and `"age"` are simple key-value pairs. Nested fields like `"contact"` and arrays like `"hobbies"` demonstrate the flexibility of the document model. The schema is not fixed, so new fields could be added to individual documents without affecting others.

- **Time-Series Database (TSDB):** A TSDB is a specialized type of database designed to store, manage, and query time-stamped data efficiently. Time-series data consists of sequences of data points indexed in time order, with each data point typically representing a measurement, observation, or event associated with a specific time. TSDBs are increasingly recognized as a distinct category due to their specialized use cases, such as IoT, system monitoring, and financial analytics. While traditionally grouped under Key-Value Stores or Column Stores because they often use similar underlying architectures—such as InfluxDB and OpenTSDB—modern classifications frequently treat TSDBs as a fifth category. This is due to their unique focus on time-based indexing, retention policies, and real-time analytics, which distinguish them from other NoSQL databases. Key features of TSDBs include their optimization for time-centric indexing, which ensures efficient storage and retrieval of sequential and historical data. They are designed for high write throughput, capable of handling large volumes of time-stamped data, such as millions of data points per second from IoT devices or system logs. TSDBs also support retention policies and aggregation, allowing automatic summarization and downsampling of older data to manage storage efficiently. Additionally, they excel in real-time analytics, providing capabilities for monitoring, visualization, and alerting. Applications of TSDBs are diverse and include managing IoT and sensor data, such as temperature readings and device states, as well as monitoring system performance metrics for servers, networks, and applications. They are also widely used in financial analytics for storing stock prices and trade data, as well as in energy management for tracking electricity usage and grid health. Examples of time-series databases include InfluxDB, a high-performance open-source TSDB with built-in visualization tools; **Example of Time-Series Data:** A time-series database monitoring temperature readings from IoT sensors might store data like this:

```
Time                 | SensorID | Temperature (°C)
-------------------------------------------------
2023-12-01T10:00Z | Sensor1   | 22.5
2023-12-01T10:00Z | Sensor2   | 21.9
2023-12-01T10:01Z | Sensor1   | 22.7
2023-12-01T10:01Z | Sensor2   | 22.0
```

  In this example, time is the primary index, ensuring that data is ordered sequentially. SensorID identifies the source of the data (e.g., IoT devices monitoring temperature). Temperature is the measurement associated with each time point. Queries might involve retrieving data for specific sensors, analyzing trends over time, or aggregating data (e.g., average temperature) for a given time range.

When Should SQL Versus NoSQL Be Used?

Choosing between SQL (relational) and NoSQL (non-relational) databases depends on the specific needs of the application and the type of data being handled. Both database types offer unique advantages, but understanding the fundamental differences between them is crucial in making the right choice. Below are some general guidelines for when to use each [2].

SQL databases are best suited for structured data with well-defined schemas, where relationships between tables are important. Consider SQL in the following scenarios [2]:

- **Building custom dashboards:** SQL excels in scenarios where structured queries and aggregations are required to create reports and visualizations.
- **Analyzing behavior-related or custom sessions:** SQL is effective when tracking and analyzing user behavior and session data within a structured environment.
- **Quick data retrieval:** SQL databases provide fast access to structured data, especially when data needs to be retrieved quickly and with a predefined schema.
- **Complex queries and joins:** SQL is ideal when the database requires sophisticated queries that involve multiple tables, joins, or relationships.
- **ACID transactions:** SQL databases are designed to handle ACID-compliant transactions, ensuring data integrity and reliability in critical applications.
- **Known and stable schema:** If the structure of the data is well-defined and unlikely to change over time, SQL is the best choice for maintaining consistent data models.

NoSQL databases are designed for flexibility, scalability, and handling large volumes of unstructured or semi-structured data. NoSQL should be used in the following cases [2]:

- **ACID support is not required:** If transactional consistency is less of a concern and the focus is on speed and scalability, NoSQL databases may be a better fit.
- **When traditional RDBMS is insufficient:** NoSQL databases excel when relational models are inadequate for certain use cases, such as with large-scale, distributed systems.
- **Flexible schema requirements:** NoSQL databases are ideal when the data schema is dynamic or likely to change over time, allowing for easier adjustments.
- **Lack of constraints or validation logic:** NoSQL is well-suited for scenarios where data validation and constraints are not a primary concern.
- **Storing data from distributed sources:** NoSQL is often used for aggregating and storing data from a variety of sources, particularly in distributed systems.
- **Temporary or session data:** For storing ephemeral data, such as logs, session information, or user wish lists, NoSQL databases provide an efficient, scalable solution.

The CAP Theorem: Trade-offs in Distributed Systems

The **CAP Theorem** is a fundamental concept in distributed systems that helps system architects choose the appropriate database platform for a data-driven solution. According to the CAP Theorem, a distributed database can satisfy only two out of the following three conditions simultaneously: **Consistency**, **Availability**, and **Partition Tolerance** (CAP). The three key components of the CAP Theorem are defined as follows [2]:

- **Partition Tolerance:** Partition tolerance ensures that the system remains operational despite network failures or delays between nodes. A partition-tolerant system can withstand any level of network partitioning, meaning that even if some nodes are disconnected, the system continues to function. Data is replicated across multiple nodes and networks, allowing the system to remain available during network outages.
- **High Consistency:** In a system with high consistency, all nodes have the same data at the same time. When a "read" operation is performed, it returns the value of the most recent "write" operation, ensuring that all nodes return the same data. A system with high consistency guarantees that any transaction will leave the system in a consistent state. If any part of a transaction encounters an error, the entire transaction is rolled back to maintain consistency.
- **High Availability:** High availability ensures that every request to the system receives a response, regardless of the state of individual nodes. In an available system, every client receives a response (either success or failure) for each request, ensuring the system remains operational 100% of the time, even if some nodes are unavailable.

The **CAP Theorem** suggests that while a distributed system can achieve consistency, availability, and partition tolerance, it can only guarantee two of these properties at the same time. Note that to offer **ACID** properties in a distributed database, the two properties from the **CAP Theorem** that should be satisfied are **Consistency (C)** and **Availability (A)**. System designers must make trade-offs depending on the specific requirements and priorities of their application.

Data Warehouse and Data Lake

A **Data Warehouse (DWH)**, also known as an **Enterprise Data Warehouse (EDW)**, has been the primary method for data integration over the past 30 years. It serves as a centralized repository for structured data, gathered from various disparate systems, and is used for running complex queries and analytical tasks. The **DWH** combines data from various **data silos** within an organization, creating a unified source of truth. For example, different departments like Marketing, Sales, and Accounting/Finance may each have their own isolated databases, which a data warehouse can consolidate to provide holistic insights. Key features of DWHs include [2]:

- **Optimized for Analytics:** Data warehouses are specifically designed to store vast amounts of historical data and support complex queries for business analysis and reporting.
- **Structured Data:** Data warehouses work exclusively with structured data and rely on **ETL** (Extract, Transform, Load) processes to clean and transform raw data into a structured format.
- **Use Case:** They are primarily used for **Online Analytical Processing (OLAP)** and decision support for business leaders, providing aggregated insights for long-term planning.

The key differences between a Data Warehouse and a Relational Database lie in their design purpose, data structure, and how they handle data processing [2]:

- **Data Optimization and Analytics:** While both are relational systems, data warehouses are optimized for **analytical queries** (OLAP), handling large volumes of historical data. In contrast, relational databases focus on **transactional queries** (OLTP), ensuring quick access to current transactional data.
- **Data Structure:** Data warehouses typically use a **denormalized** structure to support faster query performance for analytical tasks. Relational databases, on the other hand, use a **normalized** structure to minimize redundancy and maintain data integrity in operational systems.
- **Data Processing:** Data warehouses consolidate and aggregate **historical data** from multiple sources for reporting and business analysis. Relational databases focus on **real-time data processing** for day-to-day operations, such as order processing or inventory management.

A **Data Lake** addresses the limitations of traditional databases and data warehouses, providing a solution for big data. Unlike a **DWH**, which works only with structured data, a **Data Lake** can store both **structured** and **unstructured** data in its raw form until it's needed. The concept of a data lake is often compared to a large body of water where data streams in from various sources, and users can access it as needed. The Key features of a Data Lake include [2]:

- **Cost-effective Storage:** It allows for the storage of vast amounts of data at a lower cost compared to data warehouses.
- **Supports Diverse Data Types:** It can hold structured, semi-structured, and unstructured data (e.g., images, videos, logs, sensor data).
- **Query and Transformation:** Data lakes store raw data, and queries are processed and transformed when business questions arise, using a **schema-on-read** approach.
- **Flexible for Analytics:** Ideal for machine learning, advanced analytics, and big data processing, as it supports multiple workloads like batch, streaming, and interactive queries.

The key differences between a Data Lake and a Data Warehouse lie in their design purpose, data structure, and how they handle data processing [2]:

- **Data Types:** A data lake stores raw, unprocessed data in diverse formats, including structured, semi-structured, and unstructured data. A data warehouse, on the other hand, stores only structured and processed data optimized for analytics.
- **Data Use:** Data lakes are used for **data exploration** and experimentation, such as machine learning or advanced analytics, while data warehouses are used for **operational reporting** and business intelligence with curated, well-understood datasets.
- **Transaction Support:** Data lakes do not support transactional workloads, while data warehouses can handle some ACID transactions but are primarily designed

for analytical workloads. Relational databases, by contrast, excel at managing high-throughput ACID transactions in operational systems.

- **Schema:** Data lakes use a **schema-on-read** approach, where the schema is applied when the data is accessed. Data warehouses use a **schema-on-write** approach, requiring predefined schemas when data is loaded.

ETL Versus ELT

The process of **data integration** differs significantly between data lakes and data warehouses. In a **data warehouse**, the traditional process follows the **ETL (Extract, Transform, Load)** methodology, where data is extracted, transformed (cleaned, structured), and then loaded into the system for storage and analysis. In contrast, **data lakes** adopt the **ELT (Extract, Load, Transform)** approach, where raw data is extracted and loaded into the lake as-is, and transformation occurs only when the data is queried for analysis [2].

- **ETL** (in Data Warehouses): Data is extracted, cleaned, and transformed before being loaded into the system. This ensures that only structured, clean data is stored, making it ready for analysis.
- **ELT** (in Data Lakes): Data is extracted and loaded into the lake in its raw form. Transformations are performed on-demand, tailored to specific analyses or business needs, enabling the handling of diverse and large datasets.

Hybrid Storage Architecture Approach

One of the key challenges in realizing holistic IoT clouds is effective data storage. In the past, relational databases were commonly used for data storage. However, as enterprises increasingly leverage big data and IoT applications, the need for data persistence has expanded beyond traditional relational models. To address this, a hybrid architecture is required—one that supports multiple types of data repositories tailored to different access patterns and performance requirements. Ideally, the data storage architecture for any IoT environment should incorporate three distinct types of storage repositories (see Fig. 9) [2]:

- **Cold Storage:** Typically implemented using a data lake based on distributed file system technologies, cold storage is designed to handle data that is accessed infrequently. It is well-suited for storing both structured and unstructured data that does not require frequent access or real-time processing.
- **Warm Storage:** This repository, often implemented using data warehouses or relational databases, is intended for storing structured data that is accessed moderately frequently. Warm storage is appropriate for data that needs to be retrieved or analyzed periodically but does not require the high-speed access of hot storage.
- **Hot Storage:** Hot storage is used for data that is accessed very frequently and demands fast retrieval times. This type of storage is best implemented using fast in-memory databases, which provide low-latency access to real-time data needed by applications such as live analytics or monitoring systems.

By combining these different types of storage, a hybrid architecture ensures that IoT systems can efficiently handle diverse data types and access requirements, from cold, archival data to real-time, high-performance data.

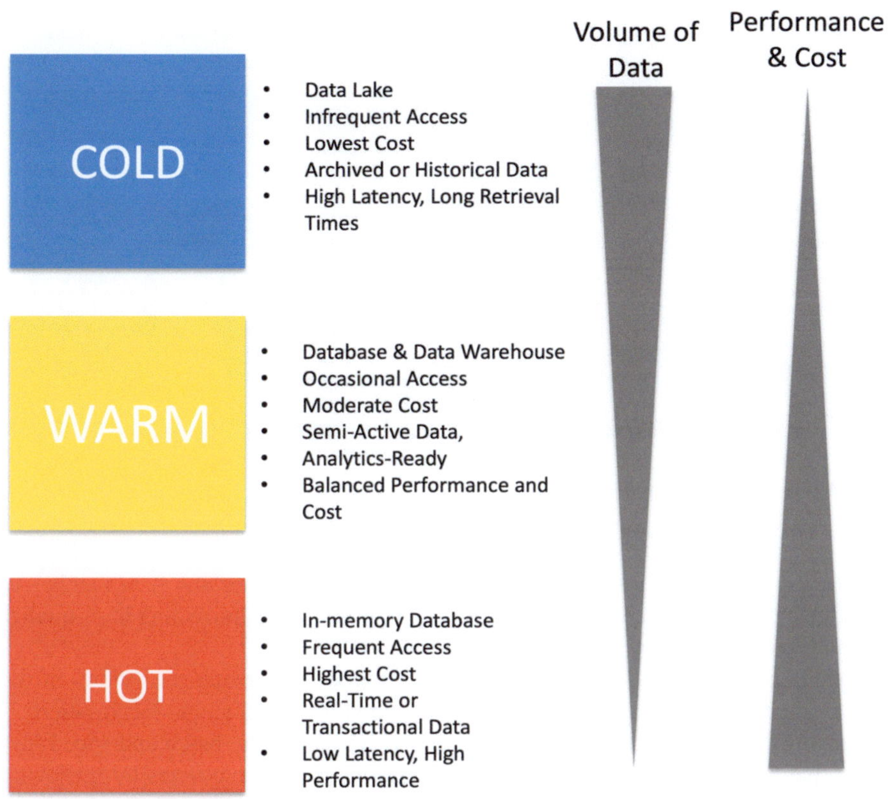

Fig. 9 A hybrid architecture to address SCH data

6.5 Data Processing Architectures in Cloud

This section provides an overview of two prominent data processing architectures: **Lambda Architecture** and **Kappa Architecture** [2].

6.5.1 Lambda Architecture

Lambda Architecture, introduced by Nathan Marz and James Warren in *Big Data: Principles and Best Practices of Scalable Real-Time Data Systems*, is a robust, fault-tolerant, and scalable framework designed to handle vast amounts of data through both batch and stream processing techniques. The architecture comprises three core layers: the **Batch Layer**, **Speed (or Real-Time) Layer**, and **Serving Layer**, as depicted in Fig. 10 [2].

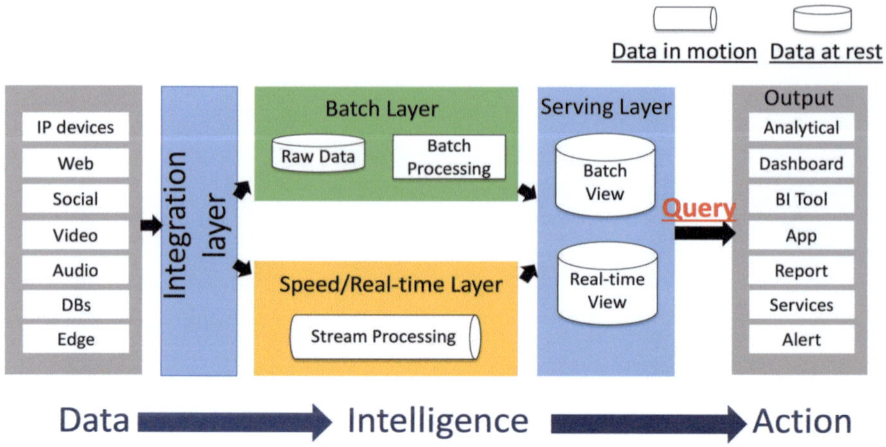

Fig. 10 Lambda architecture [2]

- **Batch Layer:** Responsible for storing raw data and processing it in batches to create batch views. Its scope can span from hours to years, providing high-quality results using complex processing rules.
- **Speed Layer:** Processes data in real-time, producing instantaneous views of the data. It enables quick insights and alerts based on the most recent information.
- **Serving Layer:** Stores and manages outputs from both the batch and speed layers (batch views and real-time views). It facilitates ad hoc queries by retrieving precomputed views or dynamically generating new ones from the stored data. By combining the strengths of batch and real-time processing, the serving layer delivers a comprehensive and holistic data view.

The key benefits of Lambda Architecture include:

- **Real-Time Insights:** The speed layer enables real-time data processing and supports simple machine learning algorithms for immediate insights and alerts.
- **Enhanced Processing Precision:** The batch layer allows for more precise, complex data processing, ensuring no loss of information or insights generated by the speed layer.
- **Optimized Storage Usage:** The batch layer minimizes the need for random write operations, reducing storage requirements.
- **Fault Tolerance and Data Integrity:** Data in the batch layer is immutable, ensuring it is safe from hardware failures or human errors. While the real-time layer is more error-prone, data loss is mitigated as new batches will eventually synchronize the results.

The challenges of Lambda Architecture include:

- **Increased Complexity:** Maintaining synchronization between the batch and speed layers requires significant effort and careful handling.

- **High Maintenance Overhead:** Supporting two distinct layers can complicate maintenance and support operations.
- **Skillset Requirements:** Building and managing Lambda architecture demands expertise in multiple technologies, which can pose recruitment and training challenges.
- **Complicated Implementation:** Deploying Lambda architecture using open-source tools or cloud/on-premises infrastructure can be complex and time-consuming.

6.5.2 Kappa Architecture

Kappa Architecture, proposed by Jay Kreps, focuses exclusively on stream processing, eliminating the batch layer found in Lambda Architecture. This architecture routes all data through a real-time processing layer, with results stored in the serving layer for querying (see Fig. 11) [2]. The central principle of Kappa Architecture is to handle real-time data processing through a single, unified stream processing engine. When updates or code changes occur, the architecture enables reprocessing by replaying the data stream. The updated processing results replace the previous outputs in the serving layer, maintaining consistency. The key benefits of Kappa Architecture include:

- **Simplicity:** By removing the batch layer, Kappa Architecture eliminates the need for managing two separate processing codes (batch and speed). Queries are simplified as they access a single serving location.
- **Streamlined Operations:** With a unified codebase, development and maintenance become less cumbersome compared to the dual-layered Lambda Architecture.

The challenges of Kappa Architecture include:

- **Stream Processing Challenges:** Tasks such as deduplication, cross-referencing events, or ensuring operation order are more complex in streaming compared to batch processing.
- **Limited Optimization:** Access to the entire dataset in the batch layer (as in Lambda) allows for higher performance and simpler algorithms, which are unavailable in Kappa Architecture.

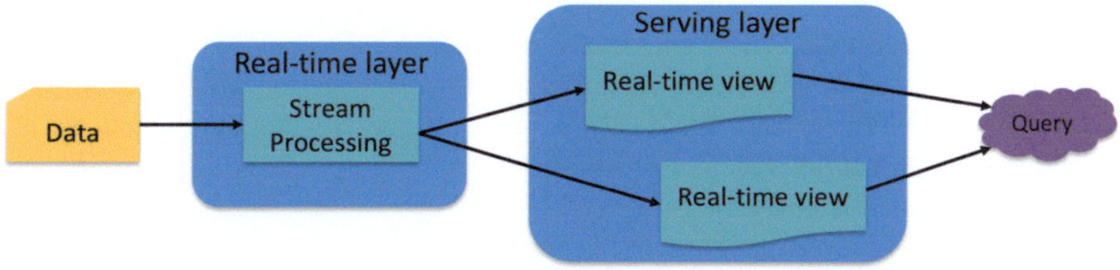

Fig. 11 Kappa architecture [2]

- **Divergent Results:** In certain cases, streaming and batch processing algorithms may produce different outcomes, complicating consistency in sophisticated scenarios.

6.5.3 Additional Data Processing Architectures

Besides the well-known **Lambda** and **Kappa Architectures**, there are several other patterns for data processing that address different challenges and use cases in modern data systems, such as the **Delta Architecture**, **Unified Batch and Stream Processing**, and **Data Mesh Architecture**, among others. Each offers unique benefits and trade-offs depending on the specific needs of an organization.

- **Delta Architecture:** Delta Architecture addresses the complexity of traditional Lambda Architecture by unifying batch and streaming workloads through a **storage-centric design**. At the core of this architecture is **Delta Lake**—an open-source storage layer developed by Databricks that runs on top of existing data lakes such as Amazon S3, Azure Data Lake Storage, or HDFS. Rather than being a standalone storage system, Delta Lake enhances these underlying object stores with key features such as *ACID transactions*, *schema enforcement*, and support for *updates and deletes*. By decoupling storage from compute, it allows various processing engines—including Apache Spark, Databricks, and Flink—to interact with the same reliable, versioned data source. This storage-level unification ensures data quality, consistency, and governance across complex data engineering pipelines.
- **Unified Processing Model (Batch + Streaming):** This approach unifies batch and streaming workloads at the *processing layer* by providing a consistent programming model for both bounded and unbounded data. Frameworks such as **Apache Beam** and **Apache Flink** are leading implementations of this paradigm, while **Kafka Streams** supports stream-first processing with limited batch capabilities. These tools enable developers to define data pipelines once and execute them seamlessly across batch and streaming inputs, using unified APIs and shared semantics. Data streams are often treated as dynamic, infinite tables, and the frameworks natively support advanced features such as *event-time processing*, and *windowing*, making them ideal for real-time analytics and low-latency decision-making. Note that unlike **Delta Architecture**, which unifies batch and stream *at the storage layer* through a transaction-capable data lake (e.g., Delta Lake), Unified Batch and Stream Processing focuses on unification within the data processing logic. It does not inherently provide ACID guarantees or schema enforcement on the underlying data store but instead relies on external storage systems. This separation of concerns allows for greater flexibility in terms of sources and sinks, making the architecture well-suited for diverse real-time applications across event-driven and analytics-driven environments.
- **Data Mesh Architecture:** Data Mesh focuses on decentralizing data ownership and management across teams. Instead of centralizing data pipelines, Data Mesh

enables domain-oriented teams to create and manage their own **data products**. These products are standardized, discoverable, and reusable across the organization. Data Mesh empowers teams to work autonomously while ensuring data governance and quality are enforced through a self-serve infrastructure. It is ideal for large organizations with diverse and siloed data domains, promoting scalability, collaboration, and domain-driven design principles.

6.6 Scaling in Cloud Computing

Modern cloud infrastructures are designed to handle dynamic workloads, often involving thousands or even millions of simultaneous requests from users or devices. Scalability ensures that resources can be dynamically added or removed to meet fluctuating demands, enhancing both performance and cost-efficiency. This adaptability is achieved through two primary strategies: *vertical scaling* (scale-up) and *horizontal scaling* (scale-out). The choice of scaling strategy depends heavily on the target—whether it is a database or an application/service—and the architectural style, workload patterns, and design assumptions of the system [2].

6.6.1 Vertical Scaling (Scale-Up)

Vertical scaling focuses on increasing the capacity of a single machine by adding more resources, such as CPU, RAM, or storage. This approach is often employed when an application/service is designed to operate on a single node and cannot easily distribute its workload across multiple nodes [2]. Common use cases include:

- **Applications/Services**: Tightly coupled applications or services that depend on a shared state or single-instance processing can benefit from vertical scaling. For instance, legacy monolithic systems often require more compute power to handle increased load.
- **Databases**: Vertical scaling is common in relational databases like MySQL or PostgreSQL to improve query performance or transaction throughput.

Design considerations and assumptions for vertical scaling include:

- **Resource Limitations**: Adding resources is constrained by the physical or virtual limits of the underlying hardware.
- **Downtime**: Upgrading a system for vertical scaling may require downtime, impacting availability.
- **Performance Gains**: Performance improvements diminish as the system approaches architectural limits.

Examples of vertical scaling include:

- **Monolithic Application**: A single-instance web application that increases allocated memory and processing power to handle higher request volumes or more complex computations.
- **Database**: A MySQL database server that is vertically scaled by increasing its memory to handle larger queries or by upgrading CPU to support more concurrent transactions.

While vertical scaling can be simpler to implement, it is less flexible for dynamic, real-time scaling demands compared to horizontal scaling.

6.6.2 Horizontal Scaling (Scale-Out)

Horizontal scaling, or scale-out, involves adding more machines (nodes) to a system to distribute workloads across multiple instances. This strategy is suitable for systems designed with scalability and fault tolerance in mind, including stateless applications, distributed databases, and services that can partition their workloads [2]. Variants and use cases of horizontal scaling include:

- **Applications/Services**:
  - **Stateless Systems**: Stateless applications or services, where each request can be handled independently, are ideal candidates for horizontal scaling. For example, web servers handling user requests can distribute workloads across multiple instances behind a load balancer.
  - **Stateful Systems**: Scaling stateful systems requires externalizing the state to a shared resource, such as a distributed cache or database. This adds complexity but enables horizontal scaling.
- **Databases**:
  - **Sharding**: Horizontal scaling for databases often involves sharding, where data is partitioned across multiple nodes based on a specific key (e.g., customer ID or geographic region).
  - **Replication**: Another approach is replication, where data is copied across multiple nodes to improve read scalability.

Design considerations and assumptions for horizontal scaling include:

- **Stateless Versus Stateful**: Stateless systems scale more easily because any node can handle any request. Stateful systems require additional effort to synchronize state or externalize it to shared resources.
- **Fault Tolerance**: Horizontal scaling inherently improves fault tolerance, as workload can be redistributed in case of node failures.

- **Consistency**: For databases, maintaining consistency across nodes in a horizontally scaled environment can be challenging, particularly in distributed systems.

Some examples of horizontal scaling include:

- **Stateless Application**: A web application deployed across multiple nodes with a load balancer distributing requests. The shared state, such as user sessions, is stored in an external cache (e.g., Redis).
- **Microservices**: A microservices-based system where each service (e.g., user authentication or inventory management) scales independently based on its specific workload, managed using a container orchestration platform like Kubernetes.
- **Distributed Database**: Horizontal scaling in databases often involves *sharding*, a technique where data is partitioned into smaller, independent segments, known as *shards*. Each shard is stored on a separate node or server, allowing the system to distribute workloads across multiple machines. Sharding typically involves selecting a *shard key*, such as a customer ID, geographic region, or timestamp, which determines how data is allocated to each shard. This approach ensures that different shards handle separate subsets of data, reducing the load on individual nodes and enabling the database to scale out effectively. While sharding improves scalability and can handle large datasets, it introduces challenges such as:

  - **Query Routing**: Application-level logic or middleware must ensure that queries are directed to the correct shard.
  - **Data Consistency**: Maintaining consistency across shards can be complex, especially in write-heavy environments.
  - **Rebalancing**: As data grows or changes, rebalancing shards to maintain even distribution can be resource-intensive.

  Examples of databases using sharding include:

  - **MySQL**: Sharding can be implemented using external tools like Vitess to manage query routing and shard rebalancing.
  - **MongoDB**: Provides built-in sharding functionality, where collections are partitioned into shards based on a user-defined *shard key*. The shard key determines how data is distributed across the cluster, and MongoDB's query router (`mongos`) ensures that queries are routed to the appropriate shard(s). This is a form of traditional sharding, with explicit data partitioning and routing.
  - **Cassandra**: Uses consistent hashing to automatically partition and distribute data across nodes, which differs from traditional sharding. In traditional sharding, data is divided into predefined shards, requiring manual management of shard boundaries and explicit query routing. By contrast, Cassandra's consistent hashing dynamically assigns data to nodes based on a partition key's hash value, eliminating the need for manual shard management and enabling seamless scaling as nodes are added or removed from the cluster.

6.7 Popular Tools for Cloud Computing

Cloud computing encompasses a vast array of tools and services tailored to various tasks. Below is a subset of popular tools, categorized by their primary use case, to serve as examples of the diverse options available [2].

6.7.1 Storage

Storage tools provide scalable solutions for storing and retrieving data. Below are common subcategories of storage:

Object Storage: A storage architecture optimized for managing unstructured data, where data is stored as discrete objects along with metadata and a unique identifier. It is ideal for storing large volumes of data such as images, videos, backups, and logs, and is commonly used in cloud environments for its scalability and durability.

- **Amazon S3**: A highly scalable object storage service by AWS.
- **MinIO**: An open-source object storage solution compatible with S3 APIs.
- **Google Cloud Storage**: A secure and durable object storage service by Google Cloud.
- **Azure Blob Storage**: A massively scalable object storage service by Microsoft Azure.
- **IBM Cloud Object Storage**: An enterprise-grade object storage solution.

Block Storage: A storage architecture that divides data into fixed-size blocks and stores them as separate pieces, allowing applications to manage data at the block level.

- **Amazon EBS**: A block storage solution for Amazon EC2 instances.
- **OpenStack Cinder**: A block storage service for OpenStack clouds.
- **Azure Disk Storage**: Block storage for Azure virtual machines.
- **Google Persistent Disks**: Reliable block storage for Google Cloud virtual machines.

File Storage: A storage architecture that manages data as files within a hierarchical directory structure.

- **Amazon EFS**: A fully-managed file storage service for EC2.
- **Azure File Storage**: A shared file storage service for Azure.
- **Ceph**: An open-source storage platform with file storage capabilities.
- **Google Filestore**: High-performance file storage for Google Cloud.

6.7.2 Databases

Databases store and manage structured or unstructured data. They can be classified into the following types:
Relational Databases: Store data in structured tables with predefined schemas.

- **MySQL**: A widely-used, open-source relational database management system known for its simplicity, reliability, and flexibility. MySQL is commonly deployed for a variety of applications due to its ease of use and robust ecosystem. MySQL uses a pluggable storage engine architecture, allowing users to choose from multiple storage engines based on their specific needs. The most commonly used engine is **InnoDB**, which provides ACID compliance, transaction support, and row-level locking for high concurrency.
  Scalability and Clustering: MySQL supports moderate scalability through its native replication features.

  - **Replication**: MySQL supports both asynchronous and semi-synchronous replication, allowing for read scalability by distributing read queries across replicas (secondary copies of the database). This setup is well-suited for read-heavy workloads, as multiple replicas can handle concurrent read operations. However, write scalability remains a limitation, as the primary server is solely responsible for processing all write operations. This can become a bottleneck under heavy write loads, and careful management of replication lag is necessary to ensure data consistency across nodes.
  - **MySQL Group Replication**: A built-in high availability solution that provides fault tolerance through synchronous replication across multiple nodes. It supports both single-primary and multi-primary modes, ensuring automatic synchronization and failover within the group. While primarily designed for consistency and fault tolerance, it also offers limited read scalability. Note that MySQL Group Replication itself does not offer direct data sharing or partitioning features. It is primarily a high-availability and fault-tolerance solution.
  - **MySQL InnoDB Cluster (Enterprise)**: A robust, fully integrated clustering solution that combines MySQL Shell, Router, and Group Replication to provide high availability, automated failover, and simplified management.
  - **MySQL Cluster (NDB Cluster)**: A technology providing shared-nothing clustering and auto-sharding for the MySQL database management system. It is designed to deliver high availability and high throughput with low latency, allowing for near-linear scalability. MySQL Cluster is implemented through the NDB (Network Database) storage engine for MySQL.
  - **Third-Party Tools**: MySQL's scalability and clustering capabilities can be enhanced with external tools like **Vitess**, which provides automatic sharding for horizontal scaling, or **Galera Cluster**, which enables synchronous multi-master replication for high availability and strong consistency.

Geospatial Support: MySQL includes native support for geospatial data through **Spatial Data Types** (e.g., `GEOMETRY`, `POINT`, `POLYGON`) and functions compatible with OpenGIS standards. These features allow for basic geospatial queries, such as proximity searches and spatial relationships (e.g., `ST_Distance`, `ST_Within`). However, MySQL's geospatial functionality is limited compared to advanced solutions like PostgreSQL with the PostGIS extension, which offers more comprehensive geospatial analysis, indexing, and topological operations.

- **MariaDB**: An open-source relational database management system that originated as a fork of MySQL, designed to enhance performance, extensibility, and compatibility. MariaDB introduces additional features such as **dynamic columns** for flexible schema design, built-in support for **JSON** and semi-structured data, and advanced storage engines like **Aria** and **ColumnStore**. These enhancements make it suitable for a wide range of use cases, from traditional transactional applications to modern analytical workloads.
- **SQL Server**: A proprietary relational database management system developed by Microsoft. SQL Server provides robust enterprise-grade features, including high availability through Always On Availability Groups, and strong support for geospatial data.
- **PostgreSQL**: A powerful, open-source relational database known for its extensibility, reliability, and advanced features. PostgreSQL supports a wide range of use cases with its rich feature set, including native support for **time-series management** through **time-based partitioning** and **range data types**. Additionally, PostgreSQL excels in handling **geospatial data** with the **PostGIS** extension, providing advanced geospatial capabilities such as spatial indexing, proximity queries, and topological operations, making it a leading choice for geospatial applications. Its strong adherence to SQL standards and support for custom data types, procedural languages, and full-text search further enhance its versatility across transactional and analytical workloads. While PostgreSQL does not natively support sharding, it provides multiple options for clustering and horizontal scaling through external tools and extensions. For instance, for distributed workloads requiring sharding, Citus is a widely-used extension for PostgreSQL that transforms it into a distributed database by automatically partitioning (sharding) data across multiple nodes. itus enables parallel query execution, making it suitable for applications that need high write scalability or large-scale analytical processing. Additionally, PostgreSQL's native partitioning features can be used to manually implement sharding by distributing data across tables in different database instances. Another approach involves using Foreign Data Wrappers (FDWs), which allow PostgreSQL to query and manage data across multiple servers, mimicking a sharding-like setup for distributed reads and writes. PostgreSQL supports clustering for high availability and fault tolerance through features like streaming replication and logical replication. Streaming replication provides synchronous or asynchronous data replication between a primary and one or more replicas, ensuring data durability and enabling read scalability by offloading read queries to replicas. Logical replication offers finer control, allowing selective replication of specific tables or rows,

which can be useful for complex migration or synchronization scenarios. While PostgreSQL lacks built-in sharding capabilities like MySQL's NDB Cluster, its flexibility and ecosystem of extensions make it a robust choice for implementing custom distributed architectures. It remains highly competitive for applications requiring advanced features such as geospatial processing (via the PostGIS extension) or time-series analytics, combining its clustering capabilities with powerful extensions for specialized workloads. Moreover, PostgreSQL's support for JSON (especially JSONB) is more powerful and efficient than MariaDB's. It is ideal for applications that require extensive use of semi-structured data or a mix of relational and NoSQL-style querying.
- **Oracle Database**: A powerful, enterprise-grade relational database designed for mission-critical applications. Oracle Database offers advanced features such as in-memory processing, sharding, and comprehensive clustering options with Real Application Clusters (RAC).
- **Google Cloud SQL**: A fully-managed relational database service offered by Google Cloud. It is compatible with MySQL, PostgreSQL, and SQL Server, providing scalability, automated backups, and seamless integration with other Google Cloud services.
- **Amazon Aurora**: A high-performance, fully-managed relational database service by AWS. Aurora is compatible with MySQL and PostgreSQL, offering features such as fault-tolerant storage, automated scaling, and faster query performance compared to traditional databases.

NoSQL Databases: Flexible schemas optimized for horizontal scaling:

- **Document-Oriented:** MongoDB, Apache CouchDB, and Amazon DynamoDB. MongoDB is a high-performance, document-oriented database that uses BSON for storage, offers strong consistency, supports advanced querying and aggregations with a rich query language, and includes MapReduce as an optional feature for processing large datasets. In contrast, CouchDB emphasizes simplicity, stores data in JSON, relies on eventual consistency, and uses MapReduce as its primary method for querying and creating views.
- **Key-Value Stores:** Examples include Redis (no longer considered open source as of March 2024), Amazon DynamoDB, Valkey, and KeyDB. Valkey, created in March 2024 as an independent initiative inspired by the open-source Redis project, emerged after Redis adopted a non-open-source license. This initiative is led by Madelyn Olson, a prominent figure in the Redis community, who sought to preserve the original open-source nature of Redis. KeyDB is an open-source and high performance fork of Redis with a focus on multithreading, memory efficiency, and high throughput. Amazon DynamoDB is a fully managed, serverless NoSQL database designed for high availability, scalability, and low-latency performance. It supports key-value functionality but extends beyond basic key-value stores by accommodating hierarchical, JSON-like data, effectively blurring the lines between key-value and document stores.
- **Column-Oriented:** Cassandra, HBase. Cassandra and HBase are both column-oriented databases, but they differ significantly in design and use cases. Cassandra

is a distributed, highly available NoSQL database optimized for scalability and fault tolerance, making it ideal for applications requiring high write throughput and multi-region setups. While its read performance is good for query-optimized schemas, it may lag in handling random or ad-hoc queries compared to other databases. In contrast, HBase, built on top of Hadoop, emphasizes strong consistency and seamless integration with the Hadoop ecosystem. While both databases support column-oriented storage, Cassandra is typically favored for operational (OLTP-like) workloads that demand continuous uptime, a distributed architecture, and high scalability. HBase, on the other hand, is better suited for analytical (OLAP-like) workloads, particularly in scenarios where deep integration with Hadoop's big data processing capabilities is crucial.

- **Graph Databases:**
  - **Neo4j**: A graph-based NoSQL database designed to represent and query highly connected data using nodes, relationships, and properties. It supports ACID transactions and uses the Cypher query language for expressive pattern matching.
  - **Amazon Neptune**: A fully managed graph database service optimized for storing and querying highly connected datasets. It supports both property graph and RDF models using SPARQL query language, and is designed for high availability, scalability, and integration with AWS services.
- **Time-Series Databases:**
  - **InfluxDB**: A time-series database optimized for metrics and real-time analytics.
  - **TimescaleDB**: A relational database for time-series data, built on PostgreSQL.
  - **Prometheus**: A time-series database focused on monitoring and alerting.
  - **ClickHouse**: ClickHouse is a high-performance, columnar OLAP (Online Analytical Processing) database primarily designed for analytics, business intelligence, and data warehousing. While not a purpose-built time-series database, it excels at time-series workloads due to its exceptional ingestion speed, advanced compression techniques, and optimized query engine. These features make it a popular choice for observability, telemetry, and event logging use cases.
  - **AWS Timestream**: A managed time-series database service by Amazon Web Services.

Data Warehouses: Designed to efficiently handle large-scale analytics, complex queries, and the consolidation of vast datasets to support decision-making and business intelligence. They can also integrate with various data sources, including operational databases, transactional systems, and external data feeds, to centralize and organize data for comprehensive analysis.

- **ClickHouse**: An open-source, columnar database optimized for real-time analytics and high-speed querying of large datasets. It is designed for high performance, scalability, and efficient storage, making it well-suited for use cases such as data warehousing, event data processing, and business intelligence.

- **Amazon Redshift**: A fully-managed, scalable data warehouse for analytics.
- **Google BigQuery**: A serverless data warehouse for large-scale SQL queries.
- **Snowflake**: A cloud-based data warehouse with high scalability and performance.
- **Azure Synapse Analytics**: A cloud-based data integration and analytics platform.
- **IBM Db2 Warehouse**: An enterprise-grade data warehouse for big data analytics.

6.7.3 Business Intelligence (BI) Tools

BI tools are designed to help organizations analyze data, generate actionable insights, and support strategic decision-making. These tools often integrate seamlessly with data warehouses and other data sources for comprehensive analytics.

- **Tableau**: A leading BI tool known for its powerful visualization capabilities, user-friendly interface, and ability to connect to a wide variety of data sources.
- **Microsoft Power BI**: A widely used BI tool that integrates tightly with Microsoft's ecosystem, including Azure Synapse Analytics. It is also recognized for its cost-effectiveness and frequent updates.
- **Looker**: A modern BI tool by Google, designed for data exploration, advanced analytics, and embedded analytics using LookML for in-depth customization.
- **Qlik Sense**: A self-service BI tool that emphasizes interactive data visualization and exploration, leveraging its unique associative engine for advanced insights.
- **SAP BusinessObjects**: An enterprise-grade BI platform for reporting, analytics, and data visualization, offering deep integration with SAP's ERP solutions.
- **IBM Cognos Analytics**: A BI solution offering reporting, data exploration, AI-driven insights, and natural language processing for intuitive data queries.
- **Metabase**: Metabase is an open-source BI platform designed to facilitate data analysis and visualization through an intuitive interface/dashboard, enabling users to generate insights without requiring advanced technical expertise.

6.7.4 Search and (Log) Analytics Engines

Search and analytics engines are designed for indexing, searching, and analyzing large volumes of structured and unstructured data. They are often used for log analysis, full-text search, and advanced data querying.

- **Elasticsearch**: A distributed search and analytics engine widely used for log and event data, as well as document-oriented storage. It is part of Elastic Stack and supports scalability and near real-time search.
- **OpenSearch**: A community-driven fork of Elasticsearch for search and log analytics.
- **Splunk**: A powerful platform for searching, monitoring, and analyzing machine-generated data, known for its real-time processing capabilities and advanced alerting features.

- **Apache Solr**: An open-source search platform built on Apache Lucene, suitable for full-text search and big data analytics.

6.7.5 Visualization Tools

Visualization tools are integral for analyzing and interpreting data. These tools provide user-friendly interfaces and graphical representations to make data insights more accessible.

- **Kibana**: A visualization tool designed to work seamlessly with Elasticsearch, offering dashboards, visualizations, and management features for the Elastic Stack.
- **Grafana**: A highly flexible visualization tool that supports a variety of data sources, including Elasticsearch, Prometheus, and InfluxDB. It is known for its interactive dashboards and robust monitoring capabilities.
- **OpenSearch Dashboards**: The visualization and user interface component of OpenSearch, allowing users to create interactive visualizations and manage search analytics.
- **Splunk Dashboards**: A built-in feature of Splunk, enabling customizable and dynamic visualizations of machine-generated data.
- **Tableau**: A leading BI tool that can be integrated with search and analytics engines to deliver rich visualizations and actionable insights.
- **Power BI**: A business intelligence platform by Microsoft that can be integrated with search and analytics data sources to create interactive dashboards, detailed reports, and data visualizations.

6.7.6 Data Processing

Data processing tools enable the transformation, enrichment, and analysis of large-scale datasets, supporting both real-time and batch processing to extract actionable insights.

- **Apache Flink**: A real-time stream processing framework for distributed computations, known for its low-latency processing and event-driven applications.
- **Apache Spark**: A unified analytics engine for large-scale data processing, supporting both batch and stream workloads efficiently.
- **Apache Storm**: A distributed real-time computation system for unbounded data streams, ideal for continuous and low-latency processing.
- **Hadoop MapReduce**: A core component of the Hadoop ecosystem, designed for distributed batch data processing using a map-reduce programming model.
- **Apache Beam**: A unified programming model for batch and stream data processing that can run on multiple execution backends, including Flink and Spark.

- **Kafka Streams**: A lightweight library for building real-time stream processing applications on top of Apache Kafka.

6.7.7 Machine Learning (ML) Tools

Machine learning tools enable the development, training, and deployment of models to uncover patterns and make predictions from data. These tools often integrate with data processing and storage systems to streamline workflows.

- **TensorFlow**: An open-source framework by Google for building and deploying machine learning and deep learning models at scale.
- **PyTorch**: A flexible and widely-used open-source deep learning framework developed by Facebook, popular for research and production use.
- **scikit-learn**: A Python library for traditional machine learning, offering a wide range of algorithms and utilities for preprocessing and model evaluation.
- **Azure Machine Learning**: A cloud-based platform by Microsoft for managing and deploying machine learning workflows.
- **Amazon SageMaker**: A managed service by AWS for building, training, and deploying machine learning models.
- **Google Vertex AI**: A unified machine learning platform for building, deploying, and scaling models, fully integrated with Google Cloud.
- **H2O.ai**: An open-source platform for machine learning and AI, providing tools for distributed and automatic machine learning.

6.7.8 Message Brokers

Message brokers facilitate asynchronous communication between distributed systems by routing messages between producers and consumers, enabling decoupled and scalable application architectures.

- **Apache Kafka**: A distributed event-streaming platform for building real-time data pipelines and applications, with robust storage and processing capabilities.
- **RabbitMQ**: A robust and lightweight message broker that supports multiple messaging protocols (e.g., AMQP, MQTT) and communication patterns such as point-to-point and publish-subscribe.
- **ActiveMQ**: An open-source message broker offering reliable and transactional communication for enterprise systems.
- **Redis Streams**: A lightweight in-memory data structure store with support for message streaming and queuing.
- **Amazon SQS**: A fully managed message queuing service by AWS, designed to decouple and scale distributed systems.

- **Google Pub/Sub**: A fully managed messaging service by Google Cloud, ideal for asynchronous communication and event-driven architectures.
- **NATS**: A lightweight, high-performance messaging system designed for simplicity, scalability, and secure communication in distributed environments. NATS supports multiple messaging patterns, including publish-subscribe, request-reply, and streaming, making it versatile for various use cases such as microservices, IoT, and cloud-native applications. With features like NATS JetStream, it provides advanced capabilities such as persistent message storage, at-least-once delivery, and historical replay.
- **MQTT-focused Brokers (e.g., Mosquitto, EMQX)**: Specialized brokers optimized for the MQTT protocol, commonly used in resource-constrained IoT environments for lightweight, low-bandwidth, publish-subscribe messaging.

6.7.9 Data Ingestion and Integration

Data ingestion tools facilitate the collection and transfer of data from diverse sources into storage or processing systems.

- **Apache NiFi**: A robust data integration tool for automating the flow of data between systems in real-time.
- **Logstash**: A data collection tool for processing and forwarding logs and events, commonly used with Elasticsearch.
- **Vector**: A lightweight and high-performance tool for data ingestion, log transformation, and forwarding.
- **Apache Flume**: A distributed system for efficiently collecting, aggregating, and moving large amounts of log data, particularly for Hadoop-based systems.
- **Amazon Kinesis**: A fully managed service for real-time data streaming and ingestion provided by AWS.
- **Google Dataflow**: A fully managed, serverless data processing service by Google Cloud for executing both batch and stream data pipelines. Built on Apache Beam, Dataflow enables complex event processing, real-time analytics, and data ingestion workflows with features like automatic scaling, windowing, and fault-tolerant execution.
- **Snowpipe (Snowflake)**: Snowflake's continuous data ingestion service for loading data automatically into Snowflake from various sources.
- **Confluent Kafka Connect**: A part of the Confluent Platform, it simplifies data integration with pre-built connectors for Apache Kafka.
- **Talend Data Integration**: A comprehensive ETL platform supporting batch and real-time ingestion workflows with built-in connectors.
- **Airbyte**: An open-source data integration tool for building ELT pipelines with customizable connectors.
- **Debezium**: A distributed platform for change data capture (CDC), streaming real-time database changes to downstream systems.

6.7.10 Monitoring, Logging, and Analytics

Monitoring, logging, and analytics tools provide critical insights into system performance, application behavior, user engagement, and overall operational health. These tools help diagnose issues, optimize user experiences, and support data-driven decision-making.

- **Graylog**: A centralized log management platform built on Elasticsearch, designed for efficiently collecting, storing, and analyzing system and application logs. It provides powerful search capabilities, alerting features, and a user-friendly interface for real-time monitoring and troubleshooting.
- **Elastic Stack (ELK)**: A stack comprising Elasticsearch, Logstash, and Kibana for centralized logging and analysis, offering robust log ingestion, storage, and visualization tools.
- **Fluentd**: An open-source log collector and processor designed for unifying log data collection and routing across systems, supporting a wide range of data sources and outputs.
- **Loki (Grafana Labs)**: A lightweight log aggregation system that integrates seamlessly with Grafana, optimized for real-time log query and exploration.
- **Loggly**: A cloud-based log management solution (part of SolarWinds) that provides centralized log aggregation, real-time search, and visualization tools for troubleshooting and analysis.
- **Splunk**: A comprehensive monitoring and log management platform offering real-time data analysis, search, and visualization capabilities.
- **Prometheus**: An open-source tool for metrics collection, alerting, and real-time monitoring, often paired with Grafana for visualization.
- **Nagios**: A powerful and flexible open-source monitoring system for networks, applications, and servers, offering extensive plugin support for tailored monitoring solutions.
- **Zabbix**: An open-source monitoring platform providing comprehensive insights into server, application, and network performance with features like trend prediction and alerting.
- **Datadog**: A cloud-based observability platform offering monitoring, log management, and analytics for applications, infrastructure, and user experience in real-time.
- **AppDynamics**: A robust performance monitoring solution that provides deep insights into application behavior, infrastructure performance, and business transaction analytics.
- **New Relic**: A cloud-based observability platform offering full-stack visibility into application performance, distributed tracing, and log management for troubleshooting and optimization.
- **Sensu**: A full-stack observability and monitoring platform supporting metrics, event processing, and alerting, with integration for modern DevOps workflows.

- **Google Analytics**: A web analytics service that tracks and reports website and app traffic, user behavior, conversion rates, and other key metrics. While not a log management tool, it provides valuable insights for website performance analysis.
- **Firebase Analytics (Google)**: A tool specifically designed for monitoring mobile apps, providing insights into user engagement, retention, and app performance.
- **Hotjar**: A popular tool for understanding user behavior on websites, providing heatmaps, session recordings, and user feedback through surveys and polls. It complements traditional monitoring tools by focusing on user experience insights.

6.7.11 Container Orchestration and Management

Container orchestration tools simplify the deployment, scaling, and management of containerized applications, making them essential for modern microservices architectures:

- **Kubernetes**: A powerful and widely adopted orchestration tool that automates the deployment, scaling, and management of containerized applications. Kubernetes offers a vast ecosystem of tools and community support, making it suitable for complex and large-scale environments.
- **Docker Swarm**: A native orchestration tool by Docker designed for managing container clusters. It prioritizes simplicity and ease of use, offering seamless integration with the Docker ecosystem for smaller-scale setups.
- **Nomad**: A flexible and lightweight orchestration tool capable of managing both containerized and non-containerized workloads. Nomad is known for its simplicity and compatibility across diverse environments.
- **Red Hat OpenShift**: An enterprise-ready Kubernetes-based platform that simplifies container orchestration and includes additional developer-friendly features such as CI/CD pipelines and a user-friendly web console.
- **Rancher**: A complete container management platform that simplifies Kubernetes deployment and management across multiple clusters, with built-in monitoring and access control.
- **AWS Elastic Kubernetes Service (EKS)**: A fully managed Kubernetes service provided by AWS. It simplifies Kubernetes cluster deployment, scaling, and management, integrating seamlessly with other AWS services.
- **Amazon Elastic Container Service (ECS)**: A container management service provided by AWS that supports Docker containers. ECS offers deep integration with the AWS ecosystem and is suitable for users who may not require Kubernetes' complexity.
- **Google Kubernetes Engine (GKE)**: A fully managed Kubernetes service by Google Cloud Platform that simplifies the deployment and scaling of containerized applications.
- **Azure Kubernetes Service (AKS)**: A fully managed Kubernetes service by Microsoft Azure, designed for simplifying Kubernetes cluster setup and management.

- **Portainer**: A lightweight, user-friendly management interface for containerized environments, compatible with Docker, Kubernetes, Nomad, and other container runtimes. Portainer provides an intuitive graphical UI for managing containers, images, volumes, networks, and orchestrator-specific workloads such as Kubernetes deployments. By abstracting the complexity of underlying orchestration systems, Portainer simplifies day-to-day operations, making container management accessible to users of all skill levels. While it does not function as a full orchestration engine, Portainer significantly streamlines cluster administration, access control, and resource monitoring across diverse platforms.

6.7.12 API Management, Documentation, and Testing

API tools address a variety of needs, from designing and documenting APIs to managing, testing, and monitoring them in production environments.

API Development and Testing: These tools assist developers in building, debugging, and validating APIs throughout the development lifecycle.

- **Postman**: A collaborative platform for designing, testing, and automating API workflows. Postman allows developers to execute HTTP requests, monitor API performance, automate testing, and generate interactive API documentation, bridging the gap between development and documentation.

API Documentation and Design:

- **Swagger**: A widely used framework for designing, documenting, and testing APIs. Swagger includes tools like Swagger Editor, Swagger UI, and Swagger Codegen, which simplify API development and documentation.
- **Redoc (Redocly)**: A tool for generating interactive API documentation from OpenAPI specifications, offering a clean and customizable user interface for API consumers.

API Management: These platforms provide end-to-end solutions for securing, deploying, monitoring, and scaling APIs, often with features like rate limiting, authentication, analytics, and developer portals.

- **AWS API Gateway**: A fully managed service by AWS for creating, publishing, and maintaining APIs at scale. It integrates seamlessly with other AWS services and offers features like traffic control, authentication, and monitoring.
- **Apigee (Google Cloud)**: A comprehensive API management platform for designing, securing, and monitoring APIs, suitable for enterprise-level applications.
- **Azure API Management**: A cloud-based API management solution by Microsoft Azure that allows developers to create, publish, secure, and monitor APIs with built-in analytics and insights.
- **Kong Gateway**: An open-source API gateway that enables API management, security, and traffic control with support for plugins and integrations.

6.7.13 Reverse Proxy and Load Balancing

Reverse proxy and load-balancing tools efficiently manage incoming traffic and distribute workloads across multiple servers or applications, ensuring high availability, scalability, and performance:

- **Nginx**: A high-performance reverse proxy, load balancer, and web server for handling HTTP and HTTPS traffic.
- **HAProxy**: A reliable open-source solution for load balancing and proxying TCP and HTTP requests.

6.7.14 Serverless Computing

Serverless computing tools provide Function-as-a-Service (FaaS) capabilities, enabling developers to run code in response to events without managing or provisioning servers. These tools support event-driven architectures and scaling based on demand:

- **AWS Lambda**: A serverless compute service by Amazon Web Services (AWS) for running code triggered by events, such as changes in data, HTTP requests, or scheduled tasks. AWS Lambda automatically scales based on the number of requests and integrates seamlessly with other AWS services.
- **Google Cloud Functions**: A fully managed serverless execution environment provided by Google Cloud. It allows developers to write and deploy event-driven code triggered by HTTP requests, Google Cloud services such as Pub/Sub and Cloud Storage, or external APIs. Google Cloud Functions supports multiple programming languages, including Python, JavaScript, and Go, and integrates seamlessly with the Google Cloud ecosystem.
- **Azure Functions**: A serverless compute service by Microsoft Azure that enables developers to execute code triggered by HTTP requests, timers, or events from Azure services like Event Grid and Service Bus. It supports a wide range of programming languages and offers robust integration with the Azure ecosystem.
- **IBM Cloud Functions**: A FaaS platform built on Apache OpenWhisk, allowing developers to run event-driven code with integration to IBM Cloud services and other APIs.
- **OpenFaaS**: An open-source serverless framework for running functions on Kubernetes or Docker Swarm. OpenFaaS provides flexibility in hosting serverless workloads within self-managed environments.

6.7.15 Communication and Messaging Platforms

Communication platforms provide APIs and services for integrating messaging, voice, video, email, and push notification functionalities into applications.

Messaging and Communication: These platforms offer APIs and services for embedding real-time communication features such as SMS, voice, and video into applications.

- **Twilio**: A cloud-based CPaaS platform offering APIs for SMS, voice calls, video conferencing, and email.
- **Vonage**: A cloud communications platform providing APIs for SMS, voice, and video services.
- **Plivo**: A communication platform offering APIs for SMS and voice calls.
- **RingCentral**: A cloud communications and collaboration platform for VoIP and video meetings.

Email Delivery: These services provide APIs and infrastructure for sending transactional, marketing, and bulk emails with high deliverability and analytics capabilities.

- **SendGrid**: An email delivery platform for transactional and marketing emails, now part of Twilio.
- **Amazon SES (Simple Email Service)**: A cost-effective email sending service for transactional and bulk emails.
- **Postmark**: An email delivery platform focused on transactional email.
- **Mailgun**: A platform for sending, receiving, and tracking emails via APIs.

Push Notifications: These platforms enable real-time communication with users by delivering targeted messages to mobile applications, enhancing engagement and user retention.

- **Firebase Cloud Messaging (FCM)**: A cross-platform messaging service for sending push notifications to mobile and web applications.
- **AWS SNS (Simple Notification Service)**: A fully managed messaging service by Amazon that enables sending notifications to various endpoints such as mobile devices, email, and SMS. SNS supports cross-platform push notifications for platforms like Apple Push Notification Service (APNs) for iOS, Firebase Cloud Messaging (FCM) for Android, and other services such as Amazon Device Messaging (ADM).
- **OneSignal**: A push notification service supporting mobile, web, and email notifications.
- **Pusher Beams**: A scalable service for delivering push notifications to mobile and web applications.
- **Airship**: A platform for sending personalized push notifications and in-app messages.

7 Software Architecture

Software architecture refers to the fundamental structure and high-level organization of a software system. It defines how components are arranged, how they interact, how data flows through the system, and how both functional and non-functional requirements are addressed [4]. Architectural decisions involve selecting specific structural options from a range of design possibilities, and these choices are often difficult and expensive to reverse once implemented. As such, software architecture represents a critical and foundational phase in the software development lifecycle. A well-crafted architecture acts as a blueprint for guiding implementation, promoting scalability, ensuring performance, and supporting long-term maintainability. It encompasses four key dimensions [4]:

- **Architecture Style:** The overarching structural model that determines how the system is organized and deployed. It establishes the foundational layout for component interactions and communication. Common styles include monolithic, microservices, layered, and event-driven architectures.
- **Architecture Characteristics:** Non-functional characteristics—commonly referred to as "-ilities"—describe how a system behaves under various conditions and inform critical architectural decisions. The following are some common examples of architectural characteristics [4]:
  - **Scalability**: The system's ability to handle increased workload or users by expanding resources without degrading performance.
  - **Elasticity**: The ability of the system to automatically scale resources up or down based on real-time demand.
  - **Efficiency**: The system's optimal use of computational and operational resources to achieve desired outcomes with minimal waste.
  - **Performance**: The system's ability to meet defined response time, throughput, and latency targets under expected workloads.
  - **Reliability**: The system's ability to consistently perform its intended functions without failure over time.
  - **Availability**: The percentage of time the system remains operational and accessible when needed.
  - **Resilience**: The capability of the system to recover from failures and continue operating, possibly in a degraded state.
  - **Recoverability**: How quickly the system can restore normal operations after a disruption or failure.
  - **Security**: The system's protection of data, functionality, and communication against unauthorized access and attacks.
  - **Authentication**: Verifying that users or systems are who they claim to be.
  - **Authorization**: Ensuring users or systems have permission to access specific resources or actions.
  - **Privacy**: The system's ability to safeguard user data from internal and external exposure.

  - **Compliance**: The system's adherence to legal, regulatory, and policy-based requirements (e.g., GDPR, HIPAA).
  - **Modularity**: The degree to which the system is composed of independent, interchangeable components with well-defined interfaces.
  - **Maintainability**: The ease with which the system can be corrected, adapted, or enhanced throughout its lifecycle.
  - **Extensibility**: The ease of adding new features or functionality without major changes to the existing architecture.
  - **Configurability**: How easily users or administrators can modify system behavior without altering the code.
  - **Upgradeability**: The ease with which the system can be updated or upgraded to newer versions without disruption.
  - **Interoperability**: The ability of the system to effectively communicate and operate with other external systems.
  - **Portability**: The ease with which the system can be deployed or moved across various environments and platforms.
  - **Testability**: How effectively the system can be validated through automated or manual testing methods.
  - **Auditability**: The system's ability to log and trace activities for compliance, debugging, and operational insight.
  - **Supportability**: The ease with which the system can be monitored, diagnosed, and maintained during operations.
  - **Usability**: The degree to which users can easily and efficiently interact with the system to achieve their goals.
  - **Accessibility**: How well the system accommodates users with disabilities or special needs.

- **Logical Components:** The core elements that define the system's behavior and business logic, including domains, entities, services, and workflows. These components reflect how the system fulfills its functional requirements [4].
- **Architecture Decisions:** The strategic design choices and guiding principles that shape the system's construction. These decisions impose constraints on how components may interact or access shared resources, effectively setting boundaries for development teams and ensuring architectural consistency. They also capture the rationale behind trade-offs made in the context of project-specific goals and limitations [4].

Together, these four dimensions form a comprehensive framework that enables the creation of robust, adaptable, and maintainable software systems while aligning technical solutions with business objectives. However, note that Software architecture is inherently contextual and governed by two fundamental principles [4]:

- *Everything is a trade-off—architectural decisions must balance competing concerns, often with no ideal solution.*
- *Why is more important than how—understanding the rationale behind decisions is more valuable than simply understanding their implementation.*

To effectively communicate and analyze software architecture, various modeling techniques are employed. One widely adopted method is the **C4 model**, which provides a hierarchical framework for visualizing software systems at different levels of abstraction. The C4 model structures architectural views into four layers [5]:

- **Context Diagrams (Level 1):** Depict the system in scope, showing how it interacts with external users and other systems.
- **Container Diagrams (Level 2):** Break down the system into containers, which represent applications, services, or data stores that execute the system's functionality.
- **Component Diagrams (Level 3):** Further decompose containers into individual components, illustrating internal structure and inter-component relationships.
- **Code Diagrams (Level 4):** Provide detailed, code-level representations using existing notations such as UML or ER diagrams.

Building on the foundational concepts of software architecture, it is important to distinguish between **architectural styles** and **architectural patterns**, as they serve complementary but distinct roles in system design. An architectural style (e.g., monolithic, layered architecture, microservices, event-driven systems, and service-oriented architecture) describes the high-level system layout, structure, and organization of a software system. It defines how components are arranged, how they communicate, and what constraints govern their interaction—providing a consistent, system-wide framework. In contrast, architectural patterns offer targeted, reusable solutions to recurring design problems that arise within or across different styles. These patterns (e.g., Saga pattern, publish-subscribe pattern, request-response, inbox, and outbox pattern) do not dictate the overall structure but instead help address specific challenges like fault tolerance or data integrity. While styles set the stage for how a system is fundamentally organized, patterns act as strategic techniques to refine and optimize behavior within that framework [4].

7.1 Architectural Styles

Software development employs a variety of architectural styles, each suited to different system requirements and contexts. From scalable microservices to streamlined monolithic designs, these styles influence how components communicate, manage data, and scale under load. Table 2 outlines the strengths, limitations, and typical use cases of five key architectural styles discussed in the following subsections [2, 4].

Before exploring these styles in detail, it is essential to understand the concept of *partitioning* in software architecture, which refers to how a system is decomposed into smaller parts. Two primary partitioning strategies are commonly used: **domain partitioning** and **technical partitioning** (See Fig. 12) [4].

Table 2 Comparison of architectural styles

Architecture	Strengths	Weaknesses	Best use cases
Microservices	Scalable; independent deployments; supports heterogeneous tech stacks (e.g., varied languages/frameworks); fault isolation; composability (reuse and combine functionality in different contexts); promotes organizational alignment (teams own specific services); easier maintenance.	Complex; requires orchestration, monitoring, and DevOps; potential inter-service latency.	Large-scale, cloud-native systems; scalable apps.
Monolithic	Simple to develop, test, and deploy; tightly integrated.	Hard to scale; difficult to maintain; single point of failure.	Small-medium apps; startups; prototypes.
Service-Oriented (SOA)	Reusable services; integrates with legacy systems; enterprise-ready.	Middleware bottlenecks; tightly coupled services; high infra cost.	Enterprise systems; reusable business logic.
Event-Driven	Real-time responsiveness; decoupled components; supports async communication; scalable.	Debugging is hard; requires async expertise; risk of message loss.	Real-time apps (e.g., chat, trading); IoT; e-commerce.
Layered (n-Tier)	Clear separation of concerns; widely used; easy to manage workflows.	Rigid; scaling individual layers is hard.	Traditional apps; predictable workflows.

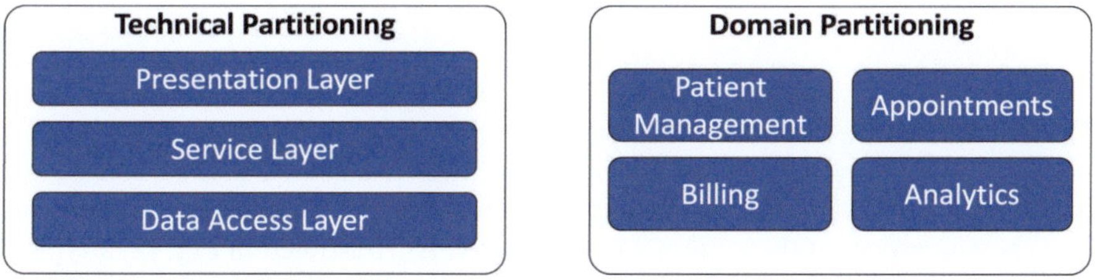

Fig. 12 Conceptual diagram illustrating technical partitioning and domain partitioning

- **Domain Partitioning**: Decomposes the system based on business capabilities or functional areas (e.g., Orders, Reporting, Payments, Inventory Management in e-commerce; or Patient Management, Analytics, Appointments, Billing in healthcare), fostering modularity and alignment with real-world business domains.

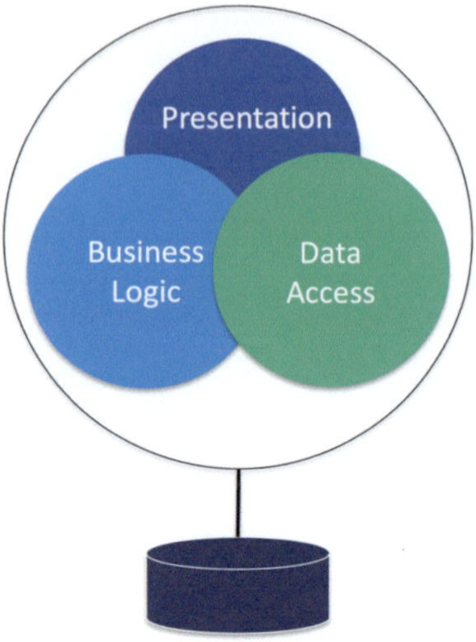

Fig. 13 Conceptual diagram of monolithic architectureconceptual diagram of monolithic architecture

- **Technical Partitioning**: Organizes the system according to technical responsibilities (e.g., Presentation Layer, Business Logic Layer, Data Access Layer), enabling clear separation of concerns across architectural layers.

7.1.1 Monolithic Architecture

Monolithic architecture organizes an application as a single, unified codebase, where all functionalities are tightly coupled and run as one process (See Fig. 13). Its simplicity makes it easier to develop, deploy, and test, particularly for small teams and startups. However, as the application grows, it becomes harder to scale and maintain because any change necessitates redeployment of the entire system. This architecture is most effective for small to medium-sized applications or projects in their early stages [2, 4].

7.1.2 Layered (n-Tier) Architecture

The layered architecture divides an application into distinct layers, each responsible for a specific aspect of the system's functionality. As illustrated in Fig. 14, a typical configuration includes the following layers [2, 4]:

- **Presentation Layer**: Manages the user interface and user interactions by collecting input and displaying results to the user.
- **Application Layer** (also known as the *Service Layer*): Coordinates the overall workflow by receiving input from the presentation layer and invoking appropriate components in the business logic layer.

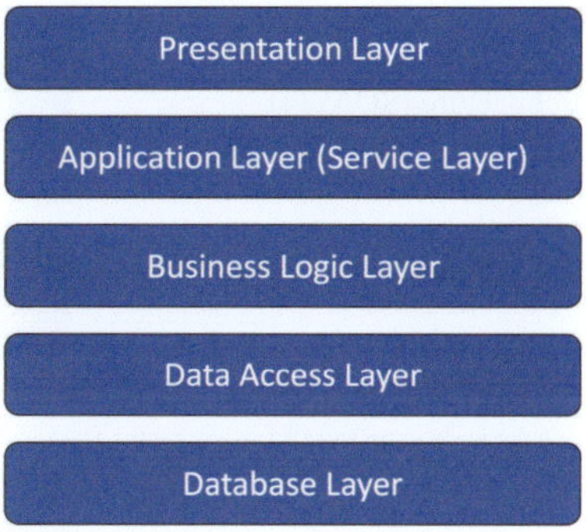

Fig. 14 Conceptual diagram of layered (n-Tier) architecture

- **Business Logic Layer** (also referred to as the *Domain Layer*): Implements the core rules, policies, and operations of the system that define how business processes behave.
- **Data Access Layer** (also referred to as the *Persistence Layer*): Handles communication with data sources by providing an interface for querying, saving, and mapping data between the business layer and the database.
- **Database Layer**: Represents the physical storage mechanism used to persist and retrieve application data.

This clear separation of concerns improves modularity and simplifies development, testing, and maintenance. In more advanced implementations, additional layers may be introduced. For example, **Infrastructure Layer** supports cross-cutting concerns such as logging, messaging, security, and integration with external systems. While layered architecture enhances structure and maintainability, its relatively rigid design can make it less adaptable in fast-evolving or highly dynamic environments. Nevertheless, it remains widely used in enterprise systems such as CRM and ERP platforms, where scalability, reliability, and standardization are critical.

7.1.3 Service-Oriented Architecture (SOA)

Service-Oriented Architecture (SOA) structures applications as a collection of modular, reusable services that communicate through a centralized middleware layer, commonly implemented via an Enterprise Service Bus (ESB) (See Fig. 15). SOA is particularly well-suited for large enterprises, as it facilitates the integration of legacy systems, encourages service reuse, and supports cross-application interoperability. However, its reliance on centralized infrastructure can introduce performance bottlenecks and tighter coupling if not carefully managed. SOA is typically composed of several key components [4]:

- **Business Services:** High-level, coarse-grained services aligned with business capabilities (e.g., *SchedulePatientVisit*, *ProcessInsuranceClaim*). Defined by busi-

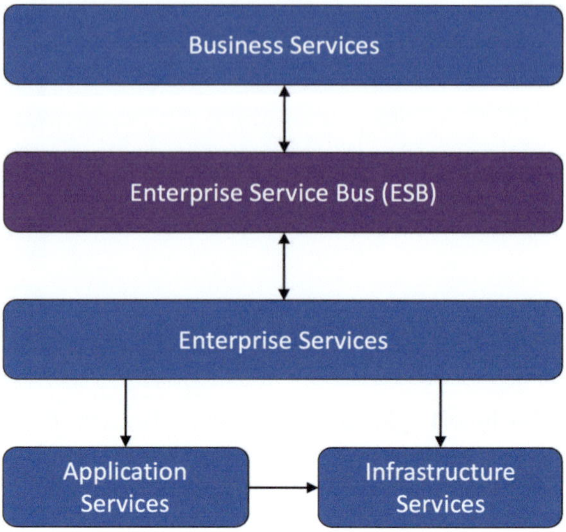

Fig. 15 Conceptual diagram of service-oriented architecture (SOA) [4]

ness users or analysts, these services serve as entry points to business processes and expose only interface definitions (inputs/outputs), not implementation details.

- **Enterprise Services:** Fine-grained, reusable components that encapsulate atomic units of functionality (e.g., *RegisterPatient*, *ValidateInsuranceCoverage*). These services are developed by technical teams and act as building blocks used by business services, enabling flexibility and reuse across workflows.
- **Application Services:** One-off or context-specific services that address the unique needs of a single application. For example, a hospital's internal dashboard may require a dedicated *GenerateWardOccupancyReport* service. These are typically not shared across the enterprise and are maintained by individual application teams.
- **Infrastructure Services:** Services that support operational and cross-cutting concerns, such as authentication, logging, monitoring, and authorization. These are usually owned by a shared infrastructure team and used throughout the SOA ecosystem.
- **Enterprise Service Bus (ESB):** The messaging and integration backbone of SOA. It handles communication between services through routing, protocol transformation, data mapping, and message mediation. In many platform-specific implementations, the ESB also includes additional capabilities, such as the **orchestration engine** and **integration hub**, providing centralized coordination of workflows and system integration.
- **Orchestration Engine:** A central component that defines, coordinates, and executes workflows involving multiple services. It manages the sequencing of service calls, transaction boundaries, message transformation, and error handling. Although conceptually distinct, orchestration logic is often implemented within the ESB in integrated platforms.

- **Integration Hub (within ESB):** A subsystem responsible for connecting heterogeneous systems, handling data format transformations, and bridging different communication protocols. The integration hub enables the ESB to link modern services with legacy applications, third-party APIs, or packaged software systems.
- **Service Registry (not shown in the diagram):** A centralized directory where services are published and registered by providers, and discovered by consumers. It enables dynamic service lookup and binding, promoting loose coupling and runtime flexibility. While often integrated with the ESB or orchestration tools, it is commonly omitted in simplified architectural diagrams.

SOA supports modular design, reuse, and scalability, but its success depends on proper abstraction, governance, and architectural discipline. When used thoughtfully, it enables organizations to evolve their systems and respond to changing business needs with greater agility.

7.1.4 Event-Driven Architecture (EDA)

Event-Driven Architecture (EDA) is a design approach in which system components interact by producing and consuming discrete events that signify state changes or completed actions (See Fig. 16). Rather than relying on direct invocations, components communicate through an intermediary, often referred to as an event broker. This model enables decoupling, enhances system scalability, and supports dynamic responsiveness, making it particularly suitable for domains requiring real-time data processing, such as IoT platforms. The communication paradigm within EDA is primarily asynchronous, although synchronous mechanisms are occasionally required depending on the use case. The communication modes can be described as follows [4]:

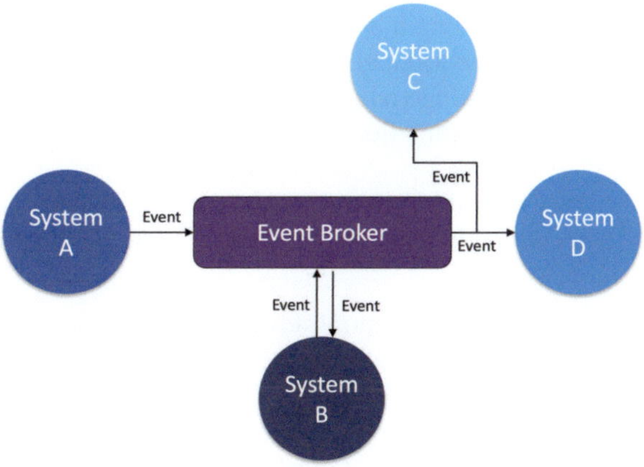

Fig. 16 Conceptual diagram of event-driven architecture (EDA)

- **Asynchronous communication:** In this default mode of operation, event producers emit signals that are captured by interested consumers without expecting any response. This one-way communication minimizes temporal dependencies between components and supports parallelism and system resilience. The key components involved include:
  - An *initiating event*, which denotes that a specific condition or activity has occurred within the system.
  - An *event broker*, a middleware component responsible for distributing events to all subscribed processors through topics, streams, or notification mechanisms.
  - One or more *event processors*, which interpret the event and perform relevant computations or business logic independently of one another.
  - Optionally, a *derived event* may be generated, indicating a secondary action triggered as a result of processing the initial event.

  It should be noted that events differ from messages in that they serve as statements of fact rather than instructions. For example, an event may convey that a patient has been discharged, whereas a message might instruct the system to initiate discharge procedures. Furthermore, events are often distributed to multiple subscribers and typically do not require acknowledgment or responses, in contrast to the directed and response-oriented nature of messages.
- **Synchronous communication (request-reply messaging):** Although EDA emphasizes loose coupling and asynchronous workflows, certain scenarios necessitate immediate data retrieval or confirmation. To address such needs, EDA can incorporate a request-reply messaging pattern that simulates synchronous interactions within an otherwise asynchronous environment.
  - The initiating service sends a request to a designated *request queue*.
  - Upon processing, the response is returned through a *reply queue*.
  - Two common implementation techniques include [4]:
    - Embedding a *correlation identifier* in the response message to associate it with the original request.
    - Generating a *temporary reply queue* specific to the request, which is discarded upon completion of the exchange.

    This hybrid approach allows systems to retain the advantages of EDA while accommodating specific business logic that requires immediate feedback.

Despite its architectural advantages, EDA introduces new complexities related to debugging, event sequencing, and system monitoring. Achieving effective implementation requires careful planning of event flows, consistent schema design, and appropriate tooling for traceability and observability. When applied judiciously, EDA fosters system robustness, adaptability, and responsiveness in highly dynamic operational environments [4].

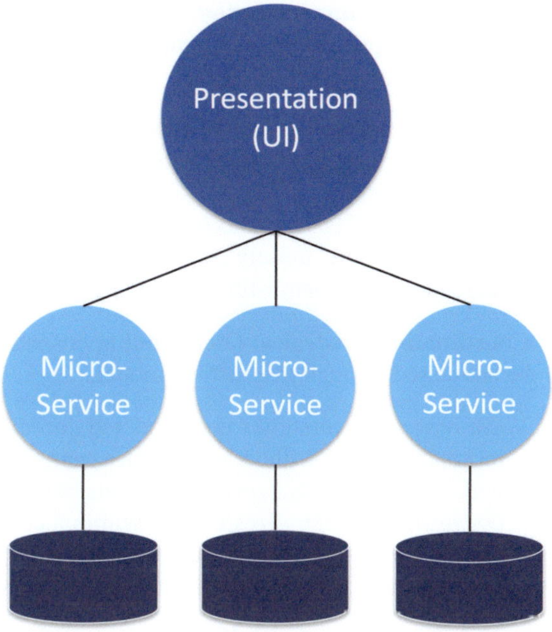

Fig. 17 Conceptual diagram of microservices architecture

7.1.5 Microservices Architecture

Microservices architecture paradigm decomposes a monolithic application into a collection of small, independently deployable services (See Fig. 17). Each microservice encapsulates a narrowly defined capability and operates autonomously within its own runtime environment. This architectural style emphasizes modularity, decentralization, and scalability, making it particularly suitable for complex, evolving systems, especially in cloud-native contexts. The core benefits of microservices include [4, 6]:

- **Autonomous scalability:** Services can be scaled independently, allowing system resources to be allocated more precisely in response to varying workload demands.
- **Technological diversity:** Teams are afforded the flexibility to select the most appropriate programming language, framework, or persistence mechanism for each service, supporting polyglot development practices.
- **Concurrent development and deployment:** Because services are loosely coupled, they can be developed, tested, and deployed independently by separate teams, which facilitates continuous delivery and accelerates development cycles.
- **Improved system resilience:** The failure of an individual service is less likely to compromise the availability of the broader system, enhancing fault tolerance and operational stability.

Despite these advantages, the transition to microservices introduces a number of architectural and operational complexities. Effective implementation requires attention to a set of supporting patterns and infrastructural considerations, such as [4]:

- **Inter-service communication:** Given that services interact over the network, selecting an appropriate communication style is essential. The two predominant approaches are:
  - *Synchronous communication*, typically realized through RESTful APIs, involves direct, blocking interactions between services, which can increase latency and interdependence.
  - *Asynchronous communication*, facilitated through *message queues* or *event streams*, supports loosely coupled interactions by decoupling service lifecycles and reducing response-time dependencies.
- **Coordination strategies:** Orchestrating business workflows across services can be addressed through two distinct models:
  - *Choreography* promotes decentralized coordination, where services emit and respond to events without relying on a central controller. This model aligns well with event-driven systems and respects bounded context boundaries.
  - *Orchestration* centralizes control within a coordinator that explicitly manages service invocation sequences. While useful for complex workflows, this model can introduce tighter coupling and a potential single point of failure.
- **Data ownership and isolation:** In alignment with domain-driven design principles, each microservice is responsible for managing its own data. This promotes decoupling but also necessitates thoughtful strategies for maintaining data consistency and supporting cross-service queries.
- **Service granularity:** Determining the appropriate scope of a microservice is a non-trivial design concern. Overly fine-grained services can result in communication overhead and operational complexity, whereas coarse-grained services may compromise modularity. Services should be scoped to reflect distinct business capabilities within well-defined bounded contexts.

While microservices architecture does not mandate the use of specific technologies or infrastructure components, its successful implementation in real-world systems often relies on a set of supporting patterns and tools. Commonly adopted practices include the use of API gateways, service discovery mechanisms, monitoring and logging systems, centralized configuration management, orchestration platforms, and CI/CD pipelines. These components are not inherent to the architecture itself but are widely recognized as essential for managing the complexity, scalability, and observability of distributed service-based systems:

- **API Gateway**: While not an absolute requirement in microservices architecture, an API Gateway is widely considered a best practice in production-grade environments. It serves as a unified entry point through which external clients access the system's distributed services. Functioning as an intermediary, the API Gateway abstracts the complexity of individual microservices by handling request routing, protocol translation, load balancing, and security concerns such as authentication and authorization. It may also perform request transformation, such as translating

RESTful HTTP calls into asynchronous messaging protocols like RabbitMQ. By centralizing these cross-cutting concerns, the API Gateway simplifies client interactions, improves scalability, and enhances system security and maintainability.

- **Load Balancer**: A load balancer ensures high availability and scalability by distributing incoming traffic evenly across multiple service instances. It prevents any single instance from being overwhelmed and improves system fault tolerance.
- **Centralized Configuration Management**: Centralized configuration management tools externalize application settings, such as database credentials, API keys, and feature flags, ensuring consistent and dynamic configuration updates without requiring service redeployment. These tools offer varying levels of compatibility depending on the programming language or framework being used, such as Spring Cloud Config (Java), Consul (language-agnostic), Etcd (Go, Python, Node.js, etc.), and AWS Parameter Store and Secrets Manager (language-independent).
- **Monitoring and Logging**: Monitoring and Logging: Observability tools like Prometheus (monitoring), the ELK Stack, and Graylog (logging) are indispensable for tracking the health, performance, and failures of microservices. These tools provide real-time insights into distributed systems, which are critical for debugging and maintaining reliability.
- **Orchestration**: A core idea of microservices architecture is that each microservice can be deployed independently, allowing teams to update, scale, or fix individual services without affecting the rest of the system. To achieve this flexibility, microservices are often packaged into containers, which provide a consistent environment for deployment across different platforms. However, managing multiple containers manually becomes complex as the number of services grows. Orchestration tools automate the deployment, scaling, and management of these containers, ensuring that resources are utilized efficiently and that services remain highly available. Tools like Kubernetes, Docker Swarm, and AWS ECS/EKS handle tasks such as scheduling, scaling, and fault recovery, enabling the lifecycle management of containerized microservices. These capabilities make orchestration essential for managing dynamic, distributed systems where reliability and scalability are key.
- **CI/CD Pipeline**: Continuous Integration and Continuous Deployment (CI/CD) pipelines are essential for managing frequent deployments in microservices architectures, where services are developed and deployed independently. Automating the build, test, and deployment processes ensures that updates can be delivered rapidly while minimizing errors and downtime. This approach enables teams to maintain high development velocity without compromising on quality or system stability. Tools like Jenkins, GitLab CI/CD, CircleCI, and AWS CodePipeline are commonly used to implement CI/CD pipelines. These tools streamline workflows by automating testing, integrating code changes, and deploying services, making them indispensable for achieving efficiency and reliability in modern distributed systems.

7.2 How to Model Microservices

Modeling microservices necessitates the identification of boundaries that ensure the system is maintainable, scalable, and conducive to independent development and deployment. A well-designed microservice encapsulates a specific domain responsibility while minimizing dependencies on other services [6].

1. Characteristics of Well-Defined Microservice Boundaries:
The establishment of effective microservice boundaries requires adherence to key principles that ensure modularity, independence, and resilience of individual services [6]:

- **Information Hiding:** Microservices should conceal their internal implementation details, exposing only the necessary interfaces through well-defined APIs. This design principle ensures that changes within a service do not propagate across the system and reduces the risk of unintended interdependencies.
- **High Cohesion:** Cohesion refers to the degree to which the responsibilities of a service are related and focused. A highly cohesive microservice should encapsulate a singular, well-defined domain concept, thereby reducing redundancy and overlap with other services.
- **Loose Coupling:** Coupling measures the extent of dependency between services. To foster independence and scalability, microservices should minimize direct dependencies on one another, thereby enabling each service to evolve independently without risking systemic failures.

2. The Interplay Between Cohesion and Coupling:
The concepts of cohesion and coupling are closely connected and are critical to system stability. High cohesion ensures that related functionality is grouped within a single boundary, while low coupling ensures changes in one boundary have minimal impact on others. Constantine's Law encapsulates this idea: "A structure is stable if cohesion is strong and coupling is low." For microservices, achieving stability is vital for independent deployability, reducing the need for constant coordination across teams. If boundaries lack stability and contracts change in a backward-incompatible way, upstream consumers will need to adapt frequently, creating instability. Balancing cohesion and coupling is context-dependent and requires thoughtful trade-offs, as no absolute "best" way exists. Over time, revisiting decisions may be necessary to align with evolving requirements and challenges [6].

3. Types of Coupling in Microservices:
Understanding the various types of coupling that may arise between microservices is essential for identifying and mitigating problematic dependencies [6]:

- **Domain Coupling:** Domain coupling occurs when one service depends on the domain-specific business rules, processes, or logic of another service. This type of dependency can reduce service autonomy and increase the risk of cascading changes across services when domain rules evolve. While some degree of domain

coupling is inevitable in complex systems, it should be minimized and abstracted through well-defined interfaces (e.g., APIs and events) to preserve service independence [6].
Example: In a healthcare system, an appointment scheduling service might depend on the patient records service not only to retrieve patient contact information but also to enforce business rules, such as checking whether the patient has active insurance coverage. This reliance on another service's domain-specific rules introduces tight coupling and reduces flexibility, as changes in the insurance validation logic within the patient records service could directly affect the appointment scheduling service.

- **Pass-Through Coupling:** Pass-through coupling occurs when a service acts merely as an intermediary, forwarding data between other services without adding any additional logic, processing, or value. This type of coupling can lead to unnecessary complexity, create tight dependencies between services, and introduce performance bottlenecks by making the intermediary service a single point of failure or latency. In a well-designed microservices architecture, such patterns should generally be avoided, favoring direct communication or event-driven decoupling where possible [6].
Example: In a healthcare system, an insurance verification service retrieves patient insurance details from the patient records service and simply forwards them to the billing service without validating the insurance coverage or adding any additional context. This design unnecessarily involves the insurance verification service in the workflow without adding value, creating tighter coupling and potential inefficiencies in the process.
- **Common Coupling:** Common coupling occurs when multiple services share a global resource, such as a database, file storage, or configuration. This practice creates tight dependencies on the shared resource, undermining service autonomy, and increases the risk of cascading failures, as changes to the resource can impact all dependent services. In a microservices architecture, this pattern should be avoided to preserve the independence and scalability of individual services [6].
Example: In a healthcare system, multiple services—such as an appointment scheduling service, a billing service, and a patient records service—directly writing to and reading from the same "PatientDetails" table in a shared database introduces tight coupling. If these services attempt to access the table simultaneously during peak usage (e.g., during a busy workday), contention can occur. *Contention* refers to a situation where multiple services compete for access to a shared resource at the same time, leading to delays, performance bottlenecks, or failures. For example, one service might lock the table for writing while others are forced to wait, delaying critical operations like retrieving patient records during emergencies and potentially causing system inefficiencies or downtime.
- **Content Coupling:** Content coupling, the most severe form of coupling, occurs when one service directly accesses or manipulates the internal implementation details (e.g., database schema, private variables, or internal methods) of another service. This type of coupling tightly binds the two services, significantly reducing modularity, scalability, and maintainability. Changes to the internal implementa-

tion of one service can directly break the dependent service, making the system fragile and difficult to evolve. Content coupling should be strictly avoided in a microservices architecture [6].
Example: In a healthcare system, consider an appointment scheduling service directly querying the internal database schema of the patient records service to retrieve patient details (e.g., contact information or medical history). If the patient records service changes its database schema (e.g., renames or restructures tables), the appointment scheduling service will break, leading to system failures. A better approach would be to define a well-structured API for the patient records service, ensuring loose coupling by allowing the appointment scheduling service to retrieve patient information through the API without directly depending on the underlying database structure.

4. Practical Guidelines for Microservice Design:
The following practical guidelines can assist in designing microservices with well-defined boundaries and minimal interdependencies [6]:

- **Apply Domain-Driven Design (DDD):** Employ *Bounded Contexts* to align microservices with distinct business capabilities, ensuring that each service is responsible for a specific domain.
- **Adhere to the Single Responsibility Principle:** Each microservice should focus on a single domain responsibility to promote modularity and simplify maintenance.
- **Emphasize Interface Design:** Microservices should expose only essential functionalities through well-defined interfaces while concealing internal implementations.
- **Prefer Asynchronous Communication:** To minimize direct dependencies, microservices should leverage asynchronous communication patterns, such as event-driven messaging or message brokers (e.g., Kafka, RabbitMQ).
- **Avoid Shared Databases:** Each microservice should maintain its own database to prevent common coupling and ensure data encapsulation.
- **Monitor Coupling and Cohesion:** Regularly evaluate the degree of coupling and cohesion in the system to ensure the continued scalability and maintainability of microservices. Tools such as distributed tracing can assist in identifying problematic dependencies.

7.3 Splitting the Monolith: Patterns and Key Considerations

The transition from a monolithic architecture to a microservices-based system represents a complex yet highly beneficial undertaking. This process necessitates meticulous decomposition into smaller, autonomous services, while simultaneously addressing both code-level and data-level concerns. The following discussion outlines common decomposition strategies, relevant architectural patterns, and key challenges associated with the transition, accompanied by practical insights [6].

Decomposition Approaches:
Several strategies can be employed to decompose a monolithic system, depending on the objectives and focus of the migration [6]:

- **Code-First Decomposition:** This strategy emphasizes the partitioning of the monolithic codebase into smaller, domain-specific services. For instance, modules such as user management and billing can be extracted into independent services. This approach is particularly effective when the monolithic application already possesses clearly defined modular boundaries or distinct domain delineations.
  Potential Challenge: Monolithic codebases often lack well-defined modularity, necessitating substantial refactoring efforts prior to decomposition.
- **Data-First Decomposition:** This approach involves prioritizing the segregation of the database layer by creating distinct, service-specific databases. Each service assumes ownership of its corresponding data, thereby ensuring independence and minimizing cross-service dependencies. This strategy is especially suitable when tightly coupled data within the monolith creates bottlenecks or operational inefficiencies.
  Potential Challenge: Migrating data to service-specific databases entails significant planning to maintain data consistency and reduce service disruption.
- **Hybrid Decomposition:** A hybrid approach combines the strengths of the code-first and data-first strategies, commencing with modularization at the code level and progressively transitioning toward service-specific data ownership. This iterative approach mitigates risks and provides a balanced pathway for achieving architectural goals.

Decomposition Patterns:
A number of architectural patterns have been identified to facilitate a smoother and incremental transition from a monolithic system to a microservices architecture [6]:

- **Strangler Fig Pattern:** This pattern derives its name from the strangler fig plant, which grows around a tree and ultimately takes over its structure, effectively killing the original tree. The Strangler Fig Pattern involves the progressive replacement of monolithic components with microservices. Initially, new microservices are introduced alongside the monolith, and requests for specific functionality are redirected to these services while the monolith continues to handle legacy functionalities. Over time, as more functionalities are migrated, the monolith is incrementally phased out. It is recommended to employ API gateways or reverse proxies (e.g., NGINX) to manage traffic routing and facilitate the transition.
- **Parallel Run:** This pattern entails developing a microservice for a specific feature and running it concurrently with the monolith. Traffic is gradually redirected to the microservice while its behavior is monitored and validated. Upon successful

validation, the corresponding functionality in the monolith is deprecated. It is recommended to utilize canary deployments or A/B testing to incrementally shift traffic, thereby minimizing potential disruptions.
- **Feature Toggles:** Feature toggles (also known as feature flags) enable or disable specific microservice implementations. This mechanism facilitates gradual rollouts and comprehensive testing while retaining the ability to revert to the monolithic implementation if required. It is important to implement feature toggles carefully to avoid long-term technical debt caused by unused or outdated toggles.

Data Decomposition Concerns:
The transition from a monolith to microservices necessitates addressing significant data management challenges [6]:

- **Performance:** Distributed systems, such as microservices, inherently introduce latency due to their decentralized nature. Strategies such as caching, asynchronous communication, and optimized APIs are essential to mitigate performance issues.
- **Data Integrity:** Maintaining data consistency across services presents considerable complexity when transitioning from tightly coupled monolithic data structures. Eventual consistency models and event-driven architectures are critical for ensuring reliable inter-service data updates.
- **Transactions:** Monolithic systems typically rely on ACID transactions within a unified database. Microservices, however, require distributed transaction patterns such as the *Saga pattern*, which divides transactions into smaller, coordinated steps across services.
- **Tooling:** Robust monitoring, logging, and debugging tools are indispensable for identifying and resolving issues within distributed microservices environments.
- **Reporting and Analytical Databases:** Centralized reporting databases are often integral to monolithic systems. In a microservices architecture, constructing a dedicated reporting layer that aggregates data from individual services into a centralized data warehouse is recommended.

7.4 Design Patterns

In microservices-based distributed systems, complex business workflows often span multiple independent services that must collaborate to deliver end-to-end functionality. For instance, in a healthcare application, processing a patient's treatment may involve services such as appointment scheduling, medical record retrieval, insurance verification, prescription management, and billing. These distributed interactions require robust coordination mechanisms to ensure consistency, reliability, and resilience. Unlike monolithic systems—where all components typically share a single database and benefit from native ACID (Atomicity, Consistency, Isolation, Durability) transaction guarantees—microservices operate with isolated data stores and decentralized control. This distributed nature introduces challenges such as partial failures, data inconsistency, latency, and coordination complexity.

To address these challenges effectively, software architects leverage **distributed system design patterns**—proven strategies that offer reusable solutions for common problems. These patterns help manage data consistency, orchestrate workflows, ensure fault tolerance, and maintain scalability across service boundaries. Key design patterns such as the Saga Pattern, CQRS, Event Sourcing, Outbox Pattern, and Two-Phase Commit provide foundational building blocks for implementing reliable and maintainable distributed systems.

7.4.1 Two-Phase Commit (2PC): A Legacy Approach

The Two-Phase Commit (2PC) protocol represents a traditional mechanism for managing distributed transactions across multiple microservices. The 2PC process is divided into two primary phases:

- **Voting Phase (Prepare Phase):** The coordinator sends a request to all participating microservices, asking if they are ready to commit the transaction. Each service evaluates the request by performing necessary checks, such as validating constraints and ensuring resource availability. During this phase, the services typically acquire locks on the relevant resources to prevent conflicting operations and ensure consistency. Based on their readiness, each service responds with either a `yes` or a `no`.

  Commit Phase (Decision Phase): If all services respond with `yes`, the coordinator issues a commit message, finalizing the transaction. Each service then applies the changes and releases the acquired locks. If any service responds with `no`, the coordinator triggers a rollback, instructing all participants to undo any changes made during the Voting Phase. Locks are released after the rollback is complete.

However, the Two-Phase Commit protocol can encounter several limitations in practical distributed environments, such as:

- **Intermediate States:** Real-world failures, such as network issues or crashes, can result in some services committing a transaction while others remain in an uncertain state. This creates temporary inconsistencies, violating the expectation of seamless transitions between valid system states.
- **Failure Handling:**
  - *Coordinator Crashes:* The failure of the coordinator during or after the voting phase can leave participants in a "prepared" state indefinitely, causing resource locks and system uncertainty. Recovery often requires manual intervention or custom logic.
  - *Unresponsive Services:* If any participating service fails or becomes unresponsive during the commit phase, the transaction may stall, increasing the risk of inconsistency.

- **Performance Bottlenecks:**
  - Resources remain locked from the *prepare phase* until the commit or abort decision is completed, preventing concurrent transactions from accessing these resources. This significantly reduces system throughput.
  - This locking mechanism introduces latency, making 2PC inefficient for high-performance, large-scale systems.
- **Scalability Issues:**
  - The centralized coordination mechanism of 2PC creates a *single point of failure* and limits the system's scalability.
  - Modern distributed architectures demand decentralized coordination, which 2PC struggles to provide.

7.4.2 The Saga Pattern

The Saga Pattern was first introduced in 1987 by Hector Garcia-Molina and Kenneth Salem in their research paper titled *"Sagas."* It was developed as a solution to address the challenges of managing long-lived transactions in distributed systems. raditional methods of ensuring data consistency relied on ACID transactions, which required a global transaction manager. However, such approaches were impractical in distributed environments due to high latency, resource locking, and scalability constraints. The Saga Pattern proposes breaking a large transaction into a sequence of smaller, independent sub-transactions, each of which can succeed or fail independently. If a sub-transaction fails, compensating actions are executed to undo the changes made by the preceding steps, ensuring the system's integrity. This approach enables eventual consistency across services without relying on a global transaction manager, which is a significant advantage in distributed systems where such centralization is infeasible. The Saga Pattern can be implemented in two primary ways:

- **Orchestration-Based Sagas:** This approach relies on a centralized controller, known as the saga orchestrator, to manage the sequence of transactions. The orchestrator coordinates saga participants by instructing them on which local operations to execute based on events. It also executes saga requests, maintains the state of each task, and handles failure recovery by initiating compensating transactions. While this structured and centralized execution flow simplifies coordination and provides clear control, it introduces a single point of control, which may reduce fault tolerance if the orchestrator is not designed with resilience in mind.
- **Choreography-Based Sagas:** In this approach, each microservice independently executes its transaction and triggers subsequent actions by emitting events. This decentralized strategy eliminates the need for a central controller, enhancing fault tolerance and scalability. However, it requires careful event design to mitigate potential coordination issues, such as event ordering conflicts and cascading failures, which can complicate the system's behavior.

The Saga Pattern addresses several key challenges faced by distributed workflows, offering solutions such as:

- **No Resource Locking:** Unlike 2PC, sagas do not require locking resources across services, which improves system throughput and scalability.
- **Eventual Consistency:** Sagas prioritize eventual consistency (where the system will eventually reach a consistent state, even in the presence of failures) over strict atomicity (where all operations succeed or fail together in a single transaction), a trade-off that is often sufficient for most business use cases.
- **Fault Tolerance:** Decentralized coordination and compensating transactions make sagas inherently more resilient to failures than 2PC.
- **Scalability:** Choreography-based sagas, in particular, allow for highly scalable architectures by removing centralized control and enabling independent service execution.

7.4.3 Command Query Responsibility Segregation (CQRS)

The Command Query Responsibility Segregation (CQRS) pattern is a design approach that separates the **write** operations (commands) from the **read** operations (queries) within a system. This separation allows each side to be independently optimized based on its specific concerns and requirements.

- **Command:** A command represents an action that changes the system's state, such as creating, updating, or deleting a resource. Commands typically involve complex business logic, validations, and interactions with multiple services or aggregates.
- **Query:** A query retrieves data from the system without modifying it. Queries are optimized for fast and efficient read access, often leveraging denormalized views, caching, or read replicas to meet performance demands.

By decoupling reads from writes, CQRS enables independent scaling of each concern—write-heavy and read-heavy workloads can be managed and optimized separately. This pattern is especially beneficial in *high-traffic systems* or *domains with complex business logic*, where the ability to evolve and tune reads and writes independently provides significant performance, maintainability, and scalability advantages.

7.4.4 Event Sourcing Pattern

The Event Sourcing Pattern ensures that all changes to the state of a system are captured and stored as a sequence of immutable events. Instead of overwriting the current state in the database, every change is recorded as a new event, creating a complete and chronological log of all modifications. The system's current state can be reconstructed at any point in time by replaying these events in the order they occurred. This approach enables auditability by maintaining a full historical record of changes, traceability to understand why and how changes were made, scalability through appending new events rather than overwriting data, and the ability to rebuild the system's state for debugging, testing, or recovery. The Event Sourcing Pattern is particularly suited for financial systems, e-commerce platforms, and other applications requiring robust traceability and the ability to replay historical actions to diagnose or recover system state.

7.4.5 Outbox Pattern

The Outbox Pattern is a design solution used to ensure reliable message delivery and maintain consistency between a database and downstream systems in distributed architectures. In this pattern, when a service updates its database as part of a transaction, it simultaneously writes an event or message to a dedicated "outbox" table in the same database. This guarantees that both the database changes and the outbox entry occur within the same transaction, ensuring atomicity. A separate process, often referred to as an "outbox relay" or "polling mechanism," periodically reads the outbox table and forwards the messages to their intended destination, such as a message broker (e.g., Kafka, RabbitMQ) or another service. Once the message is successfully delivered, it is marked as processed or removed from the outbox table to prevent duplicate processing. This approach addresses inconsistencies that may arise if a database transaction is committed but the corresponding message fails to be sent. By decoupling database transactions from message delivery, the Outbox Pattern enhances data reliability and ensures event synchronization. While a message broker is a common target for delivering outbox messages due to its scalability and durability, the pattern can also be adapted for direct communication or other delivery mechanisms. The Outbox Pattern is particularly valuable in use cases requiring strong consistency and fault tolerance in distributed systems.

8 Edge-Cloud Interplay

IoT has transformed modern systems by connecting people, processes, data, and objects into a dynamic, intelligent ecosystem. From wearable health monitors to remote patient management systems, IoT seamlessly integrates devices and applications to revolutionize healthcare delivery. Simultaneously, Cloud Computing (CC) provides virtually unlimited storage and computational resources, enabling large-scale data processing and analytics. Although IoT and cloud computing initially evolved independently, their integration into the *Cloud-IoT paradigm* has enabled advanced healthcare applications, such as medical imaging analysis, predictive diagnostics, and real-time patient monitoring [7].

Despite the capabilities of cloud computing, healthcare applications often demand low-latency, real-time data processing for critical tasks. For instance, continuous monitoring of vital signs for patients with chronic conditions or emergency response systems requires decisions to be made instantly. Relying solely on the centralized cloud can lead to delays and inefficiencies. To address these challenges, *Edge Computing* (EC) and *Fog Computing* (FC) extend computational capabilities closer to where data is generated—whether at a patient's bedside, wearable device, or hospital gateway.

Edge computing processes data locally, enabling rapid decision-making for time-sensitive tasks such as detecting irregular heart rhythms or analyzing patient vital signs. Fog computing acts as an intermediate layer, preprocessing and filtering medical data before forwarding it to the cloud for deeper analysis, such as predictive diagnostics or population health insights. Together, edge, fog, and cloud infrastructures form a hierarchical *edge-fog-cloud architecture* that balances latency, bandwidth usage, and computational load to enhance healthcare applications. This hierarchical approach ensures scalable and efficient data processing for AI-driven healthcare solutions. For example, wearable devices can perform local anomaly detection in real-time, while fog nodes at hospital networks aggregate and analyze data from multiple patients. The cloud then provides advanced capabilities such as machine learning-based disease prediction, large-scale patient data analysis, and remote collaboration across healthcare providers [7].

However, this interplay also introduces challenges. The heterogeneous nature of edge-fog-cloud infrastructures—comprising sensors, mobile devices, gateways, and servers—complicates resource management, task allocation, and data privacy. Ensuring robust security and privacy for sensitive medical data is particularly critical in healthcare applications. Furthermore, optimizing the placement of computational tasks across edge, fog, and cloud layers remains an open research problem. Addressing these challenges is essential for leveraging the full potential of the edge-fog-cloud paradigm in healthcare [7].

8.1 Edge-Fog-Cloud Architectures

Fog computing architectures are predominantly layer-based, with researchers proposing models that incorporate anywhere from three to six layers, often including one to four fog layers. These architectures aim to bridge the gap between IoT devices and the cloud by enabling efficient, distributed data processing.

Among the most prominent fog-assisted IoT architectures are [7]:

1. **Three-Layer Model** (Fig. 18a): A foundational structure with IoT devices, a fog layer, and the cloud.
    - The **IoT device layer** handles basic tasks, such as data acquisition from sensors and control of actuators.
    - The **fog layer** serves as an intermediary, performing real-time data filtering, normalization, and lightweight edge analytics.
    - The **cloud layer** manages advanced, resource-intensive tasks, including big data analysis, machine learning, and long-term storage.

This architecture reduces latency by processing data closer to its source while leveraging the cloud's computational power [8].

2. **Clustered Model** (Fig. 18b): Fog nodes are grouped into clusters to enable localized data processing and efficient resource sharing.

    - **Intra-cluster communication**: Nodes within a cluster share resources and coordinate tasks dynamically.
    - **Inter-cluster communication**: Clusters communicate globally to balance workloads and optimize overall performance.
    - This model enhances:

        - **Flexibility**: Adapts dynamically to varying workloads.
        - **Scalability**: Seamlessly integrates additional fog nodes into existing clusters.
        - **Load Balancing**: Efficiently distributes tasks across nodes to prevent bottlenecks.

    The clustered model is particularly effective in low-latency applications such as healthcare monitoring and industrial IoT [9].

3. **Tree-Based Model** (Fig. 18c): A hierarchical structure where the cloud serves as the root, branching out to intermediate fog nodes and edge devices.

    - **Edge nodes**: Located closest to IoT devices, these nodes perform data acquisition, normalization, and real-time processing.
    - **Intermediate fog nodes**: Handle data aggregation, filtering, and lightweight analytics.
    - **Higher-level fog nodes**: Positioned closer to the cloud, these nodes manage advanced computations, such as AI model training.

    This architecture organizes computation systematically across layers, ensuring a clear flow of data from edge devices to the cloud [10].

Combining available architectures and considering frequent fog computing use cases, we present a hierarchical reference model (see Fig. 19) to establish a universal computing infrastructure capable of addressing the needs of IoT applications across various domains. The architecture follows a hierarchical multi-layer structure that integrates edge, fog, and cloud layers. The fog layer is further divided into sub-layers to reflect the distribution of Fog Nodes (FNs) across the network infrastructure. Fog nodes can communicate both vertically, across different layers, and horizontally, within the same layer, to adapt to system load and application requirements. This communication is managed by Fog Orchestration Nodes (FONs), which form the control layer of the fog. In addition to FONs, the fog layer includes Fog Gateway Nodes (FGNs), serving as access points to the fog layer, and Fog Computing Nodes (FCNs), which execute tasks assigned by the FONs [7].

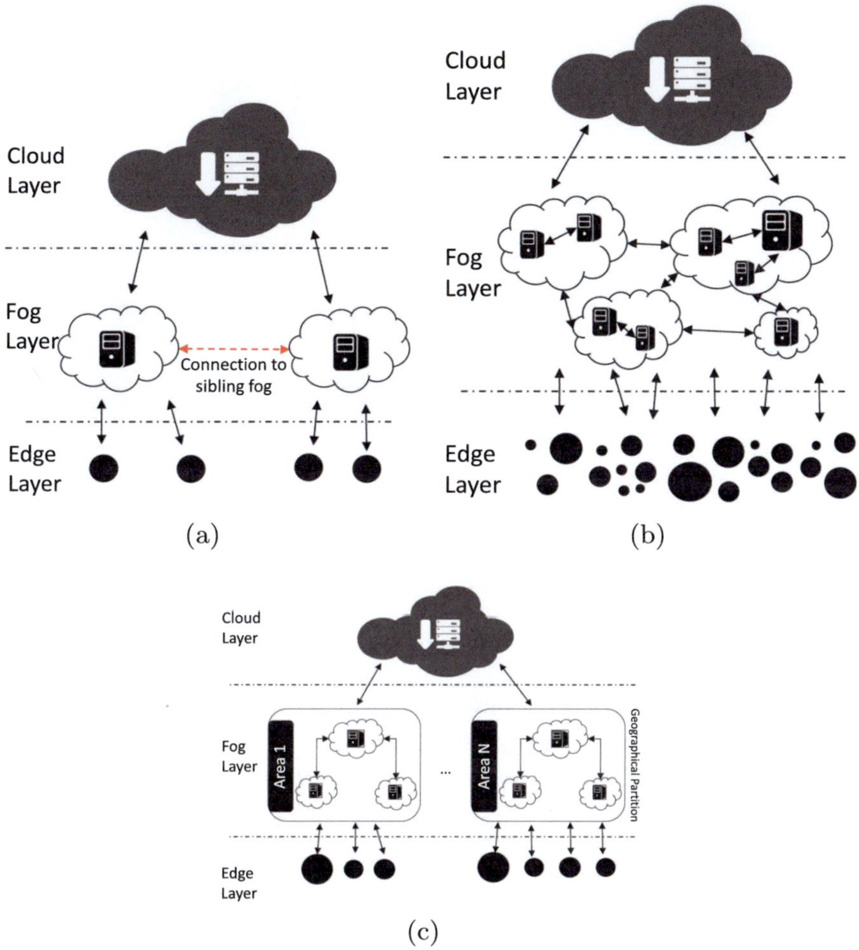

Fig. 18 Illustrations of common edge-fog-cloud architectures [1]. **a Three-Layer Model**: A basic architecture connecting IoT devices, fog nodes, and the cloud. **b Clustered Model**: Fog nodes are grouped into clusters for localized data processing and resource sharing. **c Tree-Based Model**: A hierarchical model rooted in the cloud, branching toward fog and edge nodes

The architecture allows edge devices to participate in resource orchestration, including independent service placement. Edge devices can invoke Service Placement (SP) directly to prevent starvation in multi-user scenarios where a single service is in high demand. Furthermore, edge devices can migrate tasks to nearby fog nodes using user-space containers. For example, an edge device may host part of a Convolutional Neural Network (CNN), while the remaining layers execute on a fog node. Such flexibility improves mobility by enabling edge devices to predict FN handoffs based on their movement [7].

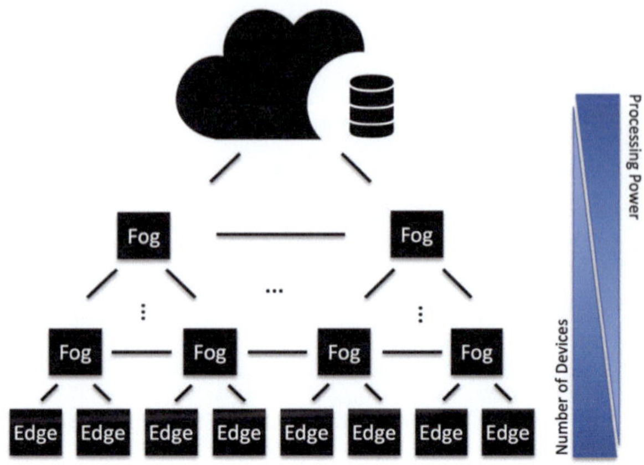

Fig. 19 Visualization of the edge, fog, and cloud computing hierarchy in a network

The reference architecture consists of two key dimensions: a vertical functional layer dimension and a horizontal cross-layer dimension [11–15].

- *Vertical Dimension:* The vertical dimension is organized into three main layers:
  - *Edge Layer:* Responsible for data acquisition, initial processing, and control of sensors and actuators.
  - *Fog Layer:* Contains multiple sub-layers, including orchestration nodes (FONs), gateway nodes (FGNs), and general-purpose computing nodes (FCNs). This layer handles intermediate tasks such as filtering, preprocessing, and lightweight analytics.
  - *Cloud Layer:* Provides large-scale data storage and advanced computations such as machine learning, predictive analytics, and big data analysis.
- *Horizontal Dimension:* The horizontal dimension represents cross-layer perspectives that apply across all layers. According to [11, 12], these perspectives include:
  - *Node Perspective:* The Node Perspective in fog computing refers to the hardware components and their roles within the edge-fog-cloud continuum. It encompasses devices such as accelerators, processors, sensors, actuators, and storage units that form the foundation of fog systems. Fog nodes, including servers, gateways, and edge devices, are deployed to handle data acquisition, storage, and computation based on their capabilities. Servers typically possess higher computational power and storage capacity to manage multiple edge devices and execute resource-intensive tasks, while gateways and edge devices operate closer to data sources, handling lightweight processing and communication. This perspective ensures that computational workloads are distributed efficiently across nodes, optimizing system performance, reducing latency, and enabling scalable and reliable operations within diverse IoT environments.
  - *Communication Perspective:* The communication perspective focuses on ensuring reliable, scalable, and high-quality communication among nodes in the edge-

fog-cloud hierarchy while meeting application-specific Quality of Service (QoS) requirements, such as low latency, high throughput, and availability. Effective communication strategies are essential to enable seamless data exchange, task orchestration, and system interoperability. Fog nodes are often embedded into network infrastructure components, such as routers, access points, or gateways, depending on their deployment and functional roles. The communication model used depends on the *node's location*, *task requirements*, and *application environment*. For instance, industrial environments may rely on wired networks, such as Controller Area Networks (CAN) buses, to enable deterministic, low-level communication with sensors and actuators. Conversely, wireless technologies, such as Wi-Fi, 5G, and LPWAN (Low Power Wide Area Networks), play a critical role in supporting downward communication (e.g., from fog nodes to IoT devices) and horizontal or upward communication between edge, fog, and cloud layers. Factors influencing the choice of communication technologies include data throughput, mobility, energy efficiency, latency constraints, spectrum licensing, and environmental conditions. For example, wireless protocols like 5G enable high-throughput, low-latency communication for real-time applications, while low-power protocols like LoRa or Zigbee are ideal for energy-constrained IoT devices in remote deployments [11, 12].

– *Control and Management Perspective:* This perspective focuses on the lifecycle management of edge, fog, and cloud nodes, ensuring seamless integration, monitoring, and operation of system components across the architecture. It encompasses various tasks, such as registration, provisioning, automated discovery, resource management, task allocation, and system monitoring.

  · *Discovery Services:* These services enable the identification, location, and integration of nodes within the system. Discovery mechanisms ensure that nodes can dynamically register, become visible to other nodes, and integrate into the network as part of the orchestration process [11, 12].
  · *Resource Management and Task Allocation:* Efficient resource management involves dynamically allocating computational, storage, and network resources across nodes based on real-time workloads and application requirements. This process minimizes latency, balances system load, and optimizes energy consumption [12]. Task allocation strategies ensure that workloads are distributed to the most appropriate nodes (e.g., edge devices, fog gateways, or servers) depending on their capabilities and proximity to data sources.
  · *System Monitoring:* Continuous monitoring of fog components enables fault detection, performance evaluation, and resource optimization. Monitoring tools collect metrics such as node utilization, task execution time, and communication latency, ensuring the system operates efficiently and reliably [11].

Note that the architecture and deployment of fog computing is highly dependent on the specific application context, leading to considerable variability in their design and implementation. However, certain core elements remain consistent across most fog computing systems. The design of a particular architecture is influenced by several key factors, such as [7]:

- **Type and scale of computational tasks**: The nature, complexity, and volume of computations to be performed at each layer.
- **Sensor and actuator characteristics**: The number, type, and data generation rate of connected devices.
- **Node capabilities**: The processing power, storage capacity, and communication bandwidth available within each layer.
- **Latency requirements**: The acceptable delay between data generation, processing, and response, both across nodes and between sensors and actuators.
- **Energy efficiency**: The power consumption of nodes, particularly for resource-constrained environments such as edge devices and battery-operated sensors.
- **Scalability requirements**: The ability of the architecture to accommodate a growing number of devices, fog nodes, and increasing data loads.
- **Fault tolerance and reliability**: Mechanisms for ensuring uninterrupted operation in the event of node failures or communication disruptions.
- **Security and privacy considerations**: Safeguarding sensitive data, especially in applications like healthcare, finance, and smart grids.
- **Network bandwidth constraints**: Limitations in communication bandwidth between IoT devices, fog nodes, and the cloud, particularly in constrained network environments.
- **Geographical distribution**: The physical dispersion of nodes and devices, which can impact latency and communication overhead in large-scale or remote deployments.
- **Real-time requirements**: The degree of responsiveness necessary for critical applications, such as autonomous vehicles, industrial automation, and healthcare monitoring.

8.2 Edge-Fog-Cloud Interplay

The widespread adoption of cloud computing has significantly reduced the need for local storage on edge devices and enabled them to offload computationally intensive tasks. Fog computing, on the other hand, allows edge devices to offload tasks with lower transmission latency and higher resilience, particularly during network backhaul congestion. While cloud computing offers virtually unlimited storage and computational power, it suffers from high latency and dependency on network connectivity. Fog computing addresses these limitations by bringing computation closer to the edge but may lack the scalability and global reach of cloud systems. Their **coordinated interplay** enables edge devices to leverage the strengths of both paradigms, leading to significant advantages [7]:

- *Optimized Task Distribution and Load Balancing:* Reduces congestion by distributing tasks dynamically across edge, fog, and cloud resources, minimizing latency and ensuring efficient use of network bandwidth. For instance, a smart factory can offload real-time monitoring tasks to fog nodes while delegating long-term analytics to the cloud [7].

- *Latency-Aware Processing and Seamless Mobility:* Ensures that time-sensitive applications, such as AR and autonomous vehicles, receive computational resources close to the source while maintaining uninterrupted service for mobile users [7].
- *Context-Aware and Adaptive Services:* Enables geographically relevant content delivery and smart decision-making based on user location and environmental data, enhancing applications such as smart cities and personalized services [7].
- *Resilient and Secure System Operations:* The decentralized nature of fog computing prevents single points of failure and allows for localized data security strategies, reducing risks from cyber threats and network disruptions. For example, in a smart grid, localized fog nodes can continue operating even if the central cloud is compromised [7].
- *Energy-Efficient and Battery-Conserving Computation:* Depending on network conditions and task complexity, offloading computationally intensive tasks can reduce the energy burden on edge devices by leveraging fog or cloud resources. However, for lightweight or latency-sensitive tasks, local processing may be more energy-efficient due to transmission overhead. Optimizing this trade-off is key to maximizing battery life and ensuring sustainable network operations [7].
- *Scalable Resource Management and IoT Expansion:* The edge-fog-cloud model dynamically orchestrates resources to handle increasing workloads, large-scale IoMT deployments, and emerging ML-driven applications. This adaptive architecture ensures efficient scaling without overwhelming local resources, enabling seamless growth in interconnected systems [7].
- *Real-Time Analytics and Decision Support:* By processing data closer to the source, the edge-fog-cloud model enables low-latency analytics and rapid decision-making for critical applications such as healthcare monitoring, industrial automation, and smart infrastructure. This localized processing delivers instant insights and facilitates proactive responses, enhancing reliability and operational efficiency [7].
- *Cost Optimization:* Reduces operational costs by leveraging fog computing for latency-sensitive tasks and cloud computing for resource-intensive workloads, minimizing unnecessary data transmission and storage expenses [7].
- *Interoperability and Standardization:* Facilitates interoperability between heterogeneous devices and platforms, enabling seamless integration across edge, fog, and cloud environments through standardized protocols and APIs [7].

8.3 Related Paradigms and Technologies

Several paradigms have emerged in parallel to the edge-fog-cloud architecture, each addressing specific challenges and use cases. These paradigms are discussed below, and a comparative overview is provided in Fig. 20 [7].

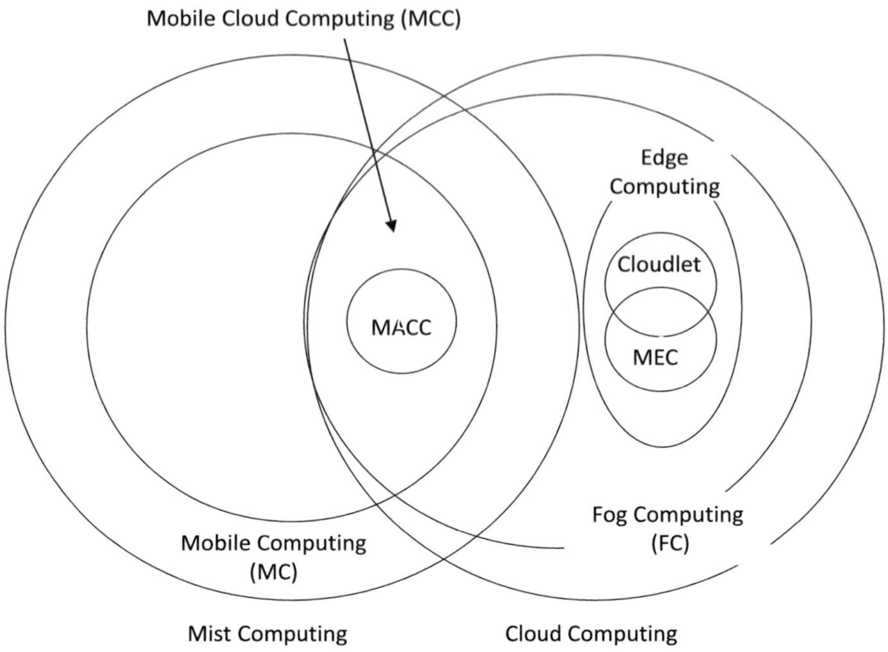

Fig. 20 Comparison of fog computing (FC) and its related computing paradigms [1, 7]

8.3.1 Multi-access Edge Computing (MEC)

A 5G-native architecture that extends cloud-computing capabilities to the network edge, minimizing latency and alleviating core network congestion. MEC nodes are strategically deployed within Radio Access Networks (RAN), Baseband Units (BBUs), Wi-Fi gateways, and enterprise premises, enabling ultra-reliable low-latency communication (URLLC) for applications such as autonomous driving, industrial automation, immersive media streaming, and smart city infrastructure. By processing data closer to the source, MEC enhances real-time analytics, mission-critical workloads, and context-aware services while reducing backhaul traffic to centralized cloud data centers. Integration with Network Function Virtualization (NFV) and Software-Defined Networking (SDN) further optimizes dynamic resource allocation and service orchestration [7].

8.3.2 Cloud-Native Edge Computing (CNEC)

A containerized and microservices-driven edge computing approach that aligns with cloud-native principles such as scalability, resilience, and automation. CNEC leverages Kubernetes-based orchestration, lightweight container runtimes (e.g., K3s, MicroK8s), and service mesh technologies to enable highly efficient edge deployments. Unlike traditional MEC, which may rely on virtualized environments, CNEC adopts serverless computing and event-driven architectures to optimize workload execution on constrained edge devices. This model is well-suited for distributed AI inference, 5G network slicing, and real-time IoT applications, enabling seamless application portability between edge and cloud environments [7].

8.3.3 Mobile Cloud Computing (MCC)

A cloud-based computational offloading model that enables resource-constrained mobile devices to execute compute-intensive tasks on remote cloud servers. MCC supports applications requiring high processing power and large-scale storage, such as deep learning model training, big data analytics, and high-performance simulations. However, it suffers from high round-trip latency and increased energy consumption due to centralized data transmission, making it unsuitable for real-time control systems, interactive applications, or URLLC workloads. Hybrid architectures combining MCC with edge computing approaches, such as MEC and cloudlets, aim to mitigate latency concerns by distributing processing workloads dynamically [7].

8.3.4 Edge AI

A computational paradigm that enables AI inference and ML models to be executed at the network edge, reducing the need for constant cloud connectivity. Edge AI is powered by hardware accelerators such as TPUs (Tensor Processing Units), NPUs (Neural Processing Units), FPGAs (Field-Programmable Gate Arrays), and dedicated AI chips like NVIDIA Jetson and Intel Movidius. This paradigm supports real-time AI-driven applications, including autonomous robotics, predictive maintenance, facial recognition, and smart surveillance. The main advantages of Edge AI include lower latency, reduced bandwidth usage, and enhanced privacy, as sensitive data is processed locally instead of being transmitted to the cloud [7].

8.3.5 Vehicular Edge Computing (VEC)

A specialized edge computing model where vehicles serve as computational nodes, either offloading tasks to nearby edge servers or collaborating with other vehicles in a V2X (Vehicle-to-Everything) ecosystem. VEC enables real-time processing for autonomous driving, smart traffic management, collision avoidance, and vehicular infotainment systems. Key technologies supporting VEC include 5G C-V2X (Cellular Vehicle-to-Everything), edge AI inference, and distributed federated learning. Challenges in VEC include dynamic network topology, latency-sensitive decision-making, and cybersecurity risks, necessitating efficient resource orchestration and AI-driven predictive analytics [7].

8.3.6 Dew Computing (DC)

An edge-extended cloud computing paradigm that facilitates local computation with intermittent cloud synchronization. Unlike continuous cloud dependency, DC ensures operational continuity even during network disruptions or intermittent connectivity, making it highly beneficial for remote monitoring systems, disaster recovery networks, and IoT-enabled environments. DC supports distributed data storage, content caching, and lightweight AI inference, optimizing bandwidth usage and energy efficiency in resource-constrained or unreliable connectivity settings [7].

8.3.7 Cloudlets

Small-scale, decentralized computing nodes that provide cloud-like processing capabilities in proximity to edge devices. Cloudlets enable low-latency computation via Virtual Machines (VMs), lightweight containers, or unikernels, supporting compute-intensive applications such as augmented reality (AR), virtual reality (VR), real-time video analytics, and mobile gaming [7].

8.3.8 Mobile Ad Hoc Cloud Computing (MACC)

A decentralized computing model in which mobile devices dynamically form ad hoc clusters to share computational workloads without requiring fixed infrastructure. Unlike traditional cloud architectures, MACC utilizes device-to-device (D2D) communication via Wi-Fi Direct, Bluetooth, ZigBee, and 5G sidelink to distribute processing tasks across multiple nodes [7].

8.3.9 Extreme Edge Computing (Mist Computing)

An ultra-distributed computing paradigm that utilizes idle computational resources from devices such as parked vehicles, drones, smart appliances, and IoT endpoints. Mist computing introduces an intermediate processing layer between traditional edge nodes and sensor networks, enabling real-time, context-aware, and energy-efficient task execution [7].

8.3.10 Gateway-Mediated Edge Connectivity (GEC)

A hierarchical networking paradigm in which edge gateways handle data aggregation, security enforcement, and protocol translation between local area networks (LANs) and wide-area networks (WANs). GEC enhances network resilience, scalability, and routing efficiency in Industrial IoT (IIoT), smart grid, and critical infrastructure deployments [7].

8.4 Critical Tasks in Edge-Fog-Cloud Computing

Efficient operation of edge-fog-cloud computing systems depends on a set of fundamental tasks that ensure optimal resource utilization, workload distribution, and system resilience. These tasks encompass a range of strategic and operational processes, from planning and provisioning to real-time allocation, workload migration, and orchestration. Addressing these challenges effectively is essential for maintaining performance, scalability, and energy efficiency in distributed computing environments. A key aspect of these tasks is resource management, which involves planning for future demands, provisioning resources in advance, dynamically allocating them based on real-time workload variations, and ensuring efficient service deployment and execution [1, 7].

- *Resource Planning:* A *proactive, high-level strategy* that anticipates future computational demands and infrastructure requirements. It uses workload forecasting, historical data analysis, and optimization models to determine the necessary provisioning of computing, storage, and network resources.
- *Resource Provisioning:* The *pre-allocation* of computing resources before they are actively used. This includes reserving capacity in advance based on demand predictions to prevent bottlenecks and underutilization.
- *Resource Allocation:* A *real-time, operational process* that assigns provisioned resources to workloads as tasks arrive. Allocation ensures efficient utilization of available resources while responding to dynamic workload fluctuations.
- *Service Placement:* The process of determining the optimal execution location—whether at the edge, fog, or cloud layer—for deploying services or computational tasks. Effective service placement minimizes latency, maximizes resource utilization, and ensures QoS compliance by strategically distributing workloads based on network conditions, resource availability, and workload requirements.
- *Task Offloading:* The process of transferring computational workloads from resource-constrained edge devices to more powerful fog or cloud nodes for execution. Offloading decisions are influenced by factors such as latency constraints, energy efficiency, network conditions, and computational demands to optimize overall system performance.
- *Service Orchestration:* A comprehensive management framework that coordinates and automates the provisioning, allocation, offloading, and execution of services across the edge-fog-cloud continuum. Orchestration strategies leverage centralized, hierarchical, or distributed control mechanisms to optimize workload distribution, improve fault tolerance, and enhance system responsiveness.

8.4.1 Capacity and Resource Planning

Resource planning is the strategic process of forecasting, estimating, and optimizing computing resources across edge, fog, and cloud environments to ensure that infrastructure meets future computational demands efficiently. Unlike resource allocation, which deals with the real-time distribution of resources, resource planning takes a predictive approach, analyzing historical trends, workload patterns, and system constraints to make informed decisions. The goal is to ensure that the computing ecosystem remains scalable, cost-efficient, and resilient, while balancing performance and energy consumption [1, 7].

Effective resource planning requires accurate demand estimation and optimization models to prevent both resource shortages and over-provisioning. Traditional deterministic models, such as Linear Programming (LP) and Mixed-Integer Programming (MIP), provide structured approaches to capacity estimation when workload behavior is predictable. However, in dynamic environments, stochastic models such as Markov Decision Processes (MDPs) and Monte Carlo simulations help account for uncertainty in demand fluctuations. Queuing theory further assists in analyzing workload arrival rates and system congestion to determine optimal resource distribution [1, 7].

To enhance accuracy and adaptability, modern resource planning integrates AI-driven predictive analytics. Time-Series Forecasting (e.g., LSTM, ARIMA) is used to analyze historical workload trends, while Reinforcement Learning (RL) dynamically learns from real-time data to optimize future resource planning decisions. Multi-Agent Learning enables collaborative decision-making across distributed edge, fog, and cloud nodes, reducing dependencies on centralized models [1, 7].

Resource planning typically faces key challenges such as workload unpredictability, latency constraints, and cost-efficiency trade-offs. Future advancements will rely on self-adaptive AI-driven models, decentralized resource planning frameworks using blockchain, and quantum-inspired optimization algorithms to improve decision-making in large-scale distributed computing environments. By integrating advanced forecasting techniques and intelligent planning models, computing infrastructures can ensure sustainable, efficient, and high-performance operations across the edge-fog-cloud continuum [1, 7].

8.4.2 2. Resource Provisioning

Resource provisioning is the process of preparing and making computing resources available—including CPU, memory, storage, and bandwidth—to support workloads. This step ensures that resources are accessible before they are needed, preventing performance bottlenecks and ensuring responsiveness. Since resources must first be provisioned before they can be allocated or assigned, provisioning is the first stage of the resource management process. Provisioning strategies are broadly categorized into two types [7]:

- *Prior Provisioning:* This proactive strategy provisions resources in advance based on predicted traffic patterns and workload forecasts. By estimating resource demands ahead of time, prior provisioning ensures immediate task processing and minimizes latency. It is particularly effective for predictable workloads, where demand fluctuations can be accurately forecasted [16].

- *On-Demand Provisioning:* In contrast, on-demand provisioning dynamically allocates resources only when a task request is received. This reactive approach optimizes resource utilization by preventing the over-provisioning of idle resources. However, since resources are allocated at the time of request, it may introduce slight provisioning delays before execution.

Each provisioning method has its advantages: prior provisioning ensures immediate resource availability but may lead to over-provisioning, while on-demand provisioning maximizes resource efficiency but may impact real-time responsiveness.

8.4.3 3. Resource Allocation

Resource allocation is the process of *assigning provisioned resources*—such as CPU, memory, storage, and bandwidth—to tasks or services. This step comes *after provisioning* and determines how these resources are distributed to ensure efficient execution of workloads. Effective resource allocation optimizes resource utilization, reduces latency, and balances system performance under dynamic workloads. It operates at two levels [7]:

- *Intra-Node Allocation:* At this level, resources are distributed within a single node, such as an edge device, fog node, or cloud instance. This fine-grained allocation determines how a node's CPU, GPU, memory, storage, or bandwidth is assigned to different tasks or processes. Efficient intra-node allocation enhances performance by ensuring that no single task monopolizes system resources while others starve.
- *Inter-Node Allocation:* This process focuses on distributing workloads *across multiple nodes* within the edge-fog-cloud infrastructure. Inter-node allocation determines which node (e.g., edge, fog, or cloud) is best suited to handle a task based on resource availability, network conditions, and QoS constraints. It works in tandem with *service placement*, ensuring that once a task is assigned to a node, resources are allocated efficiently to maintain system performance and responsiveness.

8.4.4 4. Service Placement

Service placement is the process of determining the optimal execution location—whether at the edge, fog, or cloud layer—where services or computational tasks should be deployed. Unlike resource allocation, which focuses on distributing available resources, service placement is a higher-level decision that dictates *where* a task should run based on factors such as latency, resource availability, and QoS requirements. This decision directly influences resource allocation, as once a service is placed on a specific node, its resources must be efficiently allocated to ensure smooth execution. The goals of service placement include minimizing latency, maximizing resource utilization, reducing energy consumption, and ensuring QoS for end-user applications. Placement strategies can be classified into the following categories [7]:

- *Static Service Placement:* Services are deployed at *fixed locations* without adapting to changes in network conditions or resource demands. Static placement is simple to implement but may lead to suboptimal performance in dynamic environments. For example, a service placed on a cloud node may experience high latency if network conditions degrade.
- *Dynamic Service Placement:* Placement strategies continuously monitor critical system parameters, such as end-to-end latency, bandwidth availability, and resource congestion, to *adaptively relocate* tasks and services. Dynamic placement ensures efficient resource utilization and optimized performance under fluctuating workloads. For instance, a real-time video analytics service may be migrated from a congested edge node to a nearby fog node to maintain low latency [17].

8.4.5 5. Task Offloading

Task offloading is the process of *dynamically transferring computational workloads* from *resource-constrained edge devices* to *fog or cloud nodes* for execution. This technique optimizes latency, energy efficiency, computational performance, and network adaptability, ensuring seamless operation of modern applications such as autonomous systems, real-time analytics, augmented reality (AR), and IoT deployments. Task offloading can be classified into full offloading, where an entire task is migrated, and partial offloading, where workloads are *partitioned* and distributed across multiple processing nodes. Advanced AI-driven decision-making mechanisms, such as reinforcement learning and deep neural networks, are increasingly used to predict optimal offloading decisions based on real-time conditions [7]. Task offloading strategies can be influenced by several considerations, such as [7]:

- *Latency Constraints:* Low-latency execution is critical for applications such as autonomous driving, robotics, and VR/AR systems. Offloading to proximal fog nodes reduces transmission delays and ensures real-time responsiveness, whereas cloud-based offloading may introduce higher round-trip latency [7].
- *Context-Aware Offloading Considerations:* Effective task offloading requires real-time adaptation to dynamic environmental conditions, user mobility, and heterogeneous resource availability. Decisions must consider factors such as workload priority, device energy constraints, and network stability to optimize performance while minimizing latency and resource contention. Additionally, offloading strategies should be adaptable to variations in application requirements, such as the need for low-latency execution in real-time applications or high-throughput processing for data-intensive workloads. Context-aware computing plays a crucial role in dynamically adjusting offloading policies based on situational parameters, ensuring optimal resource utilization and service quality [7].

- *Energy Efficiency:* Edge devices, which typically operate on limited battery power, conserve energy by offloading computationally intensive workloads to resource-rich fog or cloud nodes. This significantly reduces local processing power consumption, thereby extending device operational lifespan [7]. Energy-aware scheduling mechanisms further optimize offloading decisions by balancing processing energy costs and transmission overhead.
- *Network Conditions and Bandwidth Availability:* Network-aware task offloading must consider bandwidth constraints, congestion, packet loss, and link stability to minimize transmission delays [7].
- *Computational Requirements and Workload Partitioning:* Heavy computational tasks (e.g., AI model inference, large-scale data processing) are offloaded to resource-rich cloud or fog nodes, while lightweight or latency-sensitive tasks are processed locally. Partitioning strategies, such as workflow-based, dependency-aware, and energy-efficient partitioning, ensure that only compute-intensive segments are offloaded while maintaining critical components at the edge [7].
- *Security and Privacy Considerations:* Offloading computational tasks to external nodes introduces potential security and privacy risks, including data interception, unauthorized access, and integrity breaches. Ensuring secure offloading requires robust cryptographic techniques and trust mechanisms. Methods such as secure multi-party computation (MPC) and homomorphic encryption enable computation on encrypted data without exposing sensitive information, preserving confidentiality. Additionally, blockchain-based authentication and decentralized identity management enhance trust by providing immutable, tamper-resistant verification of transactions and access control. The implementation of trusted execution environments (TEE) further strengthens security by allowing sensitive computations to be executed in isolated, hardware-protected enclaves, preventing unauthorized modifications or leaks [7].
- *Reliability and Fault Tolerance:* Distributed edge-fog-cloud systems require fault-resilient offloading mechanisms to handle node failures, connectivity issues, and dynamic workload fluctuations. Redundant task replication, predictive failure detection, and checkpoint-based rollback mechanisms ensure that critical workloads continue running even in unreliable network environments [7].

8.4.6 6. Service Orchestration

Service orchestration refers to the automated and intelligent coordination of resource provisioning, allocation, and task offloading across the edge-fog-cloud continuum. It ensures that services and computational workloads are dynamically placed and optimized to meet performance, latency, and energy-efficiency constraints. Orchestration involves both decision-making (e.g., determining where services should be deployed) and execution (e.g., enforcing resource allocation and migration strategies). Orchestration strategies can be classified based on their operational mode and control architecture [7].

- *Online Versus Offline Orchestration* [7]:
  - **Offline Orchestration:** Optimization is performed before deployment, using mathematical models and solvers such as Gurobi or CPLEX to compute optimal service placement and resource allocation strategies. While this approach ensures global optimization and high QoS, it is computationally intensive and unsuitable for dynamic, time-sensitive environments.
  - **Online Orchestration:** Decision-making occurs in real-time, using heuristic, metaheuristic, and reinforcement learning-based approaches to adaptively optimize service placement and resource allocation. These methods significantly reduce computation time, making them suitable for dynamic and unpredictable workloads [18]. However, they may produce suboptimal solutions compared to offline methods.
- *Centralized Versus Hierarchical Versus Distributed Orchestration* [7]:
  - **Centralized:** A single global orchestrator—typically hosted in the cloud—has a system-wide view and makes globally optimal decisions regarding service placement, workload distribution, and resource allocation [19]. While this approach enables precise optimization, it introduces latency, scalability limitations, and a single point of failure.
  - **Hierarchical:** A multi-tier approach where lower-layer orchestrators (e.g., at fog nodes) make local optimization decisions, while higher-layer orchestrators (e.g., at the cloud) coordinate global policies. This structure balances local responsiveness with global efficiency, reducing latency while maintaining broader system-wide optimization [20].
  - **Distributed:** Multiple orchestrators, typically deployed at fog nodes, collaborate in a decentralized manner to dynamically optimize service placement and resource allocation. This approach enhances scalability, resilience, and fault tolerance, as decisions are made based on local resource conditions while leveraging inter-node cooperation to avoid suboptimal placements [21].
- *Edge-Driven Orchestration* [7]: This approach enables decentralized, real-time service placement and resource management initiated by edge devices. Unlike centralized approaches, where decisions are made at the cloud or fog level, edge-driven orchestration empowers end devices to make autonomous decisions based on real-time factors such as network conditions, resource availability, and latency constraints. This enhances system responsiveness and resilience. However, effective load balancing and coordination mechanisms are required to prevent resource congestion and suboptimal task distribution at the fog and edge layers [22].

9 Data Monetization: Where Do the Data Go

As SCH continues to evolve, the widespread adoption of wearable devices and other connected technologies has led to an overwhelming influx of data for patients, companies, and organizations. Unlocking the full potential of this data is essential for achieving success in competitive markets, as it provides immense opportunities to drive innovation, enhance decision-making, and deliver value across multiple dimensions of healthcare and business operations [23].

The monetization of SCH data has gained considerable momentum, driven by advancements in AI/ML-powered smart services and specialized platforms designed to process and analyze this data. These cutting-edge innovations empower organizations to extract actionable insights, utilize advanced analytics, and implement subscription-based models, thereby unlocking significant value from their data assets. This evolution is fueling the rise of an "intelligent and connected economy," where businesses capitalize on the seamless interconnectivity of devices, systems, and data to generate new revenue streams [23].

In this rapidly transforming landscape, the ability to efficiently and accurately monetize SCH data is emerging as a critical driver of sustainable competitive advantage in the data economy. This paradigm shift is redefining the dynamics of power between data-collecting organizations and the end users who generate the data, fostering a more collaborative and data-driven future [23].

Revenue generation from SCH data can typically be achieved through two primary approaches: **Direct Data Monetization** and **Indirect Data Monetization** [23]:

- **Direct Data Monetization:** In this approach, raw SCH data can be licensed to interested buyers through data marketplaces, offering access to valuable insights for a wide range of applications. This enables companies to efficiently unlock the financial potential of their data, generating immediate revenue while meeting the needs of diverse stakeholders.
- **Indirect Data Monetization:** SCH data can be strategically utilized to enhance business operations, create new products or services, and drive innovative business models. Optimizing IoT data typically requires the use of [2, 23]:
  - **Data-Driven Optimization:** SCH data can be harnessed to reduce costs and improve process efficiency through Big Data analytics and AI/ML technologies. These optimizations can be implemented by a single participant or through collaborative, multiparty, and privacy-saving computing techniques (e.g., Federated Learning). This approach is adaptable across various industries. For instance, SCH data can be employed to streamline healthcare workflows, shorten diagnostic timelines, or inform product development [2, 23].
  - **Data-Driven Business Models:** SCH data enables the creation of innovative services or products that drive business growth and improve customer acquisition. These models empower data owners to establish entirely new data-centric ventures rather than merely extending existing business offerings. For instance, healthcare providers are increasingly leveraging patient data to offer personalized telemedicine services, such as virtual health consultations or tailored wellness programs, that meet the specific needs of individual patients [2, 23].

9.1 Challenges of Data Silos

The historical siloing of healthcare data has left critical information uncaptured, unanalyzed, and disconnected. While the integration of SCH technologies in healthcare is accelerating, data remains fragmented across systems. Hospitals and healthcare organizations are increasingly focused on monetizing data by transforming previously siloed health information into actionable tools to reduce costs, enhance patient experiences, and support sound business decisions. However, this endeavor is fraught with challenges due to the fragmented nature of healthcare data, which is often isolated and categorized based on specific purposes, including [23]:

- **Strategic Planning Data:** Information used for long-term organizational strategies and decision-making.
- **Market Development Data:** Insights to identify new opportunities, expand services, and improve market reach.
- **Service Management Data:** Operational data for managing and optimizing healthcare services.
- **Claims Data:** Records related to insurance claims, reimbursements, and payment processes.
- **Diverse Clinical Data:** Patient information stored in electronic medical records (EMRs) and other clinical systems.
- **Medical Imaging:** This category includes both standardized formats, such as DICOM data for MRIs, X-rays, and CT scans, and non-DICOM formats, including dermatology photos, surgical videos, endoscopy recordings, pathological studies, and wound images in formats like JPEG, TIFF, and MPEG-4.
- **Patient-Generated Data:** Data collected from wearable devices, health apps, or at-home monitoring tools, providing real-time health insights.
- **Genomic Data:** Information derived from DNA sequencing or genetic studies, critical for personalized medicine and research.
- **Pharmaceutical Data:** Data related to drug prescriptions, clinical trials, pharmacovigilance, and adverse event monitoring.
- **Behavioral and Social Determinants of Health:** Information on lifestyle, socioeconomic factors, and behavioral health that impact patient outcomes.
- **Operational Data:** Data on staffing, resource allocation, and workflow management to optimize efficiency and reduce costs.
- **Financial Data:** Information on revenue cycles, billing, and cost management to support organizational profitability and sustainability.
- **Telemedicine Data:** Information from remote consultations, virtual care sessions, and telehealth devices, which has grown in importance post-pandemic.
- **Population Health Data:** Aggregated data used to track health trends and outcomes across demographics or regions, essential for public health initiatives.
- **Supply Chain Data:** Data on procurement, inventory, and logistics to ensure the availability of critical medical supplies.
- **Education and Training Data:** Information related to healthcare staff training, certifications, and performance.

9.2 *The Role of Data Sharing Across Data Silos*

Accessing healthcare data remains a significant challenge for researchers and clinicians due to its fragmentation across public and private silos. To improve patient outcomes and drive healthcare innovation, organizations must break down these silos by implementing integrated systems that can collect, process, and analyze real-time patient information. Achieving this requires expanding the scope of care, fostering collaboration among stakeholders, and providing clinicians with advanced tools to manage care within updated models.

According to Metcalfe's Law, the value of a network grows proportionally to the square of the number of its contributors. In healthcare, integrating siloed data does not result in diminishing returns; instead, it amplifies the overall value, unlocking exponential benefits. Key advantages of enhanced data sharing within healthcare systems include, but are not limited to [23]:

- **Holistic Patient Care:** Providing medical professionals with access to comprehensive health histories enables more accurate diagnoses, reduces redundant tests, and supports the development of personalized treatment plans.
- **Enhanced Business Decision-Making:** Sharing data supports strategic planning by identifying patterns and trends, such as readmissions in healthcare facilities. This enables organizations to develop strategies to improve outcomes, optimize resources, and minimize potential financial or regulatory consequences.
- **Advanced Analytics and AI Integration:** Access to integrated datasets enhances the capabilities of AI and machine learning models to recognize patterns in individual and population health, such as detecting tumors or early disease warning signs.
- **Public Health Insights:** Aggregated data from multiple healthcare systems supports monitoring of population health trends, tracking the spread of diseases, and designing effective public health policies.
- **Resource Optimization:** Data sharing helps identify gaps in care delivery, optimize staffing levels, and reduce waste in healthcare operations, leading to more efficient use of resources.
- **Collaboration and Innovation:** Shared data fosters partnerships among healthcare organizations, researchers, and technology providers, enabling the development of innovative care solutions, including new diagnostic tools and treatment protocols.

Breaking down data silos in healthcare is critical for building a more connected, efficient, and patient-centric ecosystem. By leveraging shared data, healthcare organizations can unlock unprecedented opportunities for improving care, reducing costs, and advancing medical research.

9.3 Challenges in SCH Data Monetization

Collapsing SCH data silos requires the implementation of robust legislation that promotes data sharing among private and federal healthcare organizations. Such regulations must include comprehensive security guidelines to safeguard Personally Identifiable Information (PII) and Protected Health Information (PHI). Despite advancements in legislation, significant challenges persist in areas such as data quality, information exchange, data provenance, and data mapping. Key barriers to effective healthcare data sharing include [23]:

- **Privacy Concerns:** While integrating data silos enables improved information sharing and analysis, ensuring the privacy of patient data remains paramount. Adherence to privacy regulations is critical, yet it becomes increasingly complex as the volume of shared information grows.
- **Data Security Challenges:** Protecting patient confidentiality and identity is essential during data sharing processes. The increasing reliance on virtual healthcare interactions, such as telemedicine, introduces heightened risks of data breaches due to electronic data transmission. Inadequate security infrastructures can lead to severe legal and financial repercussions, including medical identity theft.
- **Competitive Dynamics in Data Sharing:** Data sharing poses challenges due to its strategic value, which often provides a competitive edge for healthcare organizations. While data exchange between private and public entities is gradually narrowing the competitive divide, organizations must balance improving the patient experience with safeguarding their competitive positioning.
- **Workflow and Integration Barriers:** Technological barriers to integrating data silos are often overstated; instead, the primary impediment lies in workflow management. According to Soon-Shiong, a South African surgeon and researcher, "The barriers [to improving healthcare with data] technologically don't exist any longer...healthcare is falling behind other industries like banking and entertainment because it isn't using the technology properly—and that's a workflow management problem."
- **Cultural Barriers to Data Integration:** A significant obstacle to integrating siloed data is the entrenched culture within the healthcare sector, which assumes that withholding data enhances its value. In reality, siloing healthcare data diminishes value for organizations, providers, and patients alike, underscoring the need for cultural transformation.
- **Trust Deficit Among Organizations:** Trustworthy relationships between healthcare entities are critical for fostering digital communication and data sharing. While large healthcare networks often have established data-sharing systems, smaller or independent organizations frequently lack these capabilities, further exacerbating trust issues.
- **Scalability Constraints in Data Sharing:** The vast quantities of healthcare data present challenges in electronic transmission, particularly in regions with limited bandwidth or restrictive firewall settings. Scalability issues can impede data trans-

fer speeds and overall system responsiveness, particularly in rural or underserved areas.

- **Interoperability Standards and Enforcement:** The absence of enforced interoperability standards leads to inconsistencies in data formats and structures. This lack of standardization complicates data interpretation, integration, and sharing across heterogeneous systems.
- **Ownership Disputes in Healthcare Data:** Questions about who owns and controls healthcare data—patients, providers, or third-party vendors—often complicate data-sharing efforts. Clearly defining ownership is essential for fostering trust and enabling collaboration.
- **Legal and Regulatory Heterogeneity:** Different regions and countries impose varying legal and regulatory requirements for data sharing. Navigating these differences can be complex and burdensome, especially for organizations operating across borders.
- **Data Quality and Completeness Issues:** The utility of shared data often depends on its quality and completeness. Inconsistent, incomplete, or inaccurate data limits its usability, undermines trust, and reduces the effectiveness of analytics.
- **Patient-Centric Data Management Challenges:** As patient-centric data models gain traction, empowering individuals to control their data-sharing preferences introduces technical and logistical challenges, particularly in dynamically managing consent across multiple platforms.

9.4 Data Monetization Methods and Architectures in SCH

Typically, there are two primary architectures for data exchange and sharing that facilitate the transformation of SCH into a data-driven economy. These architectures form the backbone of data monetization by defining the frameworks, technologies, and processes necessary for efficient and secure data handling. The first architecture emphasizes centralized data platforms, where data is collected, stored, and monetized through a unified system. The second architecture focuses on decentralized or distributed frameworks, which leverage blockchain or peer-to-peer networks to ensure data sovereignty and enhance trust among stakeholders [23].

- **Centralized Solutions:** Centralized solutions, such as online stores, collaborative environments, data markets, or shared repositories, are often built using cloud platforms that enable the sharing, storage, and selling of data through access management systems. However, a significant challenge arises when a single organization (e.g., the market owner) collects vast amounts of data from various providers or producers. Data stored on a centralized server is inherently vulnerable to hacking, data leaks, and other security breaches.
  Moreover, in such systems, critical data is controlled by the server owner, and all data transactions flow through a single point, limiting the ability of data providers

to govern how their data is used after release. This lack of control raises concerns about misuse or unauthorized access, undermining the confidence of primary stakeholders.
Additionally, centralized solutions often deter stakeholders from sharing private or sensitive data due to risks of exposure, loss of competitive advantage, or breaches of trust. Proprietary information or strategic business data is particularly at risk in centralized setups, further diminishing their appeal to those concerned with maintaining privacy and control [23].
- **Decentralized Solutions:** Decentralized solutions aim to empower data providers by granting them control over the information they generate or share. These solutions enable data producers and consumers to interact, communicate, and exchange data in a peer-to-peer environment without relying on centralized administration. This approach is inherently fault-tolerant, eliminating the risk of a single point of failure within the system.
Typically, decentralized markets such as Enigma Data, Ocean Protocol, Streamr, or IOTA utilize secure technologies like smart contracts, DLT, and blockchain. These technologies allow data producers to directly engage in transactions with data consumers while maintaining full control over their data. This structure not only enhances security and trust but also ensures that data ownership and usage remain in the hands of the original providers [23].

Decentralized solutions have been implemented using various innovative models, each leveraging unique technologies and frameworks to empower data providers and enhance data security. Below, we discuss some of the most prominent models [23]:

1. Blockchain-Based Solutions: Data storage on blockchain can generally be implemented using two distinct approaches, each designed to address scalability and security challenges.

- **On-Chain Solutions:** In this approach, data producers store raw, encrypted data directly on the blockchain. While this method ensures strong immutability and transparency, it is typically not scalable due to the inherent limitations of blockchain networks, such as low throughput and constrained transaction processing capabilities.
- **Off-Chain Solutions:** Here, raw data is stored off-chain, while reference points (metadata or unique identifiers) for the raw data are securely stored on the blockchain. This approach is more scalable and secure, as it avoids overloading the blockchain with large volumes of data. By maintaining control over the raw data, producers can decide whether to share access by exchanging the reference points. Furthermore, off-chain solutions enhance security by preventing unauthorized access—any intercepted reference points cannot be used to access the data without proper authorization.

Off-chain solutions are widely regarded as a practical and efficient way to manage data sharing while preserving ownership and privacy for data producers. Figure 21 illustrates a conceptual model of a blockchain-based data-sharing frame-

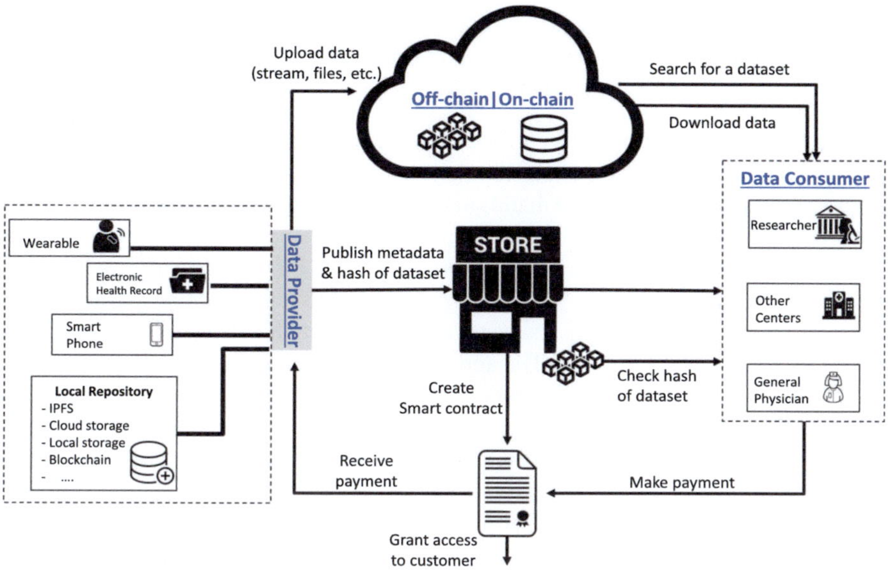

Fig. 21 A conceptual blockchain-based data marketplace, data sharing, and data monetization [23]

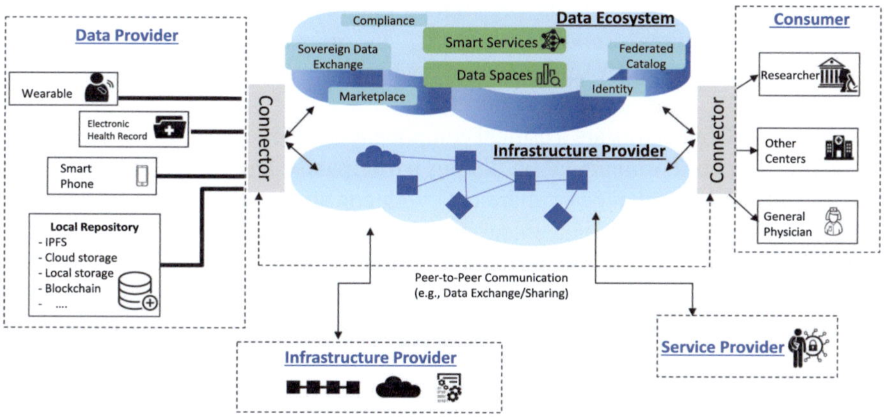

Fig. 22 A conceptual federated data ecosystem platform [23]

work, demonstrating how decentralized technologies enable secure and controlled data exchanges [24–30].

2. Federated Systems: Fig. 22 illustrates a conceptual architecture for Federated Systems, inspired by initiatives such as GAIA-X [31] and the International Data Space (IDS) [32]. These architectures facilitate sovereign data sharing through peer-to-peer connections between consumers and producers, ensuring greater autonomy and control over data usage. Federated systems go beyond data monetization, offering robust solutions for managing infrastructures and enabling advanced smart services. The architecture comprises several key components and roles, each contributing to the system's functionality and integrity [23]:

- **Data Producers/Suppliers/Vendors/Providers/Sources:** A producer refers to an entity, organization, or business that generates or legally owns and controls the data. Providers, on the other hand, act as intermediaries that collect data from multiple producers and offer those data assets to the ecosystem on behalf of the producers. It is important to note that producers can also operate as providers.
- **Service Providers:** Service providers contribute software, machine learning models, and other services to the ecosystem, enhancing its functionality and usability.
- **App (Service) Marketplace:** The app marketplace serves as a platform for application developers and service providers to design and deploy data-driven and AI-powered services. It enables developers to create innovative solutions and make them accessible to participants within the ecosystem, fostering collaboration and enhancing the overall functionality of the system.
- **Consumers (Data Sink):** Consumers are entities that utilize resources or assets (e.g., data, services) in accordance with the usage guidelines and policies defined by producers or providers.
- **Infrastructure Providers:** Also known as nodes, infrastructure providers supply computing resources to the ecosystem, supporting its operations and scalability.
- **Brokers:** Brokers facilitate the registration and discovery of available resources (e.g., infrastructures, services, datasets) by managing metadata and self-descriptions. Multiple brokers can coexist within the system, ensuring redundancy and scalability.
- **Connector:** Connectors are a critical component of the ecosystem, enabling seamless peer-to-peer communication among participants. They allow data suppliers and producers to exchange or share data with consumers efficiently and securely. Additionally, connectors can integrate with application (service) marketplaces, facilitating the purchase and download of apps or source code to enhance their functionality. For example, connectors can be utilized for tasks such as data cleaning or executing AI algorithms, including Multi-Party Computation (MPC) and Privacy-Preserving Machine Learning (PPML) models.
- **Identity Providers:** Identity providers are responsible for creating and managing the identity information of participants within the ecosystem, ensuring secure interactions and accountability.
- **Federators:** Federators enable and facilitate interactions between providers and consumers, ensuring seamless coordination and efficient resource sharing.

10 Conclusion

In this chapter, we explored the concept of SCH, starting with its definition and foundational principles. We delved into the architectural framework of SCH systems, discussing their key components, such as cloud computing, and the pivotal roles they play in enabling seamless data management and analysis. Additionally, we examined the layered structure of SCH, highlighting how each layer contributes to the overall functionality, from data collection to processing and decision-making. This comprehensive discussion provides a solid foundation for understanding SCH and its potential to revolutionize healthcare through innovation, efficiency, and improved patient outcomes.

Acknowledgements The authors acknowledge the use of Generative AI (GenAI) in editing and assisting with the development of this book chapter.

References

1. F. Firouzi, S. Jiang, K. Chakrabarty, B. Farahani, M. Daneshmand, J. Song, and K. Mankodiya, "Fusion of iot, ai, edge–fog–cloud, and blockchain: Challenges, solutions, and a case study in healthcare and medicine," *IEEE Internet of Things Journal*, vol. 10, no. 5, pp. 3686–3705, 2022.
2. F. Firouzi, K. Chakrabarty, and S. Nassif, *Intelligent Internet of Things: From Device to Fog and Cloud*. Springer, 2020.
3. B. Farahani, F. Firouzi, V. Chang, M. Badaroglu, N. Constant, and K. Mankodiya, "Towards fog-driven iot ehealth: Promises and challenges of iot in medicine and healthcare," *Future generation computer systems*, vol. 78, pp. 659–676, 2018.
4. M. Richards and N. Ford, *Fundamentals of software architecture*. "O'Reilly Media, Inc.", 2025.
5. S. Brown, "C4 model for visualising software architecture," https://c4model.com, 2024, accessed: 2025-03-25.
6. S. Newman, *Building microservices: designing fine-grained systems*. "O'Reilly Media, Inc.", 2021.
7. F. Firouzi, B. Farahani, and A. Marinšek, "The convergence and interplay of edge, fog, and cloud in the ai-driven internet of things (iot)," *Information Systems*, vol. 107, p. 101840, 2022.
8. B. Farahani, F. Firouzi, and K. Chakrabarty, "Healthcare iot," in *Intelligent Internet of Things*. Springer, 2020, pp. 515–545.
9. S. Tomovic, K. Yoshigoe, I. Maljevic, and I. Radusinovic, "Software-defined fog network architecture for iot," *Wireless Personal Communications*, vol. 92, no. 1, pp. 181–196, 2017.
10. E. Saurez, K. Hong, D. Lillethun, U. Ramachandran, and B. Ottenwälder, "Incremental deployment and migration of geo-distributed situation awareness applications in the fog," in *Proceedings of the 10th ACM International Conference on Distributed and Event-based Systems*, 2016, pp. 258–269.
11. P. Bellavista, J. Berrocal, A. Corradi, S. K. Das, L. Foschini, and A. Zanni, "A survey on fog computing for the internet of things," *Pervasive and mobile computing*, vol. 52, pp. 71–99, 2019.
12. O. Consortium *et al.*, "Openfog reference architecture for fog computing," *Architecture Working Group*, pp. 1–162, 2017.
13. F. Firouzi and B. Farahani, "Architecting iot cloud," in *Intelligent Internet of Things*. Springer, 2020, pp. 173–241.
14. J. Ren, D. Zhang, S. He, Y. Zhang, and T. Li, "A survey on end-edge-cloud orchestrated network computing paradigms: Transparent computing, mobile edge computing, fog computing, and cloudlet," *ACM Computing Surveys (CSUR)*, vol. 52, no. 6, pp. 1–36, 2019.
15. A. Yousefpour, C. Fung, T. Nguyen, K. Kadiyala, F. Jalali, A. Niakanlahiji, J. Kong, and J. P. Jue, "All one needs to know about fog computing and related edge computing paradigms: A complete survey," *Journal of Systems Architecture*, vol. 98, pp. 289–330, 2019.

16. I. Martinez, A. S. Hafid, and A. Jarray, "Design, resource management and evaluation of fog computing systems: A survey," *IEEE Internet of Things Journal*, 2020.
17. R. Cziva, C. Anagnostopoulos, and D. P. Pezaros, "Dynamic, Latency-Optimal vNF Placement at the Network Edge," in *IEEE INFOCOM 2018 - IEEE Conference on Computer Communications*. Honolulu, HI: IEEE, Apr. 2018, pp. 693–701. [Online]. Available: https://ieeexplore.ieee.org/document/8486021/
18. F. Ait Salaht, F. Desprez, A. Lebre, C. Prud'homme, and M. Abderrahim, "Service Placement in Fog Computing Using Constraint Programming," in *2019 IEEE International Conference on Services Computing (SCC)*. Milan, Italy: IEEE, Jul. 2019, pp. 19–27. [Online]. Available: https://ieeexplore.ieee.org/document/8814186/
19. L. Zhao and J. Liu, "Optimal Placement of Virtual Machines for Supporting Multiple Applications in Mobile Edge Networks," *IEEE Transactions on Vehicular Technology*, pp. 1–1, 2018. [Online]. Available: http://ieeexplore.ieee.org/document/8295232/
20. O. Skarlat, M. Nardelli, S. Schulte, and S. Dustdar, "Towards QoS-Aware Fog Service Placement," in *2017 IEEE 1st International Conference on Fog and Edge Computing (ICFEC)*. Madrid, Spain: IEEE, May 2017, pp. 89–96. [Online]. Available: http://ieeexplore.ieee.org/document/8014364/
21. Z. Nezami, K. Zamanifar, K. Djemame, and E. Pournaras, "Decentralized Edge-to-Cloud Load-balancing:Service Placement for the Internet of Things," arXiv:2005.00270 [cs], May 2020, arXiv:2005.00270. [Online]. Available: http://arxiv.org/abs/2005.00270
22. J. Tang, R. Yu, S. Liu, and J.-L. Gaudiot, "A Container Based Edge Offloading Framework for Autonomous Driving," *IEEE Access*, vol. 8, pp. 33 713–33 726, 2020. [Online]. Available: https://ieeexplore.ieee.org/document/8995560/
23. F. Firouzi, B. Farahani, M. Barzegari, and M. Daneshmand, "Ai-driven data monetization: The other face of data in iot-based smart and connected health," *IEEE Internet of Things Journal*, vol. 9, no. 8, pp. 5581–5599, 2020.
24. K. Fan, S. Wang, Y. Ren, H. Li, and Y. Yang, "Medblock: Efficient and secure medical data sharing via blockchain," *Journal of medical systems*, vol. 42, no. 8, p. 136, 2018.
25. Q. Xia, E. B. Sifah, K. O. Asamoah, J. Gao, X. Du, and M. Guizani, "Medshare: Trust-less medical data sharing among cloud service providers via blockchain," *IEEE Access*, vol. 5, pp. 14 757–14 767, 2017.
26. X. Liang, J. Zhao, S. Shetty, J. Liu, and D. Li, "Integrating blockchain for data sharing and collaboration in mobile healthcare applications," in *2017 IEEE 28th annual international symposium on personal, indoor, and mobile radio communications (PIMRC)*. IEEE, 2017, pp. 1–5.
27. Q. Xia, E. B. Sifah, A. Smahi, S. Amofa, and X. Zhang, "Bbds: Blockchain-based data sharing for electronic medical records in cloud environments," *Information*, vol. 8, no. 2, p. 44, 2017.
28. S. Wang, Y. Zhang, and Y. Zhang, "A blockchain-based framework for data sharing with fine-grained access control in decentralized storage systems," *Ieee Access*, vol. 6, pp. 38 437–38 450, 2018.
29. H. Shafagh, L. Burkhalter, A. Hithnawi, and S. Duquennoy, "Towards blockchain-based auditable storage and sharing of iot data," in *Proceedings of the 2017 on Cloud Computing Security Workshop*, 2017, pp. 45–50.
30. T. Sultana, A. Almogren, M. Akbar, M. Zuair, I. Ullah, and N. Javaid, "Data sharing system integrating access control mechanism using blockchain-based smart contracts for iot devices," *Applied Sciences*, vol. 10, no. 2, p. 488, 2020.
31. "GAIA-X, howpublished = https://www.gaia-x.eu/."
32. "International Data Space, howpublished = https://internationaldataspaces.org/."

Foundations of Machine Learning: Concepts and Algorithms

Bahar Farahani, Farshad Firouzi, and Aritra Ray

> *"Machines take me by surprise with great frequency."*
>
> *– Alan Turing*

1 Introduction to Machine Learning Models

Learning encompasses such a wide range of processes that it is difficult to define precisely. The dictionary defines learning as "gaining knowledge, skill, or understanding through study, experience, or instruction," as well as "modification of behavior through experience." While zoologists and psychologists study learning from the perspective of biological behaviors, our focus is on the learning process in machines. Many concepts in machine learning are drawn from psychological theories of human learning, which have been refined through computational models. Conversely, developments in machine learning have occasionally influenced the study of biological learning processes.

In machine learning [1], learning refers to the process by which a system refines its algorithms and patterns based on input data so it can improve its accuracy or efficiency in solving similar problems in the future. These systems learn from historical

B. Farahani (✉)
Shahid Beheshti University, Tehran, Iran
e-mail: b_farahani@sbu.ac.ir

F. Firouzi
Arizona State University, Arizona, USA

A. Ray
Duke University, Duke, USA

F. Firouzi et al. (eds.), *Smart and Connected Health: AI, IoT, and Trustworthy Technologies*, https://doi.org/10.1007/978-3-032-06286-4_2

data or in response to external information, with the goal of creating a more accurate model or prediction. Some changes, such as expanding a database by adding new information, may not strictly be considered "learning," but they can contribute to improved performance. For instance, when an image recognition system learns to differentiate between cats and dogs after processing multiple images, we can confidently say that the system has learned.

A typical machine learning process can be broken down into three main components: input, model, and output. For example, if we are given input data such as precipitation, humidity, and temperature, the task may be to predict whether the weather will be sunny or rainy. The core of this prediction process is the machine learning model, which maps the relationships between precipitation, humidity, and temperature to the actual weather conditions. Fortunately, modern machine learning techniques offer a variety of methods to achieve this, such as neural networks, logistic regression, and support vector machines. Each of these approaches provides different strategies for constructing accurate models and handling complex tasks, significantly expanding the possibilities for what machine learning systems can achieve through data-driven learning.

Machine learning models are typically categorized based on the nature of the learning process and the type of data available for training. Traditionally, these models are grouped into **three primary categories**: **supervised learning, unsupervised learning, and reinforcement learning**. However, a fourth category, **semi-supervised learning**, is often recognized as an intermediate hybrid approach that combines elements of both supervised and unsupervised learning. Additionally, **self-supervised learning** has recently gained prominence as a subcategory of unsupervised learning that enables models to learn representations from unlabeled data through automatically generated labels.

1.1 Supervised Learning

Supervised learning is one of the most intuitive and widely understood machine learning models. The term "supervised" comes from the process by which the model learns from a training dataset, much like a student learns under the guidance of a teacher. In this context, the training data serves as the "teacher" by providing both inputs and the correct outputs, allowing the model to learn through examples.

In a supervised learning model, the inputs and outputs are clearly defined, even if the internal workings of the algorithm may be complex. For instance, suppose we want to predict tomorrow's weather. The output (weather) is clearly defined—whether it will be sunny, cloudy, or rainy. However, predicting the weather involves collecting various relevant parameters, such as temperature and humidity, which indirectly influence the weather. These parameters, while not directly indicating the weather, can be used to make accurate predictions.

To train the weather prediction model, we gather historical data of weather conditions and the corresponding parameters. During the training process, the correct outcomes (weather conditions) are already known, allowing the model to iteratively make predictions and be corrected when wrong. This process continues until the model achieves a satisfactory level of accuracy, at which point it can be used to make future predictions with a certain degree of confidence.

Supervised learning models can be further divided into two main categories:

- **Classification Models**: These models are used when the output is a discrete label or class. Classification problems involve predicting categorical outcomes. For example, a model might classify weather conditions as *sunny*, *cloudy*, or *rainy*, or categorize fruits as *orange*, *apple*, or *pear*. In healthcare, a classification model could be used to diagnose a disease by classifying a patient as *healthy*, *at-risk*, or *diseased* based on their medical data.
- **Regression Models**: These models are applied when the output is a continuous numerical value. Regression problems involve predicting quantities that vary over a continuous range. For example, a regression model could be used to predict a patient's blood pressure based on factors such as age, weight, and medical history.

1.2 Unsupervised Learning

Unlike supervised learning, unsupervised learning derives insights from data without predefined labels, focusing on uncovering hidden patterns, structures, or relationships. Unsupervised learning provides a rich suite of techniques for exploring data and uncovering insights that are often inaccessible through supervised approaches. Key tasks and techniques in unsupervised learning include:

- **Clustering**: Clustering involves grouping objects such that items in the same cluster share similar characteristics, while items in different clusters are more dissimilar. For example, in healthcare, clustering can group patients based on similarities in their medical data, identifying subgroups with similar symptoms or treatment responses and enabling personalized treatment plans.
- **Dimensionality Reduction**: This technique reduces the number of features or variables in a dataset while preserving its essential information. It is particularly useful for high-dimensional data, improving visualization, computational efficiency, and model performance. Common methods include Principal Component Analysis (PCA) and t-Distributed Stochastic Neighbor Embedding (t-SNE). While most dimensionality reduction techniques are unsupervised, some, such as Linear Discriminant Analysis (LDA), use class labels and fall under supervised learning.
- **Anomaly Detection**: Anomaly detection identifies data points that deviate significantly from the majority of the dataset, often indicating rare events, fraud, or errors.

- **Self-Supervised Learning (SSL)**: Self-supervised learning (SSL) is a powerful machine learning paradigm in which models generate their own supervisory signals directly from input data, eliminating the reliance on large labeled datasets. Unlike traditional supervised learning, which depends on human-annotated labels, SSL constructs pretext tasks that enable models to discover meaningful patterns by solving self-generated challenges within the data. In natural language processing (NLP), for example, models like BERT learn by predicting masked words within a sentence, while in computer vision, SSL techniques enhance feature learning by comparing different augmented versions of the same image. The primary objective of SSL is to develop high-quality, transferable representations that require minimal additional training for downstream tasks such as classification or object detection. A major advantage of SSL is its ability to leverage vast amounts of unlabeled data, significantly reducing the costs and effort associated with manual annotation. This approach has led to groundbreaking advancements in computer vision, NLP, and speech processing, positioning SSL as a key enabler of progress in unsupervised and semi-supervised learning.

1.3 Semi-supervised Learning

Semi-supervised learning is a machine learning paradigm that effectively bridges the gap between supervised and unsupervised learning by utilizing a small amount of labeled data alongside a large pool of unlabeled data. This approach is especially valuable in scenarios where acquiring labeled data is costly, time-intensive, or laborious, but abundant unlabeled data is easily accessible. By combining the strengths of both labeled and unlabeled data, semi-supervised learning enables models to achieve better generalization and performance while minimizing the need for extensive human annotation. Techniques in this paradigm often involve leveraging the underlying structure or patterns in the unlabeled data to enhance the learning process, guided by the limited labeled examples. For example, in computer vision, semi-supervised learning is commonly used for image classification. A model may be trained on a small set of labeled images while leveraging a much larger dataset of unlabeled images. Techniques such as pseudo-labeling assign predictions to unlabeled images with high confidence and use them as additional labeled data to refine the model. This approach allows models to improve their accuracy while significantly reducing the need for manual image annotation. As a result, semi-supervised learning has found widespread application in domains such as speech recognition, computer vision, and NLP, where even a small set of labeled data can significantly improve the model's ability to learn from a much larger, unlabeled dataset. This makes it a highly efficient and practical approach for real-world machine learning challenges.

1.4 Reinforcement Learning

Reinforcement Learning (RL) [2] is the third major branch of machine learning. The core idea behind RL is that an intelligent agent interacts with a dynamic environment, learning to take actions that maximize cumulative rewards over time. In other words, RL tackles sequential decision-making problems by sampling actions, executing them, and then evaluating the outcomes. Based on feedback, the agent improves future actions, using previous experiences to inform better decisions. RL has several promising applications across industries. For example, in the field of healthcare, RL can be used in recommendation engines for personalized treatment plans, dynamic treatment regimes that adapt to a patient's evolving condition, and healthcare resource scheduling and allocation, helping optimize the use of resources in hospitals and clinics.

2 Machine Learning Pipeline

The ML pipeline is a structured sequence of stages—spanning data collection, preprocessing, model training, evaluation, deployment, and monitoring—that automates and streamlines the end-to-end process of developing, deploying, and maintaining machine learning models throughout their lifecycle.

2.1 Machine Learning Project Lifecycle

ML projects follow a structured workflow to ensure efficiency and accuracy. This workflow helps in systematically handling data-driven problems and refining models iteratively (see Fig. 1). One widely recognized framework is the **CRoss-Industry Standard Process for Data Mining (CRISP-DM)** [3], which provides a structured approach to machine learning and data mining. Rather than a strictly linear sequence, the machine learning pipeline is often **iterative**—data may be revisited, models refined, and performance metrics reassessed to guide further improvements. The main phases of this lifecycle are as follows:

- **Business and Data Understanding**: The first and most critical step in a machine learning project is **defining the problem statement** and understanding the **business objectives**. A well-defined problem statement ensures that subsequent steps align with business goals, preventing wasted efforts on irrelevant models. This phase involves **identifying key factors** that may influence the project, such as external conditions, industry-specific constraints, or user behavior patterns. Additionally, it requires **gathering relevant data** from diverse sources, such as sensor readings, customer interaction logs, transactional records, and biomedical data, ensuring that the dataset is comprehensive and suitable for analysis. Equally

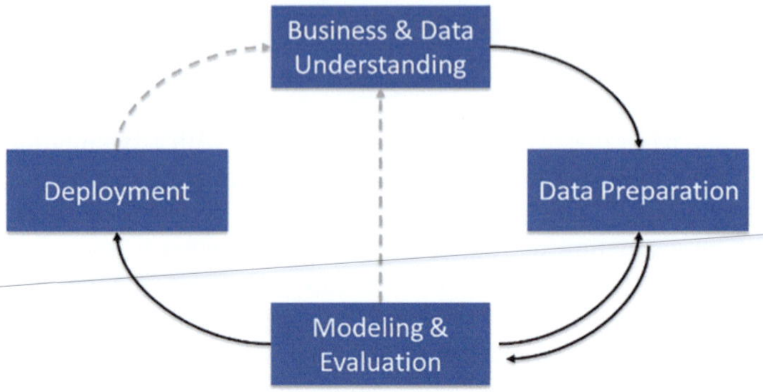

Fig. 1 Machine learning flow [4]

important is **defining success metrics and Key Performance Indicators (KPIs)** to measure the effectiveness of the model. These KPIs should align with real-world objectives, such as improving predictive accuracy, minimizing error rates, or enhancing decision-making processes. By thoroughly understanding both the **business context and the data landscape**, this phase establishes a **solid foundation** for building a machine learning pipeline that is both **relevant and impactful**.

- **Data Preparation**: Once the data is collected, it must be **preprocessed and structured** to ensure compatibility with machine learning algorithms. This phase involves **formatting** the data appropriately, handling **missing values** through techniques such as imputation or removal, and converting **categorical variables** into numerical representations using methods like **one-hot encoding** or **label encoding**. Additionally, features may need to be **normalized or standardized** to ensure consistency across different scales, preventing bias in model training. Finally, the dataset is typically **split into training and test sets** to evaluate the model's performance on unseen data. Proper data preparation is crucial for improving **model accuracy, efficiency, and generalization** in real-world applications.
- **Modeling and Evaluation**: In this phase, multiple machine learning models are trained and assessed to determine their **performance and suitability** for the given problem. This involves selecting appropriate algorithms, tuning **hyperparameters**, and optimizing model architectures to achieve the best results. Evaluation is conducted using metrics such as **accuracy, precision, recall, F1-score, or mean squared error**, depending on the task (classification, regression, etc.). The best-performing model is then chosen based on these metrics, ensuring it generalizes well to unseen data. Since machine learning is an **iterative process**, feedback from this phase often leads to refinements in **earlier steps**, such as adjusting **data preprocessing**, improving **feature selection**, or even redefining the **problem scope** to enhance overall performance.

  It should be noted that beyond accuracy and standard performance metrics, model selection and evaluation must also consider the constraints of the **target hardware** where the model will be deployed. For **edge AI and IoT devices**, factors such as **memory footprint, power consumption, and inference latency** are critical, as these devices often have limited computational resources. In such cases, model optimization and compression techniques like **quantization-aware training (QAT), pruning, and knowledge distillation** can significantly reduce model size while maintaining performance. For **TinyML applications**, where models must run on microcontrollers with extremely low power, lightweight architectures are preferred. Conversely, **cloud-based deployments** can accommodate larger, more complex models but must ensure **scalability and cost-efficiency**. Thus, during the modeling phase, trade-offs between **accuracy, computational efficiency, and hardware compatibility** must be carefully evaluated to ensure a smooth transition from training to deployment.

- **Deployment**: The final phase involves **deploying the selected model** into a production environment, where it begins making real-world predictions. However, deployment is not the end of the process—**ongoing monitoring and maintenance** are critical to ensure continued performance and adaptability to changing data patterns. Models in production may experience **concept drift**, where data distributions change over time, reducing accuracy. To address this, continuous **performance tracking**, **retraining**, and **hyperparameter tuning** may be required. Additionally, mechanisms such as **A/B testing** or **shadow deployment** can be used to validate model updates before full-scale implementation. Feedback from the deployed model's performance often leads to iterative refinements, requiring adjustments in **feature engineering**, **data preprocessing**, or even the selection of an entirely new model. A well-structured deployment strategy ensures that machine learning solutions remain **scalable, reliable, and adaptable** over time.

2.2 Data Preparation and Preprocessing

In any machine learning workflow, **data preparation and preprocessing** are essential steps that directly impact model effectiveness. Raw data is often inconsistent, incomplete, or unstructured, making it necessary to refine and transform it before feeding it into machine learning models. Without proper preprocessing, models may struggle to learn meaningful patterns, leading to poor generalization and unreliable predictions. **Data preparation** refers to the broader process of collecting, integrating, and organizing raw data, while **data preprocessing** focuses on refining and transforming it to make it suitable for learning.

A fundamental aspect of preprocessing is **data cleaning**, which ensures data quality by handling missing values, correcting errors, and removing duplicate or inconsistent records. When data is collected from multiple sources, **data wrangling** becomes necessary to restructure and integrate datasets, ensuring consistency across different formats and eliminating redundancies. These steps establish a clean and reliable foundation for further transformations.

Once data has been cleaned and structured, **normalization and feature scaling** play a critical role in ensuring that numerical variables contribute proportionally during model training. Since datasets often contain features with varying ranges, normalization ensures that all numerical values are represented on a consistent scale, preventing models from being biased toward features with larger magnitudes.

Similarly, **encoding categorical data** is essential, as machine learning models typically require numerical inputs. Since categorical variables can represent distinct groups or ordered levels, encoding methods help translate these values into numerical representations while preserving their meaning. The choice of encoding method depends on the nature of the data and the type of model being used, as some algorithms can handle categorical values directly, while others require explicit transformations. Encoding ensures that categorical information is effectively utilized without introducing unintended biases or distortions.

By ensuring that data is properly cleaned, structured, and transformed, **data preparation and preprocessing** lay the foundation for robust and efficient machine learning models. Well-prepared data reduces inconsistencies, enhances interpretability, and improves generalization, ultimately leading to more accurate and reliable outcomes.

2.2.1 Preparing the Dataset

A well-structured dataset is the foundation of any machine learning model. Before training a model, it is essential to **gather, construct, and format the data** to ensure it is clean, organized, and suitable for learning. Data can originate from various sources, including databases, APIs, sensors, web scraping, and manual collection, and it often exists in different formats that require proper structuring. Data is typically categorized into three main types:

- **Structured Data**: Data that follows a predefined schema and can be neatly arranged into rows and columns. Examples include relational databases, spreadsheets, and tabular data in CSV format, where each entry follows a fixed structure with clearly defined attributes.
- **Semi-Structured Data**: Data that does not adhere strictly to a tabular format but still contains some organizational elements that allow it to be converted into structured data. Examples include JSON, XML, and YAML files, which contain key-value pairs or hierarchical data structures that can be parsed into tabular representations.
- **Unstructured Data**: Data that lacks a predefined format, making it more challenging to process directly. This category includes text documents, images, videos, audio recordings, and sensor data. Unlike structured data, unstructured data often requires additional preprocessing steps such as feature extraction, NLP, or computer vision techniques to convert it into a usable format.

In machine learning, the representation of datasets depends on the type of data being processed. **Tabular datasets** are commonly represented in **tables**, where:

- Each **row** corresponds to an **observation** (also referred to as a **sample, data point, or instance**), representing a single record in the dataset.
- Each **column** represents a **feature** (also known as an **attribute or independent variable**), which is a measurable property of the observation that contributes to the predictive modeling process.
- In **supervised learning**, one column is designated as the **label** (also called the **target variable, response variable, or dependent variable**), representing the output that the model aims to predict based on the input features.

However, for non-tabular data such as images, audio, or graphs, representation varies:

- **Images** are typically represented as multi-dimensional arrays (tensors), where each pixel corresponds to a numerical value across different color channels.
- **Text data** is often converted into numerical form using vector embeddings or tokenized sequences before being processed by machine learning models.
- **Graph-based data** is represented using nodes and edges, where relationships between entities play a crucial role in learning.

By ensuring that the dataset is well-structured and appropriately formatted based on its nature, machine learning models can efficiently process and learn from the data, leading to **better generalization, reduced errors, and improved predictive accuracy**. Preparing the dataset correctly minimizes noise and inconsistencies, enabling models to derive meaningful patterns and insights effectively.

2.2.2 Handling Missing Values

Missing data is common in real-world datasets and can appear in various forms, such as 'NaN', 'None', or '?'. Handling missing values appropriately is crucial for building effective machine learning models. Below are key strategies for managing missing values:

- **Dropping Features (Removing Rows)**: If a feature has a high proportion of missing data and is not crucial, it may be better to drop the feature entirely.
- **Domain Knowledge**: Utilize domain expertise to interpret and handle missing data in meaningful ways.
- **Imputation**: Replace missing values with statistical measures such as the mean, median, or mode, depending on the distribution. Some of the more advanced techniques are listed below:
  - **K-Nearest Neighbors (KNN) Imputation**: Imputes missing values by identifying the most similar data points (neighbors) in the feature space. This method can be used for both numerical and categorical features.
  - **Multiple Imputation**: Generates several plausible datasets by imputing missing values multiple times and combines the results. Applicable to both numerical and categorical data.

  - **MICE (Multiple Imputation by Chained Equations)**: Iteratively imputes missing values by treating each feature (whether numerical or categorical) as the target variable in a predictive model based on the other features.

- **Interpolation**: Estimate missing values using methods like linear or polynomial interpolation based on neighboring data points.
- **Prediction**: Use regression models or other predictive algorithms to estimate missing values based on related features.
- **Frequent Category**: Replace missing values with the most frequent category within the feature.
- **Forward/Backward Filling**: Fill missing values by using the values from preceding or following observations.

2.2.3 Handling Categorical Values

Many machine learning algorithms require categorical data to be converted into numerical form to perform mathematical operations effectively. For models that require numerical input, two common encoding techniques are:

- **Label Encoding**: This method assigns a unique integer to each category. While it is simple and memory-efficient, it can potentially introduce an **ordinal relationship** (i.e., an implied ranking or order) between categories, which may lead to **ordinal misinterpretation**—when a model incorrectly assumes that the numerical values represent a meaningful hierarchy. For example, if colors are encoded as `Red = 0`, `Blue = 1`, and `Green = 2`, a model might falsely interpret "Green" as being greater than "Blue." Label encoding is best suited for **tree-based models**, which are not sensitive to ordinal relationships.
- **One-Hot Encoding**: This technique converts categorical features into multiple binary columns, where each column represents a category with a value of 1 (present) or 0 (absent). One-hot encoding avoids the issue of ordinal misinterpretation and is widely used in models that require strictly numerical input, such as neural networks. However, it can lead to a **high-dimensional feature space** when dealing with categories that have many unique values, increasing computational complexity.

  Ordinal Encoding: This method is similar to Label Encoding but is specifically used when categories have a meaningful ranking. For example, satisfaction levels such as `Poor = 0`, `Fair = 1`, `Good = 2`, and `Excellent = 3` have an inherent order, making ordinal encoding appropriate. However, this method should only be used for features with a **true ordinal relationship**.
- **Target Encoding**: This method replaces categorical values with a statistical aggregation of the target variable (label) for each category. The most common approach, known as **Mean Encoding**, assigns each category the **mean** of the target variable. However, alternative approaches such as **Median Encoding** (using the median)

or **Probability-Based Encoding** (using class likelihoods) can also be applied depending on the data characteristics.

- **Hash Encoding (Feature Hashing)**: This method applies a hash function to map categorical values to a fixed number of columns. It is particularly useful for high-cardinality features, such as words in natural language processing (NLP). However, it introduces the possibility of hash collisions, where different categories may be mapped to the same value.

  Frequency (Count) Encoding: Instead of assigning arbitrary numerical values, this method replaces each category with its frequency (or count) of occurrence in the dataset. It is useful for handling high-cardinality categorical features. However, it assumes that the frequency of a category is a meaningful predictor, which may not always be the case.
- **Binary Encoding**: This technique represents categorical variables using a combination of integer representation and binary conversion. First, each category is assigned a unique integer (similar to Label Encoding). Then, these integer values are converted into binary format, and each binary digit is placed in a separate column. This method helps reduce dimensionality compared to One-Hot Encoding while still retaining some categorical information. However, it may be less interpretable than traditional encoding methods.
- **Embedding Encoding**: This technique is used in deep learning models to represent categorical data as dense vectors in a continuous space. Unlike traditional encoding methods, embeddings are **learned during model training** and capture complex relationships between categories. Typically, categorical values are first **integer-encoded or one-hot encoded** before being mapped to a trainable **embedding layer**, which transforms them into lower-dimensional dense representations. This approach is particularly beneficial for handling **high-cardinality categorical features** (i.e., a categorical variable that has a large number of unique values) and is widely used in **NLP, recommender systems, and deep learning applications**.

2.2.4 Normalizing and Standardizing Data

Feature scaling is a crucial preprocessing step in machine learning that ensures different numerical variables contribute equally during model training. It prevents features with larger magnitudes from dominating those with smaller scales, improving model stability and convergence. The two main approaches to feature scaling are **Normalization** and **Standardization**, each serving different purposes depending on the dataset and model requirements.

Normalization rescales feature values to a fixed range, typically between [0, 1] or [−1, 1]. It is particularly useful when different features have varying units or ranges, allowing models to learn efficiently without being biased toward high-magnitude values. However, normalization is **sensitive to outliers**, as extreme values can disproportionately affect the scaling. Standardization, on the other hand, transforms data so that it has a **mean of 0** and a **standard deviation of 1**, ensuring that all features follow a comparable scale while retaining their original distributions. Common feature scaling techniques include:

- **Min-Max Normalization**: Rescales a feature to a specific range, typically [0, 1], using the formula:

$$X' = \frac{X - X_{\min}}{X_{\max} - X_{\min}}$$

This transformation ensures that all values fall within the chosen range while preserving the original distribution of the data. However, Min-Max normalization is highly **sensitive to outliers**, as extreme values can disproportionately affect the scaling.
- **Max Abs Scaler**: Normalizes each feature by dividing it by its absolute maximum value, ensuring that values remain within the range $[-1, 1]$ while preserving the sign. The transformation is defined as:

$$X' = \frac{X}{\max |X|}$$

Unlike Z-score normalization, Max Abs Scaler **does not center data at zero** but instead scales it based on absolute maximum values. This method is particularly useful for **sparse datasets** or when maintaining the original sign of the data is necessary.
- **Z-Score Normalization (Standardization)**: Also known as **Standardization**, this method transforms data so that it has a mean of 0 and a standard deviation of 1, using the formula:

$$X' = \frac{X - \mu}{\sigma}$$

Unlike Min-Max scaling, this method **does not bound values within a fixed range**, allowing data points to take any real number. Standardization is particularly effective when data follows a **Gaussian (normal) distribution** or when features have varying units and need to be scaled uniformly.
- **Robust Scaling**: Unlike Min-Max normalization, this technique is **less sensitive to outliers** as it uses the interquartile range (IQR) instead of minimum and maximum values:

$$X' = \frac{X - Q_2}{Q_3 - Q_1}$$

where Q_2 is the median, and $Q_3 - Q_1$ is the interquartile range. This method tries to ensure that extreme values do not disproportionately influence the scaling.
- **Log Scaling**: Applies a logarithmic transformation to compress large values and reduce skewness:

$$X' = \log(X + 1)$$

This method is effective when the data follows an **exponential distribution** or has **highly skewed values**.

- **Linear Scaling**: A simple transformation that applies a linear function to rescale data to a new range:

$$X' = aX + b$$

where a and b are constants chosen based on the desired range. This approach is often used when a feature needs to be shifted or stretched without distorting its distribution.

- **Clipping**: Clipping is not a feature scaling technique in the traditional sense. However, it is often used as a data preprocessing step to handle outliers before applying feature scaling:

$$X' = \begin{cases} X_{\min}, & \text{if } X < X_{\min} \\ X, & \text{if } X_{\min} \le X \le X_{\max} \\ X_{\max}, & \text{if } X > X_{\max} \end{cases}$$

This technique is useful for removing extreme outliers without affecting the rest of the distribution.

2.2.5 Dataset Partitioning for Model Training and Evaluation

Before building and evaluating machine learning models, the data must be split into separate sets for training and testing. The two most common techniques are:

- **Hold-Out Method**: This is the simplest and most commonly used approach, where the dataset is divided into three subsets:
  - **Training Set**: Typically comprises 60%–90% of the original dataset and is used to train the model.
  - **Validation Set**: A portion of the dataset (usually 10%–20%) used during training to fine-tune the model's hyperparameters and detect overfitting.
  - **Test Set**: A completely separate portion (typically 10%–20%) used to evaluate the model's performance on unseen data after training is complete.

  The hold-out method is computationally efficient and works well for large datasets but may lead to performance variation if the dataset is small.
- **K-Fold Cross-Validation**: This method provides a more robust estimate of model performance, especially when data is limited. The dataset is divided into 'k' equal-sized folds, and the model is trained on 'k-1' folds while the remaining fold is used for testing. This process is repeated 'k' times, ensuring that each fold serves as a test set once. The final model performance is averaged over all iterations. A common choice is **k = 5 or k = 10**, which balances computational efficiency and evaluation reliability.

3 Feature Extraction and Feature Selection

Feature extraction is the process of transforming raw data into a new set of variables that better represent its essential characteristics. Instead of using the original input features directly, feature extraction derives meaningful attributes that capture relevant information while reducing redundancy. This transformation can be achieved through various mathematical and statistical techniques, such as **Principal Component Analysis (PCA)** [5], **Linear Discriminant Analysis (LDA)** [6], and spectral transformations, including the **Fourier** and **wavelet transforms**.

Feature extraction is particularly useful in high-dimensional datasets, where reducing the number of features without losing critical information can improve model performance and efficiency. By emphasizing the most informative aspects of the data, feature extraction enhances pattern recognition, facilitates visualization, and minimizes the impact of noise. This process is widely applied in fields such as image processing, speech recognition, and bioinformatics, where raw data is often complex and needs to be transformed into a more structured and meaningful representation for analysis.

Feature selection, also known as variable or attribute selection, is the process of identifying and selecting a subset of the most relevant input features to build an efficient and high-performing machine learning model. The goal is to enhance model accuracy while reducing complexity by eliminating features that are irrelevant, redundant, or noisy. In supervised learning tasks, whether classification or regression, datasets often contain a large number of features, but only a subset significantly contributes to solving the problem. Including all features can increase the risk of **overfitting**, degrade generalization, and lead to higher computational costs. Feature selection algorithms help mitigate these challenges by identifying the most informative features, resulting in models that are more interpretable, efficient, and robust.

One common issue in feature selection is the presence of **irrelevant features**. These are features that provide no meaningful information about the target variable and do not contribute to learning relationships within the data. Such features can introduce noise, misleading the model and reducing predictive performance. However, in some cases, an otherwise irrelevant feature may become useful when combined with others, a phenomenon known as **feature interaction**. For example, in an XOR function, a single input feature may not provide any useful information on its own, but when considered alongside another feature, their interaction correctly determines the output. This demonstrates how individual features, even if seemingly unimportant in isolation, can form complex patterns when combined.

Another key challenge in feature selection is the presence of **highly correlated features**. When multiple features are strongly correlated, they convey similar information, making some of them redundant. These are known as **redundant features**, as they do not provide additional value beyond what is already captured by other correlated features. Including redundant features can lead to increased computational costs, unnecessary complexity, and potential issues with **collinearity**. **Collinearity** refers to a situation in which two or more features are highly correlated, making it difficult for a machine learning model to distinguish their individual contributions. This can lead to numerical instability in certain models, particularly those that rely on linear relationships, such as regression-based algorithms. By removing redundant features, models can maintain efficiency while preserving predictive power.

Feature selection provides several key advantages:

- **Reducing overfitting**: Removing irrelevant and redundant features decreases the model's sensitivity to noise, enhancing its ability to generalize to unseen data.
- **Improving accuracy**: Selecting only the most relevant features ensures that the model learns from meaningful patterns, leading to better predictive performance.
- **Enhancing efficiency**: Reducing the number of features lowers memory usage and computational costs, making model training and inference faster.
- **Simplifying model interpretation**: A smaller, more refined set of features improves model transparency, making it easier to analyze and explain.

Feature selection is particularly crucial in high-dimensional datasets, where raw data often contains thousands of variables. In these cases, selecting the most informative features is essential for improving model performance while maintaining computational feasibility. Various techniques, such as filter methods, wrapper methods, and embedded methods, can be employed to systematically identify the best subset of features, ensuring that the resulting model is both efficient and effective.

3.1 Feature Selection Techniques

Feature selection techniques can generally be categorized into two main types:

- **Univariate methods**: In this approach, input variables are evaluated individually to assess their relationship with the output. The most relevant variables—those with the strongest correlation to the output—are selected. While this method is simple and often effective, it has limitations. It ignores interactions between input variables and does not consider how features may influence each other.
- **Multivariate methods**: This approach processes the entire set of input variables simultaneously. By evaluating variables together, it accounts for their interdependencies. While more computationally demanding than univariate methods, multivariate methods tend to be more robust and accurate since they consider how features work in combination to affect the output.

To achieve effective feature selection, various techniques are employed, broadly classified into filter methods, wrapper methods, and embedded methods, each utilizing different strategies to identify relevant features:

1. Filter Methods: These are typically used in the data pre-processing phase. Filter methods [7] are often univariate and non-iterative, meaning they evaluate each feature independently without considering their interactions. The goal is to assess the predictive power of each feature based on statistical criteria, assigning each feature a score that reflects its correlation or relevance to the target variable. Examples of filter methods include:

- **Pearson Correlation**: Measures the linear correlation between a continuous feature and a continuous target variable.
- **Chi-squared test**: Measures the dependence between a categorical feature and a categorical target variable.
- **Information Gain**: Evaluates how much information a categorical feature provides about a categorical target variable.
- **ANOVA (Analysis of Variance)**: Tests whether the means of a continuous feature differ significantly across different groups or categories of a categorical target variable. ANOVA typically uses the F-test to determine statistical significance. The F-score is calculated as:

$$F = \frac{\text{Variance of the input feature between the class means}}{\text{Variance of the input feature within each class}}$$

2. Wrapper Methods: In wrapper methods, feature selection is approached as a search problem, wherein the performance of a machine learning model is utilized to evaluate and select subsets of features. These methods evaluate multiple feature subsets by training and testing the model on different combinations of features, selecting the subset that yields the best model performance. Wrapper methods tend to provide more accurate results than filter methods, but they are computationally expensive due to the iterative nature of model training. Wrapper methods can be implemented in several ways, including:

- **Forward Selection**: This iterative method starts with an empty set of features and progressively adds features that improve model performance. In each iteration, the machine learning model is run to identify the feature that, when added, provides the best performance. This process continues until the desired number of features is selected.
- **Backward Elimination**: In this iterative method, the process starts with all features and gradually removes those that have the least impact on model performance. Features are deleted one by one in each iteration until a stopping criterion is reached, often when removing further features would degrade performance.

3. Embedded Methods: Embedded methods [6] incorporate feature selection within the model training process itself. Unlike filter and wrapper methods, feature selection happens automatically as part of the learning algorithm. These methods often involve regularization techniques that impose penalties on the complexity of the model (i.e., the number of selected features). By adding constraints to the cost function, embedded methods shrink or eliminate irrelevant feature coefficients during training. Examples of embedded methods include:

- **Lasso Regression**: Uses L1 regularization, which adds a penalty equal to the absolute value of the magnitude of coefficients.
- **Elastic Net**: Combines L1 and L2 regularization.
- **Tree-based Methods**: This method is based on the idea that decision trees, Random Forests, and Gradient Boosting Trees rank feature importance by evaluating how effectively a feature can split nodes in the tree.

3.2 Chi-Square Test

The Chi-Square test (also known as the Chi-Square test of independence) is a statistical method used to determine whether there is a significant association between two categorical variables. It is important to note that the Chi-Square test is not suitable for analyzing continuous variables directly or for assessing relationships between a categorical variable and a continuous variable unless the continuous data has been discretized into meaningful categories.

To illustrate the fundamentals of the Chi-Square test, consider a simple example involving 100 people, observing their interests in the IoT and Arts. In this scenario, two categorical variables are defined: the independent variable is Gender (Male, Female) and the dependent variable (target) is Interest (IoT or Arts). These observations can be summarized in a contingency table, where rows represent Gender, columns represent Interest, and cells represent the frequency (or count) of observations. Before conducting the test, it is necessary to define two hypotheses:

- **Null Hypothesis ($\mathbf{H_0}$)**: Gender and interest (IoT or Arts) are independent (i.e., there is no correlation between the two variables).
- **Alternative Hypothesis ($\mathbf{H_1}$)**: Gender and interest are not independent (i.e., there is a correlation between the two variables).

To proceed with the Chi-Square test, the expected value for each entry in the table is calculated by multiplying the total of each column by the total of each row and then dividing by the overall total. For instance, if 60 out of 100 people are interested in IoT, and 25% of them are men, it can be expected that 15 males (25% of 60) would be represented in the cell corresponding to the Male-IoT intersection. Next, the Chi-Square statistic (χ^2) is computed using the following formula:

$$\chi^2 = \sum \frac{(observed - expected)^2}{expected}$$

In this example, the computed χ^2 value is 2. To interpret the results, it is necessary to determine where the calculated χ^2 value lies on the Chi-Square distribution curve to assess the hypotheses. This is accomplished by referring to a Chi-Square distribution table to find the probability associated with the calculated χ^2 value. For this example, the corresponding probability is 0.16. Additionally, it is important to establish a significance level (α), which is commonly set at 0.05 for the Chi-Square test. The significance level represents the probability of incorrectly rejecting the null hypothesis when it is true. In this case, with a significance level of 0.05, the p-value (0.16) is compared against α. Since 0.16 is greater than 0.05, we fail to reject the null hypothesis, indicating that there is no statistically significant association between Gender and Interest.

3.3 Pearson Correlation

The Pearson correlation method is a statistical technique used to filter features and measure the strength and direction of the linear relationship between two continuous variables. The Pearson correlation coefficient (ρ_i) can be expressed with the following formula:

$$\rho_i = \frac{cov(X_i, Y)}{\sigma_{X_i}\sigma_Y}$$

In this equation, ρ_i represents the Pearson correlation coefficient, X_i is the input feature, Y is the output variable, $cov(X_i, Y)$ denotes the covariance between the input feature and the output variable, and σ represents the standard deviation of each variable. Covariance measures how two variables change together, while standard deviation quantifies the amount of variation or dispersion in a set of values.

The Pearson correlation coefficient ranges from $[-1, +1]$. A coefficient of 0 indicates no correlation between the two variables. A value greater than 0 signifies a positive correlation, meaning that as one variable increases, the other variable also tends to increase. Conversely, a value less than 0 indicates a negative correlation, where an increase in one variable is associated with a decrease in the other variable.

3.4 Entropy and Information Gain

Entropy serves as a powerful metric for estimating correlations and selecting relevant features in datasets. It quantifies the level of **impurity** or **uncertainty** present within a group of examples. The entropy $H(X)$ can be calculated using the following equation:

$$H(X) = -\sum_i p(x_i) \log_2 p(x_i)$$

In this equation, $p(x_i)$ represents the probability of class i. To illustrate this concept with an example (See Fig. 2), consider a dataset containing 16 red circles and 14 green triangles. The probability of the circle class is calculated as $\frac{16}{16+14}$,

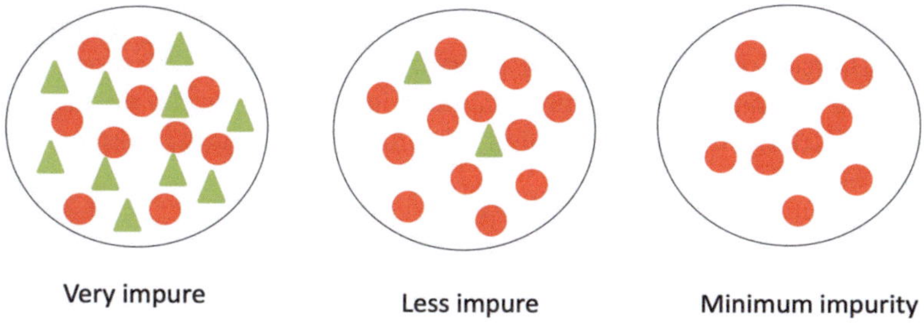

Fig. 2 Assessing the presence and distribution of various elements through the Impurity metric [4]

and the probability of the triangle class is $\frac{14}{16+14}$. Consequently, the corresponding entropy can be calculated as follows:

$$H = -\left(\frac{16}{16+14}\right)\log_2\left(\frac{16}{16+14}\right) - \left(\frac{14}{16+14}\right)\log_2\left(\frac{14}{16+14}\right)$$

The entropy of a group containing a single example class is **zero**, indicating that the group lacks information and is not a suitable training set for a learning algorithm. Conversely, a group with an equal distribution of classes (i.e., 50% of each class) has an entropy of **1**, signifying a balanced and suitable training set.

To better understand this relationship, consider Fig. 3, which depicts entropy as a function of the probability of a two-class variable (e.g., red circles and green triangles). Entropy quantifies the level of uncertainty or impurity in a system. The entropy is minimized when the probability is either 0 or 1, meaning that all instances belong to a single class (either all red circles or all green triangles). In this case, there is no uncertainty, as the outcome is completely predictable. Conversely, entropy reaches its maximum value of 1.0 when the probability is 0.5. This scenario represents the highest level of uncertainty, where the dataset is evenly split between the two classes, making it maximally impure.

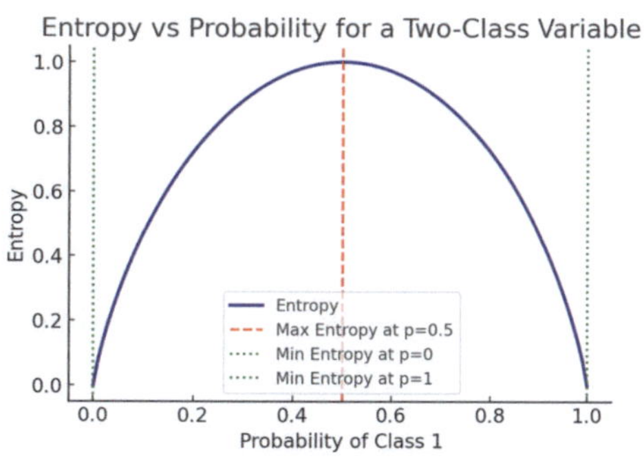

Fig. 3 Entropy as a function of probability for a two-class variable (target class is 1). Entropy is minimized at probabilities 0 and 1, while it reaches its maximum at 0.5, representing the highest uncertainty [4]

To effectively use entropy for feature selection, **Mutual Information (MI)** is defined. MI quantifies the amount of information that one variable contains about another variable. Larger mutual information values (e.g., between the target Y and feature X) indicate a stronger correlation with the target. Mutual Information can be calculated as:

$$MI(Y, X) = H(Y) + H(X) - H(Y, X)$$

In this equation, $H(Y, X)$ denotes the **joint entropy** (or **conditional entropy**) defined as:

$$H(Y, X) = -\sum_i p(y_i, x_i) \log_2 p(y_i, x_i)$$

It is important to note that Mutual Information is **zero** when there is no correlation between X and Y, indicating that they are statistically independent variables. The maximum value of Mutual Information occurs when Y is completely dependent on X.

4 Dimensionality Reduction

Dimensionality reduction refers to the process of reducing the number of features or variables in a dataset while preserving as much relevant information as possible. High-dimensional data often contain correlated or redundant features, which can negatively impact the performance and efficiency of machine learning algorithms. By transforming the data into a lower-dimensional space, dimensionality reduction helps mitigate these issues, improving computational efficiency, reducing overfitting, and enhancing data visualization. Several techniques are widely used for dimensionality reduction, including:

- **Principal Component Analysis (PCA) and Kernel PCA (KPCA)**: PCA is a linear transformation method that projects data onto a lower-dimensional subspace while retaining the maximum variance. It is particularly useful for noise reduction, feature extraction, and identifying the most informative components in high-dimensional datasets. However, PCA is limited to linear transformations and cannot capture nonlinear relationships in the data. To address this limitation, Kernel PCA (KPCA) extends PCA by using the **kernel trick**. A kernel function implicitly maps the data into a higher-dimensional space, where nonlinear patterns that are not separable in the original space can be more easily distinguished. This makes KPCA a powerful tool for capturing complex, nonlinear structures in datasets.
- **t-Distributed Stochastic Neighbor Embedding (t-SNE)**: A nonlinear dimensionality reduction technique designed primarily for visualizing high-dimensional

data. t-SNE effectively preserves local similarities and helps reveal clusters or patterns within complex datasets. However, it is computationally expensive, does not maintain global structure well, and is mainly used for visualization rather than feature selection or improving machine learning models.
- **Linear Discriminant Analysis (LDA)**: A supervised dimensionality reduction technique that maximizes class separability while reducing dimensionality. LDA is commonly used in classification tasks where distinguishing between different categories is essential. Unlike PCA, LDA considers class labels, making it suitable for supervised learning problems. However, it assumes normally distributed data with equal class covariances for optimal performance.

4.1 Principal Component Analysis (PCA)

One of the most widely used techniques for dimensionality reduction is PCA. PCA identifies the directions (principal components) along which the variance in the dataset is maximized. These principal components represent the most important underlying structure in the data. Algorithm 1 shows the details of PCA. The PCA process can be summarized in the following steps (see Fig. 4):

1. **Standardization of Features:** Standardize the dataset by subtracting the mean and dividing by the standard deviation for each feature. This ensures that all features are on the same scale and have equal weight in the analysis.
2. **Compute the Covariance Matrix:** Calculate the covariance matrix of the standardized features. The covariance matrix captures the relationships between different features and helps identify patterns of variability in the data.
3. **Decompose the Covariance Matrix:** Decompose the covariance matrix into its eigenvalues and eigenvectors. Since the covariance matrix is symmetric, this decomposition is guaranteed. The eigenvectors represent the directions (principal components) of maximum variance, and the eigenvalues indicate the magnitude of variance along those directions.
4. **Sort and Select Principal Components:** Rank the eigenvectors based on their corresponding eigenvalues in descending order. Select the top s eigenvectors corresponding to the s largest eigenvalues. These eigenvectors represent the most important directions of variance in the data.
5. **Project the Data onto a New Subspace:** Transform and project the original dataset onto the new feature subspace formed by the selected eigenvectors. This results in a reduced-dimensional representation of the data while preserving as much variance as possible.

Algorithm 1 Principal Component Analysis (PCA) Using Eigenvalue Decomposition

1: **Input:** $\mathbf{X} \in \mathbb{R}^{n \times p}$, the dataset with n observations and p features.
2: **Output:** $\mathbf{Y} \in \mathbb{R}^{n \times s}$, the transformed dataset in the principal component space with s principal components.

3: **Step 1: Compute the Mean Feature Vector:** Compute the mean feature vector for the dataset

$$\mu = \frac{1}{n} \sum_{k=1}^{n} \mathbf{X}_k$$

▷ Where $\mathbf{X}_k \in \mathbb{R}^{1 \times p}$ represents the k-th observation as a row vector. The mean vector $\mu \in \mathbb{R}^{1 \times p}$ contains the mean of each feature across all n observations.

4: **Step 2: Center the Data:** Subtract the mean feature vector from each observation to center the data

$$\mathbf{X}_{\text{centered}} = \mathbf{X} - \mathbf{1}_n \mu$$

▷ Where $\mathbf{X}_{\text{centered}} \in \mathbb{R}^{n \times p}$ is the centered data matrix, $\mathbf{1}_n \in \mathbb{R}^{n \times 1}$ is a column vector of ones, and $\mu \in \mathbb{R}^{1 \times p}$ ensures that the mean is subtracted from each feature.

5: **Step 3: Compute the Covariance Matrix:** Calculate the covariance matrix from the centered data

$$\mathbf{C} = \frac{1}{n-1} \mathbf{X}_{\text{centered}}^{\top} \mathbf{X}_{\text{centered}}$$

▷ Where $\mathbf{C} \in \mathbb{R}^{p \times p}$ is the covariance matrix. The normalization factor $\frac{1}{n-1}$ ensures that the covariance matrix is an unbiased estimator of the population covariance.

6: **Step 4: Eigenvalue and Eigenvector Decomposition:** Perform eigenvalue decomposition on the covariance matrix

$$\mathbf{C}\mathbf{w}_i = \lambda_i \mathbf{w}_i, \quad i = 1, 2, \ldots, p$$

▷ Where $\mathbf{w}_i \in \mathbb{R}^{p \times 1}$ are the eigenvectors (principal components), and λ_i are the corresponding eigenvalues. The eigenvectors form an orthonormal basis for the principal component space.

7: **Step 5: Select Principal Components:** Sort the eigenvalues in descending order and select the top s eigenvectors

$$\mathbf{W}_s = [\mathbf{w}_1, \mathbf{w}_2, \ldots, \mathbf{w}_s]$$

▷ Where $\mathbf{W}_s \in \mathbb{R}^{p \times s}$ is the matrix containing the top s eigenvectors corresponding to the largest eigenvalues. These eigenvectors are sorted in descending order of their corresponding eigenvalues.

8: **Step 6: Project Data onto Principal Component Space:** Transform the data into the new principal component space

$$\mathbf{Y} = \mathbf{X}_{\text{centered}} \mathbf{W}_s$$

▷ Where $\mathbf{Y} \in \mathbb{R}^{n \times s}$ is the reduced dataset with s principal components. The projection ensures that the new features (principal components) are uncorrelated and ordered by variance.

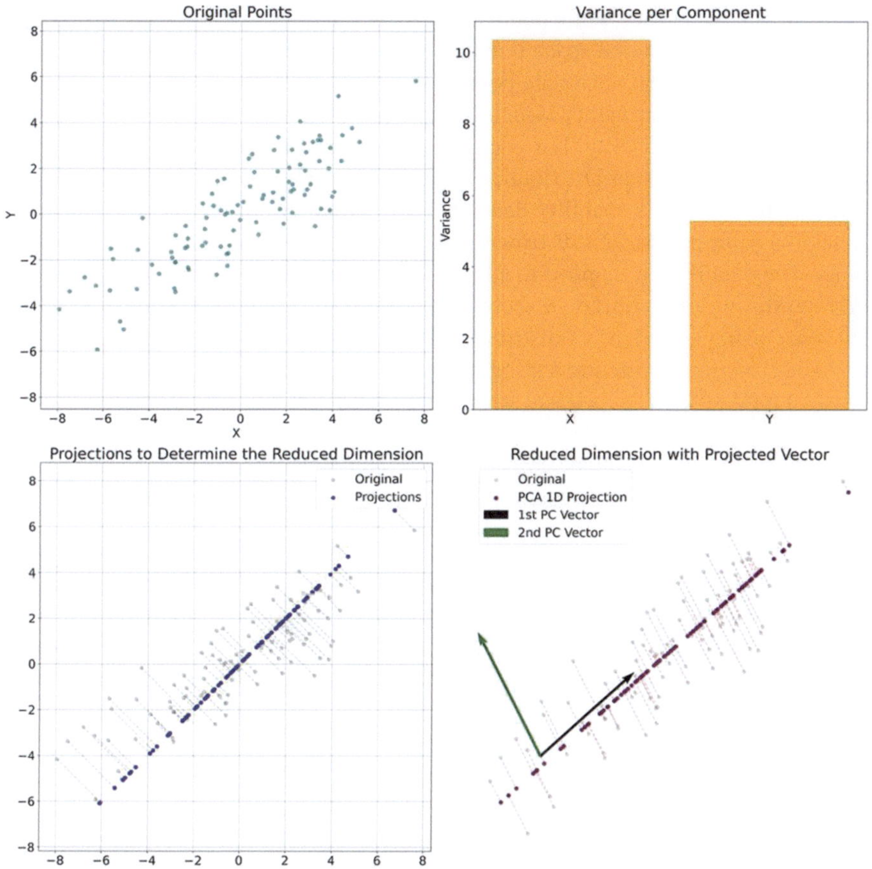

Fig. 4 Illustration of PCA: The top plot shows the original data distribution with standardized axes. The top-right plot displays the explained variance ratio for each principal component, indicating how much variance each component captures. The bottom plots visualize the data projected onto the principal components

4.2 *t-Distributed Stochastic Neighbor Embedding (t-SNE)*

t-Distributed Stochastic Neighbor Embedding (t-SNE) [8] is a **nonlinear dimensionality reduction technique** designed for visualizing high-dimensional data in a lower-dimensional space, typically two or three dimensions. Unlike PCA, which preserves the **global structure** and variance of data through linear transformations, t-SNE focuses on **retaining local relationships** between data points. This means that points that are close together in the high-dimensional space remain close in the low-dimensional representation, while distant points are pushed apart.

t-SNE works by calculating the similarity between each data point and all others in the dataset. Using these similarity measures, the algorithm projects the data into a lower-dimensional space while maintaining local relationships, ensuring that nearby points in high-dimensional space remain close in the reduced representation. t-SNE computes similarity between data points using a probabilistic approach rather than direct distance measurements, which become unreliable in high dimensions due to the curse of dimensionality. For each point, a conditional probability distribution is defined (e.g., Gaussian Distribution), where the similarity to other data points is based on the relative probability density. These probabilities are then symmetrized to create a joint probability distribution representing the relationships between all points. The data is then mapped to a lower-dimensional space, where points are initially **randomly distributed**. A second probability distribution is computed in this new space using a Student's t-distribution instead of a Gaussian, which helps maintain local structures while preventing crowding. To ensure that the low-dimensional representation reflects the original structure, t-SNE iteratively adjusts point positions by minimizing the Kullback-Leibler (KL) divergence between the two probability distributions through gradient descent. t-SNE can be summarized in the following four steps (see Fig. 5):

- **Step 1: Compute Pairwise Similarities in the High-Dimensional Space**
  The algorithm begins by measuring how similar each pair of points is in the original high-dimensional space. Given a dataset with N data points, where each point is represented as a vector x_i, the similarity between two points x_i and x_j is modeled using a **Gaussian distribution** centered at x_i with variance σ_i^2. The probability that x_i selects x_j as a neighbor is given by:

$$p_{j|i} = \frac{\exp(-\|x_i - x_j\|^2/2\sigma_i^2)}{\sum_{k\neq i} \exp(-\|x_i - x_k\|^2/2\sigma_i^2)}$$

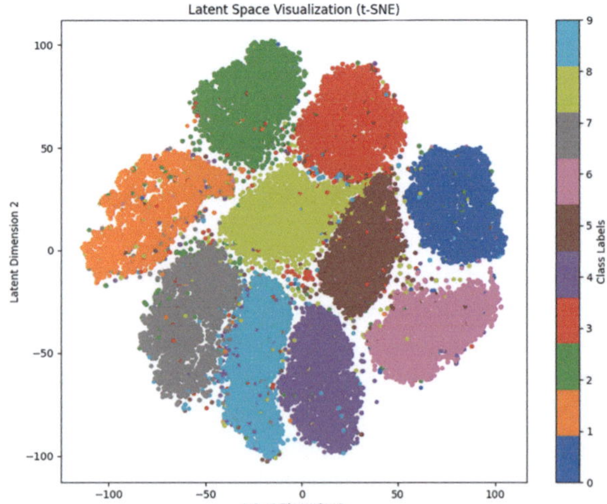

Fig. 5 t-SNE visualization of the latent space learned from the MNIST dataset, showing well-separated clusters for each digit class (0–9)

where $p_{j|i}$ represents the conditional probability of x_j being a neighbor of x_i, $\|x_i - x_j\|^2$ is the squared Euclidean distance between the points, and σ_i is the standard deviation of the Gaussian. The bandwidth of the Gaussian kernels σ_i is determined so that the entropy of the conditional distribution reaches a predefined entropy value, using the bisection method (an iterative approach that repeatedly divides an interval in half to find a root). Consequently, the bandwidth adjusts to the local density of data: smaller values of σ_i are applied in denser regions, whereas larger values are used in sparser areas. The entropy of this distribution P_i increases with perplexity, following the relationship:

$$Perp(P_i) = 2^{H(P_i)}$$

where the Shannon entropy $H(P_i)$ is given by:

$$H(P_i) = -\sum_j p_{j|i} \log_2 p_{j|i}$$

The perplexity is a user-defined parameter in t-SNE, and as the developers of t-SNE state, "perplexity can be interpreted as a smooth measure of the effective number of neighbors." The performance of t-SNE is fairly robust to changes in perplexity, and typical values range between 5 and 50. The joint probability distribution p_{ij} is computed as:

$$p_{ij} = \frac{p_{j|i} + p_{i|j}}{2N}$$

where $p_{j|i}$ and $p_{i|j}$ are the conditional probabilities computed from the high-dimensional data using Gaussian kernels. This formulation arises from the assumption that each sample is equally probable, meaning that the marginal probabilities satisfy $p_i = \frac{1}{N}$ and $p_j = \frac{1}{N}$. Additionally, it holds that:

$$\sum_{i,j} p_{ij} = 1, \quad p_{ii} = 0.$$

- **Step 2: Initialize a Random Low-Dimensional Representation**
  A set of low-dimensional points $y_1, y_2, ..., y_N$ is initialized randomly. Instead of using a Gaussian distribution like in the high-dimensional space, t-SNE models pairwise similarities in the low-dimensional space using a **heavy-tailed Student's t-distribution** with one degree of freedom. The probability that two points y_i and y_j are similar in the low-dimensional space is defined as:

$$q_{ij} = \frac{(1 + \|y_i - y_j\|^2)^{-1}}{\sum_{k \neq l}(1 + \|y_k - y_l\|^2)^{-1}}$$

where q_{ij} represents the similarity between points y_i and y_j in the low-dimensional space. The use of a **t-distribution** prevents the **crowding problem** (i.e., refers to the difficulty of accurately representing relative distances when mapping high-dimensional data into a low-dimensional space), ensuring that moderately distant points in high-dimensional space remain meaningfully separated.

- **Step 3: Minimize Kullback-Leibler (KL) Divergence**
  To align the low-dimensional representation with the original data structure and find the locations of the points y_i in the map, t-SNE minimizes the **Kullback-Leibler (KL) divergence**, which quantifies how much the probability distribution in the low-dimensional space (Q) deviates from the high-dimensional distribution (P):

$$KL(P||Q) = \sum_{i \neq j} p_{ij} \log \frac{p_{ij}}{q_{ij}}$$

  Minimizing KL divergence encourages similar points to stay close together, preserving local neighborhoods.
- **Step 4: Gradient Descent Optimization** To refine the low-dimensional embedding, the algorithm uses **gradient descent**, iteratively adjusting the positions of points. By continuously updating the positions based on the gradient, t-SNE optimizes the embedding to **preserve local relationships** while maintaining meaningful separation between distinct clusters.

4.3 Linear Discriminant Analysis (LDA)

Linear Discriminant Analysis (LDA) is a statistical method used for **dimensionality reduction** while maintaining class separability. Unlike PCA, which focuses on maximizing overall variance, LDA aims to **maximize the distance between class means while minimizing the variance within each class**. This makes LDA particularly well-suited for classification tasks, where reducing dimensionality while preserving essential class structure improves both computational efficiency and predictive accuracy.

LDA works by finding an optimal transformation that projects data into a lower-dimensional space where the differences between classes are more pronounced. This transformation is achieved by solving an optimization problem that maximizes the ratio of the variance between different class means to the variance within each class. To achieve this, LDA computes two scatter matrices: the **within-class scatter matrix**, which quantifies the spread of data within each class, and the **between-class scatter matrix**, which quantifies how much the class means differ from each other.

The **within-class scatter matrix**, denoted as S_w, measures how spread out the data points are within each class. It is defined as:

$$S_w = \sum_{c=1}^{C} \sum_{x_i \in c} (x_i - \mu_c)(x_i - \mu_c)^T \tag{1}$$

where x_i represents a sample belonging to class c, μ_c is the mean of class c, and C is the total number of classes. This matrix quantifies the variability within each class and sums the variance within all classes. By minimizing this spread during projection, LDA ensures that data points within a class remain compact.

The **between-class scatter matrix**, denoted as S_b, measures the separation between different class means. It is defined as:

$$S_b = \sum_{c=1}^{C} N_c(\mu_c - \mu)(\mu_c - \mu)^T \tag{2}$$

where N_c is the number of samples in class c, μ_c is the mean of class c, and μ is the overall mean of all samples in the dataset. This matrix captures how distinct the class means are from each other. If class means are well-separated in the transformed space, classification becomes easier.

When we project the data onto a new lower-dimensional space using a transformation matrix W, each data point is transformed as:

$$Y = W^T X \tag{3}$$

Since class means are simply averages of points, the new class means in the projected space become:

$$\mu'_c = W^T \mu_c, \quad \mu' = W^T \mu \tag{4}$$

Substituting these into the between-class scatter formula, we get:

$$S'_b = \sum_{c=1}^{C} N_c(\mu'_c - \mu')(\mu'_c - \mu')^T \tag{5}$$

Since $\mu'_c = W^T \mu_c$ and $\mu' = W^T \mu$, we can rewrite this as:

$$S'_b = W^T S_b W \tag{6}$$

Similarly, the within-class scatter matrix in the projected space is:

$$S'_w = W^T S_w W \tag{7}$$

To obtain the best projection matrix, LDA maximizes the ratio of between-class scatter to within-class scatter in the projected space. This leads to the objective function:

$$J(W) = \frac{|W^T S_b W|}{|W^T S_w W|} \tag{8}$$

where the numerator measures how well-separated the class means are after projection, and the denominator measures how much overlap exists within each class after projection. Maximizing this ratio ensures that the transformed data has **maximum class separation while reducing dimensionality**.

Since direct maximization of this ratio is challenging due to its fractional form, the problem is converted into an **eigenvalue problem**. Consider the case where the projection matrix consists of a single vector w. In this case, the objective function simplifies to:

$$J(w) = \frac{w^T S_b w}{w^T S_w w}. \tag{9}$$

The key idea is to find the vector w that maximizes this quotient. To maximize $J(w)$, we take its derivative and set it to zero:

$$\frac{d}{dw}\left(\frac{w^T S_b w}{w^T S_w w}\right) = 0 \tag{10}$$

Using the quotient rule:

$$\frac{d}{dw}\left(\frac{f(w)}{g(w)}\right) = \frac{g(w)f'(w) - f(w)g'(w)}{g(w)^2} \tag{11}$$

where:

$$f(w) = w^T S_b w, \quad g(w) = w^T S_w w \tag{12}$$

Taking the derivatives of $f(w)$ and $g(w)$:

$$\frac{d}{dw}(w^T S_b w) = 2S_b w, \quad \frac{d}{dw}(w^T S_w w) = 2S_w w \tag{13}$$

Substituting these into the quotient rule:

$$\frac{(2S_b w)(w^T S_w w) - (2S_w w)(w^T S_b w)}{(w^T S_w w)^2} = 0 \tag{14}$$

Dividing by 2:

$$S_b w(w^T S_w w) - S_w w(w^T S_b w) = 0 \tag{15}$$

Rearranging:

$$S_b w = \lambda S_w w \tag{16}$$

where:

$$\lambda = \frac{w^T S_b w}{w^T S_w w} \tag{17}$$

This equation is a **generalized eigenvalue problem**, meaning that the optimal direction w is an **eigenvector**, and the corresponding λ is an eigenvalue that quantifies how well that direction preserves class separability. Indeed, the equation:

$$S_b w = \lambda S_w w \tag{18}$$

is similar to the standard eigenvalue problem:

$$Av = \lambda v \tag{19}$$

where:

$$A = S_w^{-1} S_b, \quad v = w \tag{20}$$

Multiplying both sides by S_w^{-1}, assuming S_w is invertible, leads to:

$$S_w^{-1} S_b w = \lambda w \tag{21}$$

This confirms that the best projection directions are the **eigenvectors of** $S_w^{-1} S_b$, and the corresponding eigenvalues indicate how much separation exists along each eigenvector. The projection matrix W is formed using multiple eigenvectors. The matrix form of the eigenvalue equation is:

$$S_b W = S_w W \Lambda, \tag{22}$$

where, Λ is a diagonal matrix whose entries are the eigenvalues corresponding to the eigenvectors of $S_w^{-1} S_b$. Multiplying both sides by S_w^{-1}, the equation simplifies to:

$$S_w^{-1} S_b W = W \Lambda \tag{23}$$

Once the projection matrix W is obtained, the dataset is transformed into a lower-dimensional space using:

$$Y = XW \tag{24}$$

where X is the original dataset, W is the projection matrix formed from the top k eigenvectors, and Y is the transformed dataset.

5 Regression Analysis

Regression analysis is a subcategory of supervised machine learning that investigates the relationship between a dependent variable (often referred to as the target or output) and one or more independent variables (also known as predictors, features, or inputs). The goal of regression analysis is to understand how the independent variables influence the dependent variable. For instance, in the field of healthcare, there are several important applications of regression analysis, such as:

- **Predictive Analytics**: Regression is widely used in healthcare for predictive analytics, where historical data is analyzed to model and forecast future patient behaviors or conditions. For example, wearable IoT devices equipped with accelerometers or motion sensors can collect data on a patient's movement patterns. Next, regression models can then predict the likelihood of a fall, allowing timely interventions.
- **Operational Efficiency**: Hospitals can use regression analysis to optimize the management of resources, such as beds, medical equipment, and staffing. For instance, IoT devices and sensors tracking patient vitals or occupancy levels can be used to forecast bed availability and staffing needs.
- **Decision Support Systems**: Regression analysis supports healthcare professionals in making more informed and accurate decisions. For example, by analyzing patient data (e.g., blood pressure, glucose levels, age), regression models can help predict the risk of chronic conditions like heart disease or diabetes.
- **Treatment Optimization**: Regression analysis can help optimize treatment plans, such as by predicting the optimal dosage or treatment adjustments based on patient data.
- **New Insights**: Regression analysis can reveal hidden patterns in patient data that are difficult to detect with conventional methods. For instance, wearable IoT devices capturing patient vitals over time may reveal that fluctuations in environmental factors (e.g., temperature, air quality) correlate with changes in a patient's health status.

5.1 Formal Definition of Regression Task

In regression analysis, the goal is to model the relationship between input features (independent variables) and the output (dependent variable), typically to predict a continuous outcome. Formally, the regression task can be described as follows:

Given a training set $T = \{(x^{(1)}, y^{(1)}), (x^{(2)}, y^{(2)}), ..., (x^{(T)}, y^{(T)})\}$, where:

- $x^{(i)} = (x_1^{(i)}, x_2^{(i)}, ..., x_D^{(i)})$ represents the feature vector for the i-th sample, with D being the number of features or independent variables (also known as predictors or inputs),
- $y^{(i)}$ is the corresponding output or dependent variable for the i-th sample (also known as the target or response),
- T is the total number of training samples.

The objective of regression is to find a function $f(x)$ that best maps the input variables x to the output y, such that:

$$y = f(x) + \epsilon$$

where ϵ represents the error term, accounting for the difference between the predicted value $f(x)$ and the true value y. The goal is to minimize this error term using a loss function, such as Mean Squared Error (MSE) during the training process.

Note that, in regression analysis, the output $y^{(i)}$ is continuous, which is a key distinction from classification tasks, where $y^{(i)}$ is categorical.

5.2 Important Terminologies in Regression Analysis

Understanding key terminologies in regression analysis is essential for building reliable models and interpreting results accurately. The following concepts play a crucial role in diagnosing potential issues and improving model performance:

- **Outliers**: An outlier (See Fig. 6) is a data point that significantly deviates from the other observations in a dataset, either by being much higher or lower in value. Outliers can distort regression results, particularly when they exert high leverage on the fitted model, potentially reducing accuracy and leading to biased predictions. Handling outliers is often done during the pre-processing phase through transformation, clipping, or robust regression techniques.
- **Multicollinearity**: Multicollinearity occurs when two or more independent variables are highly correlated, leading to inflated standard errors of the regression coefficients. While this does not necessarily reduce the overall predictive accuracy of the model, it makes it difficult to assess the individual contribution of correlated predictors. In extreme cases, multicollinearity can cause some coefficients to become statistically insignificant, even if the variables are important.

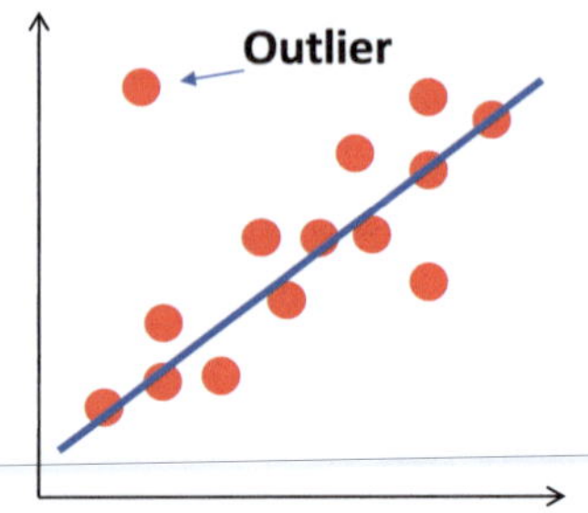

Fig. 6 An example of an outlier data point [4]

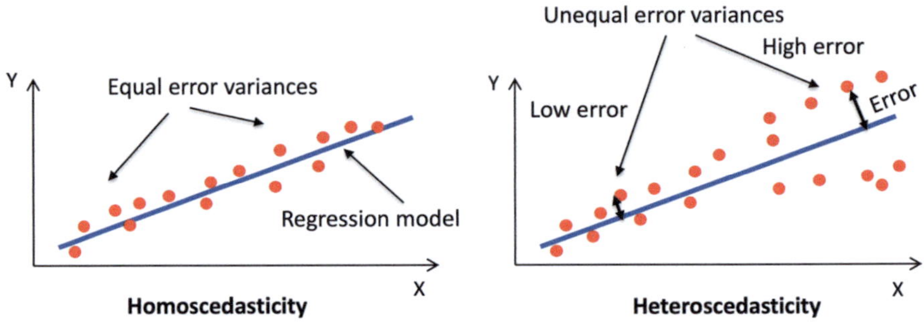

Fig. 7 Illustration of heteroscedasticity and homoscedasticity [4]

Techniques such as PCA, or ridge regression can be used to detect and mitigate multicollinearity.

- **Heteroscedasticity**: Heteroscedasticity arises when the variance of the residuals (errors) varies with the independent variable X. In regression analysis, constant variance (homoscedasticity) is assumed, meaning the spread of the residuals remains uniform across all levels of X. However, when heteroscedasticity occurs, the residuals exhibit non-constant variance, complicating the analysis and potentially undermining the reliability of the model. Figure 7 illustrates the difference: in heteroscedasticity, the variance of Y increases with X, while in homoscedasticity, the variance remains constant. Heteroscedasticity can often be detected visually or statistically. Some remedies for heteroscedasticity include using weighted regression and redefining variables.

5.3 Linear Regression

Linear regression models assume a linear relationship between the dependent variable y and the independent variables x, which can be expressed as:

$$\hat{y} = h(x) = w_0 + w_1 x_1 + \cdots + w_D x_D = w_0 + \langle w, x \rangle = w_0 + x w^T$$

where $\hat{y}$ is the predicted value, $h(x)$ is the linear function, w_0 is the intercept (bias term), and $w_1, ..., w_D$ are the coefficients (weights) corresponding to the features $x_1, ..., x_D$. The term $\langle w, x \rangle$ is the dot product of the weight vector and the feature vector.

5.3.1 Cost Functions

In linear regression, the goal is to find the optimal set of weights $w_0, w_1, ..., w_D$ that minimize the difference between the predicted values $\hat{y}$ and the actual values y. To measure this difference, we use a **cost function** (also known as a **loss function**). Several cost functions can be employed, but the most common one in linear regression is the **Mean Squared Error (MSE)**:

$$J(w) = \frac{1}{2m} \sum_{i=1}^{m} (h_w(x^{(i)}) - y^{(i)})^2$$

where, m represents the number of training examples, $h_w(x^{(i)})$ denotes the predicted value for the i-th example, and $y^{(i)}$ corresponds to the true output for the i-th example. The cost function evaluates how far off the model's predictions are from the actual values. By minimizing this function, we optimize the weights w of the model. **Mean Absolute Error (MAE)** is another widely used cost function that calculates the average of the absolute differences between predicted and actual values:

$$J(w) = \frac{1}{m} \sum_{i=1}^{m} |h_w(x^{(i)}) - y^{(i)}|$$

5.3.2 Cost Function Optimization: Gradient Descent

A common method to minimize the cost function $J(w)$ is the **gradient descent algorithm**. Gradient descent iteratively updates the weights by moving in the opposite direction of the gradient of the cost function (i.e., the direction of steepest descent):

$$w_j = w_j - \alpha \frac{\partial}{\partial w_j} J(w)$$

where: w_j is the j-th weight, α is the learning rate, which controls the size of the step taken in each iteration, $\frac{\partial}{\partial w_j} J(w)$ is the partial derivative of the cost function with respect to w_j. The following steps can summarize Gradient Descent:

1. **Initialization**: Randomly initialize the weights w_j.
2. **Calculate Gradient**: Compute the gradient of the cost function with respect to each weight.
3. **Update Weights**: Adjust each weight according to the gradient and learning rate: $W = W - \alpha G$, where $G = \frac{\partial}{\partial w_j} J(w)$.
4. **Repeat**: Continue updating the weights until convergence, i.e., until the cost function $J(w)$ stops decreasing or reaches a predefined threshold.

Gradient descent works by using the entire training dataset at once. However, it can be slow for large datasets, which is why **stochastic gradient descent (SGD)** is often preferred for large-scale problems. The core idea behind SGD is to update the model using only one randomly selected training example at each iteration rather than using the entire dataset.

5.4 Generalization and Model Capacity

In machine learning, **model capacity** refers to the ability of a model to capture and learn from the complexity of the underlying data. It is determined by factors such as the number of parameters, model architecture, and the overall flexibility of the model. A model with high capacity can capture more complex relationships and patterns in the data, while a model with low capacity may struggle to learn these patterns effectively.

Generalization refers to a model's ability to make accurate predictions on unseen data, not just on the data it was trained on. A model that generalizes well can apply learned patterns to new examples. Conversely, a model that fails to generalize is overfitting or underfitting the data:

- **Overfitting (High Variance)** occurs when a model is too complex for the amount of data available. The model fits the training data too well, learning even the noise and outliers. This results in low training error but high test error, as the model fails to generalize to new data. The model is highly sensitive to small variations in the training set.
- **Underfitting (High Bias)** occurs when a model is too simplistic to capture the underlying patterns in the data. The model does not learn enough from the training data, resulting in poor performance on both training and test datasets. This typically happens when the model's capacity is too low.

Figure 8 illustrates these concepts. The left plot demonstrates **underfitting**, where a simple linear regression model fails to capture the true quadratic nature of the data. The middle plot shows a well-fitted model, where a quadratic regression model

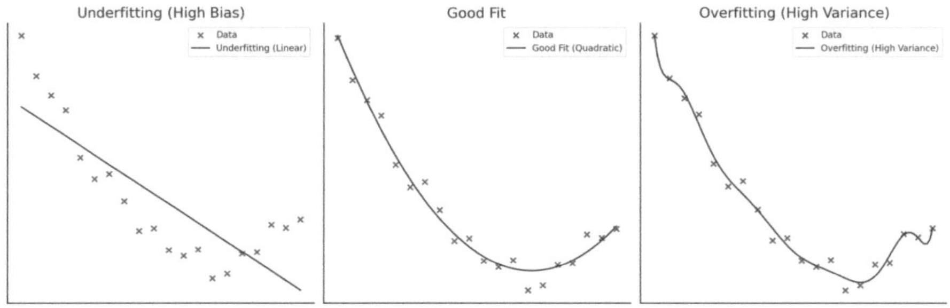

Fig. 8 Illustration of underfitting (high bias) and overfitting (high variance) in regression. The left plot represents underfitting, where a simple linear model fails to capture the underlying pattern. The middle plot shows an optimal fit, where a quadratic regression model balances complexity and generalization. The right plot demonstrates overfitting, where a high-degree polynomial regression model learns noise in the training data, leading to poor test performance. Achieving an optimal balance between bias and variance is critical for building robust regression models

balances complexity and generalization. The right plot demonstrates **overfitting**, where a higher-degree polynomial regression model fits the training data perfectly but fails to generalize, leading to poor test performance.

Model Capacity and Generalization Error: A model's generalization ability is fundamentally tied to its capacity to learn from the data while avoiding overfitting. Figure 9 demonstrates the relationship between model capacity and generalization error, highlighting the concept of the generalization gap:

- Underfitting (Left Region): When the model capacity is too low, it fails to capture key patterns in the data, resulting in high error on both training and test sets.
- Optimal Capacity: At an intermediate level of capacity, the model strikes the best balance between bias and variance, resulting in the lowest test error. This point is marked by the vertical dashed line.
- Overfitting (Right Region): Excessively complex models memorize the training data, leading to near-zero training error but significantly higher test error, as the model fails to generalize to unseen data.
- Generalization Gap: The difference between the training error (red curve) and the test error (blue curve) represents the generalization gap. A large gap indicates that the model is overfitting, performing well on training data but poorly on new data.

5.5 *Regularization*

Regularization [6] is a technique used to reduce overfitting by adding a penalty term to the cost function, which encourages the model to have smaller weights and become more generalizable. This helps the model perform better on unseen data (test data). By modifying the cost function with a regularization term, the model is trained not

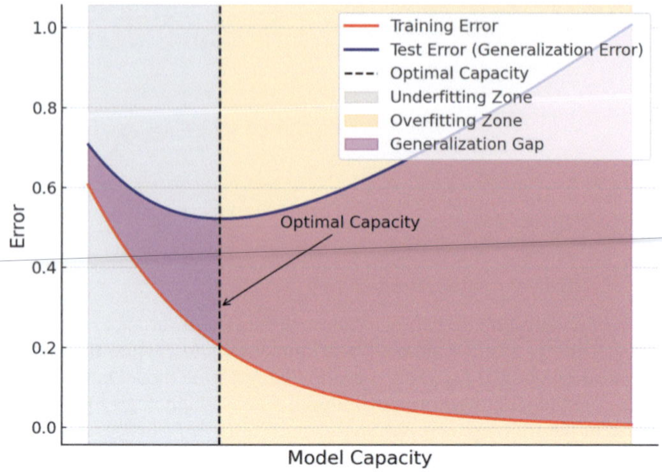

Fig. 9 Illustration of model capacity and generalization error. The red curve represents the training error (generalization error), while the blue curve represents the test error. The shaded gray region marks the underfitting zone, where the model is too simple to capture data patterns. The shaded orange region marks the overfitting zone, where model complexity leads to poor generalization. The vertical dashed line indicates the optimal capacity, where test error is minimized. The generalization gap (shaded purple region) highlights the difference between training and test error, which increases in the overfitting zone

only to minimize the error on the training data but also to avoid overfitting. The general form of a regularized cost function is:

$$\text{Cost function} = \text{Loss} + \text{Regularization term}$$

The regularization term reduces the magnitude of the model's weights, discouraging the model from fitting too closely to the training data. The two most common types of regularization methods are **L1** and **L2** regularization.

5.5.1 Norm Functions

Norm functions provide a measure of the magnitude of a vector in a given space. In general, the p-norm of a vector $v = [x_1, x_2, ..., x_n]$ is defined as:

$$\|v\|_p = \left(\sum_{i=1}^{n} |x_i|^p \right)^{\frac{1}{p}}$$

Different values of p correspond to different types of norms, each with distinct geometric properties.

- When $p = 1$, the L1 norm is obtained, which is the sum of the absolute values of the vector components:

$$\|v\|_1 = |x_1| + |x_2| + \cdots + |x_n|$$

  The set of points satisfying $\|v\|_1 = c$ forms a diamond shape (a rotated square) in two dimensions and an octahedron in higher dimensions.
- When $p = 2$, the L2 norm, also known as the Euclidean norm, is obtained. This norm measures the standard Euclidean distance:

$$\|v\|_2 = \sqrt{x_1^2 + x_2^2 + \cdots + x_n^2}$$

  The corresponding set of points where $\|v\|_2 = c$ forms a circle in two dimensions and a sphere in three dimensions.
- As $p \to \infty$, the L-infinity norm is obtained, defined as:

$$\|v\|_\infty = \max(|x_1|, |x_2|, \ldots, |x_n|)$$

  The set of points satisfying $\|v\|_\infty = c$ forms a square in two dimensions and a cube in higher dimensions.

5.5.2 L1 Regularization

L1 regularization, commonly referred to as **Lasso regression** in the context of regression models, incorporates the L1 norm as a penalty term in the cost function to encourage sparsity in the model. While Lasso is most commonly used in linear regression, L1 regularization can also be applied to other machine learning models, such as logistic regression or support vector machines, to promote feature selection and sparsity.

In the context of regularization, the cost function is modified by adding a penalty term that penalizes the absolute values of the model's weights. The updated objective function becomes:

$$\text{Cost function} = \text{Loss} + \lambda \sum \|w\|$$

Here, λ is the regularization parameter that controls the strength of the penalty, and w represents the model's coefficients.

The main advantage of L1 regularization lies in its ability to perform **feature selection**. By adding the L1 penalty, some of the model's coefficients are driven to exactly zero, effectively removing less relevant features from the model. This not only reduces the model's complexity but also improves its interpretability, as it retains only the most important features.

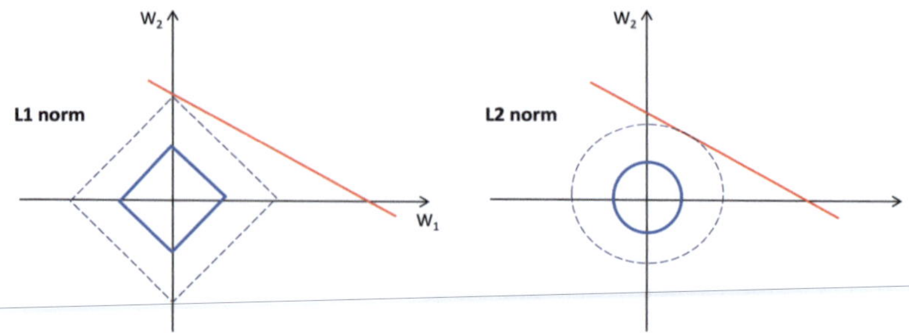

Fig. 10 Different forms of the constraint regions in Lasso and Ridge regression. The L1 norm constraint is represented as a rotated square (diamond shape) in two dimensions, which encourages sparsity by driving some coefficients to zero. The figure shows how the constraint region aligns with the axes, leading to sparse solutions. W_1 and W_2 represent the weights of regression features x_1 and x_2 [4]

The L1 norm induces sparsity by creating a constraint region in the shape of a rotated square (diamond shape) in two dimensions and an octahedron in higher dimensions. This geometric constraint plays a crucial role in guiding the model towards sparse solutions. In Fig. 10, the L1 constraint region is shown. The sharp corners of the constraint, which often align with the coordinate axes, favor solutions where at least one coefficient (e.g., W_1 or W_2) is exactly zero. When the constraint region expands to touch the optimal solution, it typically selects one of the axes, effectively removing irrelevant features. This mechanism ensures that L1 regularization performs feature selection by retaining only the most relevant features and setting others to zero.

While L1 regularization is highly effective for feature selection, it does not always lead to sparse solutions. If the optimal solution does not align with one of the tips of the constraint region, both W_1 and W_2 may remain nonzero, resulting in a model with nonzero coefficients for multiple features.

5.5.3 L2 Regularization

L2 regularization, also known as **Ridge regression** in the context of regression models, applies the L2 norm as a penalty term to the cost function, which leads to a different form of coefficient shrinkage compared to L1 regularization. The cost function in L2 regularization is modified by adding a term that penalizes the square of the weights:

$$\text{Cost function} = \text{Loss} + \lambda \sum \|w\|^2$$

In L2 regularization, λ is a hyperparameter that controls the strength of the penalty. Unlike L1 regularization, which can drive weights to exactly zero, L2 regularization **shrinks** the weights towards zero, but does not force them to become exactly zero.

This results in a model where all features are retained, but their influence is reduced. L2 regularization is also referred to as **weight decay** because it prevents any feature from becoming excessively large, thus avoiding overfitting by distributing the model's reliance on different features more evenly.

As shown in Fig. 10, the L2 constraint region is represented as a circle. The smooth nature of this constraint allows the model to distribute weight more evenly across correlated features, without driving any coefficients to zero. This property makes L2 regularization particularly effective for handling multicollinearity, where features are highly correlated, and helps prevent overfitting by ensuring that the influence of each feature is appropriately adjusted.

The L2 norm creates a smooth constraint region, forming a circular boundary in two dimensions and a spherical boundary in higher dimensions. Unlike the sharp corners of the L1 norm, the circular geometry of the L2 norm constraint does not encourage sparsity but instead shrinks the coefficients proportionally. Indeed, in contrast to L1 regularization, which can result in sparse solutions by eliminating features, L2 regularization retains all features but reduces their magnitudes. This smooth shrinking effect is beneficial in situations where all features contribute to the underlying structure of the data, but some may be less influential than others.

5.5.4 Geometric Interpretation of Regularization

To understand the geometric interpretation of regularization, consider a simple regression model with two independent variables, x_1 and x_2, and their corresponding coefficients, W_1 and W_2. The cost function, $y = f(W_1, W_2)$, could represent a loss function such as the Mean Squared Error (MSE) used in linear regression. Recall that regularization introduces an additional penalty term to the cost function, which alters the optimization process by adding a constraint that discourages overly complex models. As previously explained, the regularization terms for Lasso and Ridge regression are as follows:

- **Lasso regression (L1 regularization)**:

$$g(W) = \lambda(|W_1| + |W_2|)$$

- **Ridge regression (L2 regularization)**:

$$g(W) = \lambda(W_1^2 + W_2^2)$$

Here, λ is the regularization parameter that controls the strength of the penalty. The goal is to minimize the objective function, which combines the original loss function with the regularization term:

$$\min \left(f(W_1, W_2) + g(W_1, W_2)\right)$$

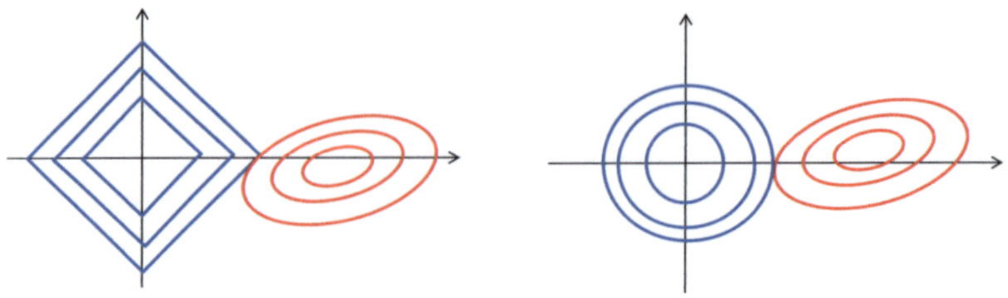

Fig. 11 Solutions of L1 and L2 norms and the effect of penalized parameters. The red contours represent the cost function $f(W_1, W_2)$, while the blue regions correspond to the constraint regions imposed by L1 and L2 regularization [4]

This process involves finding the point where the contour lines of the cost function $f(W_1, W_2)$ intersect with the constraint imposed by the regularization term $g(W_1, W_2)$. The contour plot, shown in Fig. 11, provides a visual representation of the optimization process. Contours represent regions of equal cost function values, and they form concentric curves around the minimum. In two-dimensional W_1-W_2 space, the contours illustrate how the optimization process seeks the point where the cost function is minimized. The red circles in Fig. 11 show the contours of the cost function $f(W_1, W_2)$, with the center of the circles representing the optimal values of W_1 and W_2. These values correspond to the solution that minimizes the loss function. The blue regions in Fig. 11 represent the constraint boundaries defined by L1 and L2 regularization. These constraints guide the optimization by limiting the values of the coefficients W_1 and W_2, promoting simpler models that generalize better. The intersection of the cost function's contours and the regularization constraint defines the optimal solution.

5.5.5 Lasso and Ridge: Impact on Solutions

The geometric differences between L1 and L2 regularization play a crucial role in shaping the optimization solutions. Lasso regression, which applies L1 regularization, enforces a constraint region that forms a diamond shape in two dimensions and an octahedron in higher dimensions. The sharp edges of this constraint often align with coordinate axes, increasing the likelihood that the optimization solution falls on one of these edges. As a result, Lasso frequently drives some coefficients, such as W_1 or W_2, exactly to zero. This sparsity-inducing property makes Lasso particularly effective for feature selection, as it eliminates less relevant predictors while retaining only the most significant ones.

Ridge regression, which applies L2 regularization, imposes a constraint region that is circular in two dimensions and spherical in higher dimensions. Unlike L1 regularization, the smooth, rounded boundary of the L2 constraint does not favor exact zeros. Instead, Ridge shrinks all coefficient values toward zero in a continuous manner, preventing extreme values while retaining all features in the model. This regularization technique is particularly useful when dealing with multicollinearity, as it distributes weight more evenly among correlated predictors, thereby improving model stability.

These geometric properties explain why Lasso is commonly used when feature selection is a priority, whereas Ridge is preferred for models that benefit from coefficient shrinkage without completely discarding variables. The choice between Lasso and Ridge depends on the trade-off between sparsity and robustness in predictive modeling.

5.5.6 Elastic Net Regularization

Zou and Hastie introduced the concept of the elastic net in 2005 to address the limitations of L1 (Lasso) and L2 (Ridge) regularization methods. When the number of independent variables (p) exceeds the sample size (n), i.e., when $p > n$, Lasso regression tends to select only one variable from any set of highly correlated variables, limiting the model to a maximum of n selected features. On the other hand, if the number of independent variables is less than the sample size, Ridge regression generally performs better.

Highly correlated independent variables often share similar regression coefficients, a phenomenon known as the *grouping effect*. In many real-world applications, the grouping effect is advantageous for model building. For instance, in gene identification related to diseases, researchers aim to identify all associated variables, not just one from each correlated group (as Lasso tends to do). Selecting a single variable from a group of highly correlated features can lead to a less robust model and increase the risk of precision errors. This explains why Ridge regression often performs better than Lasso in such scenarios.

Elastic net combines the strengths of both L1 and L2 regularization techniques. It shrinks some coefficients (as in Ridge regression) while setting others to zero (as in Lasso regression). The elastic net algorithm is represented as:

$$w^* = \arg\min_{w} \left(\|y - Xw\|_2^2 + \lambda_2 \|w\|_2^2 + \lambda_1 \|w\|_1 \right)$$

This method uses two regularization parameters, λ_1 and λ_2, to control the contribution of L1 and L2 norms, respectively. Elastic net is particularly useful when dealing with multiple correlated features. In cases where Lasso arbitrarily selects one feature from a pair of correlated variables, elastic net tends to include both, making the model more stable. For example, in a scenario with two highly correlated features, Lasso would randomly select one, while elastic net would account for both. Like Ridge regression, elastic net tends to be more reliable and stable for most real-world problems. Figure 12 illustrates the geometric comparison between Lasso (L1), Ridge (L2), and elastic net regularization methods.

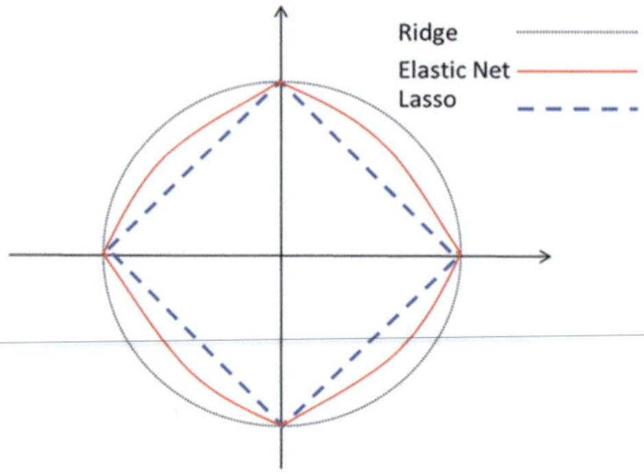

Fig. 12 Comparison of geometric interpretations of Lasso (L1), Ridge (L2), and Elastic Net regularization [4]

5.6 Bayesian Linear Regression

In standard linear regression, the output y for a given input X is modeled as a single, fixed value. However, in **Bayesian Linear Regression**, the output y is treated as a random variable that comes from a probability distribution. This reflects the Bayesian approach, which accounts for uncertainty in both the data and the model parameters. Recall that the linear regression model expresses the relationship between the input data X (features) and the output y (target) is expressed as:

$$y = \beta^T X + \epsilon$$

where β represents the model parameters (weights), X represents the input features, and ϵ is the error term, which accounts for noise or unobserved factors.In the classic **Ordinary Least Squares (OLS)** approach, the model parameters β are estimated by minimizing the sum of squared errors between the predicted and observed values of y. However, Bayesian Linear Regression takes a probabilistic approach, using probability distributions (such as a Gaussian distribution) to model the relationship between inputs (features) and output:

$$y \sim \mathcal{N}(\beta^T X, \sigma^2)$$

Here, $\beta^T X$ represents the mean of the distribution, and σ^2 represents the variance, reflecting the uncertainty due to noise in the data. The key idea in the Bayesian framework is that the model parameters β (i.e., weights) themselves are treated as random variables with their own prior distributions. In other words, in Bayesian Linear Regression, instead of assuming that we know the true relationship between inputs and outputs, we treat it as uncertain, and the ultimate goal is to find how well different possible values of the model's parameters $\beta^T X$ explain and capture the observed data y. Bayesian regression defines the following key terms:

Priors represent the initial beliefs and knowledge about the model parameters β before any data is observed. These beliefs can be based on domain expertise or, if no prior knowledge is available, a **non-informative prior** (such as a Gaussian distribution with large variance) can be used to reflect high uncertainty about the parameters. A common choice for the prior distribution of the parameters is the **multivariate Gaussian**:

$$\beta \sim \mathcal{N}(\mu_\beta, \Sigma_\beta)$$

where μ_β is the prior mean of the parameters, and Σ_β is the prior covariance matrix, which encodes the uncertainty about the parameters before observing the data.

Likelihood captures how well each possible value of β explains the observed data. For instance, if $\beta = 100$ represents a treatment effect that predicts patient recovery times close to the actual observed recovery times, the likelihood for this β value is high. In contrast, if $\beta = 500$ results in predictions that deviate significantly from the actual recovery times, the likelihood for this value of β is low. In other words, the likelihood can be viewed as a scoring mechanism. Each possible set of parameters β is assigned a score (the likelihood), which indicates how well that set of parameters explains the observed data. In Bayesian linear regression, the likelihood of the observed data y given the model parameters β is typically modeled as a Gaussian distribution as follows:

$$p(y|X, \beta) = \mathcal{N}(y|\beta^T X, \sigma^2)$$

Posterior refers to the updated belief about the parameters β after observing the data, and it is computed using **Bayes' theorem**:

$$p(\beta|X, y) = \frac{p(y|X, \beta)\, p(\beta)}{p(y|X)}$$

where $p(\beta|X, y)$ is the posterior distribution of the parameters given the data, $p(y|X, \beta)$ is the likelihood of the observed data given the parameters, $p(\beta)$ is the prior distribution of the parameters, and $p(y|X)$ is the marginal likelihood, which normalizes the posterior distribution. Bayesian regression enables us to compute the distribution of possible model parameters (**posterior**) based on the training dataset and the prior. When we have an infinite amount of data, the posterior distribution converges to the result produced by OLS regression. However, when the dataset is small or noisy, the posterior distribution remains spread out, reflecting greater uncertainty about the model parameters.

Posterior Prediction refers to making predictions ($\hat{y}$) for new data points by integrating over the posterior distribution of the model parameters β. Unlike traditional linear regression, which uses a single point estimate for the parameters, Bayesian regression accounts for uncertainty in the parameter estimates. The predictive distribution is given by:

$$p(\hat{y}|X_{\text{new}}, X, y) = \int p(\hat{y}|X_{\text{new}}, \beta)\, p(\beta|X, y)\, d\beta$$

where y represents the observed outputs and X is the input data from the training set. X_{new} is the new input for which the prediction $\hat{y}$ is made. Note that in this equation, we take into account all values of β, weighted by their posterior probabilities, rather than just using one.

6 Classification

Classification represents a category of supervised machine learning in which the output labels (dependent parameters) are categorical. This indicates that the output is partitioned into distinct groups or classes. The primary objective of classification is to train a model, referred to as a **classifier**, utilizing a training dataset, thereby enabling it to accurately predict the labels or classes of unseen data.

6.1 *Measuring Performance for Classification Problems*

Performance metrics are integral to every machine learning project, as they delineate how the performance of an algorithm is assessed and how it compares with other algorithms. In this section, we elucidate each performance metric through a straightforward classification problem: predicting whether an individual has cancer (True Positive class) or does not have cancer (False Negative class).

6.1.1 Confusion Matrix (Error Matrix)

The **confusion matrix** [9] is a tabular representation that summarizes the performance of a classification model. For a binary classifier, the confusion matrix is depicted as a 2x2 matrix, as illustrated in Fig. 13. In the context of cancer prediction, the confusion matrix encompasses two dimensions: the actual classes (true labels) and the predicted classes (predicted labels). From the confusion matrix, the following metrics can be derived:

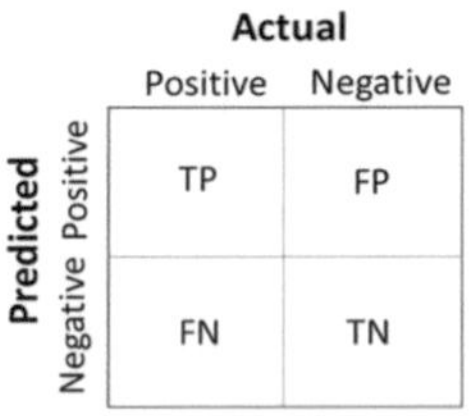

Fig. 13 Confusion matrix

- **True Positives (TP)**: The number of positive observations that are correctly predicted as positive.
- **True Negatives (TN)**: The number of negative observations that are accurately predicted as negative.
- **False Positives (FP)**: The number of actual negative observations that are incorrectly predicted as positive (Type I error).
- **False Negatives (FN)**: The number of actual positive observations that are incorrectly predicted as negative (Type II error).

The emphasis on minimizing one of these four metrics may vary based on the specific application. For instance, in **cancer prediction**, failing to identify a person with cancer (a false negative) can have serious implications, as it may delay necessary treatment. Consequently, minimizing the false negative rate is crucial in this context. Conversely, in an **email spam detection system**, true cases are identified as spam emails. If an important email is erroneously classified as spam (a false positive), it can lead to significant complications for the user awaiting that email. Thus, in this scenario, it is essential to maintain a low false positive rate to avoid missing critical communications.

6.1.2 Performance Metrics

The evaluation of classification models often relies on various performance metrics that provide insights into their predictive accuracy and effectiveness. These metrics, derived from the confusion matrix, offer a comprehensive understanding of how well a model distinguishes between different classes. Several of the most important performance metrics are presented in Fig. 14.

1. Accuracy: Accuracy computes the ratio of correct predictions (True Positives and True Negatives) to the total number of predictions made by the model:

$$\text{Accuracy} = \frac{TP + TN}{TP + FP + FN + TN}$$

Accuracy serves as a useful performance metric when the target classes are balanced. However, it becomes less reliable in cases of imbalanced datasets, where the

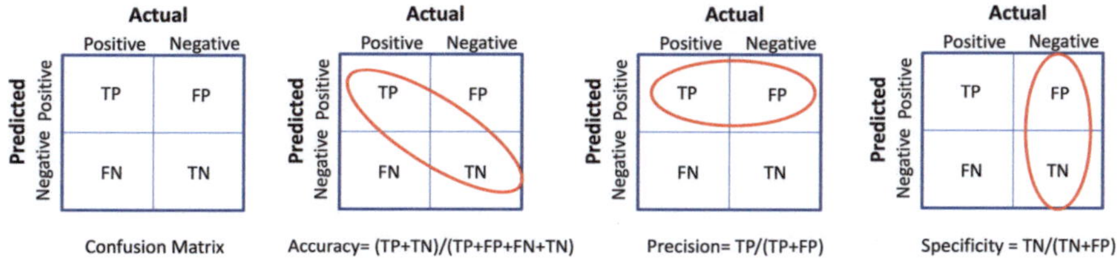

Fig. 14 Accuracy, precision, and specificity performance metrics

number of samples in each class differs significantly. In such scenarios, the probability of instances belonging to a minority class is considerably lower than that of the majority class. Consequently, the classifier may predominantly classify new observations as belonging to the majority class. For example, if there are only 5 cases of cancer out of every 100 cases, and the system predicts all 100 cases as non-cancerous, the accuracy of the model would be 95%. However, this model would perform poorly in identifying cancer cases.

2. Precision (Positive Predictive Value): Precision represents the proportion of true positive predictions among the total positive predictions made by the model. It is formally defined as the ratio of true positives to the sum of true positives and false positives:

$$\text{Precision} = \frac{TP}{TP + FP}$$

In the context of cancer detection, precision indicates the percentage of patients classified as cancerous who actually have cancer. For instance, if a model predicts only one cancerous patient and that patient is indeed cancerous, the precision is 100%.

3. Recall (Sensitivity): Recall, also known as Sensitivity, measures the model's ability to identify all relevant cases within a dataset. It is defined as the ratio of true positives to the sum of true positives and false negatives:

$$\text{Recall} = \frac{TP}{TP + FN}$$

In the cancer detection example, recall demonstrates how many of the actual cancer patients are correctly predicted as having cancer. If every patient is classified as cancerous, the recall would be 100%.

The choice between precision and recall depends on the specific problem and application context. Generally, if minimizing false negatives is a priority, efforts should focus on achieving a recall value close to 100%. Conversely, if minimizing false positives is crucial, precision should be prioritized to approach 100%.

4. Specificity (True Negative Rate): Specificity, or the true negative rate, measures the model's ability to correctly identify negative cases. In the cancer detection example, specificity represents the proportion of healthy individuals (i.e., those without cancer) correctly identified as healthy. A model achieves 100% specificity when it accurately identifies all patients without cancer.

$$\text{Specificity} = \frac{TN}{TN + FP}$$

5. F1 Score: F1 score is an optimal balance between recall and precision, calculated using the following formula:

$$F1 = \frac{2 \times \text{Precision} \times \text{Recall}}{\text{Precision} + \text{Recall}}$$

This calculation represents the harmonic mean, which, unlike a simple average, is sensitive to extreme values. For instance, a classifier with a precision of 1.0 and a recall of 0 would yield an F1 score of 0, while a simple average would result in 0.5. The F1 score is also referred to as the F Score or F Measure.

6. ROC Curve: ROC Curve (Receiver Operating Characteristic Curve) is a graphical representation that illustrates the performance of a classifier across different classification thresholds. In the ROC graph, the **True Positive Rate (TPR)** is plotted as a function of the **False Positive Rate (FPR)**. Classifiers typically generate a probability output for a given input based on its features, indicating the likelihood that the input belongs to a specific class. To classify the input, a threshold must be selected; values above this threshold indicate a positive class, while values below it indicate a negative class.

It is important to note that decreasing the classification threshold (also known as the decision threshold) results in an increase in the number of predictions classified as positive (True Positives), which simultaneously raises both False Positives and True Positives. A typical ROC curve is illustrated in Fig. 15.

7. AUC: The Area Under the ROC Curve (AUC) is a statistical parameter used for model comparison. As its name suggests, AUC measures the area underneath the entire ROC curve. This parameter aggregates the model's performance across various classification thresholds, as shown in Fig. 16. AUC provides a valuable metric for identifying which trained model best predicts the classes and assists in ranking and sorting classification models based on their predictive capabilities. Specifically, an AUC of 0.5 indicates no discriminative ability (equivalent to random guessing), values closer to 1.0 signify a strong capability to distinguish between classes, and values below 0.5 suggest performance worse than random guessing, which may indicate potential issues with the model or the data.

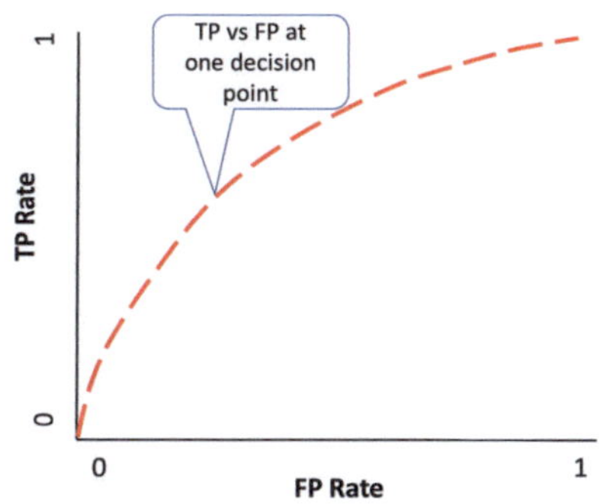

Fig. 15 True Positive (TP) and False Positive (FP) rates at different classification thresholds

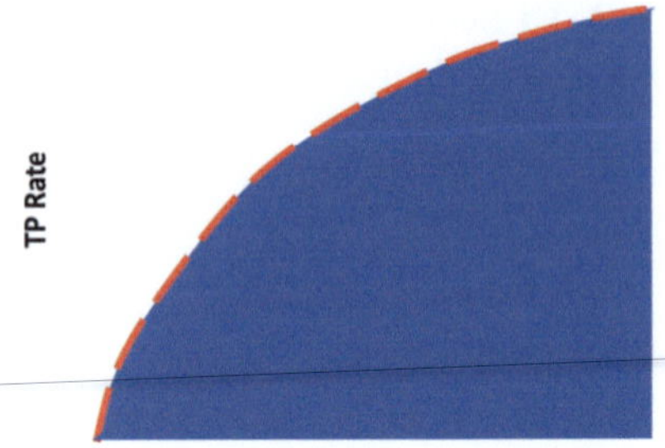

Fig. 16 Area under the ROC curve, known as AUC

6.2 *Over and Under-Sampling*

As noted in previous sections, classification metrics can be confusing and misleading, particularly when the dataset is imbalanced. Over-sampling and under-sampling are widely employed techniques to address the challenges posed by imbalanced datasets, where one class predominates compared to others. Figure 17 graphically illustrates an imbalanced dataset.

Under-sampling, as the name implies, involves selecting a subset of the majority class to match the number of data points in the minority class. This approach results in a more balanced distribution of classes, thereby improving the model's ability to learn from both classes. Conversely, **over-sampling** involves creating additional copies of the minority class to match the number of examples in the majority class. It is crucial that these copies are generated in a manner that maintains the distribution of the minority class.

6.3 *K-Nearest Neighbor (KNN)*

K-Nearest Neighbors (KNN) [10] is a simple yet popular classification technique that was first introduced in the early 1960s. The KNN algorithm can be summarized as follows (see Fig. 18):

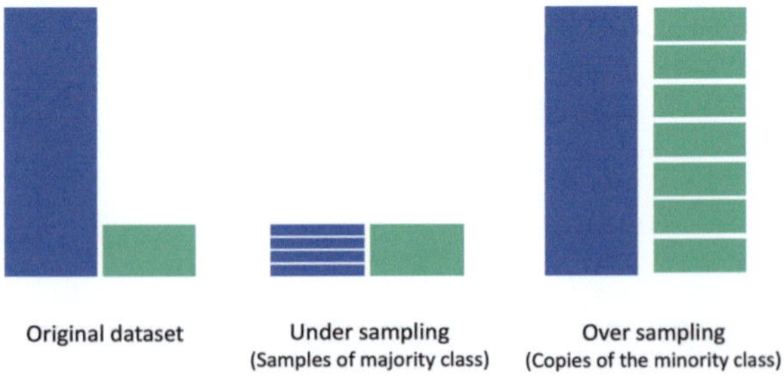

Fig. 17 Under-sampling and Over-sampling techniques to address the problem of imbalanced data sets [4]

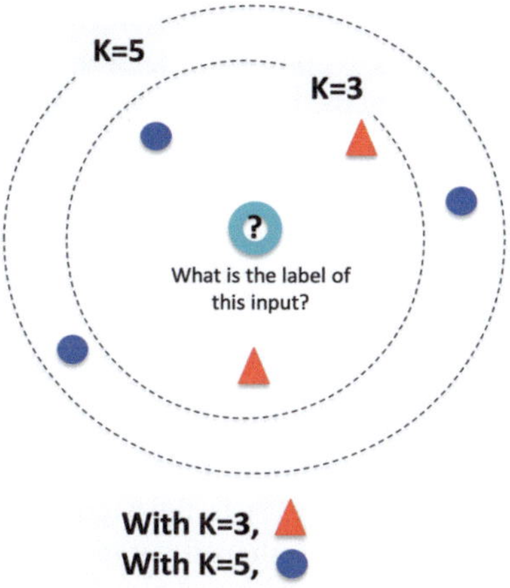

Fig. 18 A simple example of KNN classification [4]

1. Define the value of K (the number of nearest neighbors to consider).
2. Calculate the distance between the input sample (which needs to be classified) and all samples in the training set. Common distance metrics include the Minkowski distance and its special cases. In particular:
   - **Euclidean Distance**:

$$d(x, y) = \sqrt{\sum_{i=1}^{n} (x_i - y_i)^2}$$

   - **Manhattan Distance**:

$$d(x, y) = \sum_{i=1}^{n} |x_i - y_i|$$

3. Sort the distances and select the K nearest neighbors to the input sample.
4. Identify the label (class) of the input sample based on the majority label among its K neighbors. The predicted class $\hat{y}$ can be determined as follows:

$$\hat{y} = \text{mode}(y_1, y_2, \ldots, y_K)$$

   where $y_1, y_2, \ldots, y_K$ are the labels of the K nearest neighbors, and **mode** refers to the value that appears most frequently in a dataset.

6.4 Logistic Regression

Logistic regression [11] is the preferred technique for binary classification, where the dependent variable (target) has two classes. Unlike linear regression, which predicts a continuous output, logistic regression is specifically designed for categorical outcomes. It computes the probability of the default class, using a logistic function to ensure that the predicted probabilities lie within the range of 0 to 1. The mathematical representation of logistic regression is given by:

$$h_\theta(x) = g(\theta^T x)$$

where

$$g(z) = \frac{1}{1 + e^{-z}}$$

In this equation, θ represents the coefficients or weights, x denotes the input features, and $g(z)$ is the sigmoid function (also known as the logistic function), which has an S-shaped curve.

To better understand logistic regression, we need to examine the fundamentals of the Logit and Sigmoid functions:

6.4.1 Logit and Sigmoid (Logistic) Functions

Logit and Sigmoid functions are widely used in machine learning applications. For a probability p, the corresponding odds (the ratio of the probability that an event will occur to the probability that the event will not occur) can be calculated as:

$$\text{odds} = \frac{p}{1 - p}$$

The logit function is defined as the logarithm of the odds:

$$\text{logit}(x) = \log\left(\frac{x}{1 - x}\right)$$

As shown in Fig. 19, the logit function approaches positive infinity as x approaches 1 and negative infinity as x approaches 0. The logit function effectively maps probability values to the entire range of real numbers, making it useful for regression analysis.

The sigmoid function, or logistic function, is the inverse of the logit function. Therefore, for a given probability p, we have:

$$\text{sigmoid}(\text{logit}(p)) = p$$

The sigmoid function maps real values to the range of [0, 1], and larger inputs yield outputs closer to 1. It is expressed as:

$$\sigma(x) = \frac{1}{1 + e^{-x}}$$

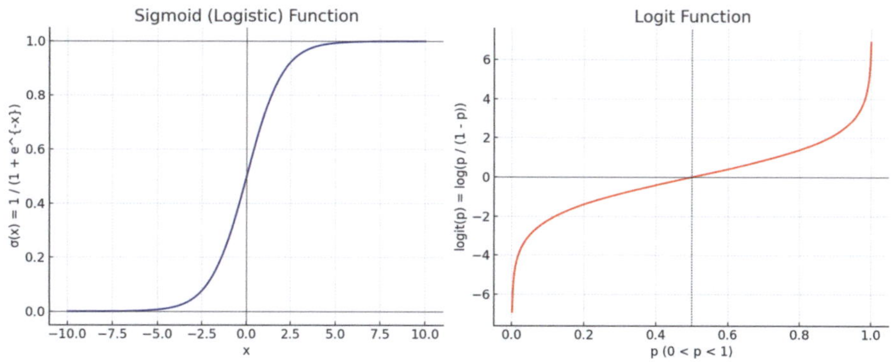

Fig. 19 Comparison of the Logit and Sigmoid (logistic) functions. The Logit function (right) maps probabilities $p \in (0, 1)$ to the entire real line via $\text{logit}(p) = \ln \frac{p}{1-p}$. Its inverse is the Sigmoid function (left), which maps real-valued inputs $x \in (-\infty, \infty)$ into probabilities in the range $(0, 1)$ according to $\sigma(x) = \frac{1}{1+e^{-x}}$

One of the most significant applications of the sigmoid function is in transforming real values into probabilities, which is commonly used in the final stages of classification algorithms.

6.4.2 Decision Boundary (Decision Surface)

To convert the logistic function's output score (a probability in the range of $[0, 1]$) into a binary class, a threshold must be defined. If the score exceeds this threshold, the output is class 1; otherwise, it is class 0. For example, when the threshold is set to 0.5, classes can be identified using the following rules (see Fig. 19):

- **Class = 1**: $h_\theta(x) \geq 0.5$
- **Class = 0**: $h_\theta(x) < 0.5$

It should be noted that the threshold can be adjusted based on the specific needs of the application, such as prioritizing precision or recall in certain contexts. To understand the meaning of the decision boundary, recall that $h_\theta(x) = g(\theta^T x)$. With a threshold value of 0.5, we conclude:

- **Class = 1**: $\theta^T x \geq 0$
- **Class = 0**: $\theta^T x < 0$

Suppose we have a training set similar to Fig. 20 and want to classify inputs into two classes (e.g., red circles and green triangles). We can represent different hypotheses about $\theta^T x$. A simple linear hypothesis might be:

$$h_\theta(x) = g(\theta_0 + \theta_1 x_1 + \theta_2 x_2)$$

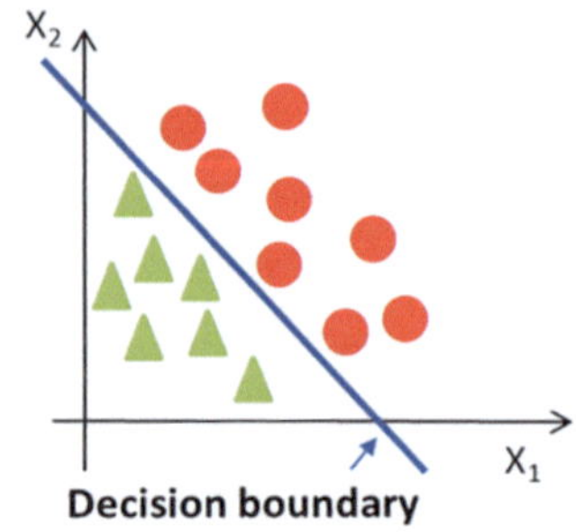

Fig. 20 Linear decision boundary [4]

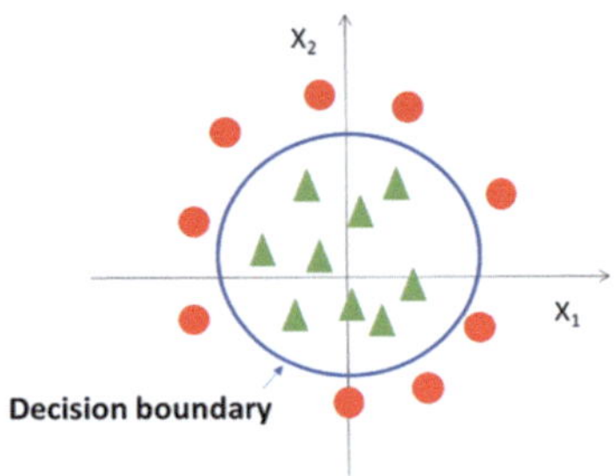

Fig. 21 Non-linear decision boundary [4]

As shown in the figure, this hypothesis corresponds to a line (in blue) that separates the inputs into two distinct classes, known as the decision boundary. Formally, a decision boundary is a hyperplane or hypersurface that divides the underlying vector space into classes.

Decision boundaries need not be linear. By incorporating higher-order polynomial terms, one can create a non-linear decision boundary. An example of a nonlinear decision boundary is depicted in Fig. 21 based on the following equation:

$$h_\theta(x) = g(\theta_0 + \theta_1 x_1 + \theta_2 x_2 + \theta_3 x_1^2 + \theta_4 x_2^2)$$

6.4.3 Cost Function in Logistic Regression

In the previous subsection, we learned about linear and nonlinear decision boundaries. The next question is how to compute the coefficients or weights of the features (inputs). Instead of using MSE as the cost function (as in linear regression), logistic regression utilizes a Cross-Entropy function, also referred to as Log Loss. This cost function is divided into two components for $y = 1$ (the first class) and $y = 0$ (the second class):

$$J(\theta) = \frac{1}{m} \sum_{i=1}^{m} \text{Cost}(h_\theta(x^{(i)}), y^{(i)})$$

Fig. 22 Cost function of logistic regression

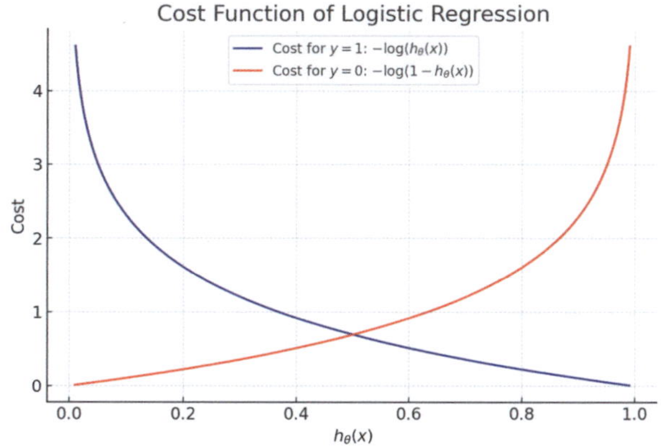

Where the cost is defined as follows:

- If $y = 1$:

$$\text{Cost}(h_\theta(x), y) = -\log(h_\theta(x))$$

- If $y = 0$:

$$\text{Cost}(h_\theta(x), y) = -\log(1 - h_\theta(x))$$

This can be combined and expressed as:

$$J(\theta) = -\frac{1}{m}\sum_{i=1}^{m}\left[y^{(i)}\log(h_\theta(x^{(i)})) + (1 - y^{(i)})\log(1 - h_\theta(x^{(i)}))\right]$$

The cost function for $y = 1$ and $y = 0$ is illustrated in Fig. 22.

6.5 *Support Vector Machine (SVM)*

The Support Vector Machine (SVM) [12] is a prominent classification method that was developed in the late 1990s. SVMs are particularly effective in identifying the optimal separation between data points belonging to two distinct classes. Figure 23a illustrates an SVM in a two-dimensional plot, displaying multiple separating hyperplanes (H1, H2, and H3), which are referred to as **decision boundaries**. These hyperplanes serve to partition the underlying feature space and facilitate the classification of inputs. A critical inquiry arises regarding the identification of the optimal hyperplane and the methodology for its computation. SVM offers a robust framework for addressing this challenge.

The primary objective of the SVM algorithm is to determine a hyperplane that maximizes the **margin** between data points of the two classes (See Fig. 23b). The

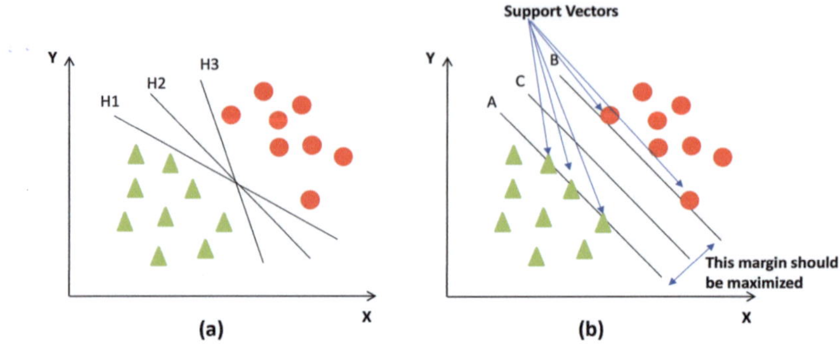

Fig. 23 A 2D support vector machine model [4]

margin is defined as the distance between the closest points of each class to the hyperplane. The hyperplane that achieves the largest margin is termed the **maximum-margin hyperplane**.

Support Vectors play a pivotal role in this context; they are the data points that lie closest to the hyperplane and define its position and orientation. The location of the hyperplane is influenced solely by these support vectors; other data points that are further away do not affect the model. The removal of a support vector typically results in a shift in the hyperplane's position, potentially altering both its orientation and margin, whereas the removal of non-support vectors has no effect on the hyperplane. For instance, Fig. 24 highlights three identified support vectors.

To formally define the problem of Linear SVM, consider a training dataset consisting of n points represented as:

$$\{(\mathbf{x}_1, y_1), (\mathbf{x}_2, y_2), \ldots, (\mathbf{x}_n, y_n)\}$$

Fig. 24 Mathematical presentation of a support vector model [4]

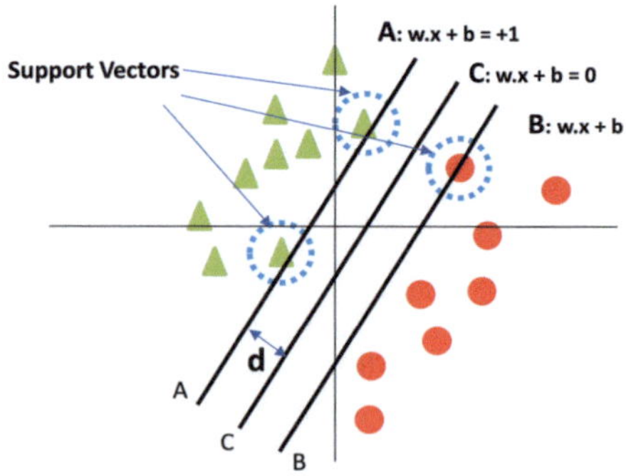

In this representation, $\mathbf{x}_i$ is an n-dimensional vector of features, and y denotes the target output (which can take values of either 1 or -1). The goal of the SVM is to maximize the margin, which can be expressed mathematically as:

$$\mathbf{w} \cdot \mathbf{x} - b = 0$$

where $\mathbf{w}$ is the weight vector, and b is the bias term. If the data points are linearly separable, two hyperplanes can be defined to bound the margin:

$$\mathbf{w} \cdot \mathbf{x} - b = 1 \quad (\text{for } y = 1)$$

$$\mathbf{w} \cdot \mathbf{x} - b = -1 \quad (\text{for } y = -1)$$

The distance between these hyperplanes is calculated as $\frac{2}{\|\mathbf{w}\|}$. Consequently, minimizing $\|\mathbf{w}\|$ will serve to maximize the margin. Constraints are imposed during optimization to ensure that data points do not encroach upon the margin:

$$y_i(\mathbf{w} \cdot \mathbf{x}_i - b) \geq 1, \quad \text{for all } i$$

This represents an optimization problem that can be solved through techniques such as the **Lagrangian multiplier** method.

In instances where the data are not linearly separable (see Fig. 25), the aforementioned constraints cannot be utilized during optimization. In such cases, it becomes necessary to identify a hyperplane that penalizes data points located on the incorrect side of the margin. This can be achieved through the application of the **hinge loss function**, which penalizes points that fall on the wrong side:

$$\max(0, 1 - y_i(\mathbf{w} \cdot \mathbf{x}_i - b))$$

This cost function yields a value of zero if the actual and predicted values reside on the same side of the margin. Conversely, for points on the wrong side, the function's value (penalty) increases in proportion to the distance of these misclassified points from the hyperplane.

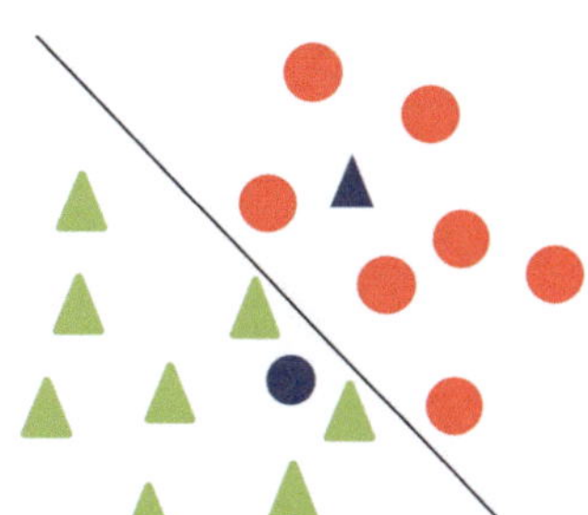

Fig. 25 An example of not linearly separable data [4]

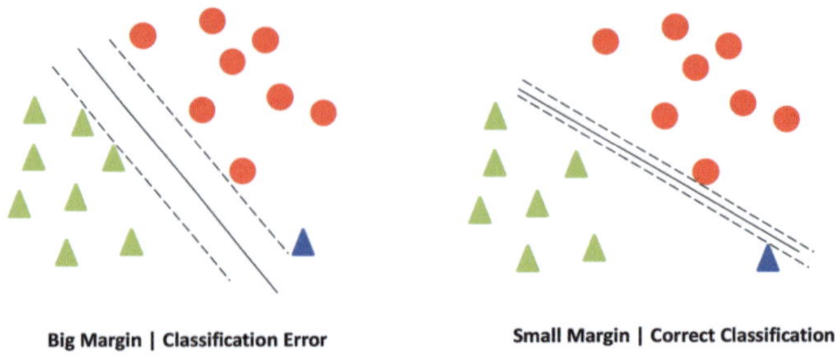

Fig. 26 Tradeoff between classification error and margin [4]

To further illustrate, consider the example shown in Fig. 26. As depicted, it is possible to extract a hyperplane with a large margin that fails to classify all data points correctly. Alternatively, one may identify a hyperplane with a very small margin (which is suboptimal for generalization) that can classify all data points accurately. In SVM, a balance must be struck between **classification error** and **margin**. This balance is facilitated by the introduction of a regularization parameter, denoted as C, leading to the formulation of the SVM cost function as follows:

$$\text{Cost} = \max(0, 1 - y_i(\mathbf{w} \cdot \mathbf{x}_i - b)) + C\|\mathbf{w}\|$$

In scenarios where the data cannot be classified by a linear hyperplane (as illustrated in Fig. 27a), a technique known as the "**Kernel trick**" can be employed. A **kernel** is defined as a function that transforms data into a higher-dimensional space without explicitly calculating the coordinates in that space. This transformation enables the SVM to identify a linear hyperplane in the higher-dimensional space that effectively separates the data points.

Commonly utilized kernels in Support Vector Machines include:

- **Linear Kernel**: This kernel applies no transformation to the data, effectively treating the original feature space as the decision boundary.
- **Polynomial Kernel**: The polynomial kernel computes a polynomial combination of the input features, expressed as $K(\mathbf{x}_i, \mathbf{x}_j) = (\mathbf{x}_i \cdot \mathbf{x}_j + c)^d$, where c is a constant and d is the degree of the polynomial.

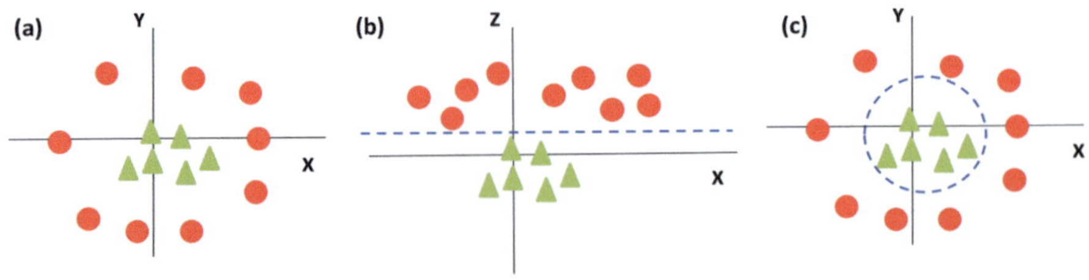

Fig. 27 Kernel trick in SVM [4]

- **Radial Basis Function (RBF) Kernel**: Also known as the **Gaussian kernel**, the RBF kernel maps data points into an infinite-dimensional space based on their pairwise distances, defined as $K(\mathbf{x}_i, \mathbf{x}_j) = e^{-\gamma \|\mathbf{x}_i - \mathbf{x}_j\|^2}$, where γ is a parameter that controls the width of the Gaussian function.

The kernel trick facilitates the application of SVMs to non-linearly separable data. For clarity, let us apply the kernel trick to our example (See Fig. 27a). In this scenario, we map our two-dimensional data points into a three-dimensional space. Suppose our mapping function is $z = x^2 + y^2$, which computes the values of data points along the z-axis. By plotting the data points using these newly computed values (in the x-z plane), we can observe a linear hyperplane effectively separating the two classes (See Fig. 27b). The final step in the kernel trick involves projecting this separation line (hyperplane) back to the original space. In our example, this transformation results in a circular boundary similar to that depicted in Fig. 27c. Finally, it is important to note that, akin to regression analysis, the SVM algorithm can also employ regularization to enhance the generalizability of the solution and mitigate the risk of overfitting.

6.6 Decision Tree Classifier

A **decision tree** [6] is a category of classification and regression algorithms characterized by a tree-like model of decisions. The classification of a new data point is accomplished by traversing down the tree, following the decisions made at each node. In decision tree algorithms, the feature space is partitioned into several regions, with each region designated by a class label (or the probability of a label), as illustrated in Fig. 28. The common terms in decision trees are defined as follows:

- **Nodes**: In nodes, the value of a particular feature (attribute) is evaluated.
- **Edges**: Edges correspond to the output of the preceding test, connecting nodes to other nodes or to leaf nodes.
- **Leaves**: Terminal nodes that represent the final prediction.

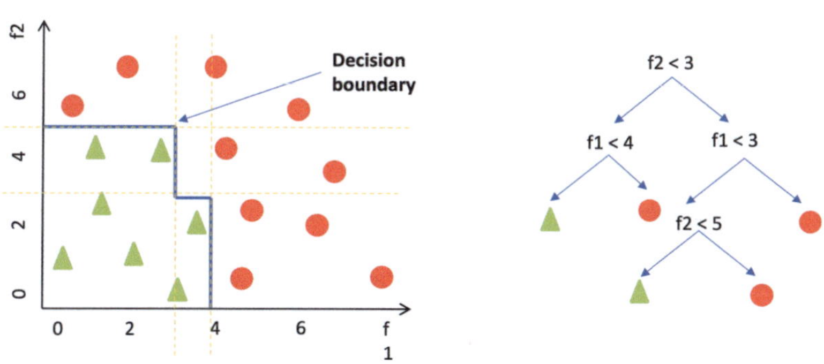

Fig. 28 A decision tree model [4]

There are several implementations of decision trees, such as C5.0, C4.5, and Iterative Dichotomiser 3 (ID3). C5.0 is the most widely recognized decision tree algorithm and has become the industry standard.

6.6.1 Recursive Divide-and-Conquer Algorithm

The fundamental **Recursive Divide-and-Conquer** algorithm for constructing a decision tree consists of the following steps:

1. **Select a Feature**: Identify the optimal feature A to serve as the root node of the tree. The selection is based on specific criteria, such as **Information Gain**.
2. **Split the Data Points**: Partition the dataset D into subsets based on the values of feature A. Let V denote the set of all possible values of feature A. For each value $v \in V$, create a corresponding subset D_v:

$$D_v = \{\mathbf{x}_i \in D \mid A(\mathbf{x}_i) = v\}$$

   In this expression, $A(\mathbf{x}_i)$ signifies the value of feature A for instance $\mathbf{x}_i$.
3. **Recursive Step**: Recursively apply the first two steps for each subset D_v generated in the previous step. This is represented as:

$$\text{TreeNode}(D_v) \rightarrow \text{Split on } A$$

   This process continues until a stopping condition is satisfied.
4. **Stopping Condition**: Terminate the recursion for a branch when all instances in that subset belong to the same class (i.e., share the same label). This can be mathematically expressed as:

$$\text{If } \forall \mathbf{x}_i, \mathbf{x}_j \in D_v, \text{ then } y_i = y_j$$

   where y_i and y_j represent the class labels corresponding to instances $\mathbf{x}_i$ and $\mathbf{x}_j$, respectively. When the stopping condition is met, specifically when $y_i = y_j$, indicating that all instances in D_v belong to the same class.

6.6.2 Choosing the Best Split

An important consideration in this algorithm is determining which feature to select as the root at each iteration. Several methods exist for selecting the best attribute for splitting at each step. Generally, a desirable attribute is one that results in successor nodes that are as pure as possible, meaning the corresponding branches predominantly contain instances of a single class.

This concept can be understood through the notion of **entropy**. The objective of the splitting procedure is to decrease the entropy, as a good attribute (node) divides

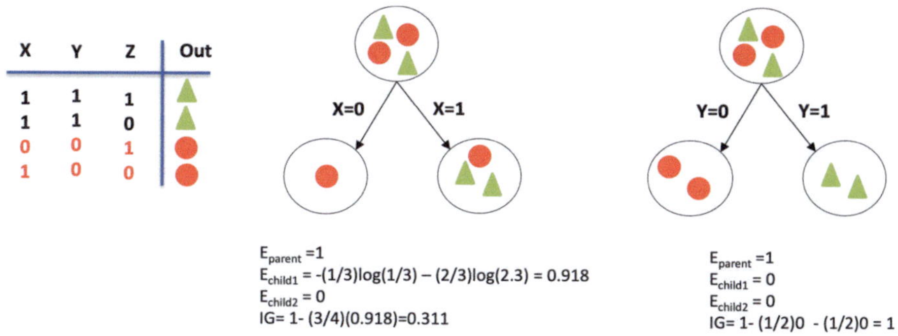

Fig. 29 A simple example of selecting the split attribute based on Information Gain (IG) [4]

a set into subsets (regions) that exhibit higher homogeneity. A higher entropy value indicates a mix of different classes within each region. To facilitate this process, we define **Information Gain (IG)** as follows:

$$\text{Information Gain} = \text{Entropy (Parent node)} - \text{Average Entropy (Child nodes)}$$

Information Gain represents the reduction in entropy following a split on an attribute. The attribute selected for the splitting procedure should be the one that yields the highest Information Gain. In other words, an attribute with the highest Information Gain leads to branches that are most homogeneous (i.e., exhibit lower entropy or are very pure subsets).

To illustrate this process, consider the example depicted in Fig. 29. Suppose we have three attributes (features) and a target with two classes. If we split based on the X attribute, the IG will be 0.31. Conversely, if we choose Y as our splitting attribute, the IG will be 1. Thus, we would split based on Y.

6.6.3 Stopping Criteria and Pruning Techniques

Another challenge in the construction of decision trees is determining the stopping criteria for the splitting procedure. Specifically, when should the splitting cease? One approach is to continue the splitting until each leaf node in the decision tree is completely pure (i.e., only instances of one class are present in the leaf). However, this method may not be efficient for all applications, as it can create numerous small regions in the feature space, leading to increased overfitting. To address the issue of overfitting in decision trees, several **pruning techniques** can be employed:

- **Pre-Pruning**: This technique halts the growth of the tree during construction when further splitting does not significantly improve the classifier's performance beyond a predefined threshold. Some of the key pruning criteria are listed below:

  - **Minimum Samples Split**: The minimum number of samples required to perform a split.
  - **Minimum Impurity Decrease**: The required decrease in impurity (e.g., entropy) for a split to be considered valid.
  - **Maximum Depth**: A limit on the tree's maximum depth.

- **Post-Pruning**: In this approach, the complete decision tree is first constructed, which may accurately classify all training data points but can lead to overfitting. After construction, an iterative pruning algorithm is applied to simplify the tree by removing certain nodes. Nodes are eliminated if their removal does not significantly impact the model performance. Examples of post-pruning include:

  - **Reduced Error Pruning**: This method evaluates the impact of removing a node based on its effect on the tree's accuracy using a validation dataset. A node is pruned if its elimination does not decrease accuracy.
  - **Cost Complexity Pruning**: This technique balances tree complexity against accuracy by tuning a hyperparameter α that penalizes overly complex trees, helping to prevent overfitting while retaining predictive power.
  - **Minimum Description Length (MDL) Pruning**: This approach minimizes the total encoded length of the tree and its classification errors, favoring simpler trees that make accurate predictions.

It should be noted that pruning techniques not only help combat overfitting but also serve as valuable tools for managing model complexity, enabling us to strike a balance between accuracy and interpretability during model exploration.

6.7 Ensembles

Ensemble machine learning combines several **weak models** (predictive models that perform slightly better than random guessing) to create a single **meta-model** (an overarching model that aggregates the predictions of multiple base models), effectively addressing issues of high bias (underfitting) and high variance (overfitting).

6.7.1 Bootstrap Aggregating (Bagging)

Bootstrap Aggregating [13], commonly referred to as bagging, is an effective technique for reducing model overfitting and handling unstable datasets. It enhances performance when training on datasets with a limited number of instances. For example, bagging can be used to combine multiple decision trees into a forest model, resulting in a stronger classifier. This method generates n small training sets from the original dataset, with each set referred to as a sample. These samples are produced by random sampling of the input dataset with replacement.

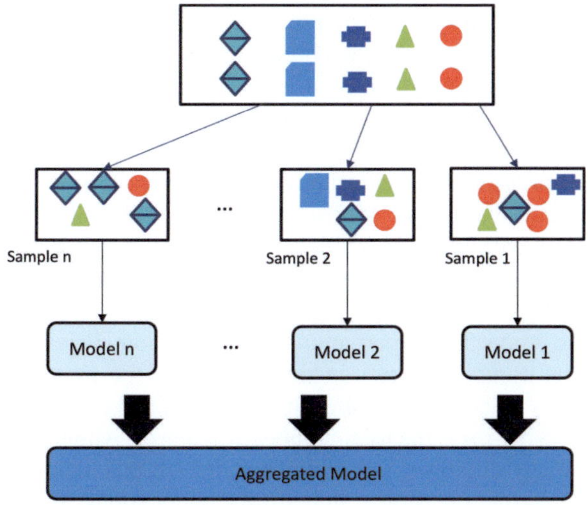

Fig. 30 An overview of Bootstrap Aggregating (Bagging) based on the random sampling with replacement technique. Bagging creates multiple subsets of the training dataset by sampling with replacement. Each subset is used to train a separate model, and their outputs are aggregated (e.g., by averaging for regression or voting for classification) to produce the final prediction [4]

To explain the concept of sampling with replacement, consider a simple example. Suppose we have three names: Farshad, Bahar, and Krish, and we need to sample two names. We place these names in a hat and randomly choose one. After selecting a name, we put it back into the hat before selecting another. The possible combinations of two sampled names include: (Farshad, Farshad), (Farshad, Bahar), (Bahar, Bahar), and so forth.

Once all samples are constructed using the random sampling with replacement technique, we independently learn and build n ensembles (i.e., individual learning models) corresponding to each sample. Each of these weak models is called an **estimator**. Finally, the outputs of these weak models are combined through voting or simple averaging to produce a **meta-model** (See Fig. 30).

Random Forest

Random Forest (RF), also known as Random Decision Forest, is a bagging ensemble learning method that builds multiple decision trees (referred to as **estimators**) and aggregates their predictions to improve accuracy. The "forest" in Random Forest refers to this collection of decision trees, which work together to enhance the predictive performance of the model. The flexibility, simplicity, and effectiveness of Random Forest have made it one of the most popular algorithms in machine learning.

RF is based on the bagging technique, where each decision tree is trained on a random bootstrap sample (i.e., a sample with replacement) of the training data. In addition to bagging, Random Forest introduces a second layer of randomness by selecting a random subset of features at each node during tree construction. This is a key difference from standard decision trees, which consider all available features when determining the best split at each node.

Specifically, instead of evaluating all p features at each node, Random Forest randomly selects m features (where m is typically much smaller than p) and uses only this subset to determine the best split. For example, if there are 10 features in the dataset ($p = 10$) and $m = 3$, the algorithm randomly selects 3 out of the 10 features at each node to decide the optimal split. This random feature selection process varies for every node and every tree, promoting diversity in the model and reducing the correlation between trees. This helps to mitigate the risk of overfitting, which is common in individual decision trees.

Therefore, two important hyperparameters in Random Forest are the **number of estimators** (the number of trees in the forest) and the **maximum number of features** m that can be considered at each split. A typical setting for classification tasks is $m = \sqrt{p}$, where p is the total number of features. Although Random Forest performs well in many situations, a potential drawback is that building a large number of trees can make predictions slower in real-time applications. Training is generally efficient, but the model's complexity can impact prediction speed when dealing with large forests.

The implementation of the Random Forest algorithm for classification can be summarized as follows:

Algorithm 2 Random Forest Algorithm for Classification

Input: Training data $D = \{(x_1, y_1), (x_2, y_2), \ldots, (x_n, y_n)\}$, number of trees T, subset size k, number of features m to sample at each node
Output: A Random Forest model consisting of T trees
for $i = 1$ to T **do**
    Sample k data points with replacement from D to create a bootstrap sample D_i
    Train a decision tree h_i on D_i using the following process:
    for each node in the decision tree **do**
        Randomly select m features from the total p features
        Select the best feature and split based on a criterion like information gain
    end for
end for
Classification Prediction:
For a new data point x, predict using majority voting:

$$H(x) = \text{mode}(h_1(x), h_2(x), \ldots, h_T(x))$$

6.7.2 Boosting

Boosting [14] is another technique for producing ensemble models, but unlike bagging, where models are created in parallel, boosting builds models sequentially. In bagging, n base models are trained independently in parallel, while in boosting, each subsequent model is built using feedback from the previous one. Specifically,

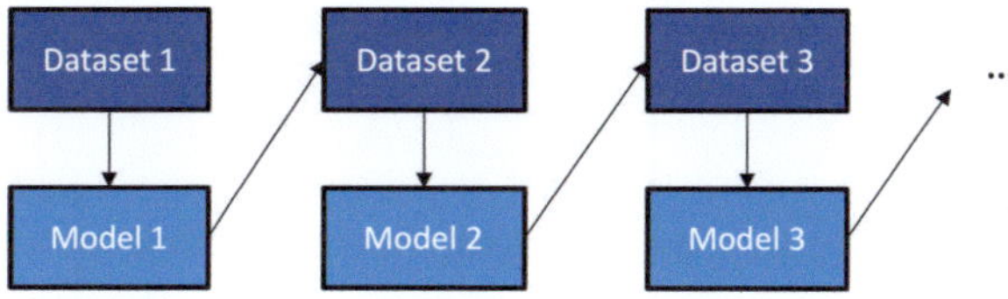

Fig. 31 Schematic presentation of the sequential steps in the boosting algorithm. The boosting algorithm builds an ensemble of weak learners sequentially, with each learner focusing on correcting the errors made by the previous ones. This iterative process improves the overall accuracy of the model [4]

boosting focuses on the training instances that were misclassified by earlier models (see Fig. 31). Compared to bagging, boosting often provides better performance for certain applications, but it can also increase the risk of overfitting.

Adaptive Boosting (AdaBoost)

AdaBoost is one of the most well-known implementations of boosting [14]. Its core concept is to sequentially improve a weak model to create a stronger classifier. AdaBoost achieves this by combining multiple weak classifiers linearly, with the final classification being a weighted sum of these classifiers. This can be expressed as:

$$f(x) = \sum_{t=1}^{T} \alpha_t h_t(x)$$

where $h_t(x)$ represents the individual weak classifiers, typically decision stumps (single-split decision trees), and α_t denotes the weight assigned to each classifier. Initially, all training instances are given equal weight. As AdaBoost progresses, it adjusts these weights, placing more emphasis on instances that are harder to classify. New weak classifiers are added sequentially, with each one focusing more on the misclassified instances from the previous model.

In each iteration, the algorithm selects a random subset of the training data and trains a weak classifier. AdaBoost then increases the weights of misclassified examples, giving them a higher probability of appearing in the next subset. This process continues, with each weak classifier honing in on the errors made by the previous models.

Figure 32 illustrates a simplified example of the AdaBoost procedure. In this figure, weights are updated at each step to construct a final model by linearly combining several weak classifiers. Larger weights are represented by bigger symbols, while smaller weights are shown by smaller symbols, indicating their relative importance in subsequent iterations. Below is a summary of the key steps involved in the AdaBoost algorithm for classification.

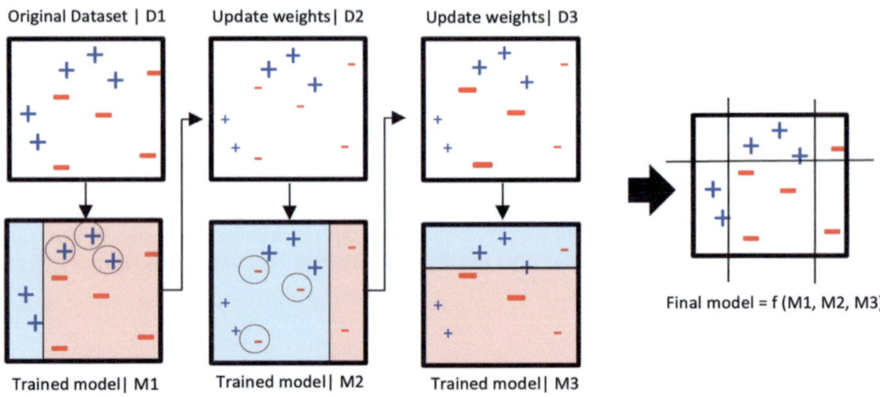

Fig. 32 The overall procedure of the AdaBoost algorithm [4]

Gradient Boosting

Gradient Boosting is a type of boosting algorithm that, unlike AdaBoost, focuses on minimizing the errors (residuals) of the current model in a sequential manner. At each iteration, a weak learner (often a decision tree) is fitted to the residual errors (also known as pseudo-residuals) of the previous model. The key difference from AdaBoost is that Gradient Boosting uses gradient descent to optimize the model. Specifically, it fits the weak learners to the negative gradients of a predefined loss function, such as the mean squared error for regression tasks or the logistic loss for classification. By iteratively adjusting the model to minimize these residuals, Gradient Boosting gradually improves its predictive performance. Below is a summary of the key steps in the Gradient Boosting Algorithm.

6.7.3 Stacking

Stacking, known as stacked generalization, is a sophisticated ensemble learning methodology utilized to enhance predictive accuracy by combining the outputs of multiple base models. The underlying principle of stacking is to exploit the complementary strengths of diverse machine learning algorithms while mitigating their individual weaknesses, thereby creating a more robust and generalized model. Unlike bagging, which reduces variance by averaging predictions from multiple models trained on different subsets of the data, stacking aims to learn an optimal combination of base models by training a meta-model—a higher-level model that aggregates

Algorithm 3 AdaBoost Algorithm for Classification

Input: Training data $D = \{(x_1, y_1), (x_2, y_2), \ldots, (x_n, y_n)\}$, number of iterations T
Output: A weighted combination of T weak classifiers
Initialize weights for all training instances: $w_1(i) = \frac{1}{n}$ for all $i = 1, 2, \ldots, n$
for $t = 1$ to T **do**
  Train a weak classifier $h_t(x)$ on the weighted training set
  Compute the error of $h_t(x)$:

$$\epsilon_t = \sum_{i=1}^{n} w_t(i) \cdot \mathbb{I}(h_t(x_i) \neq y_i)$$

▷ $\mathbb{I}(\cdot)$ is the indicator function: 1 if the condition holds, 0 otherwise.
  Compute the weight of classifier $h_t(x)$:

$$\alpha_t = \frac{1}{2} \ln \left(\frac{1 - \epsilon_t}{\epsilon_t} \right)$$

  Update the weights of the training instances:

$$w_{t+1}(i) = w_t(i) \cdot \exp(-\alpha_t \cdot y_i \cdot h_t(x_i))$$

  Normalize the weights so that they sum to 1
end for
Classification Prediction:
For a new data point x, predict using the weighted majority vote:

$$H(x) = \text{sign} \left(\sum_{t=1}^{T} \alpha_t h_t(x) \right)$$

and refines the base model predictions—on their outputs. Additionally, stacking differs from boosting, which sequentially trains models to correct the errors of previous ones, by combining models in parallel rather than in sequence. This parallel approach allows stacking to leverage the independent strengths of diverse algorithms without relying on iterative error correction, making it particularly effective in capturing complex patterns in data (See Fig. 33). Generally, the stacking methodology consists of two primary components:

- **Base Models (Level-0 Models):** These are the individual machine learning models responsible for learning patterns and generating predictions from the dataset. Base models can vary significantly, ranging from linear regression and decision trees to support vector machines and deep neural networks. Each base model operates independently and focuses on specific aspects of the data.
- **Meta-Model (Level-1 Model):** The meta-model is a higher-level model that learns to aggregate and weigh the outputs of the base models. This model is trained on the predictions generated by the base models and aims to refine the overall prediction by correcting errors or emphasizing more reliable contributors. The meta-model

Algorithm 4 Gradient Boosting Algorithm for Classification

1: **Input:** Training data $D = \{(x_1, y_1), (x_2, y_2), \ldots, (x_n, y_n)\}$, number of iterations T, differentiable loss function $L(y, f(x))$
2: **Output:** A model $f(x)$ consisting of T weak classifiers
3: Initialize the model with a constant value:

$$f_0(x) = \arg\min_c \sum_{i=1}^{n} L(y_i, c)$$

▷ Start by finding the best constant predictor, which minimizes the loss.
4: **for** $t = 1$ to T **do**
5: Compute the negative gradients (pseudo-residuals) for all data points:

$$r_t(i) = -\left[\frac{\partial L(y_i, f(x_i))}{\partial f(x_i)}\right]_{f=f_{t-1}(x)}$$

▷ $L(y_i, f(x_i))$ represents the loss function, where y_i is the true value and ▷ $f(x_i)$ is the predicted value. The loss measures the difference between them. ▷ The pseudo-residuals represent the errors the current model has not captured.
6: Fit a weak learner $h_t(x)$ to the pseudo-residuals r_t ▷ Train a weak classifier on these residuals to correct the errors.
7: Compute the step size (learning rate) γ_t by solving:

$$\gamma_t = \arg\min_{\gamma} \sum_{i=1}^{n} L\left(y_i, f_{t-1}(x_i) + \gamma h_t(x_i)\right)$$

▷ Find the optimal step size to scale the weak learner's contribution.
8: Update the model:

$$f_t(x) = f_{t-1}(x) + \gamma_t h_t(x)$$

▷ Update the model by adding the contribution of the new weak learner.
9: **end for**
10: **Classification Prediction:**
11: For a new data point x, predict the class label:

$$H(x) = \text{sign}(f_T(x))$$

is typically simpler, such as logistic regression, but its design depends on the problem context.

The stacking process can be outlined as follows:

1. **Training Base Models:** Multiple base models are trained independently on the training dataset. These models may use distinct algorithms or variations of the same algorithm with different hyperparameters.
2. **Generating Meta-Model Training Data:** The predictions of the base models are utilized as inputs (features) for training the meta-model. To ensure the meta-model generalizes effectively, predictions are often generated using cross-validation techniques.

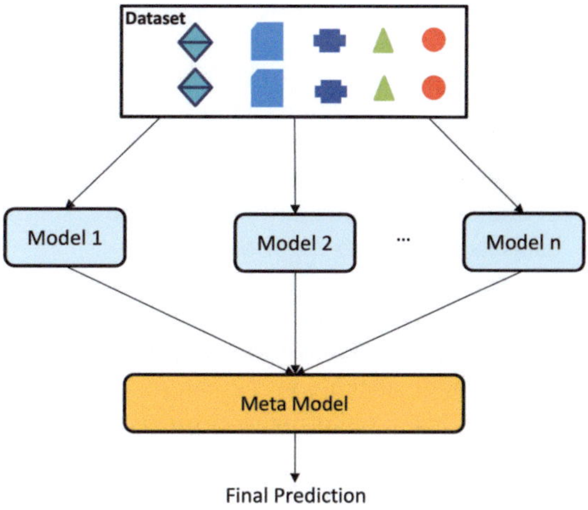

Fig. 33 An overview of the Stacking approach. In this ensemble learning method, multiple base models are trained on the same dataset, and their predictions are combined using a meta-model to produce the final output. This approach leverages the strengths of different models to improve overall performance

3. **Training the Meta-Model:** The meta-model is trained on the outputs of the base models, leveraging their predictions to learn the optimal combination or weighting.
4. **Making Final Predictions:** For unseen data, base models generate predictions that are fed into the meta-model, which produces the final prediction.

The stacking framework offers several advantages, which include:

- **Enhanced Predictive Performance:** By combining the outputs of multiple models, stacking often results in improved accuracy and robustness compared to any single model.
- **Flexibility in Model Selection:** Stacking allows for the integration of heterogeneous base models, enabling the use of diverse machine learning algorithms and techniques tailored to specific features of the data.
- **Utilization of Heterogeneous Technologies:** Each base model can be implemented using distinct programming languages or frameworks, such as employing one model in Python and another in R, thereby optimizing development workflows and computational efficiency.
- **Reduced Risk of Overfitting:** When combined with cross-validation, stacking mitigates the risk of overfitting by ensuring that the meta-model generalizes well across diverse scenarios.

Despite its advantages, stacking introduces several challenges:

- **Increased Complexity:** The requirement to manage multiple models and coordinate their outputs increases the architectural and computational complexity of the pipeline.
- **Higher Computational Costs:** Training and evaluating multiple base models, in addition to the meta-model, can be resource-intensive, particularly for large datasets or complex algorithms.
- **Overfitting Risks in the Meta-Model:** If improperly designed, the meta-model may overfit to the predictions of the base models, reducing its generalization ability.
- **Dependency on Robust Orchestration and Monitoring Tools:** Effective implementation of stacking necessitates advanced tools for orchestration, model management, and monitoring to ensure scalability and maintainability.

7 Clustering

Clustering is a well-established unsupervised learning technique that organizes data points into groups (clusters) such that points within the same cluster are more similar to each other than to those in other clusters [15]. Various approaches to clustering exist, and in this section, we review three of the most widely used techniques: K-means, Fuzzy C-Means, and Hierarchical Clustering.

7.1 K-means Clustering

K-means clustering partitions data into K clusters based on distance, typically using the Euclidean distance metric. In this method, K represents the number of clusters into which the dataset is to be divided. A fundamental limitation of K-means is that it requires the data to be continuous, as it is not inherently suitable for categorical variables. Additionally, it is advisable to normalize the data prior to applying K-means. This is necessary because distance calculations can be biased if the features have different scales or units. For example, when clustering individuals based on both weight (in kilograms) and height (in centimeters), the disparity in units may result in an undue influence of height over weight. Normalization mitigates this issue by adjusting the data to a common scale. The K-means algorithm proceeds through an iterative refinement process, as outlined in the following steps:

Algorithm 5 K-means Clustering Algorithm

1: **Input**: Dataset $X = \{x_1, x_2, \ldots, x_n\}$, where each $x_i \in \mathbb{R}^d$, the number of clusters K, and a convergence threshold ϵ (optional).
2: **Output**: Set of clusters $C_1, C_2, \ldots, C_K$ and the final cluster centroids $\mu_1, \mu_2, \ldots, \mu_K$.
3: **Step 1 (Initialization)**: Randomly select K initial centroids $\mu_1, \mu_2, \ldots, \mu_K$ from the dataset, where each centroid $\mu_j \in \mathbb{R}^d$. ▷ Initialize the centroids randomly from the data points
4: **Step 2 (Assignment)**: For each data point x_i, assign it to the closest centroid μ_j based on the Euclidean distance:

$$d(x_i, \mu_j) = \|x_i - \mu_j\|_2 = \sqrt{\sum_{k=1}^{d} (x_{ik} - \mu_{jk})^2}$$

Assign each x_i to the cluster C_j, where:

$$C_j = \{x_i : j = \arg\min_j d(x_i, \mu_j)\}$$

5: **Step 3 (Update)**: Update each centroid μ_j for cluster C_j by calculating the mean of all points assigned to that cluster:

$$\mu_j = \frac{1}{|C_j|} \sum_{x_i \in C_j} x_i$$

where $|C_j|$ denotes the number of points in cluster C_j. ▷ Recompute the centroid as the mean of the points in the cluster
6: **Step 4 (Compute the Objective Function)**: Compute the objective function J, which is the within-cluster sum of squared distances (WCSS):

$$J = \sum_{j=1}^{K} \sum_{x_i \in C_j} \|x_i - \mu_j\|_2^2$$

7: **Step 5 (Convergence)**: Repeat Steps 2 through 4 until the centroids no longer change significantly or the improvement in the objective function J between two iterations is less than a predefined threshold ϵ:

$$\Delta J = |J_{\text{old}} - J_{\text{new}}| < \epsilon$$

7.2 *Fuzzy C-Means Clustering*

C-means clustering, also known as Fuzzy C-means (FCM), is a well-known clustering algorithm that differs from K-means in a key aspect: whereas K-means assigns each data point to a single cluster, C-means allows data points to belong to multiple clusters with varying degrees of membership. These membership degrees, which are learnable parameters, indicate how strongly each point is associated with each cluster. The ultimate goal of C-means clustering is to adjust the cluster centers as well as the membership degrees in a way that the total distance between data points and their closest clusters is as small as possible while also considering the corresponding memberships with clusters. The following steps summarize the core idea behind this fuzzy algorithm:

Algorithm 6 Fuzzy C-means Clustering Algorithm

1: **Input**: Dataset $X = \{x_1, x_2, \ldots, x_n\}$, number of clusters C, fuzzification parameter $m > 1$, and convergence threshold ϵ.
2: **Output**: Fuzzy partition matrix U (membership degrees), cluster centroids $\mu_1, \mu_2, \ldots, \mu_C$.
3: **Step 1 (Initialization)**: Initialize the fuzzy partition matrix $U = [u_{ij}]$, where u_{ij} represents the membership degree of point x_i in cluster j, subject to the constraint:

$$\sum_{j=1}^{C} u_{ij} = 1 \quad \text{for all } i$$

▷ Randomly initialize the membership degrees

4: **Step 2 (Compute Centroids)**: For each cluster j, compute the centroid μ_j using the membership degrees:

$$\mu_j = \frac{\sum_{i=1}^{n} u_{ij}^m x_i}{\sum_{i=1}^{n} u_{ij}^m}$$

▷ Calculate centroids as the weighted mean of points

5: **Step 3 (Update Membership Degrees)**: For each data point x_i and cluster j, update the membership degrees u_{ij} using:

$$u_{ij} = \frac{1}{\sum_{k=1}^{C} \left(\frac{\|x_i - \mu_j\|_2}{\|x_i - \mu_k\|_2} \right)^{\frac{2}{m-1}}}$$

▷ Update membership based on the distance from centroids

6: **Step 4 (Convergence)**: Repeat Steps 2 and 3 until the change in the partition matrix U between two consecutive iterations is less than a predefined threshold ϵ:

$$\|U^{(t+1)} - U^{(t)}\| < \epsilon$$

▷ Check for convergence by measuring changes in the membership matrix

7.3 *Hierarchical Clustering*

Hierarchical clustering is a technique that builds a hierarchy of clusters, represented as a tree-like structure where clusters can differ in size and exhibit nested relationships. This method produces a hierarchical classification system and is primarily divided into two approaches: **divisive clustering** and **agglomerative clustering**. The **agglomerative approach** is a bottom-up method and follows these steps:

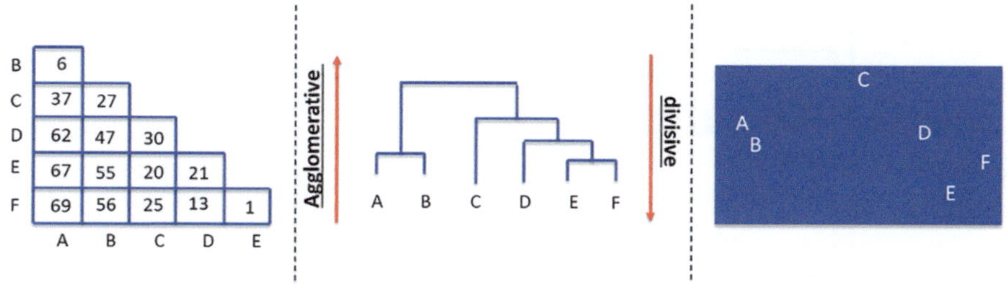

Fig. 34 A simple example of hierarchical clustering [4]

Algorithm 7 Agglomerative Hierarchical Clustering Algorithm

1: **Input**: Dataset $X = \{x_1, x_2, \ldots, x_n\}$, distance metric (e.g., Euclidean distance).
2: **Output**: Dendrogram representing the hierarchy of clusters.
3: **Step 1 (Distance Matrix)**: Construct a distance matrix D, where each entry $D(i, j)$ represents the distance between data points x_i and x_j, often computed using Euclidean distance:

$$D(i, j) = \sqrt{\sum_{k=1}^{d} (x_{ik} - x_{jk})^2}$$

4: **Step 2 (Initialization)**: Treat each data point x_i as its own cluster, so initially, there are n clusters.
5: **Step 3 (Iterative Merging)**:
6: **while** more than one cluster remains **do**
7: Identify the two clusters with the smallest distance in the distance matrix.
8: Merge the two closest clusters into a single cluster.
9: Recalculate the distances between the newly formed cluster and all remaining clusters.
10: **end while**
11: **Step 4 (Dendrogram Construction)**: The process continues until all data points are merged into one single cluster. The result is a dendrogram that illustrates the hierarchical relationships between clusters.

Figure 34 illustrates the agglomerative hierarchical clustering process. In contrast, the **divisive approach** is a top-down method, where all data points begin in a single large cluster. The algorithm then recursively splits this cluster into smaller clusters at each level, based on dissimilarity, continuing the division until the desired number of clusters is achieved or each data point forms its own cluster.

8 Reinforcement Learning

Reinforcement Learning (RL) [16] is a machine learning paradigm that enables agents to learn optimal strategies through interaction with an environment and feedback in the form of rewards. Unlike supervised learning, which relies on labeled data, RL emphasizes trial-and-error learning, making it well-suited for dynamic and

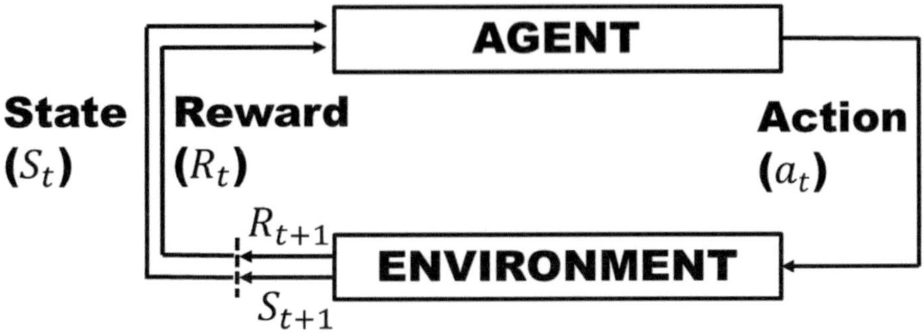

Fig. 35 Illustration of the agent-environment interaction in a Markov Decision Process (MDP). The agent observes state S_t, selects an action a_t, and transitions to a new state S_{t+1} while receiving a reward R_t. The agent's objective is to optimize cumulative rewards

complex decision-making tasks in domains such as robotics, healthcare, finance, and autonomous systems.

At its core, RL is typically formulated as a Markov Decision Process (MDP) (see Fig. 35), defined as a 5-tuple (S, A, T, R, γ), where S represents the set of possible states, A is the set of allowable actions, $T(s, a, s')$ is the transition function that determines the probability of reaching state s' from state s by taking action a, and $R(s, a, s')$ defines the immediate reward received upon transitioning to state s'. The discount factor $\gamma \in [0, 1)$ controls the trade-off between immediate and future rewards, ensuring that long-term rewards are appropriately weighted.

The goal of an RL agent is to learn an optimal policy $\pi(a \mid s)$ that dictates the best action to take in a given state to maximize the expected cumulative reward over time (considering both immediate and future rewards). This cumulative reward, also referred to as the return, is defined as:

$$G_t = \sum_{k=0}^{\infty} \gamma^k r_{t+k+1}.$$

where:

- r_{t+k+1}: The reward received at time step $t + k + 1$, indicating the immediate benefit of taking an action at a given state.
- $\gamma \in [0, 1)$: The **discount factor**, which determines how much future rewards contribute to the present value. A higher γ prioritizes long-term rewards, whereas a lower γ makes the agent more short-sighted.

A policy can be either deterministic, mapping each state to a specific action $(\pi(s) = a)$, or stochastic, defining a probability distribution over actions $(\pi(a \mid s))$, which is particularly useful in environments where exploration is crucial.

An RL agent typically comprises three key components: a value function, which estimates the expected return for states or state-action pairs to guide decision-making;

a policy representation, which maps states to actions or distributions over actions; and an environment model, which predicts state transitions and rewards to facilitate planning. Depending on how the agent interacts with the environment, RL can be categorized into model-free RL, which learns directly from experience without explicitly modeling the environment, and model-based RL, which utilizes a learned or predefined model of the environment to make decisions.

The learning process in RL can follow different paradigms. In offline learning, the agent is trained on a fixed dataset without further interaction with the environment, making it ideal for data-limited scenarios. In contrast, online learning allows the agent to continuously interact with the environment, adapting its strategy over time. Online learning is often enhanced with techniques such as experience replay, where past experiences are stored and revisited to improve sample efficiency and learning stability.

In the following sections, we discuss various types of RL approaches, including value-based methods, which focus on learning value functions to guide decision-making; policy-based methods, which directly optimize the policy without relying on value functions; actor-critic methods, which combine both value and policy-based approaches for improved stability; and model-based methods, which leverage an explicit model of the environment for planning and decision-making.

8.1 Value-Based Methods

Value-based methods focus on learning value functions that estimate the expected cumulative reward for a given state or state-action pair. These methods guide the agent to take optimal actions by leveraging the learned value functions to evaluate the long-term benefits of actions. The goal is to find a policy that maximizes the expected cumulative reward over time.

8.1.1 Q-Learning

Q-Learning is a foundational algorithm in RL and a classic example of value-based methods. It aims to learn the optimal action-value function $Q^*(s, a)$, which represents the maximum cumulative reward achievable from a state-action pair (s, a) by following the optimal policy. The Q-value update rule is based on the Bellman equation:

$$Q(s_t, a_t) \leftarrow Q(s_t, a_t) + \alpha \left[r_t + \gamma \max_{a'} Q(s_{t+1}, a') - Q(s_t, a_t) \right],$$

where:

- $Q(s_t, a_t)$: The current estimate of the Q-value for taking action a_t in state s_t. This represents the algorithm's knowledge about the value of this state-action pair before the update.
- α: The learning rate, which determines how much the Q-value is updated in response to new information. A smaller α makes updates slower and more stable, while a larger α makes the algorithm adapt faster but potentially less stable.
- r_t: The immediate reward received after taking action a_t in state s_t. This term reflects the short-term gain achieved by the agent.
- $\gamma \max_{a'} Q(s_{t+1}, a')$: The discounted estimate of the maximum future reward achievable from the next state s_{t+1}. Here:
  - γ: The discount factor, which controls how much future rewards are valued compared to immediate rewards. A higher γ places more emphasis on long-term gains, while a lower γ prioritizes immediate rewards.
  - $\max_{a'} Q(s_{t+1}, a')$: The maximum Q-value for the next state s_{t+1}, considering all possible actions a'. This term encourages the agent to seek the highest potential reward from the next state.
- $r_t + \gamma \max_{a'} Q(s_{t+1}, a')$: This represents the target value for the Q-value update. It combines the immediate reward r_t and the discounted estimate of future rewards to calculate what the Q-value should ideally be.
- $Q(s_t, a_t) + \alpha \left[r_t + \gamma \max_{a'} Q(s_{t+1}, a') - Q(s_t, a_t)\right]$: The updated Q-value is obtained by moving the current Q-value $Q(s_t, a_t)$ closer to the target value. The difference between the target and the current value, $[r_t + \gamma \max_{a'} Q(s_{t+1}, a') - Q(s_t, a_t)]$, is called the temporal difference (TD) error. The learning rate α determines how much of the TD error is incorporated in the update.

In general, the Q-Learning update rule adjusts the agent's estimate of the value of taking action a_t in state s_t based on new experiences. It balances immediate rewards (r_t) with the potential for future rewards ($\gamma \max_{a'} Q(s_{t+1}, a')$). Over time, as the agent explores the environment and updates its Q-values, it converges to the optimal Q-value function $Q^*(s, a)$, enabling it to determine the best actions to take in each state. The off-policy nature of Q-Learning means it updates Q-values assuming the optimal policy, regardless of the agent's current behavior.

8.1.2 Deep Q-Networks (DQN)

Deep Q-Networks (DQN) introduced a groundbreaking extension to Q-Learning by integrating deep neural networks to approximate the Q-function in high-dimensional environments. This innovation enabled RL to be applied to tasks with complex state spaces, such as video games and robotics.

The Q-value function in DQN is defined as:

$$Q(s, a) = \mathbb{E}\left[G_t \mid s_t = s, a_t = a\right],$$

where:

- $Q(s, a)$: The Q-value function represents the expected cumulative reward for taking action a in state s and following the current policy thereafter. DQN uses a deep neural network $Q(s, a; \theta)$ to approximate this function, where θ are the network parameters.
- $\mathbb{E}[\cdot]$: The expectation operator accounts for the stochasticity in the environment by averaging over all possible future outcomes, such as state transitions and rewards.
- G_t: The cumulative return (or discounted sum of rewards) from time t, defined as:

$$G_t = \sum_{k=0}^{\infty} \gamma^k r_{t+k+1}.$$

Here:

- r_{t+k+1}: The reward received $k+1$ steps into the future.
- $\gamma \in [0, 1)$: The discount factor, which prioritizes immediate rewards over distant rewards. A higher γ places more emphasis on future rewards.

Algorithm 8 Deep Q-Network (DQN) Algorithm

1: **Input:** Q-network with parameters θ, target network with parameters θ^-, replay buffer $\mathcal{D}$, learning rate α, discount factor γ, exploration rate ϵ
2: **Output:** Trained Q-network parameters θ
3: **Step 1 (Initialize):** Initialize environment and observe initial state s_0
4: **Step 2 (Episode Loop):** For each time step $t = 0$ to T:
5: With probability ϵ, select a random action a_t
6: Otherwise, select $a_t = \arg\max_a Q(s_t, a; \theta)$
7: Execute action a_t, observe reward r_t and next state s_{t+1}
8: Store transition (s_t, a_t, r_t, s_{t+1}) in replay buffer $\mathcal{D}$
9: **Step 3 (Training Step):** Sample a mini-batch of transitions $(s, a, r, s') \sim \mathcal{D}$
10: Compute target:

$$y = r + \gamma \max_{a'} Q(s', a'; \theta^-)$$

11: Compute loss:

$$L(\theta) = \mathbb{E}_{(s,a,r,s')} \left[(y - Q(s, a; \theta))^2 \right]$$

12: Perform a gradient descent step on $L(\theta)$ to update θ
13: **Step 4 (Target Network Update):** If update condition met, set $\theta^- \leftarrow \theta$

The key techniques used in DQN to improve stability and performance are outlined below, with references to the corresponding steps in Algorithm 8.

Target Networks: In DQN, a separate target network $Q(s, a; \theta^-)$ is maintained to compute the target Q-values. The parameters θ^- of the target network are periodically updated by copying the weights of the online Q-network θ, as shown in Step 11 of Algorithm 8. This separation of target calculation from the online Q-network helps reduce training instability caused by the frequent changes in Q-value estimates, which might otherwise lead to oscillations or divergence.

Experience Replay: To improve sample efficiency and stabilize training, DQN uses a replay buffer $\mathcal{D}$ to store past interactions in the form of transitions (s_t, a_t, r_t, s_{t+1}) (Step 7). During training, mini-batches of transitions are randomly sampled from $\mathcal{D}$ (Step 8). This random sampling de-correlates consecutive experiences, preventing the network from overfitting to recent transitions and improving generalization.

Q-Value Update Rule: The Q-value update rule in DQN is derived from the Bellman equation. The target value for the Q-update is computed as:

$$y = r_t + \gamma \max_{a'} Q(s_{t+1}, a'; \theta^-),$$

where:

- r_t: The immediate reward received after the agent takes action a_t in state s_t (Step 5).
- $\max_{a'} Q(s_{t+1}, a'; \theta^-)$: The maximum Q-value predicted by the target network for the next state s_{t+1}, considering all possible actions a' (Step 9).
- γ: The discount factor, which determines the balance between prioritizing immediate rewards and future rewards.

Using this target value, the loss function is defined as:

$$L(\theta) = \frac{1}{|\mathcal{D}|} \sum (y - Q(s, a; \theta))^2 ,$$

where $Q(s, a; \theta)$ is the predicted Q-value for the current state-action pair based on the online Q-network's parameters. The goal of the loss function is to minimize the difference between the predicted Q-values and the target values, ensuring alignment with the Bellman equation (Step 9).

The overall workflow of DQN is described in Algorithm 8. The key steps include:

- **Exploration and Exploitation (Step 2):** The agent uses an ϵ-greedy strategy to balance exploration and exploitation. With probability ϵ, it selects a random action; otherwise, it chooses the action that maximizes the Q-value predicted by the online network.
- **Updating Q-values (Step 3):** A mini-batch of transitions is sampled from the replay buffer $\mathcal{D}$, and the Q-network is updated by minimizing the loss $L(\theta)$, which measures the discrepancy between predicted Q-values and target values computed using the target network.
- **Stabilizing Training (Step 4):** The target network parameters θ^- are periodically updated to match the online network parameters θ, which helps stabilize training by providing consistent target estimates over several updates.

By integrating Q-Learning with deep neural networks, DQN enables agents to approximate the Q-value function in environments with large, high-dimensional state spaces. For example, in Atari games, the raw pixel inputs representing the game screen serve as the state. The convolutional layers of the deep neural network process these inputs to extract spatial and temporal features, allowing the agent to make informed decisions. This innovation has propelled reinforcement learning to achieve superhuman performance in complex tasks, such as mastering Atari games.

DQN's use of target networks, experience replay, and gradient-based optimization of Q-values allows it to address the instability and inefficiency of traditional Q-Learning in high-dimensional environments. These techniques ensure that DQN can generalize effectively and learn robust policies for a variety of challenging tasks.

8.1.3 SARSA

SARSA (State-Action-Reward-State-Action) is another widely used value-based method. Unlike Q-Learning, SARSA is an on-policy algorithm, meaning that it learns the action-value function based on the agent's current behavior policy. The SARSA update rule is:

$$Q(s_t, a_t) \leftarrow Q(s_t, a_t) + \alpha \left[r_t + \gamma Q(s_{t+1}, a_{t+1}) - Q(s_t, a_t) \right],$$

where a_{t+1} is the next action selected by the current policy. SARSA emphasizes learning the Q-values for the policy the agent is following, which can lead to safer behavior in some environments.

8.2 Policy-Based Methods

Policy-based methods focus on optimizing the policy π directly, without relying on value functions. These approaches parameterize the policy $\pi_\theta(a \mid s)$ and adjust its parameters θ to maximize the expected cumulative reward. Policy-based methods are particularly effective in high-dimensional or continuous action spaces, where value-based methods may struggle due to the complexity of representing the value function. Additionally, these methods naturally handle stochastic policies, making them suitable for environments requiring exploration or adaptability.

8.2.1 Policy Gradient Methods

Policy gradient methods optimize the policy by adjusting its parameters to maximize expected cumulative rewards. The policy is parameterized as $\pi_\theta(a \mid s)$, where θ represents the learnable parameters of the policy. The optimization objective is defined as:

$$J(\theta) = \mathbb{E}_{\pi_\theta}\left[\sum_{t=0}^{T} \gamma^t r_t\right],$$

where:

- $J(\theta)$: The objective function representing the expected cumulative reward.
- $\gamma \in [0, 1)$: The discount factor, which determines the relative importance of future rewards. A higher γ places more emphasis on long-term rewards, while a lower γ prioritizes immediate rewards.
- r_t: The reward received at time step t.
- T: The trajectory length or the total number of time steps in an episode.

The gradient of the objective function, known as the policy gradient, is given by:

$$\nabla_\theta J(\theta) = \mathbb{E}_{\pi_\theta}\left[\nabla_\theta \log \pi_\theta(a \mid s)\hat{A}_t\right],$$

where:

- $\nabla_\theta J(\theta)$: The gradient of the objective function with respect to the policy parameters θ, used to optimize the policy.
- $\nabla_\theta \log \pi_\theta(a \mid s)$: The gradient of the log-probability of taking action a in state s, which directs the updates toward actions that lead to higher rewards.
- $\hat{A}_t$: An estimator of the advantage function, which measures the relative value of taking action a_t in state s_t compared to the baseline (e.g., the average reward or value function). The advantage function reduces variance in policy gradient estimates.

A foundational algorithm in policy gradient methods is REINFORCE. It updates the policy parameters using the following rule:

$$\theta \leftarrow \theta + \alpha \sum_{t=0}^{T} \nabla_\theta \log \pi_\theta(a_t \mid s_t) G_t,$$

where:

- α: The learning rate, controlling the step size of updates.
- G_t: The cumulative return from time step t, defined as:

$$G_t = \sum_{k=t}^{T} \gamma^{k-t} r_k,$$

which captures both immediate and discounted future rewards.

The REINFORCE algorithm (see Algorithm 9)iteratively improves the policy by sampling actions, observing rewards, and updating the policy parameters to increase the likelihood of actions that yield higher cumulative returns. The algorithm directly optimizes the policy by increasing the probability of actions leading to higher cumulative rewards. The gradient $\nabla_\theta \log \pi_\theta(a_t \mid s_t)$ adjusts the policy to favor actions a_t that result in better outcomes, as measured by G_t. By using stochastic gradient ascent, the algorithm improves the policy iteratively. However, REINFORCE can suffer from high variance in gradient estimates, which is addressed in advanced methods using variance reduction techniques, such as advantage functions or baselines.

Algorithm 9 REINFORCE Algorithm

Input: Policy $\pi_\theta(a \mid s)$ parameterized by θ, learning rate α, discount factor γ
Output: Updated policy parameters θ
Step 1 (Initialization): Initialize environment and observe initial state s_0
Step 2 (Trajectory Generation):
while not converged **do**
    for $t = 0$ to T (trajectory length) **do**
        Sample action $a_t \sim \pi_\theta(a \mid s_t)$
        Execute action a_t, observe reward r_t, and transition to state s_{t+1}
    end for
Step 3 (Policy Update):
    for $t = 0$ to T **do**
        Compute the cumulative return:

$$G_t = \sum_{k=t}^{T} \gamma^{k-t} r_k$$

        Update the policy parameters:

$$\theta \leftarrow \theta + \alpha \nabla_\theta \log \pi_\theta(a_t \mid s_t) \cdot G_t$$

    end for
end while

In general, the REINFORCE algorithm (see Algorithm 9) works as follows:

- **Step 1 (Initialization):** The environment is initialized, and the agent observes the initial state s_0, which marks the starting point for trajectory generation.
- **Step 2 (Trajectory Generation):** The agent interacts with the environment to collect a trajectory of length T. At each time step t:

  - An action a_t is sampled from the current policy $\pi_\theta(a \mid s_t)$.
  - The action a_t is executed, and the environment returns a reward r_t and the next state s_{t+1}.

- **Step 3 (Policy Update):** After the full trajectory is collected, the agent updates the policy based on the cumulative return:
  - For each time step t, compute the return $G_t = \sum_{k=t}^{T} \gamma^{k-t} r_k$, which is the total discounted reward from time step t onward.
  - The policy parameters θ are updated using the policy gradient:

$$\theta \leftarrow \theta + \alpha \nabla_\theta \log \pi_\theta(a_t \mid s_t) \cdot G_t,$$

  which increases the likelihood of actions that yield higher returns.

- Steps 1 through 3 are repeated until the policy converges or a predefined stopping criterion is satisfied.

The REINFORCE algorithm directly optimizes the policy by associating higher probabilities with actions that yield higher cumulative rewards. The cumulative return G_t acts as a feedback signal to guide the policy updates, while the gradient $\nabla_\theta \log \pi_\theta(a_t \mid s_t)$ determines how the policy parameters should be adjusted. By iteratively sampling trajectories and updating the policy, the algorithm improves its ability to select actions that maximize long-term rewards. Despite its simplicity, the algorithm can suffer from high variance in gradient estimates, motivating the use of variance reduction techniques such as baselines and advantage functions in subsequent methods.

8.2.2 Proximal Policy Optimization (PPO)

Proximal Policy Optimization (PPO) is a state-of-the-art policy-gradient method designed to achieve a balance between simplicity, performance, and stability during training. Developed by OpenAI, PPO improves upon traditional policy-gradient algorithms by employing a clipped surrogate objective function, which ensures that policy updates remain within a safe range. This technique avoids excessive changes to the policy, leading to improved stability and performance.

The PPO objective function is defined as:

$$L^{\text{CLIP}}(\theta) = \mathbb{E}_t \left[\min \left(r_t(\theta)\hat{A}_t, \text{clip}(r_t(\theta), 1-\epsilon, 1+\epsilon)\hat{A}_t \right) \right],$$

where:

- $r_t(\theta) = \frac{\pi_\theta(a_t|s_t)}{\pi_{\theta_{\text{old}}}(a_t|s_t)}$ is the probability ratio between the new policy and the old policy.
- $\hat{A}_t$ is the advantage function, representing how much better the action a_t is compared to the policy's baseline.

- ϵ is a hyperparameter that controls the clipping range, ensuring that policy updates do not deviate significantly from the old policy.

8.2.3 Generalized Policy Optimization (GPO)

Generalized Policy Optimization (GPO) extends traditional policy optimization methods by incorporating techniques that enhance stability and sample efficiency. One notable extension is the integration of sample reuse strategies, which combine the stability benefits of on-policy algorithms with the sample efficiency of off-policy methods.

In GPO, the policy improvement step is generalized to allow leveraging information from multiple policies or experiences. This approach can lead to more efficient learning processes, especially in complex environments.

A specific instance of this is the Generalized Proximal Policy Optimization with Sample Reuse, which modifies the standard PPO algorithm to incorporate off-policy data, thereby improving sample efficiency while maintaining stability. This method has demonstrated improved performance in various decision-making tasks.

8.2.4 Generalized Regularized Policy Optimization (GRPO)

Generalized Regularized Policy Optimization (GRPO) builds upon GPO by incorporating regularization techniques into the policy optimization process. The main goal of GRPO is to improve the robustness and stability of policy updates while ensuring efficient exploration of the environment.

The GRPO objective incorporates a regularization term $R(\pi_\theta, \pi_{\text{old}})$ into the policy optimization objective:

$$L^{\text{GRPO}}(\theta) = \mathbb{E}_t \left[\min \left(r_t(\theta)\hat{A}_t, \text{clip}(r_t(\theta), 1-\epsilon, 1+\epsilon)\hat{A}_t \right) \right] - \lambda R(\pi_\theta, \pi_{\text{old}}),$$

where:

- $R(\pi_\theta, \pi_{\text{old}})$ is the regularization term, which penalizes deviations between the new policy π_θ and the old policy π_{old}. Common choices include KL divergence or entropy regularization.
- λ is a hyperparameter that controls the strength of the regularization term.

The regularization term serves two purposes:

- **Stabilization:** By limiting large changes in policy updates, GRPO enhances stability during training.
- **Exploration:** Entropy regularization encourages exploration by maintaining diversity in the action distribution.

GRPO has shown significant improvements over standard GPO and PPO in environments requiring robust and adaptive behavior, such as robotics and financial modeling.

8.3 Actor-Critic Methods

Actor-Critic methods combine the strengths of policy-based and value-based approaches by using two separate networks: a policy network (actor) and a value network (critic). The actor is responsible for selecting actions based on the current state, while the critic evaluates the selected actions by estimating the value function. This combination allows Actor-Critic methods to achieve more stable and efficient learning compared to pure policy-based or value-based approaches.

The actor updates its policy parameters to maximize the expected cumulative rewards, while the critic updates its value function to provide accurate feedback to the actor. This framework has been particularly effective in solving problems with continuous action spaces and dynamic environments.

8.3.1 Asynchronous Advantage Actor-Critic (A3C)

Asynchronous Advantage Actor-Critic (A3C) is a distributed Actor-Critic algorithm that leverages multiple agents interacting with the environment in parallel to improve sample efficiency and training stability. Each agent operates independently, interacting with its copy of the environment, while sharing a global set of parameters for the actor and critic networks.

The key features of A3C include:

- **Asynchronous Updates:** By updating the global parameters asynchronously from multiple agents, A3C reduces correlations between training samples and improves learning stability.
- **Advantage Function:** A3C uses an advantage function $\hat{A}_t = r_t + \gamma V(s_{t+1}) - V(s_t)$ to reduce the variance of policy gradient estimates, leading to more stable updates.

A3C has demonstrated strong performance in environments such as Atari games and continuous control tasks. Its ability to leverage distributed computation makes it highly scalable and efficient.

8.3.2 Proximal Policy Optimization (PPO)

Proximal Policy Optimization (PPO) is primarily a policy-gradient method, focusing on directly optimizing the policy $\pi_\theta(a \mid s)$ to maximize cumulative rewards. However, PPO also utilizes a critic network to estimate the value function $V(s_t)$, which is used to compute the advantage function:

$$\hat{A}_t = G_t - V(s_t),$$

where:

- G_t: The cumulative return from time t, representing the total discounted reward obtained from the environment starting at time step t.
- $V(s_t)$: The value function, which estimates the expected return from state s_t. This baseline helps reduce the variance of policy gradient updates.

This integration of value-based learning reduces the variance of policy updates, making PPO a hybrid actor-critic algorithm. While it is rooted in policy-based methods, its reliance on both the actor (policy) and critic (value function) networks highlights its duality as an actor-critic approach.

PPO uses separate actor and critic networks to optimize the policy while ensuring stability through its clipped surrogate objective function. The value network (critic) estimates the value function $V(s_t)$, which helps compute the advantage function $\hat{A}_t$, reducing variance in policy gradient updates.

The PPO objective function is defined as:

$$L^{\text{CLIP}}(\theta) = \mathbb{E}_t\left[\min\left(r_t(\theta)\hat{A}_t, \text{clip}(r_t(\theta), 1-\epsilon, 1+\epsilon)\hat{A}_t\right)\right],$$

where:

- $r_t(\theta) = \frac{\pi_\theta(a_t|s_t)}{\pi_{\theta_{\text{old}}}(a_t|s_t)}$: The probability ratio between the new policy and the old policy.
- $\hat{A}_t$: The advantage function, which measures how much better an action a_t is compared to the baseline value $V(s_t)$.
- ϵ: A hyperparameter that defines the clipping range, restricting policy updates to ensure stability.

By clipping the probability ratio $r_t(\theta)$, PPO prevents excessive policy updates, which could lead to divergence, ensuring stable and consistent training. PPO's simplicity, robustness, and scalability have made it widely applicable across domains such as robotics, gaming, and natural language processing.

The PPO algorithm optimizes the policy by maximizing a clipped surrogate objective, ensuring updates do not deviate significantly from the previous policy. At the same time, it updates the value function to provide more accurate feedback for computing the advantage function. These steps ensure stable and efficient learning.

Algorithm 10 Proximal Policy Optimization (PPO) Algorithm

1: **Input:** Policy $\pi_\theta(a \mid s)$, value function $V_\phi(s)$, clipping parameter ϵ, learning rates α_π, α_v, discount factor γ
2: **Output:** Optimized policy parameters θ, value function parameters ϕ
3: **Step 1 (Data Collection):** Collect trajectories $\{(s_t, a_t, r_t, s_{t+1})\}$ by interacting with the environment using the current policy π_θ
4: **Step 2 (Advantage Estimation):**
5: **for each trajectory:**
6: Compute advantage estimates:

$$\hat{A}_t = \sum_{k=t}^{T} \gamma^{k-t} \delta_k, \quad \text{where } \delta_k = r_k + \gamma V_\phi(s_{k+1}) - V_\phi(s_k)$$

7: **Step 3 (Policy Update):** Update policy parameters by maximizing the clipped surrogate objective:

$$L^{\text{CLIP}}(\theta) = \mathbb{E}_t \left[\min \left(r_t(\theta)\hat{A}_t, \text{clip}(r_t(\theta), 1-\epsilon, 1+\epsilon)\hat{A}_t \right) \right]$$

where

$$r_t(\theta) = \frac{\pi_\theta(a_t \mid s_t)}{\pi_{\theta_{\text{old}}}(a_t \mid s_t)}$$

8: **Step 4 (Value Function Update):** Update the value function by minimizing:

$$L_v(\phi) = \mathbb{E}_t \left[\left(V_\phi(s_t) - G_t \right)^2 \right]$$

9: **Step 5 (Policy Sync):** Update old policy: $\pi_{\theta_{\text{old}}} \leftarrow \pi_\theta$

The PPO algorithm (see Algorithm 10) works as follows:

- **Step 1 (Data Collection):** The agent interacts with the environment using the current policy $\pi_\theta(a \mid s)$, collecting trajectories consisting of (s_t, a_t, r_t, s_{t+1}) transitions for further training.
- **Step 2 (Advantage Estimation):** The advantage estimates $\hat{A}_t$ are computed for each time step using the temporal difference between the predicted values and actual returns. This helps assess the relative benefit of chosen actions over the baseline value $V_\phi(s_t)$.
- **Step 3 (Policy Update):** The policy is optimized by maximizing the clipped surrogate objective $L^{\text{CLIP}}(\theta)$, which prevents large updates that may destabilize training by clipping the probability ratio $r_t(\theta)$.
- **Step 4 (Value Function Update):** The value function (critic) is updated by minimizing the squared error between predicted values $V_\phi(s_t)$ and the estimated returns G_t, which serve as bootstrapped targets for value prediction.
- **Step 5 (Policy Sync):** The old policy $\pi_{\theta_{\text{old}}}$ is synchronized with the current policy π_θ to serve as a reference in the next round of updates.

PPO combines the strengths of policy-gradient methods and actor-critic frameworks by directly optimizing the policy while ensuring stability through its clipped objective. Its reliance on both the actor and critic networks enables efficient and robust learning, making it a popular choice for reinforcement learning tasks across various domains.

8.3.3 Deep Deterministic Policy Gradient (DDPG)

Deep Deterministic Policy Gradient (DDPG) is an Actor-Critic algorithm specifically designed for continuous action spaces. It extends the Deterministic Policy Gradient (DPG) algorithm by leveraging deep neural networks to approximate both the actor and critic functions. DDPG combines policy-gradient techniques with Q-learning principles to achieve stable and efficient learning in high-dimensional, continuous action spaces.

The key features of the DDPG are:

- **Deterministic Policy:** DDPG uses a deterministic policy $\mu_\theta(s)$ that directly outputs the action for a given state s, unlike stochastic policy methods which sample actions from a distribution. This makes DDPG particularly suitable for continuous action spaces.
- **Target Networks:** Separate target networks for the actor (μ_{θ^-}) and critic (Q_{ϕ^-}) are used to stabilize training. The parameters of the target networks are updated using a soft update mechanism:

$$\theta^- \leftarrow \tau\theta + (1-\tau)\theta^-, \quad \phi^- \leftarrow \tau\phi + (1-\tau)\phi^-,$$

  where $\tau \ll 1$ controls the update rate and prevents drastic changes.
- **Experience Replay:** A replay buffer $\mathcal{D}$ stores past experiences (s, a, r, s'), which are sampled randomly to break correlations between consecutive samples. This improves sample efficiency and stability.

The mathematical construct is as follows:

- **Critic Update:** The critic network $Q_\phi(s, a)$ is updated by minimizing the temporal difference (TD) error:

$$L_Q(\phi) = \frac{1}{|\mathcal{D}|}\sum \left[Q_\phi(s,a) - \left(r + \gamma Q_{\phi^-}(s', \mu_{\theta^-}(s'))\right)\right]^2,$$

where:

- r: The reward observed for taking action a in state s.
- γ: The discount factor, which weighs the importance of future rewards.
- $Q_{\phi^-}(s', \mu_{\theta^-}(s'))$: The target Q-value for the next state-action pair, estimated using the target networks.

- **Actor Update:** The actor policy $\mu_\theta(s)$ is updated to maximize the expected Q-value:

$$J(\theta) = \mathbb{E}_{s\sim\mathcal{D}}\left[Q_\phi(s, \mu_\theta(s))\right].$$

The gradient for this update is computed using the chain rule:

$$\nabla_\theta J(\theta) = \mathbb{E}_{s\sim\mathcal{D}}\left[\nabla_a Q_\phi(s, a)\nabla_\theta \mu_\theta(s)\right].$$

DDPG alternates between improving the critic, which evaluates the Q-values of state-action pairs, and the actor, which learns the policy to maximize the Q-values predicted by the critic. Target networks and experience replay are used to stabilize training and improve sample efficiency.

The workflow of the DDPG algorithm (see Algorithm 11) is as follows:

- **Step 1 (Initialization):** The target networks are initialized to match the actor and critic networks, ensuring consistent starting values. The replay buffer $\mathcal{D}$ is also initialized to store experiences.
- **Step 2 (Interaction and Training Loop):** While the training has not yet converged, the agent interacts with the environment by observing the current state s_t, selecting an action $a_t = \mu_\theta(s_t) + \mathcal{N}_t$ using the deterministic policy with added exploration noise $\mathcal{N}_t$, and storing the transition (s_t, a_t, r_t, s_{t+1}) in the replay buffer. A mini-batch of experiences is sampled from $\mathcal{D}$ for learning.
- **Step 3 (Critic Update):** The critic network is updated by minimizing the temporal difference (TD) error between its predicted Q-values and the target Q-values estimated from the target networks.
- **Step 4 (Actor Update):** The actor network is updated using the deterministic policy gradient, which maximizes the critic's predicted Q-value for the actions proposed by the actor.
- **Step 5 (Target Network Update):** The target networks are softly updated toward the current networks using the soft update rule with parameter τ, ensuring stability in training.

Algorithm 11 Deep Deterministic Policy Gradient (DDPG) Algorithm

Input: Actor network $\mu_\theta(s)$, critic network $Q_\phi(s, a)$, replay buffer $\mathcal{D}$, learning rates α_μ, α_Q, discount factor γ, soft update rate τ
Output: Optimized policy parameters θ
Step 1 (Initialization): Initialize target networks: $\mu_{\theta^-} \leftarrow \mu_\theta$, $Q_{\phi^-} \leftarrow Q_\phi$
Initialize replay buffer $\mathcal{D}$
Step 2 (Interaction and Training Loop):
while not converged **do**
    Observe current state s_t
    Select action with exploration noise:

$$a_t = \mu_\theta(s_t) + \mathcal{N}_t$$

    where $\mathcal{N}_t$ is noise (e.g., Ornstein-Uhlenbeck)
    Execute action a_t, observe reward r_t and next state s_{t+1}
    Store transition (s_t, a_t, r_t, s_{t+1}) in $\mathcal{D}$
    Sample mini-batch $(s, a, r, s') \sim \mathcal{D}$
Step 3 (Critic Update): Update critic by minimizing TD loss:

$$L_Q(\phi) = \mathbb{E}\left[\left(Q_\phi(s, a) - (r + \gamma Q_{\phi^-}(s', \mu_{\theta^-}(s')))\right)^2\right]$$

Step 4 (Actor Update): Update actor using deterministic policy gradient:

$$J(\theta) = \mathbb{E}_{s\sim\mathcal{D}}\left[Q_\phi(s, \mu_\theta(s))\right]$$

Step 5 (Target Network Update): Soft update target networks:

$$\theta^- \leftarrow \tau\theta + (1 - \tau)\theta^-, \quad \phi^- \leftarrow \tau\phi + (1 - \tau)\phi^-$$

end while

8.4 *Model-Based Methods*

Model-based methods involve learning a model of the environment's dynamics to simulate and predict outcomes. This enables efficient planning and decision-making by evaluating potential future states before interacting with the environment.

8.4.1 Monte Carlo Tree Search (MCTS)

Monte Carlo Tree Search (MCTS) is a planning algorithm that evaluates potential action sequences by iteratively building a search tree. It balances exploration (trying new actions) and exploitation (refining actions with high rewards) using a heuristic called the Upper Confidence Bound for Trees (UCT). This makes MCTS particularly useful for games and decision-making tasks where evaluating long-term strategies is crucial.

The UCT score for selecting an action a in state s is given by:

$$UCT = Q(s, a) + C\sqrt{\frac{\ln(N(s))}{N(s, a)}},$$

where:

- $Q(s, a)$: The estimated reward for taking action a in state s, calculated from past simulations.
- $N(s)$: The total number of times state s has been visited in simulations.
- $N(s, a)$: The total number of times action a has been taken in state s.
- C: The exploration constant, which controls the trade-off between exploration (trying less-visited actions) and exploitation (selecting the most promising action based on $Q(s, a)$).
- $\ln(N(s))$: Encourages exploration by increasing the score of actions that have been taken less frequently.

The UCT score combines the reward estimate $Q(s, a)$ (exploitation) with a bonus term that favors actions with fewer visits (exploration). The constant C determines the balance between these two components.

MCTS operates by simulating potential outcomes of actions and updating the search tree based on the observed rewards. The process consists of four key steps:

- Selection: Starting from the root node, traverse the tree by selecting actions that maximize the UCT score (exploration-exploitation trade-off).
- Expansion: When reaching a non-terminal node that is not fully expanded, add new child nodes representing possible actions.
- Simulation: Perform a rollout (simulate an episode) from the new node until a terminal state is reached.
- Backpropagation: Propagate the reward from the terminal state back up the tree, updating $Q(s, a)$ and visit counts.

8.5 Hierarchical Reinforcement Learning (HRL)

Hierarchical Reinforcement Learning (HRL) decomposes tasks into subtasks, allowing agents to address smaller, more manageable problems. By introducing hierarchical structures, HRL facilitates learning in environments with long time horizons and complex decision-making processes. These hierarchical structures often consist of high-level policies that make decisions over longer time scales and low-level policies that focus on executing specific actions.

8.5.1 Options Framework

The Options Framework is a formalization of HRL that introduces 'options', which represent high-level, temporally extended actions. Each option is defined by the following components:

- **Initiation Set** (I): The set of states where the option can be initiated, i.e., $s \in I$.
- **Policy** (π_ω): A policy specific to the option, governing the sequence of actions taken while the option is active. Formally, $\pi_\omega(a \mid s)$ represents the probability of selecting action a in state s under option ω.
- **Termination Condition** (β_ω): A function $\beta_\omega(s) \in [0, 1]$ specifying the probability of terminating the option in state s.

When using the Options Framework, the agent selects a high-level option $\omega \in \Omega$, which dictates the low-level actions until the option terminates. The value function for an option-based policy π can be expressed as:

$$V^\pi(s) = \mathbb{E}\left[\sum_{t=0}^{\infty} \gamma^t r_t \mid s_0 = s, \pi\right],$$

where γ is the discount factor, and r_t represents the reward at time t.

The agent's overall policy π alternates between choosing options and executing the low-level policies within each option. The Q-value for selecting an option ω in state s is given by:

$$Q_\Omega(s, \omega) = \mathbb{E}\left[r + \gamma V^\pi(s') \mid s, \omega\right],$$

where s' is the state reached after executing the option.

The options framework simplifies complex problems by allowing the agent to operate at different levels of abstraction. High-level options enable the agent to focus on long-term goals, while low-level policies handle the details of executing specific tasks.

8.5.2 Feudal Reinforcement Learning (FRL)

Feudal Reinforcement Learning (FRL) is another hierarchical approach that introduces a manager-worker hierarchy. In this framework:

- **High-Level Manager:** Sets goals or subgoals for the low-level worker, operating over an extended time horizon. Goals may be expressed as specific states to achieve or as changes in the environment.
- **Low-Level Worker:** Executes actions to achieve the subgoals provided by the manager, focusing on short-term tasks.

The manager operates on a higher-level state space S_m, where each state represents a summary of the environment. The worker operates on the original state space S_w, executing actions $a \in A$ to transition between states.

The manager's policy $\pi_m(g \mid s_m)$ selects a subgoal g in high-level state s_m, while the worker's policy $\pi_w(a \mid s_w, g)$ chooses actions to achieve the subgoal g from low-level state s_w.

The manager's value function is defined as:

$$V_m(s_m) = \mathbb{E}\left[\sum_{t=0}^{\infty} \gamma^t r_t \mid s_{m,0} = s_m, \pi_m\right],$$

where r_t includes the cumulative rewards achieved by the worker while pursuing the subgoal.

FRL leverages a hierarchy where the manager focuses on strategic planning, while the worker handles the tactical execution. By decoupling long-term planning from short-term decision-making, FRL improves learning efficiency in multi-step tasks with complex dependencies.

The options framework formalizes temporally extended actions (options) that encapsulate both policy and termination conditions. FRL on the other hand emphasizes hierarchical decision-making, with explicit roles for managers and workers, enabling coordination over different time scales. Both approaches highlight the power of HRL in addressing large-scale decision-making problems, such as robotic navigation, multi-step planning, and multi-agent coordination.

8.6 *Distributed and Multi-Agent Reinforcement Learning*

Distributed and Multi-Agent Reinforcement Learning (MARL) involves multiple agents or workers interacting with environments, often in parallel, to improve training speed, scalability, and robustness. These methods are particularly effective in complex, multi-agent systems, where agents must collaborate or compete to optimize their objectives [17, 18].

8.6.1 Asynchronous Advantage Actor-Critic (A3C)

Asynchronous Advantage Actor-Critic (A3C) is a distributed reinforcement learning algorithm that trains a global policy by asynchronously updating parameters from multiple agents. Each agent interacts with its own copy of the environment, which helps decorrelate the training data and improves stability compared to synchronous methods.

The algorithm uses an actor-critic framework, where:

- The **actor** learns a policy $\pi_\theta(a \mid s)$ to select actions.
- The **critic** estimates the value function $V(s; \phi)$, which provides a baseline for computing the advantage.

The **advantage function**, which reduces the variance of policy updates, is defined as:

$$\hat{A}_t = r_t + \gamma V(s_{t+1}) - V(s_t),$$

where:

- r_t: Reward received after taking action a_t in state s_t.
- $V(s_t)$: Predicted value of state s_t, estimating the expected cumulative rewards.
- $V(s_{t+1})$: Predicted value of the next state s_{t+1}, used to estimate future rewards.
- γ: Discount factor, balancing the importance of immediate and future rewards.

The policy parameters θ are updated using the policy gradient:

$$\nabla_\theta J(\theta) = \mathbb{E}_t \left[\nabla_\theta \log \pi_\theta(a_t \mid s_t) \hat{A}_t \right].$$

The value function parameters ϕ are updated by minimizing the mean squared error:

$$L_v(\phi) = \frac{1}{T} \sum_t \left(V(s_t; \phi) - G_t \right)^2,$$

where $G_t = r_t + \gamma V(s_{t+1})$ is the return.

A3C's key strength lies in its distributed nature, where multiple agents operate asynchronously, reducing the correlations between updates and improving training stability. This parallelism also accelerates convergence by leveraging diverse experiences from multiple agents interacting with the environment.

8.6.2 Multi-Agent Deep Deterministic Policy Gradient (MADDPG)

Multi-Agent Deep Deterministic Policy Gradient (MADDPG) extends the Deep Deterministic Policy Gradient (DDPG) algorithm to multi-agent environments. It is particularly suitable for scenarios where multiple agents must collaborate or compete, as it addresses the non-stationarity introduced by multiple learning agents.

MADDPG employs a centralized training and decentralized execution framework:

- During **training**, a centralized critic $Q_i(s, a_1, \ldots, a_N)$ is used for each agent i, where $a_1, \ldots, a_N$ are the actions of all agents. The critic evaluates the joint action space to account for interactions between agents.

- During **execution**, each agent uses its own policy $\mu_{\theta_i}(s_i)$, which depends only on its local observations s_i, enabling decentralized decision-making.

The centralized critic for agent i is trained using the following loss function:

$$L_Q(\phi_i) = \frac{1}{|\mathcal{D}|} \sum_{\mathcal{D}} \left[Q_i(s, a_1, \ldots, a_N; \phi_i) - \left(r_i + \gamma Q_i(s', a'_1, \ldots, a'_N; \phi_i^-) \right) \right]^2,$$

where:

- s: Global state observed by all agents.
- $a_1, \ldots, a_N$: Actions taken by all agents.
- r_i: Reward received by agent i.
- s': Next state after all agents take actions.
- $a'_1, \ldots, a'_N$: Actions chosen by all agents in the next state, using their respective policies.
- ϕ_i: Parameters of the centralized critic for agent i.
- ϕ_i^-: Parameters of the target critic network for agent i, updated periodically.

The actor for agent i is trained to maximize the expected Q-value:

$$\nabla_{\theta_i} J(\theta_i) = \mathbb{E}_{s \sim \mathcal{D}} \left[\nabla_{a_i} Q_i(s, a_1, \ldots, a_N) \nabla_{\theta_i} \mu_{\theta_i}(s_i) \right],$$

where $\mathcal{D}$ is the replay buffer storing experiences.

MADDPG addresses the challenges of multi-agent learning by using a centralized critic that accounts for interactions between agents during training. However, each agent's policy remains decentralized for execution, making the algorithm scalable and efficient in real-world scenarios like robotic coordination or multi-agent games.

While A3C focuses on distributed training with asynchronous updates to a shared global policy, MADDPG is specifically designed for multi-agent environments, leveraging a centralized critic for joint action evaluation. Both approaches improve scalability and learning efficiency but cater to different problem settings.

8.7 Hybrid Methods

Hybrid methods combine multiple reinforcement learning paradigms [19], such as model-based, policy-based, and value-based approaches, to leverage their complementary strengths. These methods often achieve superior performance in solving complex tasks by integrating techniques like planning, learning, and simulation.

8.7.1 AlphaGo

AlphaGo was one of the first systems to achieve superhuman performance in the game of Go. It integrates three key components:

- **Policy Network:** Predicts the probability distribution of actions, guiding the Monte Carlo Tree Search (MCTS) to focus on the most promising actions.
- **Value Network:** Estimates the probability of winning from a given state, allowing the agent to evaluate long-term outcomes.

The Q-value update in AlphaGo is governed by the Bellman equation:

$$Q(s, a) = \mathbb{E}\left[r + \gamma \max_{a'} Q(s', a')\right],$$

where:

- r: Immediate reward received after taking action a in state s.
- γ: Discount factor, controlling the trade-off between immediate and future rewards.
- $\max_{a'} Q(s', a')$: Maximum expected reward achievable from the next state s' by following the optimal policy.

The integration of deep learning (through policy and value networks) and MCTS allows AlphaGo to evaluate long-term strategies efficiently, outperforming human players in strategic board games.

8.7.2 MuZero

MuZero extends AlphaGo and AlphaZero by eliminating the need for prior knowledge of environment rules, such as the rules of a game. Instead, MuZero learns a model of the environment through three predictions:

- **Latent State Transitions:** Predicts how the environment evolves when actions are taken.
- **Rewards:** Estimates immediate rewards received after state transitions.
- **Policies and Values:** Predicts the policy $\pi(a \mid s)$ and value $V(s)$ for a given state.

This flexibility makes MuZero adaptable to diverse environments, as it combines the strengths of model-based planning and model-free reinforcement learning. Its ability to learn latent dynamics and integrate them into MCTS has enabled it to excel in tasks such as chess, Go, and Atari games.

8.7.3 Dyna-Q

Dyna-Q is a hybrid reinforcement learning algorithm that integrates direct reinforcement learning, model learning, and planning. The agent alternates between interacting with the environment and simulating experiences using a learned model of the environment dynamics. This process improves sample efficiency and enables more effective decision-making.

The Q-value update rule in Dyna-Q follows the Bellman equation:

$$Q(s_t, a_t) \leftarrow Q(s_t, a_t) + \alpha \left[r_t + \gamma \max_{a'} Q(s_{t+1}, a') - Q(s_t, a_t) \right],$$

where:

- s_t and a_t: Current state and action.
- r_t: Reward received after taking action a_t in state s_t.
- s_{t+1}: Next state after transitioning from s_t.
- α: Learning rate, determining how quickly Q-values are updated.
- γ: Discount factor, controlling the trade-off between immediate and future rewards.
- $\max_{a'} Q(s_{t+1}, a')$: Maximum estimated Q-value for the next state s_{t+1}.

Dyna-Q uses a replay-like mechanism by simulating experiences from its learned model. This allows the agent to refine its policy and value function using both real and simulated data, enhancing its ability to generalize and plan effectively.

8.7.4 AlphaZero

AlphaZero combines MCTS with deep neural networks to approximate the policy $\pi(a \mid s)$ and value $V(s)$ functions. During each iteration of MCTS, the neural network provides:

- **Policy Prediction:** $\pi(a \mid s)$, guiding MCTS to prioritize the most promising actions during exploration.
- **Value Estimation:** $V(s)$, which estimates the expected outcome of the current state and provides feedback for backpropagation.

The updated Q-value after each simulation is computed as:

$$Q(s, a) \leftarrow Q(s, a) + \alpha \left[R(s) - Q(s, a) \right],$$

where:

- $R(s)$: Rollout reward obtained from the simulation.
- $Q(s, a)$: Current estimate of the Q-value for action a in state s.
- α: Learning rate for updating the Q-value.

AlphaZero's combination of MCTS and deep learning has achieved superhuman performance in games such as chess and Go. Its ability to learn strategies through self-play and adapt to diverse tasks has set a new benchmark in reinforcement learning.

Hybrid methods like AlphaGo, MuZero, Dyna-Q, and AlphaZero demonstrate how combining multiple reinforcement learning paradigms—such as planning, value-based learning, and policy optimization—can achieve exceptional results in complex, high-dimensional tasks. By leveraging their complementary strengths, these methods have expanded the applicability and effectiveness of reinforcement learning.

Acknowledgements The authors acknowledge the use of Generative AI (GenAI) tools in the editing and development of this book chapter. They also thank Tamonash Bhattacharyya for his feedback and suggestions.

References

1. T. M. Mitchell and T. M. Mitchell, *Machine learning.* McGraw-hill New York, 1997, vol. 1, no. 9.
2. R. S. Sutton, A. G. Barto *et al.*, *Reinforcement learning: An introduction.* MIT press Cambridge, 1998, vol. 1, no. 1.
3. P. Chapman, J. Clinton, R. Kerber, T. Khabaza, T. Reinartz, C. Shearer, and R. C.-D. Wirth, "1.0 step-by-step data mining guide. crisp-dm consortium. 2000."
4. F. Firouzi, K. Chakrabarty, and S. Nassif, *Intelligent internet of things: From device to fog and cloud.* Springer, 2020.
5. I. T. Jolliffe, *Principal component analysis for special types of data.* Springer, 2002.
6. T. Hastie, "The elements of statistical learning: data mining, inference, and prediction," 2009.
7. I. Guyon and A. Elisseeff, "An introduction to variable and feature selection," *Journal of machine learning research*, vol. 3, no. Mar, pp. 1157–1182, 2003.
8. L. Van der Maaten and G. Hinton, "Visualizing data using t-sne." *Journal of machine learning research*, vol. 9, no. 11, 2008.
9. M. Sokolova and G. Lapalme, "A systematic analysis of performance measures for classification tasks," *Information processing & management*, vol. 45, no. 4, pp. 427–437, 2009.
10. T. Cover and P. Hart, "Nearest neighbor pattern classification," *IEEE transactions on information theory*, vol. 13, no. 1, pp. 21–27, 1967.
11. D. W. Hosmer Jr, S. Lemeshow, and R. X. Sturdivant, *Applied logistic regression.* John Wiley & Sons, 2013.
12. C. Cortes and V. Vapnik, "Support-vector networks," *Machine learning*, vol. 20, pp. 273–297, 1995.
13. L. Breiman, "Bagging predictors," *Machine learning*, vol. 24, pp. 123–140, 1996.
14. Y. Freund and R. E. Schapire, "A decision-theoretic generalization of on-line learning and an application to boosting," *Journal of computer and system sciences*, vol. 55, no. 1, pp. 119–139, 1997.
15. A. K. Jain, M. N. Murty, and P. J. Flynn, "Data clustering: a review," *ACM computing surveys (CSUR)*, vol. 31, no. 3, pp. 264–323, 1999.
16. R. Sutton and A. Barto, "Reinforcement learning: An introduction," *IEEE Transactions on Neural Networks*, vol. 9, no. 5, pp. 1054–1054, 1998.
17. P. Domanski, A. Ray, K. Lafata, F. Firouzi, K. Chakrabarty, and D. Pflüger, "Advancing blood glucose prediction with neural architecture search and deep reinforcement learning for type 1 diabetics," *Biocybernetics and Biomedical Engineering*, vol. 44, no. 3, pp. 481–500, 2024.

18. P. Domanski, A. Ray, F. Firouzi, K. Lafata, K. Chakrabarty, and D. Pflüger, "Blood glucose prediction for type-1 diabetics using deep reinforcement learning," in *2023 IEEE International Conference on Digital Health (ICDH)*. IEEE, 2023, pp. 339–347.
19. A. Liardo, A. Ray, F. Firouzi, K. Lafata, and K. Chakrabarty, "Neural architecture search for blood glucose prediction in type-1 diabetics," in *2024 IEEE 20th International Conference on Body Sensor Networks (BSN)*. IEEE, 2024, pp. 1–4.

Mastering Deep Learning: From Neurons to Large Language Models

Farshad Firouzi, Bahar Farahani, and Aritra Ray

> *"What we want is a machine that can learn from experience."*
>
> *– Alan Turing*

1 Deep Learning (DL) and Artificial Neural Network (ANN)

Deep Learning is a subset of machine learning that leverages artificial neural networks (ANNs) with many layers to model and learn hierarchical patterns in data. Unlike traditional machine learning methods, which often rely on hand-crafted features, deep learning enables machines to automatically extract and learn features directly from raw data. This ability to model complex, hierarchical representations has made deep learning a cornerstone of modern artificial intelligence (AI).

Neural networks, both biological and artificial, are systems designed to process and transmit information. ANNs, inspired by the structure and functionality of Biological Neural Networks (BNNs), serve as foundational tools in artificial intelligence. By mimicking the way the brain processes information, ANNs enable machines to

F. Firouzi (✉)
Arizona State University, Tempe, USA
e-mail: farshad.firouzi@asu.edu

B. Farahani
Shahid Beheshti University, Tehran, Iran

A. Ray
Duke University, Durham, USA

F. Firouzi et al. (eds.), *Smart and Connected Health: AI, IoT, and Trustworthy Technologies*, https://doi.org/10.1007/978-3-032-06286-4_3

learn from data, identify patterns, and solve complex problems, making them indispensable in fields such as healthcare.

1.1 Neural Network Fundamentals

ANN is a computational model that seeks to emulate the structure and functionality of biological neural systems. It consists of interconnected processing units called **neurons**, which transmit information through connections defined by **weights**. These weights determine the influence of one neuron's output on another.

A single neuron in an ANN computes its output by taking a weighted sum of its inputs (x_i), adding a bias (b), and applying an **activation function** (f). This process can be expressed mathematically as (see Fig. 1):

$$y = f\left(\sum_i w_i x_i + b\right)$$

.

1.2 Biological Neural Networks: The Inspiration

Biological neural networks, found in living organisms, are highly complex systems that process vast amounts of information. A **biological neuron** is a specialized cell designed to receive, process, and transmit signals. It consists of the following components:

- **Dendrites**: Tree-like extensions that receive signals from other neurons. These function as the "input" channels of the neuron.
- **Soma (Cell Body)**: The central processing unit of the neuron that integrates incoming signals and determines the appropriate output.

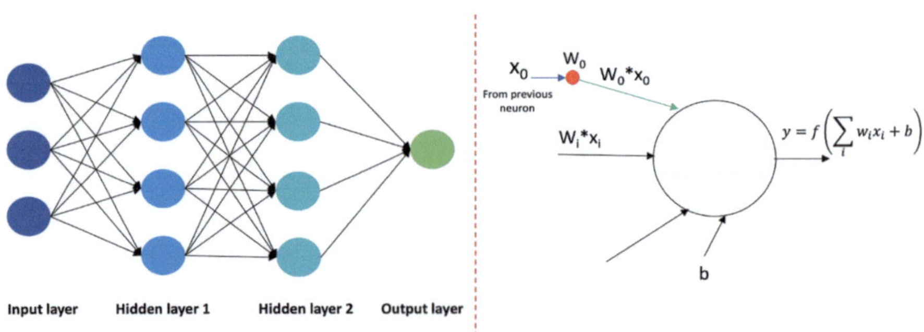

Fig. 1 Neural network architecture and neuron breakdown [1]

Table 1 Comparison between BNNs and ANNs

Biological neural network (BNN)	Artificial neural network (ANN)
Soma (cell body)	Node (Neuron)
Dendrites	Input
Synapse	Weights or Interconnections
Axon	Output

- **Axon**: A cable-like structure through which the neuron sends signals to other neurons.
- **Synapses**: The junctions between neurons where information is transmitted, often chemically or electrically.

These biological structures inspired the design of ANNs, which replicate certain aspects of their functionality in a simplified, computational framework.

While ANNs and BNNs share conceptual similarities, they differ significantly in terms of complexity, purpose, and implementation (see Table 1). ANNs are computational models inspired by the structure and function of BNNs, but they abstract and simplify the intricate mechanisms of biological systems to suit algorithmic processing.

2 Multilayer Perceptrons (MLPs): The Foundation of Deep Learning

A Multilayer Perceptron (MLP) is a type of ANN composed of multiple layers of neurons (see Fig. 1). MLPs are fundamental for understanding deep learning models, as they demonstrate how networks can learn complex, non-linear relationships by stacking layers of computational units. An MLP consists of the following components:

- **Input Layer**: Receives raw input data, such as a feature vector $\mathbf{x} = [x_1, x_2, \ldots, x_n]$, where n is the number of input features.
- **Hidden Layers**: Contain multiple neurons that process and transform the input data. Each hidden layer applies weighted transformations followed by non-linear activation functions.
- **Output Layer**: Produces the final output of the network, such as probabilities for classification tasks or continuous values for regression tasks.

Each layer contains multiple neurons, and the number of neurons in a layer determines the dimensionality of its output.

2.1 Notations and Conventions

In DL, the operations of a neural network are typically described using the following notations:

- **Superscripts** ($[l]$): Represent the layer number. For example:
  - $\mathbf{W}^{[l]}$: Weight matrix for layer l.
  - $\mathbf{b}^{[l]}$: Bias vector for layer l.
  - $\mathbf{a}^{[l]}$: Activation vector for layer l.
- **Subscripts** (i, j): Denote specific elements within a matrix or vector. For example:
  - $W_{ij}^{[l]}$: Weight connecting the j-th neuron in layer $l-1$ to the i-th neuron in layer l.
  - $a_i^{[l]}$: Activation of the i-th neuron in layer l.

2.2 Operations in a Single Neuron

Each neuron in an MLP computes its output in two steps:

1. **Pre-Activation Computation**: The neuron calculates a weighted sum of inputs from the previous layer ($\mathbf{a}^{[l-1]}$) and adds a bias term ($b_i^{[l]}$):

$$z_i^{[l]} = \sum_j W_{ij}^{[l]} a_j^{[l-1]} + b_i^{[l]}$$

   where $W_{ij}^{[l]}$ denotes the weight connecting the j-th neuron in layer $l-1$ to the i-th neuron in layer l, $a_j^{[l-1]}$ represents the activation of the j-th neuron in layer $l-1$, $b_i^{[l]}$ is the bias for the i-th neuron in layer l, and $z_i^{[l]}$ represents the pre-activation value for the i-th neuron in layer l.

2. **Activation Computation**: The neuron applies a non-linear activation function f to the pre-activation value:

$$a_i^{[l]} = f\left(z_i^{[l]}\right)$$

   where $a_i^{[l]}$ is the output (activation) of the neuron.

2.3 Forward Propagation in an MLP

Forward propagation is the process of passing input data through the layers of a MLP, where each layer applies learned weights and activation functions to transform the data, ultimately producing the final output. To compute the outputs for all neurons in a layer efficiently, we use matrix-vector notation. For layer l with $m^{[l]}$ neurons, the computations are:

$$\mathbf{z}^{[l]} = \mathbf{W}^{[l]}\mathbf{a}^{[l-1]} + \mathbf{b}^{[l]}$$

$$\mathbf{a}^{[l]} = f\left(\mathbf{z}^{[l]}\right)$$

where $\mathbf{W}^{[l]}$ is the weight matrix with dimensions $m^{[l]} \times m^{[l-1]}$, connecting all neurons in layer $l - 1$ to those in layer l, $\mathbf{b}^{[l]}$ represents the bias vector with dimensions $m^{[l]} \times 1$, containing one bias term per neuron, $\mathbf{z}^{[l]}$ is the pre-activation vector with dimensions $m^{[l]} \times 1$, containing the weighted sums for all neurons, and $\mathbf{a}^{[l]}$ represents the activation vector with dimensions $m^{[l]} \times 1$, denoting the outputs of all neurons in the layer.

Batch Processing: In practice, inputs are processed in batches for computational efficiency. If the input consists of N samples, the computations for a batch can be expressed as:

$$\mathbf{Z} = \mathbf{X}\mathbf{W} + \mathbf{b}$$

where $\mathbf{X}$ is the input matrix of size $N \times m^{[l-1]}$, where each row corresponds to a single input sample; $\mathbf{W}$ is the weight matrix of size $m^{[l-1]} \times m^{[l]}$, applied uniformly across all input samples; $\mathbf{b}$ is the bias vector of size $1 \times m^{[l]}$, broadcast across all samples in the batch; and $\mathbf{Z}$ is the output matrix of size $N \times m^{[l]}$, containing the pre-activation values for all samples and neurons in the layer.

3 Activation Functions

Activation functions play a critical role in neural networks, determining how the weighted sum of inputs is transformed into the output of a neuron. These functions introduce non-linearity, enabling the network to learn complex patterns in data. Below are the most commonly used activation functions (see Fig. 2).

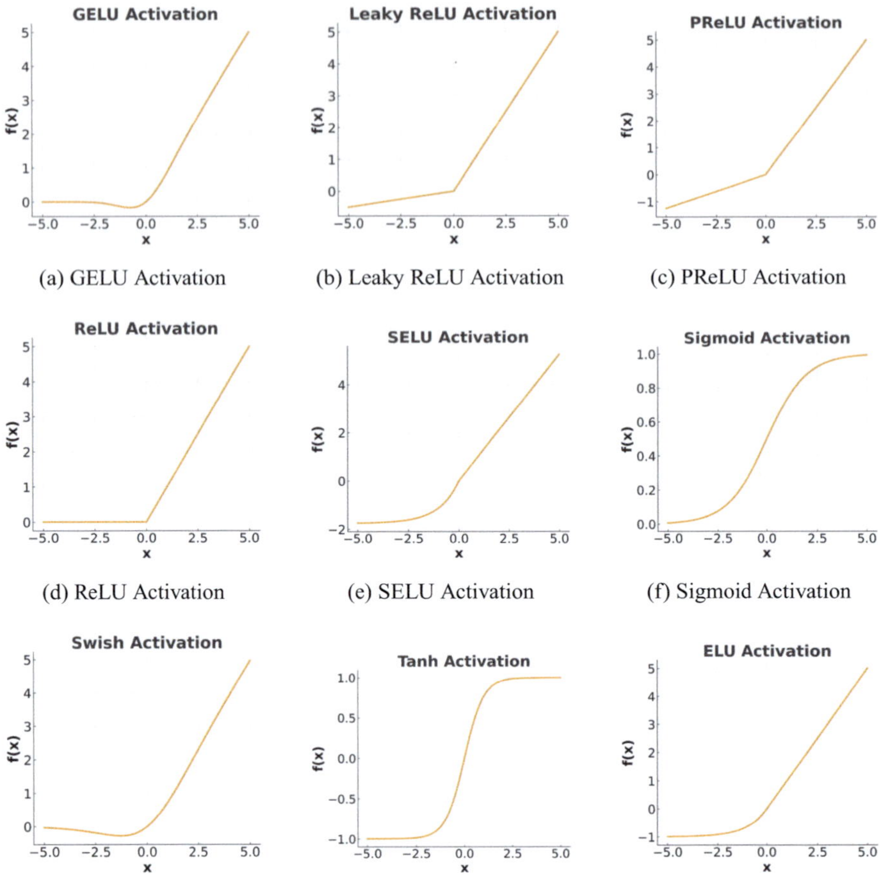

(a) GELU Activation (b) Leaky ReLU Activation (c) PReLU Activation
(d) ReLU Activation (e) SELU Activation (f) Sigmoid Activation
(g) Swish Activation (h) Tanh Activation (i) ELU Activation

Fig. 2 Visualization of different activation functions

3.1 ReLU (Rectified Linear Unit)

The Rectified Linear Unit (ReLU) is one of the most widely used activation functions in modern neural networks due to its simplicity and effectiveness. It outputs the input value if it is positive and zero otherwise. Mathematically, ReLU is defined as:

$$f(x) = \max(0, x)$$

This simple behavior introduces non-linearity while preserving computational efficiency, making ReLU an ideal choice for deep learning architectures.

ReLU offers several key advantages over other activation functions:

- **Computational Efficiency**: ReLU involves simple thresholding at zero, making it less computationally expensive compared to functions that rely on exponential operations, such as Sigmoid or Tanh.
- **Sparse Activation**: Only a subset of neurons is activated at any given time (i.e., those with positive inputs). This sparsity can help reduce overfitting and improve network interpretability.
- **Faster Convergence**: ReLU helps neural networks converge faster during training by mitigating vanishing gradient issues common with Sigmoid and Tanh functions.

Despite its widespread use, ReLU has some notable limitations:

- **Dying ReLU Problem**: Neurons can become permanently inactive (outputting zero for all inputs) if their weights are updated in a way that consistently leads to negative inputs. This issue can prevent certain neurons from learning, especially in deeper networks.
- **Unbounded Output**: For positive inputs, ReLU outputs can grow indefinitely, potentially causing instability during training if large gradients accumulate.

ReLU is the default activation function for most convolutional neural networks (CNNs) and deep feedforward networks. Its efficiency and ability to handle non-linear relationships make it particularly well-suited for large-scale datasets and architectures.

3.2 Leaky ReLU

Leaky ReLU is a variant of the Rectified Linear Unit (ReLU) designed to address the **Dying ReLU problem**, where some neurons output zero for all inputs and effectively stop learning. Unlike ReLU, which assigns zero to all negative inputs, Leaky ReLU introduces a small slope for negative values. This ensures that neurons with negative inputs can still propagate some gradient during backpropagation.

Mathematically, Leaky ReLU is defined as:

$$f(x) = \begin{cases} x, & \text{if } x > 0 \\ \alpha x, & \text{if } x \leq 0 \end{cases}$$

where α is a small positive constant (e.g., 0.01) that determines the slope for negative inputs. Leaky ReLU offers several advantages, including:

- **Gradient Propagation**: By allowing a small gradient for negative inputs, Leaky ReLU reduces the likelihood of neurons becoming inactive (dying) and ensures that all neurons contribute to learning.

- **Simplicity**: The function is computationally efficient and easy to implement, making it a practical alternative to ReLU in many scenarios.

Despite its advantages, Leaky ReLU has several limitations, including:

- **Hyperparameter Dependency**: The performance of Leaky ReLU depends on the choice of the α parameter. Improper tuning of α may result in suboptimal performance.
- **Not Zero-Centered**: Similar to ReLU, the outputs of Leaky ReLU are not zero-centered, which can sometimes slow convergence during optimization.

Leaky ReLU is widely used in deep learning architectures to enhance training stability and prevent issues associated with neuron inactivity.

3.3 Parametric ReLU (PReLU)

Parametric ReLU (PReLU) is an extension of Leaky ReLU where the slope α for negative inputs is learned during training instead of being fixed. This allows the model to adaptively optimize α based on the data, providing greater flexibility. Mathematically, PReLU is defined as:

$$f(x) = \begin{cases} x, & \text{if } x > 0 \\ \alpha_i x, & \text{if } x \leq 0 \end{cases}$$

where α_i is a learnable parameter for each neuron, with i denoting the specific neuron in the layer. PReLU offers several advantages over standard ReLU and Leaky ReLU by allowing the model to learn the negative slope parameter. Some key benefits of PReLU include:

- **Adaptive Negative Slope**: Unlike Leaky ReLU, where α is fixed, PReLU allows the model to optimize α per neuron, leading to better performance in some cases.
- **Improved Learning Capacity**: By learning α, the network can better capture complex patterns in the data.

However, PReLU also comes with challenges:

- **Increased Complexity**: Since α is learnable, it introduces additional parameters, which might slightly increase computational cost.
- **Risk of Overfitting**: If not regularized properly, PReLU can increase the risk of overfitting due to additional trainable parameters.

PReLU has been successfully applied in deep learning architectures, particularly in computer vision tasks, where adaptive activation functions help improve feature learning.

3.4 Sigmoid

The Sigmoid activation function maps inputs to a range between 0 and 1, making it particularly useful for tasks involving probabilities. It is defined mathematically as:

$$\sigma(z) = \frac{1}{1 + e^{-z}}$$

where z is the input to the neuron. The Sigmoid function is smooth, differentiable, and transforms input values into probabilities, which are useful for binary classification problems. The man advance of Sigmoir can be summarized as follows:

- **Probabilistic Output**: Sigmoid outputs can be interpreted as probabilities, making it ideal for binary classification tasks or as the activation function for the output layer in logistic regression.
- **Smooth Gradient**: The function's smooth curve ensures that small changes in input result in small changes in output, which can help stabilize training in certain scenarios.

Despite its advantages, the Sigmoid function has significant drawbacks that limit its utility in deep learning:

- **Vanishing Gradient Problem**: For very large or very small inputs, the gradient of the Sigmoid function approaches zero. This diminishes the magnitude of weight updates during backpropagation, particularly in deeper networks, and slows down learning.
- **Non-Zero-Centered Output**: The outputs of the Sigmoid function are always positive, leading to gradients that are either all positive or all negative. This imbalance can cause inefficient optimization and hinder convergence.
- **Computational Complexity**: The exponential operation in the Sigmoid function makes it slightly more computationally expensive compared to simpler functions like ReLU.

While less commonly used in modern deep learning architectures, the Sigmoid function remains relevant for:

- **Binary Classification**: It is frequently used in the output layer for binary classification problems, where the output represents the probability of a class.
- **Logistic Regression Models**: As the primary activation function in logistic regression, Sigmoid is well-suited for small-scale problems involving probabilistic outputs.

Due to its limitations, Sigmoid is often replaced by other activation functions like ReLU or Softmax in hidden layers and multi-class classification problems.

3.5 Tanh

The Tanh (hyperbolic tangent) activation function maps inputs to a range between -1 and 1, producing outputs that are zero-centered. This zero-centered property balances positive and negative activations, which helps stabilize gradients during backpropagation. It is mathematically defined as:

$$\tanh(z) = \frac{e^z - e^{-z}}{e^z + e^{-z}}$$

where z is the input to the neuron. Tanh ensures that activations are symmetrically distributed around zero, often leading to more efficient weight updates compared to the Sigmoid function. Tanh offers several advantages, including:

- **Zero-Centered Outputs**: Tanh produces outputs between -1 and 1, helping to balance positive and negative activations, which improves weight updates during backpropagation.
- **Improved Convergence**: Tanh performs better than Sigmoid in scenarios where the input data is centered around zero, as it encourages faster convergence during training.

Despite its advantages, Tanh has several limitations, including:

- **Vanishing Gradient Problem**: Like Sigmoid, Tanh suffers from vanishing gradients for inputs with large absolute values, which slows learning in deeper networks.
- **Computational Complexity**: The exponential calculations in Tanh make it slightly more computationally expensive compared to simpler functions like ReLU.

Tanh is commonly used in the following scenarios:

- **Shallow Networks**: Tanh is suitable for shallow networks with fewer layers where the vanishing gradient problem is less significant.
- **Hidden Layers**: When balanced outputs with both positive and negative values are needed, Tanh can be used in hidden layers to improve the model's performance.

While Tanh is less common in modern deep learning architectures, it remains a valuable option for specific scenarios, particularly when zero-centered outputs are beneficial.

3.6 Exponential Linear Unit (ELU)

Exponential Linear Unit (ELU) is an activation function that helps to address the vanishing gradient problem while allowing negative values, unlike ReLU, which forces all negative inputs to zero. Mathematically, ELU is defined as:

$$f(x) = \begin{cases} x, & \text{if } x > 0 \\ \alpha(e^x - 1), & \text{if } x \le 0 \end{cases}$$

where α is a positive constant that controls the smoothness of the function.

Some key benefits of ELU include:

- **Solves Vanishing Gradient Problem**: By allowing small negative outputs, ELU helps maintain gradients, improving backpropagation.
- **Better Mean Activation Centering**: ELU activations tend to be closer to zero mean, reducing bias shift during training.

3.7 Scaled Exponential Linear Unit (SELU)

Scaled Exponential Linear Unit (SELU) extends ELU by introducing scaling parameters that help achieve self-normalization in deep networks, making it particularly effective in Self-Normalizing Neural Networks (SNNs). Mathematically, SELU is defined as:

$$f(x) = \begin{cases} \lambda x, & \text{if } x > 0 \\ \lambda\alpha(e^x - 1), & \text{if } x \le 0 \end{cases}$$

where λ and α are predefined constants that ensure self-normalizing properties.

Key advantages of SELU include:

- **Self-Normalization**: Helps maintain a mean activation close to zero and a unit variance, stabilizing deep network training.
- **Reduces the Need for Batch Normalization**: Due to its self-normalizing nature, SELU reduces dependency on batch normalization layers.

3.8 Swish (Self-gated Activation Function)

Swish is a *self-gated* activation function that blends linearity and non-linearity. Unlike ReLU, which zeroes out negative values, Swish allows small negative values, improving smoothness. Mathematically, Swish is defined as:

$$f(x) = x \cdot \sigma(x) = x \cdot \frac{1}{1 + e^{-x}}$$

where $\sigma(x)$ is the sigmoid function. Key benefits of Swish include:

- **Smooth and Non-Monotonic**: Unlike ReLU, Swish is non-monotonic, leading to better gradient flow.
- **Improved Training Performance**: Swish has shown superior performance in deep networks compared to ReLU in some cases.

3.9 Gaussian Error Linear Unit (GELU)

The Gaussian Error Linear Unit (GELU) activation function is commonly used in *transformer-based models* such as BERT and GPT. It smoothly approximates the behavior of ReLU but improves gradient-based learning. Mathematically, GELU is defined as:

$$f(x) = x \cdot \Phi(x)$$

where $\Phi(x)$ is the cumulative distribution function of the standard normal distribution:

$$\Phi(x) = \frac{1}{2}\left(1 + \text{erf}\left(\frac{x}{\sqrt{2}}\right)\right)$$

Advantages of GELU include:

- **Smooth and Adaptive Activation**: Unlike ReLU, GELU smoothly scales negative inputs instead of truncating them to zero.
- **Improved Performance in Transformers**: Widely used in state-of-the-art NLP models due to its effectiveness in deep architectures.

4 Downstream Tasks

Neural networks can be applied to a variety of downstream tasks, each requiring specific configurations for the output layer. This section provides an overview of common tasks such as classification, regression, and ranking, with details about their output layer configurations.

4.1 Classification

Classification tasks aim to assign an input to one of several predefined categories. Depending on whether the classification is binary or multi-class, the choice of the output layer and activation function varies.

4.1.1 Binary Classification

For binary classification, the network predicts the likelihood of an input belonging to one of two classes. The output layer typically consists of a single neuron, and the **Sigmoid activation function** is used to map the output to a probability between 0 and 1:

$$\sigma(z) = \frac{1}{1 + e^{-z}}$$

where z is the weighted sum of inputs, defined as:

$$z = w_0 x_0 + w_1 x_1 + \cdots + w_m x_m = \sum_{i=0}^{m} w_i x_i = \mathbf{w}^T \mathbf{x}$$

The Sigmoid activation ensures that the output is interpretable as a probability, making it particularly suited for binary classification tasks.

4.1.2 Multi-class Classification

For multi-class classification tasks, the network predicts the class an input belongs to among k possible classes. The output layer consists of k neurons, each corresponding to a class, and the **Softmax activation function** is used to normalize the outputs into probabilities. The Softmax function converts raw scores (logits) into probabilities that sum to 1:

$$P(y = j|\mathbf{z}^{(i)}) = \phi_{\text{softmax}}(\mathbf{z}^{(i)}) = \frac{e^{z^{(i)}}}{\sum_{k=0}^{k} e^{z_k^{(i)}}}$$

where $P(y = j|\mathbf{z}^{(i)})$ represents the probability of class j for the i-th input, $z^{(i)}$ denotes the raw logit (pre-activation output) for the i-th sample, and k refers to the total number of classes. This ensures that the network outputs a valid probability distribution, making predictions interpretable and suitable for classification tasks.

4.2 Regression

Regression tasks involve predicting continuous numeric values. For example, a network might predict the price of a house given its features or estimate the likelihood of an event. The output layer in regression tasks typically does not use an activation function (linear activation), allowing unrestricted numeric outputs.

In regression tasks, the output layer directly produces numeric values:

$$O_i = W \cdot x_i + b$$

where O_i represents the predicted value for the i-th sample, W denotes the weights of the output layer, x_i corresponds to the input features for the i-th sample, and b refers to the bias term. The lack of activation in the output layer ensures that the network can predict values over an unrestricted range, which is essential for regression tasks.

4.3 Ranking

Ranking tasks aim to order items based on relevance or similarity. These tasks are common in recommendation systems, search engines, and similarity-based applications like face verification. The output layer in ranking tasks typically generates **embeddings** or scores that can be used to compare pairs or sets of items. For ranking tasks, the output layer produces embeddings or scalar scores:

$$f(x_i) \in \mathbb{R}^d$$

where $f(x_i)$ is the embedding vector of dimension d for input x_i. These embeddings represent items in a continuous space where geometric relationships correspond to relevance or similarity. Similar items are positioned closer together, while dissimilar items are placed farther apart.

5 Embeddings in Neural Networks

Embeddings are low-dimensional, dense representations of data points, learned by neural networks to capture meaningful relationships between inputs. These vectors map inputs, such as words, images, or users, into a continuous space where semantic or functional similarity is represented geometrically. Typically, embeddings transform high-dimensional data into a compact, dense representation in a lower-dimensional space to optimize tasks like classification, ranking, or recommendation.

Given an input data point x_i, a neural network transforms it into an embedding vector $f(x_i)$ in one of the hidden layers. The embedding vector $f(x_i)$ preserves relationships between data points geometrically. For example:

$$1/d(f(x_i), f(x_j)) \approx \text{similarity}(x_i, x_j)$$

where d is a distance metric, such as Euclidean distance or cosine similarity. In the embedding space, similar inputs (e.g., semantically related words like *king* and *queen*) are represented as vectors close together, while dissimilar inputs (e.g., *king* and *apple*) are represented as vectors farther apart. Embeddings are widely used in various tasks such as:

- **Similarity Learning**: Identifying whether two inputs are similar, as in face verification or image matching.
- **Ranking Tasks**: Learning functions to rank items based on their relevance or similarity, commonly used in recommendation systems.
- **Dimensionality Reduction**: Transforming high-dimensional data into a compact form for easier storage, visualization, or computation.
- **Natural Language Processing (NLP)**: Representing words, sentences, or documents as embeddings (e.g., Word2Vec).

6 Training Neural Networks

Training a neural network involves adjusting its parameters (weights and biases) iteratively to minimize the error between the network's predicted and true outputs. The goal is to optimize the network so it generalizes well to both the training data and unseen data. Training methods for neural networks can be categorized into two main types based on how parameter updates are performed:

- **Gradient-Based Methods**: These methods rely on gradient computations to iteratively adjust parameters. The process typically consists of two main steps:

  - **Forward Propagation**: Computes predictions by passing input data through the network. Weights, biases, and activation functions are applied sequentially layer by layer to produce an output.
  - **Backward Propagation**: Propagates errors backward through the network to compute gradients using the **chain rule of calculus**. These gradients guide updates to weights and biases, typically using optimization algorithms such as Gradient Descent or Stochastic Gradient Descent (SGD).

  Gradient-based methods are widely used due to their efficiency and compatibility with deep learning architectures.
- **Non-Gradient-Based Methods**: These methods do not rely on gradient information and are suitable for cases where the loss function is non-differentiable or

when gradient computation is computationally expensive. Forward propagation is still used to compute predictions, but parameter updates are achieved through alternative mechanisms, such as:

- **Genetic Algorithms**: Optimize parameters by simulating evolutionary processes like selection, crossover, and mutation.
- **Particle Swarm Optimization (PSO)**: Utilize swarm intelligence to iteratively explore the parameter space, based on the collective behavior of candidate solutions.
- **Forward-Forward Algorithm**: Optimizes the network by maximizing the match between activations and targets during forward passes, bypassing the need for backward propagation.

Non-gradient-based methods are valuable in specialized scenarios but are generally less efficient than gradient-based methods for deep learning tasks.

6.1 Forward Propagation: Making Predictions

Forward propagation, or inference, is the process by which input data flows through the neural network to compute predictions. The key steps are:

1. **Layer-by-Layer Calculation**: Each neuron computes a weighted sum of its inputs, adds a bias term, and applies an activation function to generate an output. This process continues sequentially from the input layer through hidden layers to the output layer.
2. **Output Generation**: The final output layer produces predictions, which are compared to the ground truth labels using a loss function to quantify the model's error.

Mathematically, for a single layer:

$$z^{[l]} = W^{[l]}a^{[l-1]} + b^{[l]}, \quad a^{[l]} = f(z^{[l]})$$

where $z^{[l]}$ represents the weighted sum of inputs in layer l, $a^{[l]}$ denotes the activations in layer l, $W^{[l]}$ refers to the weights in layer l, $b^{[l]}$ corresponds to the biases in layer l, and f is the activation function.

6.2 Backward Propagation: Learning from Errors

Backward propagation, or backpropagation, computes gradients of the loss function with respect to the network's parameters (weights and biases). These gradients are used to update the parameters, minimizing the loss function over time.

The following steps outline the core process of backpropagation.

1. **Error Calculation**: The loss function quantifies the difference between the network's predicted outputs and the true labels. Common loss functions include Mean Squared Error (MSE) for regression tasks and Cross-Entropy Loss for classification tasks. The goal is to minimize this loss over all training examples to improve the model's performance.
2. **Gradient Calculation**: Gradients of the loss function with respect to the weights and biases are computed to understand how small changes in these parameters affect the loss. This process relies on the **chain rule of calculus**, which enables efficient propagation of errors backward through the network. The chain rule states:
$$\frac{\partial E}{\partial W} = \frac{\partial E}{\partial a} \cdot \frac{\partial a}{\partial z} \cdot \frac{\partial z}{\partial W}$$
Here, E represents the loss function, a denotes the activation output of a neuron, z corresponds to the weighted input to a neuron (calculated as $z = W \cdot x + b$), and W refers to the weight parameter. This rule allows the network to decompose the gradient calculation layer by layer, starting from the output layer and propagating backward to the input layer. Each layer passes its error contribution to the previous layer, enabling efficient gradient computation across deep networks.
3. **Parameter Updates**: Using the gradients computed in the previous step, the network updates the weights and biases to reduce the loss. The update is performed using an optimization algorithm like Gradient Descent or its variants. The general update rule is:
$$W = W - \alpha \frac{\partial E}{\partial W}, \quad b = b - \alpha \frac{\partial E}{\partial b}$$
where α is the learning rate, controlling the step size of each update. This iterative process reduces the loss over time, improving the model's accuracy and performance.

6.3 Gradient Calculation in Backpropagation

Gradients are computed for all parameters to update the network during training. The **chain rule** is used to propagate gradients through multiple layers. Popular activation functions and their gradients include:

- **Sigmoid**
$$\sigma(z) = \frac{1}{1 + e^{-z}}, \quad \sigma'(z) = \sigma(z)(1 - \sigma(z))$$

- **Tanh**

$$\tanh(z) = \frac{e^z - e^{-z}}{e^z + e^{-z}}, \quad \tanh'(z) = 1 - \tanh^2(z)$$

- **ReLU (Rectified Linear Unit)**

$$\text{ReLU}(z) = \max(0, z), \quad \text{ReLU}'(z) = \begin{cases} 1 & \text{if } z > 0 \\ 0 & \text{if } z \leq 0 \end{cases}$$

- **Leaky ReLU**

$$f(z) = \begin{cases} z & \text{if } z > 0 \\ \alpha z & \text{if } z \leq 0 \end{cases}, \quad f'(z) = \begin{cases} 1 & \text{if } z > 0 \\ \alpha & \text{if } z \leq 0 \end{cases}$$

The chain rule is fundamental for computing gradients in deep networks. For instance, to compute the gradient of the total error E_{total} with respect to a weight W_5:

$$\frac{\partial E_{\text{total}}}{\partial W_5} = \frac{\partial E_{\text{total}}}{\partial \text{out}_{o1}} \cdot \frac{\partial \text{out}_{o1}}{\partial \text{net}_{o1}} \cdot \frac{\partial \text{net}_{o1}}{\partial W_5}$$

This computation enables gradients to propagate backward through the layers, updating parameters for improved performance.

6.4 *Gradient-Based Optimization Techniques*

Once gradients are computed, optimization algorithms update the weights and biases to minimize the loss function. Two commonly used methods are **Gradient Descent** and **Stochastic Gradient Descent (SGD)** (Fig. 3).

6.4.1 Gradient Descent

Gradient Descent is an iterative optimization algorithm that updates parameters (weights and biases) in the direction of the negative gradient of the loss function. The negative gradient represents the steepest descent direction, allowing the algorithm to converge toward a local or global minimum of the loss function (see Algorithm 1). The key steps in gradient descent are as follows:

1. **Initialize Parameters**: Start with random small values for weights and biases to ensure diverse activations and avoid symmetry issues.
2. **Compute Gradients**: For each iteration, calculate the gradient of the loss function with respect to all parameters (weights and biases). These gradients indicate how the loss function changes when the parameters are slightly modified.

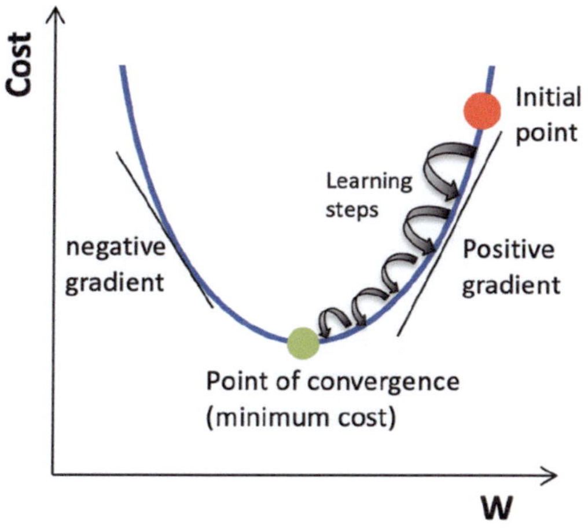

Fig. 3 Visualization of gradient descent. The process starts at an initial point (red marker) and iteratively updates the parameter W by following the negative gradient direction. The updates get smaller as the algorithm approaches the point of convergence (green marker), where the cost function reaches its minimum

3. **Update Parameters**: Adjust the weights and biases using the gradients to reduce the loss function:

$$W^{[l]} = W^{[l]} - \alpha \frac{\partial E}{\partial W^{[l]}}, \quad b^{[l]} = b^{[l]} - \alpha \frac{\partial E}{\partial b^{[l]}}$$

An intuitive explanation of gradient descent involves understanding that the parameter update formula is derived from the goal of minimizing the loss function E, which quantifies the error between the predicted and actual outputs of the neural network. Central to this process is the gradient $\frac{\partial E}{\partial W^{[l]}}$, which plays a critical role in directing the optimization by indicating the direction in which the parameters should be adjusted to reduce the error.

The gradient $\frac{\partial E}{\partial W^{[l]}}$ serves several important functions in the optimization process:

- **Measures Sensitivity**: It quantifies how sensitive the loss E is to small changes in the weight $W^{[l]}$. For example:
  - If the gradient is large, a small change in $W^{[l]}$ results in a significant change in the loss.
  - If the gradient is small, the loss is less sensitive to changes in $W^{[l]}$.
- **Indicates the Direction of Increase**: The gradient points in the direction of the steepest ascent of the loss E with respect to $W^{[l]}$. This means:
  - If $\frac{\partial E}{\partial W^{[l]}} > 0$, increasing $W^{[l]}$ will increase the loss.
  - If $\frac{\partial E}{\partial W^{[l]}} < 0$, decreasing $W^{[l]}$ will increase the loss.

To minimize E, we need to move $W^{[l]}$ in the direction that decreases E. This is achieved by updating the weight in the direction opposite to the gradient:

$$W^{[l]} = W^{[l]} - \alpha \frac{\partial E}{\partial W^{[l]}}$$

Here, the term $\frac{\partial E}{\partial W^{[l]}}$ indicates how much the loss E will increase if we make a small adjustment to $W^{[l]}$ in the positive direction of the gradient. By subtracting a fraction of this gradient ($\alpha \frac{\partial E}{\partial W^{[l]}}$) from the current weight $W^{[l]}$, we effectively reduce the value of E. This adjustment ensures that the weight $W^{[l]}$ moves closer to an optimal value that minimizes the loss. Note that the learning rate α controls the size of the step taken in the direction of the negative gradient:

- A small α results in smaller steps, leading to slower but stable convergence. This avoids overshooting the minimum but may require more iterations.
- A large α results in larger steps, enabling faster convergence. However, if α is too large, the updates may overshoot the minimum or even diverge, preventing the algorithm from converging.

In a similar manner, biases $b^{[l]}$ are updated to account for how much the loss changes with respect to the biases:

$$b^{[l]} = b^{[l]} - \alpha \frac{\partial E}{\partial b^{[l]}}$$

- The gradient $\frac{\partial E}{\partial b^{[l]}}$ measures the sensitivity of E to changes in $b^{[l]}$.
- Adjusting the biases ensures that all parameters in the model contribute to reducing the loss.

Why This Works: An Analogy
Imagine rolling a ball down a hill where the height represents the loss E, and the ball's position corresponds to the parameter $W^{[l]}$. The slope of the hill at any point is analogous to the gradient $\frac{\partial E}{\partial W^{[l]}}$:

- The steeper the slope, the larger the gradient, and the faster the ball (parameter) moves.
- The direction of the slope indicates whether $W^{[l]}$ should increase or decrease to minimize E.

By moving the ball (updating $W^{[l]}$) in the direction opposite to the slope, the ball rolls downhill toward the minimum loss.

4. **Repeat**: Perform steps 2 and 3 iteratively until a convergence criterion is met, such as a sufficiently small change in the loss function or a fixed number of iterations.

Algorithm 1 Gradient Descent

Require: Learning rate α, loss function E, initial weights W, biases b
Ensure: Optimized weights W and biases b
1: Initialize weights W and biases b randomly.
2: **while** not converged **do**
3: Compute the gradient $\nabla E(W, b)$ of the loss function with respect to W and b.
4: Update weights: $W \leftarrow W - \alpha \nabla E(W)$.
5: Update biases: $b \leftarrow b - \alpha \nabla E(b)$.
6: **end whilereturn** Optimized weights W and biases b.

Despite the effectiveness of gradient descent, it presents several challenges that can complicate the optimization process:

- **Computationally Intensive**: Gradient Descent requires calculating gradients over the entire dataset in each iteration. This involves computing the derivative of the loss function with respect to all weights and biases for every training sample, which can be computationally expensive for large datasets. The cost increases significantly as the dataset size and the number of parameters grow, making it impractical for real-world applications with massive datasets. Variants like **Stochastic Gradient Descent (SGD)** and **Mini-batch Gradient Descent** address this issue by approximating the gradient using subsets of the data, reducing computational overhead while still guiding the optimization process effectively.
- **Stuck in Local Minima**: Gradient Descent relies on the shape of the loss function to find the minimum, which works well for convex loss functions. In convex functions, any local minimum is also the global minimum, ensuring convergence to the optimal solution. However, many real-world problems involve **non-convex loss functions** that have multiple minima and saddle points. Key terms to understand this challenge are:
  - **Convex Loss Function**: A convex loss function is one where the line segment between any two points on the function lies entirely above or on the graph. Mathematically, a function $f(x)$ is convex if:

$$f(\theta x_1 + (1-\theta)x_2) \leq \theta f(x_1) + (1-\theta) f(x_2), \quad \forall \theta \in [0, 1]$$

    Convex functions are easier to optimize because they have a single global minimum, which can often be found efficiently.
  - **Non-Convex Loss Function**: In contrast, non-convex functions are more complex, with multiple minima, maxima, and saddle points. Gradient Descent can get stuck at:
    - **Local Minima**: A point where the loss is lower than the surrounding points but not the lowest possible value.

· **Saddle Points**: Points where the gradient is zero, but the function does not represent a minimum (e.g., it increases in some directions and decreases in others).

In non-convex landscapes, the initial parameter values significantly influence whether Gradient Descent converges to the global minimum or gets trapped in a suboptimal local minimum or saddle point.

- **Learning Rate Sensitivity**: The learning rate α determines the size of the steps Gradient Descent takes in the direction of the negative gradient. While the learning rate is crucial for effective optimization, choosing an appropriate value can be challenging (see Fig. 4):

  – **Small Learning Rate**: If α is too small, Gradient Descent takes very small steps toward the minimum, leading to slow convergence. This increases the number of iterations required and may result in unnecessary computational expense.
  – **Large Learning Rate**: If α is too large, the updates can overshoot the minimum, causing the algorithm to oscillate around the solution or even diverge. Divergence occurs when updates cause the loss to increase rather than decrease, preventing convergence.
  – **Dynamic Learning Rates**: To address this challenge, adaptive learning rate techniques, such as Adam, adjust the learning rate during training. These methods ensure that parameters take appropriately sized steps based on the current gradient magnitude and optimization progress, improving stability and convergence speed.

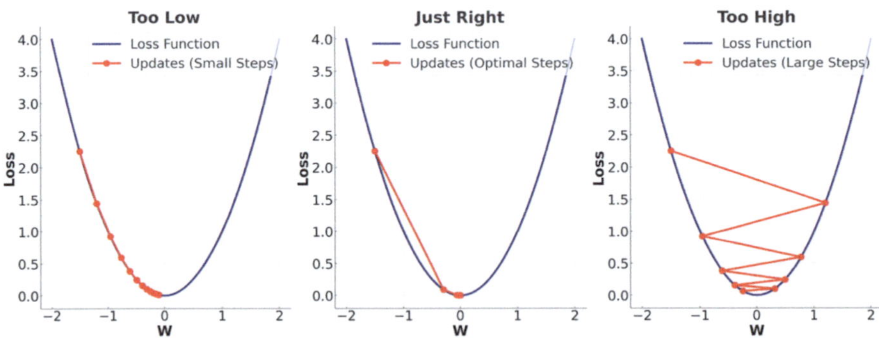

Fig. 4 Effect of learning rate on gradient descent. **Left**: A small learning rate results in slow convergence, requiring many small updates to reach the minimum. **Middle**: An optimally chosen learning rate allows efficient and stable convergence. **Right**: A large learning rate causes oscillations, preventing proper convergence and potentially leading to divergence

6.4.2 Stochastic Gradient Descent (SGD)

Stochastic Gradient Descent (SGD) is a variation of Gradient Descent that updates parameters based on the gradient computed for a single training example or a small batch of examples, rather than the entire dataset. This approach introduces randomness (stochasticity) into the optimization process, which can lead to faster updates and the ability to escape local minima. In SGD, the optimization process follows a series of steps to update the model parameters (see Algorithm 2):

1. **Initialize Parameters**: Randomly initialize weights and biases.
2. **Select Data Point or Batch**: Randomly select a single training example (x_i, y_i) or a small batch of examples $\{x_{i_k}, y_{i_k}\}$.
3. **Compute Gradients**: Calculate the gradient of the loss function for the selected data point(s):
$$\nabla E_i(W, b) = \frac{\partial E_i}{\partial W}, \quad \nabla E_i(b) = \frac{\partial E_i}{\partial b}$$
4. **Update Parameters**: Adjust weights and biases using:
$$W = W - \alpha \nabla E_i(W), \quad b = b - \alpha \nabla E_i(b)$$
5. **Repeat**: Iterate steps 2–4 for all data points in the dataset, and repeat the process for multiple epochs until the stopping criterion is met.

Algorithm 2 Stochastic Gradient Descent (SGD)

Require: Learning rate α, loss function E, initial weights W, biases b, training data $\{x_i, y_i\}$
Ensure: Optimized weights W and biases b
1: Initialize weights W and biases b randomly.
2: **while** not converged **do**
3:    **for all** training examples (x_i, y_i) or batches $\{x_{i_k}, y_{i_k}\}$ **do**
4:       Compute the gradient $\nabla E_i(W, b)$ for the selected data point or batch.
5:       Update weights: $W \leftarrow W - \alpha \nabla E_i(W)$.
6:       Update biases: $b \leftarrow b - \alpha \nabla E_i(b)$.
7:    **end for**
8: **end whilereturn** Optimized weights W and biases b.

SGD offers several notable advantages, including:

- **Faster Parameter Updates**: SGD performs updates after each data point or small batch, enabling faster convergence, which is particularly beneficial for large-scale datasets.
- **Better Generalization**: The inherent stochastic noise in SGD helps prevent overfitting by allowing the model to escape sharp local minima, promoting better generalization to unseen data.

- **Scalability**: SGD is highly scalable, as it can handle large datasets by processing data in small batches, making it suitable for online learning and large-scale machine learning problems.
- **Ability to Escape Saddle Points and Local Minima**: The randomness introduced by SGD helps the model avoid getting stuck in saddle points or local minima, which can lead to more robust optimization.
- **Memory Efficiency**: Since only a subset of the data is used for each update, SGD is more memory-efficient compared to traditional gradient descent, which requires the entire dataset to be stored in memory.

Despite its advantages, SGD also comes with several challenges, such as:

- **Noisy Updates**: The stochastic nature of the updates can introduce noise, which may prevent smooth convergence to the minimum and result in erratic progress during training. This noise can cause the algorithm to oscillate around the minimum instead of converging smoothly.
- **Learning Rate Tuning**: SGD requires careful tuning of the learning rate to strike a balance between speed and stability. An inappropriate learning rate can either lead to slow convergence or cause the optimization to overshoot the optimal solution, making convergence difficult or slower.
- **Random Sampling Variability**: The random sampling of data points introduces variability in the updates, which can lead to inconsistent results unless proper averaging or smoothing techniques, such as momentum or mini-batch averaging, are applied.
- **Convergence to Local Minima or Saddle Points**: While SGD helps in escaping some local minima, it can still get trapped in local minima or saddle points, especially in complex loss surfaces, leading to suboptimal solutions.
- **Vanishing or Exploding Gradients**: In deep networks, gradients can become very small (vanishing) or excessively large (exploding), making the optimization process difficult and potentially destabilizing the learning process.
- **Slow Convergence in Flat Regions**: In regions of the loss surface where the gradient is small (flat regions), the progress of SGD can become sluggish, leading to slower convergence during those phases of training.

6.4.3 Mini-batch Gradient Descent

While SGD updates model parameters using a single data point (x_i, y_i) at each iteration, **Mini-Batch Gradient Descent** balances the computational efficiency of SGD with the stability of full-batch gradient descent by utilizing a small subset of training data, denoted as $\{x_{ik}, y_{ik}\}$, for each update step. This approach reduces the variance in gradient estimates while maintaining computational efficiency, leading to a more stable convergence trajectory compared to standard SGD. At each iteration, a mini-batch of size B is randomly sampled from the training dataset. The gradients of the loss function E with respect to the weights W and biases b are then computed over the mini-batch:

$$\nabla E_B(W, b) = \frac{1}{B} \sum_{k=1}^{B} \nabla E_{ik}(W, b)$$

Using this batch-averaged gradient, the model parameters are updated as follows:

$$W \leftarrow W - \alpha \nabla E_B(W)$$

$$b \leftarrow b - \alpha \nabla E_B(b)$$

where α is the learning rate controlling the step size in the parameter space. The selection of an appropriate mini-batch size B influences both convergence speed and generalization. Larger mini-batches result in smoother updates but require more computational resources, while smaller mini-batches introduce controlled noise that can aid generalization. Mini-Batch Gradient Descent is widely used in deep learning due to its balance between computational efficiency and convergence stability, making it the preferred approach for large-scale datasets and modern neural networks.

6.4.4 Comparison of Batch, Mini-batch, and Stochastic Gradient Descent

Figure 5 illustrates the key differences between batch gradient descent, mini-batch gradient descent, and stochastic gradient descent. These three approaches vary in terms of computational efficiency, convergence stability, and the noise introduced in parameter updates.

Batch gradient descent (standard gradient descent) computes the gradient using the entire dataset at each iteration, leading to a smooth and stable convergence trajectory. This method ensures precise updates but becomes computationally expensive as

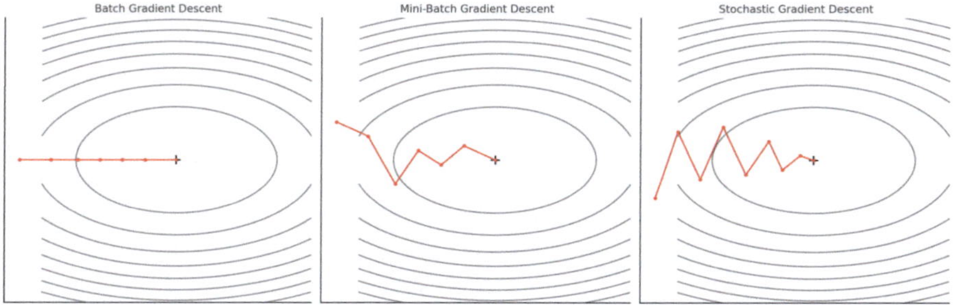

Fig. 5 Comparison of batch, mini-batch, and stochastic gradient descent

dataset size increases. Additionally, it may struggle with escaping sharp local minima due to its more deterministic nature.

Mini-batch gradient descent mitigates the computational inefficiency of batch gradient descent by using a small subset of data, referred to as a mini-batch, for each update step. This approach balances efficiency and stability, reducing memory requirements while maintaining relatively smooth convergence. The trajectory of mini-batch gradient descent is noisier than that of batch gradient descent but more stable compared to stochastic gradient descent. The choice of mini-batch size, denoted as B, affects convergence behavior. Larger mini-batches provide more accurate gradient estimates but increase computational cost, whereas smaller mini-batches introduce more variance.

SGD updates the model parameters after processing each individual data point, resulting in highly fluctuating updates. This randomness allows it to explore the loss landscape more effectively, potentially escaping saddle points and sharp local minima. However, the high variance in updates can lead to instability, requiring additional techniques such as learning rate scheduling, momentum, or adaptive optimization methods.

The selection of an appropriate learning rate is critical in gradient descent techniques. A large learning rate may cause divergence, while a small learning rate may slow down convergence. Adaptive learning rate strategies address this issue by dynamically adjusting the learning rate based on the magnitude of past gradients. The effectiveness of mini-batch gradient descent depends on two crucial factors: the mini-batch size B and the sampling strategy. Mini-batches are typically sampled randomly from the dataset to ensure diversity in gradient estimation. The selection of B depends on hardware constraints and convergence stability. While larger batches lead to smoother updates, smaller batches can introduce beneficial noise, aiding generalization. The learning rate η must also be carefully tuned to ensure a balance between convergence speed and stability.

6.4.5 Gradient Descent with Momentum

Gradient descent with momentum is an optimization technique designed to improve the efficiency and stability of gradient-based optimization. It extends the standard gradient descent methods, including batch gradient descent, mini-batch gradient descent, and SGD, by introducing a velocity term that helps smooth updates and accelerate convergence. In traditional gradient descent, updates are determined solely by the gradient at the current iteration, without any consideration of past updates. This is true for batch gradient descent, which computes updates based on the entire dataset, mini-batch gradient descent, which updates parameters using small randomly selected subsets of data, and SGD, which updates parameters using individual data points.

Momentum is particularly useful in addressing several challenges that arise in optimization, especially in deep learning applications. One such issue is **oscillations in high-curvature regions** (rapid fluctuations in update direction due to steep gradients). These oscillations occur because the gradient direction changes rapidly in such regions, causing inefficient updates that slow down convergence. Another common issue is **slow convergence in flat areas** (regions where gradients are very small, leading to minimal updates). Flat areas are often associated with **plateaus** in the loss landscape, where optimization progress is slow due to weak gradient signals.

Saddle points represent another major challenge in gradient-based optimization. **Stagnation at saddle points** (points where the gradient is near zero in multiple directions, causing the optimizer to stall) is particularly problematic in **high-dimensional spaces**, where many parameters may have near-zero gradients, making it difficult for standard gradient descent to escape. Additionally, optimization can struggle with **difficulty escaping sharp local minima** (narrow regions where the optimizer may settle instead of exploring better solutions). However, sharp local minima are generally **less common** in high-dimensional spaces compared to saddle points, making the latter a more significant concern in deep learning.

The gradient descent update rule with momentum introduces a velocity term that accumulates an exponentially weighted sum of past gradients. The velocity term is updated at each iteration by combining the previous velocity with the newly computed gradient. This update is governed by a momentum coefficient, which determines the contribution of past gradients to the current update. The weights and biases are then adjusted using the velocity term rather than the raw gradient. Mathematically, the velocity update is defined as the previous velocity scaled by the momentum coefficient, combined with the new gradient of the loss function. The weights and biases are subsequently updated by subtracting a fraction of the velocity term, weighted by the learning rate. The formulation of gradient descent with momentum is expressed as follows (see Fig. 6):

$$v^{(t+1)} = \rho v^{t} + \nabla E_{i}^{(t)}(W, b)$$

$$W^{(t+1)} = W^{t} - \alpha v^{(t+1)}$$

$$b^{(t+1)} = b^{t} - \alpha v^{(t+1)}$$

In this formulation, the velocity term at iteration $t + 1$ is computed as a combination of the **scaled previous velocity** and the **gradient of the loss function at the current iteration**. The velocity term, denoted as v, represents the accumulated momentum from previous updates. The parameter ρ, known as the momentum coefficient, controls how much of the past gradients influence the current update. Typical values for the momentum coefficient range from **0.9 to 0.99**, allowing for a balance between stability and responsiveness. The learning rate, denoted as α, determines the step size in parameter updates. By incorporating these elements, momentum-based gradient descent achieves a more stable and efficient optimization trajectory.

The effectiveness of momentum-based gradient descent stems from its ability to maintain a consistent update direction by leveraging information from past iterations. This results in smoother updates that reduce fluctuations caused by noisy gradient estimates. The accumulation of past gradients allows the optimization to move more decisively through regions of high curvature, reducing the extent of oscillations.

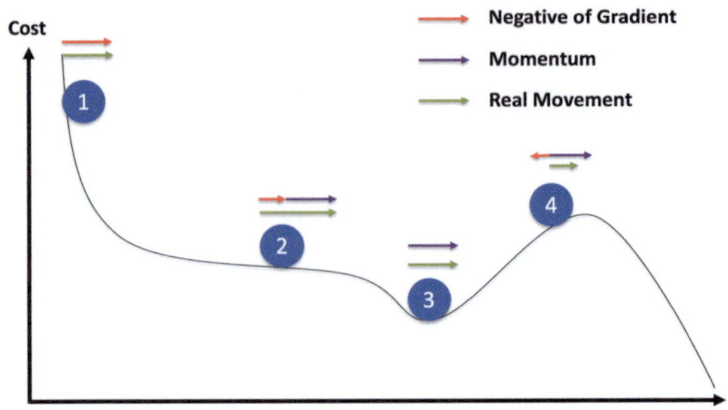

Fig. 6 Illustration of momentum in gradient descent. **Point 1**: Initial stage—the update is influenced by the gradient. **Point 2**: Momentum builds up, accelerating movement in the descent direction. **Point 3**: Momentum continues to push forward, even as the gradient weakens. **Point 4**: The gradient alone suggests moving left (backward), but accumulated momentum helps override this and keeps movement in the correct direction (right), ensuring smoother convergence without getting stuck in local minima

Additionally, the presence of momentum enables the optimizer to traverse plateaus more efficiently, preventing stagnation (in the context of optimization, stagnation refers to a situation where the training process slows down significantly or completely stops making progress due to extremely small gradient updates) in regions where the gradient is small. As a result, the overall convergence process is accelerated, leading to improved training efficiency.

6.4.6 Critical Points in Optimization

In optimization and deep learning, **critical points** are points where the first derivative (gradient) of a function equals zero. These points provide crucial insights into the behavior of loss functions and optimization dynamics. Critical points can be categorized as local minima, local maxima, or saddle points, depending on the second derivative (or the Hessian in higher dimensions), which characterizes the curvature of the function.

- **Local Minimum**: A critical point where the function reaches a lower value than its immediate surroundings. Mathematically, a local minimum at x_0 satisfies:

$$f'(x_0) = 0, \quad f''(x_0) > 0$$

  The positive second derivative indicates a convex shape, facilitating convergence when using gradient-based optimization techniques.

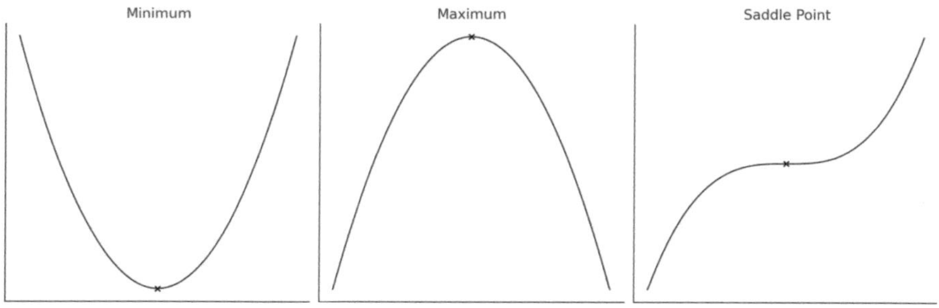

Fig. 7 Visualization of critical points in a one-dimensional function. The left plot illustrates a **local minimum**, where the function reaches a lower value than its neighbors, indicated by positive curvature ($f''(x) > 0$). The middle plot represents a **local maximum**, where the function attains a higher value than its surroundings, with negative curvature ($f''(x) < 0$). The right plot demonstrates a **saddle point**, which is neither a minimum nor a maximum, characterized by mixed curvature in different directions. These critical points are fundamental in optimization and deep learning, where saddle points frequently emerge in high-dimensional loss landscapes, affecting convergence efficiency

- **Local Maximum**: A critical point where the function attains a higher value than its neighboring points. A local maximum at x_0 satisfies:

$$f'(x_0) = 0, \quad f''(x_0) < 0$$

  Here, the function exhibits concavity, and gradient-based methods naturally avoid these points in minimization problems.
- **Saddle Point**: A critical point that is neither a minimum nor a maximum. It satisfies:

$$f'(x_0) = 0, \quad f''(x_0) \text{ changes sign in different directions}$$

  Saddle points are prevalent in high-dimensional optimization landscapes, such as those found in deep learning. They often slow down convergence, as gradient-based methods struggle to escape them efficiently.

Saddle points pose a challenge in deep learning optimization, as they are more frequent than local maxima in high-dimensional spaces. Traditional gradient descent methods may stagnate at these points due to the lack of a clear descent direction. However, modern optimizers (e.g., Adam) incorporate adaptive learning rates and second-order curvature information to mitigate the effects of saddle points and accelerate convergence (Fig. 7).

6.4.7 Curvature and Optimization

As explained in the previous sections, gradient descent optimization techniques utilize the first derivative (gradient) of the loss function $f(x)$ to iteratively update

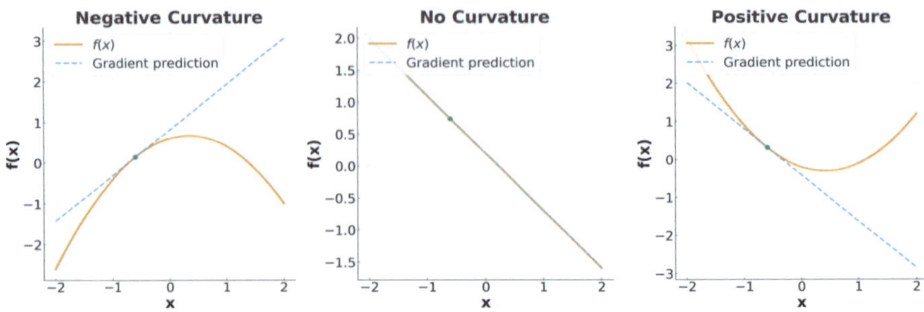

Fig. 8 Visualization of curvature effects in gradient-based optimization

model parameters. However, the efficiency and convergence behavior of gradient-based updates are significantly influenced by the second derivative $f''(x)$. The second derivative captures how the first derivative changes with respect to the input, providing crucial insight into the function's **curvature**. This curvature affects the optimization process by influencing step sizes, convergence speed, and stability. In particular, it determines whether updates lead smoothly to a minimum or result in erratic behavior, shaping the effectiveness of gradient-based methods in the following ways (see Fig. 8):

- **Negative Curvature**: When $f''(x) < 0$, the function is concave downward, meaning that the cost function decreases faster than predicted by the gradient.
- **No Curvature**: When $f''(x) = 0$, the loss function is locally linear, meaning that the gradient correctly predicts the decrease.
- **Positive Curvature**: When $f''(x) > 0$, meaning that the cost function decreases slower than expected and can even start increasing if steps are too large, leading to instability.

7 Training Challenges and Optimization Techniques

Despite the impressive successes of deep learning, the training process faces several technical challenges that must be addressed, including:

- **Overfitting and Underfitting**: Overfitting occurs when a model learns the details and noise of the training data to such an extent that it negatively impacts its performance on new, unseen data. This is typically caused by a model being too complex relative to the amount of training data. On the other hand, underfitting happens when a model is too simple to capture the underlying patterns in the data, resulting in poor performance on both the training and test sets. Balancing model complexity to avoid both overfitting and underfitting is a fundamental challenge in deep learning.

- **Vanishing and Exploding Gradients**: Vanishing gradients occur when the gradients of the loss function become extremely small as they are backpropagated through the network, particularly in deep networks. This issue can cause weights in earlier layers of deep networks to stop updating, leading to poor learning. Exploding gradients, on the other hand, occur when gradients become excessively large, causing the model's weights to grow uncontrollably, destabilizing the learning process. These issues are common in activation functions like sigmoid or tanh. Several techniques can address these challenges, including:
  - **Gradient Clipping**: This technique caps the gradients to a predefined maximum value, preventing them from growing uncontrollably during backpropagation. By ensuring stability, gradient clipping is particularly effective in training recurrent neural networks (RNNs) and other deep architectures.
  - **Residual Connections**: Residual connections, introduced in ResNets (Residual Networks), allow gradients to flow more directly through the network by introducing skip connections. A skip connection bypasses one or more layers by creating a shortcut, where the input to a layer is added directly to its output. Mathematically, this can be expressed as:

$$y = F(x, \{W_i\}) + x,$$

    where, x is the input to the residual block, $F(x, \{W_i\})$ is the output of the intermediate layers within the residual block where $\{W_i\}$ are the weights of those layers, and the addition $F(x, \{W_i\}) + x$ combines the original input x with the transformed output, effectively "skipping" the intermediate layers. This design enables the network to learn a residual mapping $F(x, \{W_i\}) = y - x$, instead of trying to learn the original mapping $y = F(x)$. By simplifying the learning objective, residual connections make it easier for the network to optimize very deep architectures.
  - **Weight Initialization Techniques**: Proper weight initialization methods, such as Xavier initialization or He initialization, can help prevent the gradients from vanishing or exploding by ensuring that weights are neither too small nor too large at the start of training.
  - **Batch Normalization**: This technique normalizes the inputs to each layer during training, reducing internal covariate shift and helping to mitigate both vanishing and exploding gradients.

- **Optimization Challenges**: Deep learning models often have a large number of parameters, making optimization complex. While methods like SGD and its variants (e.g., Adam) are commonly used, finding the right learning rate and batch size can be difficult. Furthermore, deep networks are often prone to getting stuck in local minima or saddle points, preventing the model from reaching the global minimum of the loss function.
- **Model Generalization**: Generalization remains a critical challenge in deep learning. Even with large datasets, deep models can overfit to the training data and fail

to generalize well to unseen examples. Techniques like regularization, dropout, and cross-validation are employed to improve generalization, but this remains an active area of research.

- **Data Efficiency and Labeling**: Deep learning models typically require large amounts of labeled data to achieve good performance. In many cases, especially in domains like medical imaging or niche areas, acquiring sufficient labeled data can be expensive or impractical. **Data augmentation**, **transfer learning**, and **semi-supervised learning** techniques are often used to overcome this issue.

  - **Data Augmentation**: Data augmentation involves artificially expanding the training dataset by applying various transformations to the existing data. This technique is particularly useful in domains like computer vision, where it can be applied to images. For example, transformations like flipping, rotating, cropping, scaling, and adjusting brightness or contrast can create multiple variations of each image, effectively increasing the dataset size without requiring additional data collection. In the context of NLP, techniques like paraphrasing, random deletion of words, or swapping word order can be used to augment textual data. Data augmentation helps improve the model's ability to generalize to new, unseen data by introducing more variety into the training set and preventing overfitting.
  - **Transfer Learning**: Transfer learning is a technique where a model pre-trained on a large dataset (often from a related task or domain) is fine-tuned for a specific task with less labeled data. For example, in computer vision, models like VGG, ResNet, or Inception are typically trained on large datasets like ImageNet. These models learn to extract general features such as edges, textures, and patterns, which are useful across various tasks. Once the pre-trained model is fine-tuned on a smaller, domain-specific dataset, it can perform well despite the limited data. Transfer learning is particularly useful when there is a lack of sufficient labeled data in a specialized domain (e.g., medical images), as the model can leverage learned features from more general tasks.
  - **Semi-Supervised Learning**: Semi-supervised learning combines a small amount of labeled data with a large amount of unlabeled data to improve model performance. This technique is particularly useful when labeled data is scarce or expensive to obtain, but a large pool of unlabeled data is available. By leveraging both labeled and unlabeled data, semi-supervised learning enables models to learn more generalizable patterns, improving performance while reducing the need for large labeled datasets.

- **Long Training Times**: Training deep learning models, particularly large ones, can take a considerable amount of time and computational resources. Optimizing training efficiency and ensuring convergence in a reasonable timeframe remains a significant challenge, especially for complex tasks and large datasets.
- **Interpretability and Explainability**: One of the most critical challenges is the "black box" nature of deep learning models. These models are often seen as opaque, making it difficult to understand how a model arrived at a particular decision. This

lack of interpretability is particularly concerning in high-stakes applications like healthcare and finance. Efforts like LIME, SHAP, and attention mechanisms aim to improve model transparency.
- **Bias and Fairness**: Deep learning models can inherit and even amplify biases present in the training data, leading to unfair or discriminatory outcomes. For example, facial recognition systems may show biased performance across different demographic groups. Addressing bias in deep learning models is essential to ensure fairness and ethical decision-making.

To enhance the performance and stability of DL models, the following techniques and strategies can be employed:

- **Momentum and Adaptive Learning Rates**: Incorporate advanced optimization techniques such as:
  - **Momentum**: Adds a fraction of the previous update to the current one, accelerating convergence and reducing oscillations.
  - **Adaptive Learning Rates (e.g., Adam)**: Adjust the learning rate dynamically during training based on the gradient magnitude, improving optimization efficiency and reducing the need for manual tuning.
  - **Learning Rate Schedules**: Gradually decrease the learning rate during training to balance faster convergence in the initial stages with fine-tuning near the minimum.
  - **Warm Restarts**: Periodically reset the learning rate to a high value and then gradually reduce it, allowing the model to escape local minima or saddle points. Warm restarts are particularly effective when combined with cosine annealing, as they enable more efficient exploration of the loss landscape.
- **Regularization**: Regularization techniques aim to reduce overfitting and improve generalization by either penalizing model complexity or introducing strategies to make the model more robust. These techniques can be categorized as follows:
  - **Classical Regularization Techniques**: These methods directly modify the loss function to penalize large or complex models, thereby encouraging simpler models that generalize better. Common classical techniques include:
    - **L2 Regularization**: L2 Regularization adds a penalty proportional to the square of the weights to the loss function, discouraging large parameter values and improving generalization. The modified loss function is:

$$L_{\text{total}} = L(\theta) + \lambda \sum_i \theta_i^2$$

Here, $L(\theta)$ is the original loss function, θ represents the model parameters (weights), and λ controls the strength of the regularization.

- **Weight Decay** is a technique that directly modifies the weight update rule during optimization by shrinking the weights at each step. In the context of SGD, Weight Decay is mathematically equivalent to L2 Regularization. In L2 Regularization, a penalty term proportional to the square of the weights is added directly to the loss function: $L_{\text{reg}} = L_{\text{original}} + \frac{\lambda}{2}\sum_i w_i^2$, where λ is the regularization strength. The gradient with respect to the weights is then computed with this additional penalty term: $\frac{\partial L_{\text{reg}}}{\partial w_i} = \frac{\partial L_{\text{original}}}{\partial w_i} + \lambda w_i$, which leads to an update rule for the weights: $w_i = w_i - \eta\left(\frac{\partial L_{\text{original}}}{\partial w_i} + \lambda w_i\right)$, where η is the learning rate. In the case of Weight Decay, the weight update rule is modified directly by adding a term proportional to the weights: $\theta \leftarrow \theta - \eta\left(\frac{\partial L}{\partial \theta} + \lambda\theta\right)$, where θ represents the weights. Both methods are equivalent in their effect, as they both penalize large weights and prevent overfitting, but they differ in how the regularization is incorporated—Weight Decay modifies the weight update directly, while L2 Regularization modifies the loss function.

  In **Adam**, weight decay is integrated into the gradient computation, meaning it is added directly to the gradient term. Specifically, the update rule for Adam with weight decay is $\theta \leftarrow \theta - \eta\left(\frac{\hat{m}}{\sqrt{\hat{v}}+\epsilon} + \lambda\theta\right)$, where $\frac{\hat{m}}{\sqrt{\hat{v}}+\epsilon}$ represents the **adaptive gradient update** based on Adam's first and second moment estimates, and $\lambda\theta$ is the **weight decay term**, which is added to the gradient. This means the weight decay term is influenced by the adaptive learning rates. In contrast, in optimizers like **AdamW**, weight decay is implemented as a separate step, where it is **decoupled** from the gradient computation. The weight update in AdamW is $\theta \leftarrow \theta - \eta\left(\frac{\hat{m}}{\sqrt{\hat{v}}+\epsilon}\right) - \eta\lambda\theta$. In this case, the weight decay term $\lambda\theta$ is applied **directly** to the parameters after the adaptive gradient update, independently of the gradients. This decoupling ensures that weight decay does not interact with the adaptive learning rates, as it is applied uniformly to all parameters, similar to traditional L2 regularization.
- **L1 Regularization**: L1 Regularization adds a penalty proportional to the absolute value of the weights to the loss function. This encourages sparsity in the model by driving some weights to exactly zero, effectively "turning off" certain features. The modified loss function can be expressed as:

$$L_{\text{total}} = L(\theta) + \lambda \sum_i |\theta_i|$$

  Here, λ is the regularization strength, controlling how much penalty is applied to the weights.

  Unlike L2 Regularization, which reduces the magnitude of all weights without making them exactly zero, L1 Regularization selectively eliminates less important features by driving their corresponding weights to zero.

- **Modern Regularization-Like Techniques**: In addition to classical regularization methods, there are several techniques that indirectly improve generalization and reduce overfitting by altering the training process or the dataset. These methods do not directly modify the loss function but are highly effective in stabilizing training and improving model performance. Common modern techniques include:

  - **Dropout**: Randomly ignores a subset of neurons during training, forcing the network to learn more robust and distributed representations by reducing reliance on specific neurons.
  - **Data Augmentation**: Increases the size and diversity of the training dataset by applying transformations such as flipping, cropping, scaling, or rotation, which helps the model generalize better to unseen data.
  - **Early Stopping**: Monitors the validation loss during training and stops the training process once performance on the validation set stops improving. This prevents overfitting by halting training before the model starts to memorize the training data.
  - **Batch Normalization**: Batch Normalization is a technique used to stabilize and accelerate the training of deep neural networks by normalizing the activations of each layer. During training, it computes the mean (μ_{batch}) and variance (σ^2_{batch}) of the activations within each mini-batch and normalizes them to have zero mean and unit variance using the formula $\hat{x} = \frac{x-\mu_{\text{batch}}}{\sqrt{\sigma^2_{\text{batch}}+\epsilon}}$, where x is the activation and ϵ is a small constant for numerical stability. This normalization is followed by a learnable scaling factor (γ) and a shifting factor (β) that help the network maintain flexibility and recover the original activations if necessary, with the final transformation being $y = \gamma\hat{x} + \beta$. During inference, instead of using the statistics from the current mini-batch, Batch Normalization uses the running averages of the mean and variance accumulated during training to normalize the activations, ensuring stability and consistency when making predictions on unseen data. This approach allows the model to maintain its learned behavior across different data inputs, improving generalization and making the network more robust. Additionally, because the statistics used for normalization are computed from mini-batches during training, the slight variations in these statistics introduce *stochasticity*, which acts as a form of noise and regularization, helping to prevent overfitting by reducing the model's dependency on specific data patterns.

- **Monitor Training**: Carefully monitor the training process to detect and address potential issues:

  - **Learning Rate Divergence**: Ensure the learning rate is neither too high (causing divergence) nor too low (leading to slow convergence).
  - **Overfitting**: Use validation metrics to identify overfitting early and apply techniques like early stopping or dropout as needed.

8 Loss Function

In training a neural network, the choice of a loss function plays a critical role in guiding the optimization process. The loss function quantifies the error between the network's predictions and the true outputs, providing feedback that drives parameter updates during training. Neural networks can be applied to diverse tasks, including classification, regression, and ranking, each requiring different types of loss functions.

8.1 Classification Tasks

For classification tasks, the choice of a loss function directly impacts the ability of the neural network to learn effectively. When the output layer uses the **Softmax activation function**, the **cross-entropy loss** is the most commonly employed loss function. This combination is widely used because Softmax converts raw output scores (logits) into probabilities, and cross-entropy measures the difference between these predicted probabilities and the actual target labels. The goal is to encourage the network to assign higher probabilities to the correct classes while minimizing the probabilities for incorrect ones. The cross-entropy loss function is defined as:

$$J(\mathbf{W}) = \frac{1}{n} \sum_{i=1}^{n} H(\mathbf{T}_i, \mathbf{O}_i)$$

where $\mathbf{T}_i$ represents the true target (one-hot encoded) for the i-th sample, $\mathbf{O}_i$ denotes the predicted probability distribution for the i-th sample as output by the Softmax layer, and n is the total number of training samples. The cross-entropy term $H(\mathbf{T}_i, \mathbf{O}_i)$ for each sample is computed as:

$$H(\mathbf{T}_i, \mathbf{O}_i) = -\sum_{m} T_{i,m} \cdot \log(O_{i,m})$$

where $T_{i,m}$ represents the m-th component of the true target vector for the i-th sample, and $O_{i,m}$ denotes the predicted probability for the m-th class for the i-th sample.

The cross-entropy loss is a widely used metric in classification tasks due to its strong theoretical grounding and effectiveness in guiding model optimization. The intuition behind this loss function lies in its ability to quantify the divergence between the predicted probability distribution and the true labels. It imposes a significant penalty on predictions that assign low probabilities to the correct class, leveraging the logarithmic nature of the loss to amplify errors when the model lacks confidence in its predictions. Conversely, predictions that assign high probabilities to the correct class incur a minimal penalty, thereby incentivizing both accuracy and confidence in the model's outputs. By minimizing the cross-entropy loss during training, the network iteratively adjusts its parameters to better approximate the true label distribution, ultimately improving classification accuracy and enhancing the model's ability to generalize to unseen data.

The Softmax function ensures that the network's outputs form a valid probability distribution over the classes:

$$O_{i,m} = \frac{\exp(z_{i,m})}{\sum_k \exp(z_{i,k})}$$

where $z_{i,m}$ is the raw logit for the m-th class. Softmax normalizes these logits, ensuring each output is between 0 and 1, and the total sum of probabilities equals 1.

Cross-entropy complements Softmax by quantifying how well the predicted probabilities match the true target labels. Together, they form a powerful framework for multi-class classification tasks, as Softmax provides normalized probabilities and cross-entropy evaluates the error effectively.

While Softmax and cross-entropy are the most commonly used for classification, other activation functions and loss functions may be used in specific cases:

- **Sigmoid Activation with Binary Cross-Entropy**: Used for binary classification tasks or multi-label classification.
- **No Activation with Pre-Softmax Logits**: Some implementations combine Softmax and cross-entropy into a single operation for numerical stability, allowing direct use of raw logits with cross-entropy.

In most cases, the choice of Softmax with cross-entropy remains the standard for multi-class classification due to its mathematical simplicity and effectiveness in optimizing probabilistic outputs.

8.2 *Regression Tasks*

Neural networks can also be used for regression tasks, where the output layer produces continuous values rather than probabilities. In such cases, the activation function in the output layer is typically linear (i.e., no activation) to allow unrestricted numeric outputs. Popular loss functions for regression include:

- **Mean Squared Error (MSE)**: Measures the average squared difference between the predicted and actual values, heavily penalizing large errors:

$$J = \frac{1}{n} \sum_{i=1}^{n} (y_i - \hat{y}_i)^2$$

- **Mean Absolute Error (MAE)**: Measures the average absolute difference between the predicted and actual values, making it more robust to outliers:

$$J = \frac{1}{n}\sum_{i=1}^{n} |y_i - \hat{y}_i|$$

- **Huber Loss**: Combines the advantages of MSE and MAE by being quadratic for small errors and linear for large errors:

$$J = \frac{1}{n}\sum_{i=1}^{n} \begin{cases} \frac{1}{2}(y_i - \hat{y}_i)^2 & \text{if } |y_i - \hat{y}_i| \leq \delta, \\ \delta|y_i - \hat{y}_i| - \frac{1}{2}\delta^2 & \text{if } |y_i - \hat{y}_i| > \delta. \end{cases}$$

Intuition: In regression tasks, the goal is to minimize the discrepancy between the predicted continuous values and the actual targets, and these loss functions ensure the model learns the underlying patterns in the data.

8.3 *Ranking Tasks*

In ranking tasks, the goal is to learn a function that places items in a specific order based on their relevance or similarity. Neural networks used for ranking tasks often employ loss functions like:

- **Contrastive Loss**: Encourages the network to bring embeddings of similar pairs closer together and push dissimilar pairs farther apart. It is commonly used with Siamese networks:

$$J = \frac{1}{n}\sum_{i=1}^{n} \left[y_i d^2 + (1 - y_i)\max(0, m - d)^2\right]$$

  where d is the Euclidean distance between embeddings, y_i indicates similarity, and m is the margin.
- **Triplet Loss**: Optimizes embeddings by minimizing the distance between an anchor and a positive sample while maximizing the distance to a negative sample:

$$J = \max(0, \|f(x_a) - f(x_p)\|_2^2 - \|f(x_a) - f(x_n)\|_2^2 + \alpha)$$

  Here, x_a, x_p, and x_n represent the anchor, positive, and negative samples, and α is the margin.

  Intuition: Loss functions in ranking tasks aim to create meaningful embeddings where similar items are closer together in the learned space, and dissimilar items are farther apart. These embeddings are crucial for applications, such as recommendation systems and information retrieval.

9 Self-supervised Learning (SSL)

Self-Supervised Learning (SSL) is a paradigm in machine learning wherein a model learns from data by generating its own supervisory signals, eliminating the need for externally provided labels. This approach leverages inherent structures, patterns, or transformations within the data to formulate predictive tasks, enabling the model to acquire meaningful representations. By constructing pseudo-labels from the input data itself, SSL facilitates the learning of rich feature representations that can be effectively transferred to downstream tasks, reducing dependence on large annotated datasets while maintaining competitive performance across various domains. This paradigm is conventionally structured into two fundamental stages as explained below:

- **Pretext Task (Pre-training)** :The model learns meaningful representations by solving a self-generated task using pseudo-labels derived from the data itself.
- **Downstream Task (Fine-tuning)**: The pre-trained model is adapted and refined on a specific labeled dataset to improve performance on real-world tasks.

SSL architectures can be broadly categorized into two major families: **generative architectures** and **joint embedding architectures (JEAs)**. Both classes can be trained using either contrastive or non-contrastive learning objectives.

- **Generative Architectures**: These models aim to reconstruct or predict parts of the input data, with the learning objective typically involving minimizing a reconstruction loss (e.g., mean squared error). The architecture often relies on latent variables to encode the input, which are then regularized to prevent trivial solutions. Popular techniques include Principal Component Analysis (PCA), sparse autoencoders, and *variational autoencoders (VAEs)*, where the latent space is regularized via probabilistic constraints. More advanced variants like *vector quantized VAEs (VQ-VAEs)* use discretized latent variables to enhance generative capacity and expressiveness.
- **Joint Embedding Architectures (JEAs)**: These architectures process different augmented views of an input through encoders to produce embeddings that are mutually predictable and semantically meaningful. They aim to learn view-invariant representations while avoiding representational collapse, a situation where all embeddings converge to a constant vector. This is addressed through either architectural asymmetry or regularization. Examples include Siamese networks with shared weights, as well as non-identical encoder approaches. JEAs are designed to filter out noise and irrelevant features, preserving the core informative content in the representation space.

Training Objectives. These architectures can be trained using the following strategies:

- **Contrastive Methods**: These methods learn by distinguishing between similar (positive) and dissimilar (negative) pairs. For generative models, the goal is to

assign low reconstruction error to real data samples and higher error to unlikely reconstructions. In JEAs, contrastive losses ensure embeddings of similar views are pulled closer, while dissimilar ones are pushed apart. Popular contrastive losses include the contrastive loss and triplet loss functions.
- **Non-contrastive Methods**: These approaches do not explicitly rely on negative samples. Instead, they prevent collapse by constraining the model's capacity or regularizing the latent space. For generative models, this may involve limiting the information content in the latent representation. For JEAs, maximizing the entropy or variance of the learned representations can serve as an implicit regularizer.

9.1 Contrastive Loss

The contrastive loss function aims to minimize the distance between embeddings of similar pairs (positive pairs) and maximize the distance between embeddings of dissimilar pairs (negative pairs). The contrastive loss is formulated as follows:

$$L = \frac{1}{2}\left(y \cdot D^2 + (1 - y) \cdot \max(0, m - D)^2\right)$$

where y is a binary label indicating whether the pair is similar ($y = 1$) or dissimilar ($y = 0$), D represents the distance (typically Euclidean distance) between the corresponding embeddings, and m is a margin that defines the minimum distance between dissimilar pairs to ensure they are sufficiently separated.

Triplet loss extends the concept of contrastive loss by considering triplets of data points: an anchor, a positive sample (same class as the anchor), and a negative sample (different class from the anchor). The goal is to ensure that the anchor is closer to the positive sample than to the negative sample by a specified margin. The triplet loss function is given by:

$$L = \max(d(a, p) - d(a, n) + \alpha, 0)$$

where $d(a, p)$ is the distance between the anchor a and the positive sample p, $d(a, n)$ is the distance between the anchor a and the negative sample n, and α is the margin that ensures the anchor is closer to the positive sample than to the negative sample.

9.2 Self-predictive Learning

Self-predictive learning, also referred to as **autoassociative self-supervised learning**, is a subcategory of self-supervised learning in which models are trained to infer missing portions of their input data based on the available information. This paradigm is inherently generative rather than discriminative, as it prioritizes capturing the underlying structure of data rather than categorizing distinct classes. By leveraging patterns within the data itself, self-predictive learning enables models to develop meaningful representations without the need for explicit labels. Several key methodologies are employed within this framework:

- **Autoencoders**: These neural architectures learn to map input data to a latent representation and subsequently reconstruct the original input, thereby capturing essential features. Variants such as *denoising autoencoders* introduce noise to the input to enhance robustness, while *variational autoencoders (VAEs)* extend this approach by incorporating probabilistic modeling, enabling generative capabilities.
- **Autoregressive Models**: These models predict future data points based on prior observations, making them particularly well-suited for sequential data analysis. *Autoregressive architectures*, including those used in the **GPT, LLaMa, and Claude families of large language models (LLMs)**, have demonstrated remarkable efficacy in tasks such as text generation, language modeling, and sequence prediction.
- **Masked Prediction**: This approach involves deliberately concealing segments of an input sample and training the model to reconstruct or predict the missing components. Notable implementations of this technique include **BERT (Bidirectional Encoder Representations from Transformers)** for masked language modeling and **Masked Autoencoders (MAE)** for visual representation learning.

10 Semi-supervised Learning (SSL)

Semi-Supervised Learning (SSL), also referred to as **weak supervision**, is a machine learning paradigm that combines a small amount of labeled data with a significantly larger pool of unlabeled data to improve training efficiency. SSL lies between supervised learning, which relies exclusively on labeled datasets, and unsupervised learning, which operates solely on unlabeled data. This approach has become increasingly relevant with the rise of LLMs, which demand vast training corpora. By leveraging a limited number of accurately labeled examples, SSL enables models to infer patterns from unlabeled data, thereby enhancing generalization while reducing reliance on expensive manual annotation. Several techniques have been developed to implement semi-supervised learning, as discussed next.

10.1 Foundational Assumptions in SSL

SSL relies on several core assumptions about the structure of the data to enable learning from unlabeled inputs:

- **Smoothness Assumption**: If two inputs x_1 and x_2 are close in a high-density region of the data space, their corresponding outputs y_1 and y_2 should also be similar. This implies that small changes to the input should not drastically change the output.
- **Cluster Assumption**: Data points that belong to the same cluster are likely to share the same label. This supports the idea that decision boundaries should lie in low-density areas, reducing the likelihood of misclassifying samples within the same class.
- **Manifold Assumption**: Although data may exist in a high-dimensional space, it is assumed to lie on or near a much lower-dimensional manifold. SSL methods exploit this lower-dimensional structure to improve learning efficiency.

10.2 Consistency Regularization

This technique is mainly based on two fundamental principles: the **manifold assumption** and the **smoothness assumption**. The manifold assumption states that high-dimensional data often lie on a **lower-dimensional manifold**, meaning that data points close to each other in this space should yield similar predictions. This assumption ensures that **decision boundaries** align with low-density regions, preventing them from cutting through clusters of data. The smoothness assumption further supports this idea by stating that small perturbations to an input should not drastically change the model's predictions. If two inputs are nearly identical, their corresponding outputs should also be close. In summary, the principle behind consistency regularization is that a well-generalized model should produce similar outputs for slightly different versions in the (a) **input space** and (or) **model space** (see Fig. 9) [2]. Mathematically, consistency regularization can be applied to **input variations** through the objective function [2]:

$$\min_{\theta} \sum_{x \in \mathcal{D}} d(p(y|x;\theta), \hat{p}(y|\hat{x};\theta)),$$

where $p(y|x;\theta)$ represents the model's output given an input x, while $\hat{p}(y|\hat{x};\theta)$ represents the output when a transformed version of the input, $\hat{x}$, is used. The transformed input $\hat{x} = q_x(x;\epsilon)$ is generated through a function $q_x(\cdot,\epsilon)$, where ϵ represents added noise via data augmentation and stochastic perturbation. The discrepancy function $d(\cdot,\cdot)$ quantifies the difference between the two outputs, commonly measured using **Kullback-Leibler (KL) divergence**, **cross-entropy loss**, or **mean squared error**

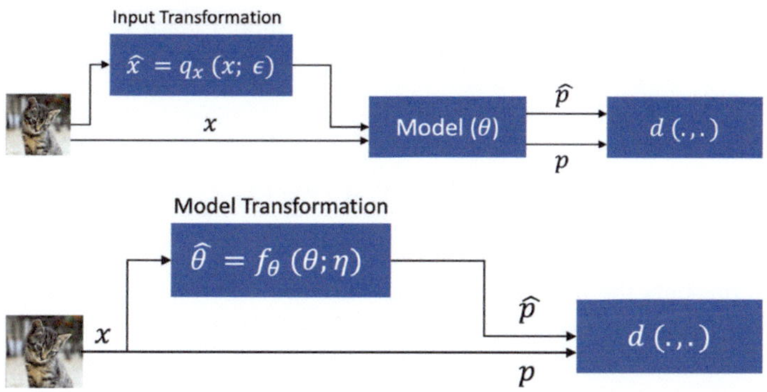

Fig. 9 Illustration of consistency regularization methods: **a** Consistency regularization under input variations, where input transformations or perturbations enforce consistency, and **b** Consistency regularization under model variations, where model parameters are perturbed to ensure stability [2]

(MSE). The goal is to minimize this discrepancy, ensuring that the model's predictions remain stable even when the input undergoes minor modifications. Consistency can also be enforced in the **model space**, where variations in the model parameters should not lead to drastically different predictions. This is expressed as [2]:

$$\min_{\theta} \sum_{x \in \mathcal{D}} d(p(y|x;\theta), \hat{p}(y|x;\hat{\theta})),$$

where $\hat{\theta} = f_\theta(\theta;\eta)$ represents a **perturbed** version of the model parameters θ, obtained through a transformation function $f_\theta(\cdot;\eta)$. Here, η accounts for randomness introduced through methods such as stochastic weight perturbation, or ensembling techniques. The function $d(\cdot,\cdot)$ again measures the discrepancy between outputs from the original and perturbed models.

10.3 Entropy Minimization

Entropy minimization is a technique in semi-supervised learning that encourages the model to make **confident predictions** by ensuring that the **class decision boundary** lies in low-density regions of the data distribution. This approach aligns with the **cluster assumption** and **manifold assumption**, which suggest that samples forming a cluster or lying on the same manifold are likely to belong to the same class [2]. By minimizing entropy, the model pushes decision boundaries away from regions of high data density where different classes overlap, leading to more reliable classification of unlabeled samples. In other words, Entropy Minimization improves model performance by encouraging the model to make confident predictions on unlabeled data. The key idea is to minimize the uncertainty (or entropy) of the model's predictions on unlabeled examples, which helps the model learn more discriminative features and improve generalization. The entropy function for a model's predictions is defined as [2]:

$$\min_{\theta} \sum_{x \in D} \left(-\sum_{j=1}^{K} p(y_j|x;\theta) \log p(y_j|x;\theta)\right) \tag{1}$$

where K represents the number of classes, and $p(y_j|x;\theta)$ denotes the probability of assigning sample x to class y_j. Lower entropy indicates greater confidence in predictions, ensuring that each unlabeled sample is strongly assigned to the most probable class.

Entropy minimization can be applied either in the **prediction space** or the **feature space**. In the prediction space, the model directly assigns class labels based on its output probabilities, enforcing confident classifications. In the feature space, class assignments are determined based on **proximity to class prototypes**, such as cluster centers, ensuring that similar data points are grouped together. By enforcing entropy minimization, the model effectively leverages **unlabeled data** to refine decision boundaries and improve performance in semi-supervised learning.

10.4 Proxy-Label Methods

Proxy-label methods generate labels for unlabeled data without supervision, providing additional training signals. These pseudo-labels, though potentially noisy, enable iterative refinement of model performance. The two principal strategies in this class are self-training and distillation.

10.4.1 Self-training

Self-training refers to an iterative process where a model trained on labeled data is used to generate **proxy labels** (or *pseudo-labels*) for unlabeled data. These proxy labels are assumed to be reliable when the model is confident in its predictions and are then used as additional training data to further refine the model. While proxy labels are often noisy, they still provide useful supervision signals.

Self-training belongs to a broader class of **proxy-label methods**, which produce labels without external supervision. These approaches include both single-view (self-training) and multi-view (co-training) techniques.

Self-training is a widely used approach in SSL that enables models to exploit unlabeled data by iteratively generating pseudo-labels for instances with high-confidence predictions. The process begins with the model being trained on a limited set of labeled data. It then predicts labels for the unlabeled samples, assuming that high-confidence predictions are more likely to be correct. These pseudo-labeled examples are subsequently incorporated into the training cycle, allowing the model to refine its performance over time. By progressively learning from both labeled and pseudo-labeled data, self-training enhances the model's ability to generalize, ultimately improving its overall effectiveness.

10.4.2 Distillation

Distillation was initially introduced as a **model compression technique**, where a large and computationally intensive **teacher model** transfers its learned knowledge to a more compact and efficient **student model**, enabling it to achieve similar performance with reduced complexity. The objective function for self-training via distillation is [2]:

$$\min_{\theta} \sum_{x \in D} d(p_s(y|x; \theta_s), z_t)$$

In this equation, θ represents the parameters of the overall model (both teacher and student parameters), θ_s refers to the parameters of the student model, $p_s(y|x; \theta_s)$ represents the student model's prediction, and z_t is the learning target generated by the teacher model. The function $d(\cdot, \cdot)$ measures the similarity between student predictions and teacher-generated targets, often using cross-entropy or Kullback-Leibler divergence. Distillation methods can be further divided into two categories:

In *model distillation*, a pre-trained teacher model generates soft pseudo-labels for the student. The teacher can be formed using a larger network or an ensemble of models. The student is trained iteratively, sometimes replacing the teacher in subsequent training cycles.

In *data distillation*, pseudo-labels are generated not from a separate teacher model but by averaging multiple forward passes of the same model on the same input under different data augmentations. Instead of training multiple networks, this method transforms input data in various ways and ensembles the resulting predictions, ensuring that pseudo-labels are more reliable.

10.5 Multi-view Training

Multi-view training (MVL) utilizes different, often complementary, views of the data—arising from various modalities or transformations. Separate models are trained on each view, and predictions are combined to enhance generalization. A common realization of MVL is co-training.

10.5.1 Co-Training

Co-training leverages multiple classifiers, each trained on a different representation of the same data. These classifiers independently generate pseudo-labels for the unlabeled data based on their respective feature views. Co-training is based on the assumption that the data can be described by multiple independent or conditionally independent views (e.g., different feature sets or modalities). For instance, in image classification, one view could be the RGB channels, and another view could be edge detection features. By exchanging pseudo-labels and learning from one another, the classifiers refine their predictions, enhancing overall model performance. The training objective for deep co-training is formulated as [2]:

$$\min_{\theta} \sum_{x \in D} d(p_1(y|x;\theta_1), z_2) + d(p_2(y|x;\theta_2), z_1) \tag{2}$$

where $p_1(y|x;\theta_1)$ and $p_2(y|x;\theta_2)$ denote the predictions of two independent classifiers, while z_1 and z_2 represent the pseudo-labels generated by the opposite classifier. The function $d(\cdot,\cdot)$ quantifies the discrepancy between predictions and pseudo-labels, commonly using cross-entropy or Jensen-Shannon divergence. This method can be extended beyond two classifiers, where pseudo-labels are assigned only if multiple models reach a consensus with high confidence. Co-training is particularly effective in scenarios where data can be represented through distinct feature spaces, such as different modalities (e.g., textual and visual features) or alternative transformations of the same input.

10.6 Graph-Based Methods

Graph-based methods in SSL leverage the structure of graphs to propagate labels from labeled to unlabeled data points. In these methods, both labeled and unlabeled data points are represented as nodes in a graph, with edges connecting nodes that are similar to each other based on some predefined metric (e.g., Euclidean distance, kernel-based similarity). The labels from the labeled nodes are then propagated to the unlabeled nodes according to these edges, with the assumption that neighboring nodes (i.e., data points that are close in the feature space) are likely to share the same label. Graph-based methods often include a regularization term to ensure that the predicted labels are smooth with respect to the graph structure. This means that if two nodes are connected by an edge with a high weight (indicating they are similar), their predicted labels should also be similar. The regularization term penalizes large differences in predicted labels between connected nodes, encouraging the model to respect the underlying graph structure. This smoothness constraint helps propagate label information effectively from labeled to unlabeled nodes, improving the model's performance on unlabeled data.

11 Convolutional Neural Networks

Convolutional Neural Networks (CNNs) are a foundational architecture in deep learning, particularly suited for tasks such as image and video recognition, classification, and natural language processing. CNNs excel at extracting hierarchical features from structured data, making them indispensable in applications ranging from computer vision to natural language processing, drug discovery, and strategy gaming.

11.1 Architecture of a CNN

CNNs consist of three main components:

1. **Input Layer**: This layer receives raw data, such as grayscale or RGB images, represented as matrices of pixel values.
2. **Hidden Layers**: These layers perform feature extraction and dimensionality reduction. Key hidden layers include:
   - Convolution (CONV) layers that detect patterns such as edges, textures, and shapes using filters.
   - Pooling (POOL) layers that reduce spatial dimensions and prevent overfitting by summarizing feature information.
   - Fully connected (FC) layers that aggregate features and produce predictions.
3. **Output Layer**: This layer generates the final predictions, such as class labels or continuous values.
   - In classification tasks, the output layer typically uses a **softmax activation function** to produce class probabilities.
   - In regression tasks, the output layer may use a **linear activation function** to generate continuous values.

The configuration and depth of these components determine the ability of CNNs to handle diverse tasks (Fig. 10).

11.1.1 Convolution Layers

Convolution layers are the backbone of CNNs, extracting features by applying filters (kernels) to the input data. Each convolution operation computes the dot product between a filter and a local region of the input, producing a feature map. For example, consider a 6×6 grayscale image and a 3×3 filter. Sliding the filter across the image with a stride of 1 produces a 4×4 feature map (Fig. 11). Stacking multiple filters allows the network to learn various features, such as edges, textures, and object shapes.

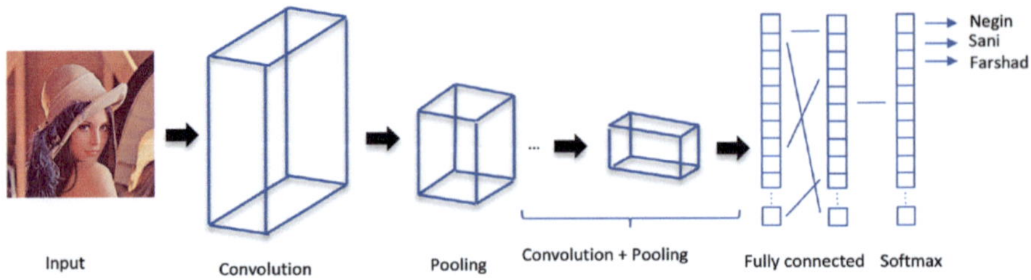

Fig. 10 Convolution neural network architecture [1]

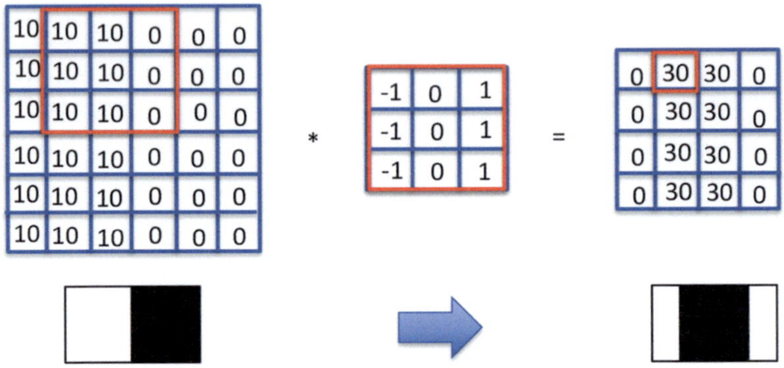

Fig. 11 Convolution computation [1]

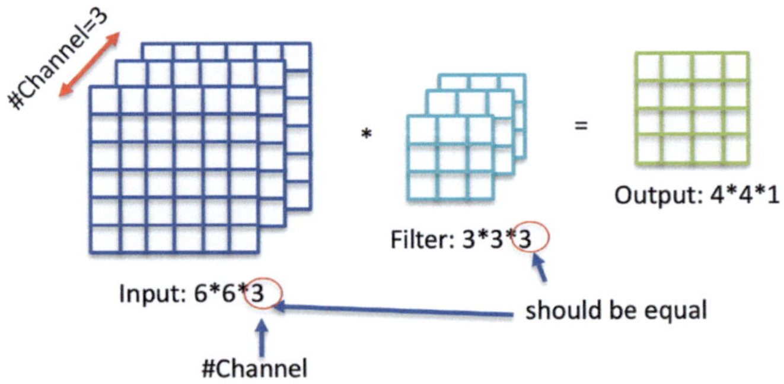

Fig. 12 Apply filter on 3-channel matrix [1]

When handling RGB images (three channels), filters have the same depth as the input ($f \times f \times 3$), and feature maps are generated for each filter. These maps are stacked to form an output volume (see Fig. 12). This approach extends to multi-channel data like satellite imagery or infrared images.

To further control the output size of convolution layers, two parameters are commonly adjusted:

- **Stride**: The stride determines how far the filter moves across the input during convolution. Larger strides reduce the output size (Fig. 13).
- **Padding**: Padding involves adding zeros (or, in some cases, other values) around the boundary of the input before applying convolution. The primary reason is to maintain the spatial dimensions of the output, so that the width and height remain the same after filtering. Equally important, padding ensures that edge features are preserved: without padding, pixels near the boundary participate in fewer convolution operations compared to central pixels (e.g., with a 3×3 kernel, a center pixel is convolved with all 8 neighbors, while an edge pixel may only have 3–5 neighbors). Zero padding addresses this imbalance by allowing the convolution kernel to fully overlap every pixel location, thereby giving equal importance to edge and central features (see Fig. 14).

In real-life applications, multiple filters are required to identify a variety of features. This necessity forms the foundation of CNNs. In such networks, each filter generates its own output, which is subsequently combined (or stacked) to create an output volume (see Fig. 15). In general, the output volume of a convolution layer can be calculated as:

$$\text{Output Volume} = \left(\frac{(n+2p-f)}{s}+1\right) \times \left(\frac{(n+2p-f)}{s}+1\right) \times n_c'$$

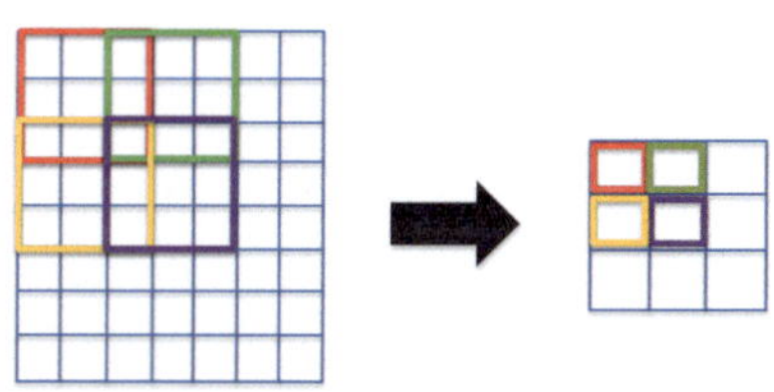

Fig. 13 Stride of 2 pixels with a filter size of 3 [1]

0	0	0	0	0	0	0	0
0	5	2	1	7	-2	1	0
0	3	3	3	5	3	8	0
0	1	1	-1	4	3	2	0
0	5	7	2	2	1	1	0

Fig. 14 Illustration of zero padding. Zero values are added around the input feature map before applying convolution. The main reason for adding zero padding is to preserve the spatial resolution of the output (i.e., maintain the same width and height as the input after convolution), and to ensure that edge pixels receive equal treatment as central pixels during filtering [1]

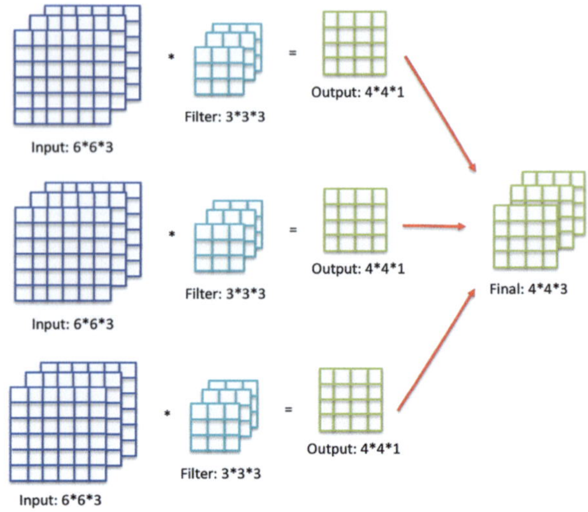

Fig. 15 Stacking outputs from different filters [1]

where n is the input size, p is the padding, f is the filter size, s is the stride, and n_c' is the number of filters.

It should be noted that Convolution and filters are long-standing concepts that have been extensively utilized in image and signal processing for decades. Filters serve various purposes, such as feature extraction and performing operations like edge detection or image smoothing (blurring). Historically, in traditional image and signal processing, filters were primarily hand-engineered, requiring significant domain expertise and manual effort. However, the advent of CNNs and advancements in deep learning have largely eliminated the need for prior knowledge and manual feature design. These modern technologies enable filters and features to be learned automatically from data, resulting in significant improvements in both efficiency and accuracy.

To illustrate this concept, we will focus on edge detection filters. Filters can be designed to detect various types of edges, such as horizontal, vertical, or diagonal lines. For instance, some filters are optimized to detect horizontal lines, while others specialize in detecting vertical or diagonal edges. Fig. 16 demonstrates different types of filters used for edge detection, and Fig. 17 provides an example where horizontal filters are applied to an image, resulting in the identification of horizontal edges.

To understand the mechanics of edge detection, let us consider vertical edge detection as an example. Vertical edge detection is achieved by applying a filter to identify vertical transitions within a 2D image. Fig. 11 illustrates this process. On the left side of the figure, a 6×6 grayscale image is shown, where the left half of the image is white and the right half is gray, creating a vertical boundary (or edge) in the middle. This boundary separates the two distinct regions.

The convolution operation plays a central role in this process. It computes the dot product of the filter and a local region of the image. Importantly, the local region must match the dimensions of the filter. By iteratively applying the filter across the

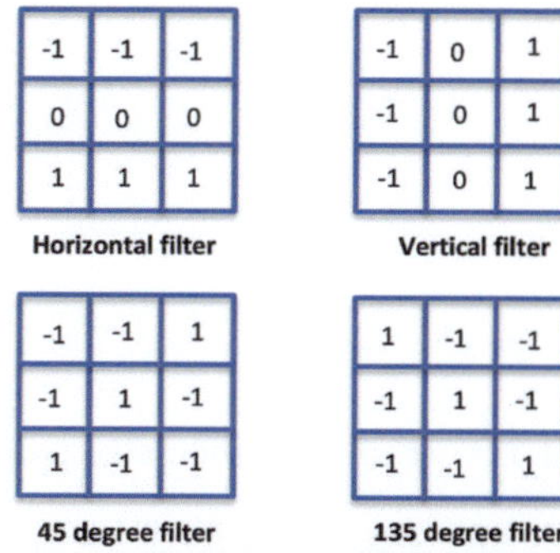

Fig. 16 Different types of filters used for edge detection [1]

Fig. 17 An example of filter application. *Left: Original image. Right: Horizontal edges produced by horizontal filters* [1]

image and shifting it one pixel at a time, the convolution operation generates an output matrix. In this case, the resulting matrix is of size 4×4 (as shown on the right side of Fig. 11). This output matrix effectively highlights the vertical edge in the middle of the image, assigning nonzero values to the edge while zeroing out the surrounding areas. This demonstrates how filters can isolate and emphasize specific features in an image.

11.1.2 Pooling Layers

Pooling layers reduce the spatial dimensions of feature maps, lowering computational costs and preventing overfitting. Two main types of pooling are commonly used:

- Max pooling retains the maximum value within the pooling region, emphasizing the most prominent features (Fig. 18).
- Average pooling computes the average of values in the pooling region, smoothing the feature map.

The output size for a pooling layer is calculated as:

$$\text{Output Size} = \frac{(N - F)}{S} + 1$$

where N is the input size, F is the filter size, and S is the stride.

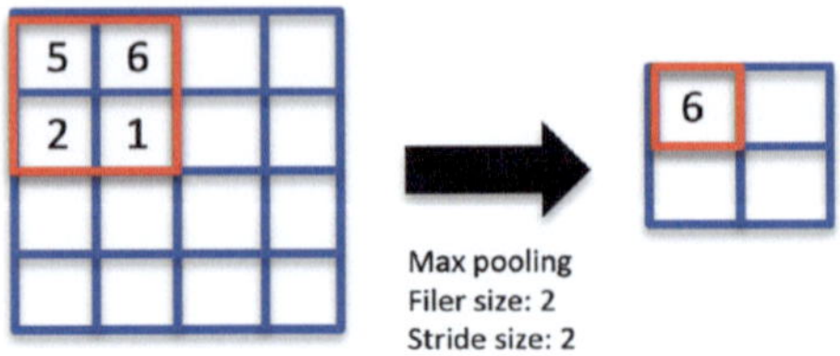

Fig. 18 Max pooling [1]

11.1.3 Fully Connected Layers

Finally, at the final stage of a CNN, fully connected layers are positioned to serve as a crucial component for synthesizing and interpreting the features extracted by earlier convolutional and pooling layers. These layers create a dense connectivity pattern, where every neuron in one layer is linked to every neuron in the subsequent layer, facilitating a complete integration of learned features. Fully connected layers translate the extracted feature representations into meaningful outputs, such as class probabilities for classification tasks or continuous values for regression tasks, ultimately delivering the final predictions of the model.

11.2 Comparison of 1D CNNs and 2D CNNs

1D CNNs are particularly effective for sequential data, where the relationships between data points are best captured along a single dimension. These networks are commonly used in tasks involving time-series data or other sequential inputs. Below are the key components and applications of 1D CNNs:

- **Input**: Sequential data, such as time-series or text.
- **Filters**: Slide along a single dimension to capture temporal patterns.
- **Applications**: Time-series forecasting, speech recognition, and text classification.

In contrast, 2D CNNs are designed to process grid-like data, such as images, where spatial relationships in both dimensions need to be captured. These networks excel at detecting patterns in 2D spaces, making them ideal for a wide range of vision tasks. Below are the key components and applications of 2D CNNs:

- **Input**: Grid-like data, such as images.
- **Filters**: Slide across two dimensions to capture spatial relationships.
- **Applications**: Image classification, object detection, and medical imaging.

The main differences between 1D and 2D CNNs lie in their input types, filter behaviors, and the types of tasks they are suited for. These differences shape how the networks are applied in real-world problems.

- **Data Type**: 1D CNNs handle sequential data, while 2D CNNs process images or grid-like inputs.

- **Filter Behavior**: Filters in 1D CNNs slide along one dimension; in 2D CNNs, they slide across two dimensions.
- **Applications**: 1D CNNs are suited for temporal or sequential tasks, whereas 2D CNNs excel at spatial pattern recognition.

While 1D CNNs are commonly used in NLP and signal analysis, 2D CNNs dominate computer vision tasks such as image classification and object detection.

12 Graph Neural Networks (GNN)

Graph Neural Networks (GNNs) are a type of neural network specifically designed to operate on graph-structured data. Unlike traditional neural networks, which work on fixed-grid structures such as images (2D grids) or sequences (1D), GNNs can effectively handle graphs, which are irregular and dynamic in nature. These models excel at capturing relationships and dependencies among entities represented in graph structures.

A graph $G = (V, E)$ is defined by:

- **Nodes** (V): Represent entities in the data (e.g., *users*, *items*, or *atoms* in a molecule).
- **Edges** (E): Represent relationships between nodes (e.g., *friendships* in a social network, *item-user interactions*, or *bonds* in a molecule).

Graphs are a fundamental representation for numerous real-world problems, capturing complex relationships and structures that other data formats cannot adequately represent. For instance:

- In **social networks**, individuals are represented as nodes, with friendships forming the edges.
- In **molecular chemistry**, atoms serve as nodes, and chemical bonds constitute the edges.
- In **transportation systems**, cities or stations are represented as nodes, and roads or tracks form the connecting edges.

These examples illustrate the versatility and prevalence of graphs across diverse domains, including social sciences, chemistry, and logistics. Beyond their general ubiquity, graphs underpin specific use cases across many fields:

1. **Social Networks**: Predict friendships, recommend connections, or classify users into communities.
2. **Molecular Graphs**: Predict molecular properties (e.g., toxicity, solubility) for drug discovery.
3. **Recommendation Systems**: Suggest products or items based on user-item interaction graphs.

4. **Knowledge Graphs**: Predict missing links or classify entities.
5. **Computer Vision**: Analyze relationships between image regions, such as for object detection or scene understanding.

Despite the importance of graphs, traditional machine learning models struggle to process graph-structured data due to its unique and irregular nature:

- Nodes often have a **variable number of neighbors**, unlike fixed input dimensions in structured data.
- Graphs **lack an inherent ordering** of nodes and edges.
- Graphs exhibit **significant variability in size**, making fixed-dimensional models ineffective.

These limitations necessitate the development of specialized models. GNNs have emerged as a powerful solution, designed specifically to handle the irregularities of graph data. By leveraging the structure of graphs, GNNs excel at learning meaningful insights from graph-based data, making them indispensable for modeling complex systems across various fields.

12.1 The Core Idea of GNNs: Message Passing

At its core, a GNN operates through a process called **message passing**, which involves the following steps:

1. Each node sends "messages" to its neighbors. These messages are typically functions of the node's features and may include learnable parameters.
2. Each node collects and aggregates the messages received from its neighbors.
3. The aggregated information is used to update the node's own features, often through a neural network or an update function.

This process enables nodes to incorporate information from their local neighborhoods, allowing GNNs to learn representations that reflect the graph's structure and the features of neighboring nodes. Below is an example of a simple graph structure:

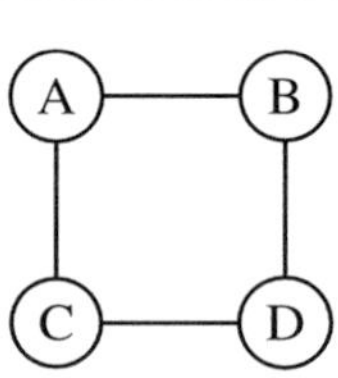

In this graph, each node starts with an initial feature vector, denoted as x_A, x_B, x_C, and x_D. During one iteration (or layer) of the GNN:

- Node A aggregates features from its neighbors B and C.
- Similarly, node B aggregates features from nodes A and D.
- This process repeats for all nodes in the graph, with each node updating its features based on the aggregated information from its neighbors.

The aggregation step is typically followed by a feature update, where the aggregated information is passed through a neural network or update function. This iterative process allows nodes to incorporate information from increasingly distant parts of the graph over multiple layers, enabling GNNs to learn rich and meaningful representations of the graph structure and its features.

12.2 Key Components and Workflow of GNNs

GNNs process information by combining node features, graph structure, and learnable transformations. Below, we discuss the essential components of GNNs.

12.2.1 Node Features

Each node in a graph is associated with a **feature vector** (x_i) that represents its attributes. These attributes vary depending on the application. For instance:

- In a social network, a feature vector may include demographic or behavioral properties such as $x_i = [\text{age, location, activity level}]$.
- In molecular graphs, the feature vector might represent physical properties like $x_i = [\text{atomic number, charge}]$.

Collectively, the features of all nodes are represented as a matrix:

$$X = \begin{bmatrix} x_1 \\ x_2 \\ \vdots \\ x_N \end{bmatrix}$$

where N is the total number of nodes, and each $x_i \in \mathbb{R}^F$, with F being the size of the feature vector.

12.2.2 Graph Structure

The structure of a graph is defined by the connectivity between nodes. This can be represented in two ways:

1. **Edge Index**: A compact representation that explicitly lists the edges in the graph. For example:

$$edge\_index = \begin{bmatrix} 0 & 1 & 1 & 2 \\ 1 & 2 & 3 & 3 \end{bmatrix}$$

   Each column represents a directed edge, such as $0 \rightarrow 1$, $1 \rightarrow 2$, $1 \rightarrow 3$, and $2 \rightarrow 3$. This format is widely used in sparse graph libraries like PyTorch Geometric.
2. **Adjacency Matrix**: A square matrix A of size $N \times N$, where $A[i, j] = 1$ if there is an edge from node i to node j, and $A[i, j] = 0$ otherwise. For example:

$$A = \begin{bmatrix} 0 & 1 & 0 & 0 \\ 0 & 0 & 1 & 1 \\ 0 & 0 & 0 & 1 \\ 0 & 0 & 0 & 0 \end{bmatrix}$$

   This representation is particularly useful for dense graphs.

12.2.3 Message Aggregation

Message aggregation is a core operation in GNNs, allowing nodes to gather contextual information from their neighbors. For a node i, the aggregation step combines feature information from its neighbors $\mathcal{N}(i)$ to compute an updated representation. Different aggregation techniques are available, each suited for specific tasks or graph structures. Some commonly used methods include:

- **Sum Aggregation**: The features of all neighbors are summed:

$$\text{Aggregate}(x_i) = \sum_{j \in \mathcal{N}(i)} x_j$$

  where $\mathcal{N}(i)$ denotes the set of neighbors of node i, and x_j is the feature vector of a neighboring node j. This method is computationally efficient and widely used in many GNN architectures.
- **Mean Aggregation**: The average of the neighbors' features is computed:

$$\text{Aggregate}(x_i) = \frac{1}{|\mathcal{N}(i)|} \sum_{j \in \mathcal{N}(i)} x_j$$

Mean aggregation normalizes the contributions of neighbors, making it less sensitive to variations in node degree.

- **Weighted Aggregation**: Each neighbor's contribution is scaled by a weight w_{ij}:

$$\text{Aggregate}(x_i) = \sum_{j \in \mathcal{N}(i)} w_{ij} x_j$$

Learnable weights, as used in Graph Attention Networks (GATs), allow the model to prioritize more important neighbors.

- **Pooling Aggregation**: Advanced techniques like attention pooling or top-k pooling aggregate only the most relevant features:

$$\text{Aggregate}(x_i) = \sum_{j \in \mathcal{N}(i)} \alpha_{ij} x_j$$

where α_{ij} is an attention coefficient, typically learned during training, that captures the importance of neighbor j to node i.

- **Concatenation**: The node's own features are concatenated with the aggregated neighbor features:

$$\text{Aggregate}(x_i) = \text{Concat}(x_i, \sum_{j \in \mathcal{N}(i)} x_j)$$

This approach retains self-information while incorporating neighborhood context, making it useful in tasks where the node's identity or initial features are critical.

As an example, consider the following graph:

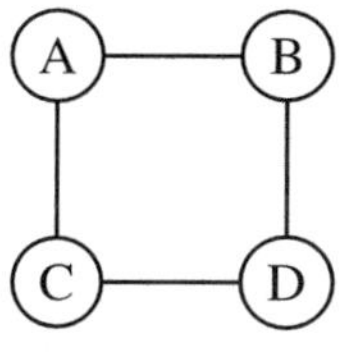

In this example, for node A, the aggregation process combines the features of its neighbors B and C:

$$\text{Aggregate}(x_A) = x_B + x_C$$

The resulting aggregated features, such as $\text{Aggregate}(x_A)$, summarize the local neighborhood information for node A. This aggregated output is passed to the next step, where it undergoes a transformation to learn higher-level representations.

By iteratively combining aggregation and transformation across multiple layers, GNNs propagate information through the graph, enabling nodes to capture both local and global structural patterns. This process forms the foundation for downstream tasks such as node classification, link prediction, and graph classification.

12.2.4 Feature Transformation

Once aggregation is complete, the features are transformed using a learnable **weight matrix** W and an activation function σ, such as ReLU:

$$h_i = \sigma(W \cdot \text{Aggregate}(x_i))$$

This transformation allows the network to identify and prioritize relevant patterns in the graph.

12.2.5 Layer Iterations

In GNNs, stacking multiple layers of aggregation and transformation allows nodes to propagate and incorporate information from increasingly distant parts of the graph. This iterative process is essential for learning rich, meaningful representations that capture both local and global structural patterns. Indeed, with each additional GNN layer, a node gathers information not only from its immediate neighbors but also from nodes that are farther away in the graph. This expansion of the **receptive field**—i.e., the set of nodes in the graph whose features directly or indirectly influence a target node's representation—enables nodes to encode contextual information from larger neighborhoods. The concept is analogous to CNNs, where deeper layers capture broader spatial relationships. In GNNs, stacked layers achieve similar effects by aggregating information across multiple "hops" in the graph:

- **Layer 1**: Each node aggregates features from its immediate neighbors (**1-hop neighbors**).
- **Layer 2**: Each node aggregates features from neighbors of neighbors (**2-hop neighbors**).
- **Layer** k: Each node aggregates features from k-hop neighbors.

As an example, consider the following simple graph:

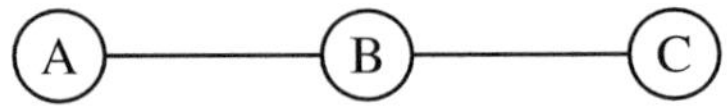

In this example, in **Layer 1**, Node A aggregates features from its 1-hop neighbor B, and in **Layer 2**, Node A further aggregates features from its 2-hop neighbor C through B. By stacking layers, Node A is able to capture long-range dependencies, accessing information from farther parts of the graph.

Stacking layers enables GNNs to capture both **local** and **global** structural patterns. For example:

- **Local Patterns**: Information from immediate neighbors is crucial for tasks such as predicting a node's category in a network.

- **Global Patterns**: Information from distant nodes is essential for understanding overall graph structure, such as in molecule classification or social network analysis.

However, care must be taken when stacking layers, as too many layers can lead to the **over-smoothing problem**, where node representations become indistinguishable. This highlights the importance of balancing depth with task-specific requirements.

For better clarification, it should also be highlighted that in MLPs, layers are structured computational components of the network, consisting of neurons with weights that transform inputs through matrix multiplications and non-linear activation functions. Each layer processes its input sequentially, applying a linear transformation followed by an activation function, with outputs dependent on the previous layer. In contrast, in GNNs, layers represent steps of message passing and aggregation, where nodes update their embeddings by collecting and combining information from their neighbors. In other words, GNN layers iteratively refine node representations based on the graph structure and the features of neighboring nodes.

12.2.6 Task-Specific Output

The final component in a GNN pipeline depends on the specific task for which the model is being applied. The GNN's ability to aggregate and transform features across multiple layers allows it to generate node-level or graph-level representations, which can then be used to perform downstream tasks:

1. **Node Classification**: The objective is to predict a label for each node in the graph. After the final GNN layer, each node retains its updated feature vector, which encodes information aggregated from the node's neighbors across multiple layers. A fully connected layer (or classifier) is applied to each node's feature vector individually to produce node-specific predictions. For example, in a social network, each node representing a user could be classified as an "influencer," a "regular user," or a "content creator."
2. **Graph Classification**: The objective is to predict a single label for the entire graph rather than for individual nodes. Features from all nodes are combined into a single graph-level feature vector using a **pooling operation**. Pooling aggregates the node-level representations into a unified representation that captures the global structure and properties of the graph. Common pooling methods include:

   - **Mean pooling**: Computes the average of all node feature vectors, capturing overall graph characteristics.
   - **Max pooling**: Takes the maximum value across all feature dimensions, focusing on the most prominent features.
   - **Sum pooling**: Sums the feature vectors of all nodes, emphasizing the cumulative contribution of nodes in the graph.

The pooled graph representation is passed through a fully connected layer to produce a graph-level prediction. For example, in molecular property prediction, a graph representing a molecule can be classified as "toxic" or "non-toxic," or assigned other chemical properties.

3. **Edge Prediction (Link Prediction)**: Predict the existence or likelihood of an edge between two nodes. For example, this can be used to predict potential friendships in a social network or interactions between proteins in a biological graph.
4. **Edge Classification**: Assign a label or property to an existing edge in the graph. For example, in a knowledge graph, relationships can be classified as "friend," "colleague," or "family." **Graph Regression**: Predict a continuous value for the entire graph. For example, this can be used to predict financial risk based on a graph's structure.
5. **Subgraph Classification**: Assign a label to a subgraph within a larger graph. For example, this can be used to detect and classify communities in a social network.
6. **Node Regression**: Predict a continuous value for individual nodes. For example, this can be used to predict the temperature of a sensor in a sensor network.
7. **Dynamic Graph Learning**: Model graphs that evolve over time. For example, this can be used to track changes in traffic flow on a road network or predict evolving interactions in a communication network.
8. **Graph Generation**: Generate new graphs with similar properties to the input data. For example, this can be used to create new molecular structures for drug discovery or generate synthetic social networks. **Graph Clustering (Community Detection)**: Group nodes into clusters based on their structural and feature similarities. For example, this can be used to identify groups of users with similar interests in a social network.

13 Recurrent Neural Networks (RNNs)

A Recurrent Neural Network (RNN) is a specialized type of neural networks designed to process **sequential data** by maintaining an internal state that captures dependencies between elements in a sequence. Unlike traditional feedforward neural networks, where inputs and outputs are treated independently, RNNs introduce **recurrence**, allowing information to persist across time steps. Some of the key features that distinguish RNNs from other neural network architectures are outlined below:

- **Sequential Processing**: RNNs process input data one element at a time, updating their state iteratively.
- **Parameter Sharing**: The same weights are used across all time steps, reducing the number of parameters and enabling generalization across sequences.

- **Stateful Design**: By maintaining a hidden state, RNNs can capture temporal dependencies, making them well-suited for tasks where context is critical.

RNNs achieve sequential dependency by sharing parameters across each time step, thereby enabling the model to generalize patterns over varying sequence lengths. In a feedforward network, each layer has distinct weight parameters, whereas in an RNN, the same set of weights is applied recurrently to process inputs across different time steps.

The most commonly used RNN architecture that follows this structure is known as an **Elman network**, introduced by Jeffrey Elman in 1990. This architecture maintains a hidden state that captures past information and updates it at each time step. However, other RNN variants exist, such as the **Jordan network**, which differs by feeding the previous output back into the next hidden layer instead of the hidden state itself. For a given input sequence $\mathbf{x} = [x_1, x_2, \ldots, x_T]$, an RNN computes a hidden state h_t at each time step t, which captures information from previous inputs. The hidden state at each time step is calculated as follows:

$$h_t = f(W_h h_{t-1} + W_x x_t + b_h)$$

where h_t represents the hidden state at time step t, summarizing information from the sequence up to that point. The term h_{t-1} denotes the hidden state from the previous time step, while x_t is the input at the current time step. The weight matrices W_h and W_x are associated with the hidden state and input, respectively, and b_h is the bias vector for the hidden state. The function f is the activation function, typically the hyperbolic tangent (tanh), which helps regulate the magnitude of the hidden state values and prevents extreme values from dominating the learning process. The output at each time step is computed using the following equation:

$$y_t = g(W_y h_t + b_y)$$

where y_t represents the output at time step t, W_y is the weight matrix for the output layer, and b_y is the bias vector for the output layer. The function g is the activation function for the output layer, which is often Softmax for classification tasks or a linear function for regression tasks (Fig. 19).

RNNs serve as a fundamental framework in deep learning for processing sequential data. Over time, they have been extended into more advanced architectures, such

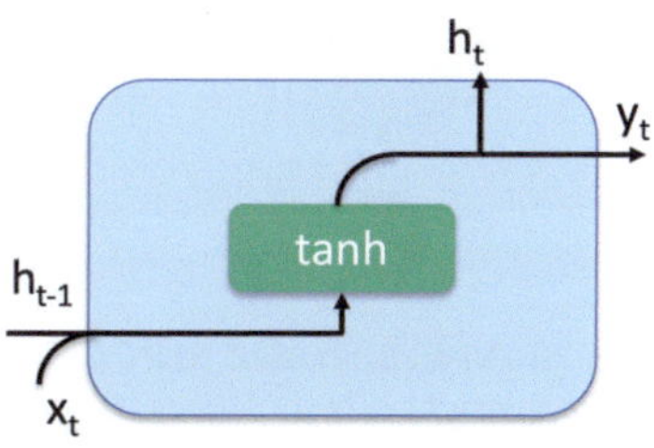

Fig. 19 Basic Recurrent Neural Network (RNN) architecture (f = tanh) based on Elman network

as **Long Short-Term Memory (LSTM)** and **Gated Recurrent Unit (GRU)**, which are specifically designed to mitigate challenges like the vanishing gradient problem, the exploding gradient problem, and short-term memory constraints, thereby enhancing the ability to capture long-range dependencies in sequential tasks.

13.1 RNN Unfolding

RNN **unfolding** refers to the process of expanding the recurrent connections of an RNN over time to represent the **sequential** nature of data processing (see Fig. 20). Unlike traditional neural networks, RNNs incorporate loops that allow information from earlier steps to be propagated forward, enabling the network to **retain** and leverage past information. This mechanism facilitates the model's ability to **comprehend context** and generate more accurate predictions by maintaining dependencies across time steps. Since RNNs store **hidden states** that carry information from previous computations, unfolding provides a structured representation that allows for the visualization and computation of gradients through time using **Backpropagation Through Time (BPTT)**. During training, this unfolded structure enhances the model's capacity to capture **temporal dependencies** by updating weights based on the entire sequence; however, it also introduces challenges such as the **vanishing** or **exploding gradient** problem. The length of the **unrolled** RNN cell is determined by the number of **time steps** in the input sequence, ensuring that each time step has a corresponding instance of the RNN cell while maintaining shared parameters across all steps. By unfolding an RNN, the cyclic nature of its architecture is transformed into a **feedforward-like** structure, facilitating the application of **optimization** techniques while preserving the **time-dependent** relationships inherent in sequential data.

13.2 Stacked RNN

A **Stacked RNN**, or **deep RNN**, consists of multiple RNN layers stacked on top of each other, where the hidden state of one layer serves as the input to the next

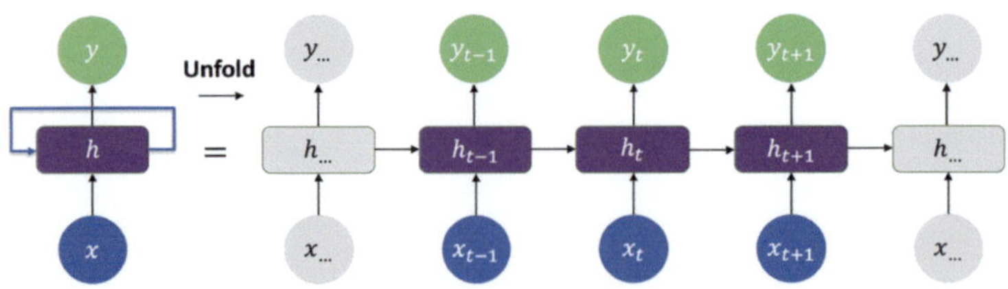

Fig. 20 Visualization of RNN unfolding. The left side represents the compact form of an RNN, while the right side illustrates its unrolled representation over multiple time steps

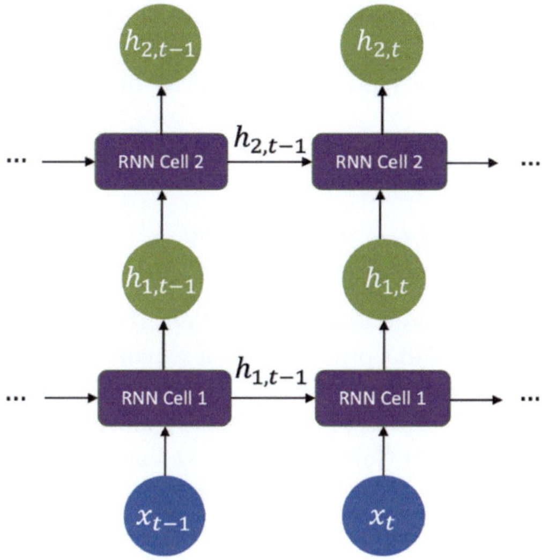

Fig. 21 Illustration of a Stacked RNN. Multiple RNN layers are stacked, where the output (hidden state) of each layer serves as the input to the next. This hierarchical structure enables deeper feature extraction and improved sequence modeling

(see Fig. 21). Each layer l has its own set of parameters (θ_l), allowing it to learn different representations at different levels. Abstractly, a stacked RNN is structured as follows:

- **Layer 1** has parameters $\theta_1 = \{W_h^{(1)}, W_x^{(1)}, b_h^{(1)}\}$ and computes its hidden state as:

$$h_{1,t} = f_{\theta_1}(W_h^{(1)} h_{1,t-1} + W_x^{(1)} x_t + b_h^{(1)})$$

  The output of this layer, $x_{1,t}$, is passed to the next layer.
- **Layer 2** has parameters $\theta_2 = \{W_h^{(2)}, W_x^{(2)}, b_h^{(2)}\}$ and processes the sequence from Layer 1:

$$h_{2,t} = f_{\theta_2}(W_h^{(2)} h_{2,t-1} + W_x^{(2)} h_{1,t} + b_h^{(2)})$$

  The transformed sequence $h_{2,t}$ is then fed to the next layer.
- **Layer n** has parameters $\theta_n = \{W_h^{(n)}, W_x^{(n)}, b_h^{(n)}\}$ and follows the same structure:

$$h_{n,t} = f_{\theta_n}(W_h^{(n)} h_{n,t-1} + W_x^{(n)} h_{n-1,t} + b_h^{(n)})$$

Finally, the **output layer** takes the hidden state from the last RNN layer to compute the final output:

$$y_t = g(W_y h_{n,t} + b_y)$$

where g is the activation function for the output layer, such as **Softmax** for classification or a **linear function** for regression. Note that each layer operates as an independent RNN, with its own parameters θ_ℓ, and its output sequence is used as the input sequence to the layer above. There is no conceptual limit to the depth of a stacked RNN, allowing for **hierarchical feature extraction** and improved sequence modeling.

13.3 Bidirectional RNN

A **Bidirectional Recurrent Neural Network (BiRNN)** is an extension of the standard RNN that processes input sequences in both **forward** and **backward** directions. Unlike a traditional RNN, which captures dependencies only from past to future, a BiRNN consists of two separate recurrent layers—one processing the sequence from **left to right (forward RNN)** and another from **right to left (backward RNN)**. At each time step, the hidden states from both directions are concatenated or combined to form a more comprehensive representation of the input. This bidirectional structure allows the network to leverage both **past and future context**, making it particularly effective for tasks such as speech recognition, machine translation, and named entity recognition. However, since BiRNNs require access to the entire input sequence before making predictions, they are less suitable for real-time applications that require streaming data processing. A bidirectional RNN consists of two RNNs: a forward RNN that processes the input sequence in one direction and a backward RNN that processes the sequence in the opposite direction (see Fig. 22). The forward RNN processes from $t = 0$ to $t = N$ using the recurrence relation

$$h_t^{\rightarrow} = f_\theta(W_h h_{t-1}^{\rightarrow} + W_x x_t + b_h)$$

$$y_t^{\rightarrow} = g(W_y h_t^{\rightarrow} + b_y)$$

where $h_t^{\rightarrow}$ is the hidden state of the forward RNN at time step t, and $y_t^{\rightarrow}$ is the corresponding output. The backward RNN processes the sequence from $t = N$ to $t = 0$ using the recurrence

$$h_t^{\leftarrow} = f_{\theta'}(W'_h h_{t+1}^{\leftarrow} + W'_x x_t + b'_h)$$

$$y_t^{\leftarrow} = g(W'_y h_t^{\leftarrow} + b'_y)$$

where $h_t^{\leftarrow}$ is the hidden state of the backward RNN at time step t, and $y_t^{\leftarrow}$ is the corresponding output. The final output at each time step is obtained by concatenating the outputs from both RNNs,

$$y_t = (y_t^{\rightarrow}, y_t^{\leftarrow})$$

or, alternatively, the hidden states can also be combined as

$$h_t = (h_t^{\rightarrow}, h_t^{\leftarrow})$$

where W_h, W_x, W_y are the weight matrices, and b_h, b_y are the bias vectors for the forward RNN, while W_h', W_x', W_y' and b_h', b_y' are the corresponding parameters for the backward RNN. The functions f_θ and $f_{\theta'}$ represent the activation functions. If the hidden states are combined instead of the outputs, the final output y_t is typically obtained by applying an additional fully connected layer (or another transformation) to the concatenated hidden state:

$$y_t = g(W_y h_t + b_y)$$

where $h_t = (h_t^{\rightarrow}, h_t^{\leftarrow})$ represents the combined hidden state from both the forward and backward RNNs.

13.4 *Long Short-Term Memory (LSTM)*

Long Short-Term Memory (LSTM) networks are an advanced variant of Recurrent Neural Networks (RNNs), specifically designed to address the challenges of vanishing and exploding gradients during training. LSTMs are particularly effective in capturing long-term dependencies in sequential data, making them well-suited for tasks such as machine translation, speech synthesis, speech recognition, and handwriting recognition.

Unlike a traditional RNN, which has a simple repeating `tanh` activation module, an LSTM consists of a more sophisticated repeating module with four interacting layers. These layers enable the network to decide what information to retain, discard,

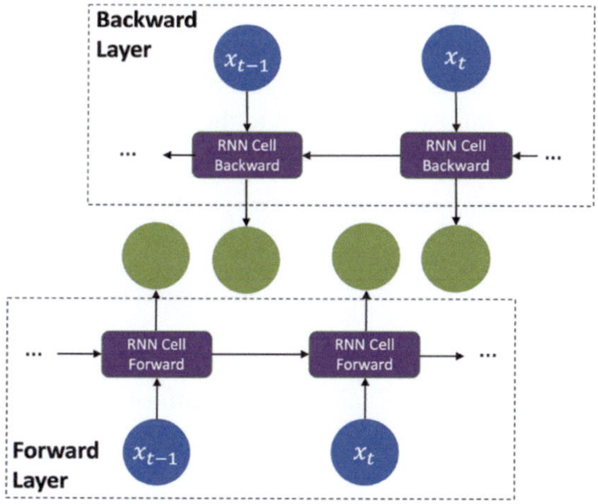

Fig. 22 Illustration of a Bidirectional Recurrent Neural Network (BiRNN). The network consists of two RNN layers: a forward RNN that processes the input sequence from left to right and a backward RNN that processes it from right to left

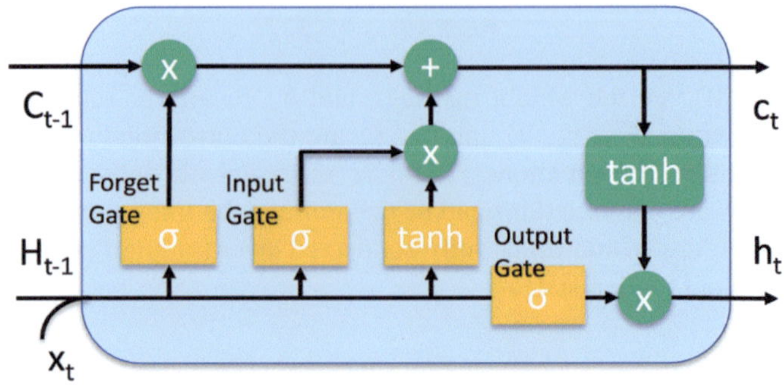

Fig. 23 LSTM cell can process data sequentially and maintain its hidden state through time

or update at each time step. The core components of an LSTM include the **forget gate**, **input gate**, **cell state**, and **output gate** (see Fig. 23).

At each time step t, the LSTM takes in three inputs:

- The current input vector x_t,
- The previous hidden state h_{t-1},
- The previous cell state c_{t-1}.

These inputs interact with the four gates to process sequential information.

The key equations in LSTMs include the following:

1. **Forget Gate** (f_t): The forget gate determines which parts of the previous cell state (c_{t-1}) to retain or forget. It uses a sigmoid activation function to produce values between 0 and 1:

$$f_t = \sigma(W_f[h_{t-1}, x_t] + b_f)$$

where W_f represents the weight matrix for the forget gate, b_f denotes the bias term for the forget gate, and σ is the sigmoid activation function.

2. **Input Gate** (i_t) **and Candidate Cell State** (g_t): The input gate decides which new information to add to the cell state. The candidate cell state (g_t) represents potential updates, regulated by the `tanh` activation function:

$$i_t = \sigma(W_i[h_{t-1}, x_t] + b_i)$$

$$g_t = \tanh(W_g[h_{t-1}, x_t] + b_g)$$

3. **Cell State Update** (c_t): The new cell state is computed by combining the previous cell state (c_{t-1}) and the input information:

$$c_t = f_t \odot c_{t-1} + i_t \odot g_t$$

Here, $\odot$ denotes element-wise multiplication.

4. **Output Gate** (o_t): The output gate determines what information to pass to the next hidden state:
$$o_t = \sigma(W_o[h_{t-1}, x_t] + b_o)$$

5. **Hidden State Update** (h_t): The new hidden state is derived from the updated cell state (c_t) and the output gate:
$$h_t = o_t \odot \tanh(c_t)$$

6. **Prediction** ($\hat{y}_t$): Finally, the LSTM generates an output prediction using the hidden state:
$$\hat{y}_t = g(W_y h_t + b_y)$$
where g is a task-specific activation function (e.g., Softmax for classification).

The intuition behind LSTMs lies in their ability to maintain a balance between remembering and forgetting information. This is achieved through several key components: the **forget gate**, which removes irrelevant or outdated information; the **input gate**, which adds relevant new information; the **cell state**, which serves as a memory pipeline regulated by the gates; and the **output gate**, which determines what information to expose to the hidden state and output. These mechanisms collectively enable LSTMs to effectively capture both short-term and long-term dependencies in sequences.

13.5 Gated Recurrent Units (GRUs): Efficient Sequence Modeling

Gated Recurrent Units (GRUs) are an advanced type of RNN designed for efficient sequence modeling (see Fig. 24). GRUs simplify the architecture of LSTM networks by reducing the number of gates and parameters while maintaining the ability to handle long-term dependencies. This makes GRUs computationally efficient and easier to train, making them particularly useful for tasks with limited resources.

Why GRU? GRUs eliminate the need for a distinct cell state (used in LSTMs) by relying entirely on the hidden state to manage information flow. They use two gates:

- **Update Gate** ($\mathbf{z}_t$): Controls how much of the hidden state is updated with new information.
- **Reset Gate** ($\mathbf{r}_t$): Determines how much of the past information is relevant for the current computation.

This simplified structure enables GRUs to retain essential functionality while reducing computational overhead, making them faster and more resource-efficient than LSTMs.

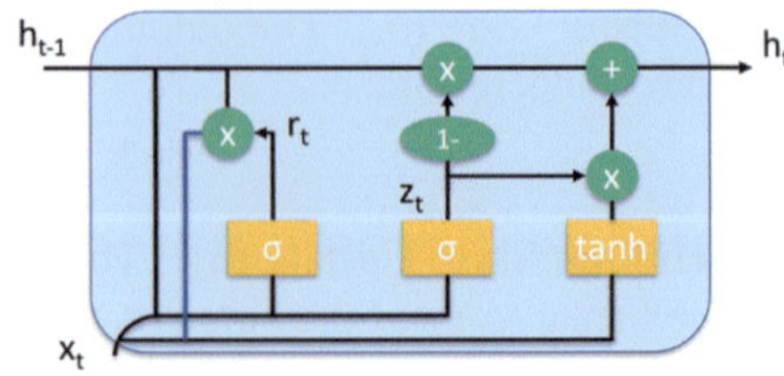

Fig. 24 Gated Recurrent Unit (GRU) architecture

Mathematical Formulation: At each time step t, GRUs take the current input $\mathbf{x}_t$ and the previous hidden state $\mathbf{h}_{t-1}$. The following equations define their operations:

1. Update Gate:

$$\mathbf{z}_t = \sigma(\mathbf{W}_z \mathbf{x}_t + \mathbf{U}_z \mathbf{h}_{t-1} + \mathbf{b}_z)$$

The update gate $\mathbf{z}_t$ determines how much of the previous hidden state $\mathbf{h}_{t-1}$ contributes to the current hidden state.

2. Reset Gate:

$$\mathbf{r}_t = \sigma(\mathbf{W}_r \mathbf{x}_t + \mathbf{U}_r \mathbf{h}_{t-1} + \mathbf{b}_r)$$

The reset gate $\mathbf{r}_t$ controls how much past information is forgotten when computing the current candidate hidden state.

3. Candidate Hidden State:

$$\tilde{\mathbf{h}}_t = \tanh(\mathbf{W}_h \mathbf{x}_t + \mathbf{U}_h (\mathbf{r}_t \odot \mathbf{h}_{t-1}) + \mathbf{b}_h)$$

The candidate hidden state $\tilde{\mathbf{h}}_t$ incorporates the reset-modified hidden state $(\mathbf{r}_t \odot \mathbf{h}_{t-1})$, capturing relevant past information.

4. Final Hidden State:

$$\mathbf{h}_t = \mathbf{z}_t \odot \mathbf{h}_{t-1} + (1 - \mathbf{z}_t) \odot \tilde{\mathbf{h}}_t$$

The final hidden state $\mathbf{h}_t$ is a weighted combination of the previous hidden state $\mathbf{h}_{t-1}$ and the candidate hidden state $\tilde{\mathbf{h}}_t$, controlled by the update gate.

5. Output:

$$\hat{\mathbf{y}}_t = g(\mathbf{W}_y \mathbf{h}_t + \mathbf{b}_y)$$

The output $\hat{\mathbf{y}}_t$ is computed from the current hidden state $\mathbf{h}_t$, often using an activation function g suitable for the task.

13.6 Architectural Configurations of Recurrent Models

Recurrent architectures (e.g., RNNs, GRUs, and LSTMs) are versatile frameworks capable of processing sequences of variable lengths. They can be configured to handle diverse tasks by adjusting the input and output sequence structures. The primary configurations include (see Fig. 25):

1. **One-to-One**: This configuration represents a simple neural network architecture where a single input maps to a single output. Although not inherently recurrent, this architecture serves as a baseline and is widely used for tasks like image classification or simple regression problems.
2. **One-to-Many**: In this configuration, a single input produces a sequence of outputs. The recurrent mechanism allows sequential generation from a fixed initial input. This is commonly used for tasks like:

    - **Image Captioning**: Generating a sequence of words describing an image.
    - **Music Generation**: Creating sequences of musical notes from an initial seed note.

3. **Many-to-One**: This configuration processes a sequence of inputs to produce a single output. It is widely used in tasks where the entire input sequence provides context for the final prediction, such as:

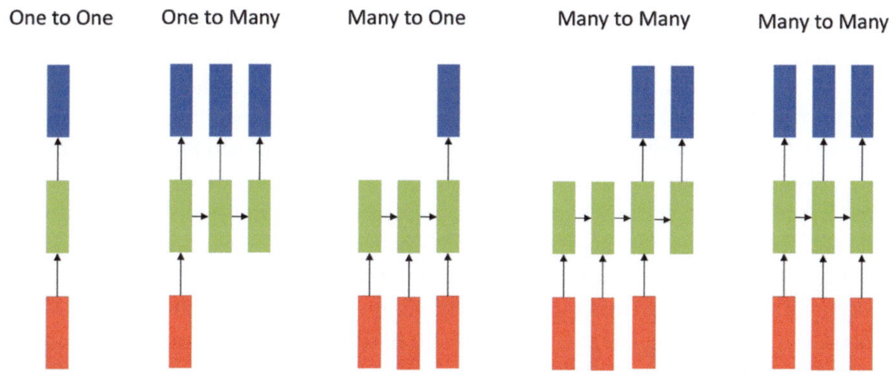

Fig. 25 Illustration of primary recurrent neural network configurations. **One-to-One** corresponds to the standard feedforward setup, such as image classification, where a single input produces a single output. **One-to-Many** generates a sequence from a single input, as in image captioning, where an image is mapped to a descriptive sentence. **Many-to-One** aggregates a sequence into a single output, as in sentiment analysis, where a text sequence is classified into a single sentiment label. **Many-to-Many** produces an output sequence aligned with an input sequence, with applications such as machine translation (translating each word in a source sentence into the target language), video-to-text tasks (mapping a sequence of frames into a textual description), or speech recognition (converting audio frames into a transcript)

- **Sentiment Analysis**: Determining the sentiment (e.g., positive or negative) of a text sequence.
- **Document Classification**: Categorizing a document based on its content.

4. **Many-to-Many**: In this configuration, both the input and output are sequences, making it the most general type of recurrent architecture. Depending on the task, the lengths of input and output sequences can be the same or different:

   - **Equal Length Sequences**: Tasks like video frame prediction or real-time time-series modeling [3], where input and output sequences are of the same length.
   - **Different Length Sequences**: Common in applications such as:

     – **Machine Translation**: Translating a sentence in one language to another, where input and output lengths often differ.
     – **Speech-to-Text**: Converting audio speech signals into text, where input audio frames are mapped to variable-length text outputs.

14 Generative Artificial Intelligence (GenAI)

Generative Artificial Intelligence (Generative AI, or GenAI) is a subfield of AI that focuses on using machine learning models to generate new, synthetic data, such as text, images, videos, and other forms of content. These models learn from large datasets to understand the underlying patterns and structures and use this knowledge to produce novel data. Generative AI has gained prominence due to its ability to produce human-like and contextually accurate outputs, making it transformative across various industries. To understand how generative AI works in practice, consider a simple scenario where we have a dataset containing real-world images of buildings, including skyscrapers, office complexes, and modern architectural designs (see Fig. 26). A generative model is trained on this dataset, allowing it to learn the underlying structures, textures, and perspectives that define architectural forms. Once trained, the model can generate new, realistic buildings that were not present in the original dataset, following a structured process [4]:

- **Training Samples (Left Side)**: The model is first exposed to real-world images of various buildings, enabling it to learn architectural patterns and styles.
- **Generative Model (Middle)**: During the generation process, the model **samples** abstract noise and progressively transforms it into structured building representations.
- **Generated Samples (Right Side)**: The final outputs are realistic building images, produced after the model has fully refined the initial noisy input.

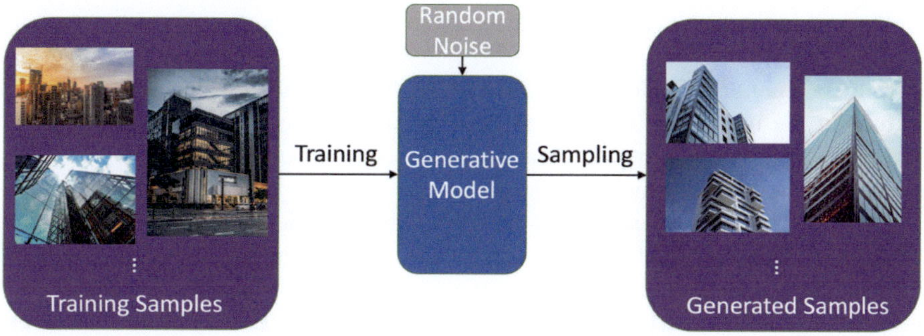

Fig. 26 Illustration of a Generative AI model generating new buildings. The model learns from real-world images (left), applies random noise (middle), and generates new synthetic buildings (right) [4]

Through this structured process, generative AI models can create an endless variety of architectural designs while maintaining realism. Some of the most widely used generative AI techniques include [4]:

- **Generative Adversarial Networks (GANs)**: A pair of neural networks—the *generator* and the *discriminator*—that work in opposition to generate high-quality synthetic data. GANs are widely used for image synthesis, data augmentation, and style transfer.
- **Variational Autoencoders (VAEs)**: These models learn to encode data into a latent space and then decode it to generate new samples, making them a fundamental technique in generative AI. They are widely used for tasks such as image generation, anomaly detection, and data augmentation.
- **Diffusion Models**: These models generate new data by gradually denoising a sample starting from random noise. Examples of popular diffusion models include **Stable Diffusion**, DALL·E 2, and Imagen, which produce highly detailed images from textual descriptions. Many diffusion models also incorporate transformer architectures to enhance their performance.
- **Transformer-Based Generative Models**: Originally developed for text processing, transformers have been adapted for image, video, and audio generation. Notable examples include MaskGIT and ViT-VQGAN, which leverage transformers for image synthesis.
- **Large Language Models (LLMs)**: These transformer-based models—such as GPT, LLaMA, and DeepSeek—are trained to generate human-like text based on natural language prompts. They excel in tasks such as text generation, summarization, translation, and conversational AI.

Generative AI operates across different **modalities** (i.e., the types of data the models are designed to generate). The key modalities include:

- **Text**: Generative AI systems trained on large textual datasets, such as **GPT-3**, **GPT-4**, **GPT-2**, **LLaMA**, and **DeepSeek**, can generate human-like text for a wide range of applications, from generating articles and blogs to creative writing and code generation.

- **Image**: Generative models such as **DALL-E**, **Midjourney**, **Stable Diffusion**, and **Imagen** are trained on vast image datasets and can generate highly realistic or artistic images based on natural language descriptions or other inputs.
- **Video**: Text-to-video models, like **Sora** by OpenAI, generate dynamic, photo-realistic video clips from textual prompts. These models enable the creation of realistic video content for various applications, including media production and entertainment.
- **Audio**: Generative AI systems like **Meta's Voicebox** can synthesize realistic speech or create music by learning from large audio datasets. These models are used for voice cloning, music generation, and text-to-speech applications.
- **3D Modeling**: AI models capable of generating 3D content, such as **AI-driven CAD** systems, produce detailed 3D models from textual descriptions or image inputs, revolutionizing industries like architecture, gaming, and product design.

In the next section of this book chapter, we will dive deeper into the specifics of each of these techniques and modalities, exploring their underlying principles, advancements, and real-world applications.

15 Transformers

The Transformer is a foundational architecture in deep learning, revolutionizing NLP and extending its impact to image processing, video understanding, and other domains. Unlike earlier models such as RNNs and LSTMs, Transformers process sequences in parallel using self-attention mechanisms, which allows them to scale efficiently and handle long-range dependencies. This ability has made them the backbone of state-of-the-art models in numerous domains.

15.1 Overall Architecture

The Transformer architecture, introduced by Vaswani et al. in *Attention Is All You Need* [5] (see Fig. 27), serves as the foundation for numerous state-of-the-art models in NLP and other domains. It is composed of two primary components: the encoder and the decoder. Each of these components consists of multiple identical layers, commonly referred to as **Transformer blocks/layer**, which facilitate efficient sequence processing and generation.

1. **Encoder**: The encoder is responsible for processing the input sequence and converting it into a high-dimensional representation that captures contextual dependencies. It consists of a stack of **Transformer blocks**, each comprising

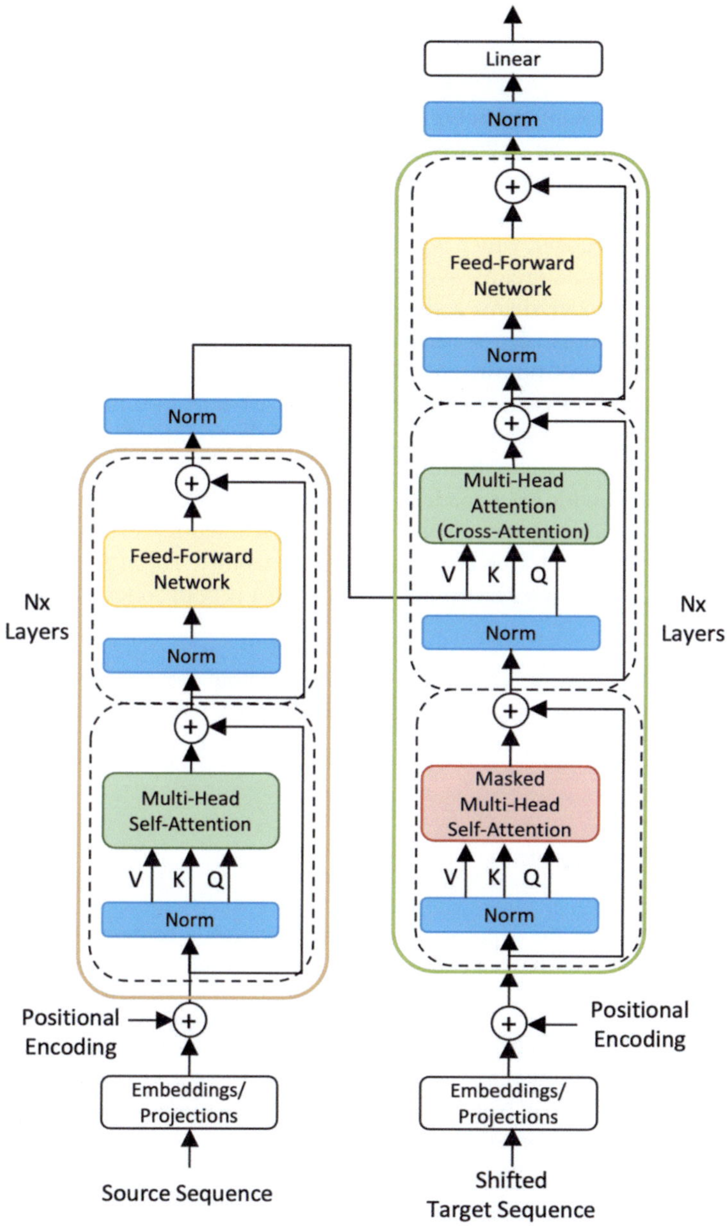

Fig. 27 A standard Transformer architecture, showing on the left an encoder and on the right a decoder [5]

three key sublayers: (i) a multi-head self-attention mechanism, which enables each token to attend to all others within the sequence, thereby capturing long-range dependencies; (ii) a feedforward network (FFN) that applies non-linearity and enhances feature extraction; and (iii) a layer normalization sublayer that stabilizes training and improves convergence. Through repeated application of these layers, the encoder constructs a context-rich representation that serves as the foundation for subsequent processing by the decoder.

2. **Decoder**: The decoder, which is also structured as a stack of Transformer blocks, is responsible for autoregressive sequence generation. While each Transformer block in the decoder shares a similar architecture with those in the encoder, it includes an additional cross-attention sublayer that allows the decoder to incorporate information from the encoder's output representations. Specifically, the decoder consists of (i) a masked multi-head self-attention mechanism, which ensures that each token attends only to previously generated tokens to maintain causality; (ii) a cross-attention mechanism, which aligns the decoder's internal representations with the encoder's output to integrate contextual information; (iii) a feedforward network, which refines the learned representations; and (iv) layer normalization, which enhances training stability. Through iterative decoding, the model generates an output sequence that is both coherent and contextually appropriate.

The modular design of Transformer blocks in both the encoder and decoder ensures scalability, efficiency, and adaptability across various tasks, making the Transformer a fundamental architecture in modern deep learning applications.

15.2 Encoder Structure

The encoder processes the input sequence and generates a context-aware representation by capturing relationships between all tokens in the sequence. Each encoder block consists of:

1. **Multi-Head Self-Attention (MHSA)**: Computes self-attention for each token relative to all other tokens.
2. **Feed-Forward Neural Network (FFN)**: Applies position-wise transformations to enhance the representation.
3. **Layer Normalization**: Stabilizes training by normalizing inputs to each sublayer.

Multiple encoder blocks are stacked to capture increasingly complex dependencies in the input sequence.

15.3 Decoder Structure

The decoder generates the output sequence by leveraging the encoded input representation and previously generated tokens. Each decoder block includes:

1. **Masked Multi-Head Self-Attention**: Ensures that predictions for each position depend only on earlier positions, preserving causal structure in the output sequence. In sequence generation tasks, the model must ensure that predictions for each position depend only on earlier positions in the sequence, preserving the causal structure. This is achieved by applying a mask to the attention mechanism, which prevents the model from accessing information about "future" tokens. The mask blocks attention scores for positions beyond the current one, ensuring that the current token is generated based only on previously seen tokens. For example, if the model is predicting the word at position t, the mask ensures that only tokens from positions 1 to $t - 1$ are considered. The attention scores for positions $t + 1$ onward are set to $-\infty$ (or a very large negative value) before applying the softmax function, effectively making their contribution zero. This ensures that the model generates the sequence one token at a time in a causally consistent manner.
2. **Encoder-Decoder Attention (Cross-Attention)**: Integrates information from the encoder's output to guide the generation process. In other words, it allows the decoder to focus on the most relevant parts of the encoded input sequence while generating each token of the output sequence. In this context, the encoder's output (E) serves as both the keys (K) and values (V) for the attention mechanism, enabling the decoder to incorporate relevant contextual information from the encoded input sequence effectively. The query (Q) is derived from the decoder's current hidden state
3. **Feed-Forward Neural Network (FFN)**: Applies further transformations to enhance the output representation.
4. **Layer Normalization**: Normalizes intermediate outputs for stability.

Similar to the encoder, the decoder consists of multiple stacked blocks.

15.4 Self-attention Mechanism

At the heart of Transformers is the **self-attention mechanism**, which allows each token in an input sequence to selectively focus on the most relevant parts when generating an output. Unlike traditional architectures that process sequences in a fixed or localized manner, self-attention dynamically adjusts the importance of different tokens, effectively capturing both *short- and long-range dependencies*. To achieve this, self-attention computes attention scores between all token pairs, determining how much influence each token should have on the others. This mechanism follows a structured process consisting of the following key steps:

1. **Projection of Input Tokens**: Each token in the sequence is **projected** into three distinct representations: **queries (Q), keys (K), and values (V)**. These representations, learned through projection matrices, form the foundation for computing attention relationships between tokens.
2. **Computing Attention Scores**: The model measures the similarity between each token's query and every other token's key, commonly using the **dot-product operation**. These raw scores determine how much influence each token exerts on others. To ensure stable gradients and effective learning, the computed attention scores are scaled and passed through the **softmax function**, converting them into a probability distribution that dictates how much attention each token allocates to others.
3. **Aggregating Information**: The attention weights obtained from the previous step are applied to the **value vectors**, producing a **weighted sum** that forms each token's final context-aware representation. This allows tokens to dynamically incorporate relevant contextual information from across the sequence.

15.4.1 Projection of Input Tokens

The self-attention mechanism first projects the input sequence into three fundamental representations—queries (Q), keys (K), and values (V)—each serving a distinct role in computing attention (see Fig. 28):

- **Query (Q)**: Represents a token's request for relevant information. Conceptually, it acts as a "question," asking which tokens in the sequence contain useful information for processing the current token.
- **Key (K)**: Represents all tokens in the sequence and serves as a "descriptor" that helps determine how much focus each token should receive when queried.
- **Value (V)**: Represents the actual content of each token that will be aggregated based on the attention scores. The values are the primary source of information used to construct the output representation.

Given an input sequence $X \in \mathbb{R}^{n \times d}$, where n is the number of tokens in the sequence and d is the dimensionality of each token's embedding, the self-attention mechanism first transforms the input into three separate subspaces:

$$Q = XW_q, \quad K = XW_k, \quad V = XW_v \tag{3}$$

where $W_q, W_k, W_v \in \mathbb{R}^{d \times d_k}$ are learnable weight matrices, and d_k is the projected dimensionality of queries and keys. These projections allow the model to learn different aspects of the token representations, enabling more flexible interactions between tokens.

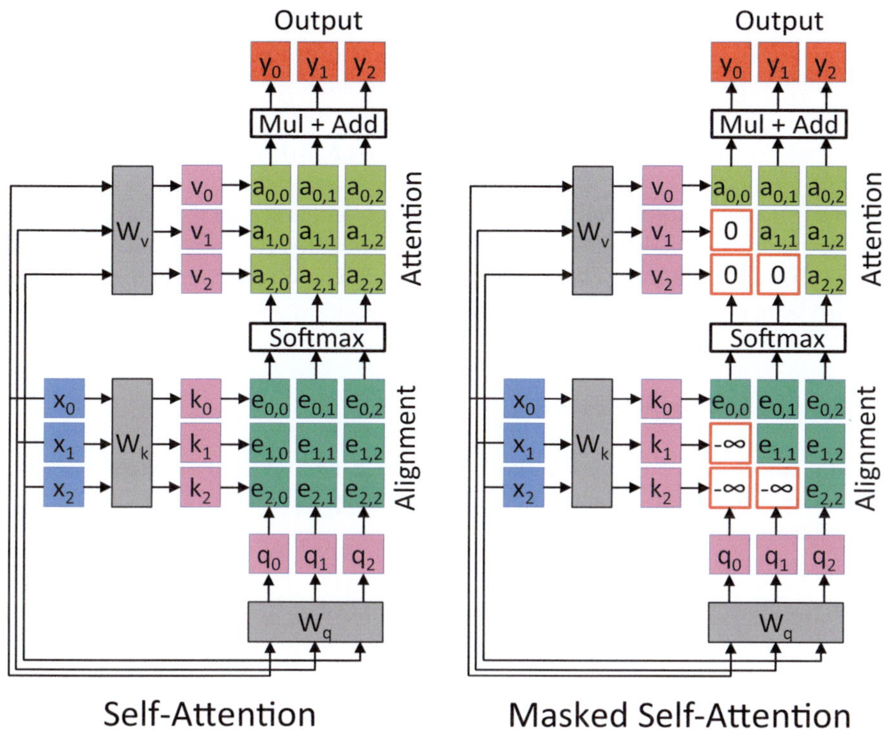

Fig. 28 Self-attention and its masked variant. Each input token x_i is projected into Query (Q), Key (K), and Value (V) vectors using learned weight matrices W_q, W_k, and W_v. The raw dot-product similarity between a query q_i and all keys k_j produces the alignment scores $e_{i,j}$. After Softmax normalization, these become attention weights $a_{i,j}$, which determine the relative importance of each token. The contextualized output y_i is then computed as the weighted sum of all value vectors, e.g., $y_0 = \sum_j a_{0,j} v_j$. In masked self-attention, a triangular mask forces alignment scores for future tokens to $-\infty$, ensuring that each token attends only to itself and preceding tokens, thereby preserving autoregressive causality

15.4.2 Computing Attention Scores

The core idea behind self-attention is that each token *attends to all other tokens* by computing similarity scores between the **query** of one token and the **keys** of all tokens in the sequence. This step determines how much focus each token should place on others. Mathematically, this is computed using the **scaled dot product**:

$$\rho(Q_{i,:}, K_{j,:}) = \frac{Q_{i,:}^{\top} K_{j,:}}{\sqrt{d_k}} \tag{4}$$

This dot product measures how similar the query of token i is to the key of token j, producing an unnormalized attention score. The division by $\sqrt{d_k}$ is a **scaling factor** that prevents large dot-product values from pushing softmax into extreme regions, ensuring stable gradients during training. To transform these raw attention scores into probabilities, a **normalization function** is applied:

$$\beta(Q_{i,:}, K_{j,:}) = \alpha(\rho(Q_{i,:}, K_{j,:})) \tag{5}$$

where $\alpha(z) = \text{softmax}(z)$ normalizes the attention scores so that they sum to 1. This ensures that the model distributes attention proportionally. It is important to note that while **softmax** is the most commonly used normalization function, other methods can also be applied depending on the use case.

15.4.3 Aggregating Information

Once the attention scores are computed, they are used to weight the **value vectors**. Each token's final representation is obtained as a **weighted sum of the value vectors** from all tokens in the sequence:

$$y_{i,:} = \sum_{j=1}^{n} \beta(Q_{i,:}, K_{j,:}) \odot V_{j,:} \tag{6}$$

where $\odot$ represents element-wise multiplication. This step ensures that information from other tokens is selectively incorporated into the current token's representation based on relevance. If a query strongly aligns with a specific key, the corresponding value will have a greater influence on the output. Conversely, if a query-key pair has low similarity, the value will contribute minimally.

15.4.4 Summary of the Self-Attention Mechanism

Summarizing the self-attention mechanism, the complete operation can be expressed as:

$$\text{Attention}(Q, K, V) = \text{softmax}\left(\frac{QK^\top}{\sqrt{d_k}}\right)V \tag{7}$$

This formulation summarizes how the attention mechanism allows each token to dynamically aggregate context from other tokens, leading to a more **context-aware representation**. By enabling each token to attend to the most relevant parts of the sequence, self-attention provides a highly effective way to model dependencies, making it a key component of modern deep learning architectures.

By dynamically adjusting attention weights based on content, self-attention allows for **efficient information flow across the entire sequence**, making it the foundation of state-of-the-art models in natural language processing, computer vision, and beyond.

15.5 Intuition Behind Attention

To illustrate how the attention mechanism works, consider the ambiguity in the word "bank" within the following two sentences:

1. *He sat by the bank of the river.*
2. *She went to the bank to deposit money.*

Traditional embedding methods would assign the same vector representation to "bank" in both sentences, failing to capture its distinct meanings: a riverside location versus a financial institution. Attention mechanisms overcome this limitation by leveraging context-specific cues.

The attention mechanism operates by defining three key components: the query (Q), the keys (K), and the values (V). These components interact to disambiguate the meaning of words within a given context. The process can be described as follows:

- **In the sentence *"He sat by the bank of the river"***:
  - The **query** (Q) corresponds to the word "bank," which seeks relevant contextual information to resolve its meaning.
  - The **keys** (K) represent the surrounding words, such as "river" and "sat," which encapsulate the contextual content.
  - The **values** (V) encode additional semantic information from these surrounding words.
  - Through the attention mechanism, higher weights are assigned to contextually relevant words such as "river." This weighting directs the representation of "bank" to align with its interpretation as a riverside location.
- **In the sentence *"She went to the bank to deposit money"***:
  - The **query** (Q) again corresponds to the word "bank," seeking disambiguation in a different context.
  - The **keys** (K) now include words such as "deposit" and "went to," which provide crucial contextual cues.
  - The **values** (V) convey semantic details of these surrounding words.
  - The attention mechanism assigns higher weights to words such as "deposit," enabling the representation of "bank" to adapt and reflect its meaning as a financial institution.

To better understand the role of attention in Transformers, consider its intuitive purpose: enabling the model to dynamically focus on the most relevant parts of the input sequence for generating each output.

- **Weighted Information Focus**: Attention assigns varying levels of importance to different tokens, ensuring the model emphasizes the most relevant parts of the input sequence while reducing the influence of less critical elements.
- **Contextual Relationship Mapping**: It establishes connections between tokens, allowing the model to capture long-range dependencies and contextual relationships across the entire sequence.
- **Dynamic Relevance Selection**: For each token, attention dynamically selects and integrates information from the sequence, enabling a flexible and context-aware processing mechanism.
- **Parallel Token Processing**: Unlike traditional sequential methods, attention computes relevance scores for all tokens simultaneously, enhancing efficiency and scalability for large sequences.
- **Capturing Semantic Similarity**: Attention leverages similarity (e.g., through dot products) between tokens to determine how much one token contributes to another, making it adept at understanding relationships in complex sequences.

15.6 Multi-head Attention

Transformers extend the capability of single-head attention by employing **Multi-Head Attention**, where multiple attention heads process the sequence in parallel. This mechanism enables the model to focus on different aspects of the input sequence simultaneously, capturing diverse relationships and patterns.

The process of Multi-Head Attention can be summarized in the following steps:

1. **Projection into Subspaces:** For each attention head, the inputs Q, K, and V are projected into smaller subspaces using learnable weight matrices $W_q^{(h)}$, $W_k^{(h)}$, and $W_v^{(h)}$:
$$Q^{(h)} = QW_q^{(h)}, \quad K^{(h)} = KW_k^{(h)}, \quad V^{(h)} = VW_v^{(h)}$$
Here, h indexes the attention head, and d_k represents the dimensionality of the projected vectors $Q^{(h)}$, $K^{(h)}$, and $V^{(h)}$. This dimensionality defines the size of the subspace where attention calculations are performed. If the input Q, K, and V have dimensions $(n \times d)$, where n is the sequence length (number of tokens) and d is the embedding size, the learnable weight matrices $W_q^{(h)}$, $W_k^{(h)}$, $W_v^{(h)}$ are of size $(d \times d_k)$. Consequently, the projected outputs $Q^{(h)}$, $K^{(h)}$, $V^{(h)}$ for each attention head have dimensions $(n \times d_k)$, facilitating efficient computation within the reduced-dimensional subspace.
These projections reduce the dimensionality of the input for each head, allowing multiple heads to compute attention efficiently and focus on different relationships in parallel.
2. **Attention Calculation for Each Head**: Each head computes scaled dot-product attention independently:

$$\text{head}_h = \text{Attention}(Q^{(h)}, K^{(h)}, V^{(h)}) = \text{softmax}\left(\frac{Q^{(h)}(K^{(h)})^\top}{\sqrt{d_k}}\right)V^{(h)}$$

Here, $\sqrt{d_k}$ is the scaling factor, which ensures that the similarity scores from the dot product $Q^{(h)}(K^{(h)})^\top$ remain numerically stable, preventing extreme values that could disrupt the softmax operation. Note that each head focuses on different parts of the input sequence or learns distinct types of relationships.

3. **Concatenation and Projection**: The outputs of all heads are concatenated and linearly transformed using a learnable projection matrix W_O:

$$\text{MultiHead}(Q, K, V) = \text{Concat}(\text{head}_1, \ldots, \text{head}_H)W_O$$

Here, $\text{Concat}(\text{head}_1, \ldots, \text{head}_h)$ has shape $\mathbb{R}^{n\times(h\cdot d_k)}$. The projection matrix W_O reduces this back to $\mathbb{R}^{n\times d}$. This step ensures that the output of the MHA mechanism aligns with the model's embedding dimension d, making it compatible with subsequent layers (e.g., the FFN or the next Transformer block).

Multi-Head Attention provides several advantages:

- **Diverse Perspectives**: Each attention head can focus on different relationships or dependencies, such as syntactic roles, semantic meanings, or positional relationships.
- **Rich Contextual Representation**: By combining the outputs of multiple heads, the model generates more comprehensive representations of each token.
- **Parallel Processing**: Multiple attention heads operate in parallel, improving efficiency and enabling the model to learn complex dependencies.

As an example, consider the sentence: "*The cat chased the mouse because it was hungry*". This sentence demonstrates how different attention heads can capture multiple relationships within a sequence:

- One attention head might focus on the relationship between "cat" and "chased," identifying the subject-action connection.
- Another attention head might resolve the pronoun "it" to refer to "cat," thereby performing coreference resolution.
- A third attention head might track the object "mouse" throughout the sequence, maintaining consistency in its representation.

Each attention head operates independently, contributing to a nuanced understanding of the sentence. By integrating the outputs of these attention heads, the model combines diverse contextual perspectives into a unified and coherent representation.

15.7 Feed-Forward Neural Network (FFN)

The output of the Multi-Head Attention (MHA) mechanism is further refined through a **position-wise Feed-Forward Neural Network (FFN)**. The FFN processes each token's vector representation independently but applies the same transformation to all tokens in parallel. The FFN typically applies two linear transformations with a non-linear activation function in between:

$$\mathrm{FFN}(x) = \phi(xW_1 + b_1)W_2 + b_2 \tag{8}$$

where:

- $x \in \mathbb{R}^{n \times d}$: Input tensor representing the sequence of token embeddings, where n is the sequence length (number of tokens) and d is the embedding dimension.
- $W_1 \in \mathbb{R}^{d \times d_{\mathrm{ffn}}}$: Weight matrix projecting the input to a higher-dimensional space.
- $b_1 \in \mathbb{R}^{d_{\mathrm{ffn}}}$: Bias term for the first linear transformation.
- $W_2 \in \mathbb{R}^{d_{\mathrm{ffn}} \times d}$: Weight matrix projecting the hidden representation back to the original embedding space.
- $b_2 \in \mathbb{R}^{d}$: Bias term for the second linear transformation.
- ϕ: Activation function, typically ReLU, defined as $\phi(z) = \max(0, z)$. ReLU is commonly used in Transformers due to its computational efficiency and ability to mitigate the vanishing gradient problem. However, alternative activation functions such as GELU (Gaussian Error Linear Unit) are sometimes employed in advanced models like BERT and GPT.

The FFN operates independently on each token's vector and ensures that the output matches the input dimensions, maintaining consistency across layers:

- The hidden layer size is typically $d_{\mathrm{ffn}} = 4 \cdot d$, though this can vary based on the model's design.
- The output of the FFN has the same shape as its input, $\mathbb{R}^{n \times d}$, ensuring compatibility with subsequent layers.

15.8 Positional Encoding

Transformers, due to their parallel processing nature, lack an inherent mechanism for capturing the sequential order of tokens in a sentence. To address this, positional encodings are added to the input embeddings, introducing information about token order. The sinusoidal positional encoding, as proposed in the original Transformer architecture, utilizes fixed, non-learnable sine and cosine functions:

$$PE(p, 2i) = \sin\left(\frac{p}{10000^{\frac{2i}{d_{\mathrm{model}}}}}\right), \quad PE(p, 2i+1) = \cos\left(\frac{p}{10000^{\frac{2i}{d_{\mathrm{model}}}}}\right)$$

Here, p denotes the token position, i represents the dimension index, and d_{model} refers to the embedding dimensionality. These encodings provide a deterministic mapping of positions to vectors, enabling the Transformer to incorporate sequence-order information without introducing additional trainable parameters. Additionally, the sinusoidal encoding generalizes well to sequences longer than those encountered during training due to its functional structure.

Although sinusoidal positional encoding has its merits, alternative approaches have been proposed to address its fixed, non-adaptive nature or enhance flexibility in sequence representation:

- **Learnable Positional Embeddings**: These embeddings are trainable vectors associated with each position in the sequence. In contrast to sinusoidal encodings, learnable positional embeddings are optimized during training, allowing the model to adapt positional representations to the specific characteristics of the task and dataset. While this approach offers greater flexibility, it introduces additional trainable parameters and may increase the risk of overfitting.
- **Relative Positional Encoding**: This method represents the relative distances between tokens, rather than their absolute positions, allowing the model to focus on contextually relevant positional relationships. By emphasizing relative rather than absolute order, this technique is particularly effective for tasks where the relationship between tokens is more critical than their fixed positions, such as dependency parsing or long-text generation.
- **Rotary Positional Embeddings (RoPE)**: This approach introduces rotational transformations to the embeddings, enabling a structured representation of relative positional information. RoPE is especially effective in capturing dependencies over long sequences, as it facilitates the modeling of relative positional relationships in a manner well-suited for tasks requiring long-range attention.

16 Large Language Models (LLMs)

Large Language Models (LLMs) are advanced generative AI systems designed to process and generate human-like text by leveraging deep architectures built upon Transformers. LLMs extend the capabilities of Transformers by scaling the number of layers, parameters, and training data, enabling them to excel in diverse language processing tasks such as translation, summarization, and dialogue generation.

16.1 What is Language AI?

Language AI refers to a subfield of AI focused on developing systems capable of understanding, processing, and generating human language. It encompasses tasks

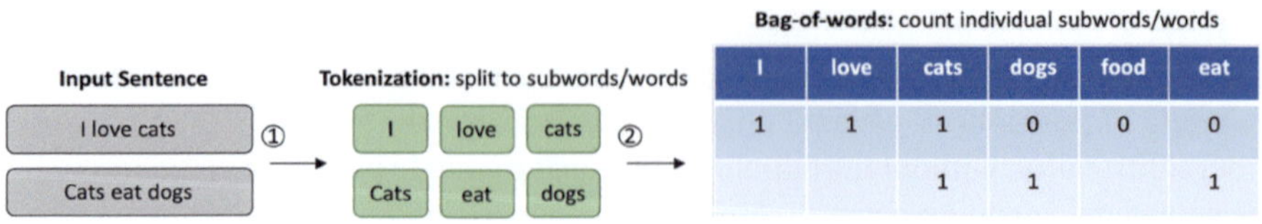

Fig. 29 Bag-of-words is created by counting individual subwords/words

such as speech recognition, machine translation, sentiment analysis, and text generation. Language AI often overlaps with NLP, leveraging statistical and machine learning methods to solve complex linguistic problems. Language AI has evolved through several stages, addressing the unique challenges of processing unstructured human language. Text, inherently unstructured, needs to be represented in forms that computers can process. Key milestones include [6]:

- **Bag-of-Words Representation** converts text into numerical data by tokenizing sentences into subwords/words and creating a vocabulary of all unique tokens in the dataset. Each sentence is represented as a vector where each dimension corresponds to a word in the vocabulary, and the value is the count of that word in the sentence. For example, the sentences "*I love cats*" and "*I love dogs*" would result in the vocabulary `["I", "love", "cats", "dogs"]` and vectors `[1, 1, 1, 0]` and `[1, 1, 0, 1]`, respectively (see Fig. 29). While simple and effective for basic tasks, Bag-of-Words ignores word order and semantic relationships, treating sentences as unordered collections of words. It also results in high-dimensional, sparse vectors, making it computationally inefficient for large vocabularies.
- **n-gram Models** extend Bag-of-Words by considering contiguous sequences of *n* tokens, thereby encoding limited word order information. For example, the sentence "*I love cats*" produces unigrams `["I", "love", "cats"]` and bigrams `["I love", "love cats"]`. By capturing local context, n-grams address one weakness of Bag-of-Words, which ignores word order entirely. However, this comes at the cost of rapidly increasing vocabulary size and sparsity: as *n* grows, the number of possible combinations explodes, leading to high-dimensional, sparse vectors. While effective for small-scale tasks and classic statistical language models, n-gram approaches become inefficient for large vocabularies and do not capture long-range semantic dependencies.
- **Dense Vector Embeddings** improve upon Bag-of-Words by capturing semantic meaning and relationships between words in dense, continuous vector spaces. Unlike Bag-of-Words, which only counts word occurrences, Dense Vector Embeddings represent words as vectors, where words with similar meanings are positioned closer together in the vector space (see Fig. 30). Models like **word2vec** learn embeddings by analyzing the contexts in which words appear. For example, word2vec may assign similar vectors to "*king*" and "*queen*" because they occur in similar contexts. A notable property of dense embeddings is their ability to capture relationships, such as `vec("king") - vec("man") + vec("woman")`

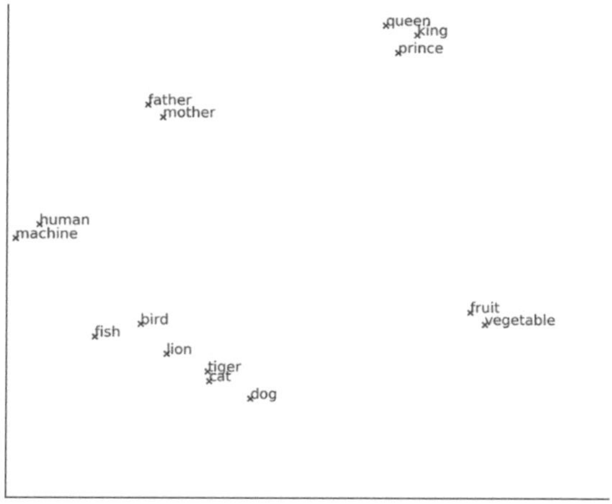

Fig. 30 A 2D scatter plot illustrating the semantic relationships between various words in the vector space. Words with similar meanings, such as "king" and "queen," "dog" and "cat," are positioned closer together

$\approx$ `vec("queen")`. This illustrates how word embeddings capture not just the meanings of individual words, but also the relationships between them in the vector space. These embeddings are trained using large corpora, employing techniques like Skip-gram and CBOW (Continuous Bag of Words). Below are the details and examples of these two techniques:

- **Skip-gram**: The Skip-gram model works by predicting surrounding words from a given target word. Essentially, for a word in a sentence, the model tries to predict the context in which it appears. The Skip-gram model works particularly well for smaller datasets and can capture rare words in the corpus better. As an example, consider the sentence "The king ruled the kingdom". If the target word is "*king*", the model tries to predict context words such as "*the*", "*ruled*", and "*kingdom*" around it. The Skip-gram model will generate predictions for each of these surrounding words based on the target word "*king*".
- **CBOW (Continuous Bag of Words)**: Unlike Skip-gram, CBOW works by predicting a target word from a given context (a set of surrounding words). In this model, instead of focusing on a target word and predicting the context, the model focuses on a surrounding window of words and attempts to predict the word that most likely fits within the window. For instance, consider the same sentence "The king ruled the kingdom". In the CBOW model, the context words (such as "*the*", "*ruled*", "*kingdom*") are used to predict the target word "*king*". So, for a given set of context words, CBOW will output the target word based on the sum or average of the surrounding context.

The word2vec model uses a shallow neural network architecture to learn word embeddings. The model is trained using either the Skip-gram or CBOW technique. The key components of this neural network are (see Fig. 31):

- **Input Layer**: Each word in the vocabulary is represented as a one-hot encoded vector, where the length of the vector corresponds to the size of the vocabulary.

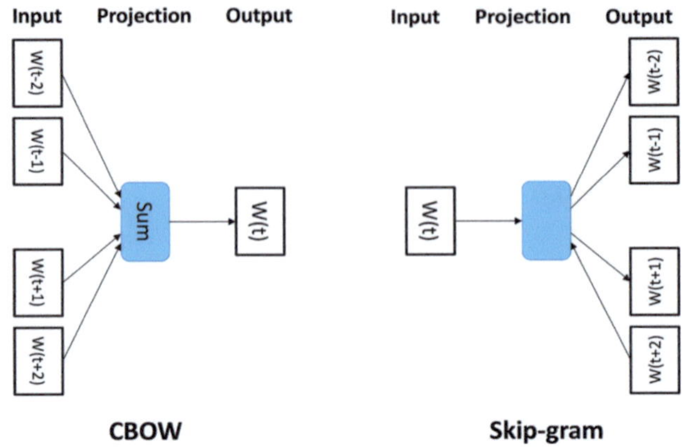

Fig. 31 The architecture of the word2vec model, consisting of an input layer with one-hot encoded vectors, a hidden layer that learns word embeddings, and an output layer that predicts the surrounding context words

For each word in a sentence, we convert it to its corresponding one-hot encoded vector. In **Skip-gram**, the target word (e.g., "king") is fed into the model. In **CBOW**, a set of context words (e.g., "the", "ruled", "kingdom") is fed into the model. Each word is processed individually as a one-hot vector, and in the case of CBOW, the context words are averaged or summed to form the input.

- **Hidden Layer**: A simple layer without any activation function. The size of the hidden layer determines the dimension of the word embeddings. The hidden layer learns to map the one-hot vectors to continuous vectors that represent the words in the vector space.
- **Output Layer**: A softmax function is applied at the output layer to generate a probability distribution over the entire vocabulary. In **Skip-gram**, the model tries to predict the context words given the target word. In **CBOW**, the model predicts the target word given the context.

• **Recurrent Neural Networks (RNNs)** introduced sequence modeling by processing text one word at a time and maintaining a hidden state to capture context. In sequence-to-sequence tasks, such as machine translation, the encoder processes the input sentence (e.g., "*I love cats*") one word at a time, updating its hidden state after each word (see Fig. 32) [6]. The final hidden state, also known as the context vector, is passed to the decoder, which generates the output sentence in an **autoregressive** manner, meaning each output word is generated based on the previous word's generation. In other words, each step in this architecture is autoregressive, where generating the next word depends on all previously generated words. For example, in the sentence "*I love cats*", the encoder starts by processing the first word, "I," and produces a hidden state that represents this word's context. Then, it processes the next word, "love," and updates the hidden state to represent the context of the sequence "*I love*". Finally, after processing the word "cats," the encoder

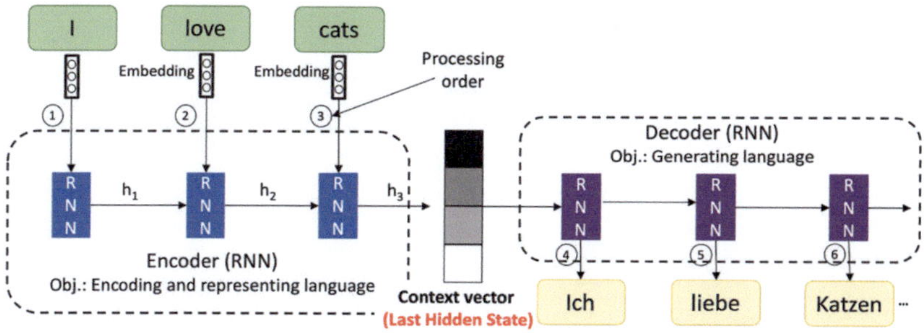

Fig. 32 Encoder–Decoder architecture without Attention. The encoder processes the input sequence and produces the final hidden state (context vector), which is passed to the decoder to generate the output sequence

updates the hidden state to represent the full context of the sentence "*I love cats*". This final hidden state (context vector) encapsulates the entire sequence's meaning and is passed to the decoder to generate the translation. However, RNNs are not ideal for processing long sequences because of the vanishing gradient problem, which makes it difficult to maintain long-range dependencies in the sequence. In other words, as the sequence length increases, the model struggles to keep track of earlier words or context, and the hidden state loses relevant information.

- **Attention Mechanisms** address the limitations of traditional encoder-decoder by allowing models to focus on the most relevant parts of an input sequence when generating output. Unlike traditional models, where the encoder outputs only the final hidden state, the attention mechanism allows the decoder to focus on different parts of the input sequence at each generation step. This is crucial when dealing with long and complex sentences, where different words may require more attention at different stages of the output generation (see Fig. 33). For example, when translating "*I love cats*" into German, the attention mechanism ensures that the word "*cats*" is aligned with "*Katzen*" in German, while "*love*" is aligned with "*liebe*". The attention mechanism dynamically assigns importance scores to different parts of the input sequence, allowing the model to focus more on the relevant parts at each step of output generation. Mathematically, the attention mechanism computes a weighted sum of all encoder hidden states at each decoding step. The weight for each hidden state is calculated using an attention score that is computed based on the decoder's current state. Let's define the attention mechanism with the following equation:

$$\text{Attention score} = \text{softmax}(\text{decoder state}_t \cdot \text{encoder hidden state}_i)$$

The decoder state at time step t represents the current state of the decoder, while the encoder hidden state at time step i is the hidden state of the encoder at that specific point in the sequence. The softmax function is applied to normalize the attention scores, ensuring that they sum to 1. This normalization indicates the

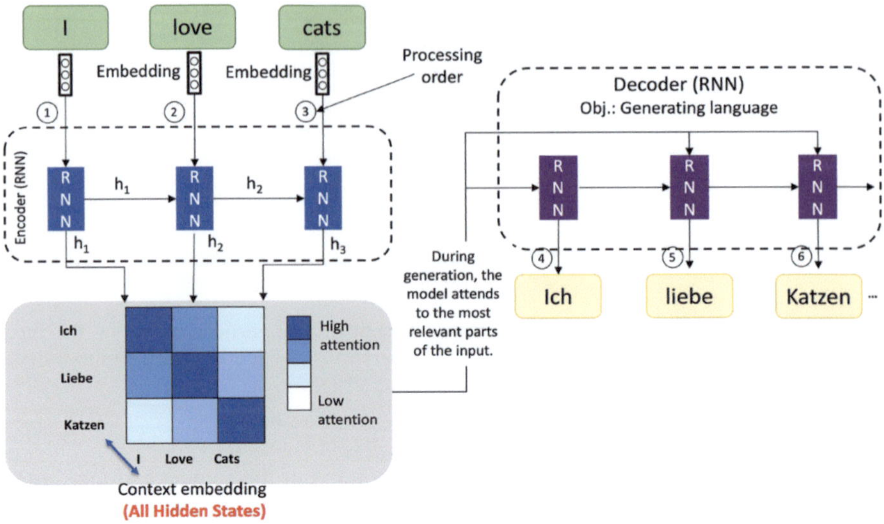

Fig. 33 Encoder–Decoder architecture with Attention. The attention mechanism is added to the decoder, allowing it to access all hidden states produced by the encoder. This enables the decoder to focus on the most relevant parts of the input sequence for each word it generates, improving the performance of the mode

relative importance of each encoder hidden state, determining how much focus the decoder should give to each part of the input sequence during the output generation process. For each output token at time t, the decoder generates a weighted sum of the encoder hidden states based on the attention scores, allowing it to focus on the most relevant parts of the input. This is a significant improvement over the simple encoder-decoder model, where the decoder only has access to the final hidden state. In sequence-to-sequence models with attention, the encoding step still generates word embeddings (e.g., using word2vec), and all the hidden states from the encoder are passed to the decoder. The attention mechanism enhances this by allowing the decoder to "attend" to the most relevant parts of the sequence, making the translation more contextually accurate.

- **Transformer Architecture** revolutionized Language AI by replacing sequential processing with self-attention, allowing models to process all words in a sequence simultaneously, rather than one by one (see Fig. 34) [6]. This parallel processing improves efficiency and enables the model to better capture relationships between words, regardless of their distance from each other. For example, in the sentence "*I love cats*", the Transformer's self-attention mechanism allows the model to understand how "*love*" is connected to both "*I*" and "*cats*", even though they are not adjacent. This ability to look at the entire sentence at once makes the model much more powerful for tasks like machine translation and text generation. In the Transformer, both the encoder and the decoder are stacked on top of each other, forming a modular architecture. The encoder takes the input sequence (e.g., "*I love*

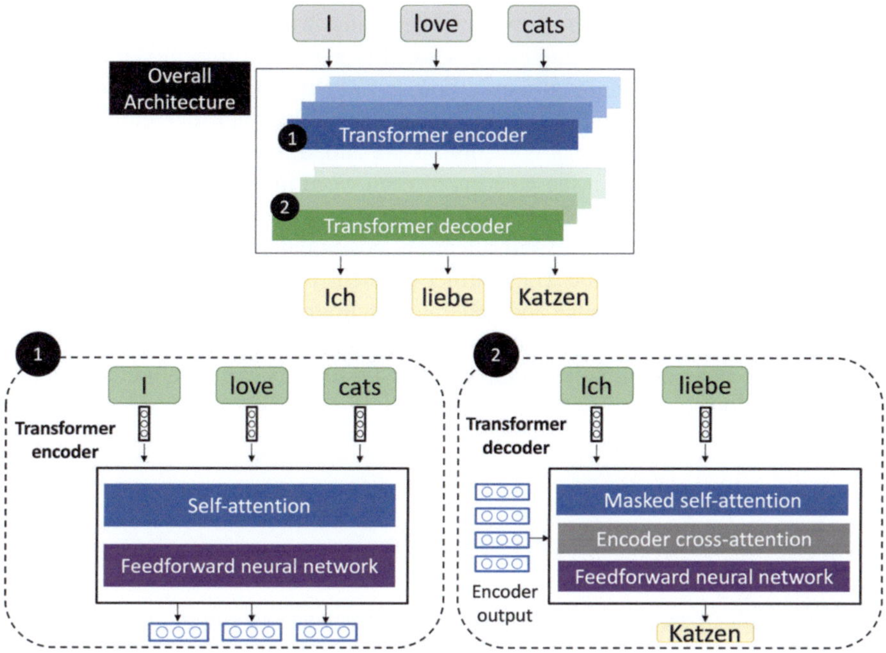

Fig. 34 The transformer architecture for language tasks

cats") and processes it through multiple layers of self-attention and feedforward networks to create a set of hidden states. These hidden states capture contextual information for each word in the sentence. In the encoder, self-attention allows each word in the input sequence to attend to every other word in the sequence. This means that, when processing a word, the encoder considers the full context of the sentence, not just the word's immediate neighbors. For example, when processing the sentence "*I love cats*", the word "*love*" will attend to "*I"* and "*cats*" to understand the overall context. Similarly, "*cats*" will attend to both "*I*" and "*love*". This allows the encoder to build a detailed representation of the entire input sequence by looking at all the words at once. The decoder also uses self-attention, but in this case, it uses masked self-attention, which ensures that the model only attends to previously generated words, not future ones. This is important for autoregressive tasks like text generation, where each word generated depends on the previously generated words. In addition to self-attention, the decoder uses cross-attention to focus on the hidden states generated by the encoder. This allows the decoder to align the relevant parts of the input sequence with the words it is generating in the output sequence. For example, when translating the sentence "*I love cats*" into German ("*Ich liebe Katzen*"), the encoder generates a set of hidden states for each word. These hidden states are passed to the decoder, which uses them to generate the translated sentence, focusing on the relevant parts of the input as needed. The decoder works in an autoregressive manner, meaning it generates one word at a time, using the previously generated words as part of the input for generating the next word. Each time a word is generated, the model attends to both the previously generated words (via self-attention) and the encoder's hidden states (via cross-attention). Since self-attention does not inherently consider the order of words, positional embeddings are added to the input embeddings to provide information about the position of each word in the sequence. This helps the model understand the word order, which is crucial for tasks like translation and text generation. By using self-attention, positional embeddings, and stacked encoder-decoder layers, the Transformer efficiently captures long-range dependencies in the input sequence and generates accurate, context-aware output. This architecture has formed the foundation for highly effective models like BERT and GPT, which have been widely adopted in tasks such as machine translation and text generation.

16.2 Architectural Variants of LLMs

LLMs are built upon the foundation of Transformer architectures by leveraging a stack of Transformer blocks, often with architectural refinements tailored to specific use cases. While Transformers serve as the underlying framework, LLMs differ in their architectural configurations, training paradigms, and scale. The key architectural variants include:

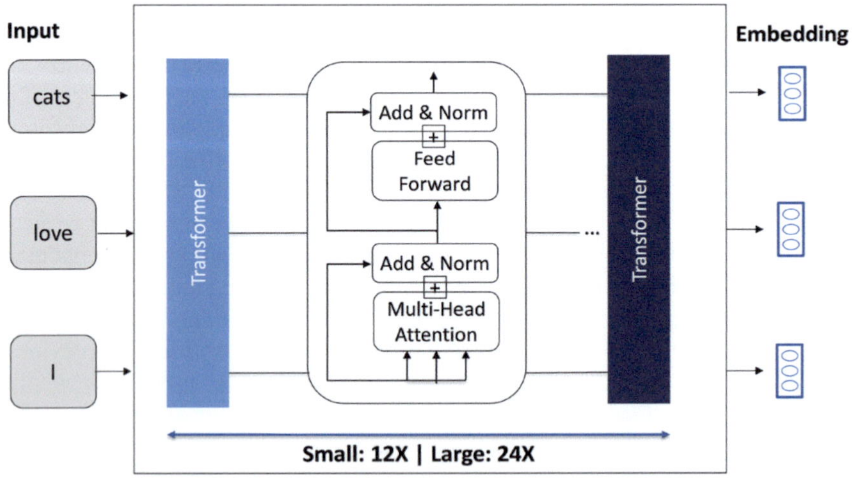

Fig. 35 Architecture of the BERT model, where BERT Base consists of 12 transformer layers (12×) and BERT Large consists of 24 transformer layers (24×)

- **Representation Models (Encoder-Only Architectures)**: Encoder-only models like BERT (Bidirectional Encoder Representations from Transformers) utilize the encoder stack of the Transformer architecture to process and understand input sequences (see Fig. 35) [6]. Representation models focus on encoding language into dense vector representations, or embeddings, that capture semantic and contextual relationships. BERT and similar models are pre-trained using techniques like **masked language modeling (MLM)**, where parts of the input text are hidden, and the model predicts the missing pieces: $P(x_{\text{masked}} \mid \text{context})$. This training enables the model to learn rich representations of language, which can then be used for downstream tasks such as text classification, semantic search, and clustering. These embeddings are generated at multiple stages of the architecture, making representation models versatile tools for feature extraction. Unlike generative models, encoder-only architectures like BERT are not designed for large-scale text generation but are optimized for understanding and analyzing textual data.
- **Generative Models (Decoder-Only Architectures)**: Generative models like GPT (Generative Pre-trained Transformer) adopt a decoder-only architecture that specializes in creating and generating coherent and contextually relevant text (See Fig. 36 [6]. These models operate in an **autoregressive** manner, predicting each token in a sequence based on all previously generated tokens:

$$P(x_t|x_{<t})$$

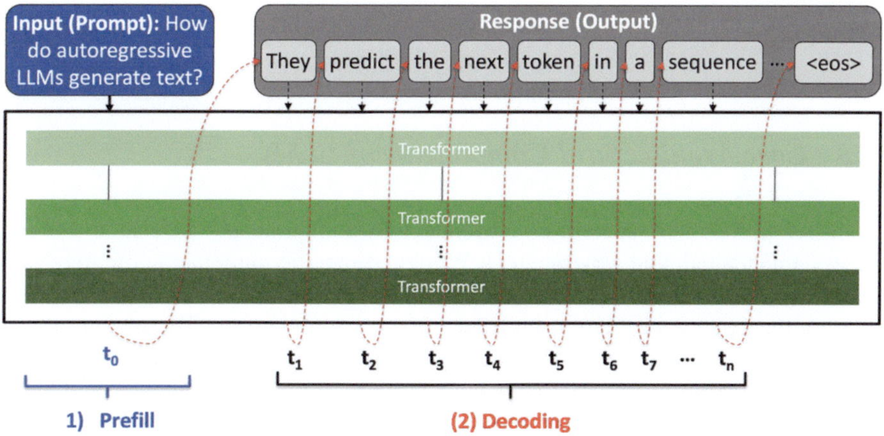

Fig. 36 The architecture of the GPT model, which follows a decoder-only transformer design and does not include an encoder component. The inference process in autoregressive LLMs typically involves two sequential phases: **prefill** and **decoding**. In the prefill phase, the model processes the entire input prompt, composed of a sequence of tokens, in parallel to generate the first token of the response in a single step. This is followed by the decoding phase, where subsequent tokens are generated sequentially in multiple steps; each decoding step produces one token conditioned on the previously generated tokens, until a termination token is reached

This design makes them well-suited for tasks like story writing, dialogue generation, summarization, and code generation. Decoder-only models are a fundamental category of LLMs, renowned for their scalability and versatility. Generative LLMs function as sequence-to-sequence systems, capable of autocompleting text based on a given input. However, their true potential is unlocked when fine-tuned for specific tasks, such as answering questions or engaging in conversational exchanges. Fine-tuned versions, often called instruct or chat models, are designed to follow user prompts and provide contextually relevant responses. An important feature of these models is the **context length** or **context window**, which defines the maximum number of tokens the model can process at once. A large context window allows these models to handle entire documents and maintain coherence across longer sequences. As tokens are generated, the context dynamically updates, enabling the model to consider both earlier inputs and its own outputs during text generation. This capability makes generative models essential tools for applications requiring adaptive and scalable AI-driven content creation.

- **Encoder-Decoder Architectures**: Models like T5 (Text-to-Text Transfer Transformer) and BART (Bidirectional and Auto-Regressive Transformers) combine both encoder and decoder stacks. The encoder processes the input sequence into a context-aware representation, while the decoder generates a target sequence based on the encoded input. Encoder-decoder architectures are highly versatile and excel in sequence-to-sequence tasks such as machine translation, summarization, and text-to-text transformations. While not traditionally labeled as LLMs, models like T5 and BART share similarities with LLMs, especially in terms of scalability and application to diverse natural language processing tasks.

16.3 Core Architectural Components of LLMs

LLMs are powered by several core architectural components that enable them to process, understand, and generate text. While the specific implementation may vary depending on the model's architecture (such as **decoder-only** or **encoder-decoder** models), the general structure is built on a few key components. These components work together to enable the model to handle complex tasks like text generation, translation, classification, and more. Below is an outline of the core components, highlighting the distinctions between **decoder-based** models (e.g., **GPT**) and **encoder-based** models (e.g., **BERT**) (see Fig. 37) [6].

- **Tokenizer**: The Tokenizer is the first component in LLMs, responsible for converting raw text into smaller units called tokens. These tokens can represent words, subwords, or characters, depending on the tokenization approach. The tokenizer breaks down the input text and assigns each token a numerical value (token ID) that the model can process. Common tokenization algorithms include Byte Pair Encoding (BPE), used in models like GPT, and WordPiece, commonly used in BERT.

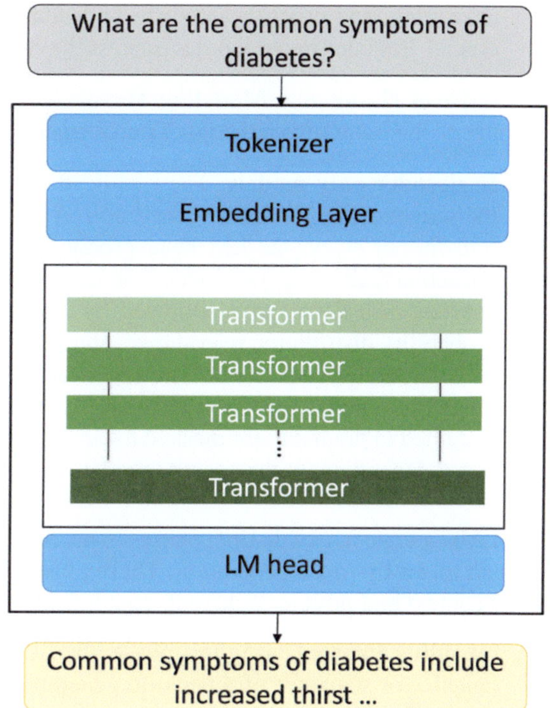

Fig. 37 A Transformer-based language model consists of a tokenizer, embedding layer, multiple layers of Transformer blocks, and a head for language modeling

- **Embedding Layer**: The embedding layer takes the numerical tokens (token IDs) produced by the tokenizer and maps them into dense, continuous vector representations in a high-dimensional space. These vectors capture the semantic meaning of each token, allowing the model to understand and process contextual relationships between tokens. The embedding layer is crucial because it transforms discrete token IDs into a continuous, meaningful representation that can be fed into subsequent layers of the model for downstream tasks. The embeddings are typically learned during the model's training process.
- **Stack of Transformers**: The Stack of Transformers refers to a series of Transformer layers that process the tokenized input. These layers are the backbone of the model, and their purpose is to capture intricate relationships and dependencies between tokens.
  - **In decoder-only models** (e.g., GPT): The stack consists of decoder layers that focus on generating text, where each layer uses self-attention to process the input and generate meaningful output. Decoder-only models are autoregressive, generating text sequentially, one token at a time.
  - **In encoder-based models** (e.g., BERT): The stack is made up of encoder layers that focus on understanding and generating contextualized representations of the entire input sequence. These models excel at tasks like classification, question

answering, and other tasks where understanding the whole input sequence is crucial. Encoder-based models process the entire input sequence bidirectionally.

- **LM Head (Language Modeling Head)**: The LM Head is the final layer that converts the output from the last Transformer layer into predictions.
  - **In decoder-only models**: The LM head typically produces the next token in the sequence, used for text generation. After processing the input through the transformer stack, the LM head produces a probability distribution over the vocabulary, allowing the model to predict the next word. The LM Head typically consists of a linear layer followed by a softmax function, which produces a probability distribution over the vocabulary for token prediction.
  - **In encoder-based models**: The LM head is often used for tasks like masked language modeling, where certain tokens in the input are masked, and the model is trained to predict those masked tokens. It might also be used for classification tasks, depending on the specific model configuration.

LLMs operate in two distinct phases to process input and generate output: encoding and token-by-token generation. These phases are integral to how LLMs understand context and produce coherent, contextually relevant responses.

1. **Encoding Phase**: During this phase, the LLM processes the input data (e.g., a sentence or sequence of tokens) to generate a context-aware internal representation. This is achieved by:
   - Parsing the input sequence into token embeddings, where each token is represented as a vector in a high-dimensional space.
   - Passing these embeddings through multiple transformer layers, which include attention mechanisms and feed-forward networks. These layers capture relationships between tokens, such as syntax and semantics, and create a unified representation of the input.
2. **Token-by-Token Generation Phase**: Once the input is encoded, the LLM generates the output sequence one token at a time. For each step:
   - The model uses the context from previously generated tokens (including the encoded input, if applicable) to predict the next token.
   - The prediction is made by computing probabilities over the vocabulary, often using a softmax function.
   - The token with the highest probability is selected (or sampled, depending on the decoding strategy) and appended to the output sequence.

   This process is repeated iteratively until a termination condition is met, such as generating an end-of-sequence token or reaching a maximum length.

This two-phase workflow enables LLMs to produce high-quality, coherent text by combining robust input understanding with iterative, context-sensitive token generation.

16.4 *Mixture of Experts (MoE)*

As the size of LLMs grows, they experience **higher inference latency**, leading to inefficiencies in real-time applications. Furthermore, their extensive parameterization results in **elevated energy consumption**, which has implications for **cost-effectiveness** and **environmental sustainability**. Addressing these challenges necessitates innovations that enable scalability without incurring excessive computational overhead. A promising solution to this issue is the concept of **sparsity**, which enables models to activate only a subset of their parameters per input, rather than utilizing the entire network as in traditional dense architectures. This selective activation significantly reduces computational complexity while preserving model efficacy. A prominent approach that leverages sparsity is the **Mixture of Experts (MoE)** framework, which incorporates two fundamental components (see Fig. 38):

1. **Sparse MoE Layers**: Unlike conventional Transformer models that employ a single, dense **MLP** in the feedforward layer, MoE introduces multiple specialized **expert networks**. At any given time, only a subset of these experts is activated, thereby optimizing computational efficiency.
2. **Router (Gate Network)**: The router dynamically identifies the most relevant expert(s) for each incoming token and directs the token accordingly.

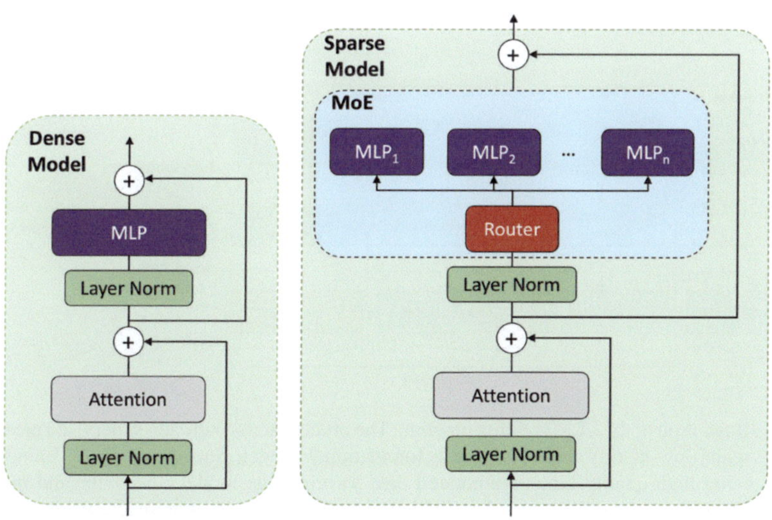

Fig. 38 Comparison of a standard Transformer block (left) and a MoE-based Transformer block (right). The left side represents the traditional Transformer architecture. The right side shows the modified MoE version, where the dense feed-forward layer (MLP) is replaced with multiple expert networks and a router. The router (gate network) dynamically selects a subset of experts for each token

16.5 Training Strategies for LLMs

LLMs are trained through a **pretraining-finetuning paradigm**, leveraging semi-supervised learning to achieve their remarkable capabilities. This paradigm consists of two main stages (see Fig. 39):

- **Pretraining**: Pretraining is a foundational stage in training LLMs, where they are trained on extensive corpora in a self-supervised manner, eliminating the need for manually labeled data. During this process, the model's parameters are iteratively updated to minimize the error on self-supervised tasks, enabling LLMs to acquire linguistic structures, contextual relationships, and patterns from diverse and massive datasets. This iterative optimization forms the basis for their broad generalization capabilities. Key aspects of pretraining include:
  - **Self-supervised Training**: LLMs generate supervision signals directly from the data without requiring manual annotations. Common tasks include predicting the next token in a sequence (Autoregressive Objective), reconstructing masked tokens (Masked Language Modeling, or MLM), and learning to map input sequences to output sequences (Sequence-to-Sequence, or Seq2Seq). These tasks allow the model to learn from diverse datasets, such as books, articles, and web pages, capturing complex linguistic structures and contextual patterns.
  - **Diversity of Data**: The pretraining phase incorporates data from a wide range of domains, such as news articles, scientific publications, code repositories, and

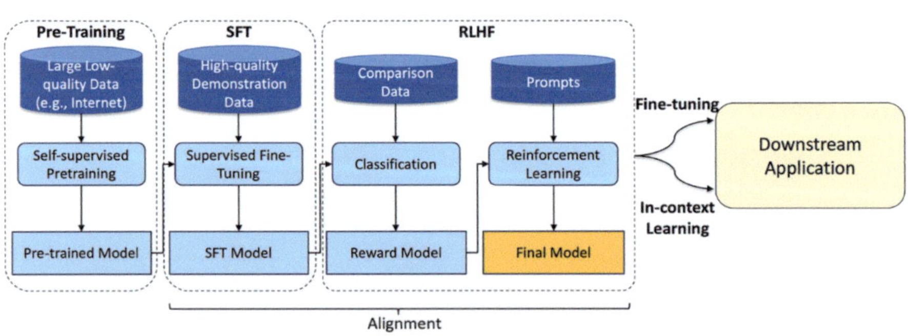

Fig. 39 Illustration of the LLM training pipeline. The process starts with self-supervised pretraining on large-scale, low-quality data, forming a foundational model. Supervised Fine-Tuning (SFT) follows, using high-quality human-annotated data to refine the model's behavior and introduce an initial level of alignment. A reward model is trained using human preference rankings to guide response selection, and Reinforcement Learning with Human Feedback (RLHF) further strengthens alignment by optimizing the model based on human feedback to ensure it is helpful, honest, and safe. The final aligned model is then deployed. Furthermore, for specific applications or domains, the model can either undergo additional fine-tuning with specialized data to enhance its performance or utilize In-Context Learning (ICL), where it adapts dynamically to new tasks based on prompts without modifying its internal parameters

conversational transcripts. This diversity enables the model to generalize effectively across different tasks and topics, enhancing its robustness and adaptability to real-world applications.
  - **Scale of Pretraining**: Pretraining operates on datasets spanning hundreds of terabytes and billions to trillions of tokens. This scale is essential for capturing subtle patterns and complex relationships in the data. Self-supervised learning facilitates efficient training on such large-scale datasets without the need for human-labeled data, making it both scalable and cost-effective.

- **Alignment**: The alignment phase optimizes a generative LLM to ensure its responses align with human-defined principles, such as helpfulness, harmlessness, safety, ethical considerations, and real-world usability. The alignment process can be performed based on these criteria through a structured approach using two main techniques

  - **Supervised Fine-Tuning (SFT)**: SFT is a process where a pretrained language model is refined using high-quality, human-annotated data to enhance its ability to generate accurate, contextually appropriate, and instruction-following responses. SFT trains the model using a next-token prediction objective, where it learns to predict the most likely next token based on a curated set of reference outputs.
  - **Reinforcement Learning from Human Feedback (RLHF)**: RLHF is a technique used to align a language model's behavior with human preferences by iteratively optimizing its responses based on structured feedback. The process begins with generating a reward model (RM), also referred to as a preference model), which is trained on human-annotated rankings of model outputs. This reward model learns to predict which responses are more aligned with desired behaviors, such as helpfulness, honesty, and safety. Once the RM is trained, reinforcement learning is applied (e.g., Policy Optimization (PPO), Generalized Proximal Policy Optimization (GPPO)), where the language model generates responses, receives reward signals from the RM, and updates its policy to favor higher-scoring outputs. By continuously refining the model based on these learned preferences, RLHF enhances the ability of language models to produce coherent, accurate, and user-aligned responses, ultimately improving their reliability and usability in real-world applications.

- **In-Context Learning (ICL)**: ICL is a method where a model learns to perform a task by conditioning on examples provided within the prompt, without updating its parameters. Unlike pretraining and fine-tuning, which require modifying the model's parameters, ICL leverages the model's existing knowledge and generalization capabilities to adapt dynamically to new tasks. By carefully structuring the prompt with task-specific instructions or examples, ICL allows models to generate more accurate and contextually relevant responses. This approach is computationally efficient and particularly useful for adapting large models to various applications without additional training. ICL is typically implemented using techniques such as Few-Shot Learning and Zero-Shot Learning:

  - **Zero-Shot Learning**: Asking the model to perform a task based solely on explicit instructions, without any prior examples.
  - **Few-Shot Learning**: Providing a few labeled examples within the prompt to help the model infer the task pattern and generate the correct output.

- **Fine-tuning**: Fine-tuning is the process of adapting pretrained LLMs to specific tasks or objectives by updating their parameters or refining their outputs. This process leverages the general linguistic knowledge acquired during pretraining (a process known as transfer learning) and applies it to specific tasks, significantly reducing the data and computational resources needed compared to training a model from scratch. Fine-tuning typically involves supervised learning with labeled data, though additional techniques may be employed for further optimization. Core approaches for fine-tuning involve two primary methods:

  - **Full Fine-Tuning**: This approach updates all model parameters using labeled datasets. It employs supervised learning steps, including forward passes, loss computation, and backpropagation, to optimize the entire model. Full fine-tuning is best suited for smaller models or scenarios where task-specific data is abundant and computational resources are sufficient. However, it is computationally intensive for very large models, as updating and storing all parameters requires significant resources, making it less practical in such cases.
  - **Parameter-Efficient Fine-Tuning (PEFT)**: To reduce the computational cost of full fine-tuning of LLMs, parameter-efficient fine-tuning methods have become increasingly popular. These techniques focus on updating only a small subset of the model's parameters or introducing lightweight adapters, making fine-tuning more accessible and scalable. Two notable methods in this category include:

    Adapter Modules: Adapter layers, first introduced in [7], offer a parameter-efficient fine-tuning method by integrating small, trainable modules into transformer models while preserving the pretrained weights (see Fig. 40). These layers function as bottleneck-style feedforward modules within each transformer block. Specifically, they (i) compress the input dimensionality using a trainable linear layer, (ii) apply a non-linearity for transformation, and (iii) restore the original dimensionality through another trainable layer. By fine-tuning only a small subset of task-specific parameters, adapter layers enable efficient and scalable adaptation of large pretrained models to diverse tasks with minimal computational cost.
    LoRA (Low-Rank Adaptation): Introduces low-rank matrices to efficiently update the model's weights with minimal computational overhead (see Fig. 41) [8]. Fine-tuning a large language model by updating all parameters is computationally expensive, requiring significant memory and processing power. To address this, LoRA introduces a more efficient approach by approximating the weight update ΔW using a low-rank decomposition.

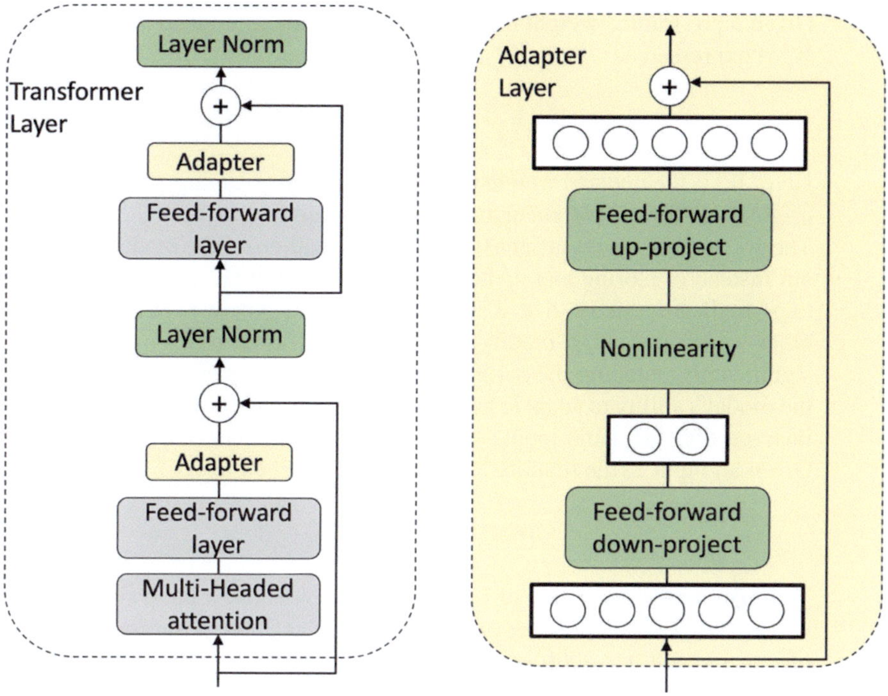

Fig. 40 Architecture of adapter layers in a transformer block [7]

Fig. 41 Low-Rank Adaptation (LoRA) introduces trainable low-rank matrices to efficiently fine-tune pre-trained models while significantly reducing the number of trainable parameters

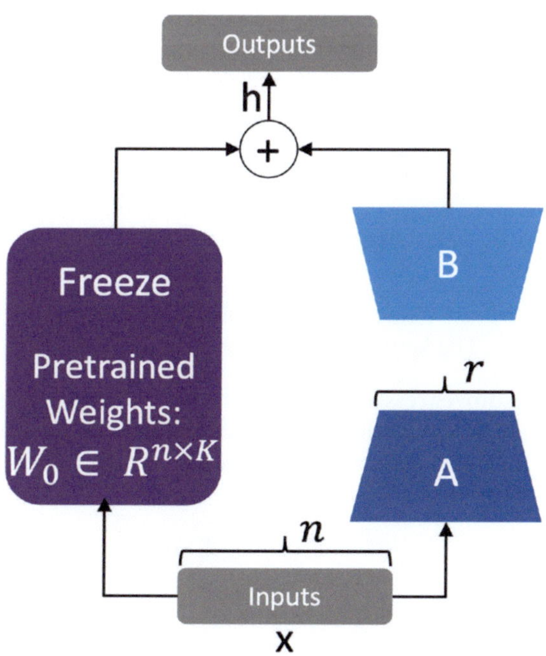

Given a pre-trained weight matrix $W_0 \in \mathbb{R}^{n \times k}$, the fine-tuned weight matrix W' is expressed as:

$$W' = W_0 + \Delta W \approx W_0 + BA$$

Here, W' is the updated weight matrix used after fine-tuning, while W_0 is the original pre-trained weight matrix, which remains frozen throughout training. The update ΔW represents the trainable modifications to the model's weights, but instead of storing and updating a full $n \times k$ matrix, it is factorized into two smaller matrices: $B \in \mathbb{R}^{n \times r}$ and $A \in \mathbb{R}^{r \times k}$, where $r \ll \min(n, k)$. In other words, instead of modifying W_0 directly, LoRA trains only A and B, significantly reducing the number of trainable parameters while preserving the model's ability to adapt to new tasks. Both W_0 and the update $\Delta W = BA$ interact with the same input, and their outputs are summed element-wise. Given an input x, the modified forward pass is:

$$h = W_0 x + \Delta W x = W_0 x + BAx$$

To illustrate the efficiency of LoRA, suppose we have a pre-trained weight matrix W_0 with dimensions $n = 2000$ and $k = 3000$, resulting in a total of $n \times k = 2000 \times 3000 = 6,000,000$ parameters. Instead of updating all $6,000,000$ parameters, LoRA introduces a low-rank decomposition with $r = 1$. The low-rank matrices A and B have dimensions $A = n \times r = 2000 \times 1 = 2000$ and $B = r \times k = 1 \times 3000 = 3000$. Thus, the total number of trainable parameters in A and B is $2000 + 3000 = 5000$, which is significantly fewer than the $6,000,000$ parameters in W_0. The resulting LoRA update is computed as BA, which has dimensions $(2000 \times 1) \times (1 \times 3000) = 2000 \times 3000$, correctly matching the shape of W_0.

16.6 Training Objectives in LLMs

LLMs are trained using specific objectives tailored to their architecture and intended tasks. The key training objectives include:

- **Autoregressive Objective**: This objective is used in decoder-only models like GPT, where the goal is to maximize the likelihood of the next token given the context of all preceding tokens:

$$\mathcal{L} = -\sum_{t=1}^{n} \log P(x_t | x_{<t})$$

Here, x_t represents the token at position t in the sequence, $x_{<t}$ refers to the sequence of tokens preceding x_t (i.e., $x_1, x_2, \ldots, x_{t-1}$), and $P(x_t|x_{<t})$ denotes the model's predicted probability of the token x_t given the preceding tokens $x_{<t}$.

Causal Masking in Autoregressive Models: To enforce the sequential prediction of tokens and prevent the model from "cheating" by looking ahead at future tokens during training, a causal mask is applied to the attention mechanism. Without such a mask, the model could attend to tokens that occur later in the sequence, learning patterns that rely on unavailable information during inference. This would compromise the model's ability to generalize in real-world text generation tasks. The causal mask ensures that the prediction of token x_t depends only on the tokens preceding it. Mathematically, the mask is represented as:

$$M_{ij} = \begin{cases} 0 & \text{if } j > i \\ 1 & \text{if } j \leq i \end{cases}$$

Here, M_{ij} is the mask applied to the attention scores. If token j occurs after token i, its attention score is set to 0 (or $-\infty$ before softmax), effectively blocking access to future tokens. For example, when predicting x_3 in the sequence "The cat chased the mouse," only "The cat" is visible to the model, while "chased the mouse" remains masked. This ensures the model learns to generate the sequence one token at a time in a causally consistent manner, enabling it to handle text generation tasks effectively.

- **Masked Language Modeling (MLM)**: This objective is used in bidirectional models like BERT. The training involves randomly masking certain tokens in the input sequence and predicting their original values based on the remaining tokens:

$$\mathcal{L} = -\sum_{t \in \text{mask}} \log P(x_t|X_{\text{masked}})$$

Here, $t \in$ mask indicates that x_t is a masked token, X_{masked} represents the input sequence with certain tokens replaced by a special `[MASK]` token, and $P(x_t|X_{\text{masked}})$ denotes the probability of recovering the original value of x_t based on the masked sequence X_{masked}. This objective allows the model to learn bidirectional dependencies, making it effective for tasks like classification, inference, and question answering.

- **Sequence-to-Sequence (Seq2Seq) Objective**: In encoder-decoder models like T5, the objective is to reconstruct a target sequence (Y) given an input sequence (X):

$$\mathcal{L} = -\sum_{t=1}^{m} \log P(y_t|y_{<t}, X)$$

Here, y_t represents the token at position t in the target sequence, $y_{<t}$ refers to the sequence of tokens preceding y_t in the target sequence, X denotes the input sequence, and $P(y_t|y_{<t}, X)$ represents the probability of predicting y_t based on

its preceding tokens and the input sequence X. This objective is commonly used in applications like machine translation, summarization, and text-to-text tasks. Unlike the autoregressive objective, the decoder in Seq2Seq models has access to the entire input sequence X in addition to the preceding tokens $y_{<t}$, enabling it to generate contextually accurate outputs.

16.7 Performance and Evaluation Metrics

Evaluating LLMs involves a variety of metrics that examine their capabilities from different perspectives. This subsection summarizes the most commonly used metrics in the field. As outlined in [9], evaluation methods can be broadly categorized into intrinsic and extrinsic approaches:

- **Intrinsic Metrics**: These assess the model's ability to generate coherent and meaningful sentences by analyzing linguistic rules and patterns.
- **Extrinsic Metrics**: These evaluate the model based on real-world applications and user feedback, focusing on how well it meets human expectations and requirements.

Evaluating the performance of LLMs requires specific metrics that capture their efficiency, accuracy, and ability to model language effectively. In the following section, we discuss the most common and important metrics used in practice.

1. **Inference Time**: Inference time measures the duration between providing input to an LLM and receiving the output. Factors influencing inference time include the model's architecture (e.g., parameter count and depth), input size, and hardware configuration.
2. **Perplexity (PPL)**: Perplexity is a standard metric for evaluating language models. It quantifies the exponentiated average negative log-likelihood of a sequence. Given a tokenized sequence $X = (x_0, x_1, \ldots, x_t)$, perplexity is calculated as:

$$\mathrm{PPL}(X) = \exp\left(-\frac{1}{t}\sum_{i=1}^{t} \log p_\theta(x_i \mid x_{<i})\right)$$

   Here, $\log p_\theta(x_i \mid x_{<i})$ represents the log-likelihood of the i-th token given its preceding tokens. A lower PPL value indicates better predictive accuracy of the LLM.
3. **Precision**: Precision is computed as the ratio of correctly generated relevant words to the total number of words generated.
4. **Recall**: Recall is calculated as the ratio of correctly identified relevant words to the total number of relevant words in the reference.
5. **F1-Score**: The F1-Score is the harmonic mean of precision and recall, providing a balanced measure of model accuracy.

6. **BiLingual Evaluation Understudy (BLEU)**: BLEU evaluates the precision of matching n-grams between the model's output and the reference output.
7. **Recall-Oriented Understudy for Gisting Evaluation (ROUGE)**: ROUGE measures the overlap of n-gram sequences, producing recall, precision, and F1-scores for generated text relative to the reference.
8. **Translation Error Rate (TER)**: TER assesses the minimum number of edits needed to convert the generated text into the reference text.
9. **BERTScore**: BERTScore compares each token in the generated text with tokens in the reference text, generating a similarity matrix. A weighted sum of these similarities is used to evaluate the quality of the output.
10. **Trustworthiness**: Trustworthiness involves evaluating aspects such as bias, privacy, safety, and interpretability. These factors assess the ethical, secure, and comprehensible nature of the generated outputs [9].
11. **Bias and Fairness**: This metric measures bias in the generated responses and evaluates the model's consistency and fairness across different user groups [9].
12. **Robustness**: Robustness examines the model's resilience to input disruptions or adversarial examples [9].
13. **Toxicity**: This metric evaluates the extent to which the model generates harmful or offensive content [9].
14. **Generalization**: Generalization assesses the model's ability to perform well on tasks it was not explicitly trained on [9].
15. **Up-to-Dateness**: This metric evaluates how current and relevant the model's outputs are to recent knowledge [9].
16. **Groundedness**: Groundedness reflects the model's ability to generate factually accurate and logically consistent outputs, avoiding hallucinations [9].

LLM-as-a-Judge: In the context of performance evaluation of LLMs, a new emerging technique known as LLM-as-a-Judge is gaining attention. This flexible evaluation approach leverages an LLM to assess the quality of outputs generated by other LLMs. Using a structured evaluation prompt, the model is guided to rate responses based on predefined criteria such as accuracy, coherence, and relevance. Rather than relying on a fixed metric, this method serves as a customizable framework that can be tailored to different evaluation needs. It supports both **pairwise comparisons**, where two outputs are assessed relative to each other, and **direct scoring**, where individual outputs are rated against specific attributes and metrics. The effectiveness of this approach depends on factors such as the clarity of the evaluation prompt, the reasoning capabilities of the LLM judge, and the complexity of the task being assessed. LLM-as-a-Judge can be applied in offline evaluations, where outputs are analyzed after generation for benchmarking, as well as in online evaluations, where assessments occur dynamically to improve real-time model performance.

16.8 Multi-modal Large Language Models (LLMs) and Vision-Language Models (VLMs)

Traditional LLMs are inherently designed to process text-based inputs. To extend their functionality to handle additional modalities such as vision, researchers have developed specialized architectures known as Multi-Modal LLMs [10–14]. Vision-Language Models (VLMs), a subset of multi-modal models, specifically focus on integrating and aligning textual and visual information for tasks like image captioning, visual question answering, and cross-modal retrieval. A general architecture of a multi-modal LLM is illustrated in Fig. 42. These architectures typically adhere to a consistent design pattern. A specialized encoder, such as ViT or CLIP for vision tasks [15–17], extracts embeddings from non-textual inputs like images or videos. Simultaneously, embeddings are derived from text inputs using a text encoder. The visual and textual embeddings are then integrated, often through concatenation, addition, or cross-modal attention. To ensure effective alignment between modalities, the combined embeddings may pass through a projection layer, refining their compatibility before being fed into the LLM for further processing. In more advanced VLMs, non-textual inputs are directly incorporated into the internal layers of the Transformer via cross-modal attention mechanisms. These mechanisms enable the model to simultaneously process and align information across modalities, resulting in better contextual understanding and synergy between text and visual data [10].

16.9 Role of RAG and KG in LLMs/VLMs

Retrieval-Augmented Generation (RAG) and Knowledge Graphs (KGs) are powerful techniques to enhance the performance and utility of LLMs. RAG combines the generative capabilities of LLMs with external knowledge retrieval, where relevant documents or data are fetched in real-time from a large corpus to improve the

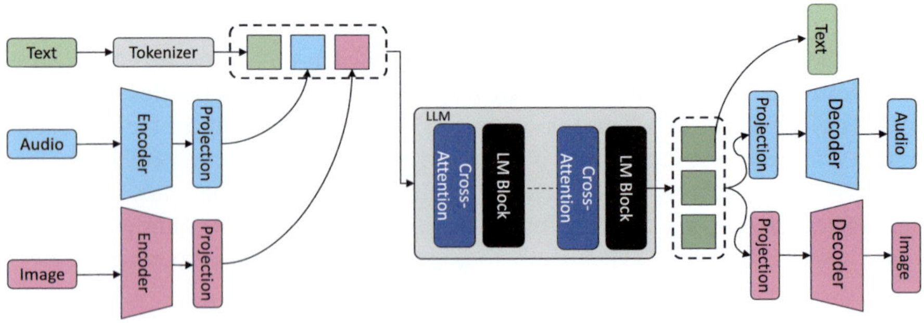

Fig. 42 Overview of the generic architecture for multi-modal LLMs, integrating text, image, and audio inputs

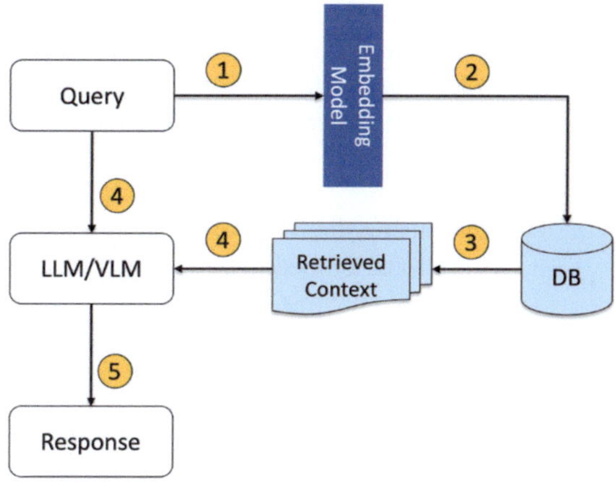

Fig. 43 Overview of the RAG/KG process. The RAG process combines external documents and user input into a prompt for an LLM to generate tailored output. The external information is retrieved from a knowledge source, which augments the LLM's response with relevant and specific information

model's accuracy and contextual understanding. This approach reduces hallucination by grounding responses in factual, up-to-date information. Knowledge Graphs, on the other hand, organize information as interconnected entities and relationships, enabling LLMs to perform structured reasoning and answer complex queries with contextually rich and precise data. By integrating KGs, LLMs gain a semantic understanding of concepts, allowing for more accurate and explainable results. The process flow for both techniques involves (see Fig. 43):

1. **Query Submission**: The user submits a query, which is directed to either the RAG or KG system based on the application logic.
2. **Knowledge Retrieval**: The system retrieves relevant knowledge from either RAG, KG, or both, depending on the use case and the nature of the query. Typically, data is stored in RAG when it is primarily unstructured, such as research articles or customer support logs. On the other hand, KG is utilized when the data is structured, such as domain-specific entities and relationships in healthcare or finance. In some cases, both RAG and KG can be used together to complement each other, where RAG handles unstructured content, and KG provides precise, structured reasoning. Table 2 summarizes the differences and use cases between RAG and KG.

   - **RAG**: The system retrieves relevant unstructured data from an external knowledge base using a vector-based semantic retrieval mechanism. For textual data, dense vector embeddings are created using models such as BERT, which capture the semantic meaning of the text. These embeddings are stored in a vector database. When multimedia documents (e.g., images, audio, or videos) are included, similar vectorization techniques are applied:

     – **Images**: Embeddings for images are generated using computer vision models such as CLIP or ResNet. These embeddings map images to a shared semantic space where they can be compared to text or other multimedia.

  - **Audio**: Audio embeddings are created using models like Wav2Vec, which convert audio signals into vector representations.
  - **Videos**: For videos, embeddings are generated by combining frame-based image embeddings (e.g., CLIP) with temporal models (e.g., 3D CNNs or transformers for video data) to encode spatial and temporal features.

  When a query is submitted, it is converted into a dense vector representation using the same embedding model (text-based for textual queries or multimodal embeddings for multimedia-aware queries). The system then performs a nearest-neighbor search in the vector database to find the top-k relevant documents, whether they are text, images, audio, or video. This approach ensures RAG can retrieve the most relevant information across multiple data modalities, making it highly adaptable for diverse use cases.

- **KG**: The system queries the KG to fetch structured data such as entities, relationships, or facts. A KG represents healthcare information as nodes (e.g., patients, diseases, treatments) and edges (e.g., `diagnosed_with`, `treated_by`) to form a connected graph structure. For example, a node representing a patient and a node representing a disease might be connected via an edge labeled `diagnosed_with`, while the disease node could be connected to a treatment node via an edge labeled `treated_by`. The data in a KG is typically stored in graph databases such as Neo4j, optimized for managing and querying large-scale, interconnected datasets. KGs often include semantic layers using standards like RDF (Resource Description Framework) or OWL (Web Ontology Language), enabling logical reasoning, such as deducing potential drug interactions or identifying relationships between symptoms and diseases. Queries to a KG use specialized languages such as SPARQL or Cypher to traverse the graph and retrieve relevant entities and relationships. For instance, a query might identify treatments for a specific disease by traversing `treated_by` edges starting from the disease node. For natural language queries, the system translates the user's input (e.g., "What are the approved treatments for diabetes?)" into graph queries, retrieves the relevant data, and passes it to the LLM for response generation. By leveraging their structured nature, KGs excel in enabling reasoning and delivering explainable answers.

3. **LLM Integration**: The retrieved data, whether from RAG or KG, is combined with the user query to form an augmented prompt, which is then passed to the LLM. This enriched input ensures the model has the necessary context and factual grounding to generate accurate and relevant responses.
4. **Response Generation**: The LLM generates a response using the provided data to ensure accuracy, relevance, and explainability.
5. **Output Delivery**: The user receives the final response, grounded in the retrieved knowledge.

Table 2 Comparison of RAG versus KG

Aspect	RAG	KG
Data type	Unstructured (text, documents)	Structured (entities, relationships)
Query complexity	Broad, general queries	Specific, reasoning-based queries
Use cases	Research papers, customer support logs	Healthcare, financial transactions, ontologies
Explainability	Moderate (based on retrieved documents)	High (based on structured, linked data)
Scalability	High (handles large unstructured corpora)	Domain-specific, smaller but highly curated

Together, RAG and KGs make LLMs more reliable, explainable, and adaptable to domain-specific tasks, using either unstructured or structured knowledge retrieval to suit the specific query requirements.

It should be noted that to achieve the best performance for a specific task, different optimization techniques offer trade-offs in accuracy, flexibility, and efficiency:

- **RAG**: Enhances responses by retrieving relevant external knowledge without modifying the model. Best for fact-based tasks requiring up-to-date information.
- **Prompt Engineering**: Adjusts inputs to steer the model's responses efficiently. Ideal for quick optimizations without additional computational costs.
- **Fine-tuning**: Trains the model on specific datasets to improve domain expertise. Suitable when deep customization and improved performance in niche areas are needed.
- **Hybrid (RAG + Fine-tuning)**: Combines external retrieval and model adaptation for optimal accuracy and flexibility. Useful when both dynamic knowledge updates and fine-tuned responses are required (Fig. 44).

16.10 LLM Agents

An agent is an entity that perceives its environment and takes actions to affect it, with its behavior shaped by the environment it operates in and the tools or actions available to it. An LLM agent is a type of agent powered by an LLM, capable of understanding and processing tasks, interacting with its **environment** (i.e., external world or system with which the agent interacts, gathers information, and performs actions to achieve its goals), and utilizing tools or executing actions to accomplish specific user-defined objectives. LLM agents are designed with four primary components—**Brain**, **Memory**, and **Tool Use**—which collectively enable them to perform diverse and complex tasks effectively (see Fig. 45).

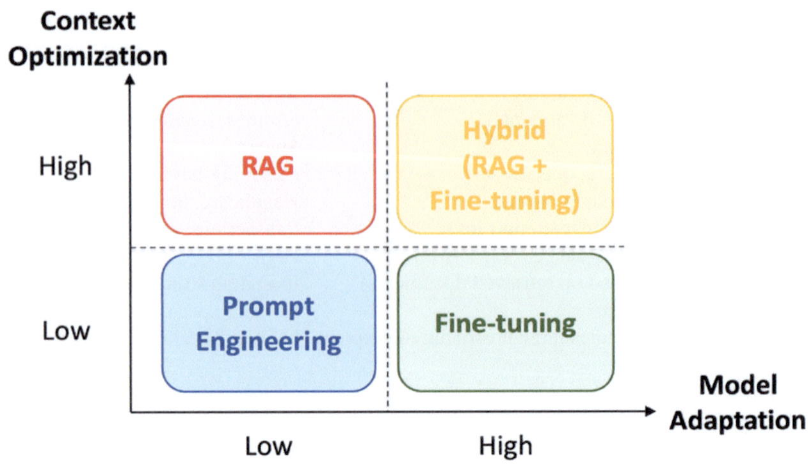

Fig. 44 Comparison of different optimization approaches for LLMs

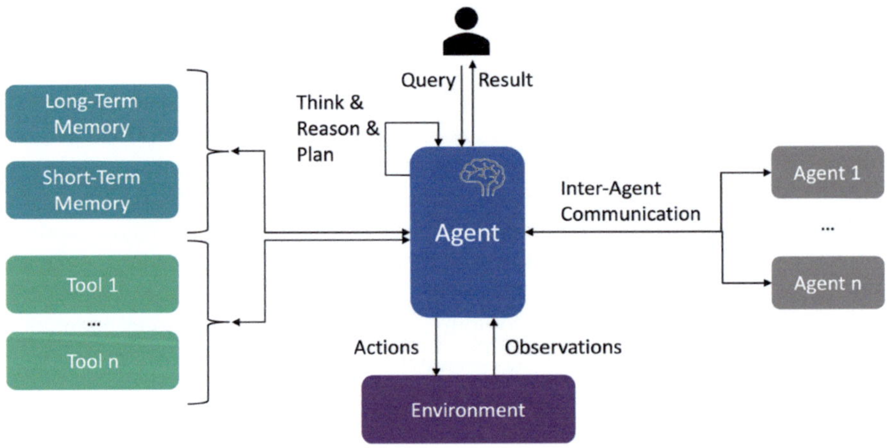

Fig. 45 Architecture of an agent-based system

- **Brain**: At the heart of an LLM agent is a language or action model trained on vast datasets to process and understand language. The agent's actions, goals, and tool usage are guided by a user-provided prompt, which serves as instructions for how the agent should respond, what tools to use, and the objectives to achieve. Additionally, LLM agents can be customized with specific personas, enabling them to adapt their responses and behavior to better suit particular tasks or interactions. This combination of advanced processing capabilities and customization makes the agent highly versatile. A fundamental role of the system's brain is to perform **planning** and **reasoning**. These capabilities enable LLM agents to effectively

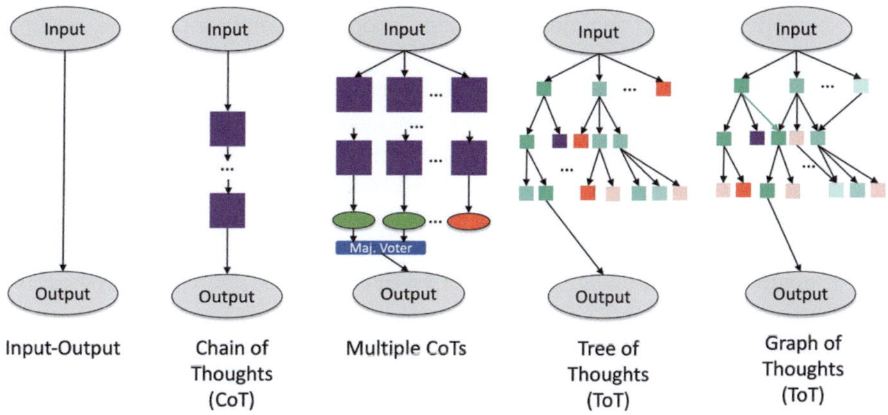

Fig. 46 Plan formulation techniques

analyze problems, decompose complex tasks into manageable subtasks, and adapt their strategies dynamically based on context and feedback:

1. **Plan Formulation**: Agents use various methods to create plans (see Fig. 46):

   – **Input–output (IO)**: The most straightforward way to map a problem input p to an output q using LLM is expressed as $q \sim L_\theta(q \mid p)$. In this context, p represents the problem input, q denotes the desired output, and θ refers to the parameters of the pre-trained LLM L_θ. The input p is directly processed by the model to generate the output q [18].
   – **Chain-of-Thought (CoT)**: Introduced by Wei et al. [19], chain-of-thought (CoT) prompting facilitates complex reasoning by incorporating intermediate steps throughout the reasoning process. CoT prompting was created to handle situations where directly linking input p to output q is challenging, such as when p is a math question and q is the resulting numerical answer [19]. The key concept involves inserting a sequence of intermediate steps $r_1, \ldots, r_k$ that connect p and q, where each r_j serves as a logical step forward in solving the problem (for example, r_j might be an intermediate equation in a math problem). In CoT-based problem-solving, each intermediate step r_j is generated sequentially following $L_\theta^{\mathrm{CoT}}(r_j \mid p, r_1 \ldots r_{j-1})$, with the final output q derived as $q \sim L_\theta^{\mathrm{CoT}}(q \mid p, r_1 \ldots r_k)$. In practice, the entire sequence $[r_1 \ldots r_k, q]$ is produced as a continuous language sequence, with the exact intermediate steps is decided by LLM [18, 19].
   – **Multiple Chain-of-Thoughts (Multiple CoT)**: Multiple Chain-of-Thought (Multiple CoT) extends CoT by exploring several parallel reasoning paths rather than relying on a single chain of steps. Each path generates its own sequence of intermediate steps, leading to a set of potential outputs. In the end, a majority voting mechanism or other aggregation strategies can be

applied to select the most consistent or optimal solution from the multiple reasoning paths.

– **Tree of Thoughts (ToT)**: ToT has been proposed to further enhance the concept of Chain of Thought (CoT) by enabling the system to explore multiple reasoning paths [18]. ToT frames any problem as a search over a tree structure, where each node represents a state $s = [p, r_1 \ldots r_i]$, with s being a partial solution that includes the input p and the sequence of thoughts $r_1 \ldots r_i$ generated so far. Within the ToT framework, two search algorithms have been mainly utilized, namely Breadth-First Search (BFS) and Depth-First Search (DFS) [18].
– **Graph of Thoughts (GoT)**: GoT extends the ToT concept by representing the reasoning process as a graph rather than a tree. In GoT, each node represents a reasoning step, and the edges represent relationships between these steps. Unlike the hierarchical structure of a tree, GoT allows for more flexibility between reasoning steps, supporting complex interactions. This graph structure provides the capability to model multiple reasoning paths that can interconnect and recombine at various stages. The GoT framework allows for richer exploration of reasoning processes and can support more dynamic problem-solving strategies.
– **Program of Thoughts (PoT)**: The main idea behind PoT is to express reasoning steps as Python programs, allowing computation to be performed by a Python interpreter [20].
– **Plan-and-Solve (PS)**: In CoT, the input typically features the trigger phrase "*Let's think step by step*." The authors of PS argued that CoT is prone to calculation errors, missing-step errors, and semantic misunderstandings [21]. Therefore, the PS method uses prompts like "*devise a plan*" and "*carry out the plan*." The authors of PS further recommend enhancing the plan-based trigger sentence with additional instructions, e.g., (1) "*pay attention to calculation*" to guide LLMs to perform calculations with greater accuracy; (2) "*extract relevant variables and their corresponding numerals*" to reduce the likelihood of skipping essential reasoning steps.
– **Reasoning and Acting in Language Models (ReAct)**: The core concept behind ReAct is to integrate reasoning traces with task-specific actions in an interleaved manner. ReAct prompts language models to simultaneously generate verbal reasoning traces and task-related actions, enabling dynamic reasoning. This approach enables the model to formulate, refine, and adapt high-level action plans (reason to act) while also interacting with external resources (e.g., Wikipedia) to enhance its reasoning process with additional information (act to reason). In this framework, an agent interacts with its environment to solve tasks. At each time step t, the agent receives an observation $o_t \in O$ from the environment and selects an action $a_t \in A$ based on a policy $\pi(a_t \mid c_t)$, where $c_t = (o_1, a_1, \ldots, o_{t-1}, a_{t-1}, o_t)$ represents the accumulated context. ReAct extends this process by expanding the agent's action space to $\hat{A} = A \cup L$, where L represents the language space. Actions within this language space, represented as $\hat{a}_t \in L$ (referred to as thoughts or

reasoning traces) do not directly impact the external environment. Instead, these thoughts are employed to reason over the current context c_t and next update the context to $c_{t+1} = (c_t, \hat{a}_t)$, supporting subsequent reasoning and actions [22].

2. **Plan Reflection**: After formulating a plan, agents assess its effectiveness through feedback mechanisms:

   - **ReAct**: Involves cycles of thought, action, and observation to refine strategies based on real-time feedback.
   - **Reflexion**: Uses user input or observations from the environment to adjust and improve planning.
   - **Memory**: Memory enables LLM agents to handle tasks effectively by retaining information about past interactions and maintaining context for ongoing tasks:

1. **Short-term Memory**: Acts as a temporary storage, capturing essential details from the current interaction. It ensures relevant responses during the task but is cleared once the task is completed.
2. **Long-term Memory**: Functions like a repository that retains patterns, insights, and user preferences from previous interactions over weeks or months. This memory allows the agent to make more tailored responses and decisions by learning from its past experiences.

The combination of short-term and long-term memory allows LLM agents to provide continuity and relevance across tasks and interactions.

- **Tool Use**: Tools significantly *extend the capabilities* of LLM agents by enabling them to *interact with external environments* and perform tasks such as **function calling**, data retrieval, and process automation. By leveraging these tools, agents can gather relevant information, execute complex workflows, and efficiently fulfill *user-defined objectives*, making them highly versatile in solving real-world problems. An important aspect of tool use is **Knowledge Augmentation**, which enhances an agent's knowledge by integrating tools that provide relevant and updated information. Examples include:

  - **RAG (Retrieval-Augmented Generation)**: Combines external retrieval systems (e.g., text retrievers, SQL executors) with the LLM to supply real-time context and improve response quality.
  - **Knowledge Graphs (KG)**: Structured repositories of interconnected information that allow agents to query and reason with organized data for decision-making and response generation.
  - **Web Browsing and APIs**: Public tools, such as web browsers or search APIs, help agents stay current by accessing real-time information like news, weather, events, and stock prices. This prevents agents from going stale due to outdated training data, reducing hallucinations and improving response relevance.

- **Inter-Agent Communication**: In multi-agent systems, an individual agent is not an isolated entity—it can communicate with other agents within the environment. This interaction enables collaboration, where agents can exchange information, coordinate strategies, and delegate subtasks. Such communication is crucial for solving complex problems that exceed the capability of a single agent, supporting scalability, distributed reasoning, and more adaptive behavior across the system.

16.11 Multi-agent Systems

Multi-agent systems offer a transformative paradigm for designing intelligent, modular, and scalable AI workflows. Instead of relying on a single monolithic model, these systems distribute intelligence across multiple specialized AI/LLM-based agents, each taking on one or more roles—such as **planner** (which decomposes high-level goals into executable steps), **reasoner** (which performs inference and contextual decision-making), **retriever** (which fetches relevant information from external sources like databases or knowledge graphs), **executor** (which carries out actions or API calls), and **evaluator** (which assesses the quality and correctness of outputs). This decomposition enables agents to communicate, collaborate, or delegate tasks according to their respective capabilities. As a result, multi-agent architectures enhance flexibility, fault tolerance, and efficiency. The topologies of multi-agent systems can be broadly categorized into several patterns, each supporting distinct coordination strategies and system-level objectives, as outlined below:

- **Single Agent**: A single LLM directly interacts with a set of predefined tools (e.g., APIs, databases). This straightforward architecture is ideal for simple pipelines where one model handles the core logic and delegates functionality to external tools.
- **Network**: Agents are interconnected in a peer-to-peer structure that allows for direct communication without a central controller. This decentralized topology enables collaborative reasoning, distributed decision-making, and robust task execution. Several specialized patterns can be seen as variants of the network topology:
  - **Sequential (Chain)**: Agents are arranged linearly, where the output of one serves as the input to the next. This topology is common in pipeline-style workflows, multi-stage reasoning, or stepwise refinement (e.g., planner $\rightarrow$ executor $\rightarrow$ verifier).
  - **Verifier Pair**: A minimal network of two agents in a bidirectional feedback loop, where one agent generates responses and the other critiques, verifies, or challenges them. This pattern is frequently used in self-consistency checking, adversarial evaluation, and red teaming.

  - **Federated**: A decentralized arrangement where agents (often distributed across devices or users) operate independently and periodically share updates or synchronize. This form supports privacy-preserving collaboration and model personalization across edge environments.

- **Supervisor**: A central agent manages a group of subordinate agents, coordinating their actions and aggregating their outputs. This centralized control structure enhances task assignment and oversight, particularly in complex multi-step workflows.
- **Hierarchical**: Agents are organized in a tiered structure, where upper-level agents delegate responsibilities to lower-level agents in a top-down manner. This design mirrors organizational hierarchies and supports scalable task decomposition and layered reasoning.
- **Custom**: A user-defined, arbitrary configuration of agents tailored to specific application requirements. This flexible topology allows for hybrid strategies, specialized interactions, and dynamic communication pathways that do not conform to standard patterns.

17 Autoencoders

Autoencoders are a class of neural networks designed for **unsupervised learning**, commonly used for dimensionality reduction, data denoising, and feature extraction. They operate by learning a compressed representation of the input data and then reconstructing the original input from this compressed representation.

An autoencoder consists of two main components (see Fig. 47):

1. **Encoder**:

   - The encoder maps the input data x to a lower-dimensional latent space representation z, also known as the **bottleneck**.
   - This step compresses the data into a smaller set of features while retaining the most important information.
   - Mathematically, the encoder is represented as a neural network f_θ parameterized by weights θ:

   $$z = f_\theta(x)$$

2. **Decoder**:

   - The decoder takes the latent representation z and reconstructs it back to the original input x'.
   - The objective is to make the reconstructed data x' as close as possible to the original input x.

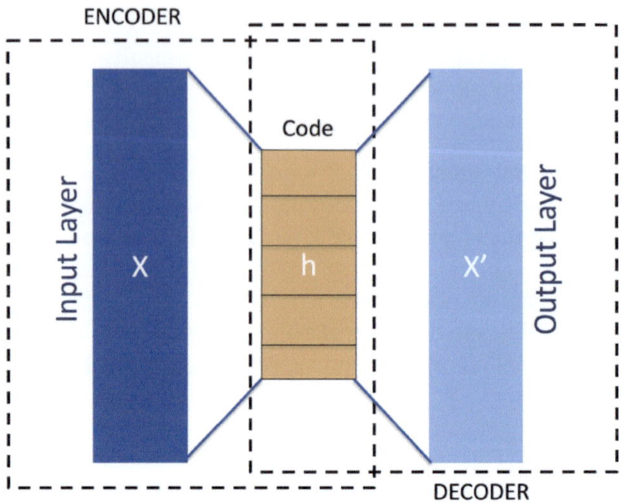

Fig. 47 A schema of an autoencoder. An autoencoder consists of two main parts: an encoder that maps the input message to a compressed representation (code), and a decoder that reconstructs the original message from the code

- Mathematically, the decoder is represented as a neural network g_ϕ parameterized by weights ϕ:

$$x' = g_\phi(z)$$

17.1 Training Objective

The training process of an autoencoder minimizes the **reconstruction error**, which quantifies the difference between the input x and the reconstructed output x'. Commonly used loss functions for this purpose include:

- **Mean Squared Error (MSE)**: Suitable for continuous input data, this loss measures the squared differences between x and x':

$$L(x, x') = \|x - x'\|^2$$

- **Binary Cross-Entropy (BCE)**: Used for binary or normalized input data, this loss evaluates the reconstruction error for each binary element:

$$L(x, x') = -\sum_i \left[x_i \log(x_i') + (1 - x_i) \log(1 - x_i')\right]$$

The optimization objective is to minimize the reconstruction loss with respect to the parameters of the encoder (θ) and decoder (ϕ):

$$\min_{\theta,\phi} L(x, x')$$

17.2 Applications of Autoencoders

Autoencoders are versatile tools with applications in a wide range of domains:

- **Dimensionality Reduction**: Extract compact features for visualization or as input for other machine learning models. **Example**: Reduce high-dimensional image data into a smaller latent space for clustering.
- **Denoising**: Learn to remove noise from input data by reconstructing clean data from noisy input. **Example**: Denoise corrupted images or audio signals.
- **Anomaly Detection**: Identify rare or abnormal patterns by observing high reconstruction error for inputs outside the training distribution. **Example**: Detect fraudulent transactions in financial data.
- **Generative Models**: Serve as a building block in more complex generative models, such as Generative Adversarial Networks (GANs). **Example**: Generate synthetic data, such as new images resembling the training data.
- **Feature Extraction**: Learn useful feature representations in an unsupervised manner for downstream tasks like classification or clustering. **Example**: Use latent space representations as inputs to a classification algorithm.

While autoencoders are powerful tools, they have several limitations:

- Autoencoders are not inherently probabilistic, which may limit their ability to model the true data distribution.
- Their performance depends heavily on the design of the architecture and the dimensionality of the latent space.
- They are prone to overfitting, particularly if the latent space is too large.

17.3 Variational AutoEncoders (VAE)

Figure 48 illustrates the concept of **Variational Autoencoders (VAE)**, a probabilistic generative model that learns to map input data x to a meaningful latent space representation z. VAE addresseses this problem using Bayesian inference. Bayesian inference provides a framework for updating beliefs about an unknown variable given

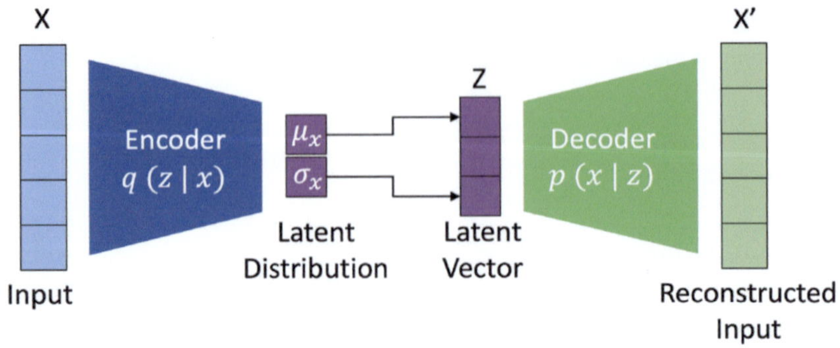

Fig. 48 Illustration of variational autoencoders (VAE)

observed data. In the context of VAEs, the objective is to infer the *posterior distribution* of the latent variable given the observed data, denoted as $p(z|x)$. This posterior is derived using Bayes' theorem:

$$p(z|x) = \frac{p(x|z)p(z)}{p(x)}$$

where $p(x|z)$ represents the *likelihood*, which describes how well a given latent variable (i.e., z) explains the observed data, $p(z)$ is the *prior* distribution over the latent variables, and $p(x)$ is the *evidence*, acting as a normalizing factor. Direct computation of $p(z|x)$ is intractable because it requires integrating over all possible values of z:

$$p(x) = \int p(x|z)p(z)dz$$

Since this integral is computationally expensive, VAEs approximate the true posterior $p(z|x)$ with a learned distribution $q(z|x)$, making inference tractable. To implement this, a VAE consists of two primary components:

- **An encoder (inference network)** that approximates the posterior distribution $p(z|x)$.
- **A decoder (generative network)** that reconstructs the input by modeling $p(x|z)$.

The encoder is responsible for approximating the posterior distribution $p(z|x)$, which is unknown and computationally intractable. Instead of computing it directly, the encoder learns a *variational approximation* $q(z|x)$, modeled as a Gaussian distribution:

$$q(z|x) = \mathcal{N}(z; \mu(x), \sigma^2(x))$$

where the mean $\mu(x)$ determines the center of the latent distribution, and the variance $\sigma^2(x)$ controls the spread of possible latent representations. The encoder network takes an input x and produces these parameters, which define a probability distribution over the latent space rather than a single deterministic point.

The decoder takes a latent variable z and reconstructs the original data by modeling the likelihood distribution $p(x|z)$. Given that the encoder maps each input to a distribution over latent variables, the decoder must learn a function that maps these latent representations back to the data space. The reconstructed output is modeled as a Gaussian distribution, where the decoder predicts both the mean and variance of the reconstructed data. The mean represents the expected value of the reconstructed output, while the variance accounts for the uncertainty in reconstruction, ensuring that the model can capture variations in the data rather than producing deterministic outputs. This formulation allows the decoder to generate realistic outputs while maintaining flexibility in modeling complex data distributions.

To ensure a structured latent space, a prior distribution $p(z)$ is imposed. This prior is typically chosen as a standard normal distribution:

$$p(z) = \mathcal{N}(z; 0, I)$$

The prior regularizes the latent space, ensuring that similar data points are mapped to nearby regions and that randomly sampled latent vectors generate meaningful outputs. The difference between the learned posterior $q(z|x)$ and the prior $p(z)$ is measured using Kullback-Leibler (KL) divergence:

$$D_{\text{KL}}(q(z|x)||p(z))$$

This regularization term prevents the model from overfitting and ensures that the latent space follows the desired probabilistic structure.

The training objective of a VAE consists of two terms:

- **Reconstruction Loss**: Ensures that the reconstructed data resembles the original input. This is typically measured using Mean Squared Error (MSE) for continuous data or Binary Cross-Entropy (BCE) for binary data.
- **KL Divergence**: Ensures that the learned posterior $q(z|x)$ remains close to the prior $p(z)$.

The final **VAE loss function** combines these two terms:

$$\mathcal{L}_{\text{VAE}} = \mathcal{L}_{\text{recon}} + \beta \mathcal{L}_{\text{KL}}$$

where β is a weighting factor introduced in *β-VAE*, which controls the trade-off between reconstruction accuracy and latent space regularization.

The structured nature of VAEs makes them particularly useful for generative modeling, data augmentation, and representation learning tasks. By ensuring that the latent space follows a well-defined probabilistic structure, VAEs enable the generation of new data points that maintain the essential characteristics of the original dataset.

This point should be emphasized that VAEs are **generative models** because they learn a structured latent space where new data samples can be generated by drawing random vectors z and passing them through the decoder. This allows VAEs to create **diverse, realistic variations** of the original dataset.

17.4 Vector Quantized Variational Autoencoder (VQ-VAE)

Vector Quantized Variational Autoencoder (VQ-VAE) is a type of generative model that learns a *discrete latent space* instead of a continuous one, as in standard VAEs. This is achieved by introducing a *codebook* of discrete latent embeddings and replacing the continuous latent variables with the nearest entry from this learned set of discrete vectors. VQ-VAE follows a typical encoder-decoder architecture but differs in the latent space representation. The model consists of:

- **Encoder**: Maps the input x to a continuous latent representation $z_e(x)$.
- **Vector Quantization (Discrete Latent Space)**: The continuous latent representation is replaced by the closest entry from a learned set of discrete embeddings (codebook). The quantized latent representation $z_q(x)$ replaces the continuous latent variable ($z_e(x)$) with a discrete embedding, where e_k is the selected codebook entry from a set of learned discrete embeddings $\{e_1, e_2, \ldots, e_K\}$. The quantization process is defined as:

$$z_q(x) = e_k \quad \text{where} \quad k = \arg\min_j \|z_e(x) - e_j\|_2$$

  where k is the index corresponding to the closest codebook entry to the encoded representation $z_e(x)$, and j represents an index iterating over all possible codebook entries. Each codebook entry e_j is the j-th embedding in the learned codebook, and the Euclidean distance $\|z_e(x) - e_j\|_2$ measures the closeness between the encoded representation $z_e(x)$ and each codebook entry, ensuring that the nearest embedding is selected.
- **Decoder**: Reconstructs the input from the quantized latent representation.

18 Diffusion Models

Diffusion models are a class of generative models that simulate a stochastic diffusion process to generate data. Particularly, diffusion models have achieved remarkable success in tasks such as image generation, text-to-image synthesis, and other generative applications. These models operate by gradually transforming data through a **forward diffusion process**, where structured data is progressively mapped into

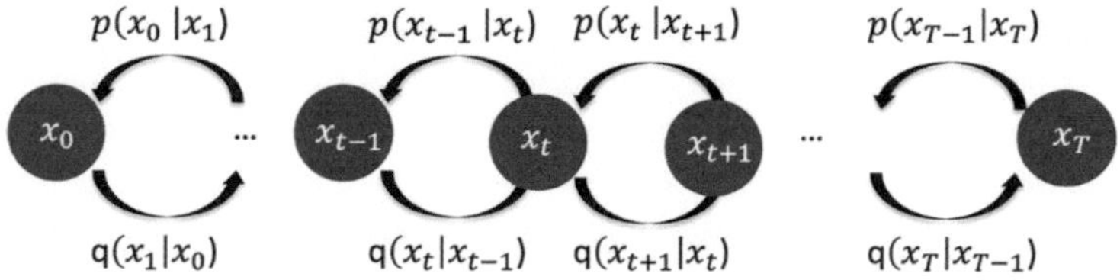

Fig. 49 Illustration of the forward and reverse diffusion processes in DDPMs. The forward process progressively adds noise to structured data, while the reverse process iteratively removes noise to reconstruct meaningful samples

a predefined prior distribution, and a **reverse diffusion process**, where the model learns to reconstruct realistic data samples from the prior.

- **Forward Diffusion**: The forward process begins with a sample drawn from the data distribution, which is subjected to progressively increasing levels of noise. Over a series of iterations, the data becomes increasingly degraded, typically culminating in a simple, tractable noise distribution (e.g., Gaussian). This process effectively transforms complex, structured data into a simpler, noise-like representation.
- **Reverse Diffusion**: To generate new data, the model reverses the forward diffusion process. Starting with a simple noise distribution, such as Gaussian noise, the model iteratively removes the noise and refines the data, gradually reconstructing it into a form that approximates the original data distribution. This reverse process is learned during training, enabling the model to produce realistic samples that closely resemble the target distribution.

Recent advancements in diffusion models revolve around several key formulations, each distinguished by its unique approach and underlying mechanisms. The most notable among these is **Denoising Diffusion Probabilistic Models (DDPMs)** [23]. DDPMs are a class of generative models that utilize a two-step Markov process: a **forward diffusion process**, which progressively corrupts structured data into noise, and a **reverse denoising process**, which reconstructs data from noise (see Fig. 49) [24].

18.1 Forward Diffusion Process

Consider an image that we progressively degrade over a sequence of steps until it becomes indistinguishable from pure noise. The goal of this process is to map the original data distribution into a standardized noise distribution while maintaining both mathematical tractability and reversibility—essential properties for generative modeling. This raises a fundamental question: *How can noise be systematically introduced in a manner that allows for both analytical tractability and an invertible generative process?* A natural approach is to define a transformation function that incrementally introduces noise at each step while controlling its variance. Iterating this process results in progressively noisier versions of the original image. A formal description of this process involves defining a Markov chain where each step depends only on the preceding step. Given an initial data distribution $x_0 \sim q(x_0)$, a sequence of progressively noisier random variables $x_1, x_2, \ldots, x_T$ is generated according to a transition kernel [23]:

$$q(x_t|x_{t-1}).$$

Since the process follows the *Markov property*, by applying the chain rule of probability, the joint distribution of the entire sequence, given x_0, can be written as:

$$q(x_1, x_2, \ldots, x_T|x_0) = \prod_{t=1}^{T} q(x_t|x_{t-1}).$$

A widely adopted transition kernel for this process is a Gaussian perturbation model, where each step x_t is obtained from x_{t-1} by adding Gaussian noise:

$$q(x_t|x_{t-1}) = \mathcal{N}(x_t; \sqrt{1-\beta_t}x_{t-1}, \beta_t I).$$

Here, β_t is a predefined **noise schedule** that controls the amount of noise introduced at each step. β_t can be learned or held constant as hyperparameters. Thereby, given a sample x_{t-1}, the next state x_t is computed as:

$$x_t = \sqrt{1-\beta_t}x_{t-1} + \sqrt{\beta_t}\boldsymbol{\epsilon}_{t-1},$$

where $\boldsymbol{\epsilon}_{t-1} \sim \mathcal{N}(0, I)$ is independent Gaussian noise. Instead of computing the diffusion process step-by-step, it is often desirable to directly obtain a noised version of x_t from x_0 without sequentially applying the transition kernel. We define $\alpha_t = 1 - \beta_t$ and $\bar{\alpha}_t$ as the cumulative product of α_i from $i = 1$ to t (i.e., $\bar{\alpha}_t = \prod_{i=1}^{t} \alpha_i$). Using these definitions, the following relationship can be established [4]:

$$\begin{aligned}
\mathbf{x}_t &= \sqrt{\alpha_t}\mathbf{x}_{t-1} + \sqrt{1-\alpha_t}\boldsymbol{\epsilon}_{t-1} \\
&= \sqrt{\alpha_t\alpha_{t-1}}\mathbf{x}_{t-2} + \sqrt{1-\alpha_t\alpha_{t-1}}\boldsymbol{\epsilon} \\
&= \ldots \\
&= \sqrt{\bar{\alpha}_t}\mathbf{x}_0 + \sqrt{1-\bar{\alpha}_t}\boldsymbol{\epsilon}
\end{aligned}$$

The second line leverages the mathematical property that the sum of two Gaussian-distributed variables remains Gaussian. This property enables a direct transition from the original data $\mathbf{x}_0$ to any point in the forward diffusion process $\mathbf{x}_t$, thereby circumventing the need to explicitly compute each intermediate step. Rather than parameterizing the diffusion process in terms of the original noise variance terms β_t, it is possible to use $\bar{\alpha}_t$, which provides a more interpretable formulation. Specifically, $\bar{\alpha}_t$ quantifies the proportion of variance attributable to the original signal ($\mathbf{x}_0$), while $1 - \bar{\alpha}_t$ corresponds to the variance contribution from the noise component ($\boldsymbol{\epsilon}$). Accordingly, the forward diffusion process q can be expressed as [4, 23]:

$$q(\mathbf{x}_t|\mathbf{x}_0) = \mathcal{N}\left(\mathbf{x}_t; \sqrt{\bar{\alpha}_t}\mathbf{x}_0, (1 - \bar{\alpha}_t)\mathbf{I}\right)$$

Choice of Diffusion Schedule. The noise schedule β_t need not be uniform across all timesteps. The function that governs how β_t (or equivalently, $\bar{\alpha}_t$) varies with t is known as the *diffusion schedule*. In the original DDPM paper, a **linear diffusion schedule** was proposed, where β_t increases linearly with t, ranging from $\beta_1 = 0.0001$ to $\beta_T = 0.02$. Later research found that a **cosine diffusion schedule** can outperform the linear schedule. The cosine schedule adjusts β_t in a way that provides smoother noise introduction, leading to enhanced generative performance [23].

18.2 Reverse Denoising Process

In DDPMs, the reverse and iterative denoising process reconstructs structured data from noise by iteratively refining a noisy sample through a learned Markov chain. The reverse process starts with a sample drawn from a Gaussian prior [23]:

$$p(x_T) = \mathcal{N}(x_T; 0, I), \tag{9}$$

and follows a learned transition model:

$$p_\theta(x_{t-1}|x_t) = \mathcal{N}(x_{t-1}; \mu_\theta(x_t, t), \Sigma_\theta(x_t, t)), \tag{10}$$

where $\mu_\theta(x_t, t)$ represents the predicted mean, and $\Sigma_\theta(x_t, t)$ denotes the predicted variance, which can be either fixed or learned. In their original formulation, DDPMs used a fixed variance $\Sigma_\theta(x_t, t) = \sigma_t^2 I$, with empirical studies showing that setting [23]:

$$\sigma_t^2 = \beta_t \tag{11}$$

or

$$\sigma_t^2 = \beta_t \frac{1 - \bar{\alpha}_{t-1}}{1 - \bar{\alpha}_t} \tag{12}$$

produced almost comparable results. A critical challenge in this process is accurately estimating $\mu_\theta(x_t, t)$, as it determines the denoising trajectory. Instead of directly predicting $\mu_\theta(x_t, t)$, the key idea of DDPM is to train and use a neural network parameterized by θ to estimate the noise component $\epsilon_\theta(x_t, t)$. This approach simplifies the learning process, as the noise is a well-behaved quantity that follows a known Gaussian distribution [4, 23].

Since the mean $\mu_\theta(x_t, t)$ can be expressed in terms of $\epsilon_\theta(x_t, t)$ as:

$$\mu_\theta(x_t, t) = \frac{1}{\sqrt{\alpha_t}}\left(x_t - \frac{1-\alpha_t}{\sqrt{1-\bar{\alpha}_t}}\epsilon_\theta(x_t, t)\right), \tag{13}$$

estimating $\epsilon_\theta(x_t, t)$ effectively allows the model to approximate $\mu_\theta(x_t, t)$. To achieve this, a U-Net architecture is employed, leveraging its ability to capture multi-scale features efficiently. The model is trained to ensure the reverse process accurately mirrors the forward diffusion process. The training procedure is detailed in Algorithm 3. The primary objective in DDPMs is minimizing [23]:

$$\mathcal{L}_{\text{MSE}} = \mathbb{E}_{x_0,\epsilon,t}\left[\|\epsilon - \epsilon_\theta(x_t, t)\|^2\right]. \tag{14}$$

By training the model to recover the added noise ϵ, it learns to denoise inputs effectively across all diffusion steps. This approach, focusing on noise prediction rather than direct data reconstruction, simplifies training while improving stability and performance [23].

Algorithm 3 Training Procedure for DDPMs

1: **repeat**
2: Sample $x_0 \sim q(x_0)$ from the data distribution.
3: Select a random timestep $t \sim \text{Uniform}(\{1, \ldots, T\})$.
4: Sample Gaussian noise $\epsilon \sim \mathcal{N}(0, I)$.
5: Construct a noisy sample:

$$x_t = \sqrt{\bar{\alpha}_t}x_0 + \sqrt{1-\bar{\alpha}_t}\epsilon.$$

6: Perform a gradient descent step to minimize the loss:

$$\mathcal{L}_{\text{MSE}} = \mathbb{E}_{x_0,\epsilon,t}\left[\|\epsilon - \epsilon_\theta(x_t, t)\|^2\right].$$

7: **until** convergence

The model employs a U-Net architecture (see Fig. 50), an encoder-decoder network with skip connections that preserve spatial details. It consists of [4]:

- **Encoder**: Composed of DownBlocks that progressively reduce spatial resolution while increasing feature channels, capturing hierarchical representations.
- **Decoder**: Composed of UpBlocks that incrementally restore spatial resolution by upsampling and decreasing the number of channels, reconstructing fine details.
- **Bottleneck (Bridge)**: The bottleneck is the lowest part of the U-Net. In a traditional U-Net, the bottleneck typically consists of two or more convolutional layers without downsampling (with ReLU activations and batch normalization) and serves as the transition between the downward and upward paths.

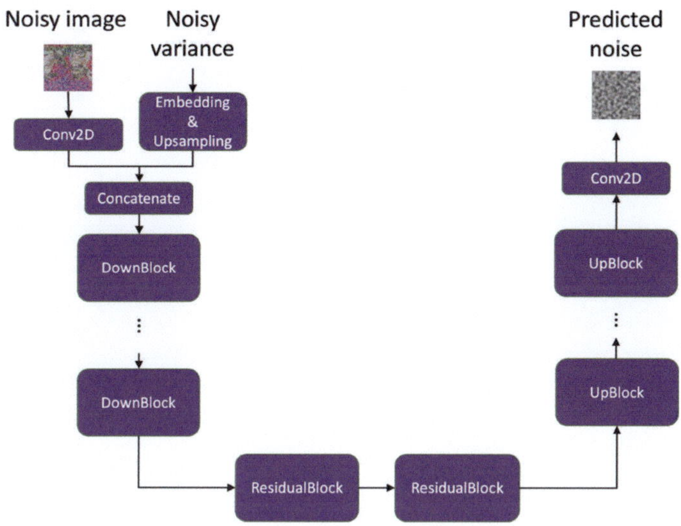

Fig. 50 An example of U-Net architecture for noise estimation in DDPMs [4]

- **Residual Blocks (Optional)**: Some modern U-Net variants incorporate Residual Blocks instead of standard convolutional layers in the bridge. This allows for better gradient flow and prevents vanishing gradients in deep architectures.
- **Skip Connections**: Directly link corresponding encoder and decoder layers, ensuring the preservation of critical spatial information and enhancing denoising accuracy.

Mathematically, the noise prediction function is:

$$\epsilon_\theta(x_t, \sigma_t) = f_{\text{U-Net}}(x_t, \sigma_t; \theta), \tag{15}$$

where $f_{\text{U-Net}}$ represents the parameterized U-Net function. The model conditions its predictions on both the noisy input x_t and the noise variance σ_t (or an equivalent derived parameter such as β_t or $\bar{\alpha}_t$), dynamically adapting to different stages of the diffusion process.

18.3 Diffusion Transformers (DiT)

After the introduction of **U-Net-based diffusion models**, researchers have explored **Transformer-based models** for diffusion processes, such as **Diffusion Transformers (DiT)** (see Fig. 51) [25]. Unlike traditional diffusion models that rely on **U-Net architectures**, DiT replaces convolutions with **self-attention mechanisms**, making it more effective for high-resolution image generation. DiT operates in **latent**

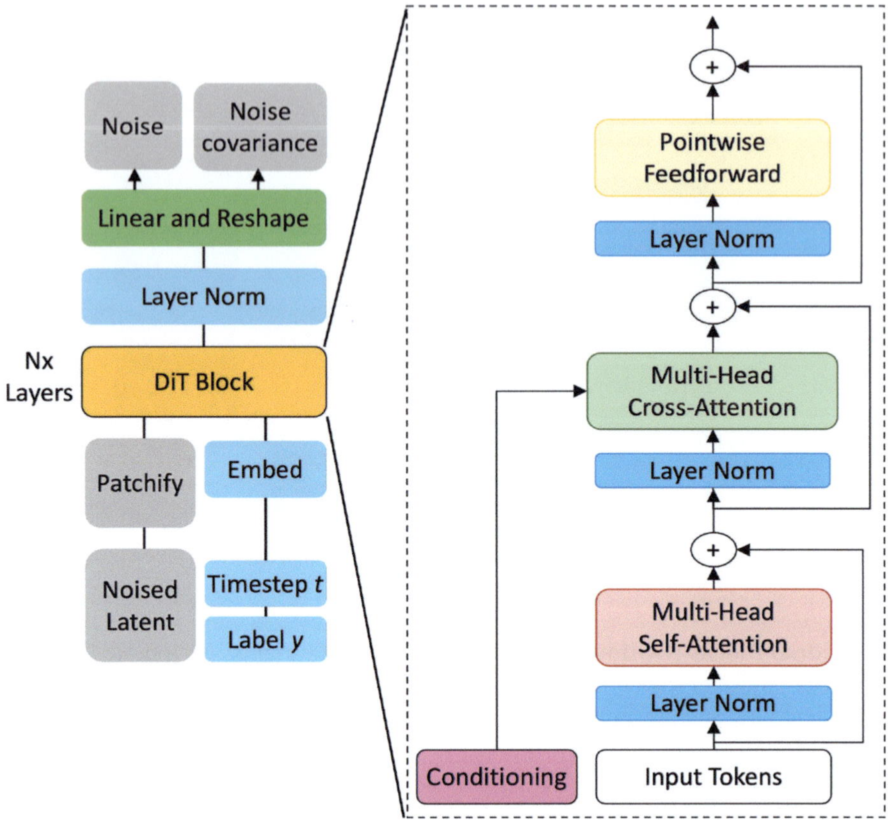

Fig. 51 Overview of the Diffusion Transformer (DiT) architecture. The model replaces U-Net with a transformer-based structure, processing noised latent representations through self-attention and cross-attention layers while incorporating conditioning information [25]

space, where the input is a **noised latent representation** of an image, divided into **patches** similar to Vision Transformers (ViTs). The **core architecture** consists of multiple **DiT blocks**, which apply **multi-head self-attention** to capture global dependencies and **cross-attention** for conditioning. Conditional diffusion models extend this approach by incorporating **additional information, such as class labels, text prompts, or other structured inputs**, alongside the noisy image at each timestep. This conditioning mechanism improves **controllability** over the generated outputs, allowing the model to synthesize images that match specific attributes or categories. For example, in **class-conditional diffusion models**, the network is trained to denoise images while considering a given **class label**, ensuring that the generated images belong to the corresponding category. The conditioning information is typically integrated through **cross-attention layers** or concatenated with the latent representation before passing through the diffusion network.

19 Generative Adversarial Networks (GANs)

Generative Adversarial Networks (GANs) are a class of **generative models** introduced in 2014. GANs consist of two neural networks, a generator and a discriminator, that compete against each other in a zero-sum game. They are widely used for generating synthetic data, such as images, audio, and text.

19.1 Architecture

A GAN consists of two primary components: the **generator** and the **discriminator**. These components work in an adversarial manner, with the generator attempting to create realistic data and the discriminator evaluating whether the data is real or fake. Their interplay forms the foundation of the GAN framework (see Fig. 52).

19.1.1 Generator

The generator in a GAN is a neural network that maps random noise vectors z (sampled from a prior distribution p_z, such as a normal or uniform distribution) to synthetic data $G(z)$. The goal of the generator is to learn the underlying data distribution $p_{\text{data}}(x)$ and produce outputs that are indistinguishable from real data samples.

- **Input: Noise Vector** (z): The generator takes as input a random noise vector $z \in \mathbb{R}^d$, sampled from a prior distribution (e.g., Gaussian or uniform distribution). The

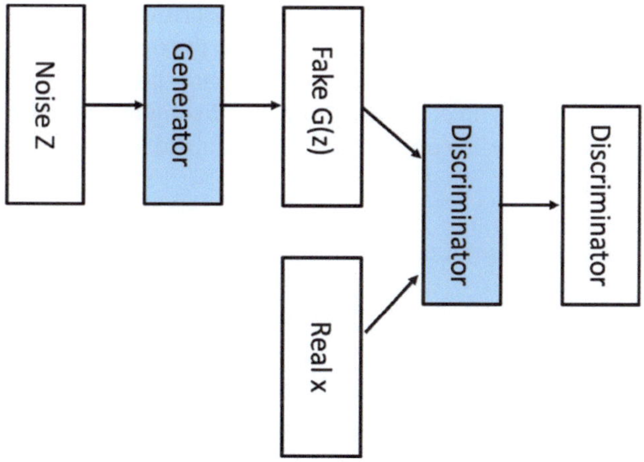

Fig. 52 An illustration of how a Generative Adversarial Network (GAN) works

noise vector acts as a seed to generate diverse outputs, encouraging the generator to capture the entire data distribution.

- **Network Structure**: The generator is typically implemented as a deep neural network, and its structure depends on the type of data being generated:
  - For low-dimensional data, fully connected (dense) networks are often sufficient.
  - For image data, convolutional networks are used, with layers designed to progressively upsample the input noise vector into a high-resolution image.
  - For sequential data like text or time series, recurrent or transformer-based architectures may be employed to handle temporal dependencies.

Several widely used generator architectures have emerged, tailored to specific tasks, such as:

- **Deep Convolutional GAN (DCGAN)**: Employs transposed convolutional layers to upsample the noise vector z, progressively creating synthetic images of higher resolutions. DCGAN's generator is particularly effective for unsupervised generation tasks, such as creating photorealistic images from random noise.
- **U-Net Generator**: Widely used in conditional GANs (cGANs) for tasks like image-to-image translation, where the generator must transform one type of input (e.g., a sketch) into another type of output (e.g., a photorealistic image). The U-Net architecture incorporates skip connections between the encoder and decoder, allowing it to retain fine-grained details from the input while generating the output.

19.1.2 Discriminator

The discriminator is a neural network that acts as a binary classifier. It takes an input (either real data x or synthetic data $G(z)$) and outputs a probability indicating whether the input is real or fake. The output of the discriminator, denoted as $D(x)$, is the probability that the input x is real. If $D(x) = 1$, the discriminator is completely confident that x is real. If $D(x) = 0$, the discriminator is completely confident that x is fake. The discriminator's objective is to correctly distinguish between real and synthetic data. Mathematically, the discriminator D is trained to maximize the following objective:

$$\max_D V(D, G) = \mathbb{E}_{x\sim p_{\text{data}}(x)}[\log D(x)] + \mathbb{E}_{z\sim p_z(z)}[\log(1 - D(G(z)))].$$

In this equation, the first term, $\mathbb{E}_{x\sim p_{\text{data}}(x)}[\log D(x)]$, represents the expected log-likelihood of the discriminator correctly classifying real data x as real. The discriminator achieves this by assigning $D(x)$, the probability it calculates for x being real, close to 1 for real data. The second term, $\mathbb{E}_{z\sim p_z(z)}[\log(1 - D(G(z)))]$, represents the expected log-likelihood of the discriminator correctly classifying synthetic data $G(z)$ as fake. In this case, the discriminator assigns $D(G(z))$, the probability it calculates for $G(z)$ being real, close to 0 for synthetic data.

19.2 Training Process

The generator and discriminator are trained simultaneously in a minimax game, where the generator tries to minimize the same objective by fooling the discriminator. The training of GANs alternates between the following steps:

- Updating the discriminator: The discriminator is trained to improve its ability to distinguish real data from synthetic data generated by the generator.
- Updating the generator: The generator is trained to improve its ability to generate data that fools the discriminator.

Mathematically, this training process is formulated as a minimax game with the following objective function:

$$\min_G \max_D V(D, G) = \mathbb{E}_{x \sim p_{\text{data}}(x)}[\log D(x)] + \mathbb{E}_{z \sim p_z(z)}[\log(1 - D(G(z)))].$$

where $x \sim p_{\text{data}}(x)$ represents samples x drawn from the real data distribution p_{data}. $z \sim p_z(z)$ represents samples z drawn from the prior distribution p_z, which is typically a simple distribution (e.g., Gaussian or uniform). z is the latent variable or noise input to the generator. $G(z)$ represents the synthetic data generated by the generator G using the noise z. $1 - D(G(z))$ shows the probability that the discriminator classifies the synthetic data $G(z)$ as fake. $\mathbb{E}_{x \sim p_{\text{data}}(x)}[\log D(x)]$ shows the expected log-likelihood of the discriminator correctly classifying real data as real. Finally, $\mathbb{E}_{z \sim p_z(z)}[\log(1 - D(G(z)))]$ captures expected log-likelihood of the discriminator correctly classifying synthetic data as fake. The adversarial nature of this training process ensures that the generator learns to produce data that closely matches the real data distribution, while the discriminator becomes increasingly robust in detecting fake data.

20 Multimodal Models

This section discusses groundbreaking multimodal models—such as CLIP, Flamingo, GLIDE, DALL-E, DALL-E 2, and Stable Diffusion—that have been integrated into other architectures and continue to be adapted as foundational building blocks for advanced AI systems.

20.1 Contrastive Language-Image Pre-training (CLIP)

Contrastive Language-Image Pre-training (CLIP) is a model developed by OpenAI that learns visual concepts from natural language descriptions [17]. CLIP is trained using **contrastive learning**, enabling it to associate images with textual descriptions without explicit supervision. Note that contrastive learning is a self-supervised learning technique that learns effective representations by distinguishing between similar and dissimilar data points. The key idea is to map data into an embedding space where similar samples are closer together, and dissimilar ones are farther apart. This process is structured around three key elements:

- **Anchor**: The reference sample whose representation is being learned; it serves as the starting point in contrastive learning, guiding how the model differentiates between similar and dissimilar examples.
- **Positive**: A sample that is semantically or structurally similar to the anchor; the model is trained to bring the positive example closer to the anchor in the learned embedding space, reinforcing similarity.
- **Negative**: A sample that is semantically or structurally different from the anchor; the model is trained to push the negative example farther from the anchor, ensuring better discrimination between different categories or concepts.

In this context, CLIP employs a **dual-encoder architecture**, consisting of separate encoders for images and text, which are trained jointly to project their respective inputs into a **shared latent space** where their relationships can be compared (see Fig. 53). The key components of CLIP's architecture include:

- **Image Encoder**: Extracts feature representations from images.
- **Text Encoder (Transformer-based)**: Converts text descriptions into high-dimensional vector representations.

CLIP is trained using a method where a batch of N (image, text) pairs is used to learn a shared multi-modal embedding space. The image encoder and the text encoder process the images and texts separately, mapping them into a common feature space. The goal is to maximize the cosine similarity of the correct (image, text) pairs while minimizing the similarity of incorrect pairings within the batch. This results in an

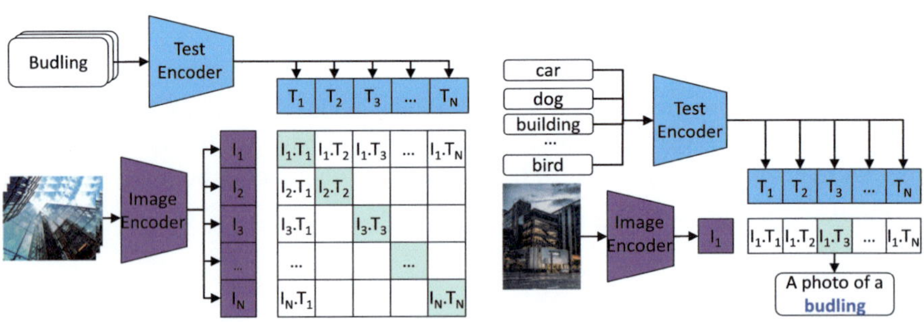

Fig. 53 Overview of CLIP's contrastive learning framework. CLIP jointly optimizes an image encoder and a text encoder to align corresponding image-text pairs. During inference, the trained text encoder constructs a zero-shot classifier by embedding the names/descriptions of the target dataset's classes [17]

$N \times N$ similarity matrix, where the diagonal contains correct matches, and the off-diagonal elements represent mismatches. Once trained, CLIP can perform a wide range of vision-language tasks, such as:

- **Zero-Shot Computer Vision Classifier**: CLIP can classify images into categories without specific training on those categories, allowing it to recognize new objects or concepts without additional labeled data.
- **Image and Text Retrieval**: CLIP enables **visual search**, allowing users to input text descriptions to find relevant images, and supports **image-to-text matching** by retrieving descriptions that best match a given image.
- **Image Captioning**: CLIP can generate descriptive captions for images, aiding accessibility for visually impaired users and improving content categorization.
- **Visual Question Answering (VQA)**: CLIP allows users to ask questions about an image (e.g., "What is the person holding?") and receive meaningful text-based answers, benefiting automation and AI-assisted interpretation.

20.2 Flamingo

Flamingo is a **Visual Language Model (VLM)** that seamlessly integrates **text, images, and videos** by combining **frozen vision encoders** with a frozen **large language model (LLM)**. To enable effective vision-language interaction, it incorporates trainable **bridging components**, specifically the **Perceiver Resampler** for compressing visual features and **Gated Cross-Attention Dense (XATTN-DENSE) layers** for integrating visual context into language processing (see Fig. 54).

20.2.1 Perceiver Resampler: Efficiently Compressing Visual Features

A key challenge in multimodal learning is handling **high-dimensional visual data** from images and videos. The **Perceiver Resampler** addresses this by **converting a variable-sized grid of vision features into a fixed set of latent tokens**, making it computationally feasible to integrate into the language model (see Fig. 55). The **vision encoder** extracts a **spatial grid of feature embeddings** from images or videos. These features are then processed by the **Perceiver Resampler**, which learns a **compact representation** using attention layers. Specifically, it takes **trainable latent vectors as queries** (Q), while the **keys** (K) and **values** (V) are derived from the extracted visual features.

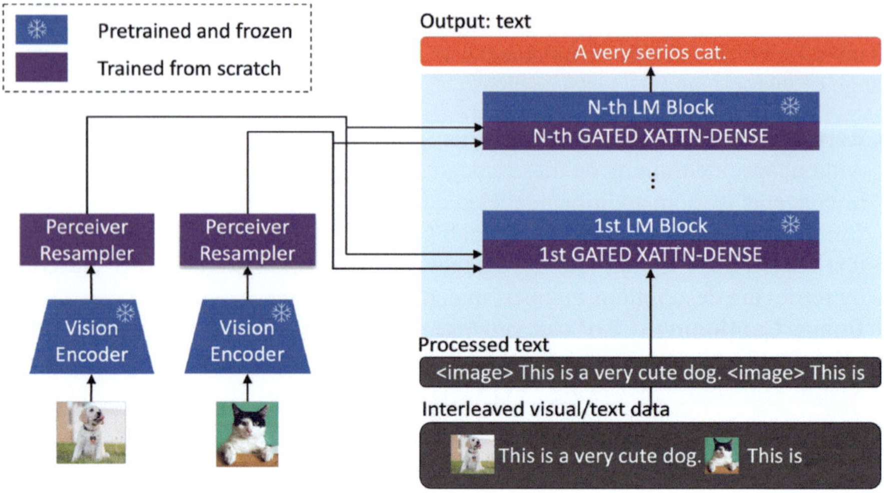

Fig. 54 Overview of Flamingo's architecture. The model processes interleaved visual and textual data, leveraging a frozen vision encoder and a frozen language model, connected via a Perceiver Resampler and Gated Cross-Attention Dense (XATTN-DENSE) layers [10]

Fig. 55 The Perceiver Resampler module converts a variable-sized grid of spatio-temporal visual features from the Vision Encoder into a fixed number of output tokens (five in this example), ensuring consistency across different input resolutions and video frame counts. It leverages a transformer mechanism, where learned latent vectors serve as queries, and the keys and values comprise both the extracted visual features and these latent vectors [10]

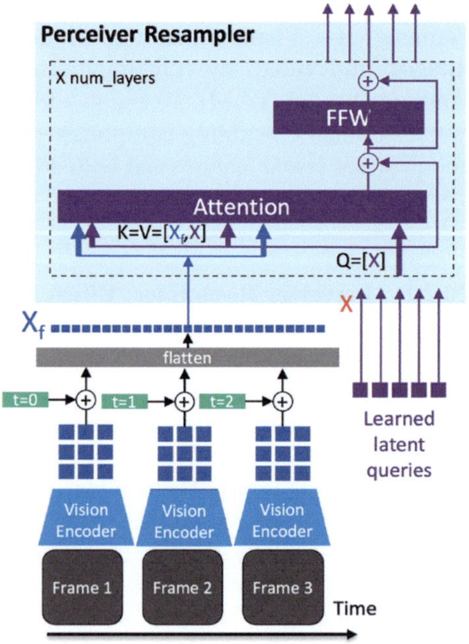

20.2.2 Gated Cross-Attention Dense (XATTN-DENSE): Injecting Vision into Language

To effectively incorporate **visual context into language understanding**, Flamingo introduces **Gated Cross-Attention Dense (XATTN-DENSE) layers** between frozen LLM layers (see Fig. 56). These layers allow the model to process language while selectively integrating relevant visual information. Each **XATTN-DENSE layer** consists of a **cross-attention mechanism**, where the **queries** (Q) originate from the language tokens, while the **keys** (K) and **values** (V) come from the visual tokens generated by the Perceiver Resampler. This enables the model to focus on the most relevant aspects of the image or video when generating text. By inserting these **vision-aware layers between the existing LM layers**, the model can process enriched language representations while preserving the original language model structure. Note that **tanh gating** is also applied at each stage of the **XATTN-DENSE module**. This mechanism **modulates the influence of visual features** and regulates how much visual context is injected into the language stream.

20.3 GLIDE

GLIDE (Guided Language-to-Image Diffusion for Generation and Editing) is a text-conditioned diffusion model developed by OpenAI for generating high-quality images from textual descriptions. Unlike traditional diffusion models, GLIDE uses

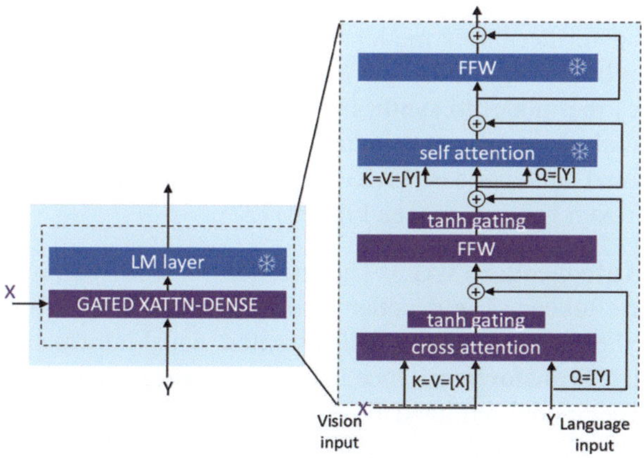

Fig. 56 To integrate visual information into the language model, new cross-attention layers are introduced between pretrained and frozen LM layers. These layers use the vision features as keys and values, while the language inputs serve as queries. The outputs are processed by dense feed-forward layers before being passed into the frozen LM layers. A gating mechanism regulates the influence of visual inputs, ensuring that the LM layers receive a stable, vision-conditioned language representation while preserving their pretrained knowledge [10]

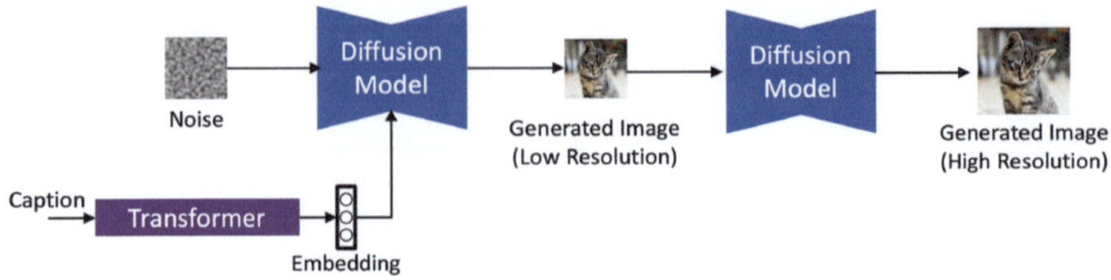

Fig. 57 Overview of GLIDE architecture: a two-stage process where a Text Conditional Diffusion Model generates a low-resolution (64×64) image, which is then upscaled to 256×256 using a Super-Resolution model [26]

natural language prompts, allowing it to generate novel and diverse images. The model consists of two main components (see Fig. 57) [26]:

- **Text Conditional Diffusion Model**: This module generates a low-resolution (64×64) image based on a given text description. It uses a **Transformer** to encode the text into embeddings, which condition the diffusion process to guide image generation.
- **Super-Resolution Model**: Once a 64×64 image is generated, this second module upscales it to 256×256 using another diffusion model trained for super-resolution.

20.4 DALL-E

DALL-E is a transformer-based model that generates images from textual descriptions by treating both text and images as a unified sequence of data tokens. This approach enables the model to synthesize highly detailed and semantically meaningful visuals purely from language inputs. The training process consists of two key stages: a variational autoencoder (VAE) for encoding images and an autoregressive transformer for token generation (see Fig. 58) [27].

- **Variational Autoencoder (VAE)**: The encoder transforms an input image into a grid of discrete tokens. The decoder reconstructs the image from these tokens, ensuring that the representation retains essential visual information.
- **Autoregressive Transformer**: Once images are represented as discrete tokens, a large transformer model is trained to predict these tokens sequentially, conditioned on text prompts. The process involves:
  - Encoding text descriptions into token sequences using a learned vocabulary.
  - Generating image tokens step by step in an autoregressive manner, ensuring consistency between text and visual content.
  - Finally, the VAE decoder is used to transform all the generated image tokens back into a complete image.

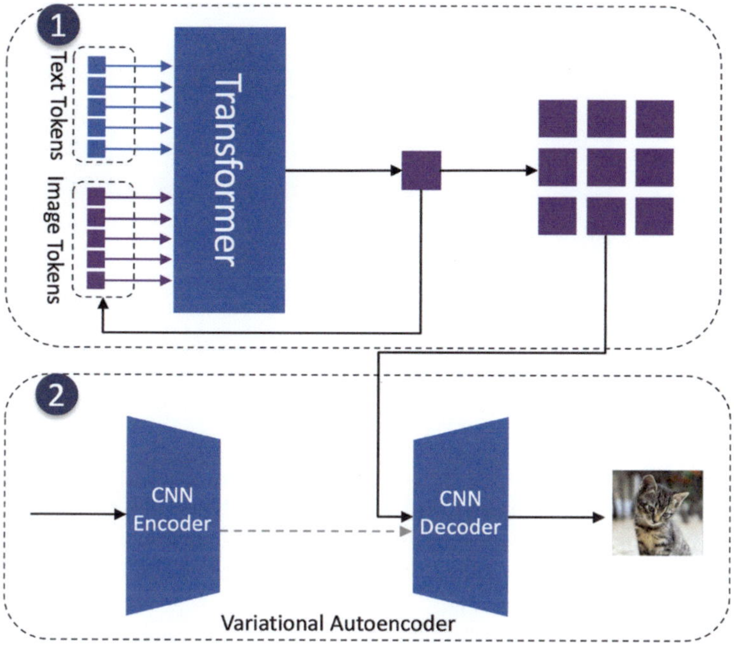

Fig. 58 Architecture of DALL·E: The model follows an autoregressive approach using a transformer. At each step, it generates an image token conditioned on the text prompt and previously generated tokens. This process continues until a full sequence of image tokens is produced. These tokens are then passed into the decoder part of a pre-trained autoencoder, which reconstructs the final image from the discrete token representation [4, 28]

20.5 DALL-E 2

DALL-E 2 operates through a structured pipeline comprising the following three key components [4, 28]:

- **CLIP**: The process begins with Contrastive Language-Image Pretraining (CLIP) to encode the input text into a numerical representation that captures its semantic content. By leveraging this pre-trained model, DALL-E 2 ensures that the generated images are well-aligned with the intended meaning of the text.
- **Prior**: The prior model acts as a bridge, converting CLIP's text embedding into its corresponding image embedding, effectively translating linguistic concepts into visual representations. This transformation is accomplished through one of two approaches:
  - **Autoregressive Model**: The autoregressive prior in DALL-E 2 is an **encoder-decoder Transformer** designed to generate a CLIP image embedding based on a corresponding CLIP text embedding (see Fig. 59). This process unfolds sequentially, with each step predicting elements of the image embedding while

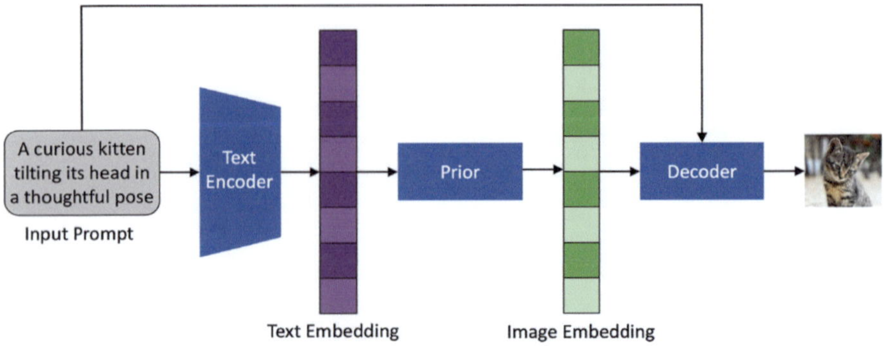

Fig. 59 DALL-E 2 architecture for text-to-image generation [4]

conditioning on previously generated outputs. The Transformer encoder first processes the text embedding, which is subsequently fed into the decoder alongside the evolving image embedding. Training utilizes the **teacher forcing** technique (i.e., the model is guided by the actual CLIP image embedding at each step) to optimize performance.

- **Diffusion Model**: The diffusion-based prior in DALL-E 2 employs a **Transformer-based diffusion model** to generate a CLIP image embedding from a CLIP text embedding. During training, paired CLIP text and image embeddings are concatenated into a single vector, and noise is progressively added to the image embedding over multiple steps until it becomes indistinguishable from random noise. The model is trained to reverse this process, using the text embedding as a conditioning signal to iteratively denoise the image embedding.

- **Decoder**: The final stage of the pipeline is the **decoder**, which is responsible for generating the image from the processed embedding. This component is adapted from **GLIDE**. However, unlike GLIDE, **DALL-E 2** leverages **CLIP-guided embeddings**, ensuring greater coherence and stronger semantic alignment between the textual prompt and the resulting image [4] (Fig. 60).

20.6 Stable Diffusion

Stable Diffusion is a **Latent Diffusion Model (LDM)** designed for efficient and high-quality image synthesis based on textual descriptions [29]. Unlike traditional diffusion models that operate in pixel space, Stable Diffusion performs the generative process in a **lower-dimensional latent space**. The architecture consists of three main components (see Fig. 61) [4, 29]: (1) a **Variational Autoencoder (VAE)**, (2) a **Denoising U-Net**, and (3) a **Text Encoder**. The **VAE encoder** compresses images

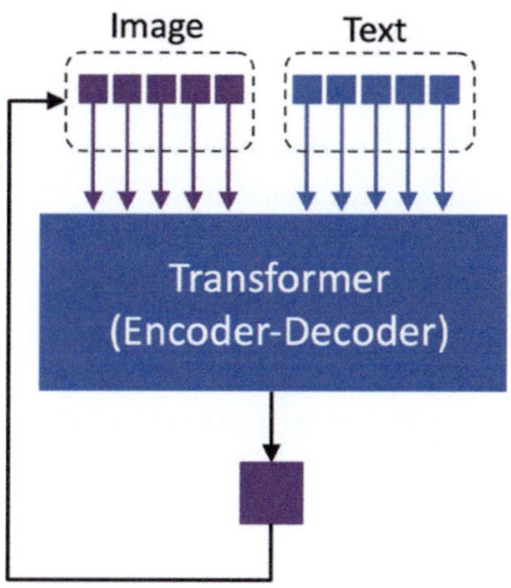

Fig. 60 Abstract representation of the autoregressive prior in DALL-E 2

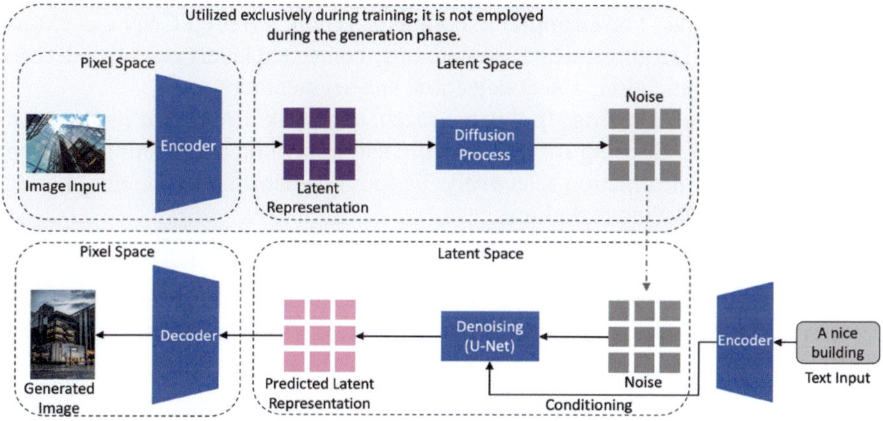

Fig. 61 Architectural diagram of the stable diffusion model [4]

into a latent space representation, where the diffusion process takes place. During training, Gaussian noise is progressively added to these latent representations, and the **Denoising U-Net** learns to reverse this process by predicting and removing noise step by step. The model is conditioned on a **text encoder**, ensuring strong semantic alignment between the generated image and the input prompt. During inference, the model starts with random noise and refines it into a meaningful image using the trained denoising process. Finally, the **VAE decoder** reconstructs the final image from the denoised latent representation.

21 Multi-task Learning

Multi-Task Learning (MTL) is a paradigm in deep learning where a single model is trained on multiple related tasks simultaneously. By leveraging shared representations across tasks, MTL improves generalization and reduces overfitting, as the model benefits from the complementary information present in the tasks. This approach is particularly effective when the tasks share commonalities, such as similar features or underlying structures, enabling the model to learn more robust and transferable representations. MTL can be implemented using two main approaches [30] (Fig. 62):

1. **Hard Parameter Sharing**: In hard parameter sharing, the majority of the model's parameters are shared across all tasks, typically in the form of a shared backbone network. Task-specific layers (heads) are added on top of the shared layers to produce outputs for each task. This approach is computationally efficient and reduces the risk of overfitting, as the shared layers are regularized by multiple tasks. For example, in computer vision, a shared CNN can extract common visual features, while separate fully connected layers can specialize for tasks like classification, object detection, and segmentation.
2. **Soft Parameter Sharing**: In this approach, each task is assigned its own dedicated network, allowing for task-specific learning while maintaining the flexibility to share information selectively. Various mechanisms enable this sharing, such as the following techniques:

- **Cross-Stitch Units**: Cross-stitch units facilitate information sharing by learning optimal linear combinations of feature representations across tasks (see Fig. 63). At each layer, the feature maps of each task are combined through learnable weights,

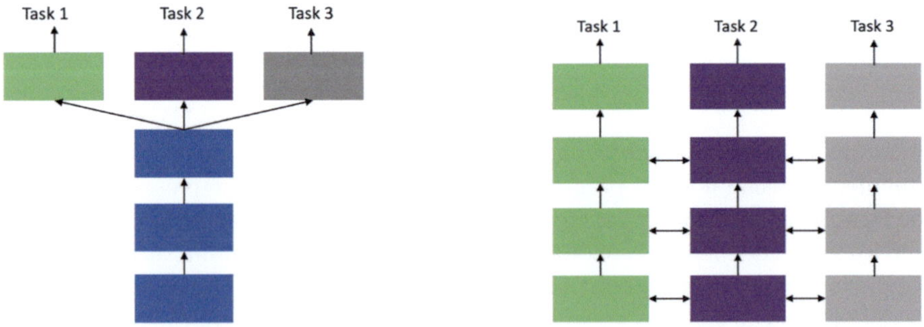

(a) Hard Parameter Sharing (b) Soft Parameter Sharing

Fig. 62 Illustration of multi-task learning approaches

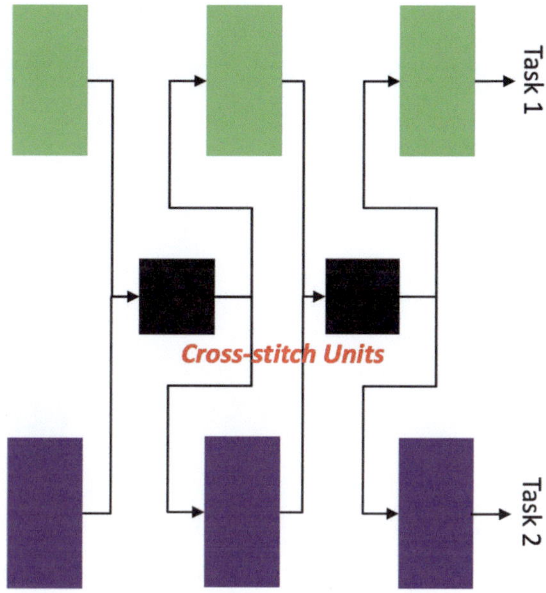

Fig. 63 Cross-stitch networks facilitate controlled feature sharing between tasks

which control how much information from one task is shared with another [31]. The updated feature maps are calculated as follows:

$$\hat{f}_A = \alpha_{AA} f_A + \alpha_{AB} f_B,$$
$$\hat{f}_B = \alpha_{BB} f_B + \alpha_{BA} f_A,$$

where f_A and f_B are the feature maps for Task A and Task B, and α values are learnable parameters optimized during training. Cross-stitch units allow tasks to balance independence and interdependence, dynamically adjusting how much information is shared or retained independently for each task.

- **Multi-Task Attention Networks (MTAN)**: The Multi-Task Attention Network (MTAN) is a multi-task learning architecture that enhances feature-sharing efficiency while preserving task-specific knowledge. MTAN consists of a single shared network that learns a global feature pool and integrates soft-attention modules for each task. These attention modules apply task-specific masks to shared features, automatically learning which features to share or specialize for a given task [32] (Fig. 64).

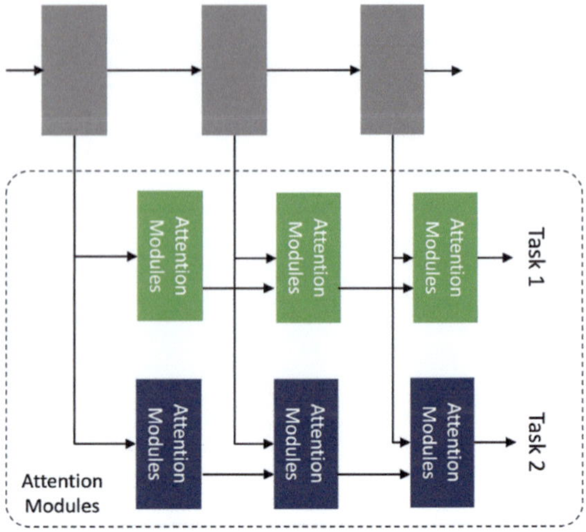

Fig. 64 MTAN architecture with task-specific attention modules (green for Task 1, blue for Task 2) interacting with a shared network (grey). Each attention module follows the same structure but learns distinct weights

22 Privacy-Preserving Machine Learning (PPML)

The exponential growth of data presents unprecedented opportunities to drive innovation and value creation. However, data across various domains is often fragmented due to concerns over data sovereignty, trust, regulatory compliance, and legal restrictions. These challenges create significant barriers to collaboration and hinder the seamless integration of data. Stringent regulations, such as the General Data Protection Regulation (GDPR) in Europe, the Health Insurance Portability and Accountability Act (HIPAA) in the U.S., and other global frameworks, further complicate the secure integration and analysis of sensitive information. As a result, these constraints hinder collaborative research and prevent organizations from fully realizing the transformative potential of data-driven applications. Privacy concerns further exacerbate these challenges, particularly in collaborative environments involving multiple organizations or departments, thereby impeding the adoption of AI/ML. Privacy-Preserving Machine Learning (PPML) techniques provide a promising solution to this dilemma. Methods such as Federated Learning (FL), Homomorphic Encryption (HE), Secure Multi-Party Computation (MPC), and Differential Privacy (DP) enable secure and privacy-compliant analysis of distributed data. By ensuring data confidentiality and building trust, these approaches (illustrated in Fig. 65) empower organizations to harness the transformative potential of AI/ML without compromising sensitive information or breaching regulatory frameworks.

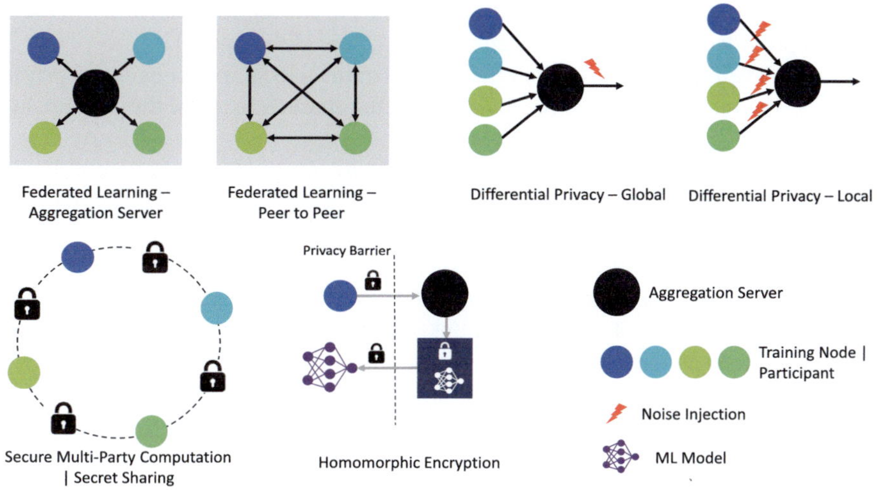

Fig. 65 Privacy-preserving machine learning techniques

22.1 Federated Learning (FL)

Federated Learning (FL) is a decentralized machine learning technique that enables multiple parties to collaboratively train a shared model without transferring their data to a central repository. This approach preserves data privacy and security by ensuring that raw data remains local to each participant while contributing to the overall model's training process. In FL, model training occurs locally on individual devices or systems, and only model updates (e.g., gradients or parameter changes) are shared with a central server, which aggregates these updates to improve the global model. The typical workflow of FL involves the following steps (see Fig. 66) [33]:

1. **Model Initialization**: A pre-trained, centralized machine learning model (referred to as the global model) and its associated parameters are initialized.

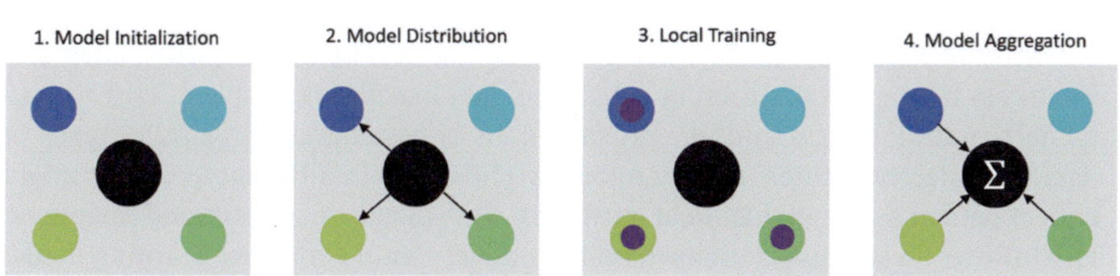

Fig. 66 Centralized federated learning workflow: model initialization, distribution, local training, and aggregation [33]

2. **Model Distribution**: A subset of participants is chosen for each training round and provided with the most recent version of the global model.
3. **Local Training**: Each selected participant trains a local model using its private data. The model is updated locally to reflect the characteristics of the participant's dataset.
4. **Model Aggregation**: Locally trained models are transmitted to the central server, where they are aggregated to update the global model. A widely used aggregation technique is Federated Averaging (FedAvg), which calculates a weighted average of the parameters from local models.
5. **Iteration and Convergence**: The updated global model is redistributed to participants for additional training iterations. This process is repeated until the model converges to an optimal state.

FL can be categorized based on the underlying network topology into three primary types [33]:

- **Centralized FL**: In this architecture, a centralized server orchestrates the entire learning process, performing key functions such as node selection, model update aggregation, and the distribution of the updated global model. The server acts as the coordinating entity, enabling participating nodes to collaboratively train a machine learning model.
- **Decentralized FL**: In contrast to the centralized approach, decentralized FL eliminates the need for a central server. Instead, participating nodes rely on a decentralized protocol to exchange and synchronize parameters, enabling collaborative training without centralized coordination. This architecture is inherently resilient to single points of failure, as the absence of a central server reduces the risk of systemic disruptions. However, the performance of decentralized FL is heavily influenced by the underlying network topology, as efficient communication and coordination between nodes are critical for convergence.

FL techniques can also be classified into three main types based on the relationship between datasets and learning objectives: Horizontal FL (HFL), Vertical FL (VFL), and Transfer FL (TFL) [33]:

- **Horizontal FL (HFL)**: Also referred to as sample-based FL or Homogeneous FL, this approach applies when the participating datasets share the same feature space but differ in sample space. In other words, the datasets include the same attributes (features) but represent different subsets of data instances (samples). HFL is particularly relevant when multiple organizations or systems collect similar types of data for different entities. For example, two hospitals in different locations might collect patient data with identical attributes, such as age, diagnosis, and treatment. HFL allows these organizations to collaboratively train a machine learning model while preserving data privacy and remaining compliant with regulatory requirements. By keeping the raw data local to each participant, HFL reduces privacy risks while achieving a common training objective.

- **Vertical FL (VFL)**: Also known as feature-based FL or Heterogeneous FL, VFL is applicable when datasets share the same sample space (e.g., overlapping user bases) but differ in feature space. For example, a healthcare organization and a social networking platform may have data on the same set of individuals, but the healthcare organization captures medical records while the social platform collects behavioral or social activity data. VFL enables these entities to collaboratively train a model by securely combining their complementary datasets.
- **Transfer FL (TFL)**: Transfer FL extends the principles of transfer learning to the federated setting. It involves leveraging a pre-trained model, initially trained on a related dataset or task, to improve training for a new task. In the context of FL, TFL is particularly useful when datasets are limited or computational resources are constrained, as the pre-trained model provides a foundation for faster and more efficient learning. For example, a general machine learning model pre-trained on healthcare data could be adapted to predict specific outcomes such as rare disease diagnoses by utilizing distributed data from multiple hospitals. By incorporating FL principles, the new model can be collaboratively fine-tuned using private, localized data from different institutions without exposing raw data. TFL significantly enhances performance and training efficiency compared to training a model from scratch in federated environments.

22.2 Homomorphic Encryption (HE)

Homomorphic Encryption (HE) is a PPML technique in which raw data is encrypted before transmission to a server for computation. Unlike traditional encryption methods, HE allows computations to be performed directly on encrypted data (non-plaintext) without requiring decryption at intermediate stages. The results of these computations remain encrypted, and when decrypted, they yield the same outputs as if the computations were performed on plaintext data [33]. This property makes HE a powerful tool for ensuring data confidentiality during ML tasks and other computations on sensitive data. HE can be broadly categorized into the following types:

- **Partially Homomorphic Encryption (PHE)**: These cryptosystems support either a multiplicative or an additive homomorphic property, but not both simultaneously.
- **Fully Homomorphic Encryption (FHE)**: These cryptosystems support both additive and multiplicative homomorphic operations on encrypted data. FHE provides greater flexibility compared to PHE and has significant potential for complex computations, including ML training and inference tasks. However, achieving both properties in a single cryptosystem introduces additional computational complexity.

Technical Limitations: Despite its theoretical advantages, HE faces significant practical challenges that limit its scalability for real-world applications. Key limitations include:

- *Performance Overhead*: HE-based computations are significantly slower compared to operations on plaintext data. The encryption, decryption, and computational processes introduce substantial latency, making HE impractical for computationally intensive tasks such as training large-scale ML models.
- *Scalability Challenges*: The resource-intensive nature of HE (e.g., memory usage and processing requirements) makes it difficult to scale for applications requiring high computational throughput or large datasets.

While HE remains infeasible for training ML models in its current state, ongoing research in cryptography and optimization techniques is expected to address these limitations. As advancements are made, HE could become a practical solution for secure and privacy-preserving ML in sensitive domains, such as healthcare, finance, and government services.

22.3 Secure Multi-party Computation (MPC)

Secure Multi-Party Computation (MPC) is a cryptographic framework that enables multiple participants to collaboratively compute a function over their private inputs while ensuring the confidentiality of these inputs. Unlike traditional cryptography, which primarily focuses on protecting data from external adversaries, MPC addresses the challenge of safeguarding each participant's data from the other participants involved in the computation. By leveraging techniques such as *secret sharing*, MPC ensures that no single party gains access to the complete dataset [34].

The concept of MPC originated in 1982 when Yao introduced secure Two-Party Computation (2PC) to solve the Millionaires' Problem—a Boolean predicate enabling two parties to securely compare their wealth. In 1986, Yao extended this idea to general computations, laying the foundation for modern MPC [35]. Subsequently, Goldreich, Micali, and Wigderson (GMW) expanded 2PC into a multi-party framework, enabling collaborative computation among more than two participants [36, 37].

In an MPC scenario, each participant (e.g., $p_1, p_2, \ldots, p_N$) holds private inputs $(d_1, d_2, \ldots, d_N)$ and collectively computes a public function $F(d_1, d_2, \ldots, d_N)$ without disclosing their individual inputs. For instance, three participants can securely determine which of them has the highest salary without revealing their actual salaries, expressed as:

$$F(x, y, z) = \max(x, y, z).$$

A cornerstone of most modern MPC protocols is the use of **secret sharing**, a technique that splits each participant's private input into multiple random "shares."

These shares are distributed among the participants, ensuring that no single party can reconstruct the original input from their shares alone. Computations are performed locally on the shares, and the results are combined to reveal the final output, maintaining the privacy of individual inputs throughout the process.

Secret sharing can take various forms, including:

- **Additive Secret Sharing**: In this approach, each private input is divided into random shares that sum to the original input. These shares are distributed among participants in such a way that no single participant can deduce the original input from the shares they receive. A simple example can illustrate this concept. Suppose three colleagues wish to compute their average salary without revealing their individual salaries. If Colleague A's salary is \$110K, this value is broken into three random shares (e.g., \$40K, \$50K, and \$20K) that sum to \$110K. Colleague A keeps one share (\$50K) and distributes the other shares to Colleague B (\$20K) and Colleague C (\$40K). Colleagues B and C repeat the same process with their own salaries, ensuring their respective shares also sum to their individual salaries. Once the shares are distributed, each participant holds one share from every other participant. These shares are independently random and uniformly distributed, such that no subset of shares smaller than the total number of participants reveals any information about the original input. To compute the result, each participant sums all the shares they hold (including their own) to produce a local total. These local totals are then aggregated to obtain the overall sum of salaries. The average is computed by dividing the total by the number of participants. If required, this process can also be implemented over a finite field $\mathbb{Z}_p$, where p is a prime larger than the largest input, to ensure modular arithmetic and avoid overflow.
- **Shamir Secret Sharing (SSS)**: Shamir Secret Sharing is based on the mathematical principle of polynomial interpolation over a finite field $\mathbb{F}_p$. A secret s is split into shares such that a minimum number of shares (the threshold k) are required to reconstruct the secret. The scheme works by constructing a degree-$(k-1)$ polynomial $f(x)$ with randomly chosen coefficients over $\mathbb{F}_p$, where $f(0) = s$ (the secret). Each share is a point on the polynomial, represented as a pair $(x, f(x) \mod p)$, where x is a unique identifier for the participant. For example, if Participant A wishes to share a secret $s = 6$ and sets a threshold $k = 2$, they might define a polynomial $f(x) = 2x + 6 \mod 11$. Shares are then computed as $(1, f(1) = 8)$, $(2, f(2) = 10)$, and $(3, f(3) = 1)$, and distributed to Participants B, C, and D, respectively. With fewer than k shares, participants cannot reconstruct the polynomial due to the underdetermined system and the exponentially large number of possible polynomials. However, with k or more shares, the polynomial can be reconstructed (e.g., using Lagrange interpolation), and the secret $f(0) = 6$ revealed.

22.4 *Differential Privacy (DP)*

Differential Privacy (DP) is a formal mathematical framework used to manage and quantify privacy risks. It is widely studied in the context of gathering, analyzing, and sharing aggregated statistics, ranging from simple averages to Machine Learning. Rather than being a standalone tool, DP serves as a criterion that data-analysis mechanisms must satisfy to ensure the protection of sensitive information. DP provides a mathematically provable guarantee of privacy, even against adversaries attempting to infer individual-specific information from aggregated data [34].

DP safeguards privacy by introducing carefully calibrated noise into sensitive data. Broadly, DP systems are categorized into **global privacy** and **local privacy** [38].

Global Privacy: In global DP systems, a trusted entity called the *curator* has access to raw, private data from multiple participants. The curator analyzes the data and injects noise into the results before sharing them. For instance, consider a researcher seeking to determine the number of hospital patients diagnosed with the Human Stigmatization Virus (HSV). A hospital administrator could count the actual number of patients and then apply the Laplace mechanism to inject noise into the count. The researcher receives a noisy estimate of the number of HSV cases, which is close to the actual value but ensures that no specific patient's diagnosis can be inferred.

Local Privacy: In local DP systems, each individual is responsible for injecting noise into their own data before sharing it with others. There is no need for a trusted curator; instead, each participant acts as the curator of their own data. For example, a political researcher could ask residents of a town if they support a particular political party. To ensure privacy, each resident could use the randomized response technique: they flip a coin secretly. If the coin lands heads, they provide a truthful answer. If it lands tails, they flip again—heads meaning "yes" and tails meaning "no." This method ensures plausible deniability for individuals while enabling the researcher to estimate aggregate support accurately.

The DP framework sets boundaries on how much information can be revealed about any individual in a dataset, typically controlled by a parameter ϵ, known as the **Privacy Budget**. This parameter quantifies the maximum privacy loss incurred by a mechanism. Mathematically, ϵ-DP is defined as follows [39–41]:

$$P[A(X) \in S] \leq e^{\epsilon} \cdot P[A(X') \in S], \tag{16}$$

where P is the probability, X and X' are two adjacent datasets differing in a single element, A is the mechanism applied to X, and S represents any possible output. A smaller ϵ indicates stronger privacy. However, repeated queries accumulate privacy loss, as ϵ values add up across queries. For example, two queries with $\epsilon = 1$ each result in an effective privacy loss of $\epsilon = 2$ [42, 43].

DP mechanisms achieve privacy through pre-defined noise-adding techniques, including the following [44–46]:

- **Randomized Response**: Originally proposed by S. L. Warner, this technique is widely used in surveys to preserve respondent privacy. For example, respondents flip two unbiased coins. If the first coin lands heads, they provide a truthful answer. If it lands tails, the second coin determines their response ("yes" for heads, "no" for tails). This method ensures plausible deniability for individuals while maintaining utility for aggregate data analysis. The mechanism satisfies differential privacy when the probability of providing a truthful answer is:

$$P = \frac{e^{\epsilon}}{1 + e^{\epsilon}}, \tag{17}$$

  where ϵ is the privacy budget, balancing privacy and utility.
- **Laplace Mechanism (LM)**: This mechanism adds noise to each data point, sampled from a Laplace distribution. The scale of the noise is determined by the sensitivity of the query (Δf) and the privacy budget (ϵ). Specifically, the noise is drawn as:

$$\text{Noise} \sim \text{Lap}\left(\frac{\Delta f}{\epsilon}\right), \tag{18}$$

  where Δf is the maximum change in the query output due to a single individual's data (L1 sensitivity). This ensures that the noise sufficiently masks individual contributions.
- **Gaussian Mechanism**: Unlike the Laplace Mechanism, this technique introduces noise sampled from a Gaussian (normal) distribution. It is often used in settings where (ϵ, δ)-DP guarantees are required, particularly for more complex mathematical properties. The scale of the Gaussian noise depends on the sensitivity of the query, the privacy budget ϵ, and the allowable δ parameter.
- **Exponential Mechanism**: This mechanism is designed for scenarios where the output is categorical rather than numeric. It selects an output probabilistically based on a scoring function, which measures the quality or desirability of each possible result. The probability of selecting an output o is proportional to:

$$\exp\left(\frac{\epsilon \cdot u(o)}{2\Delta u}\right), \tag{19}$$

  where $u(o)$ is the utility score of output o, and Δu is the sensitivity of the utility function. The Exponential Mechanism is commonly used in applications like recommendation systems and preference selection.
- **Compression Mechanism**: Inspired by information theory, this mechanism distorts or compresses data to satisfy (ϵ, δ)-DP constraints [47]. By carefully reducing the fidelity of the data, it ensures individual contributions are obscured while maintaining sufficient utility for specific tasks.

Acknowledgements The authors acknowledge the use of Generative AI (GenAI) tools in the editing and development of this book chapter. They also thank Tamonash Bhattacharyya for his feedback and suggestions.

References

1. F. Firouzi, K. Chakrabarty, and S. Nassif, *Intelligent internet of things: From device to fog and cloud.* Springer, 2020.
2. Y. Chen, M. Mancini, X. Zhu, and Z. Akata, "Semi-supervised and unsupervised deep visual learning: A survey," *IEEE transactions on pattern analysis and machine intelligence*, vol. 46, no. 3, pp. 1327–1347, 2022.
3. P. Domanski, A. Ray, K. Lafata, F. Firouzi, K. Chakrabarty, and D. Pflüger, "Advancing blood glucose prediction with neural architecture search and deep reinforcement learning for type 1 diabetics," *Biocybernetics and Biomedical Engineering*, vol. 44, no. 3, pp. 481–500, 2024.
4. D. Foster, *Generative deep learning.* " O'Reilly Media, Inc.", 2022.
5. A. Vaswani, "Attention is all you need," *Advances in Neural Information Processing Systems*, 2017.
6. J. Alammar and M. Grootendorst, *Hands-on large language models: language understanding and generation.* " O'Reilly Media, Inc.", 2024.
7. N. Houlsby, A. Giurgiu, S. Jastrzebski, B. Morrone, Q. De Laroussilhe, A. Gesmundo, M. Attariyan, and S. Gelly, "Parameter-efficient transfer learning for nlp," in *International conference on machine learning.* PMLR, 2019, pp. 2790–2799.
8. E. J. Hu, Y. Shen, P. Wallis, Z. Allen-Zhu, Y. Li, S. Wang, L. Wang, W. Chen *et al.*, "Lora: Low-rank adaptation of large language models." *ICLR*, vol. 1, no. 2, p. 3, 2022.
9. M. Abbasian, E. Khatibi, I. Azimi, D. Oniani, Z. S. H. Abad, A. Thieme, Z. Yang, Y. Wang, B. Lin, O. Gevaert *et al.*, "Foundation metrics: Quantifying effectiveness of healthcare conversations powered by generative ai," *arXiv preprint* arXiv:2309.12444, 2023.
10. J.-B. Alayrac, J. Donahue, P. Luc, A. Miech, I. Barr, Y. Hasson, K. Lenc, A. Mensch, K. Millican, M. Reynolds *et al.*, "Flamingo: a visual language model for few-shot learning," *Advances in neural information processing systems*, vol. 35, pp. 23 716–23 736, 2022.
11. H. Liu, C. Li, Q. Wu, and Y. J. Lee, "Visual instruction tuning," *Advances in neural information processing systems*, vol. 36, 2024.
12. J. Lin, H. Yin, W. Ping, P. Molchanov, M. Shoeybi, and S. Han, "Vila: On pre-training for visual language models," in *Proceedings of the IEEE/CVF Conference on Computer Vision and Pattern Recognition*, 2024, pp. 26 689–26 699.
13. S. Yin, C. Fu, S. Zhao, K. Li, X. Sun, T. Xu, and E. Chen, "A survey on multimodal large language models," *arXiv preprint* arXiv:2306.13549, 2023.
14. S. Wu, H. Fei, L. Qu, W. Ji, and T.-S. Chua, "Next-gpt: Any-to-any multimodal llm," *arXiv preprint* arXiv:2309.05519, 2023.
15. A. Dosovitskiy, "An image is worth 16x16 words: Transformers for image recognition at scale," *arXiv preprint* arXiv:2010.11929, 2020.
16. K. Han, Y. Wang, H. Chen, X. Chen, J. Guo, Z. Liu, Y. Tang, A. Xiao, C. Xu, Y. Xu *et al.*, "A survey on vision transformer," *IEEE transactions on pattern analysis and machine intelligence*, vol. 45, no. 1, pp. 87–110, 2022.
17. A. Radford, J. W. Kim, C. Hallacy, A. Ramesh, G. Goh, S. Agarwal, G. Sastry, A. Askell, P. Mishkin, J. Clark *et al.*, "Learning transferable visual models from natural language supervision," in *International conference on machine learning.* PmLR, 2021, pp. 8748–8763.
18. S. Yao, D. Yu, J. Zhao, I. Shafran, T. Griffiths, Y. Cao, and K. Narasimhan, "Tree of thoughts: Deliberate problem solving with large language models," *Advances in Neural Information Processing Systems*, vol. 36, 2024.

19. J. Wei, X. Wang, D. Schuurmans, M. Bosma, F. Xia, E. Chi, Q. V. Le, D. Zhou *et al.*, "Chain-of-thought prompting elicits reasoning in large language models," *Advances in neural information processing systems*, vol. 35, pp. 24 824–24 837, 2022.
20. W. Chen, X. Ma, X. Wang, and W. W. Cohen, "Program of thoughts prompting: Disentangling computation from reasoning for numerical reasoning tasks," *arXiv preprint* arXiv:2211.12588, 2022.
21. L. Wang, W. Xu, Y. Lan, Z. Hu, Y. Lan, R. K.-W. Lee, and E.-P. Lim, "Plan-and-solve prompting: Improving zero-shot chain-of-thought reasoning by large language models," *arXiv preprint* arXiv:2305.04091, 2023.
22. S. Yao, J. Zhao, D. Yu, N. Du, I. Shafran, K. Narasimhan, and Y. Cao, "React: Synergizing reasoning and acting in language models," *arXiv preprint* arXiv:2210.03629, 2022.
23. J. Ho, A. Jain, and P. Abbeel, "Denoising diffusion probabilistic models," *Advances in neural information processing systems*, vol. 33, pp. 6840–6851, 2020.
24. L. Yang, Z. Zhang, Y. Song, S. Hong, R. Xu, Y. Zhao, W. Zhang, B. Cui, and M.-H. Yang, "Diffusion models: A comprehensive survey of methods and applications," *ACM Computing Surveys*, vol. 56, no. 4, pp. 1–39, 2023.
25. W. Peebles and S. Xie, "Scalable diffusion models with transformers," in *Proceedings of the IEEE/CVF international conference on computer vision*, 2023, pp. 4195–4205.
26. A. Nichol, P. Dhariwal, A. Ramesh, P. Shyam, P. Mishkin, B. McGrew, I. Sutskever, and M. Chen, "Glide: Towards photorealistic image generation and editing with text-guided diffusion models," *arXiv preprint* arXiv:2112.10741, 2021.
27. A. Ramesh, M. Pavlov, G. Goh, S. Gray, C. Voss, A. Radford, M. Chen, and I. Sutskever, "Zero-shot text-to-image generation," in *International conference on machine learning*. Pmlr, 2021, pp. 8821–8831.
28. A. Ramesh, P. Dhariwal, A. Nichol, C. Chu, and M. Chen, "Hierarchical text-conditional image generation with clip latents," *arXiv preprint* arXiv:2204.06125, vol. 1, no. 2, p. 3, 2022.
29. R. Rombach, A. Blattmann, D. Lorenz, P. Esser, and B. Ommer, "High-resolution image synthesis with latent diffusion models," in *Proceedings of the IEEE/CVF conference on computer vision and pattern recognition*, 2022, pp. 10 684–10 695.
30. S. Vandenhende, S. Georgoulis, W. Van Gansbeke, M. Proesmans, D. Dai, and L. Van Gool, "Multi-task learning for dense prediction tasks: A survey," *IEEE transactions on pattern analysis and machine intelligence*, vol. 44, no. 7, pp. 3614–3633, 2021.
31. I. Misra, A. Shrivastava, A. Gupta, and M. Hebert, "Cross-stitch networks for multi-task learning," in *Proceedings of the IEEE conference on computer vision and pattern recognition*, 2016, pp. 3994–4003.
32. S. Liu, E. Johns, and A. J. Davison, "End-to-end multi-task learning with attention," in *Proceedings of the IEEE/CVF conference on computer vision and pattern recognition*, 2019, pp. 1871–1880.
33. F. Firouzi, S. Jiang, K. Chakrabarty, B. Farahani, M. Daneshmand, J. Song, and K. Mankodiya, "Fusion of iot, ai, edge–fog–cloud, and blockchain: Challenges, solutions, and a case study in healthcare and medicine," *IEEE Internet of Things Journal*, vol. 10, no. 5, pp. 3686–3705, 2022.
34. F. Firouzi, B. Farahani, M. Barzegari, and M. Daneshmand, "Ai-driven data monetization: The other face of data in iot-based smart and connected health," *IEEE Internet of Things Journal*, vol. 9, no. 8, pp. 5581–5599, 2020.
35. A. C. Yao, "Protocols for secure computations," in *23rd annual symposium on foundations of computer science (sfcs 1982)*. IEEE, 1982, pp. 160–164.
36. O. Goldreich, "Secure multi-party computation," *Manuscript. Preliminary version*, vol. 78, 1998.
37. A. Ben-David, N. Nisan, and B. Pinkas, "Fairplaymp: a system for secure multi-party computation," in *Proceedings of the 15th ACM conference on Computer and communications security*, 2008, pp. 257–266.
38. M. Yang, L. Lyu, J. Zhao, T. Zhu, and K.-Y. Lam, "Local differential privacy and its applications: A comprehensive survey," *arXiv preprint* arXiv:2008.03686, 2020.

39. A. Friedman and A. Schuster, "Data mining with differential privacy," in *Proceedings of the 16th ACM SIGKDD international conference on Knowledge discovery and data mining*, 2010, pp. 493–502.
40. M. Abadi, A. Chu, I. Goodfellow, H. B. McMahan, I. Mironov, K. Talwar, and L. Zhang, "Deep learning with differential privacy," in *Proceedings of the 2016 ACM SIGSAC conference on computer and communications security*, 2016, pp. 308–318.
41. C. Dwork, "Differential privacy: A survey of results," in *International conference on theory and applications of models of computation*. Springer, 2008, pp. 1–19.
42. A. Nadian-Ghomsheh, B. Farahani, and M. Kavian, "A hierarchical privacy-preserving iot architecture for vision-based hand rehabilitation assessment," *Multimedia Tools and Applications*, pp. 1–24, 2021.
43. Z. Ji, Z. C. Lipton, and C. Elkan, "Differential privacy and machine learning: a survey and review," *arXiv preprint* arXiv:1412.7584, 2014.
44. X. Xiong, S. Liu, D. Li, Z. Cai, and X. Niu, "A comprehensive survey on local differential privacy," *Security and Communication Networks*, vol. 2020, 2020.
45. N. Saleheen, S. Chakraborty, N. Ali, M. M. Rahman, S. M. Hossain, R. Bari, E. Buder, M. Srivastava, and S. Kumar, "msieve: differential behavioral privacy in time series of mobile sensor data," in *Proceedings of the 2016 ACM International Joint Conference on Pervasive and Ubiquitous Computing*, 2016, pp. 706–717.
46. J. Zhao, Y. Chen, and W. Zhang, "Differential privacy preservation in deep learning: Challenges, opportunities and solutions," *IEEE Access*, vol. 7, pp. 48 901–48 911, 2019.
47. A. D. Sarwate and L. Sankar, "A rate-disortion perspective on local differential privacy," in *2014 52nd Annual Allerton Conference on Communication, Control, and Computing (Allerton)*. IEEE, 2014, pp. 903–908.

Smart Sensor Technologies and Wearable IoT for Healthcare

He Sun, Rajendra P. Shukla, and Michael Daniele

An ounce of prevention is worth a pound of cure.
—Benjamin Franklin

1 Introduction

Currently, advanced sensors and connected healthcare solutions are driving innovation in the market for wearable, wireless, and connected medical devices. Smart sensors are a combination of sensing elements, microprocessors, and wireless communication that can measure, process, and transmit a variety of physiological and biological characteristics to capture vital sign data and other critical diagnostic information. The smart sensor built-in wearable Internet of Things (IoT) devices, such as smart watches, fitness bands, and smart clothing, can provide an accessible, non-invasive way of keeping track of health, making medical services more prevalent, and allowing for active participation and customization. In this section, we focus on smart sensor technologies, wearable IoT devices along with their historical development and evolution.

H. Sun · M. Daniele (✉)
Lampe Joint Department of Biomedical Engineering, North Carolina State University and University of North Carolina at Chapel Hill, Raleigh, NC, USA
e-mail: mdaniel6@ncsu.edu

R. P. Shukla · M. Daniele
Department of Electrical and Computer Engineering, North Carolina State University, Raleigh, NC, USA

F. Firouzi et al. (eds.), *Smart and Connected Health: AI, IoT, and Trustworthy Technologies*, https://doi.org/10.1007/978-3-032-06286-4_4

1.1 Smart Sensor Technologies

Healthcare practices have seen a transformation in recent decades due to the rapid advancement of mobile and data technologies. Traditionally a purview of the clinical lab or centralized facility, diagnostic medicine is most significantly impacted by the rise of smart sensor technology incorporating intelligent features such as signal processing, data analysis, and wireless communication. This advancement enables patients as well as physicians to benefit from more efficient and precise medical services. Smart sensor technologies could continuously and noninvasively record and monitor the status of multiple physiological data, giving medical professionals quick access to information about the condition of the individual and the ability to diagnose and treat conditions quickly [1]. For instance, a temperature sensor can continuously monitor core body temperature, enabling a real-time assessment of fever and prediction of infection. The heart rate sensor can give information about heart health, including support for the diagnosis and treatment of cardiac conditions like arrhythmia. To save critical rescue time for patients in need of immediate attention, blood pressure and blood oxygen sensors monitor these parameters both during and following surgery. Furthermore, glucose sensors, which have seen a transition from handheld glucometers to artificial intelligence (AI)-driven continuous glucose monitors in less than a decade, are vital tools for the management of diabetes. With real-time monitoring, patients can modify their behaving, adjust treatment, and avoid running the risks associated with both hyperglycemia and hypoglycemia. Moreover, motion sensors have emerged from playing an important role in evaluating rehabilitation training and monitoring athletic performance [2] to off-the-shelf consumer devices for continuous wellness and activity monitoring linked to subscription and even health insurance services.

Smart sensor technologies are also making a significant impact on the clinical practice. During invasive procedures, surgeons can get precise and up-to-date data, including blood pressure, pulse, and body temperature in real time with the aid of smart sensors, and they can also monitor postoperative conditions to confirm the efficacy of treatment plans and diagnoses [3]. Furthermore, smart sensors can be utilized for remote monitoring, which enables medical professionals to provide prompt medical intervention and lower hospital readmission rates. This is especially beneficial for patients with chronic illnesses and those living in remote places. However, there are still certain challenges and obstacles for sensor-based smart medical devices. With the goal to maintain accurate and comprehensive data, it is first necessary to confirm the high precision and dependability of smart sensors. The availability and expense present another difficulty. Even with ongoing advancements in sensor technology, high-precision, purpose-built smart sensor technologies remain costly. While smart sensor technologies will continue to evolve, understanding how they will transition the passive, traditional medical practice to active, preventive health care models, will be essential to realize their effective use, improve patient outcomes, and reduce medical expenses [4].

1.2 Wearable IoT Devices

Wearable IoT devices make the deployment of smart sensor technologies for medical applications more practical and user-friendly by gathering, processing, and transmitting data on the user. In addition to monitoring different physiological indicators of the users, these devices can also establish networks with other devices or systems to perform data synchronization and analysis, hence offering a range of intelligent functions and services. After the target biosignals are collected by the smart sensors, computational analysis can be done directly on-device prior to delivering a calculated biomarker, or for additional, more complex operations and analysis, data can be transmitted to external devices or cloud platforms via integrated communication interfaces such Wi-Fi, Bluetooth, Zigbee, or cellular networks [5]. Most importantly, the development of semiconductor technology and nanotechnology has made it possible to realize these kinds of measurement capabilities at the chip level, which has facilitated the process of downsizing sensors, increasing efficiency, and enabling their incorporation into both consumer and medical-grade wearable devices [6]. Consumer smartwatches and fitness trackers, for instance, are mostly used to track activity, sleep patterns, blood oxygen level (SpO_2), and heart rate. Smart textiles and skin patches mainly track physical activity, body temperature and skin conditions. Additionally, wearable medical-grade IoT devices are capable of precisely measuring blood pressure, blood glucose, and electrocardiogram (ECG) data, among numerous other functions [7].

Patients can receive real-time feedback on their health and progress from wearable IoT devices. This kind of information can encourage individuals to take preventative care, follow medical regimens, and lead healthier lifestyles. For example, consumers could be motivated to engage in more physical activity with tracking devices, and patients could benefit from smart medication reminders to keep up on treatment. In addition, through the combination of wearable IoT devices with AI and big data analysis, it is possible to offer more precise and individualized treatment options by using enormous quantities of data for pattern identification [8]. For instance, applying machine learning (ML) algorithms to wearable ECG monitor data may be able to identify arrhythmias or anticipate cardiovascular problems, allowing for proactive heart health care [9]. The success of wearable IoT devices, however, depends on user acceptability and compliance. Comfort, usability, battery life, and aesthetics all play a significant role in determining whether the patients will continue to wear and use the equipment in the future. Additionally, since these devices will gather sensitive personal health information and medical data, it is imperative to protect patient privacy and data. To that end, laws, regulations, and pertinent data privacy protection measures must be updated and put into place on a regular basis. Along with the advance of device miniaturization, integration, and proliferation, wearable IoT devices will be widespread, bringing more opportunities and challenges for the medical enterprise [10].

1.3 Historical Development and Evolution

Today, sensors play a pivotal role in monitoring vital signs, such as heart rate, blood pressure, glucose levels, oxygen saturation, and more. With improvements in sensor technology, these devices have become smaller, more accurate, and more energy-efficient, enabling continuous health monitoring. Over nearly 75 years, sensor technologies have evolved from industrial tools to personal devices (Fig. 1). Sensors were originally used heavily in industry and technology for physical measurements such as temperature, pressure, and acceleration. Electronic sensors originated in the 1960s and 1970s, and their initial integration provided a foundational platform for medical equipment. During this period, sensors were first employed to monitor physiological parameters such as heart rate and blood pressure [11]. In the 1980s, the emergence of microelectron-mechanical system (MEMs) technology marked an acceleration of sensor development. MEMs have been pivotal in miniaturizing sensors, making them more compact, energy-efficient, and cost-effective. This advancement has enabled the integration of sensors into wearable devices, facilitating continuous, non-invasive monitoring of patients' physiological parameters. In addition, wearable sensors were further developed in the 1990s with the invention of wireless communication technologies such as Bluetooth and Wi-Fi. This technology can transmit sensor data from wearable devices to remote monitoring systems, enabling real-time and continuous monitoring of patient health. The first commercial wearable medical equipment, such as in-patient ECG monitors and continuous glucose monitors, came out in the late 1990s and early 2000s [11], alongside the rapid adoption of mobile communication devices. As a highlight, in less than 100 years, the ECG monitor went from a nearly 600 lbs. instrument to a wearable mobile device. Continuing on the trend of miniaturization of the form with expansion of computing power, smartphones emerged in the 2010s, and their unprecedented mobile computing capabilities and wireless connectivity are serving as the hub for smart sensor technology information collection and transmission, seamless connections within the cloud-based Internet ecosystem, and the future large-scale integration with the current medical infrastructure [12].

Alongside the growth of the personalized and clinical medical device enterprises, the consumer-based wearable IoT industry has grown rapidly over the past decade due to the same combination of advanced sensor technology, miniaturization,

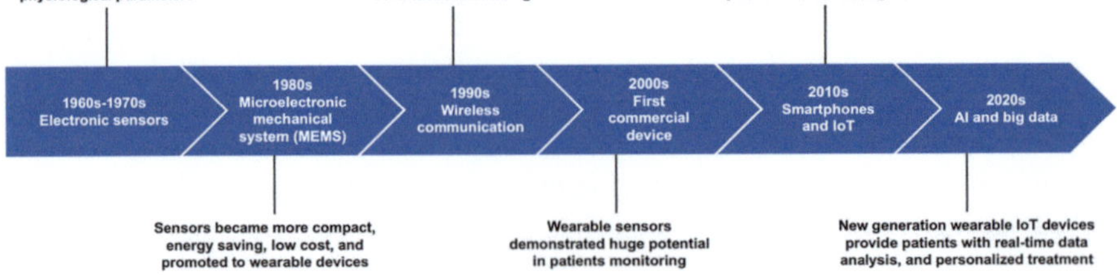

Fig. 1 Historical development and evolution of wearable IoT devices

and wireless connectivity. Smartwatches and fitness trackers are becoming increasingly common and include health monitoring features such as measuring heart rate, tracking sleep, or monitoring physical activity. Furthermore, consumer products have increased the awareness of personal health management, propelling wearable sensor markets into a multi-billion dollar (USD) market. Current research and manufacturing efforts are exploring improvements in sensor accuracy, reliability, and energy efficiency, as well as innovative designs like smart fabrics or skin patches. Yet, in the immediate future, as this is being written in 2024 and 2025, we are entering into the advent of a new AI age, in which the perfusion of large-language models and other tools will provide developers, doctors, and patients real-time data analysis, clinical support for prediction and prognosis of coming problems, and personalized treatment recommendations. Employment of AI will be a pervasive entity in the future of smart sensor technologies, and it will enable and even necessitate, for better or worse, a data-driven approach to patient care [13, 14].

2 Smart Sensors and Applications

The advancement of sensor technology has led to the emergence of intelligent systems, each with the ability to be tailored to monitor specific health factors to direct or support the provision and quality of healthcare services. Generally, there are five basic types of smart sensor systems depending on their functionality and the type of data collected. These types are biometric sensors, chemical sensors, biosensors, motion sensors, and optical sensors [15]. Each type offers unique capabilities, often suited for a select biomarker of interest, which may provide discrete or continuous measurements. This section focuses on our classification of each type of smart sensor (Fig. 2) and their applications.

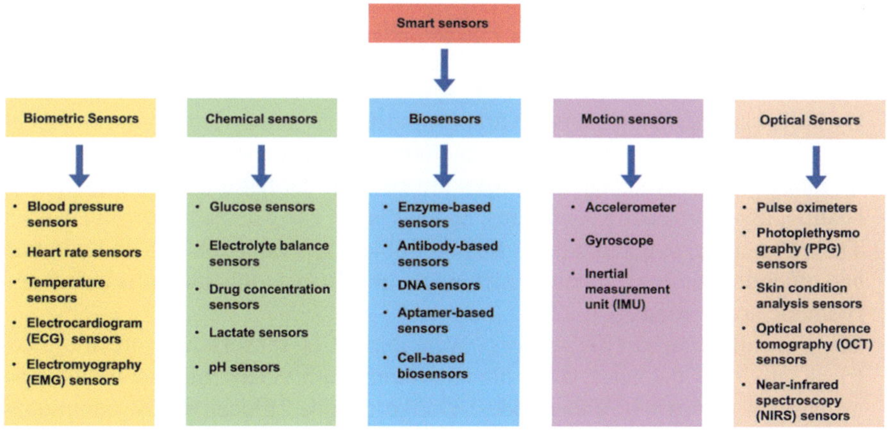

Fig. 2 Types of smart sensors

2.1 Types of Smart Sensors

2.1.1 Biometric Sensors

Biometric sensors are crucial in health monitoring as they can continuously and noninvasively measure physiological parameters and vital signs. These sensors provide real-time data, aiding early health detection and management, and can detect subtle changes before symptoms appear, offering early warning signs of potential health issues. By delivering immediate feedback, biometric sensors help improve patient outcomes. Additionally, integration with IoT platforms supports data analysis and remote monitoring, enhancing personalized therapy [16, 17]. The primary classifications of biometric sensors are as follows:

- Heart rate sensors: Heart rate monitoring typically employs both electronic and optical techniques, each with a distinct approach. Electronic sensors, which measure heart rate and monitor ECG signals, rely on electrodes to detect electrical activity in the heart. In contrast, optical heart rate monitors utilize light reflection, transmitting light through the skin and employing photoelectric plethysmography (PPG) to detect variations in blood flow [18]. These two methods offer complementary insights into cardiovascular health, each capitalizing on different physiological signals to provide accurate and real-time heart rate measurements.
- Blood pressure sensors: The management of hypertension and general cardiovascular health depends on continuous blood pressure monitoring. Traditional cuff type blood pressure sensors use mechanical occlusion combined with optical heart rate recording to measure blood pressure. Pulse transmission time is one of the novel techniques being researched to offer continuous monitoring [19].
- Temperature sensors: Temperature sensors, including those that use direct skin contact or non-contact infrared technology, are effective for monitoring body temperature. Wearable temperature sensors can track temperature fluctuations, enabling early detection of fever and assisting in the identification of potential illnesses. Additionally, these sensors can be used for monitoring menstrual cycle changes, providing valuable insights into health and well-being through continuous, non-invasive temperature tracking [20].
- Hydration and dermal sensors: Hydration and dermal sensors employ various techniques to analyze skin composition, including hydration levels, melanin concentration, and erythema. These sensors are used in dermatology to monitor skin healing, assess cosmetic conditions, and aid in the early detection of skin cancer. Their non-invasive nature allows for safe, objective evaluation of skin hydration and other dermatological issues, making them valuable tools for both clinical and cosmetic applications [21, 22].
- Electrocardiogram (ECG) sensors: Wearable ECG sensors measure the electrical activity of the heart, detecting conditions such as ischemia, infarction, and arrhythmia through skin electrodes. Unlike routine clinical ECGs, continuous monitoring with wearable sensors provides more detailed and real-time data,

enabling ongoing assessment of heart health and the early detection of potential issues [23].

- Electromyography (EMG) sensors: EMG sensors detect the electrical activity produced by skeletal muscles and are utilized in the diagnosis of neuromuscular disorders, rehabilitation, and evaluation of muscle health, activity, and exhaustion [24].

2.1.2 Chemical Sensors and Biosensors

Chemical sensors are designed to detect specific molecules or ions, providing precise data essential for accurate diagnosis and treatment. A biosensor is a hybrid sensor that combines physicochemical detectors and biological components to recognize certain biomarkers and play a critical role in the early diagnosis of disease and personalized health [25]. Many chemical sensors are minimally invasive, enhancing patient comfort and increasing compliance. For instance, wearable sweat sensors can measure markers such as lactate or sodium levels without the need for blood draws. Additionally, chemical sensors can be integrated into closed-loop drug delivery systems, where they monitor a patient's condition and automatically adjust medication doses for optimal therapeutic outcomes. A key application of this technology is in the management of conditions like diabetes or pain, where sensors help fine-tune insulin or painkiller delivery [26, 27]. Types of chemical sensors and biosensors classified by target biomarkers include:

- Glucose sensors: Glucose sensors are used mostly for the control of diabetes. They can be classified as non-enzymatic (based on the direct oxidation of glucose) or enzymatic (using glucose oxidases). Continuous glucose monitoring (CGM) systems use subcutaneous sensors to continually assess interstitial glucose levels, and some systems connected to insulin pumps to provide automatic insulin delivery. The goal of emerging technology is to use sweat analysis for noninvasive glucose monitoring. These sensors make it easier to continuously monitor blood glucose levels, which is crucial for automated insulin delivery systems and the control of diabetes [28].
- Electrolyte (i.e., ion-selective electrodes) sensors: Electrolyte sensors, often in the form of wearable sweat sensors, play a vital role in monitoring key electrolytes such as sodium, potassium, and chloride. These sensors are particularly invaluable for individuals with kidney disease, heart conditions, or those on medications that may disrupt electrolyte levels. By providing continuous, real-time data, they can help prevent dangerous imbalances—such as those that lead to arrhythmias—thus serving as an early warning system. Such sensors exemplify the potential of wearable technology in managing complex, chronic health conditions, where precise monitoring can significantly reduce the risk of acute complications [29].
- Drug sensors: Drug concentration sensors are essential tools for monitoring drug levels in the bloodstream, particularly for medications with narrow therapeutic windows. These sensors leverage electrochemical sensing technology or microanalysis techniques to provide precise measurements of drug concentrations in

real-time. By enabling the fine-tuning of individual doses, these sensors help optimize therapeutic efficacy while minimizing the risks of toxicity or underdosing [30].

- pH sensors: These sensors play a crucial role in monitoring the body's acid–base balance by continuously tracking pH levels in sweat or interstitial fluid. Such real-time monitoring is particularly valuable for patients with kidney or respiratory diseases, where maintaining the correct acid–base balance is essential for optimal health. By providing continuous data, pH sensors enable early detection of imbalances, allowing for timely interventions to prevent complications. Their ability to provide consistent, noninvasive readings makes them indispensable in managing conditions where even slight fluctuations in pH can have significant physiological consequences [31]. In addition to monitoring sweat or interstitial fluid, pH sensors are also employed in breath analysis [32, 33]. By analyzing the composition of exhaled gases, these sensors can detect changes in the concentration of volatile compounds such as carbon dioxide, which provides insights into respiratory function and metabolic processes. Breath-based pH sensors are especially useful for patients with respiratory diseases like chronic obstructive pulmonary disease (COPD) or asthma, as they offer a noninvasive method for continuous monitoring and early detection of imbalances in acid–base levels generated by organic compounds that may indicate underlying health issues.
- Lactate and Ketone sensors: Lactate sensors offer continuous, real-time monitoring of lactate levels, which serve as key indicators of tissue oxygenation and metabolic function [29]. By tracking lactate concentrations, these sensors provide valuable insights into the body's metabolic state, particularly in high-performance sports or critical care settings where precise monitoring of physical activity and metabolic status is essential. In addition to lactate, emerging sensors for wearable devices are now capable of measuring other key biomarkers, such as ketones [34, 35], which are indicative of metabolic shifts, particularly in conditions like diabetes or during periods of fasting or intense physical exertion. Together, lactate, ketone, and pH sensors form an integrated suite of tools that can offer a comprehensive picture of an individual's physiological condition, enabling personalized health management through continuous, noninvasive monitoring.

Many chemical sensors and biosensors are minimally invasive, which improves patient comfort and compliance. In addition, these sensors can be integrated into the drug release in the system of closed loop control to optimize treatment interventions. A significant advantage of chemical sensors and biosensors lies in their role in personalized medicine. By analyzing an individual's unique biochemical profile, these sensors enable the development of customized treatment plans, improving the effectiveness of interventions. For example, sensors that measure hormone levels or detect specific biomarkers can guide the adjustment of cancer therapies based on a patient's specific tumor characteristics. In summary, chemical sensors have immense potential in chronic disease management, providing accurate, real-time data that supports personalized treatment and improves patient outcomes. Their ability to monitor biochemical markers and integrate with treatment systems represents a major advancement in medical technology [36].

In complicated sample matrices, a biosensor may be a preferred technology, as the integrated bioreceptor, or biorecognition element, ideally imparts a high specificity target can accurately detect potential biological molecules while reducing the likelihood of false positives. Because of their sensitivity, these instruments can identify physiologically relevant analyte concentrations, potentially $<$ pM, which is crucial for identifying biomarkers in the initial stages of a disease or the most minute excursions from healthy concentrations. Many biosensors are designed to produce rapid results, which makes it easier to diagnose patients and make prompt clinical decisions. Furthermore, the potential for biosensor downsizing supports the development of small, portable devices and enhances their usefulness in a variety of living environments. Their expansive usefulness in various medical and environmental domains is highlighted by their versatility in detecting a broad range of biological targets. When taken as a whole, these characteristics demonstrate the critical role that biosensors play in prompt diagnosis, efficient monitoring, and customized care, greatly enhancing patient outcomes [37]. Types of biosensors classified by their bioreceptor include:

- Enzyme-based sensors: Enzyme-based sensors leverage the catalytic activity of enzymes to selectively recognize and react with target analytes. Classic examples include lactate oxidase sensors for monitoring lactate, glucose oxidase sensors for tracking glucose, and cholesterol oxidase sensors for measuring cholesterol. These devices are notable for their high sensitivity and specificity are employed to sense a relatively narrow range of biomolecules. Despite their utility, challenges such as enzyme stability and the maintenance of long-term accuracy remain critical obstacles to broader application and deployment [38].
- Antibody-based sensors: Antibody sensors, or immunosensors, exploit the highly specific binding interaction between antibodies and their corresponding antigens. This specificity enables the detection of infectious pathogens, hormone levels, and cancer biomarkers with remarkable precision. These sensors are capable of identifying target molecules at extraordinarily low concentrations, offering exceptional sensitivity and specificity. However, challenges such as ensuring consistent high-quality antibody production and mitigating potential cross-reactivity present important avenues for future research and development [39].
- DNA sensors: DNA sensors leverage the principle of complementary base pairing to detect specific DNA or RNA sequences. Their applications span pathogen detection in clinical diagnostics, personalized medicine in pharmacogenomics, and the identification of genetic abnormalities through genetic testing. While challenges such as sample preparation and the need for amplification of genetic material at low concentrations persist, these sensors exhibit exceptional specificity and accuracy in identifying genetic variations, making them indispensable tools in molecular diagnostics [40].

- Aptamer-based sensors: Aptamer-based sensors are designed to detect the full range of proteins, small molecules, and cells by utilizing synthetic oligonucleotides as recognition elements. Aptamers offer advantages over antibodies, including greater stability and the ability to be synthesized for a wide range of targets. However, developing aptamers with exceptional specificity remains a significant challenge, requiring further advances in selection methodologies and target affinity optimization [41].
- Cell-based biosensors: Cell-based biosensors use living cells as the sensing element to detect complex biological effects on cellular activity. They find applications in areas such as drug screening and toxicity testing, although they may not find practical use in mobile or wearable smart sensor technologies. While challenges like maintaining cell viability and ensuring uniformity persist, these sensors provide invaluable insights into biological responses, delivering information that extends beyond the mere presence or absence of specific compounds and thus may find some utility in distributed and environmental sensing [42].

2.1.3 Motion Sensors

Motion sensors, including accelerometers, gyroscopes, and inertial measurement units (IMUs), are crucial for tracking posture, movement, and physical activity. These sensors are widely used in applications such as activity tracking, sleep monitoring, gait analysis, and fall detection, particularly in elderly care and rehabilitation. Offering benefits like non-intrusiveness, continuous monitoring, objective data collection, affordability, and adaptability, motion sensors enable real-time responses, such as triggering alerts for fall detection [43]. In rehabilitation, they are instrumental in assessing progress, diagnosing neurological or musculoskeletal conditions, and refining treatment plans for disorders like Parkinson's disease, facilitating precise tremor evaluation and personalized care [44]. Among the examples are:

- Accelerometer: An accelerometer measures linear accelerations along one or more axes, determining the direction of acceleration relative to Earth's gravity. Its primary applications include event tracking, fall detection in the elderly, position analysis, and monitoring movements during sleep. Key advantages of accelerometers are their low power consumption and compact size, making them well-suited for continuous monitoring in wearable and portable devices [45].
- Gyroscope: A gyroscope measures angular velocity and rotational motion, providing critical information about direction and rotation. When combined with accelerometers, gyroscopes enhance the accuracy of activity recognition, enable detailed assessments of motor performance, and support balance evaluation in neurological disorders. Their key advantage lies in their ability to capture unique rotational data that other sensors, such as accelerometers, cannot provide, making them indispensable in applications requiring precise motion analysis [45].
- Inertial measurement unit (IMU): An inertial measurement unit (IMU) combines accelerometers, gyroscopes, and magnetometers to deliver comprehensive data on motion, orientation, and position. IMUs are widely used in gait analysis,

rehabilitation, elder activity recognition, and health-focused virtual reality (VR) and augmented reality (AR) applications. Their integration of multiple sensing modalities enables the most accurate and detailed representation of movement and direction, making them essential for advanced motion-tracking technologies [46]. Additionally, the contextual awareness provided by an IMU enhances smart sensing technologies by enabling adaptive responses to changes in user activity and environment.

Motion sensors offer numerous advantages, including non-intrusiveness, continuous monitoring, objective data collection, affordability, and adaptability across various settings and devices. Sensors that track acceleration and directional changes are particularly critical for fall detection in senior care and operational management, as they can trigger immediate alerts to caregivers or emergency services. In rehabilitation, motion sensors play a vital role in assessing patient progress, detecting irregularities that aid in diagnosing neurological or musculoskeletal disorders, and providing comprehensive gait analysis. Additionally, they support the evaluation and adjustment of treatment plans for conditions such as Parkinson's disease, enabling precise tremor assessment and personalized care.

2.1.4 Optical Sensors

Optical sensors are essential for measuring physiological indicators and are pivotal in both clinical and consumer health applications. Their non-invasive nature enables continuous monitoring of vital signs, making them invaluable for real-time health assessments. As these sensors have become increasingly compact and portable, their integration into smartphones and wearable devices has expanded their accessibility and utility. Capable of measuring a wide array of physiological parameters, optical sensors address diverse health monitoring needs [47]. Their rapid, reliable data provides timely clinical decision support, making them a critical tool for both routine health checks and emergency treatment scenarios [48]. Examples of optical sensors include:

- Pulse oximeters: These devices utilize dual wavelengths of light, typically red and infrared, to measure blood oxygen saturation (SpO_2), pulse frequency, and assess the ratio of oxygenated hemoglobin to total hemoglobin in the blood. Pulse oximeters are commonly employed in monitoring respiratory status, anesthesia management, sleep apnea detection, and other clinical applications. Key benefits include non-invasive measurement, continuous real-time data capture, and portability, making them essential for both patient monitoring and home care settings [49].
- Photoplethysmography (PPG) sensors: Photoplethysmogram (PPG) sensors detect fluctuations in blood volume within the microvascular bed, making them ideal for heart rate monitoring. They can also provide valuable data on blood pressure, cardiac output, and respiratory frequency. PPG sensors are integrated into devices such as smartwatches for heart rate tracking, fitness trackers for atrial

fibrillation detection, stress level assessment, and blood pressure estimation when combined with other sensors. Their compact size and energy efficiency make PPG sensors particularly well-suited for integration into wearable technologies, offering continuous, real-time health monitoring [50].

- Near-infrared spectroscopy (NIRS) sensors: NIRS sensors utilize near-infrared light to monitor tissue oxygenation and hemodynamics, offering valuable insights into deep tissue oxygen levels without the need for invasive procedures. These sensors are used in sports to track muscle oxygenation during exercise, in brain function assessments, and in monitoring wound healing, providing real-time, non-invasive data on physiological processes [51].
- Optical coherence tomography (OCT) sensors: OCT sensors provide high-resolution cross-sectional imaging of tissue microstructures, enabling detailed structural analysis without the need for invasive procedures. These sensors are applied in retinal and dental imaging, diagnosing eye diseases, and detecting skin cancer. While their non-invasive nature offers significant advantages in clinical diagnostics, the complexity and size of OCT systems currently limit their suitability for wearable smart sensor technologies. The bulky equipment and need for precise optical alignment present challenges for integration into compact, portable devices. However, with advances in miniaturization and sensor integration, OCT and other high-powered optical technology may become more feasible for wearable applications in specialized health monitoring contexts [52].

2.2 Applications for Healthcare

2.2.1 Patient Monitoring

Continuous monitoring with wearable smart sensors reveals detailed patterns and trends in physiological data that might be missed during routine clinical assessments. For instance, wearable ECG sensors can provide ongoing heart activity data, capturing arrhythmia, ischemia, and infarction that may go unnoticed during traditional clinical supervision [53]. Another key feature of smart sensor systems is real-time alarms, which notify healthcare providers when a patient's vital signs deviate from normal ranges, facilitating timely intervention in critical situations. For example, pulse oximeters can trigger an alarm if blood oxygen levels drop unexpectedly, enabling rapid response to potential respiratory distress. Advanced technologies, through trend analysis, can even predict adverse events, allowing for preventive measures to avert complications [54]. Moreover, the adoption of smart sensors for autonomous monitoring reduces the burden on medical staff. By replacing manual, periodic examinations with continuous automated data collection, clinicians can focus on more complex patient care tasks. This shift not only enhances healthcare efficiency but also has the potential to reduce overall healthcare costs through better resource allocation and decreased redundancy in tasks [55].

2.2.2 Chronic Disease Management

Smart sensors have emerged as a transformative tool in managing chronic diseases such as diabetes, hypertension, and asthma by enabling continuous tracking of relevant health indicators. This continuous monitoring empowers both patients and healthcare providers to manage these long-term conditions more effectively. By collecting real-time data, these sensors facilitate the creation of personalized treatment plans tailored to the unique needs of individual patients. The ability to customize treatment is especially critical given the dynamic nature of chronic diseases, where conditions can fluctuate over time. For example, continuous glucose monitoring enables more precise diabetes management by identifying specific glucose patterns, allowing for optimized insulin dosing and dietary adjustments. Similarly, blood pressure sensors in hypertension management provide valuable insights into how blood pressure responds to various medications, paving the way for more personalized and effective treatment strategies [56].

Improved self-management is another key benefit of smart sensors. These sensors allow patients to take an active role in the management of their ailment by giving them access to real-time data and feedback. Better patient engagement usually results in higher adherence to treatment programs and lifestyle modifications. For example, the intelligence of an asthma inhaler can detect drug use and environmental triggers, allowing users to better understand and control their own illness. Wearable sensors also provide real-time feedback on physical activity, assisting patients in aligning their activities with health goals and supporting healthy habits [57].

Smart sensors play a pivotal role in enhancing preventive care by facilitating early diagnosis of health issues. Continuous monitoring allows for the detection of emerging trends before they develop into serious problems, enabling timely interventions that can prevent complications and reduce hospitalization. For instance, sensors that track rising blood pressure can prompt early intervention, potentially averting a hypertensive crisis. This proactive approach not only improves patient outcomes by addressing health concerns before they escalate, but it also alleviates strain on the healthcare system by reducing the need for emergency care and hospital admissions [58].

2.2.3 Remote Care

Smart sensors have revolutionized telehealth by providing accurate, real-time patient data, which has become increasingly crucial in addressing global health challenges. One of their primary advantages is the reduction in the need for in-person visits, enabling patients to be monitored from home and significantly decreasing the frequency of healthcare appointments. This capability is particularly beneficial for patients with chronic conditions requiring regular monitoring, elderly individuals or those with mobility limitations, and postoperative patients who need ongoing care. For instance, wearable sensors equipped with heart rate, SpO_2, respiratory rate, and ECG monitoring capabilities enable remote tracking of COVID-19 patients, allowing healthcare providers to assess their condition without direct contact. This ability proved especially vital during the pandemic, when minimizing face-to-face interactions was critical to reducing transmission risk [59].

Increasing access to healthcare is another key advantage of smart sensors. Patients living in remote or rural areas have historically faced high travel costs when seeking specialized medical care. IoT devices help bridge these gaps by enabling the transfer of health data to doctors in different geographic locations, making specialized treatments available that would otherwise be inaccessible locally. This is especially beneficial for patients in medically underserved regions, individuals with rare conditions requiring expert care, and those with limited mobility or transportation options. For these patients, remote care facilitated by smart sensors offers a crucial avenue for receiving the care they need without the barriers posed by distance or logistical challenges.

Furthermore, remote monitoring contributes to more cost-effective practices by reducing unnecessary hospitalizations, minimizing the need for frequent in-person clinic visits, and enabling earlier interventions. These proactive measures are typically less expensive than addressing health issues at later, more critical stages, thereby lowering overall healthcare expenditures. This approach not only improves patient outcomes by addressing problems before they escalate but also enhances the efficiency of the healthcare delivery system by optimizing resource use and reducing the strain on healthcare facilities [60].

2.2.4 Rehabilitation

Smart sensors are increasingly valuable in evaluating the progress of physical therapy and rehabilitation programs, offering significant benefits for both patients and therapists. These sensors provide objective, quantitative data on motion, range of motion, and activity levels, enabling therapists to track progress and adjust treatment plans based on continuous, precise measurements. The remote monitoring capabilities of these sensors ensure patients adhere to prescribed movement guidelines, improving compliance. Real-time feedback also enhances patient motivation and engagement, fostering better adherence to rehabilitation plans. Motion sensors, such as accelerometers, gyroscopes, and IMUs, play a crucial role in rehabilitation applications by offering detailed data for gait analysis, including foot position, pressure distribution, joint angles, walking speed, and step length. This information is vital for monitoring neurological recovery, assessing rehabilitation progress after lower extremity injuries or surgeries, and optimizing athletic performance while preventing further injury [61].

Motion sensors have an essential role in fall detection and prevention, particularly in older patients or during rehabilitation for individuals with balance issues. These systems can alert caregivers in a timely manner, analyze movement patterns to detect and avoid falls, balance, and assess training and training benefits. The security is enhanced by constant monitoring during rehabilitation, which helps to ensure that patients exercise effectively and safely. Sensors can detect sudden movements that may signal pain or potential injury, as well as deviations from specific movement patterns that may result in subsequent injury. By merging smart sensors with IoT platforms and powerful data analytics, the rehabilitation process becomes more accurate, personalized, and effective, thereby improving patient outcomes and recovery paths [44].

3 Wearable IoT Devices in Healthcare

The integration of advanced sensors and wireless communication technologies into wearable devices has given rise to wearable IoT equipment, revolutionizing personal health monitoring. From smartwatches that track a variety of health metrics to intelligent clothing and rings offering more specialized monitoring, wearable IoT devices are designed to address numerous aspects of wellness and healthcare. These devices enable continuous data collection and analysis, providing valuable insights into personal health trends and patterns. The synergy between wearable technology and healthcare has led to several benefits, including enhanced data-driven insights, improved patient compliance, and valuable contributions to medical research and clinical trials. This section focuses on the integration of smart sensors with IoT in healthcare, using wearable devices as a tangible example to highlight their significant advantages in modern health management.

3.1 Integration of Smart Sensors with IoT in Healthcare

3.1.1 Interconnected Healthcare Ecosystem

The modern healthcare enterprise is undergoing a transformative change with the expansion of telemedicine, preventative care, and distributed testing, which will all be driven by advancements in smart sensor technologies, IoTs, and accessible AI tools. These advancements contribute to the development of an interconnected healthcare ecosystem that enables real-time data sharing and analysis, ultimately improving patient outcomes and enhancing clinical decision-making. The integration of smart sensors with IoT systems creates a dynamic network within healthcare, facilitating seamless communication between devices, platforms, and healthcare providers. This interconnected ecosystem involves the use of various technologies, including electronic devices, smart sensors, and IoT platforms, to collect, transmit, and analyze real-time health data. Such a system allows healthcare providers to monitor patients continuously, both remotely and at the point of care, ensuring that vital signs are tracked consistently and enabling timely intervention in medical emergencies or critical conditions. This connectivity holds great promise for the future of healthcare, offering more efficient, responsive, and personalized care. For example, wearable devices with sensors can monitor heart rate, blood glucose levels, and activity patterns, sharing this data to healthcare professionals via IoT platforms for immediate analysis and medical interventions [62]. It allows exchange of health information, medical records, and treatment strategies among healthcare settings which allows continuity of care and reducing medical errors. As technology continues to advance, the potential for innovation within the interconnected healthcare ecosystem is huge, with even better healthcare accessibility, affordability, and effectiveness worldwide.

Wearable devices equipped with sensors continuously gather real-time data, which is then transmitted to IoT platforms for processing and analysis (Fig. 3). Further, the real-time data is processed by AI/ML algorithms which enable proactive healthcare interventions. For example, wearable ECG allows early detection of cardiac irregularities, facilitating timely medical responses and improving patient outcomes. However, the deployment of IoT in healthcare comes with several challenges such as data security, device interoperability, and regulatory compliances. Robust data encryption, communication protocols, and compliance with healthcare regulations are important for protecting patient privacy in healthcare data and maintaining data integrity. Despite several challenges as mentioned above, the IOT-based healthcare ecosystem continues to grow, promising improved efficiency, effectiveness, and patient-centered healthcare delivery.

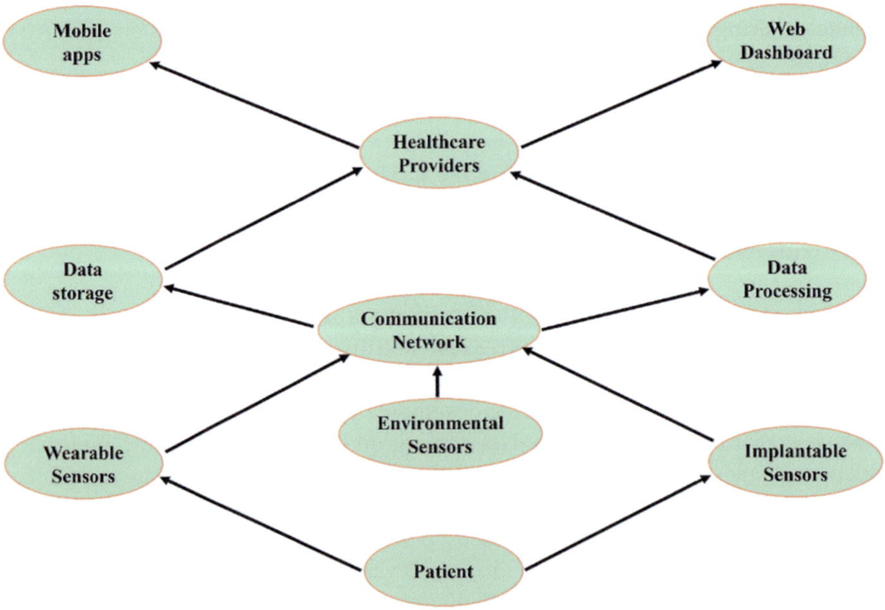

Fig. 3 Scheme of an interconnected healthcare system

3.1.2 Sensor Fusion and Data Analytics

Sensor fusion techniques enable integrated data analysis by combining information from multiple sensors, thereby significantly enhancing the accuracy, reliability, and comprehensiveness of healthcare data. This approach integrates data streams from various sensors, such as wearable devices (e.g., smartwatches and fitness trackers), implantable sensors (e.g., glucose monitors and pacemakers), and environmental sensors (e.g., air quality monitors). By merging these diverse data sets, sensor fusion offers a holistic, personalized understanding of a patient's health status. Advanced data processing algorithms analyze the combined sensor data, allowing for the detection of trends or anomalies that might be missed when considering individual data streams. For instance, combining heart rate data from a wearable device with glucose levels from an implantable sensor provides a more comprehensive picture of a diabetic patient's condition, revealing potential risk factors that could otherwise go unnoticed. This fusion of data supports improved decision-making by offering a more accurate depiction of health trends and anomalies.

In the context of chronic disease management, sensor fusion enables the early detection of exacerbations, allowing for timely interventions. Similarly, in critical care settings, these techniques allow for cross-verification of data from multiple sensors, ensuring more informed and accurate clinical decisions. Ultimately, advancements in sensor fusion and data analytics have the potential to revolutionize healthcare delivery, making it more precise, reliable, and tailored to individual patient needs [63–65].

Unlike single-mode sensing, which provides isolated data from individual sensors, sensor fusion enables the synergistic combination of diverse data sources, leading to more accurate, comprehensive, and actionable insights.

One basic example of sensor fusion in action is the work of Ishigaki et al., which investigated the increasing incidence of falls among the elderly. The study used a posture monitoring system that combined a triaxial accelerometer and a gyroscope, *i.e.,* two simple motion sensors, to evaluate gait stability through pelvic movement analysis [66]. By fusing data from both sensors, the researchers were able to capture detailed information on pelvic movement angles, angular velocities, and accelerations. This fusion provided a more complete picture of gait stability, enabling the identification of reduced pelvic movement as an indicator of gait instability, a primary risk factor for falls. The study showed that elderly individuals with gait instability demonstrated controlled walking with better stability, which could be used to assess their fall risk. This example illustrates how combining accelerometer and gyroscope data not only enhances the reliability of movement analysis but also improves fall risk predictions, leading to better care for elderly individuals.

Multimodal sensor fusion may offer even greater value. Márquez et al. developed a novel clustering algorithm called PN-EAC for improving heartbeat clustering [67]. By integrating positive and negative evidence from multiple sensor modalities, such as ECG sensors and accelerometers, PN-EAC improved the reliability of heart rhythm analysis, enabling more accurate differentiation between normal and abnormal heartbeats. This integration of heterogeneous data sources helped enhance diagnostic capabilities, allowing for better detection of heart conditions and improving patient outcomes. The use of sensor fusion techniques like PN-EAC underscores the growing potential of combining multiple sensor data types to provide deeper, more accurate insights into health conditions.

However, while sensor fusion holds significant promise for advancing healthcare, it also introduces several challenges. One of the primary challenges is the complexity of integrating heterogeneous data from different sensor types, each with its own noise levels, accuracy, and data formats. Proper calibration and synchronization of sensors are essential to ensure that data from different sources align correctly, allowing for meaningful fusion. Moreover, the computational requirements for processing and analyzing large volumes of integrated data can be demanding, requiring peripheral hardware or cloud-based computational resources, necessitating advanced data processing algorithms and ML techniques to efficiently handle and extract insights. The development of robust data fusion algorithms capable of dealing with sensor data variability and noise is essential to achieving reliable and accurate results. Nevertheless, by integrating multiplexed and multimodal sensor data, researchers and practitioners are achieving improved decision-making capabilities and operational efficiency.

3.2 Examples of Wearable Devices

Wearable devices represent a class of smart technologies seamlessly integrated into everyday life, providing continuous health monitoring, activity tracking, and interactive experiences. These devices, worn on the body, offer real-time data collection on a variety of physiological metrics, delivering personalized insights that enhance individual well-being and facilitate early detection of health conditions. Examples of wearable devices include smart watches, smart rings, smart clothing and medical wearable devices (Fig. 4).

3.2.1 Smart Watches

Smart watches are equipped with a series of smart sensors to monitor the critical health indicators, including heart rate, activity level, sleep patterns, blood oxygen level and so on. PPG sensors are commonly used to measure heart rate, while accelerometers and gyroscopes are used to track activity levels and sleep patterns. Advanced models may additionally contain an ECG function for more extensive cardiovascular monitoring. Table 1 provides a comparison of sensor technologies, their functions, and the specific applications found in different smartwatch models. For example, the Apple Watch has powerful health monitoring capabilities such as heart rate tracking, ECG features, fall detection, and blood oxygen measuring. Similarly, Fitbit Sense offers pressure monitoring, skin temperature tracking, and ECG capabilities in addition to basic health measures. The Samsung Galaxy Watch

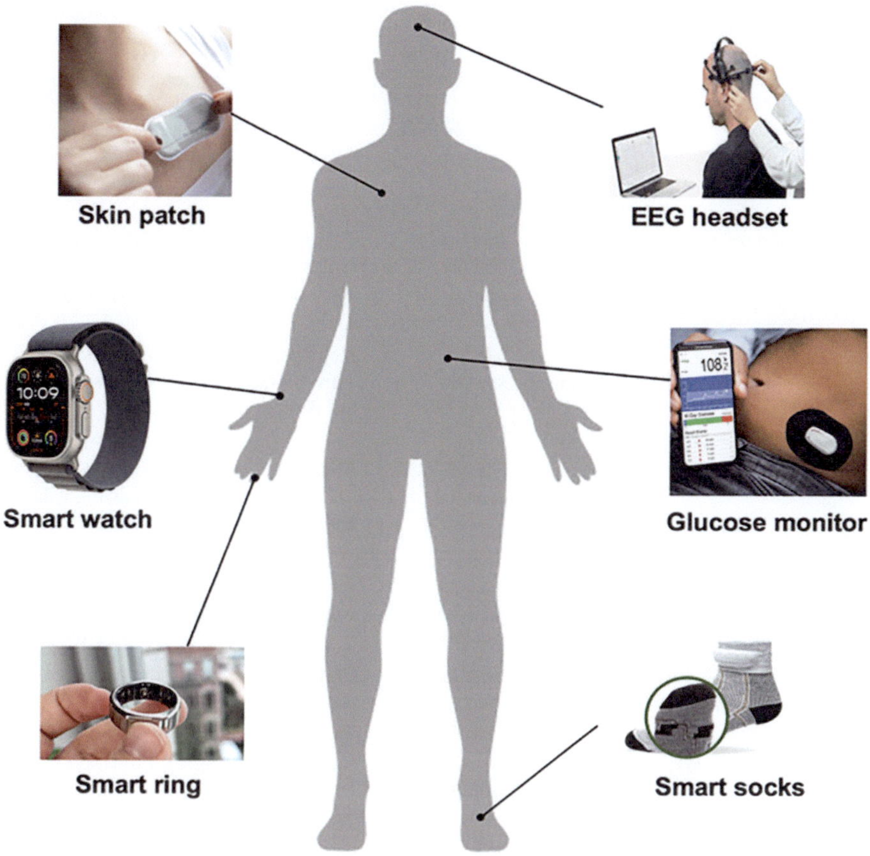

Fig. 4 Examples of wearable devices

has region-specific blood pressure monitoring in addition to heart rate, ECG, and activity tracking. Commercial smartwatches such as the Garmin HRM-Tri, Polar H7, and TomTom Spark Cardio are also examples of devices that integrate these advanced health monitoring features. In summary, smart watches have developed into incredibly sophisticated health monitoring devices that offer thorough physiology [68].

3.2.2 Smart Rings

Smart rings are compact wearables that usually include scaled-down versions of the sensors found in larger wearables, which combine sophisticated health monitoring features into a jewelry ring. They also employ accelerometers to measure exercise and PPG sensors to monitor heart rate. Nevertheless, these gadgets are a wise choice

Table 1 Sensor technologies used in various smartwatches

Sensor type	Function	Technology	Applications	Devices
Photoplethysmography (PPG)	Measures heart rate and blood oxygen levels	LEDs and photodiodes	Heart rate monitoring, SpO_2 measurement	Apple watch, fitbit, samsung galaxy watch, garmin, polar
Electrocardiogram (ECG)	Measures electrical activity of the heart	Electrodes and user interaction	Detection of arrhythmias (e.g., atrial fibrillation)	Apple watch, fitbit sense, samsung galaxy watch
Accelerometer	Tracks movement and orientation	MEMS sensors	Step counting, activity tracking, fall detection	All major smartwatches
Gyroscope	Measures rotational movement	MEMS sensors	Enhanced motion tracking, sleep analysis	Apple watch, samsung galaxy watch, fitbit, garmin
Barometer	Measures atmospheric pressure	MEMS pressure sensors	Elevation tracking, weather forecasting	Garmin, apple watch, samsung galaxy watch
Bioelectrical impedance analysis (BIA)	Measures body composition	Small electrical current	Body fat, muscle mass, and hydration analysis	Samsung galaxy watch 4, 5
Electrodermal activity (EDA)	Measures skin conductance	Sensors detecting sweat gland activity	Stress monitoring	Fitbit sense
Skin temperature sensor	Monitors skin temperature changes	Infrared or contact thermometers	Stress, illness, and ovulation tracking	Fitbit sense
Pulse wave analysis	Analyzes pulse waves to estimate blood pressure	PPG sensors and algorithms	Blood pressure monitoring	Samsung galaxy watch (region-specific)
Respiration sensor	Measures breathing rate	PPG or motion sensors	Respiration rate tracking, sleep monitoring	Garmin, polar

for health monitoring because their compact size makes them more useful for continuous usage, particularly when sleeping. Table 2 provides a side-by-side comparison to highlight the strengths and limitations of both smartwatches and smart rings. Famous examples include the Oura Ring, which is well-known for its intricate sleep tracking and monitoring functions, as well as the temperature function. The Circular

Table 2 Key differences between smart watches and smart rings

Category	Smart watches	Smart rings
Size and form factor	Larger, worn on the wrist; noticeable and tech oriented	Compact, discreet; worn on the finger like jewelry
Sensor technologies	PPG, ECG, accelerometer, gyroscope, barometer, BIA, SpO_2	PPG, accelerometer, skin temperature, SpO_2
Health tracking	Advanced: heart rate, ECG, blood oxygen, stress, fitness, GPS	Basic: heart rate, sleep, temperature, activity levels
Fitness tracking	Detailed tracking for multiple activities, GPS, personalized coaching	Basic activity tracking, no GPS, limited sports tracking
Battery life	1–7 days, depending on usage and model	4–7 days, optimized for continuous wear
User interface	Touchscreen, physical buttons, voice control	No display, data viewed via smartphone app
Comfort	Bulkier, can be uncomfortable during sleep or intense activity	Comfortable for 24/7 wear, including sleep
Data accuracy	High for a broad range of metrics, complex algorithms	Accurate for basic metrics, limited advanced capabilities
Use cases	Active users needing comprehensive health, fitness, and connectivity	Passive health monitoring, sleep tracking, minimal intrusion

Ring is a ring-sized device that monitors heart rate, blood oxygen levels, and exercise. Similarly, the BodiMetrics CIRCUL Sleep and Fitness Ring monitors heart rate continuously, tracks blood oxygen levels and analyzes sleep in depth. Overall, the smart rings are not only aesthetically attractive but also deliver useful information about sleep, exercise, heart rate, and body temperature, marking a significant advancement in wearable medical technology [69].

3.2.3 Smart Clothing

Smart clothing represents an innovative type of wearable device where sensors are directly integrated into the clothing to provide real-time health and activity monitoring. The Hexoskin Smart Shirt, for example, monitors the heart rate, breathing rate, and sleep posture. Similarly, runners might gain important insights from Sensoria Fitness Socks, which track foot landing technique and rhythm. Another application example is Nadi X Yoga Pants, which provide real-time feedback on yoga posture via subtle vibration. Recent breakthroughs include the creation of piezoelectric coating sensors that may be attached to garments to detect strain and exercise, hence improving the capacity to monitor breath and body posture (Table 3). These smart clothing technologies, which enable continuous, uninterrupted monitoring of

Table 3 Examples of smart clothing technologies

Smart clothing	Sensors and technology	Functionality	Applications
Hexoskin smart shirt	Textile-based sensors for heart rate, breathing rate, and activity	Tracks heart rate, respiratory rate, sleep, and movement	Health monitoring, fitness tracking, sleep analysis
Sensoria fitness socks	Pressure sensors, accelerometers	Tracks foot landing, running technique, cadence, and balance	Running analysis, injury prevention, gait training
Nadi X yoga pants	Haptic feedback sensors	Provides real-time feedback on yoga posture via gentle vibrations	Yoga practice, posture correction, body alignment
Piezoelectric coating	Flexible piezoelectric films	Detects strain, movement, and breathing patterns	Posture monitoring, exercise tracking, rehabilitation

vital signs, posture, and specific activities, represent an important step forward in wearable medical devices [70].

3.2.4 Skin Wearable Devices

Skin wearable devices, often known as "e-skin" or "electronic skin," are sophisticated health monitoring technology with a thin, flexible form that allows them to be directly attached to the skin. Examples of such products include the MC10 BioStamp nPoint, a multifunctional thin patch that measures exercise, muscle activity, and heart rate. Similarly, L'Oreal My Skin Track UV offers a no-battery, wearable ultraviolet sensor that may be affixed to the skin or clothing to monitor UV exposure. Recent research has highlighted prototype devices that push the boundaries of skin wearables, such as slim stretchable electronic patches that monitor vital signs, electronic devices that measure similar skin hydration levels, and temporary tattoos that continuously monitor the concentration of an alcohol sensor (Table 4). The discovery of these sophisticated materials allows flexible, elastic electrical gadgets to conform to the shape of the human body while also providing accurate sensing capabilities. Wearable equipment with smart skin highlights wearable IoT devices in the realm of medical development, providing exceptional comfort and health monitoring capability. These technologies allow for more tailored real-time health assessments, expanding the scope of routine health management beyond the typical clinical environment [71].

Table 4 Sensor technologies used in skin wearable devices

Sensor technology	Functionality	Applications	Examples of devices
Electrocardiogram (ECG) sensors	Measures electrical activity of the heart	Cardiac health monitoring, fitness tracking	MC10 BioStamp nPoint
Electromyography (EMG) sensors	Measures muscle electrical activity	Rehabilitation, sports performance, muscle analysis	MC10 BioStamp nPoint
UV sensors	Measures ultraviolet light exposure	Skin protection, UV exposure monitoring	L'Oreal my skin track UV
Hydration sensors	Measures skin hydration levels	Skincare monitoring, hydration tracking	Multifunctional electronic patches
Temperature sensors	Monitors skin and body temperature	Fever detection, health monitoring	Multifunctional electronic patches
Alcohol sensors	Measures alcohol concentration in sweat	Alcohol consumption monitoring	Temporary alcohol monitoring tattoos
Strain sensors	Detects deformation or strain in the skin	Posture monitoring, physical activity analysis	Stretchable electronic patches

3.2.5 Medical Wearables

Medical wearable devices are specifically created for certain health and precision monitoring purposes. Continuous glucose monitors, such as the Dexcom G6, allow for real-time blood sugar measurement for diabetes, making continuous blood sugar management possible. The AliveCor KardiaMobile portable ECG monitor allows for prolonged rhythm evaluation in patients with cardiac disease. Omron HeartGuide is a wearable device that can measure blood pressure at the wrist and marks a significant advancement in wearable technologies (Table 5). The examples above illustrate that wearable medical equipment can be used for a variety of functions, ranging from general health to specific medical monitoring. The integration of these devices with IoT technology provides continuous data collecting and analysis, allowing for more tailored and proactive treatment procedures. Current research focuses on enhancing the accuracy and comfort of wearable IoT devices, allowing for a more comprehensive understanding of the health conditions and increasing the potential for real-time, situation-specific management and intervention [72].

3.2.6 Other Wearables

The spectrum of wearable devices continues to expand, showcasing remarkable innovation in health and well-being monitoring. Smart hearing aids, once limited to basic amplification, have evolved into multifunctional health devices. The Starkey Livio AI, for instance, now integrates fall detection, heart rate monitoring, and activity

Table 5 Examples of medical wearable devices

Device	Sensors and technology	Functionality	Applications
Dexcom G6	Continuous glucose monitor (CGM)	Real-time blood sugar monitoring for diabetes management	Diabetes management, insulin dosing
AliveCor KardiaMobile	Portable ECG monitor	Records and analyzes heart rhythm for arrhythmia detection	Cardiac health monitoring, atrial fibrillation detection
Omron HeartGuide	Wearable blood pressure monitor	Measures blood pressure at the wrist, providing accurate readings	Hypertension management, cardiovascular health

tracking, offering a holistic approach to auditory and physiological health. Similarly, the Oticon Opn connects to the Internet, providing users with a more personalized listening experience, adapting dynamically to environmental changes. Head-mounted devices have likewise advanced, transitioning from simple smart glasses to sophisticated health monitoring tools. The Emotiv EPOC^X wireless EEG headset, for example, provides real-time monitoring of brain activity, offering valuable insights into cognitive performance and mental state. The Dreem Headband, another notable innovation, captures brain activity, heart rate, and breathing patterns, optimizing sleep evaluation by combining these metrics into a comprehensive analysis. In the realm of emotional wellness, wearables are pushing boundaries further. Devices like the Spire Stone clip, which attaches to clothing, track breathing patterns to estimate stress levels, while the Moodbeam wristband allows users to record their emotional states. By integrating these self-reported feelings with other health data, these devices create a more complete picture of emotional health, fostering a deeper understanding of the complex relationship between physical and mental well-being. These innovations underscore the rapid advancements in the Internet of Medical Things (IoMT) space. Each device uniquely contributes to personal health management, offering specialized monitoring capabilities that extend the boundaries of conventional healthcare. Technological progress in wearables strengthens the growing commitment to accessible, continuous, and individualized care, enhancing the ability to proactively manage health in real time [73].

3.3 Key Benefits

3.3.1 Data-Driven Insights

Wearable smart sensor systems can potentially usher in a transformative era in both personalized medicine and population health, enabling continuous and comprehensive data collection at an unprecedented temporal and quantitative scale, which provides valuable insights into individual health trajectories. This ongoing stream of real-time data facilitates several key advancements in medical practice, creating opportunities for early intervention and personalized care.

First, the ability to monitor long-term health trends is one of the most significant benefits of wearable smart sensor systems. By collecting data over extended periods, these devices offer a unique advantage over the occasional snapshots provided during clinical visits. This longitudinal approach allows for the identification of gradual changes in health, such as persistent increases in resting heart rate, that may signal emerging cardiovascular issues. Early detection of such trends facilitates proactive interventions that may prevent more serious conditions from developing [74].

In addition to identifying long-term health patterns, wearable smart sensor systems excel at detecting subtle, transient changes in a patient's health status. Because these devices operate continuously, they are able to flag deviations from baseline health markers that might otherwise go unnoticed. Small variations, such as alterations in sleep patterns or heart rate variability, can serve as early indicators of potential health problems, prompting timely medical attention before more significant issues arise [75].

This wealth of data gathered by wearable devices enables more informed decision-making by healthcare providers. With access to rich, continuous datasets, clinicians are better equipped to assess a patient's overall health and track changes over time. Advanced ML algorithms can analyze these datasets, predict patient outcomes, and identify anomalies, leading to more accurate diagnoses and tailored treatment plans. For example, the combination of continuous blood pressure monitoring with AI-powered analysis allows healthcare providers to adjust medication with greater precision, as opposed to relying on sporadic clinic visits for diagnosis and management [76].

The vast troves of data generated by wearable devices also unlock the potential for large-scale analysis, revealing population-level health trends and correlations. Using advanced analytical tools, such as AI/ML, researchers and clinicians can identify new risk factors, detect early warning signals, and uncover relationships between lifestyle choices and health outcomes. This ability to analyze big data on a broad scale offers unprecedented insights into public health and disease prevention, potentially transforming healthcare systems worldwide [77].

3.3.2 Improved Patient Adherence to Treatment

Smart sensor technologies are playing an increasingly significant role in enhancing patient adherence to treatment plans, a key factor that can significantly influence health outcomes. By offering continuous support, reminders, and real-time feedback, these devices are helping bridge the gap between clinical recommendations and patient compliance.

One of the most impactful features of smart sensor technologies, especially wearables, is their ability to assist with medication adherence. Through automated reminders and alerts, these devices help reduce the likelihood of missed doses. With integration into smart medication containers or dispensers, wearables can track when medications are taken, ensuring that patients follow prescribed regimens. This not only addresses forgetfulness but also minimizes issues related to incorrect timing, providing a seamless mechanism to improve compliance [78]. In addition to medication adherence, wearable device alarms or notifications are also pivotal in promoting physical activity and exercise, especially in patients undergoing rehabilitation or physical therapy. These devices can remind patients to complete their prescribed routines and even monitor the quality of their movements by analyzing motion patterns. This feedback is essential for patients recovering from surgery or injuries, ensuring that exercises are performed correctly, and progress is tracked in real-time [79].

One of the most powerful capabilities of wearables is their ability to provide immediate feedback on treatment efficacy. By continuously monitoring key health metrics, wearables create a two-way communication channel between patients and healthcare providers. This allows for more precise adjustments to treatment plans based on real-world data, and it empowers patients to see firsthand the effects of their medications or lifestyle changes. To further enhance engagement, wearables incorporate gamification elements such as goal setting, virtual rewards, and peer competition, all of which can motivate patients to stay committed to their treatment. For example, patients with hypertension may notice a reduction in their blood pressure, reinforcing their adherence to medication and encouraging them to continue their health journey [80].

Despite these advancements, the success of smart sensor technologies in improving adherence relies on patient acceptance and ease of use. Ensuring that these devices are intuitive and seamlessly integrated into daily routines is crucial for widespread adoption. Furthermore, the continuous collection of personal health data raises privacy concerns that must be addressed with robust security measures. Nevertheless, the integration of smart sensor technologies into treatment plans represents a major leap forward in healthcare, offering not only improved adherence but also enhanced health outcomes, more accurate treatment, and optimized use of healthcare resources [81].

In addition to wearable smart systems, smart pills, smart bottles, and smart labels are emerging as another innovative solution to enhance medication adherence [82–84]. These smart bottles or ingestible capsules are equipped with sensors that track when the pill is swallowed, transmitting data to a wearable device or smartphone app. This real-time tracking ensures that patients are taking their medications as prescribed, providing both patients and healthcare providers with a clear record of adherence. Smart pills and smart bottles are particularly useful for chronic conditions that require strict adherence to complex regimens, such as in the treatment of mental health disorders, organ transplant patients, or those managing conditions like tuberculosis. By incorporating this technology, healthcare providers can receive insights into patient behavior and intervene if necessary, ensuring that patients receive the full benefit of their prescribed treatment plans. Furthermore, as smart pill technology advances, it has the potential to monitor the medication's dissolution and release within the body, offering even greater insights into treatment efficacy.

3.3.3 Support for Clinical Trials and Medical Research

Wearable smart sensor systems are beginning to show potential for supporting clinical trials and medical research that provides for more adaptive and dynamic study designs. Unlike traditional clinical trials, which often rely on intermittent assessments and self-reported data, wearable technology enables ongoing, objective data collection in real-world settings. This continuous monitoring offers a much richer, more accurate picture of a patient's health status and their response to treatment [85]. With data being collected around the clock, researchers gain a more reliable understanding of how treatments are working and when they do not. As a result, the data derived from wearables leads to more trustworthy and actionable insights.

On a larger scale, smart sensor technologies have unlocked the potential for massive data collection across diverse populations. By monitoring vital signs such as heart rate, activity levels, and sleep patterns, these devices make it possible to track large cohorts over extended periods. This scale and consistency of data offer researchers the ability to detect trends and correlations that might otherwise remain hidden in smaller, less frequent datasets. The power of this data could ultimately reveal new insights into the mechanisms of disease and treatment effects, potentially sparking breakthroughs in personalized medicine [86].

Perhaps one of the most compelling aspects of wearables in research is the speed and quality they bring to medical studies. Continuous data collection allows for the rapid identification of therapeutic effects or adverse reactions, empowering researchers to make faster decisions. By analyzing data from different patient subgroups, these devices help to refine treatment plans, providing a more personalized approach. Moreover, the ability to collect data remotely lowers barriers to trial participation, which can accelerate patient recruitment and reduce dropout rates, a persistent challenge in traditional clinical trials [87].

Despite these advances, there are still hurdles to overcome. Standardizing data collection and analysis is crucial to ensure consistency across studies and research settings. As technology continues to evolve, we can expect these devices to play an increasingly central role in accelerating medical research and personalized medicine [88].

3.4 AI-Enabled Wearable Devices and Applications

3.4.1 AI/ML-Enabled Wearable Sensors

The computationally supported design and analysis of sensor development has impacted the traditional approach to developing sensor technologies. By leveraging computational tools, we gain direct insights into the parameters that require optimization or further refinement during both the design and characterization stages [89, 90]. These methods are particularly valuable in advancing miniaturization and crafting designs that are more suitable for specific experimental applications. In the realm of sensing, the most resource-intensive aspect is signal acquisition, which is highly susceptible to interference from various noise sources. This challenge is further complicated by the physiological and natural variability inherent in different user or patient populations. Factors such as molecular interactions, design and fabrication imperfections, the chemical or optical signal transduction mechanisms, and environmental variables all contribute to the overall reliability of sensors. Computational techniques address these challenges by enabling the adoption of optimal design strategies and extracting meaningful insights from otherwise raw sensor data. By combining high-quality experimental data with sophisticated simulations, we can deepen our understanding of sensor functionality and refine design processes to minimize noise, ultimately improving the accuracy and reliability of sensor systems.

Development approaches supported by advanced ML algorithms, such as random forest and support vector machine (SVM)-based algorithms have been utilized in sensing domain in various ways to get an optimal sensing performance [91]. For example, the development of cross-reactive sensor arrays combined with advance ML models, have realized "electronic noses" for detecting trace amounts of explosives and toxins in complex solutions [92–94]. ML models have also been used in diagnostics and genome data analysis, where pattern recognition in biological pathways helps in predicting diseases progression and understanding the mechanism. These algorithms allow decoding complex biological data sets, thereby shifting the attention from device design to digital data analysis [95]. Recently, deep learning and neural networks have shown enormous potential in biomedical signal analysis by capturing complex nonlinear patterns in high-dimensional data. Neural networks provide better prediction speed compared to traditional approaches like compressive sampling and signal reconstruction. These algorithms can be easily integrated into smartphones and tablets, paving the way for cost-effective, point-of-care sensing and diagnostic systems [96, 97].

Current AI/ML supported strategies follow a sequential approach where data is generated and collected by sensor hardware and then analyzed computationally. This approach overlooks the potential of embedding computational tools into the sensor design process itself. Computational design and statistical learning can be used into the sensor's development from the start with advanced parametric analysis to have idea of parameters variance and its impact on sensor functionality. A reverse engineering approach can also be followed to get desired sensing capability by computational analysis. It allows handling the noise sources and signal variation in the design phase, which can improve the sensing performance. In the perspective of Ballard et al. they emphasize "ML-inspired sensor design" as the need for the computationally enabled design of sensing systems with optimal sensor functionality [61]. This iterative design process uses initial sensor concept and computational processing, feature optimization, and statistical learning at the hardware level. Figure 5 illustrates the step-by-step design process of a smart sensing system. The process begins with an initial design phase, positioned centrally in the diagram, which is formulated through a combination of standard engineering practices or, at times, by the random selection of transduction elements or sensing components. Data collected from the initial sensors or data repository (X_i) are then used to train a ML model for

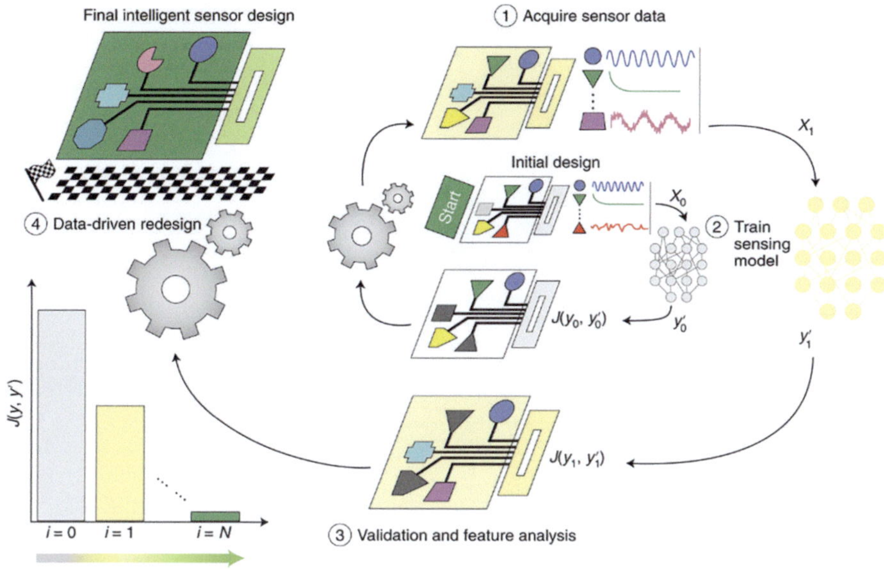

Fig. 5 Overview of ML-enabled intelligent sensor design. The subscript i = 0…N denotes the design iterations in the workflow. Reproduced with permission from Springer Nature [98]

the prediction of sensing outcomes denoted as y'_i. A cost function J (y_i, y'_i) is used to evaluate the model performance by using the ground-truth sensing information, y_i. The design iterations adopted here can be repeated to get the desired sensing performance in accordance with the user-defined cost function, which results in a final sensor design.

3.4.2 Application for Clinical Decision Making

Smart sensor systems provide access to physiological information through distributed networks of sensing units, thereby generating big datasets that range from individual metrics to population-level insights. In this complex, high-dimensional dataset, the decision-making system translates raw data into an interpretable format [99, 100]. Traditional approaches are relying on selected features for specific tasks under human guidance which has its own limitations. Whereas data-driven methods have the potential to enhance our understanding about patterns and relationships between unique features. This can be achieved by using advanced data fusion approaches by combining the data from wearable devices, *e.g.*, implantable sensors or on body area networks (BANs) [101, 102]. In an interconnected system of sensors, the data can be shared across a body area network and analyzed using these advanced data fusion approaches to reduce environmental noise in sensors. The sensing units convert physiological data into digital signals where a data pipeline is created (Fig. 6). The raw data from sensors is then processed by a data conversion unit, where digital signals

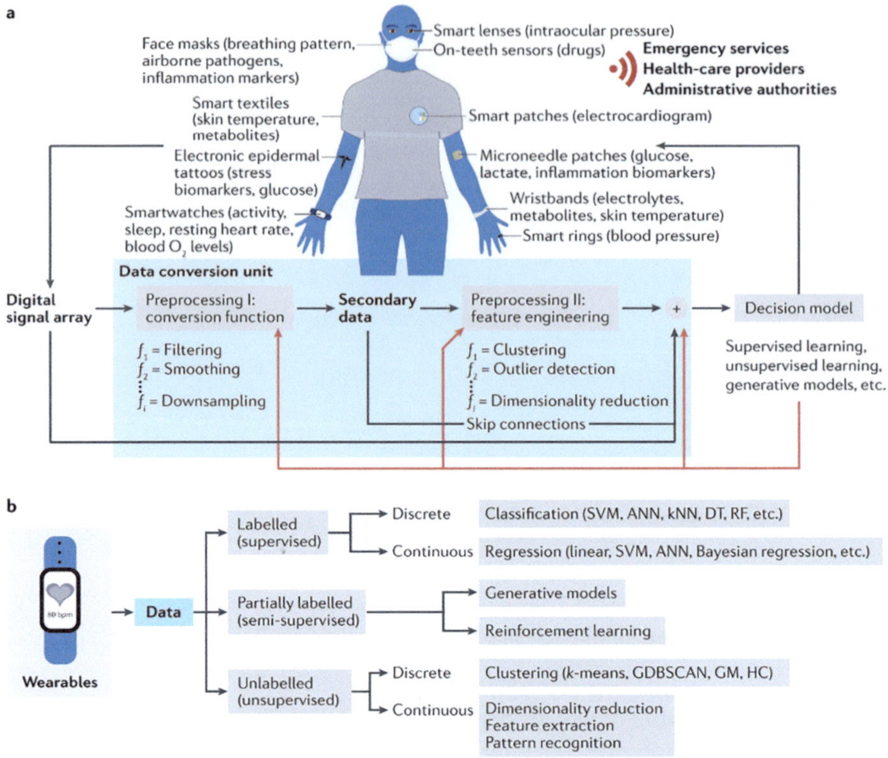

Fig. 6 Overview of the global smart sensor system and data flow across multiple sensors and modalities. Reproduced with permission from Springer Nature [101]

(such as current or voltage) are translated into secondary data (like heart rate, pH, or metabolite levels) by using specific algorithms [103]. This process is performed by a conversion function, which defines the correlation between the digital signal and the measured quantity of the sensing unit. Conversion functions are generally determined via regression analysis and require human supervision. Initial data preprocessing, such as filtering, smoothing, noise reduction, or down sampling is also used during conversion process which depends on data quality. By using fully automated data handling supported by ML models, creating an "end-to-end learnable" system [104, 105]. ML methods provide improved accuracy and computational efficiency by detecting highly nonlinear patterns of input signals and their correlation with output signals. This approach is highly recommended in case of complex physiological events, where the data is composed of multiple variables, rather than a single digital signal which may not provide sufficient information for quantitative analysis. This allows combining multiple features from different sensors to identify complex patterns more efficiently than conventional methods.

Secondary data processing involves a series of techniques designed to achieve outlier and anomaly detection, clustering, noise reduction, handling missing features, data normalization, dimensionality reduction, and baseline correction, all of which are tailored to the specific needs of the model. For instance, in datasets with high correlation, normalization becomes essential. This process scales each feature around a common mean and standard deviation, ensuring that dominant features do not skew the results. Baseline correction algorithms are employed to stabilize the data, adjusting for trends that may distort the analysis [106]. This critical stage, known as feature engineering, is where the raw data is transformed into valuable, relevant information that improves model performance. By applying appropriate feature selection algorithms, less relevant features are removed without significantly affecting the outcome, allowing the model to focus on the most informative aspects. Feature engineering also involves creating new latent variables by introducing additional features or reducing the dataset's dimensionality, preserving the data integrity [107]. This process ensures that valuable information remains intact throughout multiple data transformations. While feature engineering can be performed manually, AI now automates the identification of correlations among complex datasets, enhancing efficiency and accuracy, especially in managing sensor data from wearable networks. AI-driven methods help to optimize the entire data pipeline, ensuring that the most valuable information is consistently available for analysis.

Engineered features are further processed using data-driven models to automatically extract patterns based on specific objectives. These models perform tasks like classification, regression, clustering, and dimensionality reduction, and their efficiency is evaluated using metrics like RMSE, MAE, and Pearson correlation. In supervised learning, human-labeled data is used to predict physiological variables, though the model's performance depends on the accuracy and bias of the labeling process. To mitigate this, training data must be diverse and large enough to reduce noise and avoid overfitting. Different model complexities are applied depending on the data's volume and dimensionality, balancing the need to avoid overfitting or underfitting. Ensemble learning, which combines models based on various learning theories, helps make more robust decisions. In healthcare, where personalized strategies are often necessary, simpler models like SVM or random forest may outperform complex ones, such as deep neural networks, particularly for unique patterns like epileptic seizures. For example, if a data set consists of a unique individual patterns, such as detecting epileptic seizures from EEG signals, an ensemble of simpler models like SVM or random forest may perform better than deep neural networks, which could overfit small datasets [108]. For more stereotypical conditions, such as arrhythmia detection, larger datasets enable the use of more complex models. Ultimately, the selected data-driven model must accurately capture the physical nature of the problem at hand. For example, when predicting COVID-19 infection and disease progression, the model's design was optimized to incorporate relevant physiological markers, such as muscle degradation, respiratory tract changes, and vocal cord alterations [107, 109]. This approach can enhance model sensitivity and improve performance metrics.

3.4.3 Example Application for Stress Monitoring

Monitoring and recognizing stress in daily life is essential for maintaining both mental and physical health. The future of stress management may involve smart sensor technologies, homes, and cars that help build a comprehensive picture of a person's daily life [110]. Direct input, like active feedback from the user, or indirect data, such as digital footprints and social media activity, could provide valuable information about how individuals perceive and experience stress. Integrating advanced language models like GPT or Gemini into smart assistants and apps could enhance accessibility by using personalized chat interfaces or gamification techniques.

Figure 7 illustrates a closed-loop smart sensing system integrated with AI [111] designed to develop personalized stress management solutions [112]. Traditionally, stress has been assessed through subjective methods such as self-reporting, clinical interviews, and behavioral observations [113]. However, recent advances in stress research emphasize the need for objective physiological measurements. Key metrics include heart rate variability, hormone levels, and galvanic skin response [114, 115]. Heart rate variability reflects autonomic nervous system activity and indicates stress levels, while cortisol, a hormone that increases rapidly during stress, offers insight into the body's physiological response [116, 117]. Additionally, changes in skin conductance measured through galvanic skin response provide information on sympathetic nervous system activation. While these methods are effective for assessing stress, they often require lab environments or specialized equipment. However, wearables offer an easy to use and continuous approach to monitor physiological markers in real-life situations, making stress assessment more accessible.

Gao et al. reported the demonstration of an AI-enabled electronic skin system for monitoring stress responses [118]. Their electronic skin can measure three biometrics, skin temperature, pulse waveform, and galvanic skin response, and analyze six biomarkers in sweat, including glucose, lactate, uric acid, sodium, potassium, and ammonium ions. The system, made of composite materials, can monitor these biomarkers for over 100 h. While this work demonstrated the necessary hardware for stress monitoring, there is still a need for a database that correlates these indirect

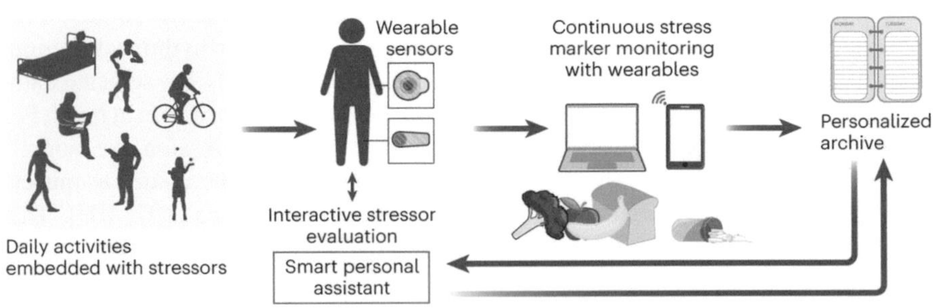

Fig. 7 AI-enabled smart sensor system for stress monitoring. Reproduced with permission from Springer Nature [112]

measurements with stress hormone level and behavioral response to better map these complex relationships.

Accordingly, to better understand, predict, or treat stress on a personal level, sensor measurements alone may be insufficient. Additional insights are required to create personalized interventions. While self-reported data have limitations, they offer valuable perspectives on how individuals perceive stress. When combined with multimodal, long-term sensor data, these reports can deepen our understanding of the relationship between physiological responses, daily activities, specific events, and self-identified stress episodes. By organizing sensor data to establish reference-state distributions, we can detect abnormal patterns of behavior. Integrating this with self-reported data during critical timeframes enables the creation of a comprehensive stress profile or physiological avatar, essentially a digital twin for each individual [119].

For example, an important aspect of self-reporting is understanding how stressors impact individuals, distinguishing between positive (eustress) and negative (distress) experiences. This distinction can be pivotal in tailoring stress detection and management to the unique needs of each person. Multimodal sensor data, organized in an unsupervised or self-supervised manner, can help establish personal baselines through correlation with data logs or diaries that denote positive or negative experiences. By mining this data, patterns can be extracted to create an individual's well-being baseline. This approach can illuminate both the identification of stressors but also analyze psychosocial and psychophysical links related to whether experiences are perceived as positive or negative, bridging the gap between objective sensor data and the subjective interpretation of an individual's experiences.

4 Challenges and Future Directions

The evolution from the IoT to the IoMT introduces unique challenges and requirements specific to the healthcare sector, including the need for cost efficiency, low computational overhead, energy conservation, and reliable wireless networking. This transition reflects a significant technological shift aimed at improving healthcare delivery and patient outcomes. However, as healthcare systems become increasingly interconnected, they also face security challenges, particularly concerning data integrity and patient safety. Blockchain technology has emerged as a promising solution, offering robust mechanisms for ensuring data integrity, transparency, and accountability in IoMT networks. Beyond technical considerations, the societal acceptance and long-term sustainability of IoMT are crucial factors influencing its successful deployment. The integration of these systems must not only address security and privacy concerns but also gain the trust of healthcare providers and patients alike to ensure their widespread adoption and continued effectiveness in improving health outcomes.

Given the sensitive nature of healthcare data, the need for robust security and privacy measures in IoMT is paramount. Cyberattacks, unauthorized access, and data breaches can compromise patient privacy, highlighting the necessity of comprehensive security protocols. Ensuring data encryption, implementing robust access controls, and securing devices against vulnerabilities are essential to safeguard against hostile entities exploiting weaknesses in IoMT systems. Furthermore, adherence to healthcare regulations, such as the Health Insurance Portability and Accountability Act (HIPAA) in the United States, is mandatory to uphold patient rights and maintain the trust of both patients and healthcare providers. Ethical considerations, including obtaining informed consent and practicing transparent data handling, are fundamental to respecting individual autonomy and preserving the integrity of healthcare data. Thus, security and privacy are not only regulatory requirements but also the foundations for safe and effective healthcare delivery.

4.1 Data Security and Encryption

Data security and encryption are crucial in IoMT networks and devices to ensure the integrity and confidentiality of sensitive medical information. As medical devices within the IoMT ecosystem collect and process private health data, robust security measures are essential. However, the miniaturized nature of many IoMT devices presents challenges due to their limited computational power, complicating the implementation of advanced security schemes. Researchers have proposed various solutions to address these challenges.

Recent studies have emphasized the critical need for robust security and encryption mechanisms to protect sensitive data in the IoMT ecosystem. Sun et al. reviewed cutting-edge security and privacy approaches for IoMT-enabled healthcare systems, underscoring the importance of encryption in safeguarding medical information [120]. In a similar vein, Masud et al. proposed a lightweight, secure key establishment protocol tailored for IoMT applications in COVID-19 patient care, focusing on ensuring the confidentiality and integrity of data exchanged between IoT nodes [121]. However, the limited computational power of IoMT devices presents significant challenges in implementing such security measures, as highlighted by Hamza et al. [122] who explored the vulnerabilities of IoT devices in applying security algorithms. They recommended solutions like the Key Management Scheme (KMS) and Public Key Infrastructure (PKI) to address concerns related to confidentiality and data integrity. Additionally, the increasing use of blockchain technology in IoMT applications has raised new security concerns, as explored by Li et al., who surveyed the security of blockchain systems and identified common threats and attacks. Their work emphasized the need for further research to enhance blockchain security in healthcare settings [123].

Lastly, Sadeghi et al. extended the conversation to the broader IoT landscape, focusing on industrial IoT systems with lessons that are directly applicable to IoMT, and stressing the importance of a comprehensive security framework to mitigate cyberattack risks. Collectively, the ongoing challenges and the need for continuous innovation to ensure the security and privacy of IoMT systems will be a major consideration for all smart sensor technologies. The development in the general and industrial IoT sectors [124] may provide preliminary guidance to serve the next-generation of security frameworks, that address the unique challenges of IoMT devices and networks, ensuring patient privacy and data integrity in an increasingly connected healthcare ecosystem.

4.2 Authentication and Access Control

Effective authentication and access control mechanisms are essential for ensuring the security of IoMT systems. The proliferation of IoMT devices complicates the management and maintenance of these systems' security. Adversaries can exploit system vulnerabilities, necessitating the implementation of robust mutual authentication and access control schemes to protect IoMT networks from adversarial threats. Masud et al. highlighted the absence of robust mutual authentication and key establishment schemes as key factors attracting adversaries to IoMT networks. Existing mutual authentication schemes are often computationally and communication-expensive, which can drain the energy reserves of IoT sensor nodes. To overcome these challenges, Masud et al. suggested a mutual authentication and secret key establishment technique for IoMT networks that is both lightweight and physically secure [121]. This protocol utilizes Physical Unclonable Functions (PUF) to verify the legitimacy of doctors and sensor nodes before establishing a session key, ensuring authentication, confidentiality, integrity, and anonymity in IoMT networks. Hasan et al. reviewed various methods, including security threats, data vulnerabilities, and countermeasures for 5G-enabled IoMT, underlining the importance of encryption techniques, access control, verification, and data encryption in improving IoMT device security and reliability [125]. The lack of reliable authentication mechanisms and network access controls in IoMT devices is a major vulnerability. Gupta et al. addressed privacy and security issues, such as confidentiality, entity authentication, and integrity, using lattice-based data authentication and access control protocols applicable in resource-constrained quantum environments [126].

4.3 Considering Patient Privacy

Ensuring patient privacy is paramount in IoMT environments due to the sensitive nature of personal health data. Researchers such as Oh et al. and Sahi et al. have stressed the need for strong privacy-preservation mechanisms to protect patient information, emphasizing the importance of secure sessions and data protection techniques. IoMT systems are vulnerable to various cybersecurity threats, including weak passwords, insufficient encryption, and denial-of-service (DoS) attacks, highlighting the need for comprehensive security frameworks. Blockchain technology offers benefits such as immutability, transparency, and secure data management, as shown by Ktari et al.'s e-health monitoring platform [127]. Additionally, zero-trust frameworks bolster security by requiring continuous verification of trustworthiness and monitoring network traffic. Secure bootstrapping and firmware verification help maintain device integrity by preventing unauthorized access and tampering. Anonymizing patient data remains crucial for privacy, with several methods proposed to balance privacy protection and data utility. Ongoing innovation and robust security solutions are essential to address these challenges and strengthen the security and privacy of IoMT systems.

4.4 Regulatory Considerations

The deployment and use of smart sensors in the IoMT are governed by a complex framework of regulatory requirements and standards designed to ensure patient safety, data integrity, and system security. Adhering to these regulations is essential for maintaining trust in healthcare systems and ensuring their secure operation without compromising patient privacy or safety [128]. In United States, the HIPAA sets national standards for protecting health information, including the Privacy Rule for medical records and the Security Rule for electronic Protected Health Information (PHI). In the EU, the General Data Protection Regulation (GDPR) mandates lawful, fair, and transparent data processing and requires timely breach notifications. The Medical Device Regulation (MDR) and In Vitro Diagnostic Regulation (IVDR) govern the safety and performance of medical devices, while the U.S. FDA regulates medical devices through Premarket Approval and 510(k) Clearance. Key ISO standards, such as ISO 13485 for quality management, ISO 27001 for information security, and ISO 14971 for risk management, ensure exacting standards of quality and safety. Compliance with these regulations is crucial to protecting patients and their data, preserving trust, and avoiding legal and financial penalties. Healthcare organizations must remain vigilant and up-to-date with the latest standards to safeguard IoMT systems and provide secure healthcare services [129].

References

1. Srikrishnarka, P.; Haapasalo, J.; Hinestroza, J. P.; Sun, Z.; Nonappa, Wearable Sensors for Physiological Condition and Activity Monitoring. *Small Science* **2024**, 2300358.
2. Zeng, X.; Deng, H.-T.; Wen, D.-L.; Li, Y.-Y.; Xu, L.; Zhang, X.-S., Wearable multi-functional sensing technology for healthcare smart detection. *Micromachines* **2022,** *13* (2), 254.
3. Thai, M. T.; Phan, P. T.; Hoang, T. T.; Wong, S.; Lovell, N. H.; Do, T. N., Advanced intelligent systems for surgical robotics. *Advanced Intelligent Systems* **2020,** *2* (8), 1900138.
4. Staszewski, W.; Boller, C.; Tomlinson, G. R., *Health monitoring of aerospace structures: smart sensor technologies and signal processing*. John Wiley & Sons: 2004.

5. Baskar, S.; Shakeel, P. M.; Kumar, R.; Burhanuddin, M.; Sampath, R., A dynamic and interoperable communication framework for controlling the operations of wearable sensors in smart healthcare applications. *Computer Communications* **2020,** *149*, 17-26.
6. Nasiri, S.; Khosravani, M. R., Progress and challenges in fabrication of wearable sensors for health monitoring. *Sensors and Actuators A: Physical* **2020,** *312*, 112105.
7. Bayo-Monton, J.-L.; Martinez-Millana, A.; Han, W.; Fernandez-Llatas, C.; Sun, Y.; Traver, V., Wearable sensors integrated with Internet of Things for advancing eHealth care. *Sensors* **2018,** *18* (6), 1851.
8. Ha, N.; Xu, K.; Ren, G.; Mitchell, A.; Ou, J. Z., Machine learning-enabled smart sensor systems. *Advanced Intelligent Systems* **2020,** *2* (9), 2000063.
9. Krittanawong, C.; Rogers, A. J.; Johnson, K. W.; Wang, Z.; Turakhia, M. P.; Halperin, J. L.; Narayan, S. M., Integration of novel monitoring devices with machine learning technology for scalable cardiovascular management. *Nature Reviews Cardiology* **2021,** *18* (2), 75-91.
10. Mbunge, E.; Muchemwa, B.; Batani, J., Sensors and healthcare 5.0: transformative shift in virtual care through emerging digital health technologies. *Global Health Journal* **2021,** *5* (4), 169–177.
11. McGrath, M. J.; Scanaill, C. N., *Sensor technologies: healthcare, wellness, and environmental applications.* Springer Nature: 2013.
12. Ometov, A.; Shubina, V.; Klus, L.; Skibińska, J.; Saafi, S.; Pascacio, P.; Flueratoru, L.; Gaibor, D. Q.; Chukhno, N.; Chukhno, O., A survey on wearable technology: History, state-of-the-art and current challenges. *Computer Networks* **2021,** *193*, 108074.
13. Ali, F.; El-Sappagh, S.; Islam, S. R.; Ali, A.; Attique, M.; Imran, M.; Kwak, K.-S., An intelligent healthcare monitoring framework using wearable sensors and social networking data. *Future Generation Computer Systems* **2021,** *114*, 23-43.
14. Sony, S.; Laventure, S.; Sadhu, A., A literature review of next-generation smart sensing technology in structural health monitoring. *Structural Control and Health Monitoring* **2019,** *26* (3), e2321.
15. Javaid, M.; Haleem, A.; Rab, S.; Singh, R. P.; Suman, R., Sensors for daily life: A review. *Sensors International* **2021,** *2*, 100121.
16. Jain, A.; Hong, L.; Pankanti, S., Biometric identification. *Communications of the ACM* **2000,** *43* (2), 90-98.
17. Awad, A.; Trenfield, S. J.; Pollard, T. D.; Ong, J. J.; Elbadawi, M.; McCoubrey, L. E.; Goyanes, A.; Gaisford, S.; Basit, A. W., Connected healthcare: Improving patient care using digital health technologies. *Advanced Drug Delivery Reviews* **2021,** *178*, 113958.
18. He, R.; Shen, L.; Wang, Z.; Wang, G.; Qu, H.; Hu, X.; Min, R., Optical fiber sensors for heart rate monitoring: A review of mechanisms and applications. *Results in Optics* **2023,** *11*, 100386.
19. Kim, J.; Chou, E. F.; Le, J.; Wong, S.; Chu, M.; Khine, M., Soft wearable pressure sensors for beat-to-beat blood pressure monitoring. *Advanced healthcare materials* **2019,** *8* (13), 1900109.
20. Zhao, Y.; Bergmann, J. H., Non-contact infrared thermometers and thermal scanners for human body temperature monitoring: a systematic review. *Sensors* **2023,** *23* (17), 7439.
21. Madhvapathy, S. R.; Wang, H.; Kong, J.; Zhang, M.; Lee, J. Y.; Park, J. B.; Jang, H.; Xie, Z.; Cao, J.; Avila, R., Reliable, low-cost, fully integrated hydration sensors for monitoring and diagnosis of inflammatory skin diseases in any environment. *Science advances* **2020,** *6* (49), eabd7146.
22. Schiavoni, R.; Monti, G.; Piuzzi, E.; Tarricone, L.; Tedesco, A.; De Benedetto, E.; Cataldo, A., Feasibility of a Wearable Reflectometric System for Sensing Skin Hydration. *Sensors* **2020,** *20* (10).
23. Ramasamy, S.; Balan, A., Wearable sensors for ECG measurement: a review. *Sensor Review* **2018,** *38* (4), 412-419.
24. Gohel, V.; Mehendale, N., Review on electromyography signal acquisition and processing. *Biophysical reviews* **2020,** *12* (6), 1361-1367.

25. Ramesh, M.; Janani, R.; Deepa, C.; Rajeshkumar, L., Nanotechnology-enabled biosensors: A review of fundamentals, design principles, materials, and applications. *Biosensors* **2022,** *13* (1), 40.
26. Banica, F.-G., *Chemical sensors and biosensors: fundamentals and applications.* John Wiley & Sons: 2012.
27. Booth, M. A.; Gowers, S. A.; Leong, C. L.; Rogers, M. L.; Samper, I. C.; Wickham, A. P.; Boutelle, M. G., Chemical monitoring in clinical settings: recent developments toward real-time chemical monitoring of patients. *Anal. Chem* **2018,** *90* (1), 2-18.
28. Rodbard, D., Continuous glucose monitoring: a review of successes, challenges, and opportunities. *Diabetes technology & therapeutics* **2016,** *18* (S2), S2–3-S2–13.
29. Seshadri, D. R.; Li, R. T.; Voos, J. E.; Rowbottom, J. R.; Alfes, C. M.; Zorman, C. A.; Drummond, C. K., Wearable sensors for monitoring the physiological and biochemical profile of the athlete. *NPJ digital medicine* **2019,** *2* (1), 72.
30. Teymourian, H.; Parrilla, M.; Sempionatto, J. R.; Montiel, N. F.; Barfidokht, A.; Van Echelpoel, R.; De Wael, K.; Wang, J., Wearable electrochemical sensors for the monitoring and screening of drugs. *ACS sensors* **2020,** *5* (9), 2679-2700.
31. Brown, D.; Wagner, C. A., Molecular mechanisms of acid-base sensing by the kidney. *Journal of the American Society of Nephrology* **2012,** *23* (5), 774-780.
32. Tsai, M. C.; Wang, C. C.; Chung, H. H.; Hsieh, P. H.; Tsai, H. Y., Development of a pH-strip-based sensor for the detection of ammonia in exhaled breath as a promising method for the diagnosis of infections. *J Chin Chem Soc-Taip* **2024,** *71* (9), 1103–1110.
33. Vaughan, J.; Ngamtrakulpanit, L.; Pajewski, T. N.; Turner, R.; Nguyen, T.; Smith, A.; Urban, P.; Hom, S.; Gaston, B.; Hunt, J., Exhaled breath condensate pH is a robust and reproducible assay of airway acidity. *Eur Respir J* **2003,** *22* (6), 889-894.
34. Moonla, C.; Reynoso, M.; Casanova, A.; Chang, A. Y.; Djassemi, O.; Balaje, A.; Abbas, A.; Li, Z. X.; Mahato, K.; Wang, J. S., Continuous Ketone Monitoring via Wearable Microneedle Patch Platform. *Acs Sensors* **2024,** *9* (2), 1004-1013.
35. Del Caño, R.; Saha, T.; Moonla, C.; de la Paz, E.; Wang, J. S., Ketone bodies detection: Wearable and mobile sensors for personalized medicine and nutrition. *Trac-Trend Anal Chem* **2023,** *159.*
36. Yao, J.; Yang, M.; Duan, Y., Chemistry, biology, and medicine of fluorescent nanomaterials and related systems: new insights into biosensing, bioimaging, genomics, diagnostics, and therapy. *Chemical reviews* **2014,** *114* (12), 6130-6178.
37. Chadha, U.; Bhardwaj, P.; Agarwal, R.; Rawat, P.; Agarwal, R.; Gupta, I.; Panjwani, M.; Singh, S.; Ahuja, C.; Selvaraj, S. K., Recent progress and growth in biosensors technology: A critical review. *Journal of Industrial and Engineering Chemistry* **2022,** *109*, 21-51.
38. Wilson, G. S.; Hu, Y., Enzyme-based biosensors for in vivo measurements. *Chemical reviews* **2000,** *100* (7), 2693-2704.
39. Holford, T. R.; Davis, F.; Higson, S. P., Recent trends in antibody based sensors. *Biosensors and Bioelectronics* **2012,** *34* (1), 12-24.
40. Paludan, S. R.; Bowie, A. G., Immune sensing of DNA. *Immunity* **2013,** *38* (5), 870-880.
41. Willner, I.; Zayats, M., Electronic aptamer-based sensors. *Angewandte Chemie International Edition* **2007,** *46* (34), 6408-6418.
42. Gupta, N.; Renugopalakrishnan, V.; Liepmann, D.; Paulmurugan, R.; Malhotra, B. D., Cell-based biosensors: Recent trends, challenges and future perspectives. *Biosensors and Bioelectronics* **2019,** *141*, 111435.
43. BASSETT, D. R., Validity of four motion sensors in measuring moderate intensity physical activity. *Medicine & Science in Sports & Exercise* **2000,** *32* (9), S471-S480.
44. Stavropoulos, T. G.; Papastergiou, A.; Mpaltadoros, L.; Nikolopoulos, S.; Kompatsiaris, I., IoT wearable sensors and devices in elderly care: A literature review. *Sensors* **2020,** *20* (10), 2826.
45. Jain, A.; Kanhangad, V., Human activity classification in smartphones using accelerometer and gyroscope sensors. *IEEE Sensors Journal* **2017,** *18* (3), 1169-1177.

46. Ahmad, N.; Ghazilla, R. A. R.; Khairi, N. M.; Kasi, V., Reviews on various inertial measurement unit (IMU) sensor applications. *International Journal of Signal Processing Systems* **2013,** *1* (2), 256-262.
47. Santos, J. L.; Farahi, F., *Handbook of optical sensors.* Crc Press: 2014.
48. Kaur, B.; Kumar, S.; Kaushik, B. K., Novel wearable optical sensors for vital health monitoring systems—A review. *Biosensors* **2023,** *13* (2), 181.
49. Baker, S. B.; Xiang, W.; Atkinson, I., Internet of things for smart healthcare: Technologies, challenges, and opportunities. *Ieee Access* **2017,** *5*, 26521-26544.
50. Castaneda, D.; Esparza, A.; Ghamari, M.; Soltanpur, C.; Nazeran, H., A review on wearable photoplethysmography sensors and their potential future applications in health care. *International journal of biosensors & bioelectronics* **2018,** *4* (4), 195.
51. Wang, D.; Xie, L.; Yang, S. X.; Tian, F., Support vector machine optimized by genetic algorithm for data analysis of near-infrared spectroscopy sensors. *Sensors* **2018,** *18* (10), 3222.
52. Aumann, S.; Donner, S.; Fischer, J.; Müller, F., Optical coherence tomography (OCT): principle and technical realization. *High resolution imaging in microscopy and ophthalmology: new frontiers in biomedical optics* **2019**, 59–85.
53. Pramanik, P. K. D.; Upadhyaya, B. K.; Pal, S.; Pal, T., Internet of things, smart sensors, and pervasive systems: Enabling connected and pervasive healthcare. In *Healthcare data analytics and management*, Elsevier: 2019; pp 1-58.
54. Li, Y.; Liu, C.; Zou, H.; Che, L.; Sun, P.; Yan, J.; Liu, W.; Xu, Z.; Yang, W.; Dong, L., Integrated wearable smart sensor system for real-time multi-parameter respiration health monitoring. *Cell Reports Physical Science* **2023,** *4* (1).
55. Tavakoli, M.; Carriere, J.; Torabi, A., Robotics, smart wearable technologies, and autonomous intelligent systems for healthcare during the COVID-19 pandemic: An analysis of the state of the art and future vision. *Advanced intelligent systems* **2020,** *2* (7), 2000071.
56. Kim, B. Y.; Lee, J., Smart devices for older adults managing chronic disease: a scoping review. *JMIR mHealth and uHealth* **2017,** *5* (5), e7141.
57. Zheng, H.; Nugent, C.; McCullagh, P.; Huang, Y.; Zhang, S.; Burns, W.; Davies, R.; Black, N.; Wright, P.; Mawson, S., Smart self management: assistive technology to support people with chronic disease. *Journal of telemedicine and telecare* **2010,** *16* (4), 224-227.
58. Formica, D.; Schena, E., Smart sensors for healthcare and medical applications. MDPI: 2021; Vol. 21, p 543.
59. El-Miedany, Y., Telehealth and telemedicine: how the digital era is changing standard health care. *Smart Homecare Technology and Telehealth* **2017**, 43–51.
60. Chandrasekaran, V.; Babu, D. V.; Thanikaiselvan, V.; Somasundaram, K. In *Designing of Tele-Health Smart Sensor Device to assist home care staff*, Journal of Physics: Conference Series, IOP Publishing: 2021; p 062106.
61. Postolache, O.; Hemanth, D. J.; Alexandre, R.; Gupta, D.; Geman, O.; Khanna, A., Remote monitoring of physical rehabilitation of stroke patients using IoT and virtual reality. *IEEE Journal on Selected Areas in Communications* **2020,** *39* (2), 562-573.
62. Viswanadham, N., Ecosystem model for healthcare platform. *Sādhanā* **2021,** *46* (4), 188.
63. Gravina, R.; Alinia, P.; Ghasemzadeh, H.; Fortino, G., Multi-sensor fusion in body sensor networks: State-of-the-art and research challenges. *Information Fusion* **2017,** *35*, 68-80.
64. Muhammad, G.; Alshehri, F.; Karray, F.; El Saddik, A.; Alsulaiman, M.; Falk, T. H., A comprehensive survey on multimodal medical signals fusion for smart healthcare systems. *Information Fusion* **2021,** *76*, 355-375.
65. Harb, H.; Mansour, A.; Nasser, A.; Cruz, E. M.; de la Torre Diez, I., A sensor-based data analytics for patient monitoring in connected healthcare applications. *IEEE Sensors Journal* **2020,** *21* (2), 974-984.
66. Ishigaki, N.; Kimura, T.; Usui, Y.; Aoki, K.; Narita, N.; Shimizu, M.; Hara, K.; Ogihara, N.; Nakamura, K.; Kato, H., Analysis of pelvic movement in the elderly during walking using a posture monitoring system equipped with a triaxial accelerometer and a gyroscope. *Journal of biomechanics* **2011,** *44* (9), 1788-1792.

67. Márquez, D. G.; Félix, P.; García, C. A.; Tejedor, J.; Fred, A. L.; Otero, A., Positive and negative evidence accumulation clustering for sensor fusion: An application to heartbeat clustering. *Sensors* **2019,** *19* (21), 4635.
68. Chandel, R. S.; Sharma, S.; Kaur, S.; Singh, S.; Kumar, R., Smart watches: A review of evolution in bio-medical sector. *Materials Today: Proceedings* **2022,** *50*, 1053-1066.
69. Poongodi, M.; Hamdi, M.; Malviya, M.; Sharma, A.; Dhiman, G.; Vimal, S., Diagnosis and combating COVID-19 using wearable Oura smart ring with deep learning methods. *Personal and ubiquitous computing* **2022**, 1–11.
70. Muhammad Sayem, A. S.; Hon Teay, S.; Shahariar, H.; Luise Fink, P.; Albarbar, A., Review on smart electro-clothing systems (SeCSs). *Sensors* **2020,** *20* (3), 587.
71. Soni, M.; Dahiya, R., Soft eSkin: distributed touch sensing with harmonized energy and computing. *Philosophical Transactions of the Royal Society A* **2020,** *378* (2164), 20190156.
72. Wang, P.; Kricka, L. J., Current and emerging trends in point-of-care technology and strategies for clinical validation and implementation. *Clinical chemistry* **2018,** *64* (10), 1439-1452.
73. Khan, Y.; Ostfeld, A. E.; Lochner, C. M.; Pierre, A.; Arias, A. C., Monitoring of vital signs with flexible and wearable medical devices. *Advanced materials* **2016,** *28* (22), 4373-4395.
74. Vijayan, V.; Connolly, J. P.; Condell, J.; McKelvey, N.; Gardiner, P., Review of wearable devices and data collection considerations for connected health. *Sensors* **2021,** *21* (16), 5589.
75. Muthu, B.; Sivaparthipan, C.; Manogaran, G.; Sundarasekar, R.; Kadry, S.; Shanthini, A.; Dasel, A., IOT based wearable sensor for diseases prediction and symptom analysis in healthcare sector. *Peer-to-peer networking and applications* **2020,** *13* (6), 2123-2134.
76. Kario, K., Management of hypertension in the digital era: small wearable monitoring devices for remote blood pressure monitoring. *Hypertension* **2020,** *76* (3), 640-650.
77. Mamdiwar, S. D.; Shakruwala, Z.; Chadha, U.; Srinivasan, K.; Chang, C.-Y., Recent advances on IoT-assisted wearable sensor systems for healthcare monitoring. *Biosensors* **2021,** *11* (10), 372.
78. Aldeer, M.; Javanmard, M.; Martin, R. P., A review of medication adherence monitoring technologies. *Applied System Innovation* **2018,** *1* (2), 14.
79. Greiwe, J.; Nyenhuis, S. M., Wearable technology and how this can be implemented into clinical practice. *Current allergy and asthma reports* **2020,** *20*, 1-10.
80. Lu, L.; Zhang, J.; Xie, Y.; Gao, F.; Xu, S.; Wu, X.; Ye, Z., Wearable health devices in health care: narrative systematic review. *JMIR mHealth and uHealth* **2020,** *8* (11), e18907.
81. Amin, T.; Mobbs, R. J.; Mostafa, N.; Sy, L. W.; Choy, W. J., Wearable devices for patient monitoring in the early postoperative period: a literature review. *Mhealth* **2021,** *7*.
82. Utilization of patient reported outcomes generated by electronic medical record and smart pill bottles with follow up telehealth encounters to improve adherence to adjuvant endocrine therapy in breast cancer patients. *Cancer Res* **2022,** *82* (4).
83. Shanmugam, D. B.; Dhilipan, J.; Sivasankari, A.; Munusamy, S.; Karpagam, S., Smart Pill Box using IoT and Cloud. *Biosci Biotech Res C* **2020,** *13* (4), 139-145.
84. Mauro, J.; Mathews, K. B.; Sredzinski, E. S., Effect of a Smart Pill Bottle and Pharmacist Intervention on Medication Adherence in Patients with Multiple Myeloma New to Lenalidomide Therapy. *J Manag Care Spec Ph* **2019,** *25* (11), 1244-1254.
85. Canali, S.; Schiaffonati, V.; Aliverti, A., Challenges and recommendations for wearable devices in digital health: Data quality, interoperability, health equity, fairness. *PLOS Digital Health* **2022,** *1* (10), e0000104.
86. Lomotey, R. K.; Pry, J.; Sriramoju, S., Wearable IoT data stream traceability in a distributed health information system. *Pervasive and Mobile Computing* **2017,** *40*, 692-707.
87. Guk, K.; Han, G.; Lim, J.; Jeong, K.; Kang, T.; Lim, E.-K.; Jung, J., Evolution of wearable devices with real-time disease monitoring for personalized healthcare. *Nanomaterials* **2019,** *9* (6), 813.
88. Izmailova, E. S.; Wagner, J. A.; Perakslis, E. D., Wearable devices in clinical trials: hype and hypothesis. *Clinical Pharmacology & Therapeutics* **2018,** *104* (1), 42-52.
89. Khoshbin, Z.; Housaindokht, M. R.; Izadyar, M.; Bozorgmehr, M. R.; Verdian, A., Recent advances in computational methods for biosensor design. *Biotechnology and Bioengineering* **2021,** *118* (2), 555-578.

90. Zhang, J. Z.; Nguyen, W. H.; Greenwood, N.; Rose, J. C.; Ong, S.-E.; Maly, D. J.; Baker, D., Computationally designed sensors detect endogenous Ras activity and signaling effectors at subcellular resolution. *Nature Biotechnology* **2024**, 1–11.
91. Shang, C.; Yang, F.; Huang, D.; Lyu, W., Data-driven soft sensor development based on deep learning technique. *Journal of Process Control* **2014,** *24* (3), 223-233.
92. Wang, W.; Xue, C.; Zhao, J.; Yuan, C.; Tang, J., Machine learning-based field geological mapping: A new exploration of geological survey data acquisition strategy. *Ore Geology Reviews* **2024**, 105959.
93. Wasilewski, T.; Kamysz, W.; Gębicki, J., Bioelectronic tongue: Current status and perspectives. *Biosensors and Bioelectronics* **2020,** *150*, 111923.
94. Wasilewski, T.; Gębicki, J.; Kamysz, W., Bioelectronic nose: Current status and perspectives. *Biosensors and Bioelectronics* **2017,** *87*, 480-494.
95. Mazafi, A.; Shukla, R. P.; Shukla, S. K.; Lavon, A.; Ben-Yoav, H.; Silue, A.; McDowell, B.; Peixoto, N. In *Intelligent multi-electrode arrays as the next generation of electrochemical biosensors for real-time analysis of neurotransmitters*, 2018 IEEE International Symposium on Medical Measurements and Applications (MeMeA), IEEE: 2018; pp 1–5.
96. Singh, S.; Varma, P.; Sreelekha, G.; Adak, C.; Shukla, R. P.; Kamble, V. B., Metal oxide-based gas sensor array for VOCs determination in complex mixtures using machine learning. *Microchimica Acta* **2024,** *191* (4), 196.
97. Shukla, R. P.; Aroosh, M.; Matzafi, A.; Ben-Yoav, H., Partially functional electrode modifications for rapid detection of dopamine in urine. *Advanced Functional Materials* **2021,** *31* (17), 2004146.
98. Ballard, Z.; Brown, C.; Madni, A. M.; Ozcan, A., Machine learning and computation-enabled intelligent sensor design. *Nature Machine Intelligence* **2021,** *3* (7), 556-565.
99. Verma, S.; Shukla, R. P.; Dutta, G., Machine learning-enabled biosensors in clinical decision making. In *Next-Generation Nanobiosensor Devices for Point-Of-Care Diagnostics*, Springer: 2022; pp 163-194.
100. King, R. C.; Villeneuve, E.; White, R. J.; Sherratt, R. S.; Holderbaum, W.; Harwin, W. S., Application of data fusion techniques and technologies for wearable health monitoring. *Medical engineering & physics* **2017,** *42*, 1-12.
101. Ates, H. C.; Nguyen, P. Q.; Gonzalez-Macia, L.; Morales-Narváez, E.; Güder, F.; Collins, J. J.; Dincer, C., End-to-end design of wearable sensors. *Nature Reviews Materials* **2022,** *7* (11), 887-907.
102. Al-Zaiti, S. S.; Alghwiri, A. A.; Hu, X.; Clermont, G.; Peace, A.; Macfarlane, P.; Bond, R., A clinician's guide to understanding and critically appraising machine learning studies: a checklist for Ruling Out Bias Using Standard Tools in Machine Learning (ROBUST-ML). *European Heart Journal-Digital Health* **2022,** *3* (2), 125-140.
103. Faes, L.; Liu, X.; Wagner, S. K.; Fu, D. J.; Balaskas, K.; Sim, D. A.; Bachmann, L. M.; Keane, P. A.; Denniston, A. K., A clinician's guide to artificial intelligence: how to critically appraise machine learning studies. *Translational vision science & technology* **2020,** *9* (2), 7-7.
104. Cui, F.; Yue, Y.; Zhang, Y.; Zhang, Z.; Zhou, H. S., Advancing biosensors with machine learning. *ACS sensors* **2020,** *5* (11), 3346-3364.
105. Witt, D. R.; Kellogg, R. A.; Snyder, M. P.; Dunn, J., Windows into human health through wearables data analytics. *Current opinion in biomedical engineering* **2019,** *9*, 28-46.
106. Benvenuto, D.; Giovanetti, M.; Vassallo, L.; Angeletti, S.; Ciccozzi, M., Application of the ARIMA model on the COVID-2019 epidemic dataset. *Data in brief* **2020,** *29*, 105340.
107. Chakraborty, T.; Ghosh, I., Real-time forecasts and risk assessment of novel coronavirus (COVID-19) cases: A data-driven analysis. *Chaos, Solitons & Fractals* **2020,** *135*, 109850.
108. Huang, G.; Liu, Z.; Van Der Maaten, L.; Weinberger, K. Q. In *Densely connected convolutional networks*, Proceedings of the IEEE conference on computer vision and pattern recognition, 2017; pp 4700–4708.
109. Laguarta, J.; Hueto, F.; Subirana, B., COVID-19 artificial intelligence diagnosis using only cough recordings. *IEEE Open Journal of Engineering in Medicine and Biology* **2020,** *1*, 275-281.

110. Samson, C.; Koh, A., Stress monitoring and recent advancements in wearable biosensors. *Frontiers in bioengineering and biotechnology* **2020,** *8*, 1037.
111. Zhang, Y.; Hu, Y.; Jiang, N.; Yetisen, A. K., Wearable artificial intelligence biosensor networks. *Biosensors and Bioelectronics* **2023,** *219*, 114825.
112. Ates, H. C.; Ates, C.; Dincer, C., Stress monitoring with wearable technology and AI. *Nature Electronics* **2024,** *7* (2), 98-99.
113. Ates, H. C.; Alshanawani, A.; Hagel, S.; Cotta, M. O.; Roberts, J. A.; Dincer, C.; Ates, C., Unraveling the Impact of Therapeutic Drug Monitoring via Machine Learning. *medRxiv* **2023,** 2023.10. 23.23297424.
114. González Ramírez, M. L.; García Vázquez, J. P.; Rodríguez, M. D.; Padilla-López, L. A.; Galindo-Aldana, G. M.; Cuevas-González, D. In *Wearables for Stress Management: A Scoping Review*, Healthcare, MDPI: 2023; p 2369.
115. Ernst, H.; Scherpf, M.; Pannasch, S.; Helmert, J. R.; Malberg, H.; Schmidt, M., Assessment of the human response to acute mental stress–An overview and a multimodal study. *PLoS One* **2023,** *18* (11), e0294069.
116. Dalmeida, K. M.; Masala, G. L., HRV features as viable physiological markers for stress detection using wearable devices. *Sensors* **2021,** *21* (8), 2873.
117. Torrente-Rodríguez, R. M.; Tu, J.; Yang, Y.; Min, J.; Wang, M.; Song, Y.; Yu, Y.; Xu, C.; Ye, C.; IsHak, W. W., Investigation of cortisol dynamics in human sweat using a graphene-based wireless mHealth system. *Matter* **2020,** *2* (4), 921-937.
118. Xu, C.; Song, Y.; Sempionatto, J. R.; Solomon, S. A.; Yu, Y.; Nyein, H. Y.; Tay, R. Y.; Li, J.; Heng, W.; Min, J., A physicochemical-sensing electronic skin for stress response monitoring. *Nature Electronics* **2024,** *7* (2), 168-179.
119. Jin, C.; Kim, S.-J.; Kim, M.; Guha, A.; Lee, D.; Esenkov, O. E.; Xu, W.; Smith, B.; Ryu, S. In *Digital Twin Method for Global Motion and Stress Monitoring of a Steel Lazy Wave Riser*, OCEANS 2022, Hampton Roads, IEEE: 2022; pp 1–7.
120. Sun, Y.; Lo, F. P.-W.; Lo, B., Security and privacy for the internet of medical things enabled healthcare systems: A survey. *IEEE Access* **2019,** *7*, 183339-183355.
121. Masud, M.; Gaba, G. S.; Alqahtani, S.; Muhammad, G.; Gupta, B. B.; Kumar, P.; Ghoneim, A., A lightweight and robust secure key establishment protocol for internet of medical things in COVID-19 patients care. *IEEE Internet of Things Journal* **2020,** *8* (21), 15694-15703.
122. Hamza, M.; Akbar, M. A.; Shafiq, M.; Kamal, T.; Baddour, A. M., Identification of privacy and security risks of Internet of Things: An empirical investigation. *Review of Computer Engineering Research* **2019,** *6* (1), 35-44.
123. Li, X.; Jiang, P.; Chen, T.; Luo, X.; Wen, Q., A survey on the security of blockchain systems. *Future generation computer systems* **2020,** *107*, 841-853.
124. Sadeghi, A.-R.; Wachsmann, C.; Waidner, M. In *Security and privacy challenges in industrial internet of things*, Proceedings of the 52nd annual design automation conference, 2015; pp 1–6.
125. Hasan, M. K.; Ghazal, T. M.; Saeed, R. A.; Pandey, B.; Gohel, H.; Eshmawi, A. A.; Abdel-Khalek, S.; Alkhassawneh, H. M., A review on security threats, vulnerabilities, and counter measures of 5G enabled Internet-of-Medical-Things. *IET communications* **2022,** *16* (5), 421-432.
126. Gupta, D. S.; Mazumdar, N.; Nag, A.; Singh, J. P., Secure data authentication and access control protocol for industrial healthcare system. *Journal of Ambient Intelligence and Humanized Computing* **2023,** *14* (5), 4853-4864.
127. Ktari, J.; Frikha, T.; Ben Amor, N.; Louraidh, L.; Elmannai, H.; Hamdi, M., IoMT-based platform for E-health monitoring based on the blockchain. *Electronics* **2022,** *11* (15), 2314.
128. Stein, W.; Boswell, M., The current ethical and regulatory status of the internet of medical thing (IoMT) and the need of a new IoMT law. *Journal of Healthcare Ethics & Administration* **2018,** *4* (2), 32-38.
129. Ravizza, A.; De Maria, C.; Di Pietro, L.; Sternini, F.; Audenino, A. L.; Bignardi, C., Comprehensive review on current and future regulatory requirements on wearable sensors in preclinical and clinical testing. *Frontiers in bioengineering and biotechnology* **2019,** *7*, 313.

Computer Vision for Healthcare and Medicine: Unlocking Insights from Visual Data

Aritra Ray, Farshad Firouzi, Bahar Farahani, and Krishnendu Chakrabarty

> *"The eye sees only what the mind is prepared to comprehend."*
>
> *– Henri Bergson*

1 Introduction

The integration of Computer Vision (CV) into healthcare has revolutionized the way medical images are analyzed, diagnoses are made, and treatment plans are developed. The journey began with early attempts at automated image analysis, relying on rule-based systems and handcrafted features [1]. These systems, popular from the 1970s to the 1990s, used sequential applications of edge detection, region growing, and mathematical modeling to analyze medical images. Despite their success, the process of designing handcrafted features was resource-intensive, time-consuming, and often lacked the adaptability needed to address diverse medical imaging tasks.

By the late 1990s, the paradigm shifted towards supervised learning approaches, where training data was used to guide the development of algorithms. Techniques like active shape models for segmentation [2], atlas-based methods [3], and statistical classifiers [4] became the foundation of many computer-aided detection and

A. Ray (✉) · F. Firouzi
Department of Electrical and Computer Engineering, Duke University, Durham, USA
e-mail: aritra.ray@duke.edu

F. Firouzi · K. Chakrabarty
School of Electrical and Computer and Energy Engineering, Arizona State University, Tempe, USA

B. Farahani
Shahid Beheshti University, Tehran, Iran

F. Firouzi et al. (eds.), *Smart and Connected Health: AI, IoT, and Trustworthy Technologies*, https://doi.org/10.1007/978-3-032-06286-4_5

diagnosis systems. These systems relied heavily on human-designed features, where researchers extracted discriminative attributes from images to train models.

The advent of deep learning marked a significant leap forward in medical image analysis. Deep learning models, particularly Convolutional Neural Networks (CNNs) [5], eliminated the dependency on handcrafted features by enabling computers to learn directly from data. These models, composed of multiple layers, transform raw inputs into high-level representations, allowing them to classify diseases, segment structures, and detect anomalies with unprecedented accuracy. Although CNNs had been explored as early as the 1980s, their widespread adoption followed breakthroughs in computational efficiency and model optimization, exemplified by AlexNet's success in the 2012 ImageNet competition [6].

Since then, CNNs and related architectures have become the cornerstone of medical image analysis, demonstrating human-level performance in tasks like retinal image analysis and skin lesion classification [7]. The rapid growth in this field is evident from the increasing number of publications and the inclusion of deep learning topics in major conferences and journals.

This chapter explores the evolution of CV in healthcare, focusing on its transition from rule-based systems to deep learning-driven approaches. It provides a comprehensive overview of methodologies, applications, and the pivotal role of CV in medical imaging, pathology, ophthalmology, and patient monitoring. Furthermore, it examines the challenges confronting the domain, including the scarcity of annotated data, issues with generalization, and the need for privacy-preserving solutions. By addressing these challenges, CV can continue to enhance diagnostics, personalize treatment, and improve healthcare accessibility in an era of precision medicine.

2 Fundamentals of CV

2.1 *Object Detection*

Object detection is a critical aspect of CV, focused on identifying and pinpointing objects of specific categories in digital images. This task aims to answer two fundamental questions for CV applications: *What objects are present?* and *Where are they located?* Broadly speaking, the key performance metrics in this regard include classification accuracy, and localization accuracy, alongside detection speed.

Object detection serves as a foundational component for broader CV applications like instance segmentation [8], image captioning [9], and object tracking [10]. However, the complexity of object detection varies based on the specific task. Challenges include recognizing objects under varying conditions, such as different viewpoints, lighting (including exposure, contrast), and within-class diversity. Additional challenges involve object rotation, scaling issues based on object dimensions, and accurate localization amid crowded scenes or occlusions. Speed optimization also remains crucial for real-time applications.

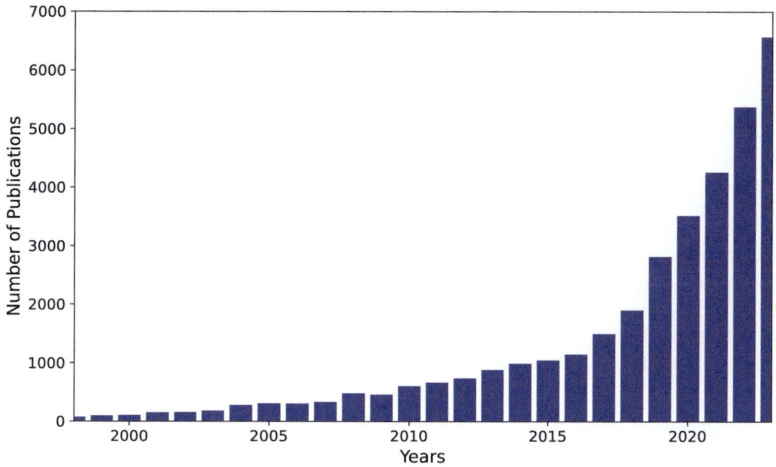

Fig. 1 The number of publications on object detection has steadily increased from 1998 to 2023, according to data from Google Scholar's advanced search using the terms allintitles: "object detection" or "detecting objects"

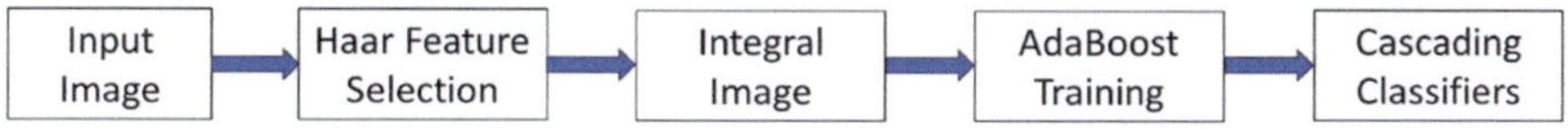

Fig. 2 Schematic workflow of the Viola-Jones detector for object detection

This subsection explores the evolution of object detection, examining the key techniques, recent innovations, and effective methods for accelerating detection processes. It also addresses available datasets, evaluation metrics, and technical advancements that have shaped the domain of object detection over the past decades. Figure 1 highlights the steady increase in interest among researchers in this field.

Key Techniques in Object Detection. We discuss next some of the most influential approaches from traditional methods to deep learning-driven techniques that defined the domain of object detection.

Handcrafted Feature-Based Techniques. Early object detection methods relied on manually engineered features, aiming to capture critical image characteristics. The Viola-Jones detector [1], introduced in 2001, was a pioneering face detection method that introduced a sliding window approach, scanning images at multiple scales to detect objects. Key innovations included the integral image for rapid feature computation, feature selection using Adaboost, and a cascade of classifiers for real-time performance. The schematic workflow is illustrated in Fig. 2.

Introduced in 2005, Histogram of Oriented Gradients (HOG) [11] improved object detection by calculating edge directions within localized regions. This approach provided robustness against variations in illumination and object scale, making it popular for tasks like pedestrian detection. Figure 3 illustrates the HOG for a sample image, with directional gradients.

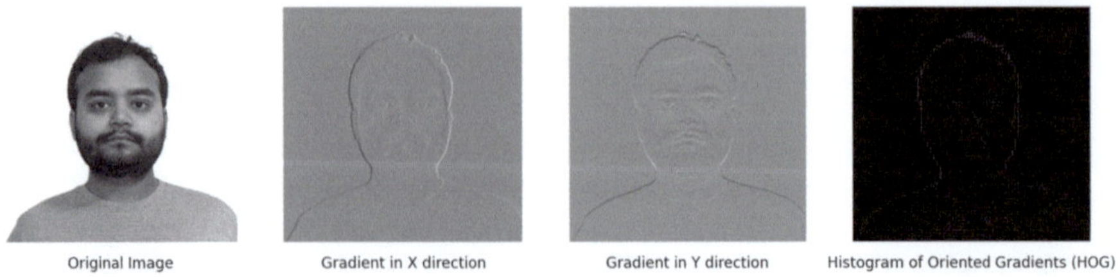

Fig. 3 Illustration of the Histogram of Oriented Gradients (HOG) for a sample image (human face), showing directional gradients. The original grayscale image is processed to extract gradients in the x-direction and y-direction, highlighting horizontal and vertical edge information, respectively. The HOG visualization combines these directional gradients, providing a representation of oriented gradients that captures structural features essential for image recognition and classification tasks

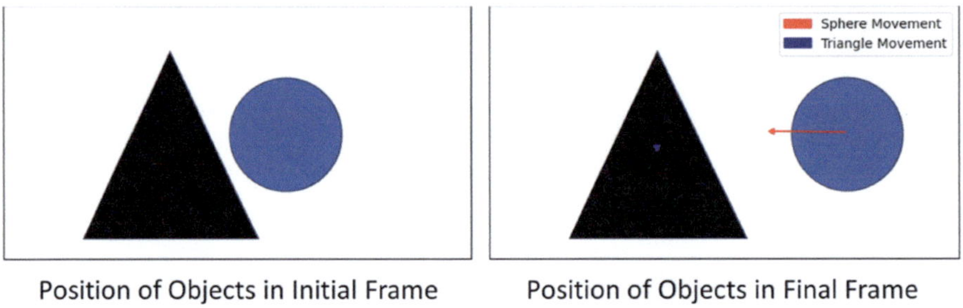

Fig. 4 Illustration of a Deformable Part-Based Model (DPM) applied to detect and track the movement of object parts. The example shows a blue sphere and a black triangle as separate 'parts' within the images. The left panel displays the initial positions of each part, while the right panel shows their final positions with arrows indicating movement. The red arrow represents the displacement of the sphere, and the blue arrow shows the movement of the triangle. This approach demonstrates the DPM's ability to capture and model deformations or positional changes in individual parts, which is useful for detecting subtle shifts in object configurations

Deformable Part-Based Models (DPM). In the late 2000s, object detection transitioned to more sophisticated approaches like the DPM [12]. DPM represented objects as combinations of parts, each with its own spatial configuration. This 'divide and conquer' strategy enabled better handling of occlusions and variations in object pose. By decomposing complex objects into simpler components, DPM introduced the idea of mixture models to capture diverse object appearances and used bounding box regression for localization improvement. Figure 4 illutrates the DPM model.

Convolutional Neural Networks (CNNs) for Two-Stage Detection. The advent of deep learning represented a pivotal advancement in the field of object detection. CNN-based models revolutionized the field of object detection by learning features directly from training data. Proposed in 2014, Regions with CNN Features (R-CNN) [13] combined region proposals with CNNs to classify objects, leading to a substantial boost in accuracy. However, the model was computationally intensive due to redundant feature extraction for each proposal, making it unsuitable for real-world applications. Fast R-CNN [14] and Faster R-CNN [15] variants improved efficiency

Table 1 Comparison of R-CNN and Faster R-CNN

Feature	R-CNN	Faster R-CNN
Region proposal	Uses selective search to generate around 2000 region proposals, which is slow	Introduces a Region Proposal Network (RPN) to generate proposals, making the process faster and more efficient
Inference speed	Slow, as it processes each region independently in a separate forward pass	Faster, as the RPN generates proposals on feature maps, reducing redundant calculations
Training complexity	Requires multi-step training (e.g., pre-training, fine-tuning) making it cumbersome	Allows end-to-end training, simplifying the process and making it more robust
Object detection time	High latency due to separate processing of each proposal for feature extraction, classification, and bounding box regression	Lower latency as it combines region proposal and detection stages in one pass
Accuracy	Accurate but slow due to separate processing of proposals	Comparable accuracy with much faster inference, suitable for real-time or near real-time applications
Suitability for real-time applications	Not suitable for real-time applications because of slow processing	Faster R-CNN's RPN makes it feasible for real-time applications, though it is still slower than models like YOLO

by sharing convolutional computations across proposals. Faster R-CNN introduced a Region Proposal Network (RPN), generating region candidates directly within the network for faster processing. A comparison between R-CNN and Faster R-CNN is illustrated in Table 1. Later on, to address scale variability another significant milestone included the Feature Pyramid Networks (FPNs) [16] which utilized a top-down architecture with lateral connections, leveraging feature maps from multiple layers to detect objects at different scales. Figure 5 illustrates object detection results using Faster R-CNN with a ResNet-50 backbone and FPN.

One-Stage Detection Techniques. To enhance speed and simplify the detection pipeline, one-stage detectors like YOLO (You Only Look Once) [18], and Single Shot Multibox Detector (SSD) [19] were introduced, eliminating the region proposal phase and significantly reducing inference latency. YOLO treats object detection as a single regression problem, predicting bounding boxes and object classes simultaneously for the entire image. This approach enables real-time detection, although it may struggle with small objects. SSD introduced multi-scale feature extraction, predicting objects of different sizes at various layers. This innovation improved accuracy, particularly for small objects, without sacrificing speed.

Fig. 5 Object detection results using Faster R-CNN with a ResNet-50 backbone and Feature Pyramid Network (FPN) on an image of a person riding a bicycle. The model, pre-trained on the COCO [17] dataset, detects various objects with differing confidence levels. At a high confidence threshold ($\geq$0.8), the model identified 'person' and 'bicycle' with 100% confidence and 'truck' with 80% confidence. At a lower confidence threshold ($\geq$0.3), additional objects were detected, including a 'car' (49% confidence) and two instances of 'bench' (43% and 37% confidence). Bounding boxes highlight each detected object, demonstrating Faster R-CNN's ability to detect objects accurately even with complex backgrounds. The total inference time for processing this image was 7.29 s, showcasing the model's efficiency

Anchor-Free Detection. Modern trends in object detection focus on anchor-free methodologies, simplifying the design and reducing dependency on predefined reference boxes. CornerNet [20] and CenterNet [21] models approach object detection as a keypoint prediction problem, identifying specific points like corners or centers to form bounding boxes. This method reduces the complexity of traditional anchor-based methods, streamlining the detection process. Detection transformers, like DETR3D [22] leverages transformers to discards traditional CNNs for a set-based prediction framework using self-attention mechanisms. This architecture removes the need for anchor boxes, achieving a global context understanding and enabling end-to-end object detection.

Object Detection Datasets. The development of sophisticated object detection algorithms relies heavily on comprehensive datasets that are both diverse and unbiased. Next we discuss some of the benchmark datasets for object detection.

- PASCAL VOC [23]: The PASCAL Visual Object Classes (VOC) Challenge, held from 2005 to 2012, was a significant milestone in early object detection research. Datasets like VOC2007 and VOC2012 included thousands of annotated images with objects spanning 20 everyday categories, such as 'person', 'dog', and 'bicycle'. These datasets have been instrumental in evaluating early detection methods, providing a standard for comparison.
- ImageNet Large Scale Visual Recognition Challenge (ILSVRC) [24]: Starting in 2010, the ILSVRC became a driving force for innovation. The ImageNet dataset features millions of images across 200 object categories, with annotations and bounding boxes that set a high standard for generic object detection.

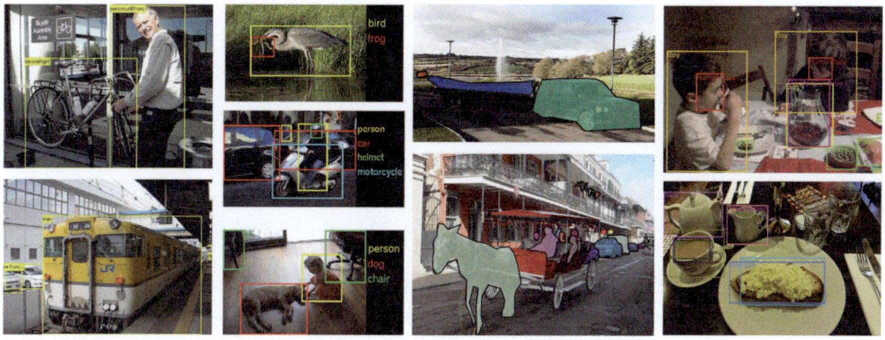

(a) (b) (c) (d)

Fig. 6 Examples of images and annotations from **a** PASCAL-VOC07, **b** ILSVRC, **c** MS-COCO, and **d** Open Images dataset

- MS-COCO [17]: Launched in 2015, MS-COCO (Microsoft Common Objects in Context) is considered one of the most challenging and comprehensive datasets. It contains over 160,000 images and nearly a million annotated instances across 80 object categories. Beyond bounding boxes, MS-COCO includes segmentation masks, which offer fine-grained localization and are particularly useful for detecting small and densely packed objects.
- Open Images [25]: Released in 2018, the Open Images Dataset is one of the largest and most diverse object detection datasets available. It features over 1.9 million images and 15 million bounding box annotations for 600 object categories. Unlike earlier datasets, Open Images includes a visual relationship detection task, identifying how objects interact with each other within an image.

Figure 6 illustrates some of the images and annotations from PASCAL-VOC07 [23], ILSVRC [24], MS-COCO [17], and Open Images [25] dataset.

2.1.1 Metrics for Evaluating Object Detection

Evaluating the performance of object detection models requires reliable metrics to assess both classification and localization accuracy. Over time, the evaluation methodologies have evolved to better capture detection nuances:

The Average Precision (AP) metric, introduced by PASCAL VOC [23], often calculated as the area under the precision-recall curve, can be represented as follows:

$$AP = \int_0^1 p(r)\, dr$$

where $p(r)$ represents precision as a function of recall r; $\int_0^1 p(r)\, dr$ is the integral over the recall range [0, 1], summing up the precision values at each level of recall.

For Mean Average Precision (mAP), which averages AP across all object classes, the equation is:

$$mAP = \frac{1}{N}\sum_{i=1}^{N} AP_i$$

where N is the total number of object classes; AP_i is the Average Precision for the i-th object class.

Precision (p) is the ratio of true positive detections (correctly identified objects) to the total detections (true positives and false positives). It indicates how accurate the detections are among all detected instances. Recall (r) is the ratio of true positives to the total ground truth instances (true positives and false negatives). It measures the coverage or completeness of the detection model in identifying all objects of interest. Average Precision (AP) measures the area under the precision-recall curve for a specific object class, providing a single precision value averaged over different recall levels. A higher AP score indicates better model performance for that class. Mean Average Precision (mAP) is the average of AP scores across all object classes in a dataset, providing an overall measure of model performance for multi-class detection tasks. It is particularly useful for evaluating the model's effectiveness across varied object classes in a detection dataset.

Intersection over Union (IoU) [26] is a metric for assessing localization accuracy. It calculates the overlap between the predicted bounding box and the ground truth bounding box, as shown below:

$$\text{IoU} = \frac{\text{Area of Overlap}}{\text{Area of Union}} = \frac{B_p \cap B_{gt}}{B_p \cup B_{gt}}$$

where B_p represents the area of the predicted bounding box; B_{gt} represents the area of the ground truth bounding box; $B_p \cap B_{gt}$ is the area of overlap between the predicted and ground truth bounding boxes; $B_p \cup B_{gt}$ is the combined area covered by both the predicted and ground truth bounding boxes.

A predicted object is considered 'correctly detected' if its IoU with the ground truth exceeds a predefined threshold (e.g., 0.5). Higher IoU thresholds, as used in datasets like MS-COCO, encourage more precise localization, which is essential for accurate object positioning.

Unlike earlier benchmarks, MS-COCO [17] employed a more rigorous evaluation system, averaging AP over a range of IoU thresholds (from 0.5 to 0.95 in increments of 0.05). This multi-threshold approach ensures that models are not only accurate in broad detection but also precise in localizing object boundaries, making it a comprehensive measure of detection performance.

Evolution of Multiscale Detection. Detecting objects of varying sizes and aspect ratios has been a key challenge in object detection. We next discuss how the landscape has transformed for handling scale variations.

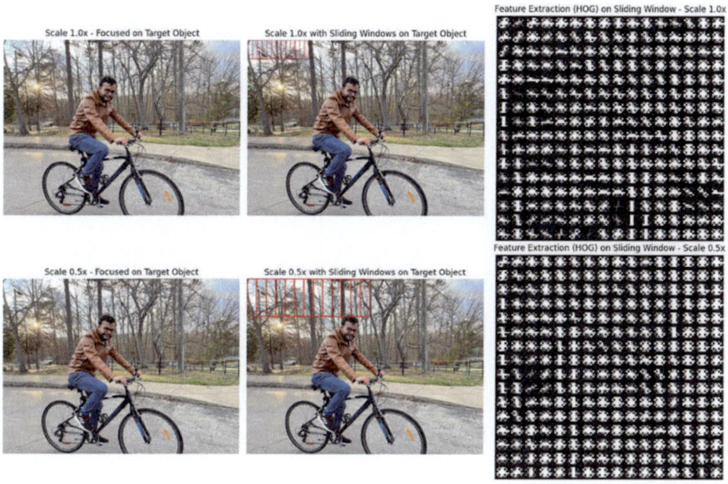

Fig. 7 Feature pyramid and sliding window approach for object detection. Each row shows the image at a different scale, enabling detection of objects at various resolutions. Sliding windows are focused around the 'person' and 'bicycle', targeting relevant regions. For each window, HOG feature extraction highlights edges and gradients, helping models like DPM identify object-like patterns

Feature Pyramids and Sliding Windows. In the early days, object detection relied on feature pyramids combined with a sliding window approach, using fixed-size windows to search for objects [27]. Models like HOG and DPM used this paradigm, often treating objects with different aspect ratios as separate categories or using multiple detectors for different shapes (Fig. 7).

Object Proposals. The concept of object proposals was introduced to avoid exhaustive searches [28]. These methods generated candidate regions likely to contain objects, reducing computational complexity. With the rise of deep learning, top-down, learning-based methods for generating proposals became dominant, eventually giving way to anchor-free models.

Deep Regression and Anchor-Free Detection. Recent methods [18] simplify multiscale detection using deep learning, predicting bounding box coordinates. Some techniques now focus on keypoint detection, identifying object corners or centers, and using these points to determine object properties like size and orientation.

Multireference and Multiresolution Detection. Modern detectors employ a set of reference anchors at various image locations or perform detection at multiple network layers to capture objects of different sizes, making these strategies foundational for contemporary detection systems [15, 19].

Context Utilization for Detection. Context plays a crucial role in accurately identifying objects, using surrounding information to improve detection [27].

Local Context. Leveraging nearby visual information can enhance object detection. For example, including some background context has been shown to improve detection accuracy in early face and pedestrian detection models [29].

Global Context. Integrating scene-wide context allows for understanding the broader environment, aiding in object identification [30, 31]. Modern techniques, including deep convolutional and attention-based methods like transformers, have enabled networks to capture global context efficiently, leading to improved detection in complex scenes.

Context Interaction. Exploring relationships between objects, and between objects and their environment, has enhanced detection accuracy [32]. This approach utilizes dependencies among visual elements, integrating relational context to refine predictions.

Evolution of Sample Selection and Training Strategies. Training object detectors involves overcoming challenges like data imbalance, optimizing classification, and accurately predicting object locations.

Hard Negative Mining (HNM). Detecting rare objects often involves an imbalanced dataset, with far more background examples than objects of interest. Techniques like HNM focus on selecting challenging examples that are most informative for training, improving model robustness [33].

Loss Function Innovations. Loss functions are critical for guiding model training. Early models used simple loss metrics like L2 loss, but newer loss functions, like cross-entropy for classification and IoU-based losses for localization, offer enhanced performance by addressing issues like class imbalance and outlier sensitivity [34].

Evolution of Inference Techniques. Research demonstrated that object localization can be improved under with probabilistic frameworks [35]. Another popular approach involves Non-Maximum Suppression (NMS) based inference strategies in refining predictions during object detection, as discussed next.

Greedy Selection for NMS. This traditional method [36] selects the highest-confidence bounding box and eliminates overlapping candidates based on a threshold. Although effective, it struggles with suppressing false positives and nearby objects (Fig. 8).

Bounding Box Aggregation. A more refined approach [37] that combines overlapping boxes into a single detection, considering object relationships and spatial layout. This method enhances detection in crowded scenes.

Learning-based NMS. Newer strategies [38] integrate NMS into the learning pipeline, using a network to predict detection scores and suppress redundant detections in an end-to-end fashion. These methods improve the handling of occlusions and dense scenes.

NMS-Free Detectors. Recent developments [39] aim to eliminate the need for NMS altogether, using one-to-one label assignment to ensure each object is detected only once, similar to human visual perception.

Inference Speed of Detection. Accelerating object detection inference has been a persistent challenge. Speedup techniques focus on optimizing three main areas: the detection pipeline, the network backbone, and numerical computation. We discuss the key strategies next.

Fig. 8 Demonstration of Greedy Non-Maximum Suppression (NMS) for object detection. The faster R-CNN model initially detects multiple overlapping bounding boxes with associated confidence scores. Greedy NMS selects the highest-confidence box and suppresses overlapping boxes based on an IoU threshold of 0.3, retaining high-confidence detections. This approach effectively reduces redundant detections but may struggle with closely positioned objects, as seen in traditional object detection pipelines

Feature Map Shared Computation. In object detection, feature extraction dominates computational demands. A common optimization is to compute the feature map for the entire image only once, significantly reducing redundant computations and accelerating the overall detection process [14].

Cascaded Detection. Cascaded detection [1] employs a step-by-step refinement strategy, starting with simple computations to eliminate obvious background regions. Complex regions are refined in later stages, enhancing detection accuracy while saving computational resources. This is particularly useful for detecting small objects within large scenes, like faces or pedestrians.

Network Pruning and Quantification. Network pruning [40] involves trimming unnecessary weights and layers, while quantification reduces the precision of computations, converting them to simpler forms. These strategies aim to maintain detection accuracy while minimizing computational overhead. Binarization is an example, where weights are reduced to binary values, simplifying calculations.

Lightweight Network Design. Designing efficient and smaller networks is another tactic for speeding up inference. Techniques include:

- Factorizing Convolutions [41]: Break large filters into smaller, more manageable ones to save resources.
- Group Convolution [42]: Dividing channels into groups for independent processing reduces computational costs.
- Depthwise Separable Convolutions [43]: Separate spatial and depth computations to significantly reduce the number of operations.

- Bottleneck Layers [44]: Using layers with fewer parameters to compress the data flow without compromising the capacity of the model.

Numerical Computation Optimization. Numerical efficiency at low-level implementation has a significant impact on detection speed:

- Integral Image [45]: Accelerates image processing by rapidly adding regions.
- Frequency Domain Operations [46]: Fourier transforms are used to speed up convolutions, allowing quick computations of inner products through fast algorithms like FFT.
- Vector quantization [47]: This technique compresses data into prototype vectors, making computations more efficient.

2.2 Segmentation

Image segmentation involves dividing an image into meaningful segments, typically corresponding to distinct objects or regions. This process can be approached as a classification task, where each pixel is labeled with a category in semantic segmentation, or as a more complex problem in instance segmentation, where individual objects are distinguished within the same category. Applications span a variety of fields, from medical imaging, enabling precise tumor identification, to autonomous driving, where accurate segmentation of road environments is critical for navigation and safety.

2.2.1 Fully Convolutional Networks (FCNs) for Semantic Segmentation

The introduction of Fully Convolutional Networks (FCNs) [48] marked a significant advancement in semantic segmentation, transforming the way deep learning models handle pixel-wise predictions. One of the pioneering works [48] modified traditional Convolutional Neural Networks (CNNs) like VGG16 and GoogLeNet, eliminating the fully connected layers and replacing them with convolutional layers. This allowed the model to accept images of varying dimensions and generate segmentation maps of the same size, facilitating end-to-end training for dense predictions (Fig. 9).

FCNs utilize skip connections to merge high-level semantic information from deeper layers with fine-grained spatial details from shallower layers. This combination enhances segmentation accuracy by incorporating both contextual and appearance data, leading to more detailed results. Tested on benchmarks like PASCAL VOC [23], NYUDv2 [49], and SIFT Flow [50], FCNs achieved leading performance, establishing deep learning as a powerful tool for image segmentation.

Despite their impact, FCNs have limitations, including slow real-time performance and insufficient global context integration. To address these gaps, methods like ParseNet [51] were developed. ParseNet extends FCNs by pooling feature maps

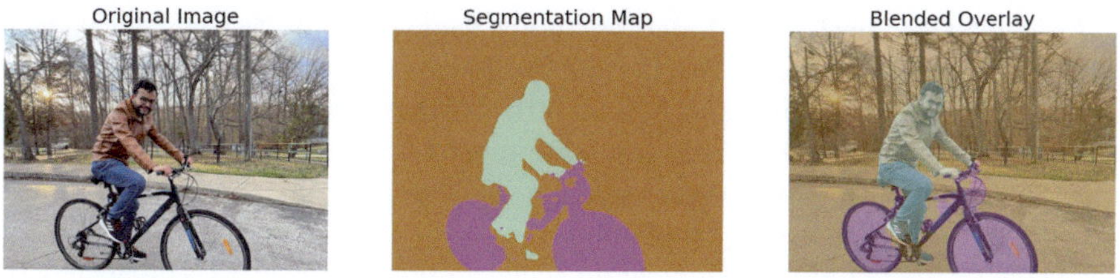

Fig. 9 Example of semantic segmentation using a Fully Convolutional Network (FCN) with a ResNet-50 backbone. The model generates a pixel-wise segmentation map, with each color representing a different class (e.g., person, road, sky). The visualization includes the original image, the segmentation map, and a blended overlay that combines both. This demonstrates the FCN's ability to perform dense predictions, capturing object boundaries and distinguishing between different regions in the scene, leveraging skip connections to enhance spatial details and contextual understanding

over the entire image to capture global context, thereby improving segmentation quality. This strategy smooths segment boundaries and enhances accuracy for complex scenes.

FCNs have been applied to a wide array of segmentation tasks, from medical imaging like brain tumor detection to specific challenges such as skin lesion and iris segmentation, demonstrating their versatility and importance in modern CV applications.

Convolutional Models with Graphical Models Fully Convolutional Networks (FCNs) are limited in their ability to incorporate broader scene-level context, which is critical for enhancing semantic segmentation. To address this, several methods integrate probabilistic graphical models, such as Conditional Random Fields (CRFs) and Markov Random Fields (MRFs), within deep learning architectures. These models aid in refining spatial details, improving segmentation accuracy, especially at object boundaries.

Chen et al. [52] proposed a method combining CNNs with fully connected CRFs to improve localization precision in segmentation. Recognizing that CNNs, by design, often lack the fine-grained localization needed for accurate segmentation, they used CRFs to sharpen boundaries and enhance object separation. Schwing and Urtasun [53] extended this by creating a framework that jointly trains CNNs with CRFs, achieving strong results on challenging datasets like PASCAL VOC 2012.

Zheng et al. [54] further advanced this idea by directly integrating CRFs into CNNs, forming an end-to-end trainable system. Similarly, Lin et al. [55] developed a contextual CRF-based algorithm, using both patch-to-patch and patch-to-background context to refine segmentations with additional relational information.

Lastly, Liu et al. [56] introduced a Parsing Network using MRFs with high-order relationships and contextual label combinations, effectively bypassing the iterative optimization usually required in MRFs. This approach allows deterministic, single-pass computation, combining context integration with computational efficiency.

Encoder-Decoder Based Models for Image Segmentation. The encoder-decoder architecture has become a prominent framework in deep learning for image segmentation, comprising a downsampling encoder for feature extraction and an upsampling decoder for reconstruction. These models are used across applications and can be divided into two main categories: those for general segmentation tasks and those tailored for medical image segmentation.

Encoder-Decoder Models for General Segmentation. Noh et al. [57] pioneered the use of deconvolutional networks for semantic segmentation, employing VGG-based convolutional layers as an encoder and deconvolution layers as a decoder to produce pixel-wise predictions. This approach achieved impressive accuracy on datasets like PASCAL VOC 2012 by enabling precise segmentation mask generation. Similarly, SegNet proposed by Badrinarayanan et al. [58], built on this concept by introducing max-pooling indices from the encoder for efficient upsampling, reducing parameter count while maintaining performance. SegNet's Bayesian variant later introduced a probabilistic framework to model uncertainty in segmentation tasks.

The High-Resolution Network (HRNet) [59] advanced the field by preserving high-resolution representations throughout the encoding process, connecting multiple resolution streams in parallel, and allowing effective contextual modeling. Recent segmentation architectures often utilize HRNet as a backbone due to its capability of retaining fine-grained details and effectively handling context with self-attention.

Several other models, including Stacked Deconvolutional Network (SDN) [60], and Linknet [61] have explored transposed convolutions to enhance general segmentation. However, these models typically struggle with high-resolution detail loss during encoding, a limitation addressed by architectures like HRNet.

Encoder-Decoder Models for Medical and Biomedical Image Segmentation. Medical image segmentation has benefited greatly from encoder-decoder architectures designed to handle limited data and 3D structures. U-Net, introduced by Ronneberger et al. [62], became a foundational model in this area. It features a symmetric encoder-decoder structure with skip connections that transfer spatial information from encoder layers to their corresponding decoder layers, enabling high-precision segmentation. U-Net's architecture allows it to effectively learn from small datasets through data augmentation, making it especially valuable for biological and medical imaging tasks. Numerous extensions, like 3D U-Net [63] and nested U-Net [64], have adapted this model for volumetric data and complex hierarchical segmentation (Fig. 10).

V-Net, another influential model, was developed by Milletari et al. [65] for 3D segmentation tasks. This network introduced a Dice coefficient-based loss function to manage class imbalance in medical images, specifically optimizing the segmentation of MRI volumes. Extensions of V-Net and other 3D-CNN-based architectures have been designed for various medical applications, including lesion and lobe segmentation in radiology.

Multi-Scale and Pyramid Network-Based Models for Image Segmentation. Multi-scale analysis, a longstanding concept in image processing, has been effectively incorporated into deep learning architectures for segmentation. One influential model in this area is the FPN by Lin et al. [16], originally developed for object detection. FPN leverages the natural multi-scale hierarchy in CNNs, using a combination

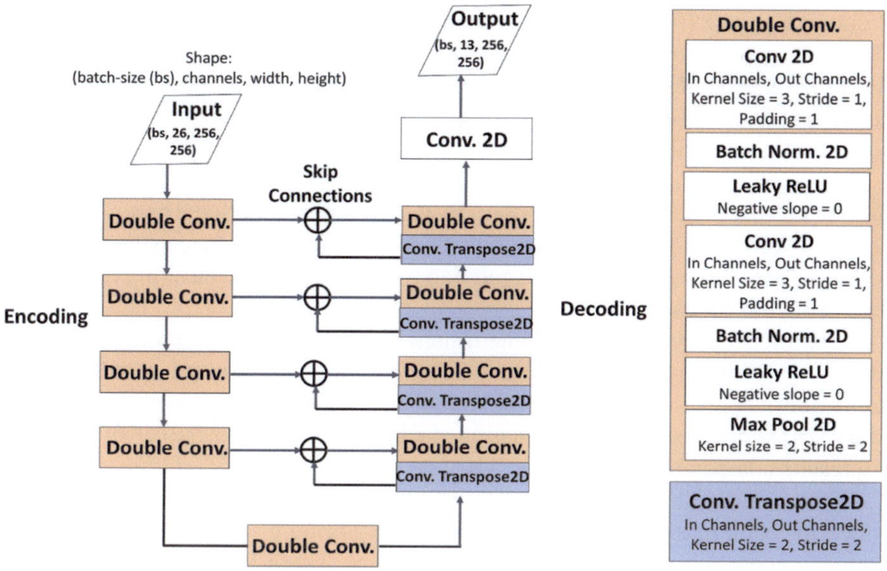

Fig. 10 Schematic of U-NET architecture

of bottom-up and top-down pathways with lateral connections to merge features of varying resolutions. This approach enables effective detection across scales, where segmentation predictions are generated by processing the integrated feature maps through multi-layer perceptrons.

The Pyramid Scene Parsing Network (PSPN) [66], is another prominent model for multi-scale segmentation. PSPN enhances scene understanding by applying a pyramid pooling module that pools features at multiple scales. These pooled features capture both global context and fine-grained details, which are then combined with initial feature maps to improve segmentation accuracy.

Other notable approaches include the Laplacian Pyramid Network [67], which refines segment boundaries using a multi-resolution architecture with skip connections. Additionally, models such as Dynamic Multi-scale Filters Network (DM-Net) [68], Context Contrasted Network (CCN) [69], and Adaptive Pyramid Context Network (APC-Net) [70] employ dynamic multi-scale filters and gated multi-scale mechanisms to capture contextual information across different resolutions, contributing to advancements in segmentation precision.

2.2.2 R-CNN Based Models for Instance Segmentation

The Regional Convolutional Neural Network (R-CNN) and its variants, including Fast R-CNN [14], Faster R-CNN [15], and Mask R-CNN [71], have played a pivotal role in advancing instance segmentation. Faster R-CNN [15] introduced a Region

Proposal Network (RPN) to generate bounding box candidates, enhancing detection precision by predicting object location and class within Regions of Interest (RoIs). Building on this, Mask R-CNN [71] added a third branch for pixel-level segmentation masks, enabling simultaneous object detection and instance segmentation, significantly outperforming earlier models on COCO benchmarks.

Path Aggregation Network (PANet) [72], an extension of Mask R-CNN and FPN, further improved segmentation accuracy by integrating an additional bottom-up pathway to enhance low-layer feature propagation. In PANet, each stage in the bottom-up pathway fuses its output with corresponding top-down features, ensuring rich feature representation for each instance.

Other models have sought to refine instance segmentation with specialized tasks. MaskLab [73], for example, integrates Faster R-CNN with semantic and directional predictions, delivering precise localization and segmenting each object. TensorMask [74] addresses dense segmentation via a unique 4D tensor framework, achieving results comparable to Mask R-CNN.

Further innovations include models like DeepMask [75] and PolarMask [76], which focus on mask proposals, and methods exploring grouping cues for instance segmentation, such as Deep Watershed Transform. These R-CNN extensions have driven significant improvements in instance segmentation, allowing models to achieve precise segmentation for complex scenes.

Dilated Convolutional Models and the DeepLab Family. Dilated convolutions are a key advancement in segmentation models [77], expanding the receptive field without increasing computational cost. By introducing a dilation rate, these convolutions allow networks to capture larger spatial information in fewer operations. For example, a 3×3 kernel with a dilation rate of 2 captures the same area as a 5×5 kernel, while using fewer parameters. This technique has become popular for real-time segmentation in models such as the DeepLab series [77], which has shown impressive results in handling complex scene segmentation.

The DeepLab models [77] are among the most influential in this category. DeepLabv1 [52] and DeepLabv2 [77] introduced Atrous Spatial Pyramid Pooling (ASPP) [78], a method of applying dilated convolutions at different rates to extract features at multiple scales, effectively capturing both object and contextual information. This design improved boundary localization and achieved state-of-the-art performance on benchmarks like PASCAL VOC [23] and Cityscapes [79].

DeepLabv3 [80] further refined the model by integrating parallel modules of dilated convolutions, concatenating their outputs for enhanced feature representation. To improve this even further, DeepLabv3+ [81] adopted an encoder-decoder structure with atrous separable convolutions, combining spatial depthwise convolutions and 1×1 pointwise convolutions, which reduced computational overhead while improving segmentation accuracy. DeepLabv3+ set new standards in semantic segmentation, achieving high mean Intersection over Union (mIoU) scores on PASCAL VOC and other datasets.

The DeepLab family's use of dilated convolutions and ASPP has become a benchmark in multi-scale and context-sensitive segmentation, influencing numerous subsequent models in the field.

Recurrent Neural Network-Based Models for Image Segmentation While CNNs are widely used in segmentation tasks, recurrent neural networks (RNNs) offer unique advantages by capturing sequential dependencies across pixels, thereby enhancing the contextual understanding of an image. RNN-based models enable global context modeling, which is crucial for precise segmentation.

Visin et al. introduced ReSeg [82], an RNN-based model adapted from their earlier ReNet model, which was designed for image classification. In ReSeg, RNN layers process features from VGG-16 convolutional layers, encoding pixel dependencies by sweeping across the image both horizontally and vertically. Using Gated Recurrent Units (GRUs), ReSeg achieves a balance between memory efficiency and computation, generating segmentation maps through upsampling layers.

Byeon et al. [83] took a different approach with 2D Long Short-Term Memory (LSTM) networks to manage spatial dependencies in natural scenes, integrating classification and segmentation in one model. In another innovation, Liang et al. [84] proposed the Graph LSTM network, where superpixels are treated as graph nodes connected by spatial edges, effectively capturing structural relationships. This graph-based model enhances segmentation by adding global context to convolutional features and updates node representations using a confidence-driven approach.

Expanding the use of RNNs, Data Associated RNNs (DA-RNNs) by Xiang et al. [85] integrate scene mapping with semantic labeling in RGB-D videos, feeding semantic data into 3D scene reconstruction. Additionally, Hu et al. [86] presented a model combining CNNs and LSTMs to produce segmentations based on natural language descriptions, enabling complex, context-specific queries like 'the red umbrella' for precise segmentation (Fig. 11).

2.2.3 Attention-Based Models for Semantic Segmentation

Attention mechanisms have become increasingly influential in semantic segmentation by allowing models to focus on relevant image features. Chen et al. [87] introduced an attention mechanism that adaptively weights multi-scale features at each pixel, surpassing traditional pooling methods. This approach enables the model to prioritize important features across scales and spatial locations, improving segmentation accuracy.

Reverse Attention Network (RAN), proposed by Huang et al. [88], explores reverse attention to identify non-target class features, thus enhancing the model's discriminative ability. This three-branch architecture simultaneously learns direct and reverse attention, broadening the model's semantic understanding. Similarly, Li et al. [89] developed a Pyramid Attention Network that combines attention with spatial pyramids to effectively capture dense features across multiple scales without relying on dilated convolutions.

Expanding on this, Fu et al. [90] introduced Dual Attention Network captures contextual dependencies using self-attention, employing two attention modules to model spatial and channel dependencies independently, enriching scene segmentation.

Fig. 11 Simulated natural language-based segmentation for the query 'person' using Mask R-CNN. In this demonstration, we utilized a pre-trained Mask R-CNN model to detect and segment instances within the image. A simple query filter ('person') was applied to the model's output, allowing us to isolate the person in the image while ignoring other objects. The result highlights the detected person with a bounding box and a semi-transparent segmentation mask overlay. This example illustrates how a segmentation model can respond to natural language queries by selectively displaying relevant parts of the image, even though further enhancements would be needed for more complex, context-specific queries

Additional attention-based approaches have emerged, such as OCNet (object context pooling with self-attention) [91], Expectation-Maximization Attention (EMANet) [92], Criss-Cross Attention Network (CCNet) [93], and Discriminative Feature Network (DFN) [94]. These models illustrate the versatility and power of attention in refining semantic segmentation, advancing the field by enhancing feature selection, boundary clarity, and spatial awareness (Fig. 12).

Generative Models and Adversarial Training for Image Segmentation Generative models, particularly Generative Adversarial Networks (GANs), have been successfully adapted for image segmentation by leveraging adversarial training to improve segmentation accuracy and quality. In this approach, a segmentation model generates predictions that are refined by an adversarial network, which discriminates between ground-truth segmentation maps and model-generated maps, promoting more realistic outputs.

Luc et al. [95] pioneered this by using an adversarial network alongside a segmentation network, leading to improved results on datasets like Stanford Background and PASCAL VOC 2012. Souly et al. [96] furthered this by introducing semi-weakly supervised segmentation with a GAN-based generator creating additional training examples, enhancing model robustness.

Hung et al. [97] developed a semi-supervised framework that uses a discriminator to compare predicted probability maps against ground-truth distributions. They employed a composite loss function that combines cross-entropy, adversarial, and confidence-based losses, encouraging better spatial coherence.

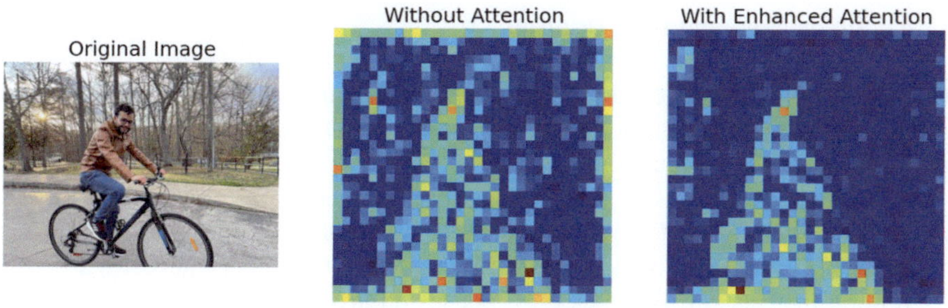

Fig. 12 Illustration of enhanced spatial attention applied to feature maps. In this example, a spatial attention mechanism is applied to feature maps extracted from a pre-trained ResNet model. The attention mechanism intensifies focus on key regions within the feature map by amplifying important features through a non-linear scaling operation. The original feature map (center) is shown alongside the attention-enhanced map (right), where the high-attention areas are more prominently highlighted, demonstrating the impact of spatial attention on feature refinement. Quantitative metrics highlight the difference: Mean Squared Error (MSE) increased to 0.0152, Structural Similarity Index (SSIM) dropped to 0.7228, and Cosine Similarity decreased to 0.9281, indicating substantial alterations in both pixel intensity and structure as a result of the enhanced attention

Adversarial training has also been adapted for specialized applications, such as cell segmentation and generating occluded object parts, underscoring its versatility in diverse segmentation challenges. These generative models have introduced powerful strategies for improving segmentation quality, even in limited data scenarios.

CNN Models with Active Contour Models for Image Segmentation Integrating CNNs with Active Contour Models (ACMs) [98] has shown promise for enhancing segmentation tasks by leveraging ACM principles, such as contour energy minimization. Some approaches formulate ACM-inspired loss functions; for instance, Chen et al. [99] developed a supervised loss layer that includes area and size constraints, applied to ventricle segmentation in cardiac MRI.

Other approaches combine ACMs as post-processing or co-learning components. For instance, Le et al. [100] used level-set ACMs as recurrent neural networks (RNNs) to refine CNN outputs, while Rupprecht et al. [101] introduced Deep Active Contours to improve contour accuracy. In medical imaging, Hatamizadeh et al. [7] proposed the Deep Active Lesion Segmentation (DALS) model, in which CNNs predict parameter functions for locally-parameterized ACMs.

Structured prediction methods have also emerged, such as Marcos et al. [102] Deep Structured Active Contours (DSAC) for building instance segmentation in aerial images and Cheng et al.'s Deep Active Ray Network (DarNet) [103], which employs a polar coordinate ACM formulation to avoid contour overlap.

2.2.4 Image Segmentation Datasets

This section provides an overview of widely used datasets for image segmentation, grouped by 2D images, 2.5D RGB-D images, and 3D images, each offering pixel-wise annotations for evaluation.

2D Datasets. 2D datasets are prominent in segmentation research due to their accessibility. PASCAL VOC [23] offers annotations for various tasks and 21 classes, while MS COCO [17] provides complex scenes with 91 object categories and over 2.5 million instances. Cityscapes [79] focuses on urban street scenes with detailed pixel-level annotations across 30 classes. ADE20K [104] contains extensive scene annotations, and SiftFlow [50] covers diverse outdoor environments. Datasets like Stanford Background [105] and Berkeley Segmentation Dataset (BSD) [106] support semantic understanding and boundary detection, while YouTube-Objects [107] and KITTI [108] aid in object detection for dynamic scenes.

2.5D Datasets. RGB-D [109] datasets are popular in applications requiring depth information. NYU-D V2 [49] provides RGB and depth images for various indoor scenes, while SUN-3D [110] and SUN RGB-D [111] offer detailed annotations across multiple environments. UW RGB-D [112] includes household objects, and ScanNet [113] supplies millions of RGB-D views with semantic segmentations.

3D Datasets. 3D datasets are essential in fields like robotics and medical analysis. Stanford 2D-3D [114] contains multiple registered modalities for indoor spaces. ShapeNet Core [115] offers a broad collection of 3D models across common categories, and the Sydney Urban Objects Dataset [116] includes scans of urban objects like vehicles and trees, contributing to diverse 3D segmentation applications.

Data augmentation is often applied to expand labeled samples, especially in small datasets, enhancing model training by transformations like translation, rotation, and color shifts. This process supports improved generalization and faster convergence, especially in limited data scenarios such as medical imaging.

Performance Metrics for Segmentation Models Evaluating segmentation models typically focuses on accuracy, speed, and memory requirements. We next discuss the primary accuracy metrics for segmentation models.

Pixel Accuracy (PA) measures the ratio of correctly classified pixels to the total number of pixels across all classes:

$$\text{PA} = \frac{\text{Number of Correctly Classified Pixels}}{\text{Total Number of Pixels}}$$

or, equivalently,

$$\text{PA} = \frac{\sum_{i=1}^{N} \text{True Positive}_i}{\sum_{i=1}^{N} \text{Total Pixels}_i}$$

where N is the total number of classes; True Positive_i is the count of correctly classified pixels for class i; Total Pixels_i is the total number of pixels in class i.

Mean Pixel Accuracy (MPA) is calculated by taking the average of the pixel accuracy across all classes:

$$\text{MPA} = \frac{1}{N}\sum_{i=1}^{N}\frac{\text{True Positive}_i}{\text{Total Pixels}_i}$$

where N is the total number of classes; True Positive_i is the count of correctly classified pixels for class i; Total Pixels_i is the total number of pixels in class i.

Intersection over Union (IoU), also known as the Jaccard Index, measures the overlap between predicted and ground-truth areas, divided by their union. is given by:

$$\text{IoU} = \frac{\text{Area of Overlap}}{\text{Area of Union}} = \frac{\text{True Positive}}{\text{True Positive} + \text{False Positive} + \text{False Negative}}$$

Mean IoU (mIoU) is the average of IoU across all classes:

$$\text{mIoU} = \frac{1}{N}\sum_{i=1}^{N}\text{IoU}_i$$

where N is the total number of classes; True Positive represents pixels correctly classified as the target class; False Positive represents pixels incorrectly classified as the target class; False Negative represents pixels of the target class incorrectly classified as other classes.

Precision, Recall, and F1 Score: These metrics calculate accuracy in terms of true positives (TP), false positives (FP), and false negatives (FN). The F1 Score, the harmonic mean of precision and recall, balances both metrics.

Precision and Recall are defined as follows:

$$\text{Precision} = \frac{\text{TP}}{\text{TP} + \text{FP}}$$

$$\text{Recall} = \frac{\text{TP}}{\text{TP} + \text{FN}}$$

The F1 Score, the harmonic mean of Precision and Recall, is:

$$\text{F1 Score} = 2 \cdot \frac{\text{Precision} \times \text{Recall}}{\text{Precision} + \text{Recall}}$$

where TP is the count of True Positives; FP is the count of False Positives; FN is the count of False Negatives.

Dice Coefficient: Especially popular in medical imaging, it measures the overlap between prediction and ground truth, similar to IoU. When applied to binary data, it aligns with the F1 Score, offering insight into the model's performance in segmenting critical regions.

Dice Coefficient is given by:

$$\text{Dice} = \frac{2 \times \text{True Positive}}{2 \times \text{True Positive} + \text{False Positive} + \text{False Negative}}$$

where True Positive is the count of correctly segmented pixels; False Positive is the count of incorrectly segmented pixels; False Negative is the count of missed pixels from the target class.

These metrics collectively assess the effectiveness of segmentation algorithms, with IoU and Dice Coefficient particularly favored in fields requiring detailed overlap analysis.

2.3 *Classification*

Image classification involves assigning one or more labels to an image and is foundational in CV and pattern recognition, with applications spanning from content analysis to biometrics. Traditional methods rely on manually defined low-level features (e.g., color, texture) or mid-level features derived through algorithms like bag-of-visual-words, followed by classifiers such as SVMs.

In contrast, deep learning integrates feature extraction and classification within a single network, learning complex, high-level representations through layered architectures, improving upon traditional approaches. This end-to-end approach enhances accuracy and adaptability across various classification tasks.

Network Architecture: The architecture of CNNs advances traditional Artificial Neural Networks (ANNs) by incorporating layers tailored for feature extraction and classification. Typically, CNNs include convolutional layers for extracting image features, pooling layers for dimensionality reduction, and fully connected layers for final classification.

Convolution Layer: The convolution layer applies filters across an image, transforming pixel data into feature maps. This process extracts key image characteristics by sliding filters over the input matrix, generating a new representation focused on essential visual features.

Pooling Layer: Pooling layers, often inserted between convolution layers, reduce the spatial size of feature maps, cutting down computational complexity. Max pooling, a popular approach, downsamples by taking maximum values from sub-regions, preserving the most prominent features and preventing overfitting.

Fully Connected Layer: The fully connected layer, integral to classification tasks, takes the flattened outputs from convolution and pooling layers to classify images. Each output is mapped to the desired classification categories based on the learned weights, producing the final label prediction.

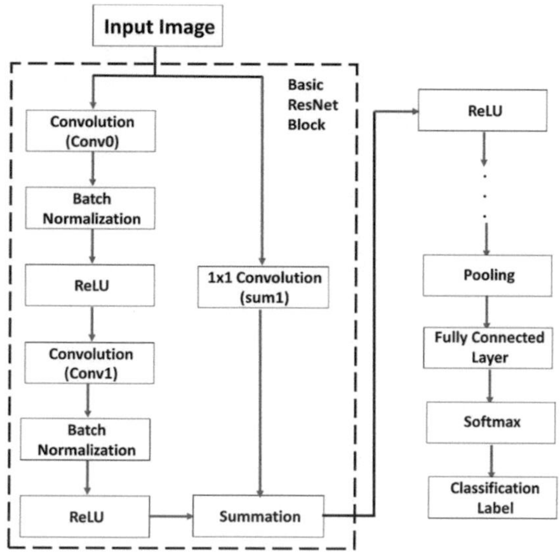

Fig. 13 Schematic of ResNet architecture

Loss Function: Loss functions, such as mean squared error and cross-entropy, quantify the prediction error during training. Optimizing this loss through backpropagation enhances the model's predictive accuracy.

AlexNet [6], a pioneering deep CNN with eight layers, achieved notable success in 2012, setting benchmarks in image classification. With five convolutional layers and three fully connected layers, AlexNet demonstrated the power of deep learning for feature-rich, high-accuracy classification tasks.

ResNet (Residual Network) [117] introduced residual connections to address network depth challenges and prevent performance degradation. By leveraging shortcut connections that pass information across layers, ResNet models are more efficient, effectively managing deeper architectures and achieving high classification accuracy, particularly in medical image analysis contexts (Fig. 13).

2.3.1 Image Classification Datasets

Image classification has significantly benefited from large-scale annotated datasets, which serve as benchmarks for developing and evaluating machine learning models. We next discuss some of the most widely used image classification datasets that have shaped advancements in the field (Fig. 14).

ImageNet [118], introduced for the ImageNet Large Scale Visual Recognition Challenge (ILSVRC), contains over 14 million images across 1,000 classes. It covers a broad spectrum of object categories, making it essential for training deep convolutional neural networks (CNNs) and for advancing transfer learning in image classification.

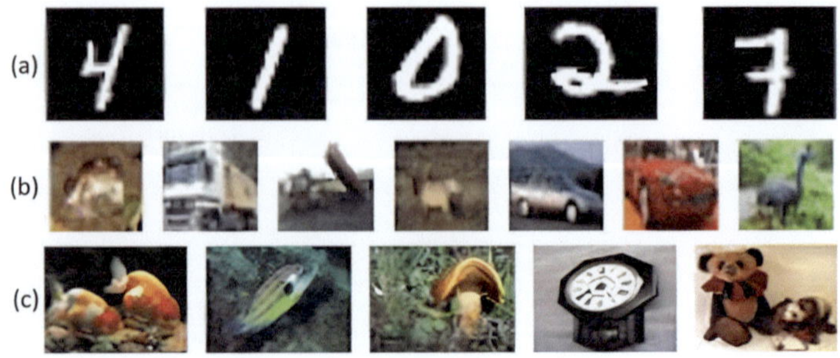

Fig. 14 Datasets examples for classification, **a** MNIST, **b** CIFAR, **c** ImageNet dataset

The CIFAR datasets [119] contain 60,000 32×32 color images. CIFAR-10 covers 10 classes with 6,000 images per class, while CIFAR-100 expands to 100 classes with 600 images each. These datasets are particularly useful for evaluating model performance on smaller, diverse datasets and are commonly used in early-stage neural network experiments.

MNIST[1] contains 70,000 grayscale images of handwritten digits, with 10 classes (0–9). The dataset, with its simplicity and ease of use, remains popular for benchmarking classification algorithms, especially in tasks requiring digit recognition and for testing novel architectures.

Created as a drop-in replacement for MNIST, Fashion-MNIST [120] has 70,000 grayscale images across 10 categories of clothing items. It provides a more challenging benchmark than MNIST, enabling researchers to test algorithms on a dataset similar in structure but with more complex visual patterns.

Stanford Cars dataset [121] contains 16,185 images of 196 classes of car models, with detailed annotations that include make, model, and year. It is particularly useful for fine-grained image classification tasks, requiring the model to differentiate between subtle differences within a single category.

The Flowers102 dataset [122] has 102 flower categories, each with between 40 and 258 images. Its varied classes and high inter-class similarity make it suitable for evaluating algorithms on fine-grained classification problems in natural image categories.

2.4 *Mapping and Tracking*

Camera-based mapping and tracking methods in Medical CV (MCV) are crucial for observing environments, estimating motion, and developing spatial representations of areas such as tissues and instruments. Tracking focuses on identifying the position and movement of objects, including cameras, while mapping creates a persistent

---

[1] https://yann.lecun.com/exdb/mnist/index.html.

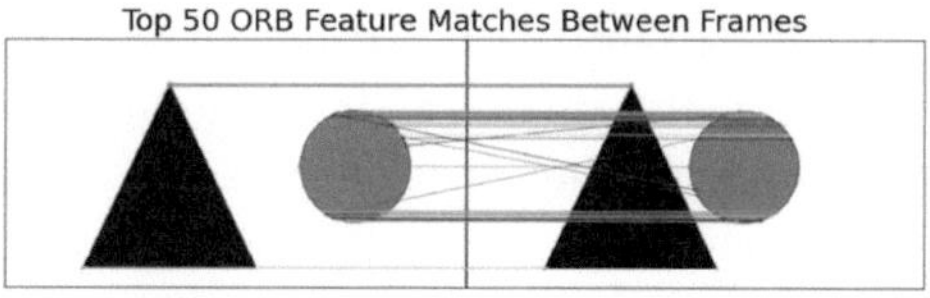

Estimated Homography Matrix

	Column 1	Column 2	Column 3
Row 1	0.99495	0.00298	-129.43339
Row 2	-0.00091	1.00022	0.19171
Row 3	-1e-05	1e-05	1.0

Fig. 15 Feature detection and matching using ORB for MCV tracking and mapping. This example demonstrates the detection and matching of Oriented FAST and Rotated BRIEF (ORB) features between two consecutive frames, a key component in ORB-SLAM for real-time localization and mapping in Medical CV. Keypoints detected by ORB (visualized as circles) are matched between frames using the Brute Force Matcher (BFMatcher) to track motion, and a homography matrix is estimated to simulate a mapping stage. This approach is effective in endoscopic mosaicking, where continuous frames are combined to produce a larger map of the environment

representation or 'memory state' of the environment. For example, in mosaicking, the map is an image, while in Simultaneous Localization and Mapping (SLAM) [123], it often takes the form of a point cloud (Fig. 15).

MCV methods, commonly using endoscopes, bronchoscopes, and similar devices, face unique challenges. In clinical settings like endoscopy and minimally invasive surgery, accurate localization of tissue and instruments aids in procedures like tumor resection and cancer detection. However, issues such as low texture in organs, light reflections from fluid-covered surfaces, and image blurring from smoke or blood can impact accuracy. Deformation of organs or environmental changes further complicate the generation of reliable maps and tracking data.

Addressing these challenges can lead to significant improvements in patient outcomes, ease clinicians' tasks, and potentially lower healthcare costs. Unlike non-medical CV tasks, MCV applications must account for the unique textures and appearance of medical data. Integrating domain-specific loss functions, specialized training on medical datasets, and model architectures designed with these unique challenges in mind has proven crucial. Consequently, while insights from broader CV fields like SLAM and neural rendering are valuable, MCV requires adaptations that address the specific demands and visual characteristics of the medical field.

Key Algorithms in MCV Tracking and Mapping

Scale-Invariant Feature Transform (SIFT): Detects local features by capturing scale and rotation invariance, effective in diverse MCV environments but computationally intensive.

Speeded-Up Robust Features (SURF): Uses an approximation of the Hessian matrix for faster detection of keypoints, providing a balance of speed and accuracy for MCV use.

Oriented FAST and Rotated BRIEF (ORB): Combines the FAST keypoint detector with the BRIEF descriptor, providing a low-cost, effective alternative to SIFT and SURF, suitable for real-time applications.

SuperGlue and LightGlue: Neural network-based matchers; SuperGlue uses graph neural networks to enhance point matching, integrating motion and contextual information, making it more robust for dynamic and deforming environments.

Grid-Based Motion Statistics (GMSMatch): A heuristic-based approach that helps reduce mismatches in environments with large visual changes by considering the consistency of surrounding points' motion.

CNN-based Homography Estimation: Uses CNNs to estimate the transformation between image pairs, creating smooth and robust mosaics suitable for endoscopic mosaicking.

Deep Learning for Fetoscopic Mosaicking: Combines segmentation and homography estimation using CNNs tailored for fetoscopy, enhancing robustness against specularities and other artifacts common in MCV environments.

Stereo Depth Mapping (GA-Net [124], RAFT-Stereo [125]): Estimates depth using stereo images by computing disparity maps, ideal for rigid environments but adaptable to endoscopic tasks.

Monocular Depth Methods [126]: Estimate relative depth using only single images, useful for constrained environments like bronchoscopy where dual-camera setups are impractical.

Structure from Motion (SfM) [127] Generates 3D reconstructions from 2D images in rigid settings, valuable in mapping for surgical planning.

Non-Rigid SfM (NRSfM) [128]: Extends SfM to handle deformable objects by incorporating motion constraints, critical in flexible tissue tracking.

Shape-from-Template (SfT) [129]: Uses a learned template to match the 3D form of deforming objects in real-time, supported by neural networks that account for variability in tissue appearance.

ORB-SLAM [130]: A widely adopted approach for mapping with feature matching and loop closure; optimized for real-time performance in MCV applications.

EKF-SLAM [131]: Uses an Extended Kalman Filter for mapping and tracking in minimally textured environments, adjusting positions based on expected positions over time.

DefSLAM [132]: Incorporates non-rigid deformation into mapping using triangular mesh structures, often used for organ or tissue mapping.

E-DSSR [133]: A deep learning-based approach that combines CNNs with SLAM for improved tissue and instrument tracking in surgery.

Super [134]: Uses embedded deformation for tracking tissues in MCV environments, optimizing for changes in soft tissue dynamics in real-time (Fig. 16).

2.4.1 Mapping and Tracking Datasets

Datasets play a crucial role in advancing mapping and tracking technologies, particularly in applications like autonomous driving, robotics, and medical imaging. These datasets provide annotated sequences, including spatial, temporal, and sensory data, which are essential for training and benchmarking algorithms. Below are some prominent datasets widely used in mapping and tracking:

Fig. 16 Datasets examples for tracking and mapping, **a** KITTI, **b** Cityscapes, **c** Oxford RoboCar

The KITTI dataset [108] is one of the most comprehensive resources for autonomous driving research, offering real-world data captured from a moving vehicle. It includes stereo camera, LiDAR, and GPS data, covering tasks like depth estimation, object detection, tracking, and visual odometry. KITTI has become a standard benchmark for evaluating mapping and tracking performance in dynamic urban environments.

The Cityscapes dataset [79] is designed for semantic urban scene understanding, with 5,000 finely annotated images and 20,000 coarsely annotated ones. It captures scenes from 50 different cities, providing pixel-level annotations for 30 classes. Cityscapes is widely used for semantic and instance segmentation in urban settings, making it valuable for evaluating algorithms in mapping static objects in complex scenes.

TUM RGB-D dataset[2] contains RGB-D data sequences from an RGB-D camera, commonly used for evaluating visual odometry and SLAM algorithms. TUM RGB-D provides ground truth trajectories from a motion capture system, offering benchmarks for indoor mapping and tracking with real-world depth information. Its diversity of scenes (static and dynamic) and inclusion of loop closures are particularly useful for indoor navigation systems.

Oxford RobotCar[3] captured across a year, the Oxford RobotCar dataset offers extensive, time-variant data from a self-driving vehicle in Oxford, UK. It includes stereo images, LiDAR, GPS, and inertial data. The repeated visits across changing weather, lighting, and seasonal conditions make it ideal for testing the robustness of mapping and tracking algorithms under diverse environmental conditions, especially in outdoor urban scenarios.

ApolloScape[4] provides 147,000 frames of high-resolution street scenes with 2D and 3D annotations. It supports various tasks, including semantic segmentation, 3D car instance tracking, and road scene parsing, with an emphasis on tracking

---

[2] https://cvg.cit.tum.de/data/datasets/rgbd-dataset.

[3] https://robotcar-dataset.robots.ox.ac.uk/.

[4] https://apolloscape.auto/.

moving objects in urban settings. This dataset helps evaluate mapping and tracking for autonomous navigation in dense urban traffic.

Mapillary Vistas[5] is a large-scale dataset offers high-quality, street-level imagery from diverse environments worldwide. With over 25,000 images annotated with 66 object categories, it is particularly valuable for semantic segmentation and mapping in varied geographic locations and environmental conditions. It aids in developing robust mapping and tracking systems that generalize across global datasets.

EuRoC MAV[6] targeted at visual SLAM and drone applications, the EuRoC MAV dataset includes stereo images, IMU data, and accurate ground truth from a motion capture system. Captured in indoor environments, it is frequently used to test tracking and mapping algorithms for MAVs (Micro Aerial Vehicles) navigating in small, cluttered spaces, making it essential for indoor autonomous drone navigation.

ETH3D[7] provides high-resolution images and accurate ground truth for 3D reconstruction and visual odometry. Its benchmarks cover both indoor and outdoor scenes, supporting evaluation of dense and sparse reconstruction techniques. ETH3D is valuable for mapping applications in robotics and augmented reality, where accurate 3D scene representation is critical.

SLAMBench [135] is a performance evaluation framework with various SLAM algorithms, including KinectFusion and ElasticFusion. It provides RGB-D datasets with ground truth for benchmarking SLAM algorithms, focusing on real-time performance, accuracy, and computational efficiency. SLAMBench aids in evaluating SLAM systems across hardware platforms, with relevance for both indoor and outdoor environments.

Segmentation Metrics.

Intersection over Union (IoU): Measures overlap between predicted and ground truth segmentation, assessing accuracy for applications like tissue segmentation.

Structural Similarity Index (SSIM): SSIM is calculated by comparing three components between two images: luminance (l), contrast (c), and structure (s). The SSIM index between images x and y is given by:

$$\mathrm{SSIM}(x, y) = \frac{(2\mu_x\mu_y + C_1)(2\sigma_{xy} + C_2)}{(\mu_x^2 + \mu_y^2 + C_1)(\sigma_x^2 + \sigma_y^2 + C_2)}$$

where μ_x and μ_y are the mean values of images x and y; σ_x^2 and σ_y^2 are the variances of x and y; σ_{xy} is the covariance of x and y; C_1 and C_2 are constants to stabilize the division.

Endpoint Error (EPE): EPE measures the Euclidean distance between the estimated endpoint $p = (x_p, y_p)$ and the ground truth endpoint $q = (x_q, y_q)$ and is calculated as:

[5] https://www.mapillary.com/dataset/vistas.

[6] https://projects.asl.ethz.ch/datasets/doku.php?id=kmavvisualinertialdatasets.

[7] https://www.eth3d.net/datasets.

$$\text{EPE} = \sqrt{(x_p - x_q)^2 + (y_p - y_q)^2}$$

where x_p and y_p are the coordinates of the estimated endpoint; x_q and y_q are the coordinates of the ground truth endpoint.

Forward-Backward Error (FBE): FBE is calculated by comparing the position of a tracked point in the forward direction with its backward-projected position. Given a point tracked from t to $t + \Delta t$ and then back to t, the FBE is:

$$\text{FBE} = \| p_t - p_{t+\Delta t \to t} \|$$

where p_t is the original position of the point at time t; $p_{t+\Delta t \to t}$ is the position after tracking forward and then backward to time t; $\| \cdot \|$ represents the Euclidean distance.

Absolute Trajectory Error (ATE): ATE measures the Euclidean distance between the estimated trajectory and the ground truth trajectory across all time steps. It is given by:

$$\text{ATE} = \sqrt{\frac{1}{n} \sum_{i=1}^{n} \| \mathbf{p}_i^{\text{est}} - \mathbf{p}_i^{\text{gt}} \|^2}$$

where $\mathbf{p}_i^{\text{est}}$ is the estimated position at time step i; $\mathbf{p}_i^{\text{gt}}$ is the ground truth position at time step i; n is the total number of time steps.

Relative Pose Error (RPE): RPE evaluates the frame-to-frame error in camera movement by comparing the difference between the estimated and ground truth transformations between consecutive frames. It is calculated as:

$$\text{RPE} = \frac{1}{n-1} \sum_{i=1}^{n-1} \| (\mathbf{p}_{i+1}^{\text{est}} - \mathbf{p}_i^{\text{est}}) - (\mathbf{p}_{i+1}^{\text{gt}} - \mathbf{p}_i^{\text{gt}}) \|$$

where $\mathbf{p}_{i+1}^{\text{est}} - \mathbf{p}_i^{\text{est}}$ is the estimated transformation between frames i and $i + 1$; $\mathbf{p}_{i+1}^{\text{gt}} - \mathbf{p}_i^{\text{gt}}$ is the ground truth transformation between frames i and $i + 1$.

Root Mean Square Error (RMSE): RMSE is a common metric for depth estimation accuracy, measuring the square root of the average squared differences between the estimated depth d_i^{est} and the ground truth depth d_i^{gt}:

$$\text{RMSE} = \sqrt{\frac{1}{n} \sum_{i=1}^{n} (d_i^{\text{est}} - d_i^{\text{gt}})^2}$$

where d_i^{est} is the estimated depth at point i; d_i^{gt} is the ground truth depth at point i; n is the total number of depth points.

Mean Absolute Error (MAE): MAE measures the average absolute difference between the estimated depth and the ground truth depth. It is expressed as:

$$\text{MAE} = \frac{1}{n}\sum_{i=1}^{n} |d_i^{\text{est}} - d_i^{\text{gt}}|$$

where the terms are defined as in the RMSE equation.

Peak Signal-to-Noise Ratio (PSNR): PSNR is a measure of the ratio between the maximum possible power of a signal and the power of distorting noise, used to assess the quality of depth predictions. It is calculated as:

$$\text{PSNR} = 10 \cdot \log_{10}\left(\frac{L^2}{\text{MSE}}\right)$$

where L is the maximum possible value of the depth data (e.g., 255 for 8-bit depth); MSE is the Mean Squared Error between the estimated and ground truth depths.

2.5 *Explainability in Computer Vision*

CNNs have achieved remarkable success across various tasks in CV, including image classification, object detection, and semantic segmentation. However, their complex architecture makes it challenging to interpret model decisions, especially in high-stakes applications like autonomous driving and healthcare. As a result, explainable AI (XAI) methods in CNNs have become essential to enhancing transparency, trust, and usability. The current XAI literature on CNNs can be categorized into two main models, decision models [136], and architecture models [137].

Decision Models. These models interpret CNN decisions by analyzing input-output mappings and can be classified into feature relevance [136], and visual explanation [138]. Feature relevance [139] focuses on ranking feature importance by identifying parts of the input that most significantly contribute to a model's decision. Visual explanation [140] provides visual saliency maps to highlight the spatial importance of input regions.

Some of the visual explanation techniques include:

- Class Activation Maps (CAM): Replaces the fully connected layers in CNNs with a global pooling layer to produce visual heatmaps of relevant features, though it may reduce accuracy [141].
- Grad-CAM and Variants (Grad-CAM++, Smooth Grad-CAM++): Computes pixel-level gradients to produce heatmaps without altering CNN structure, with extensions for more refined interpretations [142].
- Score-CAM: Combines forward pass and confidence measures to generate activation maps, bypassing the gradient-related issues of earlier methods [143].

Some of the feature relevance techniques include:

- Layer-wise Relevance Propagation (LRP): This technique backpropagates the prediction to assign relevance scores to each neuron, producing heatmaps that highlight relevant image areas [144].

- DeepLIFT: Overcomes gradient saturation by selecting reference inputs to compute contributions, offering better clarity on CNN decision-making processes [139].
- Integrated Gradients: Calculates attribution scores by analyzing gradients with respect to model predictions, providing robust visual explanations for model outputs [145].

Architecture Models. These models alter CNN structures to improve interpretability and are classified into architecture modification and architecture simplification.

Architecture modification involves modifying CNN layers or loss functions to enhance interpretability. Architecture simplification employs techniques like pruning or distillation to simplify CNNs, thereby improving their transparency.

Some of the architecture modification techniques include:

- Attention Mechanisms: Models such as GALA (Global-and-Local Attention) and DomainNet incorporate attention mechanisms to focus on critical parts of an input image, guiding the CNN's focus for accurate feature extraction [137].
- Residual Attention Networks: By adding attention modules at different stages, these networks produce feature maps that improve classification accuracy with a cost of increased computational load [146].
- Loss-based Attention: Instead of additional attention layers, this method incorporates attention into the loss function, leveraging inherent CNN features for interpretability without extra parameters [147].

Some of the architecture simplification techniques include:

- Network Pruning: Methods like CAR [148] compress CNNs by removing unimportant filters to improve interpretability while preserving classification accuracy [149].
- Decision Tree Integration: For CNN interpretation, decision trees connected to high-level CNN features decompose complex decisions into a rule-based format, making CNN outputs easier to understand [150].
- Subnetwork Extraction: Creates smaller, interpretable subnetworks focused on specific classes or samples, allowing more straightforward interpretations of CNN predictions [149].

The interpretability of CNN models can be evaluated using Model-centric Metrics (e.g., visualization quality, object localization via IoU, and robustness against noise and adversarial attacks) and Human-centric Metrics (e.g., trust assessments through human subject experiments). The models are assessed by visual clarity, faithfulness to predictions, and classification accuracy, aiming to balance transparency with CNN performance.

3 Applications of CV in Healthcare

3.1 *Medical Imaging Analysis*

The utilization of deep learning in medical imaging has revolutionized several key areas, allowing for improved accuracy and efficiency in diagnosis and analysis. Here, we outline prominent applications, with specific emphasis on their clinical relevance and technical methodologies.

3.1.1 Classification

Image and Exam Classification. Deep learning has been widely utilized in medical imaging for exam classification, where an exam typically comprises of one or more images analyzed to produce a single diagnostic output, such as determining disease presence. Given that medical datasets are typically smaller than those in general CV, transfer learning has become a pivotal technique. It often employs pre-trained models on natural images, either as feature extractors or through fine-tuning. Studies, such as [151] have examined these strategies, finding contrasting results that highlight how transfer learning strategies may vary depending on the specific medical application.

In recent years, CNNs have become the dominant architecture for image classification tasks in medical imaging, demonstrating human-level performance in some cases, as exemplified by fine-tuning Google's Inception v3 on medical data for skin and retinal images [152].

Object and Lesion Classification. Object classification in medical imaging focuses on the identification and categorization of specific lesions or features, such as lung nodules or skin abnormalities. Multi-scale or multi-stream CNN architectures [153] are often employed to process images at various resolutions, providing a robust understanding of both local and global contexts, which is essential for accurately distinguishing small lesions.

Detection. Organ, Region, and Landmark Localization. Localization tasks in medical imaging, such as identifying anatomical landmarks, have benefited greatly from CNN-based approaches. Traditional methods relied on segmenting 3D volumes, but advancements now allow for the parsing of 3D space by employing orthogonal 2D planes [154]. More recent methods utilize CNNs directly to predict landmarks in 3D space, a significant step towards real-time diagnostic tools.

Object or Lesion Detection. Object detection aims to localize and identify specific regions within an image that contain lesions or abnormalities. CNN-based detection systems often perform pixel-wise classifications, with methods such as fully convolutional networks (fCNNs) used to optimize this task by reducing computational redundancies. Weakly-supervised approaches and multi-stream CNNs have also been developed to manage class imbalances that often arise in lesion detection [155].

Segmentation. Organ and Substructure Segmentation. Segmentation is critical for analyzing anatomical features, particularly for volumetric or morphological studies in the brain or heart. The U-net architecture [156] is a cornerstone in medical segmentation tasks due to its unique design that combines upsampling and downsampling paths with skip connections, enabling a holistic view of the image. Adaptations of U-net, such as V-net and variants with ResNet-like skip connections, have further optimized 3D medical image segmentation.

Lesion Segmentation. Lesion segmentation is a more complex variant of segmentation, requiring multi-scale or multi-stream networks to accommodate the variability in lesion shapes and sizes. Techniques such as loss function adjustments and data augmentation have been utilized to handle class imbalance issues commonly seen in lesion segmentation tasks [157].

Registration. Image registration in medical imaging is the spatial alignment of images, often applied iteratively to determine an optimal transformation. Deep learning has enabled two main strategies: using networks to estimate similarity metrics to guide registration algorithms; or directly predicting transformation parameters. For example, Yang et al. [158] demonstrated the potential for using U-net-inspired architectures for significantly accelerated registration, showcasing the potential of deep learning in optimizing computationally intensive processes like medical image registration.

Advanced Applications in Medical Imaging. Content-Based Image Retrieval (CBIR). CBIR enables the retrieval of similar medical cases, potentially aiding clinicians in diagnosing rare conditions. CNNs are commonly used in CBIR to extract features that serve as the basis for similarity assessments. Applications in retinal imaging and X-rays have shown promise, yet challenges remain due to the high variability and specificity of medical image data.

Image Generation and Enhancement. Generative models in deep learning have been applied to improve medical image quality, generate missing data, and enhance image resolution. Techniques like multi-stream CNNs allow for the synthesis of high-resolution images from multiple low-resolution sources, aiding in clinical interpretation and diagnostic accuracy.

Combining Image Data with Reports. Integrating textual reports with image data is a growing area, aiming to improve the contextual understanding of images and automate report generation. This task typically employs CNNs for image representation and recurrent neural networks for generating descriptive labels or full reports, inspired by advancements in natural image captioning [159].

3.2 Pathology

In recent years, deep learning has revolutionized digital pathology by providing advanced techniques for analyzing histopathological images, enabling higher accuracy, consistency, and speed in diagnostic processes. Digital Pathology (DP) focuses on the analysis of digital images derived from tissue samples to support disease diagnosis and prognosis. Within this domain, deep learning models, particularly CNNs, have proven effective in various tasks, including classification, segmentation, and detection, by identifying complex patterns that traditional methods might miss.

Key Applications of Deep Learning in Digital Pathology. Image Preprocessing and Stain Normalization. The variability in staining intensity and color, resulting from differences in scanner types, staining protocols, and environmental conditions, can impact the consistency of pathology results. Stain normalization [160] is crucial in digital pathology to ensure uniformity across images, as it adjusts the color profiles to reduce these inconsistencies. Deep learning-based stain normalization approaches leverage color deconvolution and style transfer techniques, where generative adversarial networks [161] have recently been employed to achieve high-quality, consistent stain normalization.

Cell-Level Analysis. Cellular analysis is central in pathology, as nuclear features like spatial arrangement and morphological characteristics are essential biomarkers for cancer diagnosis. Tasks like cell segmentation, detection, and instance segmentation facilitate the automatic identification and classification of cellular objects, which traditionally required labor-intensive manual examination by pathologists. In cell segmentation, CNNs [162] and FCNs [163] are commonly used, with architectures like U-Net and variants designed for precise localization of cellular boundaries, aiding in the quantification of attributes such as nuclear pleomorphism [164] and mitotic count [165].

Mitosis Detection. Mitosis detection plays a critical role in assessing tumor aggressiveness, requiring accurate identification of dividing cells. This is a complex task due to the diverse shapes and growth stages of mitotic figures. Advanced deep learning approaches, such as multi-scale CNNs and regression-based FCNs [166], have demonstrated improved detection accuracy by distinguishing mitotic figures from apoptotic cells and other similar structures, addressing both the need for precision and computational efficiency.

Gland-Level Analysis. Glands are key structures in organs like the colon, pancreas, and breast, where the degree of glandular differentiation is a strong indicator of cancer progression. Gland instance segmentation enables the assessment of these structural characteristics, with DL models handling the varying shapes, sizes, and arrangements of glands. Approaches like Deep Contour-Aware Network (DCAN) [167] and Holistically-Nested Edge Detection (HED) [168] models leverage both semantic and contour information to enhance segmentation accuracy, even in cases where gland morphology is distorted due to disease.

Region-Level Analysis. Region-level tasks extend deep learning applications to larger structures in whole-slide images (WSIs), facilitating tasks such as tumor identification, cancer subtype classification, and the prediction of cancer severity. Deep learning models are employed to process large-scale WSIs through patch-based analysis, with techniques like Multiple Instance Learning (MIL) [169] supporting weakly supervised learning where exact labels for subregions are unavailable. For example, deep learning models trained on small patches extracted from WSIs can predict cancer presence and subtype, even when annotations are limited to whole regions rather than specific details within the tissue [169].

Prognosis and Cancer Grading. Deep learning models contribute significantly to prognosis by extracting features linked to disease progression and recurrence risk. In cancer grading [170], DL-based models utilize histopathological patterns to assign cancer grades, which are strongly correlated with survival rates [171]. For instance, models like ResNet and Inception networks have been adapted to identify genetic markers and predict gene mutations, providing insights into tumor severity and aiding in personalized treatment planning. This application is particularly impactful in Non-Small Cell Lung Cancer (NSCLC) [172] and gastrointestinal cancers, where genetic and morphological features drive treatment options.

3.3 Ophthalmology

Deep learning has significantly impacted ocular imaging by providing automated methods for accurate segmentation, diagnosis, and disease progression analysis. Various advanced imaging technologies, including Optical Coherence Tomography (OCT), Magnetic Resonance Imaging (MRI), and Computed Tomography (CT), generate high-resolution images essential for examining ocular structures. Deep learning models have been adapted to these imaging modalities to enhance the diagnosis and treatment of ocular diseases, such as glaucoma, diabetic retinopathy, and age-related macular degeneration.

Retinal Disease Detection and Analysis. Retinal imaging is crucial in diagnosing conditions such as diabetic retinopathy, which can lead to vision loss if untreated. OCT, a non-invasive imaging technique, captures detailed cross-sectional images of the retina. Deep learning algorithms, particularly CNNs, effectively analyze these images, enabling automated segmentation of retinal layers and detection of microaneurysms, exudates, and other retinal abnormalities [173]. For instance, CNN-based models can distinguish between healthy and diseased retinal tissues with high accuracy, facilitating early diagnosis and intervention [174].

Glaucoma Screening and Progression Monitoring. Glaucoma, a leading cause of irreversible blindness, requires early detection and monitoring of the Optic Nerve Head (ONH) and retinal nerve fiber layer (RNFL) for effective management. Deep learning models applied to OCT images of the optic disc can segment ONH and RNFL regions, allowing accurate assessment of glaucoma progression [175]. These models help quantify structural changes in the ONH and RNFL, providing clinicians with reliable tools for monitoring disease progression and tailoring treatment.

Cataract Detection and Classification. Cataract detection has also benefited from the integration of deep learning with ocular imaging [176]. Automated systems using deep learning algorithms on slit-lamp or fundus images can classify cataract severity based on the density and opacity of the lens [177]. These systems support clinicians in diagnosing cataracts at different stages, allowing for more personalized treatment plans, whether for non-surgical management or preparation for surgical intervention.

Ocular Tumor Segmentation and Analysis. Deep learning has enhanced the analysis of ocular tumors, such as retinoblastoma and ocular melanoma, using MRI and CT imaging [178]. Automated segmentation methods identify tumor boundaries, assess tumor volume, and predict tissue characteristics. This segmentation helps in determining the extent of tumor spread and guides surgical planning and radiation therapy, improving patient outcomes. For example, U-Net-based architectures have been effectively applied to segment tumors from MRI scans [179], offering clinicians detailed insights into tumor size and location.

Corneal Disease Diagnosis. Corneal imaging is essential for diagnosing keratoconus and other corneal diseases. Deep learning methods, particularly CNNs and transformer-based models, are applied to corneal topography and tomography scans to detect subtle shape changes and abnormalities in the corneal structure [180]. These algorithms enable automated detection and grading of corneal disease severity, assisting ophthalmologists in early intervention strategies.

Multi-Modal Image Fusion for Comprehensive Diagnosis. Combining different ocular imaging modalities enhances the diagnostic accuracy of deep learning models. For instance, fusing OCT and fundus photography data enables a more comprehensive analysis of both retinal and optic nerve health [181]. Deep learning models designed for multi-modal image fusion can integrate data from multiple imaging sources, enabling simultaneous assessment of multiple ocular structures and conditions. This multi-modal approach is particularly valuable in complex cases where different modalities provide complementary information.

3.4 Monitoring and Rehabilitation

Deep learning has significantly advanced the field of monitoring and rehabilitation, particularly by enhancing the precision and effectiveness of continuous patient monitoring and physical rehabilitation therapies. These applications leverage a variety of deep learning techniques for analyzing sensor data, automating patient assessments, and personalizing therapeutic interventions.

Continuous Health Monitoring. In continuous monitoring, deep learning enables real-time analysis of physiological data collected from wearable sensors, such as heart rate, respiratory rate, and muscle activity. These systems often involve complex, multi-sensor networks, where deep learning models analyze bioelectric and movement data for applications in cardiac health, respiratory monitoring, and fall detection. By interpreting data from body-worn sensors, these models detect anomalies that indicate potential health risks, such as arrhythmias or respiratory irregularities, allowing for early intervention. For example, CNNs have shown efficacy in ECG data analysis, identifying irregular patterns that signify cardiac abnormalities with high accuracy [182].

The integration of wearable sensors with cloud and mobile health (mHealth) systems also allows data to be accessed remotely by healthcare providers. Deep learning algorithms deployed on these platforms can assess data in real time, enabling remote patient monitoring and facilitating the development of Health 4.0 applications, where patient data is collected, processed, and analyzed continuously. This integration of deep learning with IoT infrastructure has revolutionized health monitoring, making it feasible to conduct remote diagnostics and improving accessibility to healthcare services.

Physical Rehabilitation Systems. In rehabilitation, deep learning applications support patient progress assessment and feedback, often using movement data collected from sensors such as inertial measurement units (IMUs) and surface electromyography (sEMG). Recurrent neural networks (RNNs) and Long Short-Term Memory (LSTM) networks are particularly effective in these applications as they handle temporal data, allowing precise tracking of joint angles, limb positions, and muscle activation patterns [183]. Such models are crucial for designing adaptive rehabilitation programs, where exercises are dynamically adjusted based on real-time feedback.

Deep learning-driven analysis has facilitated automated assessments in various rehabilitation exercises, such as gait analysis for patients with neurological impairments like stroke and Parkinson's disease. Wearable sensors provide data on body movements, which are analyzed by machine learning models to monitor progress and detect deviations from normal movement patterns. In some cases, these systems are integrated with robotic exoskeletons or VR environments, enhancing the interaction between the patient and therapeutic equipment [184]. These hybrid systems employ deep learning models to adapt movements and provide feedback, enhancing patient engagement and improving recovery outcomes.

Assistive Technologies in Rehabilitation. Deep learning has also enhanced the capabilities of assistive technologies, particularly for individuals with severe motor limitations. Brain-computer interfaces (BCIs), for example, leverage deep learning models to interpret EEG signals, enabling patients to control external devices such as wheelchairs or robotic limbs with minimal physical effort [185]. Additionally, facial recognition and gesture-based interfaces, powered by CNNs and other CV models, are employed for communication in cases of speech impairment. These models help recognize facial expressions, eye movements, or hand gestures, translating them into actionable commands for assistive devices.

Furthermore, virtual reality (VR) and serious games in rehabilitation often use deep learning to provide realistic, interactive training environments. VR-based systems help patients practice specific movements, while deep learning models ensure the accuracy of simulated responses. For example, in stroke rehabilitation, patients use VR systems to retrain motor functions with real-time feedback on their performance, enabling therapists to modify treatment protocols based on data from previous sessions [186].

4 Challenges and Limitations

CVhas made significant strides in medical imaging and diagnostics, yet several challenges continue to hinder its broader and effective adoption in healthcare settings. These limitations encompass technical, ethical, and practical constraints that must be addressed to realize CV's full potential in enhancing healthcare outcomes.

Annotation Requirements and Safety-Critical Nature. The development of reliable CV models in healthcare depends on extensive annotated datasets, where images must be meticulously labeled by domain experts. This process is time-intensive, costly, and requires specialized medical expertise, such as radiologists or pathologists, who are often limited in availability. Furthermore, healthcare is a safety-critical domain; inaccuracies in CV predictions can lead to severe consequences, underscoring the need for exceptionally high standards of precision and reliability in these models.

Lack of Model Resilience and Generalization. CV models in healthcare face challenges in generalizing across different patient populations, imaging modalities, and clinical environments. A key issue is model resilience, or the capacity to handle fluctuations and uncertainties in input data, such as noise, artifacts, or distortions commonly encountered in medical imaging. Models that overly rely on patterns learned from a specific dataset may exhibit overfitting, limiting their effectiveness when applied to novel data or unexpected scenarios. This issue is especially pertinent in medical contexts where data variability is high, as a lack of generalization can undermine a model's applicability to diverse patient populations and healthcare settings.

Sensitivity to Environmental and Lighting Variations. Unlike human observers, CV models often struggle with changes in illumination, which can affect the visibility and contrast of anatomical structures in medical images. Lighting variations, caused by different imaging devices or environmental factors, can alter the appearance of tissues, lesions, and other critical features, leading to potential misinterpretation by CV models. This limitation can impair the performance of CV systems in real-world settings, where conditions often vary significantly from the controlled environments in which models were initially trained.

Limited Performance in Rare and Complex Cases. CV models frequently show a drop in performance when dealing with uncommon conditions or complex cases. These scenarios are often the most challenging to diagnose and are particularly important in healthcare, as clinicians rely on precise insights to handle rare conditions effectively. If a CV system underperforms in such cases, healthcare professionals may be reluctant to adopt the technology due to concerns about reliability and trustworthiness, impacting its potential for widespread use.

Privacy and Ethical Considerations. Data privacy is a significant barrier to implementing CV in healthcare. Medical imaging data is inherently sensitive, containing information that, if mishandled, could compromise patient confidentiality. As digital health systems and CV applications become more prevalent, ensuring data privacy through robust frameworks and adhering to regulatory standards becomes crucial. This challenge is particularly pressing in cases involving facial recognition or patient monitoring, where data misuse could lead to privacy violations.

High Dependency on Specialized Expertise. Deploying and maintaining CV systems in healthcare necessitates expertise in artificial intelligence and machine learning. Professionals with a deep understanding of these technologies, combined with medical knowledge, are essential to optimize CV models for specific healthcare applications. However, the scarcity of such interdisciplinary expertise is a bottleneck that can limit the effective development and integration of CV solutions in clinical workflows.

Susceptibility to Adversarial Attacks. CV models in healthcare are vulnerable to adversarial attacks, where intentional modifications to input images may deceive the model into making inaccurate predictions. These attacks expose a significant security risk, especially in safety-critical environments like healthcare. Ensuring robustness against adversarial manipulations is essential to build trust and ensure reliable performance under diverse conditions.

Extensive Training Data Requirements. CV models require large volumes of annotated data for effective training. For healthcare applications, creating these datasets involves meticulous tagging by experts to ensure model accuracy in detecting specific clinical features. The dependence on annotated data not only extends development timelines but also limits the scalability of CV systems, as obtaining high-quality annotated medical images is a substantial resource investment.

5 Conclusion

This chapter has explored the fundamental concepts and applications of CV in healthcare, highlighting its transformative impact on medical imaging, pathology, ophthalmology, monitoring, and rehabilitation. CV, powered by deep learning and other advanced techniques, has greatly enhanced the accuracy and efficiency of clinical diagnostics, personalized patient care, and therapeutic interventions. Through a range of applications—such as segmentation, detection, and classification—CV enables faster, more accurate analyses, empowering clinicians with data-driven insights and automated support in decision-making.

Despite these advances, challenges remain in implementing CV at scale within healthcare. Issues related to model generalization, privacy, and annotation requirements illustrate the complex ethical and technical considerations in deploying these tools in clinical environments. Moreover, the unique characteristics of healthcare data, including high variability, safety-critical implications, and privacy requirements, demand tailored solutions to meet clinical standards.

Ongoing research to address these limitations is crucial, as overcoming them will facilitate more robust, secure, and trustworthy CV systems. With interdisciplinary efforts spanning medical and technical domains, the future of CV in healthcare holds the promise of more accessible, accurate, and comprehensive patient care, ultimately improving healthcare delivery and patient outcomes worldwide. This chapter highlights the need for continuous innovation, and ethical diligence to realize the full potential of CV in revolutionizing healthcare.

Acknowledgements The authors acknowledge the use of GenAI for editing this chapter.

References

1. P. Viola and M. Jones. Rapid object detection using a boosted cascade of simple features. *Proc. IEEE Computer Society Conference Computer Vision Pattern Recognition (CVPR)*, 2001.
2. Bram Van Ginneken, Alejandro F Frangi, Joes J Staal, Bart M ter Haar Romeny, and Max A Viergever. Active shape model segmentation with optimal features. *IEEE transactions on medical imaging*, 21(8):924–933, 2002.
3. Mariano Cabezas, Arnau Oliver, Xavier Lladó, Jordi Freixenet, and Meritxell Bach Cuadra. A review of atlas-based segmentation for magnetic resonance brain images. *Computer methods and programs in biomedicine*, 104(3):e158–e177, 2011.
4. Buket D Barkana, Inci Saricicek, and Burak Yildirim. Performance analysis of descriptive statistical features in retinal vessel segmentation via fuzzy logic, ann, svm, and classifier fusion. *Knowledge-Based Systems*, 118:165–176, 2017.
5. Shervin Minaee, Yuri Boykov, Fatih Porikli, Antonio Plaza, Nasser Kehtarnavaz, and Demetri Terzopoulos. Image segmentation using deep learning: A survey. *IEEE transactions on pattern analysis and machine intelligence*, 44(7):3523–3542, 2021.
6. Alex Krizhevsky, Ilya Sutskever, and Geoffrey E Hinton. Imagenet classification with deep convolutional neural networks. *Advances in neural information processing systems*, 25, 2012.
7. A. Hatamizadeh, A. Hoogi, D. Sengupta, W. Lu, B. Wilcox, D. Rubin, and D. Terzopoulos. Deep active lesion segmentation. In *Proceedings of the International Workshop on Machine Learning in Medical Imaging*, volume 11861 of *Lecture Notes in Computer Science*, pages 98–105. Springer, 2019.
8. Abdul Mueed Hafiz and Ghulam Mohiuddin Bhat. A survey on instance segmentation: state of the art. *International journal of multimedia information retrieval*, 9(3):171–189, 2020.
9. Taraneh Ghandi, Hamidreza Pourreza, and Hamidreza Mahyar. Deep learning approaches on image captioning: A review. *ACM Computing Surveys*, 56(3):1–39, 2023.
10. Wenhan Luo, Junliang Xing, Anton Milan, Xiaoqin Zhang, Wei Liu, and Tae-Kyun Kim. Multiple object tracking: A literature review. *Artificial intelligence*, 293:103448, 2021.
11. N. Dalal and B. Triggs. Histograms of oriented gradients for human detection. *Proc. IEEE Computer Society Conference Computer Vision Pattern Recognition (CVPR)*, 2005.
12. P. F. Felzenszwalb, R. B. Girshick, and D. McAllester. Cascade object detection with deformable part models. In *Proceedings of the IEEE Computer Society Conference on Computer Vision and Pattern Recognition (CVPR)*, pages 2241–2248, Jun. 2010.
13. R. Girshick, J. Donahue, T. Darrell, and J. Malik. Rich feature hierarchies for accurate object detection and semantic segmentation. In *Proceedings of the IEEE Conference on Computer Vision and Pattern Recognition (CVPR)*, pages 580–587, Jun. 2014.
14. R. Girshick. Fast R-CNN. In *Proceedings of the IEEE International Conference on Computer Vision (ICCV)*, pages 1440–1448, Dec. 2015.
15. S. Ren, K. He, R. Girshick, and J. Sun. Faster R-CNN: Towards real-time object detection with region proposal networks. In *Proceedings of the Advances in Neural Information Processing Systems (NeurIPS)*, pages 91–99, 2015.
16. T.-Y. Lin, P. Dollar, R. Girshick, K. He, B. Hariharan, and S. Belongie. Feature pyramid networks for object detection. In *Proceedings of the IEEE Conference on Computer Vision and Pattern Recognition (CVPR)*, pages 2117–2125, Jul. 2017.

17. Tsung-Yi Lin, Michael Maire, Serge Belongie, James Hays, Pietro Perona, Deva Ramanan, Piotr Dollár, and C Lawrence Zitnick. Microsoft coco: Common objects in context. In *Computer Vision–ECCV 2014: 13th European Conference, Zurich, Switzerland, September 6-12, 2014, Proceedings, Part V 13*, pages 740–755. Springer, 2014.
18. J. Redmon, S. Divvala, R. Girshick, and A. Farhadi. You only look once: Unified, real-time object detection. In *Proceedings of the IEEE Conference on Computer Vision and Pattern Recognition (CVPR)*, pages 779–788, Jun. 2016.
19. Wei Liu, Dragomir Anguelov, Dumitru Erhan, Christian Szegedy, Scott Reed, Cheng-Yang Fu, and Alexander C Berg. Ssd: Single shot multibox detector. In *Computer Vision–ECCV 2016: 14th European Conference, Amsterdam, The Netherlands, October 11–14, 2016, Proceedings, Part I 14*, pages 21–37. Springer, 2016.
20. H. Law and J. Deng. Cornernet: Detecting objects as paired keypoints. In *Proceedings of the European Conference on Computer Vision (ECCV)*, pages 734–750, Sep. 2018.
21. K. Duan, S. Bai, L. Xie, H. Qi, Q. Huang, and Q. Tian. Centernet: Keypoint triplets for object detection. In *Proceedings of the IEEE/CVF International Conference on Computer Vision (ICCV)*, pages 6569–6578, Oct. 2019.
22. Y. Wang, V. C. Guizilini, T. Zhang, Y. Wang, H. Zhao, and J. Solomon. DETR3D: 3d object detection from multi-view images via 3d-to-2d queries. In *Proceedings of the Conference on Robot Learning (CoRL)*, pages 180–191, 2022.
23. M. Everingham, L. Van Gool, C. K. I. Williams, J. Winn, and A. Zisserman. The PASCAL visual object classes (VOC) challenge. *International Journal of Computer Vision*, 88(2):303–338, Jun. 2010.
24. Olga Russakovsky, Jia Deng, Hao Su, Jonathan Krause, Sanjeev Satheesh, Sean Ma, Zhiheng Huang, Andrej Karpathy, Aditya Khosla, Michael Bernstein, et al. Imagenet large scale visual recognition challenge. *International journal of computer vision*, 115:211–252, 2015.
25. Alina Kuznetsova, Hassan Rom, Neil Alldrin, Jasper Uijlings, Ivan Krasin, Jordi Pont-Tuset, Shahab Kamali, Stefan Popov, Matteo Malloci, Alexander Kolesnikov, et al. The open images dataset v4: Unified image classification, object detection, and visual relationship detection at scale. *International journal of computer vision*, 128(7):1956–1981, 2020.
26. H. Rezatofighi, N. Tsoi, J. Gwak, A. Sadeghian, I. Reid, and S. Savarese. Generalized intersection over union: A metric and a loss for bounding box regression. In *Proceedings of the IEEE/CVF Conference on Computer Vision and Pattern Recognition (CVPR)*, pages 658–666, Jun. 2019.
27. P. F. Felzenszwalb, R. B. Girshick, D. McAllester, and D. Ramanan. Object detection with discriminatively trained part-based models. *IEEE Transactions on Pattern Analysis and Machine Intelligence*, 32(9):1627–1645, Sep. 2010.
28. J. R. R. Uijlings, K. E. A. van de Sande, T. Gevers, and A. W. M. Smeulders. Selective search for object recognition. *International Journal of Computer Vision*, 104(2):154–171, Apr. 2013.
29. C. Chen, M.-Y. Liu, O. Tuzel, and J. Xiao. R-cnn for small object detection. In *Proceedings of the Asian Conference on Computer Vision (ACCV)*, pages 214–230, Cham, Switzerland, 2016. Springer.
30. Songtao Liu, Di Huang, et al. Receptive field block net for accurate and fast object detection. In *Proceedings of the European conference on computer vision (ECCV)*, pages 385–400, 2018.
31. S. Bell, C. L. Zitnick, K. Bala, and R. Girshick. Inside-outside net: Detecting objects in context with skip pooling and recurrent neural networks. In *Proceedings of the IEEE Conference on Computer Vision and Pattern Recognition (CVPR)*, pages 2874–2883, Jun. 2016.
32. Q. Chen, Z. Song, J. Dong, Z. Huang, Y. Hua, and S. Yan. Contextualizing object detection and classification. *IEEE Transactions on Pattern Analysis and Machine Intelligence*, 37(1):13–27, Jan. 2015.
33. J. Hosang, R. Benenson, P. Dollár, and B. Schiele. What makes for effective detection proposals? *IEEE Transactions on Pattern Analysis and Machine Intelligence*, 38(4):814–830, Sep. 2016.

34. G. Cheng, P. Zhou, and J. Han. RIFD-CNN: Rotation-invariant and fisher discriminative convolutional neural networks for object detection. In *Proceedings of the IEEE Conference on Computer Vision and Pattern Recognition (CVPR)*, pages 2884–2893, Jun. 2016.
35. S. Gidaris and N. Komodakis. Locnet: Improving localization accuracy for object detection. In *Proceedings of the IEEE Conference on Computer Vision and Pattern Recognition (CVPR)*, pages 789–798, Jun. 2016.
36. J. Hosang, R. Benenson, and B. Schiele. Learning non-maximum suppression. In *Proceedings of the IEEE Conference on Computer Vision and Pattern Recognition (CVPR)*, pages 6469–6477, Jul. 2017.
37. R. Rothe, M. Guillaumin, and L. Van Gool. Non-maximum suppression for object detection by passing messages between windows. In *Proceedings of the Asian Conference on Computer Vision (ACCV)*, pages 290–306, Cham, Switzerland, 2014. Springer.
38. C. Desai, D. Ramanan, and C. C. Fowlkes. Discriminative models for multi-class object layout. *International Journal of Computer Vision*, 95(1):1–12, Oct. 2011.
39. X. Zhou, D. Wang, and P. Krähenbühl. Objects as points, 2019.
40. S. Han, H. Mao, and W. J. Dally. Deep compression: Compressing deep neural networks with pruning, trained quantization and huffman coding, 2015.
41. Z. Li, C. Peng, G. Yu, X. Zhang, Y. Deng, and J. Sun. Light-head r-cnn: In defense of two-stage object detector, 2017.
42. X. Zhang, X. Zhou, M. Lin, and J. Sun. Shufflenet: An extremely efficient convolutional neural network for mobile devices, 2017.
43. F. Chollet. Xception: Deep learning with depthwise separable convolutions, 2016.
44. B. Wu, A. Wan, F. Iandola, P. H. Jin, and K. Keutzer. Squeezedet: Unified, small, low power fully convolutional neural networks for real-time object detection for autonomous driving. In *Proceedings of the IEEE Conference on Computer Vision and Pattern Recognition Workshops (CVPRW)*, pages 446–454, Jul. 2017.
45. P. Simard, L. Bottou, P. Haffner, and Y. LeCun. Boxlets: A fast convolution algorithm for signal processing and neural networks. In *Proceedings of the Advances in Neural Information Processing Systems (NeurIPS)*, pages 571–577, 1999.
46. M. Mathieu, M. Henaff, and Y. LeCun. Fast training of convolutional networks through ffts, 2013.
47. M. A. Sadeghi and D. Forsyth. Fast template evaluation with vector quantization. In *Proceedings of the Advances in Neural Information Processing Systems (NeurIPS)*, pages 2949–2957, 2013.
48. J. Long, E. Shelhamer, and T. Darrell. Fully convolutional networks for semantic segmentation. In *Proceedings of the IEEE Conference on Computer Vision and Pattern Recognition (CVPR)*, pages 3431–3440, 2015.
49. Nathan Silberman, Derek Hoiem, Pushmeet Kohli, and Rob Fergus. Indoor segmentation and support inference from rgbd images. In *European Conference on Computer Vision (ECCV)*, pages 746–760. Springer, 2012.
50. Ce Liu, Jenny Yuen, and Antonio Torralba. Sift flow: Dense correspondence across scenes and its applications. *IEEE transactions on pattern analysis and machine intelligence*, 33(5):978–994, 2010.
51. Gopal Sharma, Difan Liu, Subhransu Maji, Evangelos Kalogerakis, Siddhartha Chaudhuri, and Radomír Měch. Parsenet: A parametric surface fitting network for 3d point clouds. In *Computer Vision–ECCV 2020: 16th European Conference, Glasgow, UK, August 23–28, 2020, Proceedings, Part VII 16*, pages 261–276. Springer, 2020.
52. L.-C. Chen, G. Papandreou, I. Kokkinos, K. Murphy, and A. L. Yuille. Semantic image segmentation with deep convolutional nets and fully connected crfs, 2014.
53. A. G. Schwing and R. Urtasun. Fully connected deep structured networks, 2015.
54. S. Zheng, S. Jayasumana, B. Romera-Paredes, V. Vineet, Z. Su, D. Du, C. Huang, and P. H. Torr. Conditional random fields as recurrent neural networks. In *Proceedings of the IEEE International Conference on Computer Vision (ICCV)*, pages 1529–1537, 2015.

55. G. Lin, C. Shen, A. Van Den Hengel, and I. Reid. Efficient piecewise training of deep structured models for semantic segmentation. In *Proceedings of the IEEE Conference on Computer Vision and Pattern Recognition (CVPR)*, pages 3194–3203, 2016.
56. Z. Liu, X. Li, P. Luo, C.-C. Loy, and X. Tang. Semantic image segmentation via deep parsing network. In *Proceedings of the IEEE International Conference on Computer Vision (ICCV)*, pages 1377–1385, 2015.
57. H. Noh, S. Hong, and B. Han. Learning deconvolution network for semantic segmentation. In *Proceedings of the IEEE International Conference on Computer Vision (ICCV)*, pages 1520–1528, 2015.
58. V. Badrinarayanan, A. Kendall, and R. Cipolla. Segnet: A deep convolutional encoder-decoder architecture for image segmentation. *IEEE Transactions on Pattern Analysis and Machine Intelligence*, 39(12):2481–2495, 2017.
59. Y. Yuan, X. Chen, and J. Wang. Object-contextual representations for semantic segmentation, 2019.
60. J. Fu, J. Liu, Y. Wang, J. Zhou, C. Wang, and H. Lu. Stacked deconvolutional network for semantic segmentation. *IEEE Transactions on Image Processing*, 2019.
61. A. Chaurasia and E. Culurciello. Linknet: Exploiting encoder representations for efficient semantic segmentation. In *2017 IEEE Visual Communications and Image Processing (VCIP)*, pages 1–4. IEEE, 2017.
62. Olaf Ronneberger, Philipp Fischer, and Thomas Brox. U-net: Convolutional networks for biomedical image segmentation. In *Medical image computing and computer-assisted intervention–MICCAI 2015: 18th international conference, Munich, Germany, October 5-9, 2015, proceedings, part III 18*, pages 234–241. Springer, 2015.
63. Ö. Çiçek, A. Abdulkadir, S. S. Lienkamp, T. Brox, and O. Ronneberger. 3d u-net: Learning dense volumetric segmentation from sparse annotation. In *International Conference on Medical Image Computing and Computer-Assisted Intervention (MICCAI)*, pages 424–432. Springer, 2016.
64. Z. Zhou, M. M. R. Siddiquee, N. Tajbakhsh, and J. Liang. UNet++: A nested u-net architecture for medical image segmentation. In *Deep Learning in Medical Image Analysis and Multimodal Learning for Clinical Decision Support*, pages 3–11. Springer, 2018.
65. F. Milletari, N. Navab, and S.-A. Ahmadi. V-net: Fully convolutional neural networks for volumetric medical image segmentation. In *2016 Fourth International Conference on 3D Vision (3DV)*, pages 565–571. IEEE, 2016.
66. H. Zhao, J. Shi, X. Qi, X. Wang, and J. Jia. Pyramid scene parsing network. In *Proceedings of the IEEE Conference on Computer Vision and Pattern Recognition (CVPR)*, pages 2881–2890, 2017.
67. G. Ghiasi and C. C. Fowlkes. Laplacian pyramid reconstruction and refinement for semantic segmentation. In *European Conference on Computer Vision (ECCV)*, pages 519–534. Springer, 2016.
68. J. He, Z. Deng, and Y. Qiao. Dynamic multi-scale filters for semantic segmentation. In *Proceedings of the IEEE International Conference on Computer Vision (ICCV)*, pages 3562–3572, 2019.
69. H. Ding, X. Jiang, B. Shuai, A. Qun Liu, and G. Wang. Context contrasted feature and gated multi-scale aggregation for scene segmentation. In *Proceedings of the IEEE Conference on Computer Vision and Pattern Recognition (CVPR)*, pages 2393–2402, 2018.
70. J. He, Z. Deng, L. Zhou, Y. Wang, and Y. Qiao. Adaptive pyramid context network for semantic segmentation. In *Proceedings of the IEEE Conference on Computer Vision and Pattern Recognition (CVPR)*, pages 7519–7528, 2019.
71. K. He, G. Gkioxari, P. Dollár, and R. Girshick. Mask R-CNN. In *Proceedings of the IEEE International Conference on Computer Vision (ICCV)*, pages 2961–2969, 2017.
72. S. Liu, L. Qi, H. Qin, J. Shi, and J. Jia. Path aggregation network for instance segmentation. In *Proceedings of the IEEE Conference on Computer Vision and Pattern Recognition (CVPR)*, pages 8759–8768, 2018.

73. L.-C. Chen, A. Hermans, G. Papandreou, F. Schroff, P. Wang, and H. Adam. Masklab: Instance segmentation by refining object detection with semantic and direction features. In *Proceedings of the IEEE Conference on Computer Vision and Pattern Recognition (CVPR)*, pages 4013–4022, 2018.
74. X. Chen, R. Girshick, K. He, and P. Dollár. Tensormask: A foundation for dense object segmentation, 2019.
75. P. O. Pinheiro, R. Collobert, and P. Dollár. Learning to segment object candidates. In *Advances in Neural Information Processing Systems (NeurIPS)*, pages 1990–1998, 2015.
76. E. Xie, P. Sun, X. Song, W. Wang, X. Liu, D. Liang, C. Shen, and P. Luo. Polarmask: Single shot instance segmentation with polar representation, 2019.
77. L.-C. Chen, G. Papandreou, I. Kokkinos, K. Murphy, and A. L. Yuille. Deeplab: Semantic image segmentation with deep convolutional nets, atrous convolution, and fully connected crfs. *IEEE Transactions on Pattern Analysis and Machine Intelligence*, 40(4):834–848, 2017.
78. M. Yang, K. Yu, C. Zhang, Z. Li, and K. Yang. Denseaspp for semantic segmentation in street scenes. In *Proceedings of the IEEE Conference on Computer Vision and Pattern Recognition (CVPR)*, pages 3684–3692, 2018.
79. M. Cordts, M. Omran, S. Ramos, T. Rehfeld, M. Enzweiler, R. Benenson, U. Franke, S. Roth, and B. Schiele. The cityscapes dataset for semantic urban scene understanding. In *Proceedings of the IEEE Conference on Computer Vision and Pattern Recognition (CVPR)*, pages 3213–3223, 2016.
80. L.-C. Chen, G. Papandreou, F. Schroff, and H. Adam. Rethinking atrous convolution for semantic image segmentation, 2017.
81. L.-C. Chen, Y. Zhu, G. Papandreou, F. Schroff, and H. Adam. Encoder-decoder with atrous separable convolution for semantic image segmentation. In *Proceedings of the European Conference on Computer Vision (ECCV)*, pages 801–818, 2018.
82. F. Visin, M. Ciccone, A. Romero, K. Kastner, K. Cho, Y. Bengio, M. Matteucci, and A. Courville. Reseg: A recurrent neural network-based model for semantic segmentation. In *Proceedings of the IEEE Conference on Computer Vision and Pattern Recognition Workshops*, pages 41–48, 2016.
83. W. Byeon, T. M. Breuel, F. Raue, and M. Liwicki. Scene labeling with lstm recurrent neural networks. In *Proceedings of the IEEE Conference on Computer Vision and Pattern Recognition (CVPR)*, pages 3547–3555, 2015.
84. X. Liang, X. Shen, J. Feng, L. Lin, and S. Yan. Semantic object parsing with graph lstm. In *European Conference on Computer Vision (ECCV)*, pages 125–143. Springer, 2016.
85. Y. Xiang and D. Fox. Da-rnn: Semantic mapping with data associated recurrent neural networks, 2017.
86. R. Hu, M. Rohrbach, and T. Darrell. Segmentation from natural language expressions. In *European Conference on Computer Vision (ECCV)*, pages 108–124. Springer, 2016.
87. L.-C. Chen, Y. Yang, J. Wang, W. Xu, and A. L. Yuille. Attention to scale: Scale-aware semantic image segmentation. In *Proceedings of the IEEE Conference on Computer Vision and Pattern Recognition (CVPR)*, pages 3640–3649, 2016.
88. Q. Huang, C. Xia, C. Wu, S. Li, Y. Wang, Y. Song, and C.-C. J. Kuo. Semantic segmentation with reverse attention, 2017.
89. H. Li, P. Xiong, J. An, and L. Wang. Pyramid attention network for semantic segmentation, 2018.
90. J. Fu, J. Liu, H. Tian, Y. Li, Y. Bao, Z. Fang, and H. Lu. Dual attention network for scene segmentation. In *Proceedings of the IEEE Conference on Computer Vision and Pattern Recognition (CVPR)*, pages 3146–3154, 2019.
91. Y. Yuan and J. Wang. Ocnet: Object context network for scene parsing, 2018.
92. X. Li, Z. Zhong, J. Wu, Y. Yang, Z. Lin, and H. Liu. Expectation-maximization attention networks for semantic segmentation. In *Proceedings of the IEEE International Conference on Computer Vision (ICCV)*, pages 9167–9176, 2019.
93. Z. Huang, X. Wang, L. Huang, C. Huang, Y. Wei, and W. Liu. Ccnet: Criss-cross attention for semantic segmentation. In *Proceedings of the IEEE International Conference on Computer Vision (ICCV)*, pages 603–612, 2019.

94. C. Yu, J. Wang, C. Peng, C. Gao, G. Yu, and N. Sang. Learning a discriminative feature network for semantic segmentation. In *Proceedings of the IEEE Conference on Computer Vision and Pattern Recognition (CVPR)*, pages 1857–1866, 2018.
95. P. Luc, C. Couprie, S. Chintala, and J. Verbeek. Semantic segmentation using adversarial networks, 2016.
96. N. Souly, C. Spampinato, and M. Shah. Semi-supervised semantic segmentation using generative adversarial network. In *Proceedings of the IEEE International Conference on Computer Vision (ICCV)*, pages 5688–5696, 2017.
97. W.-C. Hung, Y.-H. Tsai, Y.-T. Liou, Y.-Y. Lin, and M.-H. Yang. Adversarial learning for semi-supervised semantic segmentation, 2018.
98. M. Kass, A. Witkin, and D. Terzopoulos. Snakes: Active contour models. *International Journal of Computer Vision*, 1(4):321–331, 1988.
99. X. Chen, B. M. Williams, S. R. Vallabhaneni, G. Czanner, R. Williams, and Y. Zheng. Learning active contour models for medical image segmentation. In *Proceedings of the IEEE Conference on Computer Vision and Pattern Recognition (CVPR)*, pages 11632–11640, 2019.
100. T. H. N. Le, K. G. Quach, K. Luu, C. N. Duong, and M. Savvides. Reformulating level sets as deep recurrent neural network approach to semantic segmentation. *IEEE Transactions on Image Processing*, 27(5):2393–2407, 2018.
101. C. Rupprecht, E. Huaroc, M. Baust, and N. Navab. Deep active contours, 2016.
102. D. Marcos, D. Tuia, B. Kellenberger, L. Zhang, M. Bai, R. Liao, and R. Urtasun. Learning deep structured active contours end-to-end. In *Proceedings of the IEEE Conference on Computer Vision and Pattern Recognition (CVPR)*, pages 8877–8885, 2018.
103. D. Cheng, R. Liao, S. Fidler, and R. Urtasun. Darnet: Deep active ray network for building segmentation. In *Proceedings of the IEEE Conference on Computer Vision and Pattern Recognition (CVPR)*, pages 7431–7439, 2019.
104. B. Zhou, H. Zhao, X. Puig, S. Fidler, A. Barriuso, and A. Torralba. Scene parsing through ade20k dataset. In *Proceedings of the IEEE Conference on Computer Vision and Pattern Recognition (CVPR)*, 2017.
105. S. Gould, R. Fulton, and D. Koller. Decomposing a scene into geometric and semantically consistent regions. In *2009 IEEE 12th International Conference on Computer Vision (ICCV)*, pages 1–8. IEEE, 2009.
106. D. Martin, C. Fowlkes, D. Tal, and J. Malik. A database of human segmented natural images and its application to evaluating segmentation algorithms and measuring ecological statistics. In *Proceedings of the 8th International Conference on Computer Vision (ICCV)*, volume 2, pages 416–423, July 2001.
107. A. Prest, C. Leistner, J. Civera, C. Schmid, and V. Ferrari. Learning object class detectors from weakly annotated video. In *2012 IEEE Conference on Computer Vision and Pattern Recognition (CVPR)*, pages 3282–3289. IEEE, 2012.
108. A. Geiger, P. Lenz, C. Stiller, and R. Urtasun. Vision meets robotics: The kitti dataset. *The International Journal of Robotics Research*, 32(11):1231–1237, 2013.
109. S. Song, S. P. Lichtenberg, and J. Xiao. Sun rgb-d: A rgb-d scene understanding benchmark suite. In *Proceedings of the IEEE Conference on Computer Vision and Pattern Recognition (CVPR)*, pages 567–576, 2015.
110. J. Xiao, A. Owens, and A. Torralba. Sun3d: A database of big spaces reconstructed using sfm and object labels. In *Proceedings of the IEEE International Conference on Computer Vision (ICCV)*, pages 1625–1632, 2013.
111. S. Song, S. P. Lichtenberg, and J. Xiao. Sun rgb-d: A rgb-d scene understanding benchmark suite. In *Proceedings of the IEEE Conference on Computer Vision and Pattern Recognition (CVPR)*, pages 567–576, 2015.
112. K. Lai, L. Bo, X. Ren, and D. Fox. A large-scale hierarchical multi-view rgb-d object dataset. In *2011 IEEE International Conference on Robotics and Automation*, pages 1817–1824. IEEE, 2011.
113. A. Dai, A. X. Chang, M. Savva, M. Halber, T. Funkhouser, and M. Nießner. Scannet: Richly-annotated 3d reconstructions of indoor scenes. In *Proceedings of the IEEE Conference on Computer Vision and Pattern Recognition (CVPR)*, pages 5828–5839, 2017.

114. I. Armeni, A. Sax, A. R. Zamir, and S. Savarese. Joint 2d-3d semantic data for indoor scene understanding, Feb. 2017. ArXiv e-prints.
115. A. X. Chang, T. Funkhouser, L. Guibas, P. Hanrahan, Q. Huang, Z. Li, S. Savarese, M. Savva, S. Song, and H. Su et al. Shapenet: An information-rich 3d model repository, 2015.
116. M. De Deuge, A. Quadros, C. Hung, and B. Douillard. Unsupervised feature learning for classification of outdoor 3d scans. In *Proceedings of the Australasian Conference on Robotics and Automation*, volume 2, page 1, 2013.
117. Kaiming He, Xiangyu Zhang, Shaoqing Ren, and Jian Sun. Deep residual learning for image recognition. In *Proceedings of the IEEE conference on computer vision and pattern recognition*, pages 770–778, 2016.
118. Jia Deng, Wei Dong, Richard Socher, Li-Jia Li, Kai Li, and Li Fei-Fei. Imagenet: A large-scale hierarchical image database. In *2009 IEEE conference on computer vision and pattern recognition*, pages 248–255. Ieee, 2009.
119. Alex Krizhevsky, Geoffrey Hinton, et al. Learning multiple layers of features from tiny images. 2009.
120. Han Xiao, Kashif Rasul, and Roland Vollgraf. Fashion-mnist: a novel image dataset for benchmarking machine learning algorithms. *arXiv preprint* arXiv:1708.07747, 2017.
121. Jonathan Krause, Michael Stark, Jia Deng, and Li Fei-Fei. 3d object representations for fine-grained categorization. In *2013 IEEE International Conference on Computer Vision Workshops*, pages 554–561, 2013.
122. Maria-Elena Nilsback and Andrew Zisserman. Automated flower classification over a large number of classes. In *2008 Sixth Indian conference on computer vision, graphics & image processing*, pages 722–729. IEEE, 2008.
123. Hugh Durrant-Whyte and Tim Bailey. Simultaneous localization and mapping: part i. *IEEE robotics & automation magazine*, 13(2):99–110, 2006.
124. F. Zhang, V. Prisacariu, R. Yang, and P. H. Torr. Ga-net: Guided aggregation net for end-to-end stereo matching. In *2019 IEEE/CVF Conference on Computer Vision and Pattern Recognition (CVPR)*, pages 185–194. IEEE, 2019.
125. L. Lipson, Z. Teed, and J. Deng. Raft-stereo: Multilevel recurrent field transforms for stereo matching. In *2021 International Conference on 3D Vision (3DV)*, pages 218–227, 2021.
126. M. Visentini-Scarzanella, T. Sugiura, T. Kaneko, and S. Koto. Deep monocular 3d reconstruction for assisted navigation in bronchoscopy. *International Journal of Computer Assisted Radiology and Surgery*, 12:1089–1099, 2017.
127. J. L. Schonberger and J. M. Frahm. Structure-from-motion revisited. In *Proceedings of the IEEE Conference on Computer Vision and Pattern Recognition (CVPR)*, pages 4104–4113, 2016.
128. A. Sengupta and A. Bartoli. Colonoscopic 3d reconstruction by tubular non-rigid structure-from-motion. *International Journal of Computer Assisted Radiology and Surgery*, 16:1237–1241, 2021.
129. A. Malti, A. Bartoli, and T. Collins. Template-based conformal shape-from-motion-and-shading for laparoscopy. In *Lecture Notes in Computer Science (LNCS)*, volume 7330, 2012.
130. R. Mur-Artal, J. M. M. Montiel, and J. D. Tardos. Orb-slam: A versatile and accurate monocular slam system. *IEEE Transactions on Robotics*, 31:1147–1163, 2015.
131. P. Mountney, D. Stoyanov, A. Davison, and G. Z. Yang. Simultaneous stereoscope localization and soft-tissue mapping for minimal invasive surgery. In *Lecture Notes in Computer Science (LNCS)*, volume 4190, 2006.
132. J. Lamarca, S. Parashar, A. Bartoli, and J. Montiel. Defslam: Tracking and mapping of deforming scenes from monocular sequences. *IEEE Transactions on Robotics*, 37:291–303, 2021.
133. Y. Long, Z. Li, C. Yee, C. Ng, R. H. Taylor, M. Unberath, and Q. Dou. E-dssr: Efficient dynamic surgical scene reconstruction with transformer-based stereoscopic depth perception. In *Proceedings of the Medical Image Computing and Computer-Assisted Intervention (MICCAI)*, 2021.
134. Y. Li, F. Richter, J. Lu, E. Funk, R. Orosco, J. Zhu, and M. Yip. Super: A surgical perception framework for endoscopic tissue manipulation with surgical robotics. *IEEE Robotics and Automation Letters*, 5:2294–2301, 2020.

135. Luigi Nardi, Bruno Bodin, M Zeeshan Zia, John Mawer, Andy Nisbet, Paul HJ Kelly, Andrew J Davison, Mikel Luján, Michael FP O'Boyle, Graham Riley, et al. Introducing slambench, a performance and accuracy benchmarking methodology for slam. In *2015 IEEE international conference on robotics and automation (ICRA)*, pages 5783–5790. IEEE, 2015.
136. Francesco Ventura and Tania Cerquitelli. What's in the box? explaining the black-box model through an evaluation of its interpretable features. *arXiv preprint* arXiv:1908.04348, 2019.
137. Drew Linsley, Dan Shiebler, Sven Eberhardt, and Thomas Serre. Learning what and where to attend. *arXiv preprint* arXiv:1805.08819, 2018.
138. Karen Simonyan. Deep inside convolutional networks: Visualising image classification models and saliency maps. *arXiv preprint* arXiv:1312.6034, 2013.
139. Avanti Shrikumar, Peyton Greenside, and Anshul Kundaje. Learning important features through propagating activation differences. In *International conference on machine learning*, pages 3145–3153. PMIR, 2017.
140. Bolei Zhou, Aditya Khosla, Agata Lapedriza, Aude Oliva, and Antonio Torralba. Learning deep features for discriminative localization. In *Proceedings of the IEEE conference on computer vision and pattern recognition*, pages 2921–2929, 2016.
141. Mohammed Bany Muhammad and Mohammed Yeasin. Eigen-cam: Visual explanations for deep convolutional neural networks. *SN Computer Science*, 2(1):47, 2021.
142. Aditya Chattopadhay, Anirban Sarkar, Prantik Howlader, and Vineeth N Balasubramanian. Grad-cam++: Generalized gradient-based visual explanations for deep convolutional networks. In *2018 IEEE winter conference on applications of computer vision (WACV)*, pages 839–847. IEEE, 2018.
143. Haofan Wang, Zifan Wang, Mengnan Du, Fan Yang, Zijian Zhang, Sirui Ding, Piotr Mardziel, and Xia Hu. Score-cam: Score-weighted visual explanations for convolutional neural networks. In *Proceedings of the IEEE/CVF conference on computer vision and pattern recognition workshops*, pages 24–25, 2020.
144. Sebastian Bach, Alexander Binder, Grégoire Montavon, Frederick Klauschen, Klaus-Robert Müller, and Wojciech Samek. On pixel-wise explanations for non-linear classifier decisions by layer-wise relevance propagation. *PloS one*, 10(7):e0130140, 2015.
145. Mukund Sundararajan, Ankur Taly, and Qiqi Yan. Axiomatic attribution for deep networks. In *International conference on machine learning*, pages 3319–3328. PMLR, 2017.
146. Fei Wang, Mengqing Jiang, Chen Qian, Shuo Yang, Cheng Li, Honggang Zhang, Xiaogang Wang, and Xiaoou Tang. Residual attention network for image classification. In *Proceedings of the IEEE conference on computer vision and pattern recognition*, pages 3156–3164, 2017.
147. Xiaoshuang Shi, Fuyong Xing, Kaidi Xu, Pingjun Chen, Yun Liang, Zhiyong Lu, and Zhenhua Guo. Loss-based attention for interpreting image-level prediction of convolutional neural networks. *IEEE Transactions on Image Processing*, 30:1662–1675, 2020.
148. Reza Abbasi-Asl and Bin Yu. Structural compression of convolutional neural networks. *arXiv preprint* arXiv:1705.07356, 2017.
149. Yulong Wang, Hang Su, Bo Zhang, and Xiaolin Hu. Interpret neural networks by extracting critical subnetworks. *IEEE Transactions on Image Processing*, 29:6707–6720, 2020.
150. Quanshi Zhang, Yu Yang, Haotian Ma, and Ying Nian Wu. Interpreting cnns via decision trees. In *Proceedings of the IEEE/CVF conference on computer vision and pattern recognition*, pages 6261–6270, 2019.
151. Joseph Antony, Kevin McGuinness, Noel E O'Connor, and Kieran Moran. Quantifying radiographic knee osteoarthritis severity using deep convolutional neural networks. In *2016 23rd international conference on pattern recognition (ICPR)*, pages 1195–1200. IEEE, 2016.
152. Andre Esteva, Brett Kuprel, Roberto A Novoa, Justin Ko, Susan M Swetter, Helen M Blau, and Sebastian Thrun. Dermatologist-level classification of skin cancer with deep neural networks. *nature*, 542(7639):115–118, 2017.
153. Jeremy Kawahara and Ghassan Hamarneh. Multi-resolution-tract cnn with hybrid pretrained and skin-lesion trained layers. In *International workshop on machine learning in medical imaging*, pages 164–171. Springer, 2016.

154. Dong Yang, Shaoting Zhang, Zhennan Yan, Chaowei Tan, Kang Li, and Dimitris Metaxas. Automated anatomical landmark detection ondistal femur surface using convolutional neural network. In *2015 IEEE 12th international symposium on biomedical imaging (ISBI)*, pages 17–21. IEEE, 2015.
155. Mark JJP Van Grinsven, Bram van Ginneken, Carel B Hoyng, Thomas Theelen, and Clara I Sánchez. Fast convolutional neural network training using selective data sampling: Application to hemorrhage detection in color fundus images. *IEEE transactions on medical imaging*, 35(5):1273–1284, 2016.
156. Nassir Navab, Joachim Hornegger, William M Wells, and Alejandro Frangi. *Medical Image Computing and Computer-Assisted Intervention–MICCAI 2015: 18th International Conference, Munich, Germany, October 5-9, 2015, Proceedings, Part III*, volume 9351. Springer, 2015.
157. Tom Brosch, Lisa YW Tang, Youngjin Yoo, David KB Li, Anthony Traboulsee, and Roger Tam. Deep 3d convolutional encoder networks with shortcuts for multiscale feature integration applied to multiple sclerosis lesion segmentation. *IEEE transactions on medical imaging*, 35(5):1229–1239, 2016.
158. Xiao Yang, Roland Kwitt, and Marc Niethammer. Fast predictive image registration. In *Deep Learning and Data Labeling for Medical Applications: First International Workshop, LABELS 2016, and Second International Workshop, DLMIA 2016, Held in Conjunction with MICCAI 2016, Athens, Greece, October 21, 2016, Proceedings 1*, pages 48–57. Springer, 2016.
159. Thomas Schlegl, Sebastian M Waldstein, Wolf-Dieter Vogl, Ursula Schmidt-Erfurth, and Georg Langs. Predicting semantic descriptions from medical images with convolutional neural networks. In *International Conference on Information Processing in Medical Imaging*, pages 437–448. Springer, 2015.
160. Adnan Mujahid Khan, Nasir Rajpoot, Darren Treanor, and Derek Magee. A nonlinear mapping approach to stain normalization in digital histopathology images using image-specific color deconvolution. *IEEE transactions on Biomedical Engineering*, 61(6):1729–1738, 2014.
161. Aïcha BenTaieb and Ghassan Hamarneh. Adversarial stain transfer for histopathology image analysis. *IEEE transactions on medical imaging*, 37(3):792–802, 2017.
162. Dan Ciresan, Alessandro Giusti, Luca Gambardella, and Jürgen Schmidhuber. Deep neural networks segment neuronal membranes in electron microscopy images. *Advances in neural information processing systems*, 25, 2012.
163. Youyi Song, Ee-Leng Tan, Xudong Jiang, Jie-Zhi Cheng, Dong Ni, Siping Chen, Baiying Lei, and Tianfu Wang. Accurate cervical cell segmentation from overlapping clumps in pap smear images. *IEEE transactions on medical imaging*, 36(1):288–300, 2016.
164. Mitko Veta, Paul J Van Diest, and Josien PW Pluim. Cutting out the middleman: measuring nuclear area in histopathology slides without segmentation. In *Medical Image Computing and Computer-Assisted Intervention–MICCAI 2016: 19th International Conference, Athens, Greece, October 17-21, 2016, Proceedings, Part II 19*, pages 632–639. Springer, 2016.
165. Mitko Veta, Paul J van Diest, and Josien PW Pluim. Detecting mitotic figures in breast cancer histopathology images. In *Medical Imaging 2013: Digital Pathology*, volume 8676, pages 70–76. SPIE, 2013.
166. Shadi Albarqouni, Christoph Baur, Felix Achilles, Vasileios Belagiannis, Stefanie Demirci, and Nassir Navab. Aggnet: deep learning from crowds for mitosis detection in breast cancer histology images. *IEEE transactions on medical imaging*, 35(5):1313–1321, 2016.
167. Hao Chen, Xiaojuan Qi, Lequan Yu, Qi Dou, Jing Qin, and Pheng-Ann Heng. Dcan: Deep contour-aware networks for object instance segmentation from histology images. *Medical image analysis*, 36:135–146, 2017.
168. Saining Xie and Zhuowen Tu. Holistically-nested edge detection. In *Proceedings of the IEEE international conference on computer vision*, pages 1395–1403, 2015.
169. Yan Xu, Tao Mo, Qiwei Feng, Peilin Zhong, Maode Lai, I Eric, and Chao Chang. Deep learning of feature representation with multiple instance learning for medical image analysis. In *2014 IEEE international conference on acoustics, speech and signal processing (ICASSP)*, pages 1626–1630. IEEE, 2014.

170. Nathan Ing, Zhaoxuan Ma, Jiayun Li, Hootan Salemi, Corey Arnold, Beatrice S Knudsen, and Arkadiusz Gertych. Semantic segmentation for prostate cancer grading by convolutional neural networks. In *Medical Imaging 2018: Digital Pathology*, volume 10581, pages 343–355. SPIE, 2018.
171. Kunal Nagpal, Davis Foote, Yun Liu, Po-Hsuan Cameron Chen, Ellery Wulczyn, Fraser Tan, Niels Olson, Jenny L Smith, Arash Mohtashamian, James H Wren, et al. Development and validation of a deep learning algorithm for improving gleason scoring of prostate cancer. *NPJ digital medicine*, 2(1):48, 2019.
172. Hidetaka Uramoto and Fumihiro Tanaka. Recurrence after surgery in patients with nsclc. *Translational lung cancer research*, 3(4):242, 2014.
173. Muhammad Zeeshan Tahir, Muhammad Nasir, and Sanyuan Zhang. Advances in retinal microaneurysms detection, segmentation and datasets for the diagnosis of diabetic retinopathy: a systematic literature review. *Multimedia Tools and Applications*, pages 1–39, 2024.
174. Qaisar Abbas, Imran Qureshi, Junhua Yan, and Kashif Shaheed. Machine learning methods for diagnosis of eye-related diseases: a systematic review study based on ophthalmic imaging modalities. *Archives of Computational Methods in Engineering*, 29(6):3861–3918, 2022.
175. Mark Christopher, Christopher Bowd, Akram Belghith, Michael H Goldbaum, Robert N Weinreb, Massimo A Fazio, Christopher A Girkin, Jeffrey M Liebmann, and Linda M Zangwill. Deep learning approaches predict glaucomatous visual field damage from oct optic nerve head en face images and retinal nerve fiber layer thickness maps. *Ophthalmology*, 127(3):346–356, 2020.
176. Hongyan Zhang, Kai Niu, Yanmin Xiong, Weihua Yang, ZhiQiang He, and Hongxin Song. Automatic cataract grading methods based on deep learning. *Computer methods and programs in biomedicine*, 182:104978, 2019.
177. Tiarnan DL Keenan, Qingyu Chen, Elvira Agrón, Yih-Chung Tham, Jocelyn Hui Lin Goh, Xiaofeng Lei, Yi Pin Ng, Yong Liu, Xinxing Xu, Ching-Yu Cheng, et al. Deeplensnet: deep learning automated diagnosis and quantitative classification of cataract type and severity. *Ophthalmology*, 129(5):571–584, 2022.
178. Victor IJ Strijbis, Christiaan M de Bloeme, Robin W Jansen, Hamza Kebiri, Huu-Giao Nguyen, Marcus C de Jong, Annette C Moll, Merixtell Bach-Cuadra, Pim de Graaf, and Martijn D Steenwijk. Multi-view convolutional neural networks for automated ocular structure and tumor segmentation in retinoblastoma. *Scientific reports*, 11(1):14590, 2021.
179. Mayank Goswami. Deep learning models for benign and malign ocular tumor growth estimation. *Computerized Medical Imaging and Graphics*, 93:101986, 2021.
180. Yan Zhu, Rebecca Salowe, Caven Chow, Shuo Li, Osbert Bastani, and Joan M O'Brien. Advancing glaucoma care: integrating artificial intelligence in diagnosis, management, and progression detection. *Bioengineering*, 11(2):122, 2024.
181. Yongli Xu, Run Sun, Man Hu, and Hui Zeng. A dual-modal fusion network using optical coherence tomography and fundus images in detection of glaucomatous optic neuropathy. *Current Eye Research*, pages 1–7, 2024.
182. Sanjeev Kumar Saini and Rashmi Gupta. Artificial intelligence methods for analysis of electrocardiogram signals for cardiac abnormalities: state-of-the-art and future challenges. *Artificial Intelligence Review*, 55(2):1519–1565, 2022.
183. Arvind Gautam, Madhuri Panwar, Dwaipayan Biswas, and Amit Acharyya. Myonet: A transfer-learning-based lrcn for lower limb movement recognition and knee joint angle prediction for remote monitoring of rehabilitation progress from semg. *IEEE journal of translational engineering in health and medicine*, 8:1–10, 2020.
184. Omar Mubin, Fady Alnajjar, Nalini Jishtu, Belal Alsinglawi, and Abdullah Al Mahmud. Exoskeletons with virtual reality, augmented reality, and gamification for stroke patients' rehabilitation: systematic review. *JMIR rehabilitation and assistive technologies*, 6(2):e12010, 2019.
185. Wireko Andrew Awuah, Arjun Ahluwalia, Kwadwo Darko, Vivek Sanker, Joecelyn Kirani Tan, Tenkorang Ohenewaa Pearl, Adam Ben-Jaafar, Sruthi Ranganathan, Nicholas Aderinto, Aashna Mehta, et al. Bridging minds and machines: the recent advances of brain-computer interfaces in neurological and neurosurgical applications. *World Neurosurgery*, 2024.

186. Faisal Amin, Asim Waris, Javaid Iqbal, Syed Omer Gilani, M Zia Ur Rehman, Shafaq Mushtaq, Niaz Bahadur Khan, M Ijaz Khan, Mohammed Jameel, and Nissren Tamam. Maximizing stroke recovery with advanced technologies: A comprehensive review of robot-assisted, emg-controlled robotics, virtual reality, and mirror therapy interventions. *Results in Engineering*, page 101725, 2023.

TinyML: Applications, Algorithms, Co-design and Implementations

Omar Ghazal, Tousif Rahman, and Rishad Shafik

Big things have small beginnings
—Lawrence of Arabia (1962)

1 The TinyML Ecosystem

The rise of Internet-of-Things (IoT) based sensors and wearable devices has dramatically increased the ability to collect data at any time. Machine Learning (ML) provides a generalized processing method for this data to develop classification or prediction applications. This has had a profound impact in areas such as personalized healthcare, smart home monitoring, and predictive maintenance (Fig. 1).

However, one of the caveats to the successful development of the aforementioned applications is translating the ML components into the IoT sensors and wearable devices themselves. The advantages of doing this are significant: The ML application can process potentially user-sensitive data in situ instead of transferring to a more powerful computing device through the network. This in situ approach is typically faster and more energy efficient and is known as ML on the *edge* (on edge devices). It can alleviate both network security concerns and bandwidth contention.

O. Ghazal (✉) · T. Rahman · R. Shafik
Newcastle University, Newcastle upon Tyne (GB), Tyne, United Kingdom
e-mail: o.g.g.awf2@newcastle.ac.uk

T. Rahman
e-mail: s.rahman@newcsatle.ac.uk

R. Shafik
e-mail: rishad.shafik@newcastle.ac.uk

F. Firouzi et al. (eds.), *Smart and Connected Health: AI, IoT, and Trustworthy Technologies*, https://doi.org/10.1007/978-3-032-06286-4_6

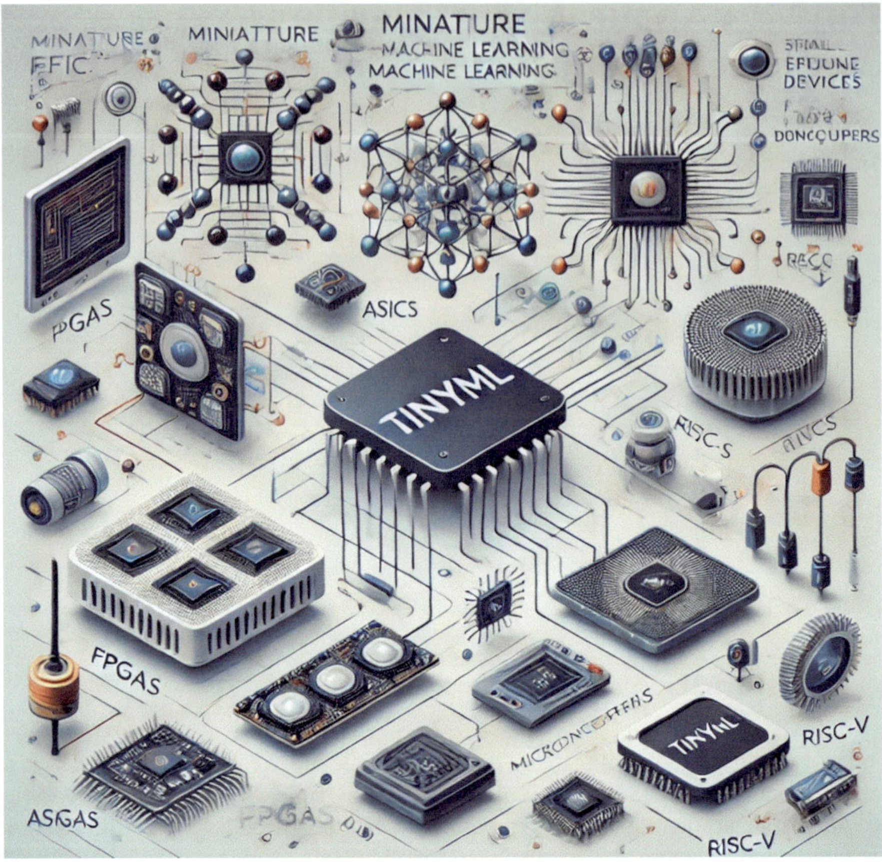

Fig. 1 A DALL-E 3 generated image for TinyML when given the prompt: "*Generate a picture of TinyML suitable as the introduction page of a book chapter*". It also produces the following accompanying caption: "This image visually symbolizes the integration of machine learning into small, resource-constrained devices like microcontrollers and smart sensors"

The disadvantage of keeping the application within the sensor or the wearable itself comes from the resource limitations of these devices. This means that the ML model must fit into the edge device memory, the computation time must meet the application's latency requirement, and the model must provide adequate performance in its classification/prediction task. The trade-off between balancing the effectiveness of the ML application and whether it can be incorporated into the edge sensor or wearable is the principle crux of *Tiny Machine Learning* (*TinyML*). TinyML aims to bring intelligence to IoT devices, wearables, and other edge endpoints, enabling them to perform tasks autonomously without relying on cloud connectivity [1, 2].

The significance of TinyML in modern technology cannot be overstated. It offers a platform where machine learning models can be pushed to user devices, enhancing user experience across diverse applications while addressing concerns such as energy

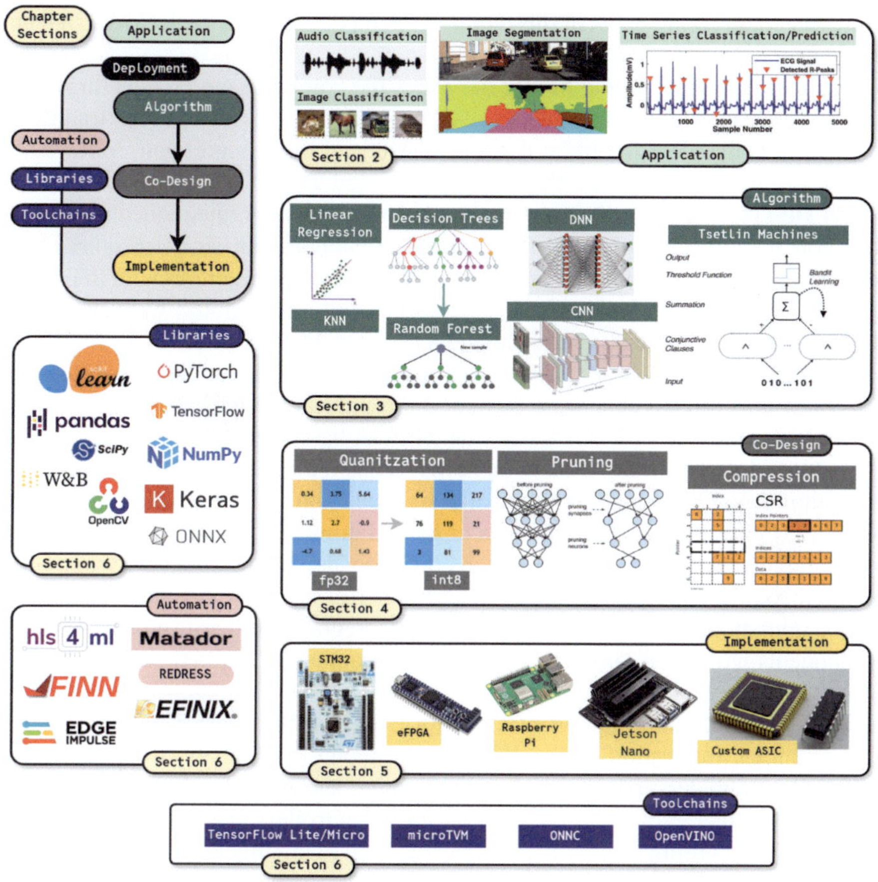

Fig. 2 A visual overview of the TinyML ecosystem of algorithms, co-design methods, implementations and the development tools and automation flows that link them together. This image also forms the organization of the sections presented in this chapter

efficiency, reduced costs, data security, and low latency prevalent in contemporary cloud computing technology [3]. By enabling the deployment of machine learning models on tiny devices, TinyML contributes to widening access to applied machine learning, fostering the development of real-world applications and ethical considerations in technology through hands-on programming and deployment experiences [4]. Furthermore, TinyML plays a crucial role in the development of energy-efficient IoT sensors for computer vision applications, demonstrating its potential to improve the sustainability and effectiveness of IoT devices [5].

To further unpack these ideas, this chapter examines fundamental concepts laced with the latest developments in TinyML by presenting a pipeline that one might use when deploying a TinyML application. This is seen in Fig. 2. The left corner shows the main components of this deployment pipeline: Algorithm, Co-Design and Implementation, we briefly define them now:

Algortihms: This section briefly describes the compute, and architecture of the core and emerging algorithms that are used for TinyML. This includes the traditional ML methods like decision trees, random forests and linear regression, followed by deep learning approaches and the recently proposed Tsetlin Machine [6].

Co-Design: The goal of co-design is to translate the aforementioned algorithms into more memory and compute efficient implementations that can be accommodated into the resource constrained edge nodes that are typically ubiquitous with IoT sensor systems, this includes small RISC processors with very limited flash and RAM packaged as development boards like the STM32 or ESP32 series.

Implementation: This section focuses on architectures that are used to implement the condensed models that were designed through the co-design methods. These architectures include CPU, GPU, TPU and FPGA based accelerators.

Automation, Libraries and Toolchains: Finally, the full flow of transition from algorithm to implementation is usually wrapped into an automation flow. This section explores the main components of these automation flows. By utilizing these lightweight frameworks, developers can efficiently deploy machine learning models on microcontrollers, allowing edge devices to perform complex AI tasks with minimal computational overhead [7]. The seamless integration of hardware and software components within the TinyML ecosystem is essential for enabling intelligent edge computing solutions.

The synergy of the components above form the TinyML *Ecosystem*. This ecosystem promotes innovation and collaboration among researchers, developers, and industry stakeholders, advancing edge computing and AI technologies. By democratizing machine learning on ubiquitous microcontrollers, TinyML has expanded access to AI capabilities, fostering the development of intelligent edge devices across industries. Collaborative efforts within the TinyML ecosystem have led to the emergence of novel approaches and solutions, such as Tiny Deep Learning Architectures and Physics-Enhanced TinyML, catering to specific application domains with unique requirements [8, 9]. This collaborative spirit and innovation within the TinyML ecosystem drive advancements in edge AI and the evolution of intelligent IoT solutions.

Chapter Organization: Figure 2 also contains the organization of this chapter. The following section will discuss some suitable applications for TinyML and the type of ML task that they require. Section 3 examines the properties of commonly used ML algorithms in the context of TinyML. Section 4 presents optimization methods for these algorithms to transition them to TinyML devices. Section 5 explores the actual implementation of the ML model and the application on the mentioned devices. The remaining section explores how the previous four sections can be tied together for an easier transition or end-to-end automation.

2 Applications

The TinyML ecosystem extends to various use cases in wearables, smart home devices, and industrial IoT applications, demonstrating the versatility and applicability of TinyML technology [10]. In wearables, TinyML enables the implementation of AI algorithms on compact devices, enabling them to perform tasks such as user recognition and activity monitoring efficiently. Smart home devices benefit from TinyML by incorporating intelligent features like anomaly detection and energy optimization, improving functionality and user experience. Moreover, in industrial IoT settings, TinyML is applied in smart factories and predictive maintenance systems, enabling real-time data analysis and decision making at the edge [11, 12]. TinyML finds applications in wearable and ambulatory systems, enabling embedded machine learning on microcontrollers for health and care applications [13].

Table 1 summarizes some of these applications of TinyML across different domains, providing an overview of its diverse implementations.

3 Algorithms (Refresher)

Before discussing the theoretical concepts behind TinyML, it is important to quickly outline the fundamentals of widely used ML algorithms, this time from the perspective of TinyML. While reading this section, it is worth thinking about the following questions about each of the algorithms presented.

Table 1 Applications of TinyML in various industrial sectors

Sector	Applications
Industrial	Predictive maintenance (PdM) [14], Enhancing reproducibility [11], Gas leakage detection [15], Energy efficiency [16], Object detection [17], Obstacle detection [18]
Other	Smart greenhouse [12], Smart farming [19], Environmental monitoring [20], Adaptive resolution inference [21]
Consumer	Smart home devices [22], Safe lock entry [23], Voice command recognition [24], Multisensor fusion framework [25], Gas leakage detection [15], Diabetes detection [26]
Health	Healthcare [13], LPWAN wearable devices [27], COVID-19 disease detection [28], Air pollution prediction [29], Implantable devices [30]
Finance	Fraud detection [31], Credit risk assessment [32], Algorithmic trading [33]
Retail	Retailer supplier switching [34], Inventory optimization [35]
Government	Big data management [10], CCTV object detection [36], People counting [37], Hydraulic infrastructure health [38]
Energy	Green AI [39], Demand response [40], Energy conservation [41]
Transport	Mobility vehicles fall detection [42], Crowd monitoring [5]

- How intensive is the computation—how much hardware would be required?
- How large is the model—how easily would it be accommodated into memory constrained implementations?
- How fast is the algorithm (this will be a combination of the first two points)?
- What are the TinyML performance metrics to determine whether a particular algorithm is most suited to be the classifier or regressor in the ML application?

Overview

This section provides a brief summary of traditional ML approaches, deep learning, and emerging methods in chronological order. Traditional ML uses statistical approaches to make predictions (e.g. linear regression, k-Nearest Neighbors). Deep learning allows for application development without as much design effort in feature engineering (e.g. Convolutional Neural Networks). For each algorithm, the model computation, size, speed, and metrics are given to determine their suitability for TinyML. Later sections will explore how the most suitable algorithms can be further developed into implementations.

This chapter presents a very high-level summary of these algorithms, where appropriate, more in-depth explanations have been referenced. The next three subsections will explore the algorithms presented in Fig. 3. The same chronology presented in the timeline in the figure will be used to explore how ML algorithms have transitioned from traditional statistical models to more intricate non-linear and compute heavy approaches to solve more complex problems. Finally, there will be an examination of one particular emerging algorithm called the *Tsetlin Machine (TM)*.

3.1 Traditional ML Algorithms

Machine Learning algorithms find their roots in statistical methods. Typically, datasets were constructed using domain-specific knowledge for feature extraction—this extraction process was crucial to the learning efficacy of traditional models [43]. This data would then be used in statistical analysis. With respect to Fig. 3, these are the early works of 1805–1901.

3.1.1 Linear Regression

One of the simplest algorithms, it models the relationship between a dependent variable and one or more independent variables by fitting a linear equation. The method of least squares, which is the foundation of linear regression, was first introduced by Adrien-Marie Legendre in 1805. Later, Carl Friedrich Gauss expanded on it in 1821,

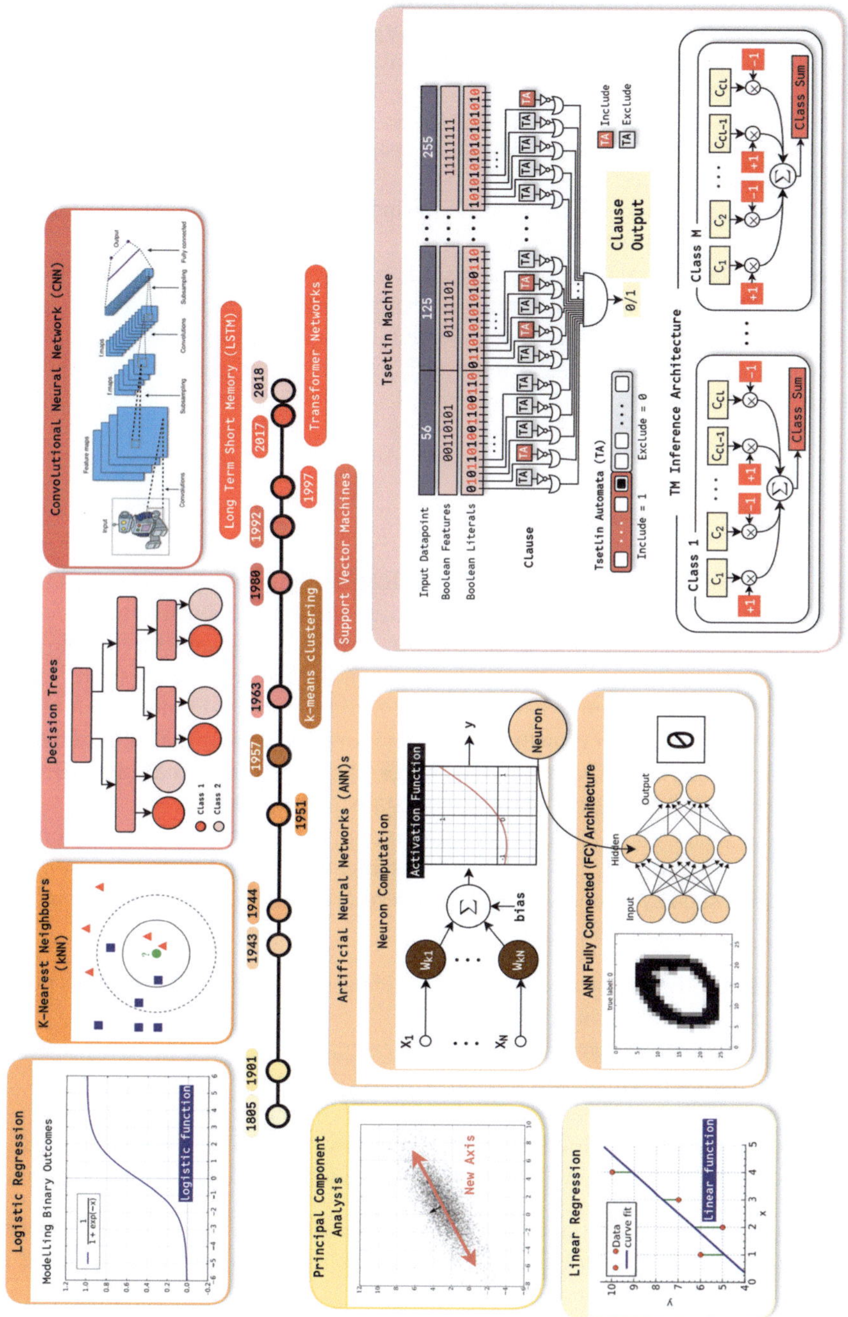

Fig. 3 Timeline showing the invention of core algorithms used in ML applications. The selected algorithms have been visually depicted in a stylized manner. Color coding is used to associate the algorithm with its place in the timeline

using the same principles for astronomical data fitting. Linear regression is highly suitable for TinyML. It requires minimal computational resources and can be easily deployed on microcontrollers. Its small memory footprint and fast inference times make it a great option for tasks such as sensor data analysis or predictive maintenance in IoT devices.

- **Computation**: Minimal computation is required, making it one of the most lightweight algorithms.
- **Model Size**: Extremely small, often requiring only coefficients that fit easily in memory-constrained environments.
- **Speed**: Very fast to train and predict, as it is based on simple matrix operations.
- **Metrics**: The most common metrics to evaluate regression tasks are the mean squared error (MSE), the root mean squared error (RMSE) and the R-squared. It is suitable when there is a linear relationship between the input features and the output.

3.1.2 Logistic Regression

Despite its name, logistic regression is a classification algorithm. It is used for binary classification problems. Logistic regression applies the logistic function to model the probability of a binary outcome, using the log-odds of the dependent variable as a linear function of the independent variables. Joseph Berkson coined the term "logit" in 1944, but the logistic function itself was used earlier by Pierre-François Verhulst in 1838 to describe population growth. The logistic function was formalized for binary classification by David Cox in 1958 in his paper The Regression Analysis of Binary Sequences published in Journal of the Royal Statistical Society.

Like linear regression, logistic regression is lightweight and easy to implement on memory-constrained devices. Its binary classification capability makes it suitable for applications such as anomaly detection, where decisions need to be made quickly and with minimal processing power.

- **Computation**: Slightly more complex than linear regression, as it calculates probabilities, but still very lightweight.
- **Model Size**: Small, requiring storage for weights and bias terms, easily accommodated in memory-constrained environments.
- **Speed**: Fast, with performance comparable to linear regression.
- **Metrics**: Key metrics include accuracy, precision, recall, and F1-score. Logistic regression works well when there's a linear decision boundary between classes.

These ideas are further discussed in "Machine Learning: A Probabilistic Perspective" (Murphy, 2022), which explores logistic regression in depth and compares it with other probabilistic models.

3.2 Principal Component Analysis

PCA is used for dimensionality reduction, identifying the axes of greatest variance in the data and projecting it onto fewer dimensions while retaining most of the original information. PCA was introduced by Karl Pearson as a technique for dimensionality reduction by finding the directions (principal components) that capture the most variance in data. Pearson's original work On Lines and Planes of Closest Fit to Systems of Points in Space was published in Philosophical Magazine in 1901.

PCA can be beneficial in TinyML when used for dimensionality reduction, especially for preprocessing large datasets on constrained devices. Reducing the number of features helps minimize the computational and memory overhead of downstream tasks.

- **Computation**: Relatively lightweight but can become computationally expensive for very high-dimensional data due to eigenvalue decomposition.
- **Model Size**: Small, as PCA reduces dimensionality, producing fewer components than the original features.
- **Speed**: Fast for low-dimensional data, but slows with increased dimensionality.
- **Metrics**: Explained variance ratio is used to determine how much variance is captured by each principal component. PCA is ideal for dimensionality reduction when feature correlation is high.

3.3 Decision Trees

Decision Trees are a tree-like model that recursively splits the data into subsets, making decisions based on the feature that provides the best split until it arrives at a final classification or prediction. The foundation for decision trees was laid by William Belson in 1963 when he developed a method for creating binary trees. Later, the Classification and Regression Trees (CART) algorithm was introduced by Breiman et al. in 1984.

Decision trees, especially shallow trees, are well-suited for TinyML. They can perform fast inference with low computational overhead. However, large or deep trees may require more memory and processing power, which might limit their use in very constrained environments.

- **Computation**: Training can be more computationally intensive, especially with deep trees, but remains manageable compared to deep learning models.
- **Model Size**: Decision trees can grow large depending on the depth and number of features, potentially becoming memory-intensive.
- **Speed**: Relatively fast for prediction once trained, though training time increases with data complexity.

- **Metrics**: Gini impurity, entropy, and accuracy are used for classification, while MSE and R-squared are common for regression. Decision trees are a good choice when interpretability is key.

3.4 k-Nearest Neighbors (kNN)s

kNN is a non-parametric algorithm used for classification and regression. For a given input, it looks at the k closest data points in the feature space and classifies the input based on the majority class among its neighbors or averages the values for regression. Evelyn Fix and Joseph Hodges introduced the kNN algorithm as a non-parametric classification method. The original paper Discriminatory Analysis: Non-parametric Discrimination: Consistency Properties was published in 1951 as part of the USAF School of Aviation Medicine Project RAND.

kNN is generally unsuitable for TinyML due to the memory-intensive nature of the algorithm, which requires storing the entire dataset and performing real-time distance calculations for each new prediction. However, optimizations such as reducing the number of stored data points or applying appropriate nearest-neighbor techniques could make it feasible for limited applications.

- **Computation**: Training is nearly instantaneous (just storing the dataset), but prediction is computationally intensive, as it involves distance calculations with the entire dataset.
- **Model Size**: Memory requirements are high, as the entire dataset must be stored.
- **Speed**: Slow for prediction due to the need to compute distances between the query and all data points.
- **Metrics**: Accuracy, precision, recall, and F1-score are common for classification. Euclidean distance is typically used to measure similarity.

3.4.1 K-Means Clustering

This unsupervised learning algorithm groups data points into k clusters by minimizing the distance between each data point and the centroid of its assigned cluster. The centroids are updated iteratively until the clusters converge. K-means can be adapted to TinyML for lightweight clustering tasks.

Since K-means only need to store cluster centroids rather than the entire dataset, it can be used for simple classification or clustering in resource-constrained environments. Its computational efficiency makes it a good candidate for on-device anomaly detection. The K-means clustering algorithm was first introduced by Stuart Lloyd in 1957 during his work at Bell Labs. However, it was not widely known until it was rediscovered and published in 1982. Lloyd's 1957 work remained internal until it was republished as Least Squares Quantization in PCM in 1982 in IEEE Transactions on Information Theory.

- **Computation**: K-means is computationally efficient for small-to-medium datasets but can become slow with a large number of clusters or features.
- **Model Size**: Very small, as the model only stores the cluster centroids.
- **Speed**: Training can be slow when recalculating centroids iteratively, but the overall computational cost is moderate.
- **Metrics**: Inertia (sum of squared distances to cluster centers) and silhouette score are common metrics to evaluate clustering performance. K-means is best suited for spherical clusters.

3.5 Support Vector Machines (SVMs)

SVMs attempt to find the best boundary (or hyperplane) that separates data into different classes, maximizing the margin between data points of different categories. Vladimir Vapnik and Alexey Chervonenkis first introduced the theory of Support Vector Machines (SVMs) in the 1960s, but the modern version of the algorithm was developed in 1992 by Bernhard E. Boser, Isabelle Guyon, and Vladimir Vapnik. The seminal work A Training Algorithm for Optimal Margin Classifiers was presented at the Fifth Annual Workshop on Computational Learning Theory (COLT) in 1992.

- **Computation**: Can be computationally expensive, particularly with large datasets, as it requires solving a quadratic optimization problem.
- **Model Size**: The model's size depends on the number of support vectors, which can increase with large datasets.
- **Speed**: Training can be slow, especially with complex kernels, but prediction is generally fast.
- **Metrics**: Accuracy, precision, recall, and F1 score are standard for classification. SVMs are ideal when there is a clear margin of separation between classes.

For a more in-depth explanation, refer to "Support Vector Machines Explained" [44], which details the computational and theoretical aspects of SVMs.

3.6 Limitations of Traditional ML Algorithms

Although traditional machine learning has been the workhorse of AI for decades, it is not without limitations. Reliance on feature engineering means that the quality of the model is often limited by the expertise of the practitioner. In fields like image recognition and natural language processing, traditional algorithms struggle because manually identifying useful features is incredibly complex. Enter deep learning, a transformative approach that addresses these challenges.

3.7 Deep Learning Methods

Deep learning is inspired by the structure of the human brain and its neural networks. Unlike traditional ML algorithms, deep learning automatically discovers representations from raw data. This leap in capability is enabled by neural networks with many layers, hence the term "deep" learning. The success of deep learning is largely attributed to the availability of large datasets, powerful computing hardware, and advancements in optimization techniques. These ideas will be covered in later sections of this chapter. For an in-depth overview of neural networks, see "Deep Learning" by Ian Goodfellow, Yoshua Bengio, and Aaron Courville (2016) .

3.7.1 Artificial Neural Networks (ANN)

At the heart of deep learning is the neural network, composed of layers of interconnected nodes (neurons) that process data through weighted connections. ANNs are the simplest form of neural networks. They consist of an input layer, hidden layers, and an output layer. Each layer transforms the input, passing information forward and adjusting weights to minimize errors. Warren McCulloch and Walter Pitts developed the first artificial neuron model, laying the foundation for ANNs. This model was inspired by how biological neurons might function. Their paper A Logical Calculus of Ideas Immanent in Nervous Activity was published in Bulletin of Mathematical Biophysics in 1943.

Basic ANNs with a small number of layers and neurons can be adapted for TinyML, though deep networks with many layers are impractical without significant optimization. ANN models can be compressed to reduce their size, but their performance is often limited compared to simpler traditional ML models.

- **Computation**: Computationally intensive, especially as the number of layers and neurons increases.
- **Model Size**: Moderate to large depending on the number of layers and parameters. Memory requirements grow as the network depth and width increase.
- **Speed**: Slower to train compared to traditional models but can be fast for prediction if optimized.
- **Metrics**: Accuracy, precision, and recall are commonly used for classification, while MSE is used for regression tasks. ANNs excel in scenarios where manual feature engineering is difficult.

3.7.2 Convolutional Neural Networks (CNN)

A CNN uses a series of filters that slide across the input image, extracting features such as lines and corners. These features are passed through layers of pooling and activation functions to learn hierarchical representations from simple shapes to complex objects. CNNs are designed specifically for processing structured grid-like data, such as images. CNNs use convolutional layers that automatically detect patterns like edges, textures, and shapes. They have revolutionized fields such as image recognition, object detection, and even video analysis.

Kunihiko Fukushima introduced the concept of a CNN in 1980 with his Neocognitron model, designed to recognize visual patterns. This architecture was further developed by Yann LeCun in the 1990s with the introduction of CNNs for digit recognition. LeCun's seminal work Gradient-Based Learning Applied to Document Recognition was published in the Proceedings of the IEEE in 1998.

Although standard CNNs are too computationally demanding for most TinyML applications, lightweight versions like MobileNet and SqueezeNet have been developed specifically for resource-constrained environments. These models use fewer parameters and computational resources, which makes them more suitable for TinyML.

- **Computation**: Highly computationally intensive due to convolution operations, especially with deep architectures like VGG or ResNet.
- **Model Size**: Large, as each layer adds numerous parameters. CNNs require substantial memory, particularly for high-resolution images.
- **Speed**: Slower to train but relatively fast for inference if implemented on specialized hardware like GPUs.
- **Metrics**: Accuracy, precision and IoU (Intersection over Union) for tasks such as image classification or object detection. CNNs are best suited for vision tasks and large-scale image datasets.

3.7.3 Recurrent Neural Networks (RNNs) and Long Short-Term Memory (LSTM)

RNNs are used for sequential data, where previous outputs inform future predictions, making them ideal for tasks like time series forecasting and natural language processing. LSTM is a type of RNN that addresses the vanishing gradient problem by using memory cells to store and retain information over longer sequences. It is effective for tasks requiring long-term dependencies, such as speech recognition and language translation. RNNs and LSTMs, due to their sequential processing and high memory requirements, are generally not well-suited for TinyML. However, small RNN models optimized through pruning or quantization can be used for limited tasks such as basic time series forecasting or voice recognition.

RNNs were introduced by David Rumelhart, Geoffrey Hinton, and Ronald J. Williams in 1986. Later, Sepp Hochreiter and Jürgen Schmidhuber developed LSTM networks to address the vanishing gradient problem in RNNs in 1997. The original RNN work appeared in Learning Representations by Back-Propagating Errors (1986), while the LSTM paper, Long Short-Term Memory, was published in Neural Computation (1997).

- **Computation**: Computationally intensive due to the need to process sequences of data.
- **Model Size**: Large, especially with long sequences, as the model must retain information over time.
- **Speed**: Slower than CNNs due to sequential processing, though faster variants like GRU improve performance.
- **Metrics**: Accuracy and BLEU score for NLP tasks, and MSE for time-series prediction. LSTMs are preferred when capturing long-range dependencies in sequential data.

3.7.4 Transformer Networks

Transformers use self-attention mechanisms to process sequential data in parallel, allowing them to capture long-range dependencies without relying on recurrence. They have revolutionized tasks in natural language processing and are the basis for models like BERT and GPT. The Transformer architecture was introduced by Vaswani et al. in 2017. This new model revolutionized natural language processing by introducing the attention mechanism, allowing for parallelization during training, unlike RNNs. The paper Attention is All You Need was published at the Advances in Neural Information Processing Systems (NeurIPS) conference in 2017.

Transformers, such as GPT and BERT, are known for their heavy computational and memory demands. As a result, they are not practical for TinyML applications unless significantly scaled down. Some research has been conducted on simplifying transformers for constrained environments, but the trade-offs in performance make them less appealing for TinyML.

3.8 Emerging ML Methods: Tsetlin Machines

Deep learning introduces complexity in architectures for ML applications. The trend seems to be added complexity with each new algorithm i.e. fully connected layers to convolution to transformers. The recently proposed Tsetlin Machine (TM) [6] offers one potential alternative to this direction of development. The TM is an intrinsically logic based algorithm that relies only on bitwise operations and integer addition for inference. This makes them extremely well suited for TinyML applications in terms of both compute and memory.

The core components of the TM are presented in Fig. 3 and a more detailed description is offered in [6, 45]. Following Fig. 3, the TM must convert the inputs into Boolean Literals for the inference computation. The example shows how an input datapoint is converted into these Boolean Literals. This process is referred to as *Booleanization*. In the example in Fig. 3, the Booleanization is a conversion of the integer value into its binary representation. However, other approaches are available based on the precision that the user would like to use for their learning problem. This is discussed further in [46].

Each Boolean Literal has its own learning element called the Tsetlin Automata (TA). The TA is a finite state machine with two actions `Include = 1` and `Exclude = 0`. Each TA will have an equal number of states corresponding to each action. The clause is the bitwise operation that links each TA to its respective Boolean Literal to generate a 1 bit clause output.

The 1 bit clause output forms the basis of the TM inference architecture. For a multi-class classification problem, each clause in each class contributes a vote to form a class sum. The clause may have either a positive polarity ("multiply" the clause output by `+1`) or a negative polarity ("multiply" the clause output by `-1`). The class sum from each of these clause votes will go through an argmax function to produce the final classification.

- **Computation**: The use of bitwise operations for the clause computation and the integer-based class sum compute makes the TM one of the simplest algorithms for computation.
- **Model Size**: Relatively large, the TM typically requires a large number of clauses to solve complex problems. Therefore, memory compression approaches are paramount to porting TMs to TinyML devices.
- **Speed**: The simplicity of the operations can be exploited to yield fast speeds.
- **Metrics**: Test accuracy (%) is the only important metric as precision is set intrinsically to a single bit.

Summarizing: There are limitations to all mentioned algorithms that prevent them from being easily deployed on the edge. The main issues are the floating-point computations and the memory footprints. Addressing these two points is the central idea for the co-design of algorithms for TinyML and is explored in the following section.

4 Co-design

This section condenses the theoretical concepts presented by seminal papers dealing with co-design of ML algorithms. The following overview summarizes the most effective concepts.

Overview
This section is concerned with how the trained ML models from the previous section can be *compressed* or *distilled* into smaller and less compute intensive versions while retaining as much as the original trained model's performance characteristics. These methods include *quantization* of the precision of learning elements, *pruning* of redundant or insignificant components for *compression* into more memory-frugal data structures suitable for edge nodes.

4.1 Compression

The compression of models is central to fitting a trained model into the TinyML device. Compression typically involves reduction of the precision of the learning elements and data (referred to as Quantization), removal of redundancy or less important model components (referred to as Pruning) or exploiting model sparsity to store the model in smaller data structures. These three ideas are discussed next.

4.1.1 Quantization

Quantization involves reducing the precision of the elements used in the parameters of a model. Instead of storing and high-precision 32-bit floating point numbers (FP32), we use 16-bit or 8-bit integers (INT8). This drastically reduces the model's memory needs and accelerates the inference computations as the compute intensity is now reduced. Of course, there is a trade-off: lowering precision can sometimes reduce accuracy, but with careful tuning, the impact is often minimal. This process can happen after (post-training quantization) the model training or during (quantization aware training). Both of these are now considered in turn:

Post-training quantization: Once the model is trained, its parameters such as weights and activations are reduced to a lower precision. Typically this comes with a small reduction in accuracy. Contoli et al. [47] provide an insightful example of Post-Training Quantization, particularly in the context of Human Activity Recognition tasks on low-power devices. The impact was examined for three compression techniques lite conversion, dynamic quantization, and full-integer quantization on accuracy and energy consumption when deploying models on an ESP32 microcontroller. Their results show that dynamic quantization offers the best balance, with

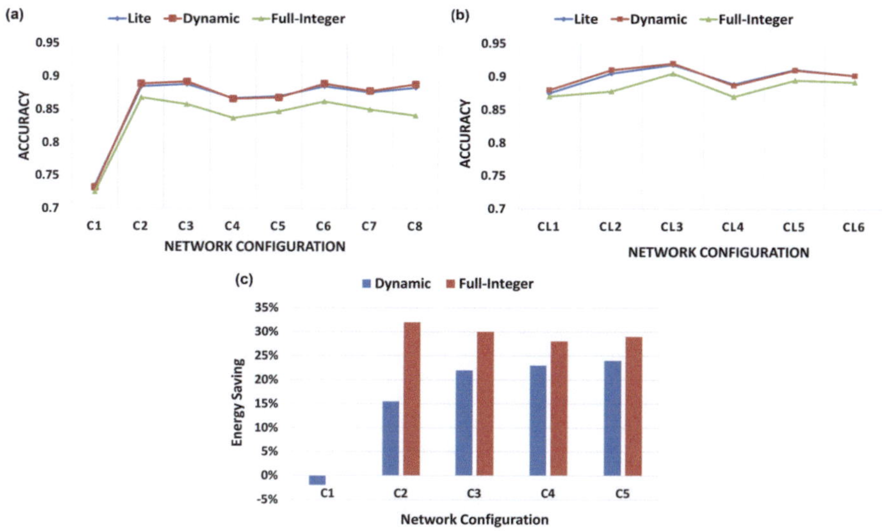

Fig. 4 **a** and **b** Classification accuracy of the 1D CNN (one-dimensional convolutional neural network) and 1D CNN LSTM (long short-term memory) models as a function of network complexity. **c** Classification accuracy of the 1D CNN (one-dimensional convolutional neural network) and 1D CNN LSTM (long short-term memory) models as a function of network complexity. Reprint from [47]

negligible accuracy loss (less than 0.1%), while achieving up to 30% energy savings, as shown in Fig. 4. On the other hand, full-integer quantization slightly outperforms dynamic quantization in energy savings but causes a more significant 2-3% drop in accuracy, particularly for larger models. Moreover, the accuracy changes as network complexity increases (from C1 to C8 for 1D CNN and CL1 to CL6 for 1D CNN LSTM), demonstrating that dynamic quantization retains high accuracy even for complex models.

Quantization-aware training (QAT): In this approach, the model is trained with quantization in mind, simulating the lower precision during training. This helps the model better adapt to being compressed, resulting in less loss of accuracy compared to post-training quantization. Krishnamoorth [48] used QAT, where quantization is incorporated during the training process, resulting in higher accuracy compared to post-training quantization. QAT allows both weights and activations to be quantized during training, and the model learns to adjust to the quantization-induced noise. QAT reduced the accuracy gap to floating-point models to within 1% at 8-bit precision, and at 4-bit precision, the accuracy loss ranges from 2 to 10%, depending on network size as seen in Fig. 5. Additionally, the results demonstrate that QAT offers significant improvements at lower bitwidths, making it particularly effective for resource-constrained environments.

Dynamic quantization: The model maintains high precision when stored but uses lower precision computations during inference. It is a balance between performance

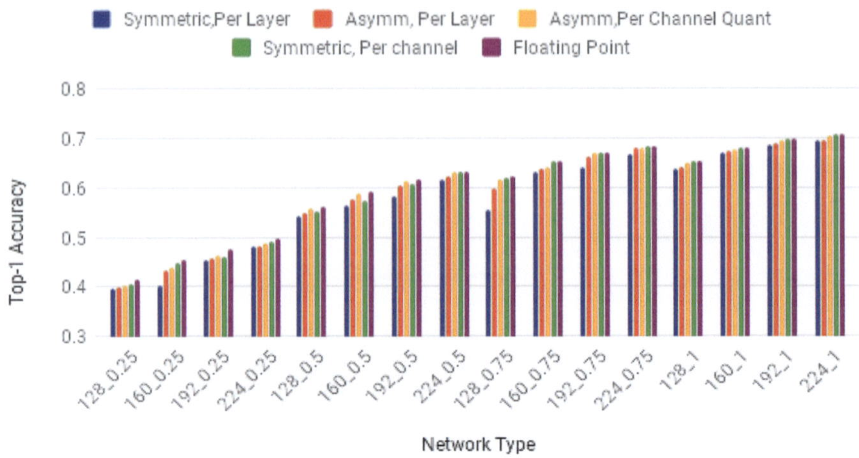

Network	Asymmetric, per-layer (PTQ)	Symmetric, per-channel (PTQ)	Asymmetric, per-layer (QAT)	Symmetric, per-channel (QAT)	Floating Point
Mobilenet-v1_1_224	0.001	0.591	0.70	0.707	0.709
Mobilenet-v2_1_224	0.001	0.698	0.709	0.711	0.719
Nasnet-Mobile	0.722	0.721	0.73	0.73	0.74
Mobilenet-v2_1.4_224	0.004	0.74	0.735	0.745	0.749
Inception-v3	0.78	0.78	0.78	0.78	0.78
Resnet-v1_50	0.75	0.751	0.75	0.75	0.752
Resnet-v2_50	0.75	0.75	0.75	0.75	0.756
Resnet-v1_152	0.766	0.762	0.765	0.762	0.768
Resnet-v2_152	0.761	0.76	0.76	0.76	0.778

PTQ: Post-Training Quantization. QAT: Quantization-Aware Training.

Fig. 5 The comparison of quantization-aware training schemes for Mobilenet-V1. The graph shows how QAT closes the accuracy gap to floating-point performance, even at 4-bit quantization. The table: comparison of top-1 accuracy for different quantization schemes shows that QAT reduces the accuracy gap compared to PTQ, particularly at lower precisions. Reprint from [48]

and accuracy. For TinyML, quantization is one of the go-to strategies because it can dramatically reduce the model's footprint, making it feasible to run on small devices without much sacrifice in performance. Angelo et al. [49] proposed the DARK-SIDE architecture for Extreme Edge On Chip DNN Inference and Training was proposed. Employing the dynamic quantization technique achieved an impressive energy efficiency of 835 GOPS/W for 2/4/8-bit integer kernels with peak performance of 65 GFLOPS and 300 GFLOPS/W for FP16 workloads with 18.2 GFLOPS. When applied to the Bottleneck layer of MobileNetV2, the system shows a 1.85× improvement in performance using the RVNN cores (RISC-V Neural Network) with M&L (fused MAC-load instruction) and the specialized Depthwise Engine. These optimizations, along with the DataMover for efficient data organizing, significantly reduce execution time and boost energy efficiency.

4.1.2 Pruning

This technique involves removing less important weights or neurons from a neural network, which can lead to significant reductions in the model's memory footprint and computational requirements and minimal impact on the model's output. This can be done in various ways.

Weight pruning: Focuses on removing individual weights from the model. Weights that have minimal impact on the output are identified and pruned. Molchanov et al. [50] proposed an approach to pruning convolutional kernels in neural networks to enhance inference efficiency. The main focus is on pruning entire feature maps rather than individual weights, and the method interleaves greedy criteria-based pruning with fine-tuning by back-propagation using Taylor expansion to approximate the cost function. This ensures the pruned networks retain good generalization abilities. Their method demonstrated that pruning could be effectively integrated into the training process, resulting in models that are both lightweight and efficient. The method starts by fine-tuning the network, followed by iterative pruning and further fine-tuning until the desired balance between accuracy and efficiency is achieved. Taylor expansion, which is based on first-order gradient information, is used to approximate the impact of pruning on the cost function. The proposed pruning outperforms other criteria, such as weight magnitude pruning, activation statistics, Optimal Brain Damage (OBD), and mutual information, in terms of maintaining accuracy while reducing the number of parameters and the reduced GFLOPs (giga floating point operations per second). It encapsulates the effectiveness of the pruning method on a real-world transfer learning task. Moreover, A recurrent 3D-CNN was pruned for dynamic hand gesture recognition using the Taylor criterion, as shown in Fig. 6. The pruning process reduces the required GFLOPs from 37.8 to 3.0, while the accuracy drops by 2.5%, which is largely recovered with additional fine-tuning. Significant speedups were achieved using pruned networks on hardware. For example, with the pruned AlexNet/Flowers-102, VGG-16/ImageNet a 2.1x to 3.4x speedup where achieved compared with various GPUs and CPUs.

Neuron Pruning: This technique involves removing entire neurons from the network. As a result, this will lead to a more substantial reduction in model size compared to weight pruning. For example, Chen et al. [51] explored how pruning can be effectively applied to Vision Transformers (ViTs) to reduce computational costs while maintaining or even improving model accuracy. The used pruning techniques are one-shot weight magnitude pruning and gradually magnitude pruning, which seamlessly integrates pruning into the training process as shown in Fig. 7a. Through sparse Vision Transformer Exploration (SViTE), the unimportant parameters were dynamically pruned during training, reducing the model size and FLOPs, as shown in Fig. 7b. An extended version, S2ViTE, introduces structured sparsity by pruning self-attention heads, making the model more hardware-friendly. Furthermore, SViTE+ co-explores data and architecture sparsity by using a token selection mechanism that identifies the most important image patches, allowing the model to operate more efficiently. A prim example of the effectiveness of pruning is seen in SViTE+-Small, which, at 50% unstructured sparsity combined with 5% data sparsity, achieves

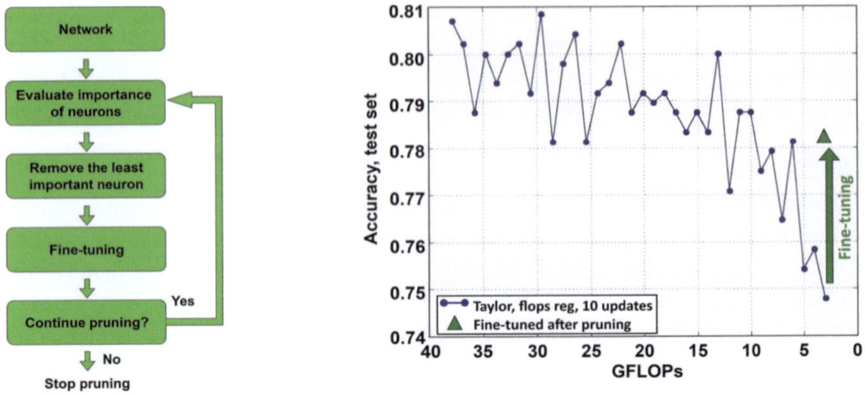

(a) The network pruning approach

(b) Pruning of a recurrent 3D-CNN for dynamic hand gesture recognition.

Fig. 6 The weight pruning process [50]

49.32% FLOPs savings and even improves the top-1 accuracy to 80.18%. This result showcases that by selectively pruning less important components, the model not only becomes more efficient but can also slightly enhance its performance, highlighting the power of pruning as an essential technique for modern machine learning models. The models used in the study are based on DeiT (Data-efficient Image Transformers), which are specialized ViTs designed for image classification with enhanced training efficiency.

Structured Pruning: Instead of pruning individual weights, structured pruning removes entire structures, such as filters or channels, from convolutional layers. This method can lead to more efficient implementations on hardware. For instance, Zhang et al. [52] discussed AKECP (Adaptive Knowledge Extraction for Channel Pruning), a method designed to efficiently prune DNNs by extracting effective knowledge from feature maps (FM). AKECP addresses the limitations of previous FM-based pruning techniques by using fewer images, sometimes just one, to establish pruning criteria, significantly improving the speed and efficiency of pruning. It prunes all layers simultaneously, avoiding the slow layer-by-layer process, and achieves substantial compression without significant accuracy loss. For instance, on ResNet-110, it reduces parameters by 59.9% and FLOPs by 59.8% on CIFAR-10, and on ResNet-50, it cuts 40.5% of memory and reduces FLOPs by 44.1% on ImageNet with only a 0.32% drop in Top-1 accuracy.

Dynamic Pruning: Adapts the pruning process based on the model's performance during training or inference. For example, Pronoet et al. [53], showcased dynamic pruning in the context of IoT applications, where models were pruned based on real-time performance metrics. The dynamic pruning utilized a novel Multiply-And-Max/min (MAM) map-reduce neuron paradigm, enabling substantial pruning of DNNs with minimal accuracy loss. Unlike traditional MAC neurons, MAM neu-

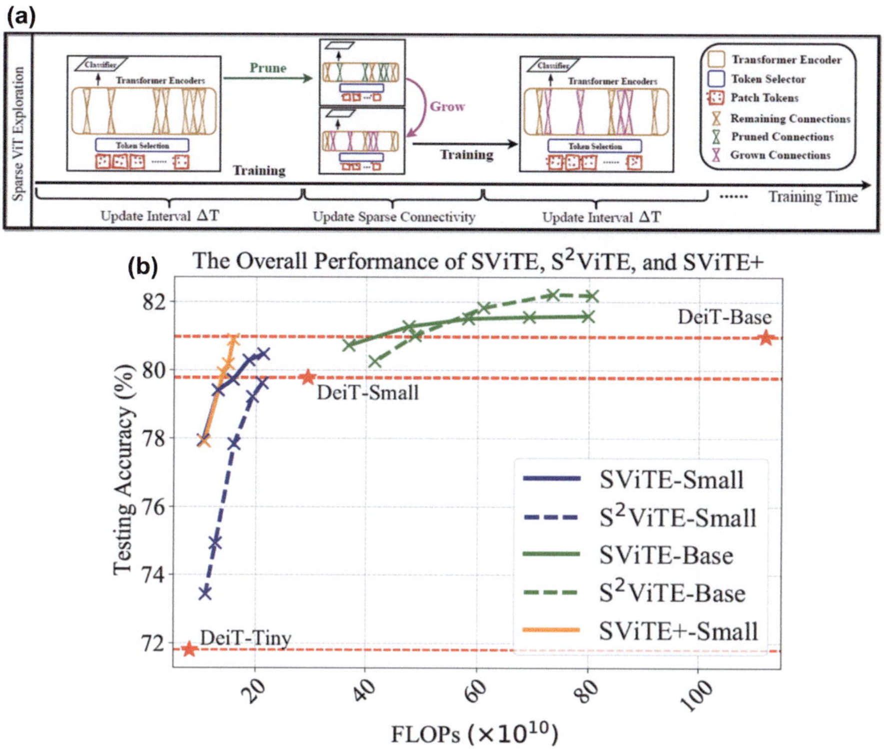

Fig. 7 **a** SViTE process to explore critical sparse connectivities, **b** Top-1 accuracy versus FLOPs trade-off for different sparse models (SViTE, S2ViTE, SViTE+) on ImageNet, demonstrating the efficiency improvements through pruning [51]

rons enhance interconnection sparsity, allowing for much more aggressive pruning. The MAMs demonstrated effectiveness on several DNN architectures, including AlexNet, VGG-16, and Vision Transformers (ViT-B/16), using datasets like CIFAR-10, CIFAR-100, and ImageNet-1K. Notably, MAM neurons retained 98% of the model's accuracy while reducing interconnections to just 0.04% on the ViT-B/16 model, compared to MAC layers requiring 56.16% of interconnections to achieve the same accuracy. Moreover, using AlexNet, MAM-based models maintained 87.34% accuracy on CIFAR-10 with only 0.03% remaining weights, while MAC-based models needed 0.48% of the original weights. For CIFAR-100, MAM maintained 62.52% accuracy with 0.26% weights kept compared with 25.01% weights kept for MAC. The MAM's ability to dramatically reduce the model size and FLOPs while preserving performance makes it a powerful technique for dynamic pruning. Mathematically ,as shown in Fig. 8, the output z_i of the MAM neuron for the i-th neuron would be:

$$z_i = \max(v_{i1}, v_{i2}, \ldots, v_{iN}) + \min(v_{i1}, v_{i2}, \ldots, v_{iN}) + b_i$$

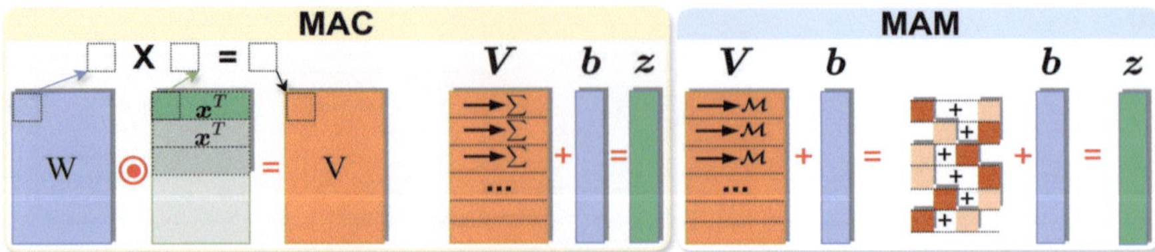

Fig. 8 The MAC and the MAM neuron paradigm in reducing the number of interconnections

Where v_{ij} is the weighted input (the product of the jth input x_j and the corresponding weight w_{ij}). $\max(v_{i1}, \ldots, v_{iN})$ is the largest value from the weighted inputs, $\min(v_{i1}, \ldots, v_{iN})$ is the smallest value from the weighted inputs and b_i is the bias term added to the final output for the i-th neuron.

Iterative Pruning: This method involves repeatedly pruning the model and then fine-tuning it to recover any lost accuracy. Disabato et al. [54] highlighted the importance of iterative pruning in TinyML, where models are pruned and fine-tuned incrementally to adapt to the limited resources of embedded systems. The technique leverages transfer learning combined with a kNN classifier, allowing on-device learning without the need for off-device training. This incremental approach supports gradual model adjustments, making the system more adaptable to new data and environmental changes. Their experiments in image classification and speech command recognition demonstrated that as more samples are introduced, classification accuracy improves iteratively. For instance, in the image classification scenario, the kNN classifier's accuracy plateaus after 20–30 samples per class, while the Support Vector Machine (SVM) model performs slightly better with a gap of 0.05–0.10 in accuracy where the SVM classifier used as a comparison is not incremental. Although SVM performs better, it requires more complex training, which is impractical to perform the training phase for embedded devices such as MCUs. Furthermore, by employing dimensionality reduction, computation time is significantly reduced, especially in the feature extraction and classification stages. where, the feature extraction on the STM32F7 microcontroller took 2.099 seconds per image, compared to 0.154 seconds on the more powerful Raspberry Pi 3B+. Similarly, kNN inference time increased from 0.004 seconds on the Raspberry Pi to 0.009 seconds on the STM32F7 with 10 neighbors, showcasing the effectiveness of pruning techniques even in constrained hardware environments.

4.1.3 Sparsity Based Compression for Tsetlin Machines

Certain models have intrinsic sparsity. This is true for Tsetlin Machine where the number of excludes far outnumbers the number of includes, as seen in Fig. 9. This kind of sparsity leads to more than 95% compression of the model and is explored in [55] where these compressed TMs are compared against embedded binary neural networks (eBNN). Notice in the Extract TAs graph for a Keyword Spotting

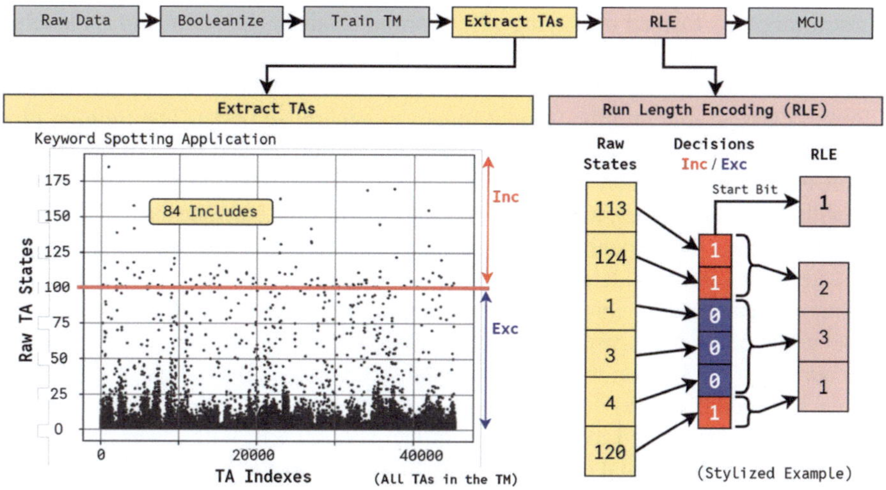

Fig. 9 One possible approach to exploiting sparsity in Tsetlin machine models. There is a large number of exclude TAs vs include TAs. This allows trained TM models to be represented in compressed data structures such as Run Length Encoded arrays. This approach is used by [55]

application, there are only 84 includes in a model with 45K TAs. If the TA include and exclude decisions are flattened into a 1D array, there will be large runs of 0s before a 1 appears. The strategy used in this figure is to use Run Length Encoding (RLE) to compress the large runs of excludes. This style of encoding is suitable for TinyML implementations where the compressed model will be iterated one element of the RLE array at a time. This is very suitable for microcontrollers.

4.1.4 Knowledge Distillation

Knowledge distillation is a powerful model compression technique that involves transferring knowledge from a larger, more complex model ("teacher") to a smaller, simpler model ("student"). This process allows the student model to achieve performance levels comparable to the teacher model while significantly reducing its size and computational requirements. The fundamental idea behind knowledge distillation is that the teacher model has learned rich representations and decision boundaries during training, which can be effectively utilized to guide the training of the student model, which results in a lightweight model that's efficient enough for TinyML devices, yet still performs admirably in the real world. The process of knowledge distillation generally involves two phases: training the teacher model and training the student model. In the teacher training phase, a high-capacity model, often a deep neural network with multiple layers and parameters, is trained on a dataset to achieve the best performance. The model's complexity allows it to learn complex patterns within the data. In the student training phase, the outputs of the teacher model serve as "soft targets." Instead of relying solely on hard labels (i.e., actual class labels), the student model learns from the probability distributions generated by the teacher model. These soft targets provide richer information about the relationships between different classes, enabling the student model to generalize better. The training process involves minimizing the difference between its outputs and those of the teacher model, typically by employing a loss function that combines traditional cross-entropy loss with a distillation loss, which quantifies the divergence between the teacher's and student's outputs [56]. Shahraki et al. [57] explored the use of knowledge distillation to create lightweight models suitable for IoT traffic classification. It demonstrated that the student model achieved almost comparable accuracy to the larger teacher model while being more resource-efficient. Specifically, the student model achieved an accuracy of 98.31% on the Cambridge dataset, which is only slightly lower than the teacher model's 98.51%. The size of the student model was 768 KB, a significant reduction compared to the teacher model's 9,060 KB, with fewer parameters and reduced computational cost. The study highlights that knowledge distillation is effective even with reduced training data and shows robust performance when some traffic classes are excluded, emphasizing the value of soft labels for training student models under limited data scenarios.

4.1.5 Weight Clustering

Weight clustering (or weight sharing) is a compression technique that reduces the number of unique weight values of the Neural Network to be stored for efficient compression as well as representation. It groups the weights of a neural network layer into clusters and shares the centroid value of each cluster [58]. The process of weight clustering involves the following:

- **Clustering Weights**: divide the weights of each layer into an X number of clusters, where each weight is allocated to the closed cluster.
- **Weight Sharing**: the centroid of each cluster will be stored in the model rather than saving the individual weights, where all weights that belong to a cluster are replaced by the centroid value. This method reduces the number of unique weight values and allows the model to be compressed without significant loss in performance.
- **Centroids tune**: the cluster of the centroids can be fine-tuned to reduce the error by initializing it randomly, linearly, or density-based. The linear technique yielded superior results. Because large weights are more significant, utilizing linear distribution increases their chances of generating a large centroid.

4.1.6 Filter Bank Design

Filter bank is a crucial aspect of signal processing in TinyML, particularly for applications involving audio, image, and sensor data. A filter bank is a collection of band-pass filters that divide the input signal into multiple components, each representing a specific frequency range. These filters can be used to extract temporal and frequency-domain features, depending on the type of sensor data. This division allows for more efficient processing, analysis of signals, and image compression making filter banks essential for various machine learning tasks in resource-constrained environments. Some types of filter banks are

- Quadrature Mirror Filter Banks (QMF) [59]
- DFT Modulated Filter Banks [60]
- Multi-Band Filtering (MBF) [61]
- Frequency Response Masking (FRM) [62]
- Non-Uniform Filter Banks [63]
- Gammatone Filter Banks [64]
- Wavelet Packet Transform Filter Banks [3]
- Adaptive Filter Banks [65].

5 Implementation

Selecting the *Right* Existing Hardware... or Developing *New* Hardware

Overview

The key focus of this chapter is about taking the optimized model described in the previous section and understanding the rationale for *mapping* it to the most suitable the hardware implementation. Following the decision process presented in Fig. 11, the core *metrics* are derived that form the direction for whether the model is deployed on generalized hardware such as CPUs or GPUs or whether specialized hardware is required (see Fig. 10).[1] CPU-based implementations have constraints on hardware resources and development support that should be considered. Specialized hardware is underpinned by core design principles that emerge from the question of whether a system is bounded by compute or memory. Ultimately, the specialized implementation is either a custom ASIC, an FPGA, or a Tensor Processing Unit (TPU) [70]. An exploration of recent research works and commercial platforms offers further insight into the direction and trends for TinyML *implementation*.

[1] It is assumed for this section that the reader has some notion already about computer architecture and how CPUs, FPGAs and GPUs work. Otherwise, a quick refresher on the latest functionality and design trends in CPUs and GPUs can be seen here: [67]. Harris and Harris offer a good summary on the fundamentals [68]. For FPGAs, [69] presents a detailed guide to development using Xilinx (AMD) platforms.

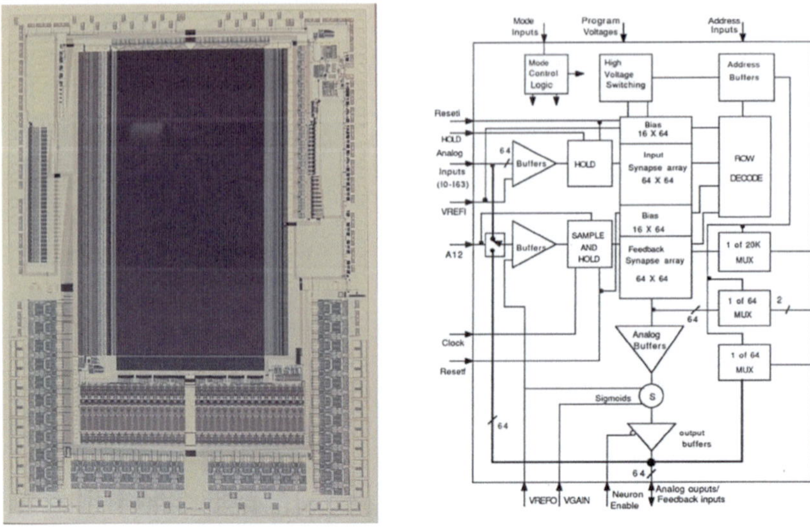

Fig. 10 The Intel ETANN 80170 is widely regarded as the first ML accelerator implementation (1989). ETANN stands for Electrically Trainable *Analog* Neural Network. The block diagram shows that this accelerator only incorporated 64 neurons—truly TinyML! (see [66] for more details on how the US Naval Air Warfare Center used this implementation for missile seeker applications)

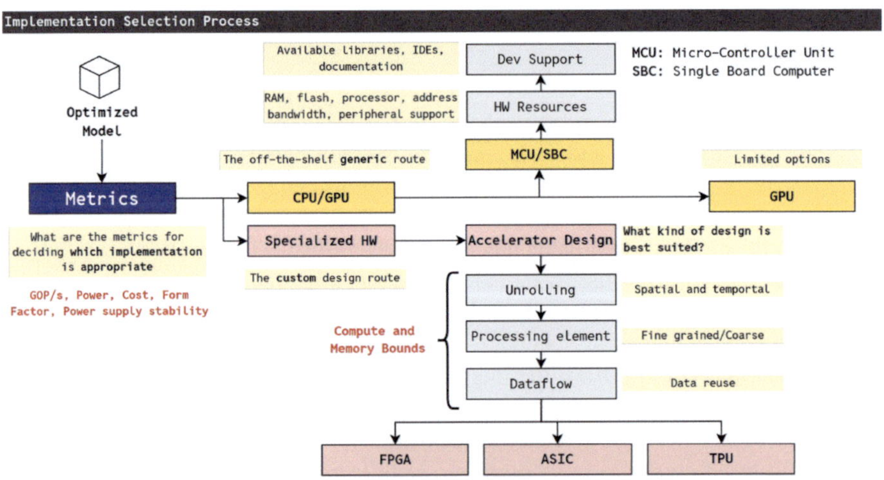

Fig. 11 Visual overview of this section. The focus is selecting suitable hardware for the optimized model. This can be either a general off-the-shelf CPU or GPU or a more specialized architecture

5.1 Metrics Underpin Implementation Choice!

The new metrics: Before discussing the different implementations that the ML algorithm can be deployed on, it is worth understanding a bigger picture of what the implementation is supposed to offer. In many ways, implementation is the convergence point of the TinyML pipeline, it is the culmination of the previous two sections of algorithm and co-design now mapped to a hardware platform. There are certain metrics that are important for TinyML—from the learning efficacy perspective there is accuracy—e.g. test accuracy or F1 score. It could also be latency and model size.

Figure 11 demonstrates that implementations begin with metrics. When mapping the algorithm to a platform, newer metrics emerge in addition to what was seen in the previous sections; these metrics are concerned with how the hardware platform can leverage better energy efficiency from its algorithm workload (i.e. the inference computation using the optimized model). The cost of and the development time required for implementation as well as the form factor (size) also play a crucial part.

For example, ASIC implementation is more costly and time consuming than developing on a FPGA.However the FPGA designs can only be mapped onto existing fabrics and typically of a larger form factor and are less energy efficient. Similarly CPU good generalization but there is a plethora of choice when it comes to MCUs and Single Board Computers (SBCs). GPUs typically offer better throughput than CPUs for tensor based models but at the cost of higher power consumption.

Figure 11 organizes the implementation choice into two routes. Either a generic route and a specialized route. The reminder of this section will examine both these route in terms of decision decisions, available platforms, recent research and most importantly how they address the implementation metrics.

The two routes that can be taken for TinyML implementation:

- **The generic route**:Deploy the optimized algorithm on a existing piece of hardware like a RISC based CPU (typically in a MCU development board), a single board computer (SBC)—or small GPU.
- **The specialized hardware route**: Design an *accelerator*. This is a custom hardware implementation to deploy the optimized algorithm—this could be an ASIC or FPGA or a commercially available TPU.

Making the decision between the routes: The key to making the decision of which path to take is determined by how much computation should be completed within a given power budget for the application.

The Big Picture: Figure 12 shows three graphs that explain how custom implementations should be measured, not only for TinyML implementations but across the range of all ML accelerators. These implementations have focused solely on Neural Network based algorithms. The axis of the graph show the two main metrics: Number of Giga Operations (GOPs) and the Power (W). These graphs are from the Nanoscale Integrated Circuits and Systems Lab, Department of Electronic Engi-

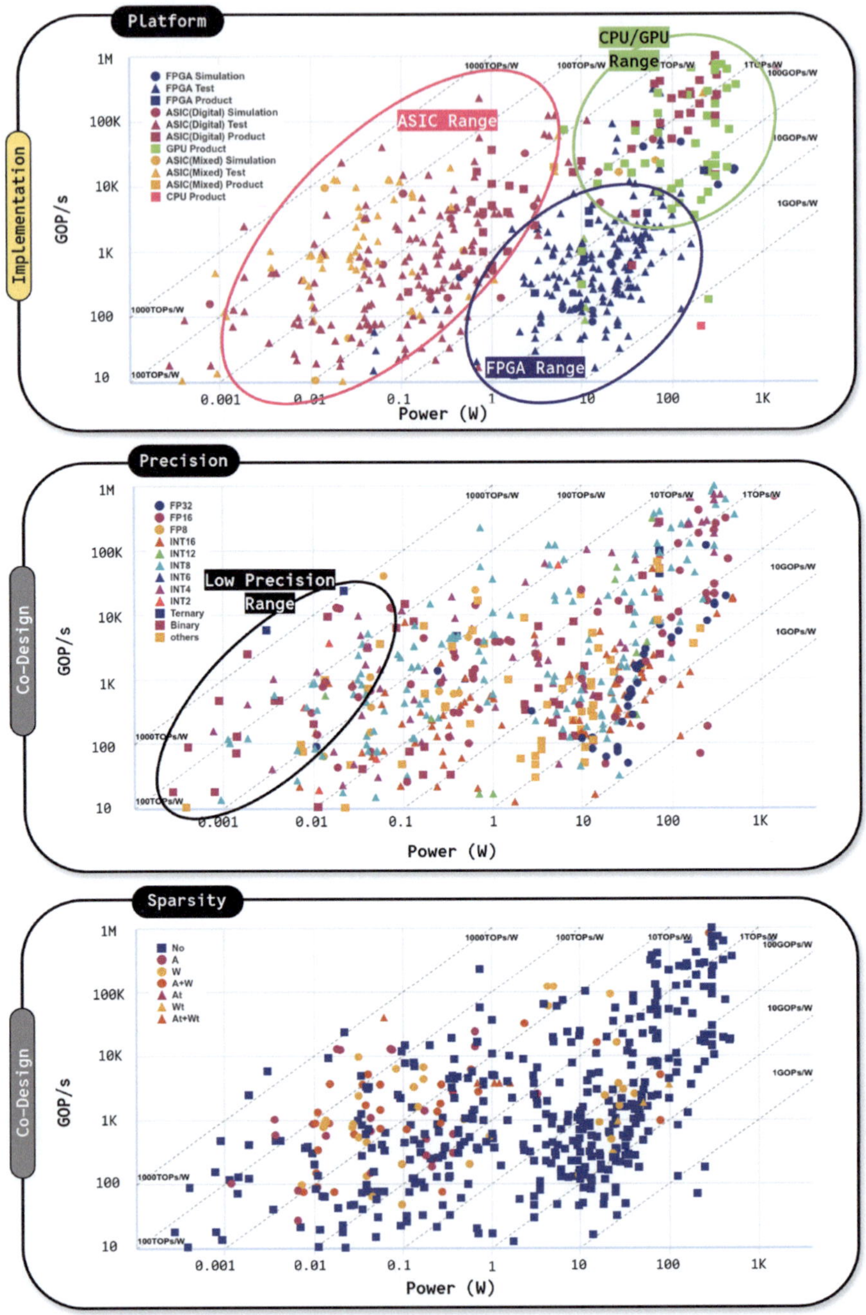

Fig. 12 Neural network implementation map adapted from data presented in [71]

neering, Tsinghua University, Energy Efficient Computing Group (NCIS-EFC). They show the most up-to-date hardware custom platforms from industry and research as reported in IEEE Transactions on Circuits and Systems I, Journal of Solid State Circuits and other prestigious publication channels. The key point to note is how the GOPs/W pareto frontiers constrain the different implementation platform types. For example the ASIC implementations are more power efficient while the FPGA ans GPU can achieve greater GOPs at the expense of more power—going well beyond what is appropriate for TinyML.[2]

Therefore, choice of implementation is centred around how to maximise the GOP/s within the power budget of the TinyML application. One important finding from the top graph is the groupings formed by the different hardware implementations. ASICs offer the most energy efficient option while FPGAs, CPUs, and GPUs take up more power. Additional nuances to this will be explored later.

The two graphs below the Platform graph in Fig. 12 shows the different co-design techniques that have been used by each of the implementations. Many have utilized some degree of quantization but far fewer have looked at sparsity. For the sparsity key, **A** refers to activation function and **W** refers to the weights.

A caveat when making direct comparisons between implementations:

Notice the GOP/s in the top graph do not account for precision. The precision of the model can play a huge part in the number of operations that can be performed and the power budget as seen in the second graph (this will be explored in detail later). Therefore, it is important to look at GOP/s and GOPs/W with respect to *precision*.

The next two subsections explore the two routes seen in Fig. 11 of either a generic platform or a specialized platform. The focus of these subsections is to understand how the specific hardware components in each platform type (CPU, GPU, ASIC, and FPGA) can be harnessed to optimize one or more of the metrics encountered so far. But also to understand how intrinsic properties of both the algorithm and the hardware can also lead to distinct clusters as seen in Fig. 12.

5.2 *The Specialized HW Route*

Taxonomy and Design Principles for Accelerators

This subsection explores the specialized hardware route presented in Fig. 11. The rationale for going down this route of custom hardware is the freedom in design capability when mapping the optimized model to either an ASIC, FPGA or TPU.

[2] These graphs are interactive—a specific point of interest (i.e. a accelerator implementation) can be examined if interested—see here: `https://nicsefc.ee.tsinghua.edu.cn/project.html`. These accelerators are not limited to TinyML applications but are useful for demonstrating the overarching trend.

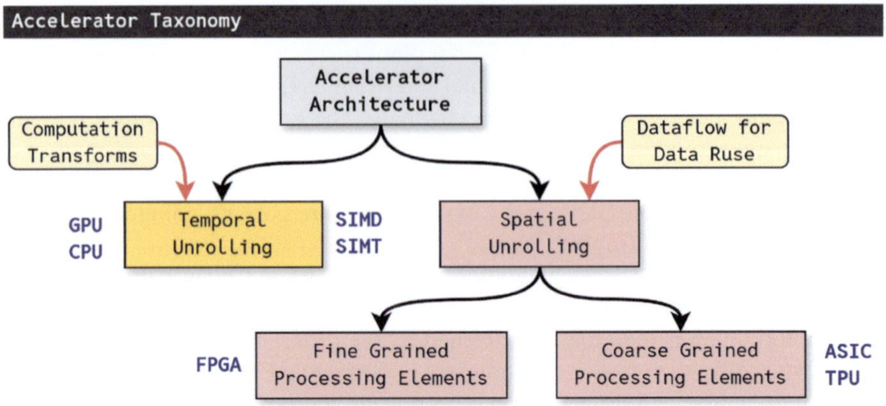

Fig. 13 Generalizing accelerator design approaches into a taxonomy. We will refer to this diagram frequently throughout this section

Therefore, it is important to understand the core principles used in this mapping (rolling, processing element design, and data flow).[3]

5.2.1 The Taxonomy for Accelerator Designs

At the heart of the ML algorithm is a computation. Typically for Neural Network based approaches these are General Matrix Multiplications (GEMM). However, any algorithm that contains `for` loops are also applicable (e.g. TMs). The role of the accelerator is to maximise the GOPs through highly parallel structures. However, these structures cannot be infinitely parallel as there are always restrictions in compute and memory. Therefore, the design must carefully balance the memory and compute *bounds* of the platform. For the remainder of this section, a particular focus is placed on DNNs. The GEMM operations in DNNs expose opportunities to discuss all the main design decisions expressed in Fig. 11. It is worth noting that the taxonomy is extended to all ML accelerators, not just for TinyML implementations.

Spatial and Temporal Unrolling: Figure 13 shows how accelerator design approaches can be organized into a taxonomy. The first branch of the taxonomy occurs between *Temporal Unrolling* and *Spatial Unrolling*. Typically, accelerators will use a combination of both temporal and spatial unrolling. The choice of how much of each to use depends on whether the implementation is more compute-bound or memory-bound. These ideas will form the core focus of this subsection.

Temporal unrolling is common in CPU and GPU architectures as well where they may use Single Instruction Multiple Data (SIMD) or Single Instruction Multiple

[3] Given the plethora of options that are available in this area, only the most state-of-the-art is considered when reviewing implementations, and only the most widely used commercial platforms are selected for review.

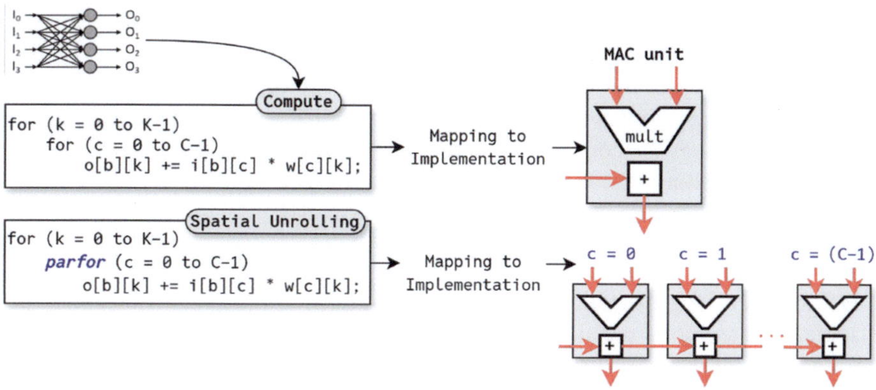

Fig. 14 Visualizing spatial unrolling. The processing elements here are the MAC units. The first nest `for` loop box (Compute) represents a temporal unrolling, the same MAC unit is used again and again for all the iterations of the loop

Thread (SIMT) [72]. Typically implementations like this are very resource heavy e.g. many ALUs and not so appropriate for TinyML. In comparison spatially unrolled architectures are more common for DNN based TinyML implementations. They key design principles for each unrolling is either computational transforms (loop reordering or breaking) or maximising data reuse [72].

Focusing on spatial unrolling, this approach involves the parallel execution of multiple operations by allocating distinct hardware resources (such as multiple processing units (PEs)) for each operation. Instead of reusing a single computational unit, multiple units are deployed simultaneously to perform the same or different tasks in parallel. This is a key strategy in compute-bound systems, where the bottleneck is the processing capability rather than memory access. By increasing the number of PEs, accelerators can achieve higher throughput, executing multiple operations in parallel. In Fig. 14, the `for` loop in the first compute box is replaced with a `parfor` loop. The compute here represents a fully connected layer; however, any loop iteration compute applies here. In mapping terms for this compute, it is the equivalent of replacing a single MAC unit that is iterated for the two nest for loops with a parallel version. Note that this `parafor` has unrolled the full loop, however this does not have to be the case. The amount of rolling possible depends on the hardware resources available.

Compute Bounded Implementations: In compute-bound systems, the bottleneck is the availability of compute resources. Here, memory access is fast enough to keep up with computational demands, and the performance is limited by the number of operations the hardware can perform. Spatial unrolling is ideal in these scenarios because it maximizes parallelism, allowing more computations to occur simultaneously and thus speeding up execution.

Memory Bounded Implementations: In these systems, the performance bottleneck lies in how quickly data can be fetched from memory. If the bandwidth or latency of the memory is the limiting factor, increasing computational resources (as in spatial unrolling) will not yield significant performance improvements, as the computational units will often be idle, waiting for data. Temporal unrolling is more effective in such systems, as it conserves resources and balances memory access with computation cycles.

Before examining the state-of-the-art works in research and industry that make use of spatial and temporal unrolling it is useful to summarize the main characteristics, pros and cons of these type of design approaches. Most of the reviewed works will incorporate a balance between the two of these.

The Characteristics of Spatial Unrolling:

- **Maximizing Parallelism**: Spatial unrolling is optimal for compute-bound systems, where the primary bottleneck is the availability of compute resources rather than memory latency or bandwidth. Each operation runs in parallel, allowing for significant performance gains by increasing the number of PEs in the design.
- **Latency Reduction**: Spatial unrolling can significantly reduce the latency of a task, as more operations are performed simultaneously.
- **Limitations of Spatial unrolling:**
  - High demand for hardware resources (e.g., area on FPGAs or logic gates on ASICs).
  - Can lead to inefficiencies if the system is memory-bound, as the parallel PEs may become underutilized waiting for data.

The Characteristics of Temporal Unrolling:

- **Resource Conservation**: Temporal unrolling uses fewer hardware resources at a time but requires more cycles to complete a task.
- **Efficiency in Memory-bound Systems**: Since data needs to be fetched and processed over multiple cycles, temporal unrolling is useful when the system is memory-bound—that is, when the latency and bandwidth of memory access are bottlenecks in the design. By reusing the same computational elements over time, temporal unrolling can balance the slower memory access rates.

- **Limitations of Temporal unrolling:**
  - Longer execution time due to sequential processing.

Dataflow Considerations for Accelerator Design: Different dataflow architectures focus on maximizing various types of data reuse, optimizing for specific workloads like convolution, matrix multiplication, or fully connected layers in neural networks. These architectures include:

- **Weight-Stationary Dataflow**: Weights remain in local memory while inputs are streamed through, maximizing weight reuse. This is ideal for convolutional layers where filters are applied repeatedly to different inputs.
- **Output-Stationary Dataflow**: Intermediate results (e.g., partial sums) are accumulated and reused locally, reducing the need for memory writes. This approach is common in matrix multiplication.
- **Row-Stationary Dataflow**: Optimizes the reuse of both input and weight data by focusing on rows of the input matrix and weights. It balances the reuse of multiple types of data.

By designing custom dataflow architectures for specific operations in neural networks, accelerators can optimize for memory bandwidth, minimize energy consumption, and enhance overall performance.

Data reuse is a crucial aspect of efficient accelerator design, particularly for applications like TinyML, where power and memory constraints are significant. By maximizing input reuse, output reuse, and weight and activation reuse, dataflow architectures can deliver higher performance with lower energy consumption. Optimizing dataflow for specific tasks such as convolutions or matrix multiplications ensures that FPGAs, ASICs, and other accelerators can achieve the necessary throughput and efficiency for edge inference and other real-time applications.

Continuing with Fig. 13, the next branching on the Spatial Unrolling branch introduces the idea of granularity. Granularity here refers to the compute capabilities of the processing element. These could be MAC units or they could be simpler logic elements. This is another factor to consider when designing custom implementations. First we discuss fine grained processing elements.

Fine Grained Processing Elements (FPGAs): Field-programmable gate arrays, abbreviated as FPGAs, offer custom hardware design capabilities through a reconfigurable implementation based on routing small fine-grained processing elements together. Figure 15 shows a stylized architecture of a FPGA, typically there are columns of soft logic blocks, memory blocks (Block RAMs (BRAMs)), and arithmetic blocks (usually Digital Signal Processing (DSP) blocks) that can be connected together. The soft logic blocks typically contain Look-up Tables LUTs as seen through the OpenFPGA soft logic block (see [73] for details). These LUTs are used to model Boolean logic functions; for example, an LUT-6 block will take 6 inputs and map them to one output. Different vendors use different LUT and soft logic processing elements to get a good trade-off between compute elements (i.e. logic

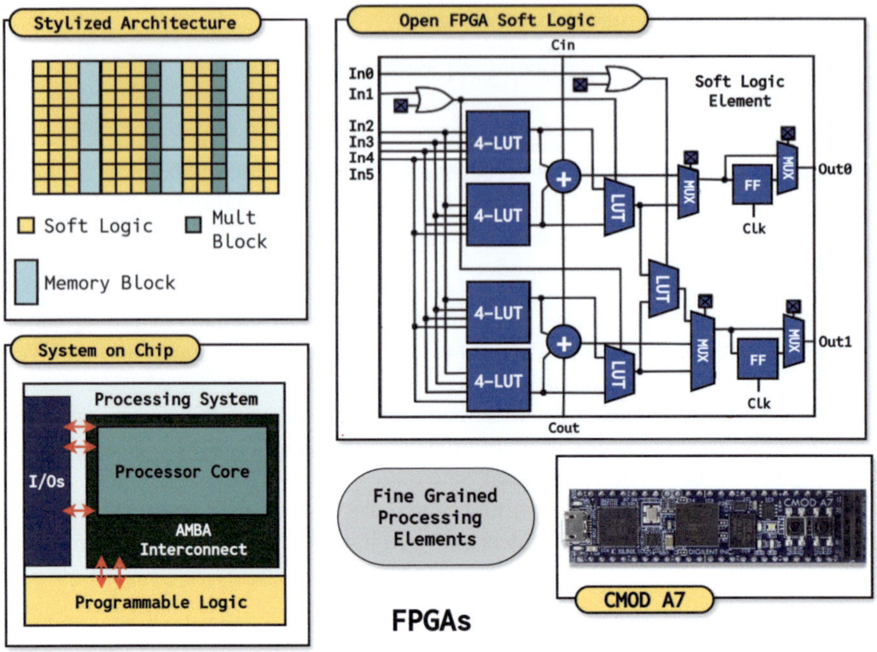

Fig. 15 Fine grained spatial unrolling through FPGAs. The figure shows a picture of a FPGA board available from diligent (CMOD A7), A stylized architecture showing FPGA composition, an expanded view of one possible soft logic block and finally now FPGAs can be developed into System on Chip (SoC) styled implementations

density) and routing. These simple logic-based processing elements make FPGAs fine-grained.

FPGA Vendors: Xilinx (acquired by AMD) and Intel (through its acquisition of Altera) dominate the commercial FPGA market, providing high-performance devices that are widely used. These companies have continually pushed the boundaries of FPGA architecture by enhancing their devices with high-speed transceivers, large amounts of logic elements, and embedded resources such as DSP slices and memory blocks. In the next section, we will discuss the toolchains that these vendors offer to translate from algorithm to implementation. Recently, open source toolchains for FPGA development have allowed the development of custom FPGA fabrics offering even more customization [73, 74].[4] In recent years the main vendors, along with emerging vendors like Lattice, have started offering smaller form factor, lower power high density FPGAs. Often these are packaged as a System on Chip (SoC), these are referred to as embedded FPGAs (eFPGAs). A good example is the Xilinx Zynq7020. In addition, FPGAs like the Lattice iCE40 series are specifically designed for always-on, low-power inference.

[4] The next section will discuss more about toolchains for TinyML deployment.

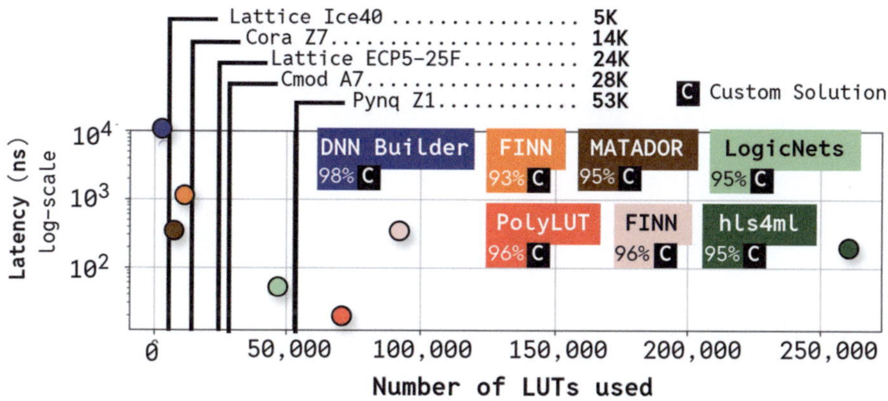

Fig. 16 Figure showing a general trend among FPGA implementations in terms of latency and throughput. The vertical lines indicate the kind of FPGA and eFPGA platforms that are available and their respective LUT availability. A key point to note is that maximally unrolling the model is not so suitable for TinyML—even for a simple application like MNIST that was used for the implementations in this figure. Only three approaches fall into the TinyML territory

FPGA Design Trends: FPGA implementations that may be regarded as TinyML typically can be divided into two categories; parameterized compute engines that use a combination of spatial and temporal unrolling to allow for greater control in resource usage (FPGA Route 1) [75–78], or alternatively, custom implementations that maximally unroll the full optimized model into a wired logic solution (FPGA Route 2). These directions have been visualized in Fig. 16. Notice that most of the solutions are well beyond the LUT availability of embedded FPGA platforms seen in the vertical lines. The selected solutions all use an automated pipeline to deploy models on the FPGA. This will be elaborated on further in the next section. For now, the focus is the two types of implementation.

FPGA Route 1 (Parameterized Compute Engine): One of the most popular implementations on this route is FINN from Xilinx [75, 76]. By having design knobs that control the amount of spatial versus temporal unrolling, it is possible to create a more resource-fugal design most implementations refer to this as the *reuse factor* [78, 79]. In all cases, the implementations use a row-stationary data flow design. This design is essentially a matrix-vector multiplication kernel that can be modified per model. This design thinking is shared by other implementations such as DNN Builder [78], hls4ml [79] and FracBNN [77]. All these works have favored either binary quantization or 2bit quantization of the weights and activation functions.

To understand the impact of quantization more quantitatively, Fig. 17 shows the roofline diagram for a particular FPGA platform. The x-axis of the graph represents the operations performed per byte of data retrieved from memory. For low quantization, the majority of the data is retrieved off-chip. For single bit quantization, all the data can be kept on chip. Notice the dashed red line; where the line is diagonal, the implementation is memory bound; however, where the line is horizontal, the

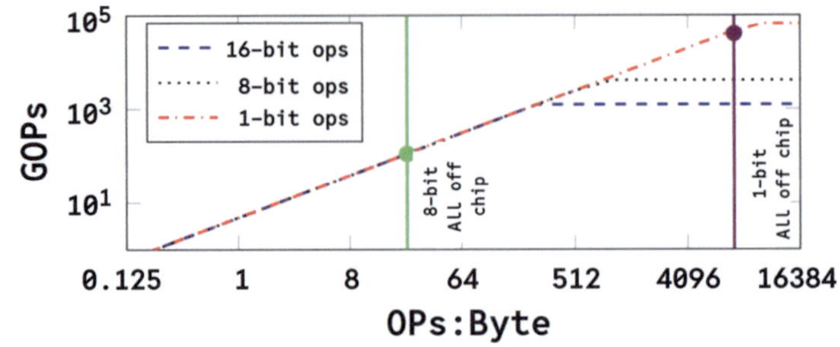

Fig. 17 Figure showing the roofline diagram for the ZU19EG FPGA adapted from [75]. Quantization plays a significant role in whether the implementation is compute or memory bound

implementation is compute bound. Therefore, the binary implementations are better for achieving better throughput but return poorer performance from the model—notice how the DNN Builder (12bit quantization) gives much poorer latency but with much lower LUT usage. This is an important trade-off to consider for FPGA implementations.

FPGA Route 2 (Maximal Spatial Unrolling): The alternative to finding the optimal trade-off between spatial and temporal unrolling is to maximally unroll the model. The key is to map the computation efficiently to the fine-grained LUTs. This is the approach taken by PolyLUT [80] and the works it builds from like LogicNets [81] and LUTNets [82]. The approach here is mainly concerned with maximizing throughput over LUT usage. Typically this results in deployment on larger and more power hungry FPGA platforms.

MATADOR [83] explores the maximal unrolling of TMs. The difference is the sparsity that is seen in the TMs allows for both fast throughput but also more frugal LUT usage meaning the implementations can fit on resource constrained eFPGAs. These maximally unrolled solutions are custom for each model. Therefore it is important that they come with an automation flow that can easily convert the co-designed model into implementation. This is explored further in the following section focusing on automation toolchains.

Summarizing FPGAs for TinyML: FPGAs offer a highly versatile and efficient platform for accelerating machine learning workloads, particularly for TinyML and edge computing applications. With innovations in spatial unrolling and systolic array designs, FPGAs can achieve low-latency and high-throughput inference, even in power-constrained environments [84]. However, the key is to use parameterized compute engines like FINN and FracBNN or implementations that support extreme sparsity MATADOR. As a result, both commercial FPGAs (e.g., Xilinx, Lattice, and Intel) and open-source solutions are becoming critical tools in the development of custom accelerator implementations for TinyML.

Coarse Grained Processing Elements: The multiply accumulate operations for Neural Network architectures can be implemented into more coarse grained processing elements where each element will be a MAC unit. ASIC implementations allow for greater flexibility in the size and functionality of these units.

The Google Edge TPU (Tensor Processing Unit) is a hardware accelerator designed to enhance machine learning inference at the edge from Google. It is incorporated into Google's Coral platform, which aims to enable efficient AI applications on resource-constrained devices. The Edge TPU is optimized for running TensorFlow Lite models, making it particularly suitable for TinyML applications where low power consumption and high performance are critical. TensorFlow Lite models enable developers to leverage pre-trained models for various applications—this will be covered more in the following section.

Edge TPU provides a performance throughput of up to 4 trillion operations per second (TOPS) while maintaining low power consumption. Its architecture is designed to efficiently execute neural network operations, particularly those commonly used in deep learning models. The Edge TPU employs a systolic array architecture, which allows for efficient data flow and parallel processing of matrix operations. This design is particularly beneficial for convolutional neural networks (CNNs), which rely heavily on matrix multiplications [85]. It features a significant amount of on-chip memory, which reduces the need for external memory access and minimizes latency by processing data locally, reducing the need for cloud-based inference. This is crucial for real-time applications requiring quick responses [86]. The Edge TPU is designed to operate within a power envelope suitable for battery-powered devices, making it ideal for IoT applications. Its efficiency allows it to perform complex computations without draining the device's battery, typically around 2 watts, which is essential for deployment in resource-constrained environments [87].

The main challenge is model size limitation, which is when the model size does not meet the limits of its on-chip memory. Additionally, although the Edge TPU primarily supports TensorFlow Lite, developers using other frameworks may need to convert their models to ensure compatibility, which can introduce additional complexity [88].

Apple Neural Engine (ANE) is a dedicated neural processing unit (NPU) integrated into Apple's custom silicon, such as A-series and M-series chips. It is designed to accelerate machine learning tasks on iOS and macOS devices. ANE has developed significantly through its generations, with each providing considerable improvements in throughput and efficiency for machine learning tasks, from basic facial recognition in the A11 Bionic to advanced real-time processing in the A18 chip as listed in Table 2. ANE has become a critical component in enabling intelligent applications across Apple's ecosystem. As the demand for on-device AI continues to grow, the ANE plays a substantial role in enhancing its capabilities like privacy-preserving by enabling on-device processing of models without needing cloud computation; tightly coupled with the CPU GPU on Apple SoCs (System on Chips) facilitates sharing of resources. Furthermore, ANE supports the bfloat16 (brain floating point 16-bit) format, making it useful for training large models where memory efficiency is critical. The multi-core nature of the ANE accelerates complex neural networks by parallelizing operations like convolutions, matrix multiplications, and activation functions, thereby offering

Table 2 Development of Apple Neural Engine used in iPhones [89, 90]

Chipset	Peak TOPS	Technology (nm)	Core count	Other features	Used in
A18 Pro	35	Enhanced 3 (N3E)	16	20% faster and 2x faster ray tracing than A17	iPhone 16 Pro, 16 Pro Max
A17 Pro	35	3 (N3B)	16	20% faster and 4x faster ray tracing than A16	iPhone 15 Pro, 15 Pro Max
A16 Bionic	17	4 (N4P)	16	CPUs with 20% less power, 50% more GPU memory bandwidth than A15	iPhone 14 Pro, 14 Pro Max, 15, 15 Plus
A15 Bionic	15.8	5	16	40% faster CPU A12	iPhone 13, 13 Pro, 14, 14 Plus
A14 Bionic	11	5	16	40% faster CPU, 30% faster GPU compared to A12	iPhone 12 series, iPad Air (4th gen)
A13 Bionic	6	7	8	20% faster CPU and 15% lower power	iPhone 11 series, iPhone SE (2nd gen)
A12 Bionic	5	7	8	9x faster, 90% lower power consumption	iPhone XS, XR, iPad Pro (3rd gen)
A11 Bionic	0.6	10	2	First Neural Engine, Real-time ML tasks, like Animoji and Face ID	iPhone 8, iPhone X

Note N3E, N3B, and N4P refer to TSMC's advanced semiconductor technology nodes. N3E is an enhanced 3nm node, N3B is the baseline 3nm node, and N4P is an optimized 4nm node, offering different trade-offs in power efficiency, performance, and cost

beneficial features for TinyML applications and edge computing, as well as mobile and wearable technologies. Various Applications leveraging the ANE from Real-time Image Processing (e.g., object detection, facial recognition), Natural Language Processing on voice-enabled devices (Siri), and Health Monitoring through real-time sensor analysis in devices like Apple Watch and Augmented Reality (AR) applications. ANE works seamlessly with Apple's Core ML framework, making it easy for developers to deploy machine learning models on Apple devices. The main limitation is that unlike traditional GPUs or CPUs, which can be programmed for a wide range of tasks, the ANE is highly specialized for neural networks. For applications outside ML, it does not provide as much flexibility as a general-purpose processor. ANE is only available in Apple's ecosystem, limiting its accessibility to developers who work with non-Apple hardware. Core ML is Apple's machine learning framework that allows developers to integrate pre-trained models into their apps. The framework decides whether to run the model on the CPU, GPU, or ANE based on the model's structure and the available hardware. ANE accelerates specific operations such as CNNs, and RNNs. Matrix Multiplications are common in fully connected networks. Core ML can automatically optimize models for further improving performance by quantizing models into low-precision formats, which the ANE can process faster.

Qualcomm's Hexagon DSP (Digital Signal Processor) is a specialized hardware accelerator designed to optimize performance for various computational tasks, particularly in mobile, embedded systems, and IOT. It specifically focusing on real-time, low-power AI tasks, making it ideal for applications in tinyML (tiny Machine Learning) where resource constraints are a significant concern [91, 92]. The Hexagon DSP architecture is integrated into Qualcomm's Snapdragon SoCs. The architecture includes features like vector processing units (VPUs), supported by Hexagon Vector Extensions (HVX), to enable more efficient vector processing capabilities. These VPUs are designed to handle multiple data points simultaneously, leveraging the principles of Single Instruction Multiple Data (SIMD) architecture. This capability allows the Hexagon DSP to execute operations on large data sets efficiently, which is particularly beneficial for applications requiring real-time processing, such as audio, video processing, and sensor data, as well as machine learning inference tasks [92–94]. Additionally, it supports a range of data types and operations, including fixed-point arithmetic, which is often used in machine learning models to reduce computational load and power usage [93]. The Hexagon DSP's architecture allows for the efficient execution of neural networks and other machine learning algorithms, enabling on-device inference without the need for cloud processing.

5.3 Application-Specific Integrated Circuit (ASIC)s

Offer the most freedom when developing coarse-grained processing elements. Designers have control over the precision of the computation and its hardware composition. As shown in Fig. 12, the ASIC has the smallest power consumption of the platform types, but can offer competitive GOP/s for some implementations.

Generally speaking, the advantages of ASICs mentioned above must be considered against their high cost (most expensive route compared with CPUs/GPUs/FPGAs) and longest time to market. Shuvo et al. offer a summary of current ASIC implementations for Neural Networks [95]. Their review indicates that ASIC designs are focused on being able to service many different NN structures, most commonly focusing on VGG-16, AlexNet. Almost all of them focus on image classification, segmentation, or analysis. One of the key design aspects here is the different modalities; designs now offer variable precision MAC units and as such variable GOP/W. They also have different architectural modules that may be switched off depending on the application.

5.4 Compute-in-Memory (CiM)

CiM is gaining traction as an effective approach for accelerating TinyML applications by addressing the memory wall challenge inherent in traditional computing systems, called Von Neumann. CiM offers system enhancements by reducing the energy and inference latency required for data movement between memory and processing units. These environments are promising solutions for systems typically operating on resource-constrained devices such as TinyML. CiM architectures mitigate this bottleneck by performing computation directly within the memory arrays; the in-situ data processing allows the running of complex machine-learning tasks on devices that operate on battery power [96, 97]. CiM further reduces the latency issue by allowing for parallel data processing directly within memory cells [98]. Low latency and parallelism are sectionals, especially in applications requiring real-time analysis. In addition, IMC reduces the exposure of TinyML applications to data corruption and noise, which are common challenges in resource-constrained environments as the computations are directly within memory [99, 100]. CiM can be implemented using digital or analog approaches, each leveraging different memory technologies as described in the following sections.

5.5 Digital in-Memory Computing (DiMC)

DiMC relies on binary arithmetic and logic operations, such as AND, OR, and NOR, to perform basic computational tasks directly within memory arrays. The DiMC architectures typically utilize standard memory technologies based on charge storage, such as static random access memory (SRAM), dynamic random access memory (DRAM), and Flash memory as storage mediums. The memory architecture in DIMC generally follows a row/column organization, where memory cells perform computations in parallel across rows, resulting in energy-efficient bulk operations like matrix-vector multiplication (MVM). The binary matrices and vectors are stored in memory, and MAC operations are conducted using the memory's inherent properties. In SRAM arrays capacitors can be added to SRAM arrays to accumulate MVM charge, enabling efficient matrix computations. Similar operations can be performed in DRAM by redistributing charge across multiple rows or columns.

5.5.1 Static Random Access Memory

SRAM plays a pivotal role in DiMC due to its fast access times and high cycling endurance. Each SRAM cell is typically composed of six transistors (6T) that form a bistable latch, enabling rapid switching between binary states. With sub-nanosecond access times (< 1 ns) and high endurance ($> 10^{16}$ cycles), SRAM is well-suited for

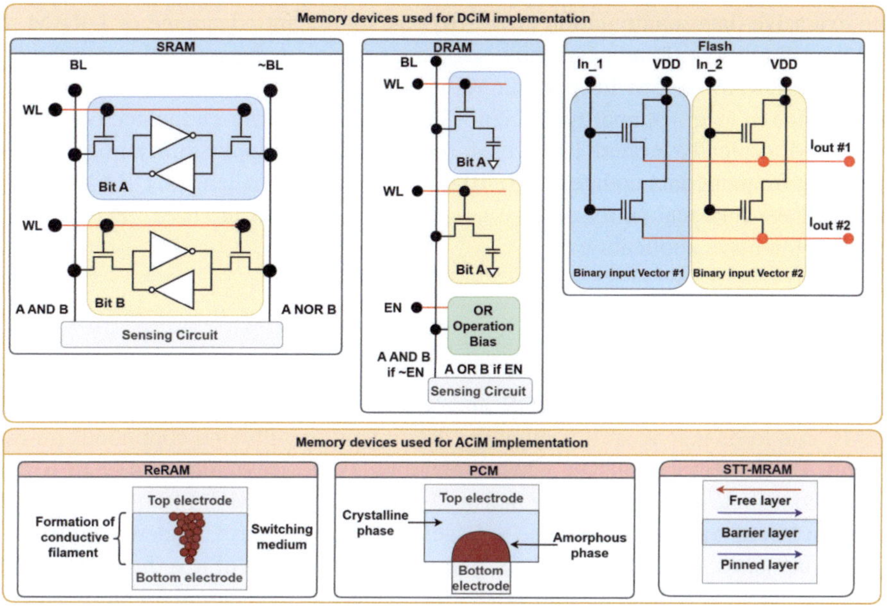

Fig. 18 CMOS-based digital memory devices and emerging analog memory devices used in compute-in-memory architectures for machine learning applications

DiMC applications that demand frequent data reads and writes, such as real-time AI inference and deep learning accelerators [101, 102].

One of the critical benefits of SRAM in DiMC is its ability to perform MVM and logical operations directly within the memory, as shown in Fig. 18, significantly reducing data movement and thus improving energy efficiency. However, due to the need for continuous power to maintain data, SRAM's high power consumption is a notable drawback. Additionally, SRAM's large cell size limits its scalability compared to other memory types [102]. Despite these challenges, SRAM remains a critical component in DiMC systems where low latency and high computational throughput are priorities.

5.5.2 Dynamic Random Access Memory

DRAM is another important memory type used in DiMC, primarily due to its high density and low cost per bit. DRAM cells consist of a capacitor and a transistor, enabling a smaller cell size around $6F^2$, which allows for higher storage density compared to SRAM [101, 103]. In DiMC, DRAM can execute basic logic operations such as AND and OR by manipulating charge redistribution across capacitors, as shown in Fig. 18. Although DRAM's access times are longer than SRAM's (around 10 ns), its capacity for large-scale data storage makes it suitable for applications involving extensive data manipulation [101, 102]. A significant advantage of DRAM in DiMC is its ability to handle bulk bitwise operations, enhancing parallel computation. However, DRAM's need for periodic refreshing to maintain data integrity presents a challenge in terms of energy efficiency. Moreover, its lower endurance compared to SRAM, typically around 10^9 cycles, further limits its applicability in scenarios requiring frequent data updates [101, 102]. Despite these challenges, DRAM's cost-effectiveness and scalability make it a valuable component in DiMC architectures for data-intensive applications.

5.5.3 Flash Memory

Flash memory, particularly in NOR and NAND configurations, is widely used in DiMC due to its non-volatility, allowing data retention without a continuous power supply. Flash memory cells are smaller, enabling a cell size of around $4F^2$ to $10F^2$, which contributes to high storage density [101, 102]. In DiMC, Flash memory can perform MVM by modulating gate voltages and storing matrix elements as charges on floating gates, as shown in Fig. 18, with typical read times of less than 100 ns and write latencies exceeding $\sim 10\mu$ s [101, 102]. NOR Flash is favored for DiMC due to its ability to support parallel access, which is critical for high-throughput operations. One of Flash's key advantages is its scalability, making it ideal for large-scale storage applications. However, Flash memory has lower endurance, typically around 10^5 cycles, and higher write energy requirements, often requiring voltages exceeding 10 V [101, 102]. Despite these limitations, the non-volatility and high density of Flash memory make it an attractive choice for DiMC applications that require long-term data storage and infrequent writes, such as machine learning inference and data archiving.

In summery DiMC is advantageous for applications requiring high accuracy and reliability, as digital circuits are less susceptible to noise and variability than their analog counterparts. Furthermore, DIMC can be integrated with existing digital processing architectures, making it easier to adopt in current systems.

5.5.4 Case Study of DiMC: XNOR-SRAM

The XNOR-SRAM is a 65-nm CMOS ML accelerator based on the DiMC concept [104], as shown in Fig. 19a. XNOR-SRAM offers efficient integration of computation within the memory array to enhance performance in binary and ternary DNNs. This architecture embeds ternary-XNOR-and-accumulate (XAC) operations directly into a 256-by-64 SRAM bit-cell memory array. By enabling simultaneous activation of all rows, XNOR-SRAM achieves analog accumulation of bitwise ternary-XNOR outputs on the read bit-line (RBL), which are then digitized using an 11-level flash ADC. Operating at a supply voltage of 0.6V, this direct in-memory computation eliminates the need for row-by-row memory access, reducing energy consumption and computation latency in conventional digital implementations.

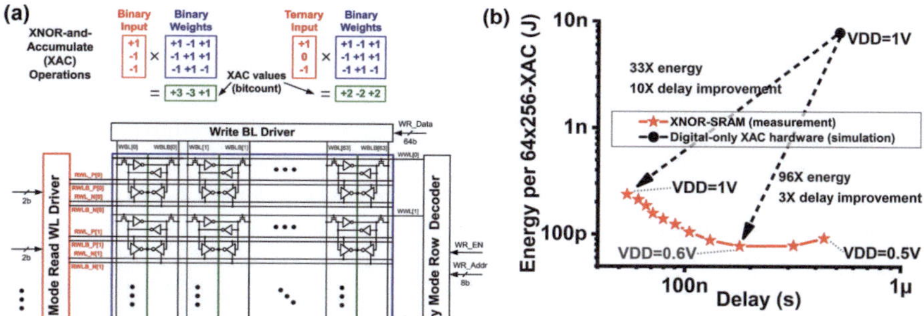

Dataset	MNIST		CIFAR-10			
DNN Model	MLP		VGG-like CNN		ResNet-14 CNN	
Activation Precision	Binary	Ternary	Binary	Ternary	Binary	Ternary
SW Baseline Accuracy	98.77%	99.07%	88.60%	90.70%	89.61%	90.80%
HW Measured Accuracy	98.65% (-0.12%)	98.84% (-0.23%)	87.23% (-1.37%)	88.78% (-1.92%)	85.72% (-3.89%)	87.05% (-3.75%)

Fig. 19 XNOR-SRAM overview: **a** Architecture: A 256-by-64 SRAM array performing in-memory ternary-XNOR-and-accumulate (XAC) operations with analog accumulation and flash ADC digitization. **b** Energy and Delay: XNOR-SRAM achieves 33× higher energy efficiency and 300× better energy-delay product than a digital baseline, with 403 TOPS/W at 0.6V. **c** Accuracy: Achieves 98.8% on MNIST and 88.8% on CIFAR-10 using ternary precision, combining efficiency and precision

Figure 19b shows that the XNOR-SRAM achieves a 33× improvement in energy efficiency and a 300× better energy-delay product while performing ternary-XAC operations with a 403 TOPS/W performance. This efficiency is accompanied by high test accuracy, reaching 88.8% for the CIFAR-10 dataset and 98.8% for MNIST when using ternary precision, as detailed in Fig. 19c. These results demonstrate the balance between energy efficiency and computational accuracy in this DiMC architecture.

The architecture supports binary weights and activations (+1, −1) and ternary inputs (+1, 0, −1), making it highly versatile for DNN applications. The linear quantization scheme incorporated into the flash ADC enhances CIFAR-10 accuracy by optimizing voltage intervals, improving the result from 85.7% to 88.8%. This improvement reflects the robustness of the XNOR-SRAM design in preserving precision despite the challenges of analog processing.

5.6 Analog in-Memory Computing (AiMC)

AiMC extends the conventional digital computing paradigm by capitalizing on continuous, non-binary resistive, or charge states in memory devices to perform complex arithmetic operations, such as MVM and finite-state machine (FSM) processing. Unlike digital approaches that rely on sequential processing, AiMC executes these operations directly within the memory array, enabling highly parallel and energy-efficient computation. The core principle of AiMC lies in leveraging Ohm's law and Kirchhoff's Law to perform in-situ arithmetic. In this approach, trained model parameters (such as weights or states) are encoded as resistive values within the memory array, while input data (such as encoded features) is applied as voltage signals. The interaction between these two operands generates accumulative currents at the array's output nodes, representing the computed results of MVM or other tasks. Resistive or memristive memory technologies, including Resistive RAM (ReRAM), Phase Change Memory (PCM), and Magnetoresistive RAM (MRAM), are essential to AiMC as these cells represent the learning elements. These technologies provide non-volatile storage, high density, and analog programmability, making them ideal for encoding model parameters and executing computations in memory. The analog outputs from AiMC can either be processed further in the analog domain or converted to the digital domain for subsequent processing. For conversion, techniques such as Analog-to-Digital Converters (ADCs), Current Sense Amplifiers (CSAs), or Transimpedance Amplifiers (TIAs) are employed.

AiMC has advantages over DiMC, particularly for machine learning applications and workloads that demand high throughput through massively parallel operations within memory arrays and low power consumption. Moreover, it leverages the physical properties of memory devices for computation, allowing higher computational density and faster execution of tasks like MVM. By integrating storage and computation, AiMC offers a promising solution to overcome the memory-wall bottleneck in traditional computing. In the following sections, we delve deeper into the operational principles and unique advantages of ReRAM, PCM, and MRAM in AiMC applications.

5.6.1 ReRAM Memory

ReRAM is a non-volatile memory technology that has emerged as a key enabler for AiMC due to its inherent resistive switching properties. ReRAM operates by modulating the resistance of a memory cell between a Low Resistance State (LRS) and a High Resistance State (HRS) using electrical SET and RESET pulses. Additionally, certain ReRAM devices can stabilize intermediate resistance states, enabling them to store analog data, an essential feature for performing in-memory computations like MVM [105].

Structurally, ReRAM consists of a simple yet versatile metal-insulator-metal (MIM) stack, where the insulator layer, typically made of transition metal oxides like hafnium oxide (HfO_2) [106] or titanium dioxide (TiO_2) [107], governs the resistive switching behavior. The switching mechanism relies on the formation and dissolution of conductive filaments (CFs) within the insulator, which are driven by the migration of oxygen vacancies or metallic ions. During the SET process, CFs bridge the electrodes, creating an LRS, as shown in Fig. 18. Conversely, the RESET process disrupts these pathways, transitioning the cell to the HRS. This robust switching mechanism enables ReRAM cells to act as tunable resistors, capable of encoding both binary weights by switching between the LRS and HRS and analog weights by leveraging the ability to stabilize resistance levels within the intermediate states between LRS and HRS [108].

For AiMC applications, ReRAM arrays are typically organized in dense two-dimensional crossbar architectures. This configuration enhances computational throughput by facilitating parallel operations across the ReRAM array. Each ReRAM cell in the crossbar represents a synaptic weight, and by applying input voltages (V_i) to the rows of the array, the cumulative current (I_i) measured at the columns represents the computational result. This process directly leverages physical laws such as Ohm's Law and Kirchhoff's Current Law to perform MVM efficiently:MVM $= \sum_{i=1}^{m}(G_i V_i)$ where G_i is the conductance of the ReRAM cell [109]. This in-memory approach eliminates the overhead associated with data movement between memory and processing units, offering significant energy and speed advantages. ReRAM's non-volatile feature eliminates the need for refresh cycles, significantly reducing power. Its ability to stack multiple layers of crossbar arrays further enhances storage density and computational capacity, making it well-suited for AI workloads and neuromorphic computing [109, 110]. Despite its promising features, ReRAM faces challenges such as device variability, limited write endurance, and non-linear switching characteristics [111, 112]. Addressing these challenges will require advancements in materials engineering, device design, and fabrication techniques.

5.6.2 PCM Memory

PCM is another promising candidate for AiMC, leveraging the unique properties of phase change materials such as $Ge_2Sb_2Te_5$ to achieve non-volatile storage. PCM exploits Joule heating to induce rapid transitions between amorphous (HRS) and crystalline (LRS) states, as shown in Fig. 18. By carefully tuning the energy of these pulses, intermediate resistance states can be achieved, allowing for precise analog storage. This characteristic is particularly beneficial for AiMC, where the continuous resistance range enables accurate representation of model parameters [113, 114].

PCM cells often feature a mushroom-shaped architecture, with a bottom electrode designed to confine heat and current for energy-efficient operations. In AiMC applications, PCM can be arranged in dense two-dimensional crossbar architectures, similar to ReRAM. PCM arrays are used to execute MVM directly in memory by encoding weights as conductance values and applying input voltages across rows of the array. The resulting cumulative currents at the output nodes correspond to the weighted sums, calculated using Ohm's and Kirchhoff's laws. PCM's accumulative behavior further supports iterative adjustments during tasks like neural network training, where conductance can be progressively tuned [115]. However, challenges such as high programming energy, thermal crosstalk, and material fatigue due to repeated phase transitions must be addressed. Despite these limitations, PCM remains a powerful tool for AiMC due to its precision, stability, and scalability [116].

5.6.3 Spin-Transfer Torque Magnetic RAM

STT-MRAM is an advanced non-volatile memory technology that leverages spintronics to provide high-speed, low-power, and high-density memory solutions. STT-MRAM operates by manipulating the spin of electrons to switch the magnetic state of a Magnetic Tunnel Junction (MTJ), which serves as the core building block of this memory technology. The MTJ consists of two ferromagnetic layers separated by a thin insulating barrier. One layer is fixed (the reference layer), while the other (the free layer) can change its magnetization direction, as shown in Fig. 18. When a current is passed through the MTJ, the spin-polarized electrons exert a torque on the magnetic moments of the free layer, enabling it to switch between parallel (LRS) and anti-parallel (HRS) configurations. This mechanism allows for binary data storage [117].

For AiMC applications, STT-MRAM can be arranged in dense two-dimensional crossbar architectures, similar to other in-memory computing technologies such as ReRAM. This structure allows for efficient MVM operations, where each cell represents a synaptic weight in a neural network. The rows and columns of the array can be activated simultaneously to perform computations, enabling efficient and rapid processing of large data volumes in learning applications. The resulting cumulative currents at the output nodes correspond to the computed results, leveraging Ohm's Law for direct in-memory processing [118].

STT-MRAM offers several advantages. Firstly, its non-volatility allows it to retain data even when power is removed. Secondly, STT-MRAM provides high-speed performance, with write speeds in the nanosecond range, faster than traditional Flash memory and comparable to SRAM. Additionally, STT-MRAM offers high density, achieving high integration densities that allow more data to be stored in a smaller physical area. Finally, it has high endurance, being capable of withstanding many write cycles without degradation [119–121].

However, STT-MRAM also faces several challenges that need to be addressed. One such challenge is write energy, as the energy consumption during write operations can be higher than that of SRAM, which may pose challenges in energy-sensitive applications. Another issue is process variations, as STT-MRAM devices are sensitive to variations in manufacturing processes, which can lead to inconsistencies in performance and reliability [119]. Thermal stability is also a concern, as elevated temperatures can affect the reliability and performance of STT-MRAM cells [122, 123]. Finally, the complexity of control circuits is another challenge, as precise control of the write and read operations in STT-MRAM can complicate the design of the associated circuitry, potentially increasing the overall system complexity [124, 125].

5.6.4 Case Study of AiMC: The IMBUE

The IMBUE in [126] introduces a novel accelerator design aimed at improving the energy efficiency and performance of Tsetlin Machines using in-memory computing

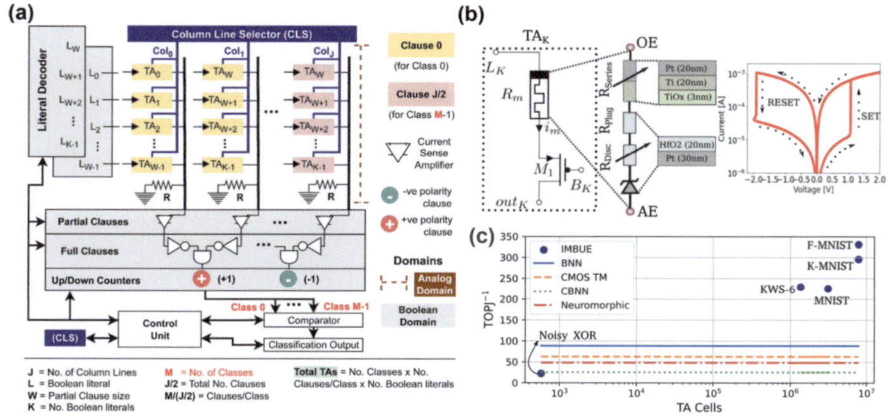

Fig. 20 **a** TM inference architecture using IMBUE. **b** An equivalent circuit for a TA is represented by a 1T1R cell, and **c** IMBUE energy efficiency for TM inference is applied to various datasets

principles as shown in Fig. 20. The IMBUE architecture leverages ReRAM technology, integrating computation and memory to address von Neumann bottlenecks in traditional machine learning ML systems. The IMBUE accelerator interacts directly with the input Boolean feature as digital voltages by translating them into current paths and exploiting the simplicity of Tsetlin Automata for logic inference tasks. This Boolean-to-Current mechanism enabled the streamlining of the inference process and eliminated the requirement for high precision analog to digital converters, thus significantly reducing circuit overheads [98, 127]. The design is organized into partial clauses to handle the inherent non-linearity in ReRAM devices and to mitigate sneak currents. IMBUE minimizes power consumption by exploiting the sparse activation of TAs (mostly "exclude" actions), leading to a remarkable up to 12.99x improvement over state-of-the-art binarized convolutional neural networks and up to 5.28x improvement over digital TM implementations. Critical results include energy efficiency evaluations across datasets like MNIST and Fashion-MNIST, demonstrating up to 5.283x energy reduction. By addressing the challenges associated with traditional computing architectures and leveraging the unique properties of in-memory computing, It is expected to see increasingly sophisticated implementations that push the boundaries of what is possible in TinyML and artificial intelligence.

5.7 *Logic-Based ASICs: Tsetlin Machine Accelerators*

The design in [128] presented a custom ASIC accelerator to achieve energy-efficient machine learning tailored for IoT applications. The ASIC utilizes the Tsetlin Machine, a logic-based algorithm that replaces computationally intensive neural network operations with propositional logic, significantly reducing energy consumption

Table 3 Comparison of ASIC and FPGA implementations for noisy XOR and binary Iris datasets [128]

Feature	Noisy XOR (ASIC)	Noisy XOR (FPGA)	Binary Iris (ASIC)
# Inputs, Clauses, Classes	12, 10, 1	12, 10, 1	16, 20, 3
ASIC area (mm^2)	0.246	–	0.386
Frequency (MHz)	118	110	33.3
Training time (ms)	4.24	4.55	1.80
Avg. power (training)	16.8 μW	1.10 W	1.85 mW
Energy/datapoint (training)	142 fJ	10.0 nJ	55.6 pJ
Energy/datapoint (inference)	–	–	30.6 pJ

and hardware complexity. The binarized data encoding process reduces the feature space while maintaining classification accuracy. For example, feature encoding was reduced by 2.67x in the binary Iris dataset, enabling a compact and efficient hardware implementation. The architecture further supports on-chip learning, which allows real-time adaptability to changing environments, and uses an FPGA-based platform for rapid hyperparameter tuning.

The ASIC performance on the Iris dataset is a major highlight. Using the TM with the configuration of 16 inputs, 100 clauses, and 3 classes, the design achieved an accuracy of 96.3% for training and 97.0% for testing, comparable to state-of-the-art solutions. The implementation demonstrated an area of 0.386 mm^2, operating at 33.3 MHz, and a training power of 1.85 mW, as shown in Table 3. Energy consumption per datapoint was measured at 55.6 pJ for training and 30.6 pJ for inference, underscoring its suitability for low-power IoT environments. The ASIC leverages 65 nm technology, operating at a voltage of 1.0 V, with an architecture that prioritizes efficiency through logic-based clauses and power-gated reinforcement blocks, achieving performance of 62.7 Top/J, surpassing other accelerators such as BNNs. By integrating energy-efficient hardware with adaptable logic-based learning, this ASIC sets a new prototype for edge AI in IoT applications.

This section examined **ASIC** advancements for efficient computing, highlighting Compute-in-Memory approaches such as **DiMC** and **AiMC**, which utilize technologies like SRAM, DRAM, FLASH, ReRAM, STT-MRAM, and PCM for energy-efficient processing. Case studies, including XNOR-SRAM, IMBUE, and logic-based designs, illustrated the practical benefits of these accelerators in optimizing ML performance and power usage.

5.8 *The Generic Route: CPUs and GPUs*

5.8.1 RISC-V Accelerators

RISC-V (Reduced Instruction Set Computing) is an open-source instruction set architecture (ISA) that has gained significant traction in the TinyML community and various computing domains [129]. One of the most compelling features of RISC-V is its modularity. The architecture consists of a base instruction set that can be extended with optional features tailored to specific applications. This design flexibility and The open nature of RISC-V encourage the developers to create customized processors that meet the unique requirements for low-power embedded systems such as CNN accelerators, edge intelligence applications, and network packet processing. The following are the key features of RISC-V:

- **Modularity and Extensions** RISC-V is built around a core integer instruction set that can be enhanced with optional extensions. These modules add specialized functionalities for different use cases. For instance, M (Multiplication/Division) adds support for integer multiplication and division. A (Atomic) enables atomic operations for multi-threaded environments.F/D/Q (Floating Point) supports single, double, and quad-precision floating-point arithmetic. C (Compressed) reduces instruction size to save memory and bandwidth. V (Vector) provides vector-based computation for high-performance and AI workloads.
- **Base ISA Flexibility** RISC-V offers three major base ISAs are RV32I (32-bit), RV64I (64-bit), and RV128I (128-bit) these frozen standards ensure long-term stability and backward compatibility.
- **Flexible Register Design** RISC-V utilizes 32 general-purpose registers (x0–x31). The zero register (x0) is hardwired to 0, simplifying certain operations. the remaining registers can be customized based on Application Binary Interface (ABI) requirements.
- **Instruction Formats** RISC-V uses six main instruction formats (R, I, S, U, B, and J), all of which are 32-bit wide. This consistency simplifies instruction decoding and improves pipeline efficiency, making RISC-V easier to work with than ARM or x86 architectures.

RISC-V has been implemented in various cores, processors, and SoC platforms, demonstrating its versatility across numerous applications. The PULP(Parallel Ultra-Low-Power Processing) platform is a silicon-proven framework that provides highly energy-efficient solutions for IoT and embedded applications [130]. Developed by ETH Zurich and the University of Bologna, PULP leverages clusters of RISC-V cores with shared tightly coupled data memory. It offers scalability for a wide range of applications, from simple microcontroller systems to advanced multi-cluster accelerators. PULP's architecture is implemented using RISC-V-compatible cores written in SystemVerilog, and it integrates a variety of peripherals, custom accelerators, and interconnect solutions. PULP features several processors, including CV32E40P (RI5CY), a 32-bit, 4-stage core with DSP-specific extensions; Ibex (Zero-riscy), a 2-bit, 2-stage core optimized for control applications and area efficiency; Micro-risky provides a 32-bit, 2-stage core with 16 registers, implementing RV32-EC without a hardware multiplier; CVA6 (Ariane), a 64-bit,6-stage core supporting Unix-like operating systems offering three privilege levels (M, S, U) for effective OS management; and Another notable core is Snitch, a single-stage, 32-bit, 1 stage core optimized for energy efficiency. By externalizing the FPU and focusing on simple, single-issue operations, Snitch aims to maximize computational throughput while keeping power consumption low. The simpler PULP-based microcontroller systems, PULPino and PULPissimo. PULPino is a minimal single-core RISC-V SoC that has attracted attention for its simplicity and effectiveness in embedded applications. PULPissimo, a more advanced version, introduces a logarithmic interconnect that allows multiple access ports between the core and the memory subsystem. It is integrated with a micro-DMA (μDMA), enabling peripherals to interconnect to the memory [130].

5.8.2 Single Board Computer (SBC)

SBC is a compact computing device that integrates all the essential components of a computer onto a single circuit board. SBC has a CPU, memory RAM, and various I/O interfaces for peripheral connectivity. The compact nature of SBCs and the advancements in technology enable them to address increasingly complex computations, including machine learning tasks particularly suitable for at the edge, where limited resources (capacity and power) and real-time efficiency are critical [131, 132].

Raspberry Pi: is among the most popular SBCs is the Raspberry Pi, developed by the UK-based Raspberry Pi Foundation, established in 2009 with the primary mission of advancing computer science education in schools [133]. The first iteration, the Raspberry Pi Model B, debuted in February 2012, featuring 256MB of RAM and a single-core ARM11 processor at 700 MHz. This initial model was widely utilized in academic and hobbyist projects due to their versatility, affordability,and a rapidly expanding support community [134]. Since then, Raspberry Pi has experienced several iterations, each improving in terms of performance, connectivity, and capabilities. The Raspberry Pi 2 2015 upgraded with a more powerful quad-core ARM Cortex-A7 processor and 1GB of RAM, allowed for more challenging applications [135]. To enhance usability in IoT applications, built-in Wi-Fi and Bluetooth were added to the Raspberry Pi 3, followed in 2016 [133]. The Raspberry Pi 4, released in 2019, marked a significant advancement, with enhanced graphics capabilities, support for dual 4K HDMI outputs, USB 3.0, and a memory capacity of up to 4GB of RAM, establishing it as a practical SBC option for lightweight computing applications [135]. The most recent model, the Raspberry Pi 5, was released in October 2023. It was equipped with a quad-core ARM CPU of 2.4 GHz speed and 8 GB RAM. The Raspberry Pi 5's compatibility with running the in-house Raspberry Pi OS as well as Android 15 makes it well-suited for demanding TinyML applications.

Raspberry Pi is A Pioneering SBC in TinyML:The integration of Raspberry Pi in TinyML applications presents several advantages that enhance the efficiency and effectiveness of deploying tinuML models on edge devices.

- **The versatility and Low cost** One of the primary benefits of using Raspberry Pi in TinyML is its affordability and accessibility. The low cost of Raspberry Pi devices allows for widespread adoption in various TinML applications due to its ability to interface with various sensors, and actuators, which enhance the Raspberry Pi's versatility. The Raspberry Pi's GPIO (General Purpose Input/Output) pins facilitate multiple sensor connections, allowing concurrent data processing, a vital feature for real-time applications that gather data from diverse environmental sources [136]. This local processing reduces latency and minimizes bandwidth usage associated with cloud computing, making it especially valuable for TinyML applications in remote or power-limited environments [137].
- **Small form factor** Another significant advantage of utilizing Raspberry Pi in TinyML is its processing power relative to its small form factor. It supports the efficient execution of lightweight machine-learning models in space-constrained environments [138]. For instance, in a study focused on air quality prediction, the Raspberry Pi was employed for real-time inference to run CNNs for accurate hourly forecasts, demonstrating its ability to handle complex computations at the edge [136]. Moreover, Creating clusters of SBCs further enhances their computational capabilities, allowing for distributed processing and improved performance in more complex ML tasks [139].
- **Libraries** Because of the Raspberry Pi's flexibility, developers may take advantage of pre-existing TinyML frameworks and tools, which makes model deployment and prototyping easier[o25]. It has broad support for a variety of IoT and TinyML applications.The main operating system, Raspberry Pi OS is equipped with pre-installed libraries and programming tools, making it the perfect environment for developers working with Python, C++, and Java. furthermore, a vast array of libraries and frameworks that support development across all areas enhance the Raspberry Pi ecosystem, including OpenCV [140], TensorFlow Lite [141], and the Node-RED platform [142]. Additionally, The Raspberry Pi community plays a crucial role in supporting developers through forums, tutorials, and documentation such as GitHub and Stack Overflow.

The Raspberry Pi's processing capability allows it to implement lightweight RNNs or LSTM networks [143]. These scenarios, where high-frequency data streams are involved, typically require more memory and processing power; in return, it may limit Raspberry Pi's effectiveness in handling complex models. Nonetheless, optimizations such as quantization and pruning can be employed to reduce the model size and computational requirements, making it feasible to be on the Raspberry Pi and underscoring its suitability for TinyML deployments [144].

Microcontroller Units (MCUs): Are another example of SBC. MCUs are designed to govern operations in embedded systems, where TinyML systems can input direct data from various sensors. It can use a micro, nano level neural network. The system can accommodate an MCU with or without hardware accelerators.Additionally, MCUs allow for real-time data processing and decision-making without relying on cloud resources. They are also supported by an extensive ecosystem of libraries and development tools. One prominent MCU vendor is **STMicroelectronics**, which uses Arm Cortex-M processors as the core CPUs in its STM32 series of MCUs [145]. The STM32 included different models to meet the most demanding balance between the computation performance and the Power requirements for tinyML applications. For instance, the STM32H7 series offers high performance based on the 32-bit Arm Cortex-M7 core, running at up to 600 MHz, embedded flash memory of up to 2 Mbytes, and up to 1.4 Mbytes of SRAM. STM32U5 is another series optimized for ultra power-saving down to 16uA/MHz in the active mode and 110nA in the low power mode. It was built based on an Arm Cortex-M33 of 160MHz speed and equipped with up to 4 Mbytes of flash memory and 3Mbytes of SRAM [145].

Microchip is another vendor offering a wide range of MCUs that span from 8-bit to 32-bit architectures [146]. The 8-bit-based MCUs, such as PIC16 and AVRa6 series, are suitable for basic control and data acquisition applications. The 16-bit-based MCUs, like the 16-bit PIC24F MCU series, are affordable solutions for many applications. These MCUs are equipped with processing up to 70 MHz, nonvolatile memory 1024Kbyte, and SRAM up to 52 Kbyte. For applications demanding higher performance, microchip assembled the 32-bit MCUs, like the PIC32 family, operating up to 300 MHz, with memory of 8192 KB and 1024 KB RAM. These 32-bit based MCUs provide the performance and functional capabilities to meet the requirements of modern electronics from general-purpose embedded control to secured IoT solutions. **Texas Instruments (TI)** is another manufacturer supplying the 16-bit MSP430 MCUs to provide affordable solutions for all applications [147]. MSP430 is processing up to 25MHz and has installed 512Kbyte and up to 66Kbyte flash and SRAM memory onboard, respectively as shown in Fig. 21. MSP430 is supported by a design resources software called MSP430Ware software and is often used in sensor-based TinyML tasks like anomaly detection [148].

Single Board Computers represent a significant advancement in computing, combining all essential components on a single board for versatile and cost-effective applications. As technology advances, the capabilities of SBCs continue to expand, reinforcing their pivotal role in modern computing. The evolution of the Raspberry Pi reflects the impact of accessibility and innovation on computing. It has transcended its initial function in education to evolve a highly flexible platform for various applications, including TinyML. Moreover, the various vendors offer specialized MCUs tailored to different applications to be implemented efficiently on-device processing. The persistent advances in MCU technology are paving the way for more sophisticated and capable TinyML devices.

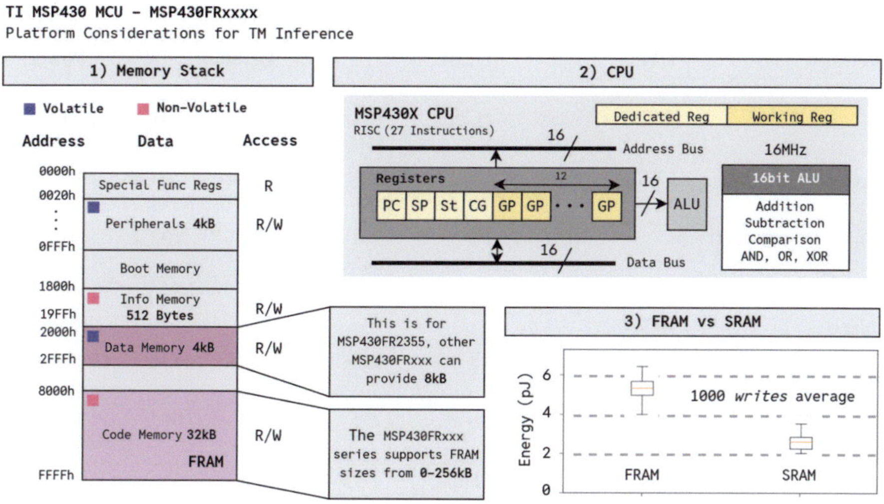

Fig. 21 Stylized MSP board. Stylized overview of the compute and memory capabilities of the MSP430FRxxxx series (FR indicates it uses FRAM). **(1) Memory Stack** shows the contiguous memory segments that can be traversed through the CPU's stack pointer (SP). **(2) CPU** shows the MSP430X CPU, a 16bit CPU with limited ALU capability running at around 16MHz. It is connected to 16bit address and data busses. It has 4 dedicated registers: Program Counter (PC), SP, Status Register (St) and Constant Generator (CG). It also has 12 General Purpose registers (GP). **(3) FRAM vs SRAM** shows the average energy expense of writing to FRAM vs SRAM memory

5.8.3 GPU Accelerators (NVIDIA Jetson)

The NVIDIA Jetson platform is a series of embedded computing devices designed to accelerate AI and ML applications at the edge. These devices integrate powerful GPUs with CPUs, enabling efficient processing of complex neural networks through featuring multi power modes, which allow to switch between performance and energy-saving modes, making it adaptable for TinyML use. The Jetson family includes several models, such as Nano, TX2, and Xavier NX, each tailored for different performance and application requirements, as shown in Table 4. The Jetson platform leverages NVIDIA's CUDA architecture, allowing for parallel processing of neural network computations. Jetson supports popular deep learning frameworks such as TensorFlow, PyTorch, and Caffe, facilitating the development and deployment of machine learning models. Its specialized architecture, support for deep learning frameworks, and high performance make it an ideal choice for a wide range of applications, including robotics, healthcare, environmental monitoring, and real-time object detection and tracking to enable to navigate and interact with their environment effectively [149].

Table 4 Jetson series specifications [149]

Jetson series	AI (TOPS)	GPU		CPU		Memory		Power
		Core	Max GHz	Core	Max GHz	GB	GB/s	(W)
AGX Orin	200–275	1792–2048	0.93–1.3	8–12	2.0–2.2	32–64	204.8	15–75
Orin NX	70–100	–	0.765–0.918	6-8	2.0	8-16	102.4	10–25
Orin Nano	20–40	512–1024	0.625	6	1.5	4-8	34-68	7–15
AGX Xavier	30–32	512	1.211–1.377	8	2.0–2.2	32–64	136.5	10–40
Xavier NX	21	384	1.1	6	1.9	8–16	59.7	10–20
TX2	1.26–1.33	256	1.12–1.3	Dual Denver2 + quad Cortex-A57	1.95–2.0	4–8	51.2–59.7	7.5–20
Nano	472	128	0.921	Quad Cortex-A57	1.43	4	25.6	5–10

6 Automation, Libraries, and Toolchains

6.1 MATADOR

The MATADOR is another tool that introduces a solution for automating the design and deployment of Tsetlin machine based accelerators on System-on-Chip FPGA (SoC-FPGAs), addressing the constraints faced in edge computing [150]. The MATADOR framework automates the translation of TM models into Register Transfer Level (RTL) designs. Moreover, it significantly facilitates the creation of machine learning inference accelerators for resource constrained environments. One of the key contributions of MATADOR is its ability to outperform state-of-the-art accelerators like Quantized and Binary Neural Networks (QNN/BNN), achieving up to 13.4x faster performance, up to 7x better resource efficiency and 2x lower power consumption. This is made possible by using the inherent sparsity and logic sharing characteristics of TM models, which allows MATADOR to generate compact designs without sacrificing performance. Running on the MNIST dataset, MATADOR achieves a throughput of 3.85 million inferences per second compared to the 954,000 inferences per second of a state-of-the-art BNN, with a corresponding latency reduction to 0.32 μs from 1.047 μs. Additionally, MATADOR reduces resource usage on FPGAs, using only 8,709 Look-up Tables (LUTs) compared to

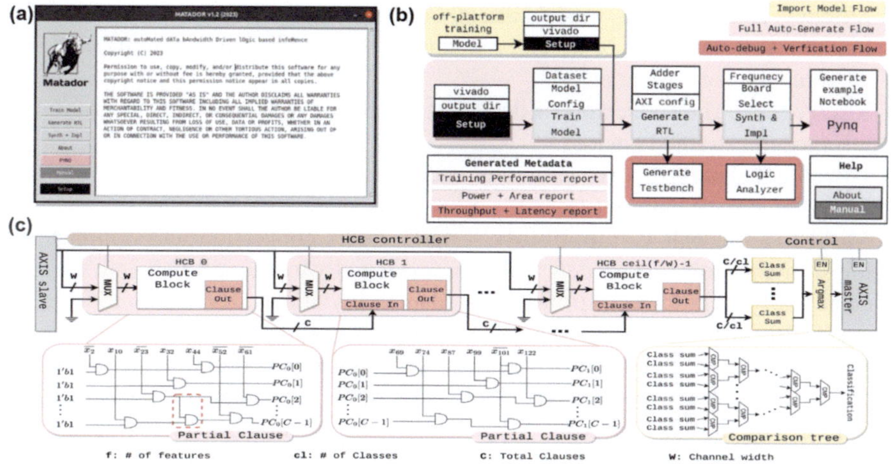

Fig. 22 Matador design flow. **a** Matador's GUI. **b** the design generating flow. **c** Block diagram of the generated MATADOR inference accelerator architecture

the framework FINN's 11,622 while also lowering dynamic power consumption to 1.29W versus FINN's 1.458W. The MATADOR inference architecture is explained through a block diagram, see Fig. 22, demonstrating how the accelerator efficiently handles large-scale data inputs by splitting computations into Hard Coded Clause Blocks (HCBs). This framework offers an excellent automated solution for deploying a resource frugal ML models at the edge accelerators.

6.2 *LiteRT: The Next Evolution of TensorFlow Lite*

LiteRT (short for Lite Runtime) is the new name for TensorFlow Lite (TFLite), marking an expanded vision for Google's high-performance runtime for on-device applications [151]. By supporting multiple frameworks and focusing on the constraints of edge computing, LiteRT enables developers to create sophisticated ML applications that are fast, private, and efficient. LiteRT supports models created not only in TensorFlow but also in PyTorch, JAX, and Keras. This versatility aligns with LiteRT's commitment to enabling developers to deploy models across various platforms while maintaining exceptional performance. LiteRT has been specifically tailored for microcontrollers, enabling machine learning on devices with very limited resources. With a core runtime that fits within 16 KB of memory on an ARM Cortex M3, LiteRT facilitates the deployment of basic models without needing extensive operating system support or dynamic memory allocation. This capability opens the door to smart appliances and IoT devices that can operate intelligently in real-time, all while preserving user privacy by keeping data local. LiteRT is optimized for on-device machine learning by addressing the following constraints:

- **Latency**: Executing ML models directly on the device.
- **Privacy and Security**: User data remains on the device, where LiteRT operates without an internet connection.
- **Binary Size**: The framework is designed to reduce both model and data sizes on resource-constrained devices.
- **Power Efficiency**: LiteRT promotes efficient inference, conserving power and extending device battery life.
- **Multi-Framework Compatibility**: LiteRT offers conversion and optimization tools for transforming models from TensorFlow, PyTorch, and JAX into the efficient FlatBuffers format (.tflite). This adaptability allows developers to leverage state-of-the-art models regardless of their initial training framework.
- **Integrate the Model**: LiteRT provides APIs in multiple programming languages (Java/Kotlin, Swift, Objective-C, C++, Python) for easy integration into applications across various platforms.
- **Pre-Trained Models**: LiteRT offers a rich library of pre-trained models, enabling quick deployment of machine learning functionalities for tasks like Vision APIs (ImageClassifier, ObjectDetector, ImageSegmenter, ImageSearcher, ImageEmbedder), Natural Language (NL) APIs (NLClassifier, BertNLClassifier, BertQuestionAnswerer, TextSearcher, TextEmbedder), Audio APIs (AudioClassifier), and more. Furthermore, LiteRT utilizes delegates to enhance performance through hardware acceleration, leveraging on-device accelerators such as GPUs, CPUs, NPU, and DSPs.

6.3 Edge Impulse

Edge Impulse is a powerful cloud-based platform designed to facilitate the development of machine learning models specifically for TinyML-targeted edge devices. With support for various sensors and hardware platforms, Edge Impulse enables developers to create intelligent applications that can process data directly on edge devices, ensuring low latency and real-time responses. The platform features a user-friendly interface that guides users through the entire machine learning pipeline, from data collection and preprocessing to model training and deployment. Thousands of embedded developers rely on Edge Impulse to run critical ML projects across millions of data samples. The platform addresses the significant challenge of limited memory on Arm Cortex-M MCUs, which typically offer less than 128K of RAM, making it difficult to build, test, and deploy embedded ML models. Edge Impulse has introduced the Edge Optimized Neural (EON) Compiler to overcome this limitation. The EON Compiler accelerates embedded ML code to operate with 25-55% less RAM and up to 35% less flash memory, with the same accuracy as TensorFlow Lite for Microcontrollers. The EON Compiler compiles neural networks to C++. This process simplifies the code, reduces device power consumption, and saves valuable development time [152].

6.4 uTensor

uTensor is an open-source, lightweight machine learning inference framework optimized for resource-constrained environments, particularly microcontrollers, and IoT-edge devices. Built on top of TensorFlow and Mbed, uTensor is designed to enable rapid prototyping and deployment of machine learning models in devices with minimal computational resources. It converts the pre-trained model from frameworks like Keras into compact, self-contained C++ code suited for deployment on embedded platforms such as Arm-based microcontrollers, where the framework requires as little as 2 KB of memory for its runtime. uTensor's architecture supports convolutional neural networks and pooling layers and utilizes quantization techniques to reduce model size. Its lightweight and flexible design plays a role in advancing the TinyML ecosystem [153].

6.5 The Embedded Learning Library (ELL)

ELL is a framework developed by Microsoft that enables the design and deployment of intelligent machine learning models on resource-constrained platforms such as Raspberry Pi and Arduino. ELL allows models to run on devices without requiring cloud access, making it ideal for applications that demand real time and low latency processing. It serves as a powerful tool for developers, empowering them to create AI-driven solutions on small, single-board computers. ELL functions as a cross compiler, written in C++ with an optional Python interface to generate machine code on a PC that can be executed on embedded devices. It currently supports tasks such as image and audio classification. Though still in active development, ELL is its focus on model optimization and size reduction ELL toward AI for embedded systems [154].

6.6 STM32Cube.AI

STM32Cube.AI is an optimization tool to deploy AI models into resource constrained devices. It automatically converts the pre-trained neural networks and classical machine learning models into optimized C code for STM32 microcontrollers. STM32Cube.AI supports deep learning frameworks like Keras and TensorFlow Lite, as well as models from PyTorch and MATLAB using the ONNX format, as shown in Fig. 23. It supports various scikit-learn models such as (isolation forest, support vector machine (SVM), and K-means) as well as quantized neural networks. Moreover, the relocatable option for model updates makes STM32Cube.AI an essential resource-efficient tool for developers. STM32Cube.AI's optimizations ensure that

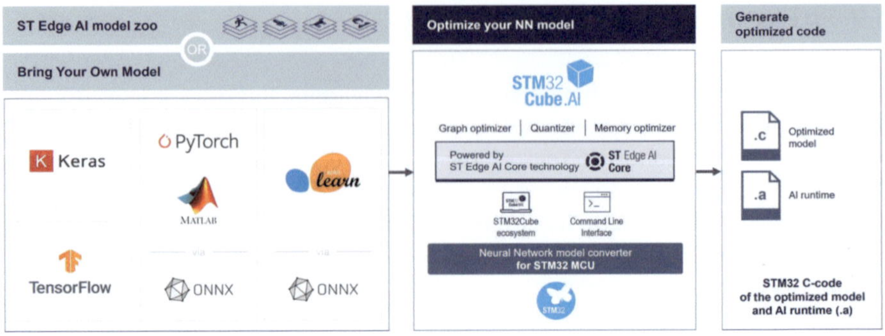

Fig. 23 STM32Cube.AI framework for deploying AI on STM32 MCUs [155]

complex neural networks can run efficiently, making it suitable for real-time applications in areas such as IoT, predictive maintenance, and autonomous systems [155].

6.7 *Other Tools*

Include NanoEdge AI Studio [156], which facilitates optimal library selection and performance testing before edge deployment, supporting STM32 Nucleo-32 and Arduino Nano 33 IoT boards for anomaly detection and classification tasks. PyTorch Mobile [157] enables executing ML models on edge devices, from the training to deployment within the PyTorch ecosystem, while supporting privacy features like federated learning. Currently in beta, it provides APIs for preprocessing, tracing, and scripting, and integrates XNNPACK and QNNPACK for optimized performance on iOS, Android, and Linux. μTVM extends TVM's compiler for ARM microcontroller targets, enabling on-device autotuning and supporting standalone model inference on various hardware, including STM Nucleo-F746ZG and nRF 5340 Development Kit [158]. μTVM optimizes deep learning workloads across diverse hardware, while MinUn addresses memory management and efficiency in TinyML applications [159]. Additional tools to explore include Keras [160], OpenVINO [161], ONNX Runtime [162], SODA [163], and Cainvas [164].

7 Conclusion

TinyML is an innovative approach for enabling the execution of machine learning algorithms on devices with restricted resources, such as microcontrollers, which are frequently used in IoT applications. TinyML has a lot of potential, but in order for it to be widely used and be effective, a lot of challenges must be overcome. In this section, we look at some of the challenges, identify future directions, and provide a roadmap.

7.1 Challenges of TinyML

Resource Constraints: One of the most significant challenges in TinyML is the limited computational power, memory, and energy resources available on edge devices. These constraints necessitate the development of lightweight and highly efficient algorithms that can operate within the confines of microcontroller units. These devices typically run at clock speeds as low as 40-100 MHz and often have less than 1MB of flash memory and SRAM. The trade-off between model complexity and performance is a critical consideration, as more sophisticated models often require more resources than are available on typical IoT devices.

Model Compression and Optimization: To fit machine learning models into the constrained environment of TinyML devices, techniques such as model pruning, quantization, and knowledge distillation are widely used. However, these methods often lead to reduced model accuracy. Achieving the right balance between accuracy and model size is still a major area of research. While advances in hardware (e.g., Tensor Processing Units and Neural Processing Units) have helped, the optimization of models for TinyML still presents significant challenges.

Benchmarking and Standardization: To advance the field of TinyML, establishing a comprehensive benchmarking framework is essential. TinyML models, often deployed in dynamic environments such as wearable devices or remote sensors, face challenges in performance evaluation due to the need for standardized benchmarks. Current benchmarks, designed for more traditional ML models, need to be adapted to TinyML's constraints. These benchmarks must be designed to support multilevel quantization and accuracy and have enough flexibility to accommodate the heterogeneity of accelerators and platforms. Furthermore, standards are crucial to ensure model interoperability, enabling models to be reused across different platforms. It also promotes scientific comparisons and innovation while reducing development time.

Energy Efficiency: Energy consumption is one of the critical bottlenecks for TinyML, especially in battery-powered edge devices. TinyML system must accommodate energy efficiency in its model's features, where it is designed to operate on minimal energy for extended periods. Despite advancements in energy-efficient algorithms and hardware, achieving a balance between low power consumption and high computational performance is still a major challenge. TinyML models require a certain level of energy to maintain the accuracy level of the algorithms. Hence, It is challenging to specify a universal power-management module for various platforms and applications. Research into energy harvesting for low-power accelerators is crucial for making TinyML applications more possible in the real world.

Security and Privacy: Due to the distributed nature of TinyML, where models operate on decentralized devices that often handle sensitive data, security and privacy are major concerns. TinyML systems must implement secure protocols for communication, data handling, and inference without overburdening the device's limited resources. Privacy preserving techniques, such as on-device processing and encrypted model inference, are increasingly necessary to safeguard data and prevent vulnerabilities. However, the limited resources available for this technology are an obstruction.

7.2 Future Directions

Advanced algorithms for more sophisticated algorithms that can operate efficiently on low resource devices should be in future research. This includes exploring alternative ML architectures specifically designed for TinyML applications, such as lightweight CNNs, RNNs, and SNNs in analog computation that are under research for tinyML applications. These architectures should balance between performance and energy efficiency.

Enhancing the capabilities of TinyML through leveraging methods like federated learning and collaborative approaches. Through Federated learning, multiple devices collaboratively learn from distributed data while maintaining privacy, where only the updates are shared while the raw data never leaves the local device. This helps mitigate concerns about data privacy and security breaches. TinyML can not only improve the model accuracy by implementing collaborative learning but also enhance the model robustness when the data may be heterogeneous and noisy.

Additionally, the integration of TinyML with edge computing can expand its capabilities further. Edge computing enables data processing and decision-making to occur close to the source of data, which is crucial for applications that require real time responses, such as smart home devices, autonomous vehicles, and health monitoring systems. By offloading tasks from the cloud to edge devices, this integration reduces latency, improves bandwidth efficiency, and ensures faster response times, which leads to more complex and sophisticated applications.

ML inference demands significant memory support, where memory facing challenges similar to those in general computing. Memristors and emerging non-volatile memory technologies present promising solutions. Despite the inefficiency of Von Neumann architectures for ML, cost constraints might limit us to using small, inexpensive microcontrollers based on this architecture. This requireing enhancing design through improved training methods, datasets, optimizations techniques . Additionally, exploring new computing paradigms, such as in-memory computing, could benefit ML applications but requires more research and development for incorporation into tinyML systems.

Furthermore, interdisciplinary collaboration is critical to unlocking the full potential of TinyML. Bringing together experts in computer science, electrical engineering, and data science can spark innovative solutions to the challenges TinyML faces and broaden its applications across different fields. To broaden the adoption of TinyML, educational initiatives aimed at increasing awareness and understanding of this technology are essential. Developing accessible resources and training programs can empower a new generation of developers and researchers to contribute to the TinyML ecosystem.

References

1. W. Shi, J. Cao, Q. Zhang, Y. Li, and L. Xu, "Edge computing: vision and challenges," *IEEE Internet of Things Journal*, vol. 3, pp. 637–646, 2016.
2. T. Meuser, L. Lovén, M. Bhuyan, S. G. Patil, S. Dustdar, A. Aral, S. Bayhan, C. Becker, E. d. Lara, A. Y. Ding, J. Edinger, J. Gross, N. Mohan, A. D. Pimentel, E. Rivière, H. Schulzrinne, P. Simoens, G. Solmaz, and M. Welzl, "Revisiting edge ai: Opportunities and challenges," *IEEE Internet Computing*, vol. 28, no. 4, pp. 49–59, 2024.
3. T. S. Ajani, A. L. Imoize, and A. A. Atayero, "An overview of machine learning within embedded and mobile devices–optimizations and applications," *Sensors*, vol. 21, p. 4412, 2021.
4. V. J. Reddi, B. Plancher, S. Kennedy, L. Moroney, P. Warden, L. Suzuki, A. Agarwal, C. Banbury, M. Banzi, M. R. Bennett, B. Brown, S. Chitlangia, R. Ghosal, S. Grafman, R. Jaeger, M. Lam, D. Leiker, C. Mann, M. Mazumder, D. Pajak, D. Ramaprasad, J. Smith, M. Stewart, and D. Tingley, "Widening access to applied machine learning with tinyml," *Harvard Data Science Review*, 2022.
5. M. E. Monteiro and A. Nardi, "Evaluation of the energy viability of smart iot sensors using tinyml for computer vision applications: a case study," *International Robotics & Automation Journal*, vol. 9, pp. 78–85, 2023.
6. O.-C. Granmo, "The tsetlin machine – a game theoretic bandit driven approach to optimal pattern recognition with propositional logic," 2021. [Online]. Available: https://arxiv.org/abs/1804.01508
7. J. Morales-García, A. Bueno-Crespo, R. Martínez-España, J. Posadas, P. Manzoni, and J. M. Cecilia, "Evaluation of edge computing platforms through tinyml workloads," 2022.
8. A. K. Nejad, M. Jamshidi, and S. B. Shokouhi, "A super-efficient tinyml processor for the edge metaverse," *Information*, vol. 14, p. 235, 2023.
9. G. Donati, F. Zonzini, and L. D. Marchi, "Tiny deep learning architectures enabling sensor-near acoustic data processing and defect localization," *Computers*, vol. 12, p. 129, 2023.
10. A. Karras, A. Giannaros, C. Karras, L. Theodorakopoulos, C. S. Mammassis, G. A. Krimpas, and S. Sioutas, "Tinyml algorithms for big data management in large-scale iot systems," *Future Internet*, vol. 16, p. 42, 2024.
11. H. Ren, D. Anicic, and T. A. Runkler, "How to manage tiny machine learning at scale: an industrial perspective," 2022.
12. J. Morales-García, A. Bueno-Crespo, R. Martínez-España, J. Posadas, P. Manzoni, and J. M. Cecilia, "Evaluation of low-power devices for smart greenhouse development," *The Journal of Supercomputing*, vol. 79, pp. 10 277–10 299, 2023.
13. M. S. Diab and E. Rodriguez-Villegas, "Embedded machine learning using microcontrollers in wearable and ambulatory systems for health and care applications: a review," *IEEE Access*, vol. 10, pp. 98 450–98 474, 2022.
14. E. Njor, J. Madsen, and X. Fafoutis, "A primer for tinyml predictive maintenance: input and model optimisation," *IFIP Advances in Information and Communication Technology*, pp. 67–78, 2022.
15. V. Tsoukas, A. Gkogkidis, E. Boumpa, S. Papafotikas, and A. Kakarountas, "A gas leakage detection device based on the technology of tinyml âŁ ," *Technologies*, vol. 11, p. 45, 2023.
16. W. Raza, A. Osman, F. Ferrini, and F. G. B. D. Natale, "Energy-efficient inference on the edge exploiting tinyml capabilities for uavs," *Drones*, vol. 5, p. 127, 2021.

17. S. Jang, J. Park, E. Lee, and S. Lee, "Lightweight and energy-efficient deep learning accelerator for real-time object detection on edge devices," *Sensors*, vol. 23, p. 1185, 2023.
18. S. E. ARSLAN, "Tiny machine learning model for obstacle detection with multi-zone time of flight sensors," *Recent Advances in Science and Engineering*, pp. 9–14, 2023.
19. A. M. Hayajneh, S. A. Aldalahmeh, F. Alasali, H. Al-Obiedollah, S. A. Zaidi, and D. McLernon, "Tiny machine learning on the edge: a framework for transfer learning empowered unmanned aerial vehicle assisted smart farming," *IET Smart Cities*, vol. 6, pp. 10–26, 2023.
20. A. Hayajneh, S. Aldalahmeh, S. A. R. Zaidi, D. McLernon, H. Obeidollah, and R. Alsakarnah, "Channel state information based device free wireless sensing for iot devices employing tinyml," *2022 4th IEEE Middle East and North Africa COMMunications Conference (MENACOMM)*, 2022.
21. Z. Wang, P. Reviriego, F. Niknia, J. Conde, S. Liu, and F. Lombardi, "Adaptive resolution inference (ari): energy-efficient machine learning for internet of things," *IEEE Internet of Things Journal*, vol. 11, pp. 14 076–14 087, 2024.
22. O. Atanane, A. Mourhir, N. Benamar, and M. Zennaro, "Smart buildings: water leakage detection using tinyml," *Sensors*, vol. 23, p. 9210, 2023.
23. O. O. Martins, A. A. Aribisala, M. M. Abdulhamid, I. O. Joshua, O. J. Oluwatobi, and O. R. Samuel, "Development of a facial recognition model using optimized convolution neural network with tiny-ml and fingerprint authentication for safe lock entry," *Dutse Journal of Pure and Applied Sciences*, vol. 10, pp. 311–321, 2024.
24. N. A. Abbas and M. R. Ahmad, "Keyword spotting system with nano 33 ble sense using embedded machine learning approach," *Jurnal Teknologi*, vol. 85, pp. 175–182, 2023.
25. O. D'Souza, S. C. Mukhopadhyay, and Q. Z. Sheng, "Health, security and fire safety process optimisation using intelligence at the edge," *Sensors*, vol. 22, p. 8143, 2022.
26. A. Gudiño-Ochoa, J. A. García-Rodríguez, R. Ochoa-Ornelas, J. I. Cuevas-Chávez, and D. A. Sánchez-Arias, "Noninvasive diabetes detection through human breath using tinyml-powered e-nose," *Sensors*, vol. 24, p. 1294, 2024.
27. R. Sánchez-Iborra, "Lpwan and embedded machine learning as enablers for the next generation of wearable devices," *Sensors*, vol. 21, p. 5218, 2021.
28. S. Chakkor, M. Baghouri, Z. Cheker, A. E. Oualkadi, A. J. E. Hangouche, and J. Laamech, "Intelligent network for proactive detection of covid-19 disease," *2020 6th IEEE Congress on Information Science and Technology (CiSt)*, 2020.
29. I. N. K. Wardana, S. A. Fahmy, and J. W. Gardner, "Tinyml with meta-learning on microcontrollers for air pollution prediction," *Eurosensors 2023*, 2024.
30. B. Krivokapic, S. Tomovic, I. Radusinovic, and A. Jovanovic, "Implementation and performance evaluation of convolutional neural network models for low-power microcontrollers with constrained resources," *2024 28th International Conference on Information Technology (IT)*, 2024.
31. R. M. M. B. P. R. M. N. Pranav Kumar Chaudhary, Aakash Kishore Chotrani, "Ai in fraud detection: evaluating the efficacy of artificial intelligence in preventing financial misconduct," *Journal of Electrical Systems*, vol. 20, pp. 1332–1338, 2024.
32. X. Wang, Y. Tang, and A. G. Quaranta, "Machine learning-driven lending decisions in bank consumer finance," *International Journal of Information Systems and Supply Chain Management*, vol. 17, pp. 1–19, 2024.
33. R. Agrawal, "Effectiveness of machine learning in stock price prediction," *International Journal for Research in Applied Science and Engineering Technology*, vol. 11, pp. 256–262, 2023.
34. L. Liang, J. Chen, and D. Yao, "Switching to profitable outside options under supplier encroachment," *Production and Operations Management*, vol. 32, pp. 2788–2804, 2023.
35. C. Eroglu, B. D. Williams, and M. A. Waller, "Using the pack-and-a-half rule to eliminate backroom inventories in retail operations," *Journal of Business Logistics*, vol. 39, pp. 164–181, 2018.
36. H. Turtiainen, A. Costin, T. Lahtinen, L. Sintonen, and T. Hamalainen, "Towards large-scale, automated, accurate detection of cctv camera objects using computer vision. applications and implications for privacy, safety, and cybersecurity. (preprint)," 2020.

37. M. Piechocki, M. Kraft, T. Pajchrowski, P. Aszkowski, and D. Pieczyński, "Efficient people counting in thermal images: the benchmark of resource-constrained hardware," *IEEE Access*, vol. 10, pp. 124 835–124 847, 2022.
38. C. Huang, X. Sun, and Y. Zhang, "Tiny-machine-learning-based supply canal surface condition monitoring," *Sensors*, vol. 24, p. 4124, 2024.
39. T. Li, J. Luo, K. Liang, C. Yi, and L. Ma, "Synergy of patent and open-source-driven sustainable climate governance under green ai: a case study of tinyml," *Sustainability*, vol. 15, p. 13779, 2023.
40. S. Ye, J. Luo, H. Liang, and S. Zhao, "The optimal control algorithm of the integrated energy system based on demand response," *Journal of Physics: Conference Series*, vol. 2636, p. 012015, 2023.
41. D. Piątkowski, T. Puślecki, and K. Walkowiak, "Study of the impact of data compression on the energy consumption required for data transmission in a microcontroller-based system," *Sensors*, vol. 24, p. 224, 2023.
42. R. Sánchez-Iborra, L. Bernal-Escobedo, J. Santa, and A. F. Skarmeta, "Tinyml-based fall detection for connected personal mobility vehicles," *Computers, Materials & Continua*, vol. 71, pp. 3869–3885, 2022.
43. C. M. Bishop, *Pattern Recognition and Machine Learning (Information Science and Statistics)*. Berlin, Heidelberg: Springer-Verlag, 2006.
44. Z.-H. Zhou, "Support vector machine", in *Machine Learning*, pp. 129–153, 2021. Springer, Singapore. https://doi.org/10.1007/978-981-15-1967-3_6
45. S. Maheshwari, T. Rahman, R. Shafik, A. Yakovlev, A. Rafiev, L. Jiao, and O.-C. Granmo, "Redress: Generating compressed models for edge inference using tsetlin machines," *IEEE Transactions on Pattern Analysis and Machine Intelligence*, vol. 45, no. 9, pp. 11 152–11 168, 2023.
46. T. Rahman, A. Wheeldon, R. Shafik, A. Yakovlev, J. Lei, O.-C. Granmo, and S. Das, "Data booleanization for energy efficient on-chip learning using logic driven ai," in *2022 International Symposium on the Tsetlin Machine (ISTM)*, 2022, pp. 29–36.
47. C. Contoli and E. Lattanzi, "Energy efficiency of deep learning compression techniques in wearable human activity recognition," *IFIP Advances in Information and Communication Technology*, pp. 102–113, 2023.
48. R. Krishnamoorthi, "Quantizing deep convolutional networks for efficient inference: a whitepaper," 2018.
49. A. Garofalo, Y. Tortorella, M. Perotti, L. Valente, A. Nadalini, L. Benini, D. Rossi, and F. Conti, "Darkside: a heterogeneous risc-v compute cluster for extreme-edge on-chip dnn inference and training," *IEEE Open Journal of the Solid-State Circuits Society*, vol. 2, pp. 231–243, 2022.
50. P. Molchanov, S. Tyree, T. Karras, T. Aila, and J. Kautz, "Pruning convolutional neural networks for resource efficient inference," 2016.
51. T. Chen, Z. Gan, Y. Liu, L. Zhang, and Z. Wang, "Chasing sparsity in vision transformers: an end-to end exploration," 2021.
52. H. Zhang, L. Liu, H. Zhou, W. Hou, H. Sun, and N. Zheng, "Akecp: adaptive knowledge extraction from feature maps for fast and efficient channel pruning," 2021.
53. L. Prono, P. Bich, C. Boretti, M. Mangia, F. Pareschi, R. Rovatti, and G. Setti, "A multiply-and-max/min neuron paradigm for aggressively prunable deep neural networks," 2024.
54. S. Disabato and M. Roveri, "Incremental on-device tiny machine learning," *Proceedings of the 2nd International Workshop on Challenges in Artificial Intelligence and Machine Learning for Internet of Thi*, vol. 46, pp. 7–13, 2020.
55. A. Bakar, T. Rahman, A. Montanari, J. Lei, R. Shafik, and F. Kawsar, "Logic-based intelligence for batteryless sensors," in *Proceedings of the 23rd Annual International Workshop on Mobile Computing Systems and Applications*, ser. HotMobile '22. New York, NY, USA: Association for Computing Machinery, 2022, p. 22–28. [Online]. Available: https://doi.org/10.1145/3508396.3512870

56. R. K. Kushawaha, S. Kumar, B. Banerjee, and R. Velmurugan, "Distilling spikes: knowledge distillation in spiking neural networks," 2020.
57. M. Abbasi, A. Shahraki, J. Prieto, A. G. Arrieta, and J. M. Corchado, "Unleashing the potential of knowledge distillation for iot traffic classification," *IEEE Transactions on Machine Learning in Communications and Networking*, vol. 2, pp. 221–239, 2024.
58. TensorFlow Team, "Weight clustering: Model optimization guide," 2024, accessed: 2024-10-17. [Online]. Available: https://www.tensorflow.org/model_optimization/guide/clustering
59. S. Dhabal, S. Chakraborty, and P. Venkateswaran, "An efficient quadrature mirror filter design and its applications in audio signal processing," *2011 International Conference on Communication and Industrial Application*, 2011.
60. J. Jiang, F. Zhou, S. Ouyang, and G. Liao, "Efficient design of very large-scale dft modulated filter banks using m th band condition," *IET Signal Processing*, vol. 8, pp. 381–391, 2014.
61. Y. Lu, Z. Song, Q. Gao, D. Zhu, and D. Sun, "Bearing fault diagnosis based on multi-band filtering," *IET Science, Measurement & Technology*, vol. 16, pp. 101–117, 2021.
62. T. Devis and M. Manuel, "Hardware-efficient auto-reconfigurable hearing aids using 3-level octave interpolated filters for auditory compensation applications," *Physical and Engineering Sciences in Medicine*, vol. 44, pp. 785–798, 2021.
63. S. Patel, R. Dhuli, and B. Lall, "Analysis of signals via non-maximally decimated non-uniform filter banks," *IEEE Transactions on Circuits and Systems I: Regular Papers*, vol. 66, pp. 3882–3895, 2019.
64. U. H. Jaid and A. AbdulHassan, "End-to-end speaker profiling using 1d cnn architectures and filter bank initialization," *International Journal of Online and Biomedical Engineering (iJOE)*, vol. 19, pp. 65–81, 2023.
65. S. M. M. Kim and S. Bleeck, "An open development platform for auditory real-time signal processing," *Speech Communication*, vol. 98, pp. 73–84, 2018.
66. L. Kern, "Design and development of a real-time neural processor using the intel 80170nx etann," in *[Proceedings 1992] IJCNN International Joint Conference on Neural Networks*, vol. 2, 1992, pp. 684–689 vol.2.
67. Y. Sun, N. B. Agostini, S. Dong, and D. Kaeli, "Summarizing cpu and gpu design trends with product data," 2020. [Online]. Available: arXiv:1911.11313
68. S. Harris and D. Harris, *Digital Design and Computer Architecture: ARM Edition*, 1st ed. San Francisco, CA, USA: Morgan Kaufmann Publishers Inc., 2015.
69. S. Churiwala, *Designing with Xilinx FPGAs: Using Vivado*, 1st ed. Springer Publishing Company, Incorporated, 2016.
70. S. Hosseininoorbin, S. Layeghy, B. Kusý, R. Jurdak, and M. Portmann, "Exploring deep neural networks on edge tpu," 2021.
71. "Project—nicsefc.ee.tsinghua.edu.cn," https://nicsefc.ee.tsinghua.edu.cn/project.html, [Accessed 09-10-2024].
72. V. Sze, Y.-H. Chen, T.-J. Yang, and J. Emer, "Efficient processing of deep neural networks: A tutorial and survey," 2017. [Online]. Available: arXiv:1703.09039
73. X. Tang, E. Giacomin, B. Chauviere, A. Alacchi, and P.-E. Gaillardon, "Openfpga: An open-source framework for agile prototyping customizable fpgas," *IEEE Micro*, vol. 40, no. 4, pp. 41–48, 2020.
74. D. Koch, N. Dao, B. Healy, J. Yu, and A. Attwood, "Fabulous: An embedded fpga framework," in *The 2021 ACM/SIGDA International Symposium on Field-Programmable Gate Arrays*, ser. FPGA '21. New York, NY, USA: Association for Computing Machinery, 2021, p. 45–56. [Online]. Available: https://doi.org/10.1145/3431920.3439302
75. Y. Umuroglu, N. J. Fraser, G. Gambardella, M. Blott, P. Leong, M. Jahre, and K. Vissers, "Finn: A framework for fast, scalable binarized neural network inference," in *Proceedings of the 2017 ACM/SIGDA International Symposium on Field-Programmable Gate Arrays*, ser. FPGA '17. ACM, Feb. 2017. [Online]. Available: https://doi.org/10.1145/3020078.3021744
76. M. Blott, T. B. Preußer, N. J. Fraser, G. Gambardella, K. O'brien, Y. Umuroglu, M. Leeser, and K. Vissers, "Finn-r: An end-to-end deep-learning framework for fast exploration of quantized neural networks," *ACM Trans. Reconfigurable Technol. Syst.*, vol. 11, no. 3, Dec. 2018. [Online]. Available: https://doi.org/10.1145/3242897

77. Y. Zhang, J. Pan, X. Liu, H. Chen, D. Chen, and Z. Zhang, "Fracbnn: Accurate and fpga-efficient binary neural networks with fractional activations," in *The 2021 ACM/SIGDA International Symposium on Field-Programmable Gate Arrays*, ser. FPGA '21. New York, NY, USA: Association for Computing Machinery, 2021, p. 171–182. [Online]. Available: https://doi.org/10.1145/3431920.3439296
78. X. Zhang, J. Wang, C. Zhu, Y. Lin, J. Xiong, W.-m. Hwu, and D. Chen, "Dnnbuilder: an automated tool for building high-performance dnn hardware accelerators for fpgas," in *2018 IEEE/ACM International Conference on Computer-Aided Design (ICCAD)*, 2018, pp. 1–8.
79. "hls4ml: An open-source codesign workflow to empower scientific low-power machine learning devices," 2021. [Online]. Available: arXiv:2103.05579
80. M. Andronic and G. A. Constantinides, "Polylut: Learning piecewise polynomials for ultra-low latency fpga lut-based inference," in *2023 International Conference on Field Programmable Technology (ICFPT)*. IEEE, Dec. 2023. [Online]. Available: https://doi.org/10.1109/ICFPT59805.2023.00012
81. Y. Umuroglu, Y. Akhauri, N. J. Fraser, and M. Blott, "Logicnets: Co-designed neural networks and circuits for extreme-throughput applications," 2020. [Online]. Available: arXiv:2004.03021
82. E. Wang, J. J. Davis, P. Y. K. Cheung, and G. A. Constantinides, "Lutnet: Rethinking inference in fpga soft logic," in *2019 IEEE 27th Annual International Symposium on Field-Programmable Custom Computing Machines (FCCM)*, 2019, pp. 26–34.
83. T. Rahman, G. Mao, S. Maheshwari, R. Shafik, and A. Yakovlev, "Matador: Automated system-on-chip tsetlin machine design generation for edge applications," in *2024 Design, Automation & Test in Europe Conference & Exhibition (DATE)*, 2024, pp. 1–6.
84. A. Shawahna, S. M. Sait, and A. El-Maleh, "Fpga-based accelerators of deep learning networks for learning and classification: A review," *IEEE Access*, vol. 7, pp. 7823–7859, 2019.
85. A. Yazdanbakhsh, S. Kiran, B. Akin, J. Laudon, and R. Narayanaswami, "An evaluation of edge tpu accelerators for convolutional neural networks," 2021.
86. S. Hosseininoorbin, S. Layeghy, M. Sarhan, R. Jurdak, and M. Portmann, "Exploring edge tpu for network intrusion detection in iot," 2021.
87. H. Kwon, C. Hegde, Y. Kiarashi, V. S. K. Madala, R. Singh, A. Nakum, R. Tweedy, L. M. Tonetto, C. M. Zimring, M. Doiron, A. D. Rodriguez, A. I. Levey, and G. D. Clifford, "A feasibility study on indoor localization and multiperson tracking using sparsely distributed camera network with edge computing," *IEEE Journal of Indoor and Seamless Positioning and Navigation*, vol. 1, pp. 187–198, 2023.
88. L. Klein, P. Žmij, and P. Krömer, "Partial discharge detection by edge computing," *IEEE Access*, vol. 11, pp. 44 192–44 204, 2023.
89. Apple, "Machine learning research at apple: Neural engine and transformers," 2024, accessed: 2024-09-19. [Online]. Available: https://machinelearning.apple.com/research/neural-engine-transformers
90. A. Wiki, "Neural engine," 2024, accessed: 2024-09-19. [Online]. Available: https://apple.fandom.com/wiki/Neural_Engine
91. Z. Fredin, J. Zemanek, C. Blackburn, E. Strand, A. Abdel-Rahman, and P. Rowles, "Discrete integrated circuit electronics (dice)," *2020 IEEE High Performance Extreme Computing Conference (HPEC)*, 2020.
92. D. Xu, M. Xu, Q. Wang, S. Wang, Y. Ma, H. Kang, G. S. Huang, X. Jin, and X. Liu, "Mandheling: mixed-precision on-device dnn training with dsp offloading," 2022.
93. M. Abeysinghe, J. Villarreal, T. T. Hahn, and J. D. Bakos, "Automated loop fusion for image processing," 2024.
94. C. Yang, S. Chen, Y. Wang, and J. Zhang, "The evaluation of dcnn on vector-simd dsp," *IEEE Access*, vol. 7, pp. 22 301–22 309, 2019.
95. M. M. H. Shuvo, S. K. Islam, J. Cheng, and B. I. Morshed, "Efficient acceleration of deep learning inference on resource-constrained edge devices: A review," *Proceedings of the IEEE*, vol. 111, no. 1, pp. 42–91, 2023.

96. Z. I. Chowdhury, S. K. Khatamifard, Z. Zhao, S. Resch, M. Razaviyayn, J. P. Wang, S. S. Sapatnekar, and U. R. Karpuzcu, "Spintronic in-memory pattern matching," *IEEE Journal on Exploratory Solid-State Computational Devices and Circuits*, vol. 5, pp. 206–214, 2019.
97. R. Srinivasagan, M. Mohammed, and A. Alzahrani, "Tinyml-sensor for shelf life estimation of fresh date fruits," *Sensors*, vol. 23, p. 7081, 2023.
98. O. Ghazal, W. Wang, S. Kvatinsky, F. Merchant, A. Yakovlev, and R. Shafik, "Impact: Inmemory computing architecture based on y-flash technology for coalesced tsetlin machine inference," *arXiv preprint* arXiv:2412.05327, 2024.
99. M. L. Gallo, A. Sebastian, R. Mathis, M. Manica, H. Giefers, T. Tůma, C. Bekas, A. Curioni, and E. Eleftheriou, "Mixed-precision in-memory computing," *Nature Electronics*, vol. 1, pp. 246–253, 2018.
100. A. Haj-Ali, R. Ben-Hur, N. Wald, and S. Kvatinsky, "Efficient algorithms for in-memory fixed point multiplication using magic," *2018 IEEE International Symposium on Circuits and Systems (ISCAS)*, pp. 1–5, 2018.
101. A. Sebastian, M. Le Gallo, R. Khaddam-Aljameh, and E. Eleftheriou, "Memory devices and applications for in-memory computing," *Nature nanotechnology*, vol. 15, no. 7, pp. 529–544, 2020.
102. O. Krestinskaya, M. E. Fouda, H. Benmeziane, K. El Maghraoui, A. Sebastian, W. D. Lu, M. Lanza, H. Li, F. Kurdahi, S. A. Fahmy *et al.*, "Neural architecture search for in-memory computing-based deep learning accelerators," *Nature Reviews Electrical Engineering*, pp. 1–17, 2024.
103. S. E. Kim, J. Y. Sung, J. D. Jeon, S. Y. Jang, H. M. Lee, S. M. Moon, J. G. Kang, H. J. Lim, H.-S. Jung, and S. W. Lee, "Toward advanced high-k and electrode thin films for dram capacitors via atomic layer deposition," *Advanced Materials Technologies*, vol. 8, no. 20, p. 2200878, 2023. [Online]. Available: https://doi.org/10.1002/admt.202200878
104. S. Yin, Z. Jiang, J.-S. Seo, and M. Seok, "Xnor-sram: In-memory computing sram macro for binary/ternary deep neural networks," *IEEE Journal of Solid-State Circuits*, vol. 55, no. 6, pp. 1733–1743, 2020.
105. G. W. Burr, R. S. Shenoy, K. Virwani, P. Narayanan, A. Padilla, B. Kurdi, and H. Hwang, "Access devices for 3d crosspoint memory," *Journal of Vacuum Science & Technology B*, vol. 32, no. 4, 2014.
106. S. S. Kim, S. K. Yong, J. Kim, J. M. Choi, T. W. Park, H. Kim, H. J. Kim, and C. S. Hwang, "Fabrication of a hole-type vertical resistive-switching random-access array and intercell interference induced by lateral charge spreading," *Advanced Electronic Materials*, vol. 9, 2022.
107. A. Hazra, A. Tripathi, A. Jan, S. Kundu, and P. K. R. Boppidi, "Multiple nano-filaments based efficient resistive switching in tio2 nanotubes array influenced by thermally induced self-doping and anatase to rutile phase transformation," *Nanotechnology*, vol. 32, p. 115201, 2020.
108. M. Sivan, Y. Li, H. Veluri, Y. Zhao, B. Tang, X. Wang, E. Zamburg, J. F. Leong, J. X. Niu, U. Chand, and A. Thean, "All wse2 1t1r resistive ram cell for future monolithic 3d embedded memory integration," *Nature Communications*, vol. 10, 2019.
109. Z. Sun, G. Pedretti, E. Ambrosi, A. Bricalli, W. Wang, and D. Ielmini, "Solving matrix equations in one step with cross-point resistive arrays," *Proceedings of the National Academy of Sciences*, vol. 116, no. 10, pp. 4123–4128, 2019. [Online]. Available: https://doi.org/10.1073/pnas.1815682116
110. Q. Wang, Y. Kim, and P. Li, "Neuromorphic processors with memristive synapses: Synaptic interface and architectural exploration," *J. Emerg. Technol. Comput. Syst.*, vol. 12, no. 4, May 2016. [Online]. Available: https://doi.org/10.1145/2894756
111. S. Lee, S. Lee, J. Lee, J. Choi, D. Kwon, S. Hong, and K. Kwon, "Architecture-accuracy co-optimization of reram-based low-cost neural network processor," *Proceedings of the 2020 on Great Lakes Symposium on VLSI*, pp. 427–432, 2020.
112. W. Chen, Z. Qi, Z. Akhtar, and K. Siddique, "Resistive-ram-based in-memory computing for neural network: a review," *Electronics*, vol. 11, p. 3667, 2022.

113. S. Song, S. Zhang, P. Cheng, Y. Gu, Z. Zhang, Y. Lv, D. Yao, L. Wu, and B. Liu, "Performance improvement of phase-change memory cell using alsb3te and atomic layer deposition tio2 buffer layer," *Nanoscale Research Letters*, vol. 8, 2013.
114. Y. Wang, T. Guo, G. Li, T. Li, S. Lv, S. Song, Y. Cheng, W. Song, K. Ren, and S. Zhang, "Sc-centered octahedron enables high-speed phase change memory with improved data retention and reduced power consumption," *ACS Applied Materials & Interfaces*, vol. 11, pp. 10 848–10 855, 2019.
115. H. Sun, L. Chen, X. Hao, C. Liu, and M. Ni, "An energy-efficient and fast scheme for hybrid storage class memory in an aiot terminal system," *Electronics*, vol. 9, p. 1013, 2020.
116. M. Kim, H. Lee, H. Kim, and H. Lee, "Wl-wd: wear-leveling solution to mitigate write disturbance errors for phase-change memory," *IEEE Access*, vol. 10, pp. 11 420–11 431, 2022.
117. X. Fong, S. H. Choday, and K. Roy, "Design and optimization of spin-transfer torque mrams," *More Than Moore Technologies for Next Generation Computer Design*, pp. 49–72, 2015.
118. Y. Park, G. Kil, and Y. Song, "A novel sensing algorithm for spin-transfer-torque magnetic ram (stt-mram) by utilizing dynamic reference," *IEICE Electronics Express*, vol. 9, pp. 153–159, 2012.
119. M. Mayahinia, M. B. Tahoori, M. Perumkunnil, H. Zahedmanesh, K. Croes, T. Marinelli, J. I. Gomez Perez, T. Evenblij, G. S. Kar, and F. Catthoor, "Time-dependent electromigration modeling for workload-aware design-space exploration in stt-mram," *Ieee Transactions on Computer-Aided Design of Integrated Circuits and Systems*, 2022.
120. S. Yang and Y. Ryu, "A buffer management for stt-mram based hybrid main memory in sensor nodes," *Proceedings of the International Conference on Computer, Networks and Communication Engineering (ICCNCE 2013)*, 2013.
121. A. Shafaei, Y. Wang, and M. Pedram, "Low write-energy stt-mrams using finfet-based access transistors," *2014 IEEE 32nd International Conference on Computer Design (ICCD)*, 2014.
122. X. Fong, R. Venkatesan, A. Raghunathan, and K. Roy, "Non-volatile complementary polarizer spin-transfer torque on-chip caches: A device/circuit/systems perspective," *Ieee Transactions on Magnetics*, 2014.
123. R. Gupta, H. Park, Mojumder, and S. Roy, "Layout-aware optimization of stt mrams," 2012.
124. G. Verma, S. Soni, A. Nisar, and B. K. Kaushik, "Multi-bit mram based high performance neuromorphic accelerator for image classification," *Neuromorphic Computing and Engineering*, vol. 4, p. 014008, 2024.
125. Y. Umeki, K. Yanagida, S. Yoshimoto, S. Izumi, M. Yoshimoto, H. Kawaguchi, K. Tsunoda, and T. Sugii, "Stt-mram operating at 0.38v using negative-resistance sense amplifier," *IEICE Transactions on Fundamentals of Electronics, Communications and Computer Sciences*, vol. E97.A, pp. 2411–2417, 2014.
126. O. Ghazal, S. Singh, T. Rahman, S. Yu, Y. Zheng, D. Balsamo, S. Patkar, F. Merchant, F. Xia, A. Yakovlev, and R. Shafik, "Imbue: In-memory boolean-to-current inference architecture for tsetlin machines," in *2023 IEEE/ACM International Symposium on Low Power Electronics and Design (ISLPED)*, 2023, pp. 1–6.
127. O. Ghazal, G. Maot, T. Lan, J. Ojukwu, F. Xia, A. Yakovlev, and R. Shafik, "Asynchronous control for tsetlin machine with binary memristor-transistor array," in *2023 IEEE International Symposium on Circuits and Systems (ISCAS)*, 2023, pp. 1–5.
128. A. Wheeldon, R. Shafik, T. Rahman, J. Lei, A. Yakovlev, and O.-C. Granmo, "Learning automata based energy-efficient ai hardware design for iot applications," *Philosophical Transactions of the Royal Society A: Mathematical, Physical and Engineering Sciences*, vol. 378, no. 2182, p. 20190593, 2020. [Online]. Available: https://doi.org/10.1098/rsta.2019.0593
129. S. Kalapothas, M. Galetakis, G. Flamis, F. Plessas, and P. Kitsos, "A survey on risc-v-based machine learning ecosystem," *Information*, vol. 14, p. 64, 2023.
130. P. Platform, "Pulp: Parallel ultra low power," https://pulp-platform.org/, 2024, accessed: 2024-10-02.
131. J. Lei, T. Rahman, R. Shafik, A. Wheeldon, A. Yakovlev, O.-C. Granmo, F. Kawsar, and A. Mathur, "Low-power audio keyword spotting using tsetlin machines," *Journal of Low*

Power Electronics and Applications, vol. 11, no. 2, 2021. [Online]. Available: https://www.mdpi.com/2079-9268/11/2/18
132. M. A. Al Jbaar and S. A. Dawwd, "Simd implementation of deep cnns for myopia detection on a single-board computer system," *Eastern-European Journal of Enterprise Technologies*, vol. 5, pp. 98–108, 2023.
133. H. K. Kondaveeti, S. Raman, and P. Raj, "Prototyping with raspberry pi in healthcare domain," *International Journal of Advanced Computer Science and Applications*, vol. 11, 2020.
134. E. Gamess and S. P. Hernandez, "Performance evaluation of different raspberry pi models for a broad spectrum of interests," *International Journal of Advanced Computer Science and Applications*, vol. 13, 2022.
135. Y. Afrianto, V. Mardiansyah, Ritzkal, F. S. Ramadhan, A. D. Batistuta, B. Wulandari, and W. T. Atmojo, "Enhanced π approximation through mimd parallel computing: an efficiency analysis utilizing raspberry pi," *Mathematical Modelling of Engineering Problems*, vol. 10, pp. 1657–1664, 2023.
136. I. N. K. Wardana, J. W. Gardner, and S. A. Fahmy, "Optimising deep learning at the edge for accurate hourly air quality prediction," *Sensors*, vol. 21, p. 1064, 2021.
137. S. R. Manalu, J. V. Moniaga, D. A. Hadipurnawan, and F. Sahidi, "Obd-ii and raspberry pi technology to diagnose car's machine current condition: study literature," *Library Hi Tech News*, vol. 34, pp. 15–21, 2017.
138. D. P. Hausherr and D. Berben, "(all-in-one) power supply system for mobile and network-wired raspberry pi-based internet of things applications," *Hardware*, vol. 1, pp. 54–69, 2023.
139. D. Fernández-Cerero, J. Y. Fernández-Rodríguez, J. A. Álvarez García, L. M. S. Morillo, and A. Fernández-Montes, "Single-board-computer clusters for cloudlet computing in internet of things," *Sensors*, vol. 19, p. 3026, 2019.
140. K. H. Fen, "Development of a portable community video surveillance system," *International Journal of Electrical and Computer Engineering (IJECE)*, vol. 9, p. 1814, 2019.
141. I. S. M. Isa, C. J. Yeong, and N. L. Azyze, "Real-time traffic sign detection and recognition using raspberry pi," *International Journal of Electrical and Computer Engineering (IJECE)*, vol. 12, p. 331, 2022.
142. M. S. R. Krishna, K. Dinesh, and N. S. Shanbog, "Low cost remote monitoring of solar plant through rs485 communication," *International Journal of Innovative Technology and Exploring Engineering*, vol. 8, pp. 3034–3037, 2019.
143. A. Hassan, H. Nahar, W. M. Shah, A. Abd-Aziz, S. A. Sahiran, N. Bahaman, M. R. Ahmad, I. R. A. Hamid, and M. A. B. Sidik, "Performance evaluation of raspberry pi as an iot edge signal processing device for a real-time flash flood forecasting system," *International Journal of Advanced Computer Science and Applications*, vol. 13, 2022.
144. D. Salami, R. Hasibi, S. Palipana, P. Popovski, T. Michoel, and S. Sigg, "Tesla-rapture: a lightweight gesture recognition system from mmwave radar sparse point clouds," *IEEE Transactions on Mobile Computing*, vol. 22, pp. 4946–4960, 2023.
145. STMicroelectronics, "Stm32 32-bit arm cortex mcus," 2024, accessed: 2024-10-17. [Online]. Available: https://www.st.com/en/microcontrollers-microprocessors/stm32-32-bit-arm-cortex-mcus.html
146. Microchip Technology Inc., "Microchip microcontrollers (mcus)," 2024, accessed: 2024-10-17. [Online]. Available: https://www.microchip.com/en-us/products/microcontrollers-and-microprocessors
147. Texas Instruments, "Msp430 microcontrollers," 2024, accessed: 2024-10-17. [Online]. Available: https://www.ti.com/design-development/embedded-development/msp430-mcus.html
148. A. Bakar, T. Rahman, R. Shafik, F. Kawsar, and A. Montanari, "Adaptive intelligence for batteryless sensors using software-accelerated tsetlin machines," in *Proceedings of the 20th ACM Conference on Embedded Networked Sensor Systems*, 2022, pp. 236–249.
149. NVIDIA, "Nvidia jetson for next-generation robotics," 2024, accessed: 2024-10-16. [Online]. Available: https://www.nvidia.com/en-gb/autonomous-machines/embedded-systems/
150. T. Rahman, G. Mao, S. Maheshwari, R. Shafik, and A. Yakovlev, "Matador: Automated system-on-chip tsetlin machine design generation for edge applications," in *2024 Design, Automation & Test in Europe Conference & Exhibition (DATE)*, 2024, pp. 1–6.

151. G. AI, "Tensorflow lite for microcontrollers," 2024, accessed: 2024-12-19. [Online]. Available: https://ai.google.dev/edge/litert
152. E. Impulse, "Eon compiler," https://docs.edgeimpulse.com/docs/edge-impulse-studio/deployment/eon-compiler, 2024, accessed: 2024-10-02.
153. uTensor, "Hands on embedded machine learning," https://utensor.github.io/website/, 2024, accessed: 2024-10-02.
154. Microsoft, "Embedded learning library (ell): An open source library for embedded ai and machine learning," https://microsoft.github.io/ELL/, 2024, accessed: 2024-10-02.
155. STMicroelectronics, "Free ai model optimizer for stm32," https://stm32ai.st.com/stm32-cube-ai/, 2024, accessed: 2024-10-02.
156. STMicroelectronics, "Nanoedge ai studio: Automated machine learning (ml) tool for stm32 developers," 2024, accessed: 2024-10-17. [Online]. Available: https://www.st.com/en/development-tools/nanoedgeaistudio.html
157. PyTorch, "Pytorch mobile," 2024, accessed: 2024-10-17. [Online]. Available: https://pytorch.org/mobile/home/
158. Apache Software Foundation, "Tvm: Open deep learning compiler stack," 2024, accessed: 2024-10-17. [Online]. Available: https://tvm.apache.org/
159. S. Jaiswal, R. K. K. Goli, A. Kumar, V. Seshadri, and R. Sharma, "Minun: Accurate ml inference on microcontrollers," ser. LCTES 2023. New York, NY, USA: Association for Computing Machinery, 2023, p. 26–39. [Online]. Available: https://doi.org/10.1145/3589610.3596278
160. Keras Team, "Keras: The python deep learning api," 2024, accessed: 2024-10-17. [Online]. Available: https://keras.io/
161. Intel Corporation, "Openvino toolkit," 2024, accessed: 2024-10-17. [Online]. Available: https://www.intel.com/content/www/us/en/developer/tools/openvino-toolkit/whats-new.html
162. ONNX Community, "Onnx: Open neural network exchange," 2024, accessed: 2024-10-17. [Online]. Available: https://onnx.ai/
163. N. B. Agostini, S. Curzel, A. Limaye, V. Amatya, M. Minutoli, V. G. Castellana, J. Manzano, F. Ferrandi, and A. Tumeo, "Soda: An end-to-end open-source hardware compiler for machine learning accelerators."
164. AI Tech Systems, "Cainvas platform," 2024, accessed: 2024-10-17. [Online]. Available: https://cainvas.ai-tech.systems/accounts/login/

Cloud Resource Orchestration: Scaling Healthcare in the Digital Age

Aritra Ray, Farshad Firouzi, and Krishnendu Chakrabarty

"The cloud is a democracy."

– Marc Benioff

1 Introduction

DNNs and GenAI models are increasingly leveraged across various domains, including computer vision, natural language processing, recommendation systems, and robotics, among others [1, 2]. Training and inference of DNN and GenAI models for healthcare applications are resource-intensive and necessitate access to hardware accelerators. Consequently, users often offload their machine learning models for training and inference to compute clusters in public, private, or hybrid cloud environments [3].

Cloud infrastructures designed for DNN training promote multi-tenancy of compute resources. Efforts to enhance the multi-tenancy of hardware accelerators, particularly GPUs, have focused on service-level orchestration [4], runtime-level scheduling [5], and resource-level management [6, 7]. Studies [8, 9] aim to improve Job

This chapter is adapted from A. Ray et al., 'Job Recommendation Service for GPU Sharing in Kubernetes,' in IEEE Cloud Summit 2023; and A. Ray et al., 'Inference Serving System for Stable Diffusion as a Service' in IEEE Cloud Summit 2024.

A. Ray (✉) · K. Chakrabarty
Department of Electrical and Computer Engineering, Duke University, Durham, NC, USA
e-mail: aritra.ray@duke.edu

K. Chakrabarty
e-mail: krishnendu.Chakrabarty@asu.edu

F. Firouzi · K. Chakrabarty
Arizona State University, Tempe, AZ, USA
e-mail: farshad.firouzi@asu.edu

F. Firouzi et al. (eds.), *Smart and Connected Health: AI, IoT, and Trustworthy Technologies*, https://doi.org/10.1007/978-3-032-06286-4_7

Completion Time (JCT) of user-defined tasks, enhance kernel performance [10], and reduce latency [5], while adhering to Service Level Objectives (SLOs) [6, 7]. Service-level orchestration often involves designing query schedulers with heuristic-based preemption [9], whereas resource orchestration typically focuses on multi-process scheduling [6], adaptive batching [6], and kernel tuning [10]. However, GPU schedulers [11, 12] do not address the need to preserve user privacy when scheduling user-defined ML training jobs. Preserving privacy is essential as users' health data and hyperparameters, such as model architecture, batch size, and training datasets, are often sensitive and proprietary, containing valuable intellectual property or confidential information that should not be accessed during the scheduling process.

Moving on to inference serving in traditional DNN deployments in the cloud, models are usually served on CPUs, with each input query triggering a single forward pass to generate a classification label. Optimization in this context often revolves around model switching [13], ease of use [14], and reducing inference latency [15]. However, inference in GenAI applications, such as Stable Diffusion (SD) for healthcare applications differs significantly [16]. Instead of a single forward pass, SD maps input text into token embeddings and starts with a noisy latent image array. Through multiple diffusion steps, the model iteratively refines the latent array, progressively adding details until the image decoder produces the final image. The increasing adoption of GenAI models, particularly for text-to-image generation in healthcare applications, and their deployment in cloud environments [17, 18] highlight the need for efficient inference approaches. These differences in inference processes, compared to traditional DNNs, motivate the design of a model-less, privacy-preserving, low-latency inference framework specifically tailored for Stable Diffusion as a Service (SDaaS), presented in this chapter.

The rest of this chapter is organized as follows: Sect. 2 presents the GPU sharing job recommendation tool for Kubernetes. Section 3 presents a model-less, privacy-preserving, low-latency inference framework for SDaaS. Section 4 concludes the discussion on resource orchestration for Machine Learning as a Service (MLaaS).

2 Job Recommendation Tool for GPU Sharing in Kubernetes

Kubernetes (K8S) is a widely used framework for the auto-deployment and management of containerized applications, providing production-ready container orchestration [19]. The kube-scheduler, a key component of K8S, assigns pods to nodes based on resource requirements and system-level constraints, functioning in a coarse-grained manner (checks for valid placement of pods to nodes). K8S makes scheduling decisions based on overall resource availability rather than fine-tuning decisions for optimizing workload placement or addressing specific job-level requirements. K8S lacks tools to intelligently decide which workloads or pods should be mapped to worker nodes. In general, user-defined ML training jobs are placed on pods and executed on a container runtime as directed by the kube-scheduler.

We present a fine-grain, online, privacy-preserving job scheduler that integrates with Argo[1] workflow on top of K8S. Our design includes a job profiler on worker nodes to extract features from user-defined ML training jobs. We categorize ML workloads into clusters and assess inter- and intra-cluster task interference, leveraging feature extraction and clustering to guide scheduling decisions that satisfy scheduler-level objectives. Throughout this process, we maintain black-box access to the user-defined ML training jobs. We refrain from accessing any sensitive job-specific parameters such as model architecture, batch size, datasets, and the number of epochs.

The remainder of this sub-section is organized as follows: In Sect. 2.1, we discuss the methodology, while we present the evaluation results in Sect. 2.2.

2.1 Design Methodology

In the traditional K8S architecture, the kube-scheduler assigns pods to available nodes based on their resource requirements while preserving system constraints. As a coarse-grain scheduler, it determines the appropriate pod placement for each node, considering all system-level constraints. In K8S, user-defined ML training workloads are deployed within pods with a randomly selected competitor job and executed on assigned nodes with available computing resources. We present a fine-grain, online, privacy-preserving job scheduler that enhances multi-tenancy by intelligently selecting competing jobs for concurrent execution within each pod.

We introduce two components in the control plane of K8S: the *Knowledge Base* (*KB*) and the *Job Recommendation Service* (*JRS*), along with a *Job Profiler* (*JP*) located in the worker node, as depicted in Fig. 1. We assume the presence of n user-defined ML training workloads, denoted as j_1, j_2, ..., j_n, in the job queue. With black-box access, *JP* profiles each job j to extract relevant features while preserving user privacy. The *KB* component maintains a record of the profiling results from *JP* and clusters the jobs based on their extracted features. It assigns an uncertainty score to each job, determined by its distance from the assigned cluster center. The *JRS* component conducts an online analysis of inter- and intra-cluster task interference using the defined SLO metrics to generate relative rankings, which guide the selection of competitor jobs.

Considering ⓐ the available number of jobs to be scheduled, ⓑ the task interference analysis, and ⓒ the multi-tenancy objective prioritizing either the user, the cloud provider, or a combination of both, the *JRS* recommends two or more competitor jobs to the kube-scheduler for concurrent scheduling in the same pod. Guided by the recommendations from the *JRS*, the kube-scheduler schedules the competitor jobs within the same pod, leveraging the Argo workflow for concurrent execution.

[1] https://argoproj.github.io/workflows/.

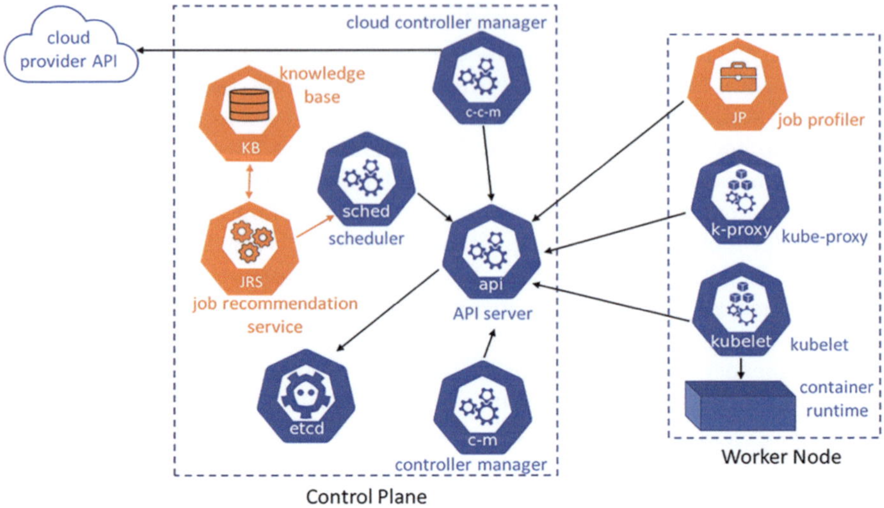

Fig. 1 Design of the fine-grain, online, privacy-preserving job scheduler built on top of the Kubernetes platform. The components outlined in blue represent the traditional elements of a Kubernetes cluster. In the control plane, we introduce two new components: *KB* (Knowledge Base) and *JRS* (Job Recommendation Service), along with a *JP* (Job Profiler) located in the worker node, all deployed as pods. For each user-defined ML training workload scheduled within the compute cluster, *KB* logs the job profile obtained from *JP* while maintaining user privacy via black-box access. Through an online analysis of inter- and intra-cluster task interference, *JRS* recommends the most suitable competitor job to the kube-scheduler, satisfying the multi-tenancy SLO at the scheduler level. (Adapted from [20].)

The outcomes are then fed back to the *JRS* in a feedback loop to dynamically update the task interference analysis.

The following sub-sections provide a detailed exploration of workload characterization and task interference analysis, followed by an in-depth discussion of the *JRS* component, which recommends competitor jobs to the kube-scheduler to enhance multi-tenancy while adhering to scheduler-level objectives.

2.1.1 Workload Characterization and Task Interference Analysis

We analyze ML training workloads on standard benchmark architectures and create a dataset with privacy-preserving features. We perform exploratory data analysis on the collected dataset and generate synthetic workloads to augment the dataset size. Using unsupervised clustering techniques, we cluster the ML workloads while preserving user privacy.

- **Characterizing ML Training Workloads**: Our job profiler, *JP*, characterizes user-defined ML training workloads across various attributes, including network parameters, GPU and CPU utilization metrics, GPU and disk memory usage, interrupts, system calls, runtime environment, and context switching. We profile jobs on standard benchmark architectures using publicly available datasets from categories such as image classification, image segmentation, reinforcement learning, generative adversarial networks, and natural language processing.

- **Feature Selection**: To ensure privacy, we exclude sensitive hyper-parameters such as job ID, batch size, epochs, dataset used, and model architecture. For our target variables—maximum GPU utilization percentage ($Max.GPU_{util}$) and maximum GPU memory allocation percentage ($Max.GPU_{memAlloc}$)—which act as potential bottlenecks to meet scheduler-level objectives, we conduct exploratory data analysis using Pearson's correlation coefficient to assess their relationship with other features. Additionally, we analyze feature interdependence with the target variables using random forests, ridge regression, and lasso regression.
- **Synthetic Data Generation**: We generate synthetic ML training workloads for our target and predictor variables to gain additional insights during clustering. Synthetic workloads with varying batch sizes are generated from the ground truth data (GT_{real}) using interpolation techniques, including barycentric, Krogh, pchip, spline, and linear methods. To evaluate the quality of synthetic samples, we mask the original ground truth GT_{real}, use the synthetic workload as pseudo-ground truth, and regenerate the ground truth ($ReGen_{GT}$). We assess the quality of synthetic data by estimating the mean absolute difference between $ReGen_{GT}$ and GT_{real}, selecting the best interpolation technique for each workload.
- **Clustering**: We cluster both GT_{real} and synthetic workloads using k-means based on the target and predictor variables, assigning each ML training job to a cluster. We conduct extensive scheduling of inter- and intra-cluster training workloads on the same GPU to analyze task interference, logging relative task interference rankings based on mean values. For online scheduling decisions, *JRS* recommends jobs based on the updated relative ranking of inter- and intra-cluster task interference to achieve scheduling objectives, with rankings refreshed after each execution.

2.2 Job Recommendation Service

We define two metrics, Individual Slowdown (Ind_{SD}) and Packing Saving (Pac_{Sav}) to analyze task interference. The Individual Slowdown (Ind_{SD}) measures the percentage difference in the JCT of a job when executed alone on a hardware accelerator compared to when it runs concurrently with a competitor job. Equation (1) defines Ind_{SD} for a job, j, where J_c represents the JCT when j is executed alongside one or more competitor jobs, and J_a is the completion time when j is executed alone on the same GPU, with all other computing resources kept constant.

Packing Saving (Pac_{Sav}) quantifies the percentage difference in JCT when multiple jobs are executed concurrently on the same GPU compared to when each job is executed sequentially. Equation (2) defines Pac_{Sav} for n jobs, j_1, ..., j_n, executed concurrently on the same GPU, where M denotes the maximum function.

$$Ind_{SD}(j)\% = \frac{J_c(j) - J_a(j)}{J_a(j)} \tag{1}$$

$$Pac_{Sav}(j_1, .., j_n)\% = \frac{\sum_{x=j_1}^{j_n} J_a(x) - M[J_c(j_1), ..., J_c(j_n)]}{\sum_{x=j_1}^{j_n} J_a(x)} \tag{2}$$

When a new user-defined job becomes available in the job queue, the *JP* profiles the job by extracting the target and predictor variables, and the *KB* component assigns it to a cluster. The *JRS*, leveraging prior knowledge of inter- and intra-cluster task interference results, recommends the most suitable competitor job from the job queue that aligns with the desired scheduler-level objectives. The *JRS* also considers the uncertainty scores of jobs within the cluster, giving selection precedence to those with lower uncertainty scores.

The scheduler-level objectives are tailored based on the prioritization of user and cloud provider demands. Typically, users aim to minimize Individual Slowdown (Ind_{SD}), while cloud providers seek to maximize Packing Saving (Pac_{Sav}). To address these conflicting objectives and promote multi-tenancy for GPU sharing in compute clusters, we present three algorithms: ⓐ one that balances Ind_{SD} and Pac_{Sav}, ⓑ one that focuses on minimizing Ind_{SD}, and ⓒ one that aims to maximize Pac_{Sav}.

The *JRS* provides scheduling recommendations to the kube-scheduler, facilitating the concurrent execution of two (or more) competitor jobs that best satisfy the defined scheduler-level objectives. These recommended jobs are then scheduled using the K8S Argo workflow framework. The specific scheduler-level objectives are defined as follows:

- **Balancing** $Ind_{SD}(j)$ **and** Pac_{Sav}: The *JRS* utilizes the relative rankings of inter- and intra-cluster task interference for $Ind_{SD}(j)$ and Pac_{Sav}, which are dynamically updated after each concurrent execution. From the user's perspective, the primary objective is to minimize $Ind_{SD}(j)$. Conversely, from the cloud provider's perspective, scheduling multiple jobs concurrently on the same GPU enhances multi-tenancy, optimizing the utilization of available computing resources. The goal of the presented algorithm is to optimize the trade-off between maximizing Pac_{Sav} while concurrently executing n user-defined ML workloads and minimizing Ind_{SD} across all jobs. To address these conflicting objectives, the *JRS* recommends two competitor jobs based on the updated relative rankings of inter- and intra-cluster task interference, balancing the interests of both users and cloud providers.
- **Emphasizing** $Ind_{SD}(j)$: Ideally, scheduling each user-defined ML training job on a dedicated GPU would reduce $Ind_{SD}(j)$ to zero. However, this approach significantly increases computing costs for cloud providers. Multi-tenancy, which involves scheduling multiple training jobs concurrently on the same computing

resources, helps mitigate these costs. Therefore, the selection of competitor jobs becomes critical, as the Ind_{SD} of the user-defined job j is highly influenced by the competing jobs. Using the relative ranking from inter- and intra-cluster task interference analysis, the *JRS* intelligently selects competitor jobs that minimize Ind_{SD} for the user-defined job j and recommends them to the kube-scheduler to achieve this scheduler-level objective.

- **Emphasizing** Pac_{Sav}: From the cloud provider's perspective, maximizing Pac_{Sav} is achieved by scheduling multiple ML training jobs concurrently. However, the relative ranking of inter- and intra-cluster task interference for packing saving has a high variance. To address this, we present a bin-packing strategy to minimize the variability in Pac_{Sav} analysis. For each user-defined ML training job j, the *JP* extracts the $Max.GPU_{util}$ and $Max.GPU_{memAlloc}$. Jobs from the job queue are packed based on $Max.GPU_{util}$ or $Max.GPU_{memAlloc}$ until the GPU's full capacity is utilized. The remaining jobs in the queue are then scheduled based on their relative inter- and intra-cluster ranking of the Pac_{Sav} metric.

For each of the three scheduler-level objectives, we preserve user privacy by excluding sensitive hyper-parameters of the user-defined ML training task, j, such as batch size, epochs, dataset used, and model architecture. Although these parameters are influential in determining computing resource requirements, they are not considered in the competitor job selection process to maintain privacy. Our design prioritizes jobs with lower uncertainty scores within each cluster to guide scheduling decisions.

Additionally, we introduce controlled randomized intra-cluster exploration through an exploration factor, ϵ, which represents the probability of overriding the precedence given to lower uncertainty scores during scheduling. This approach allows for a balance between exploitation and exploration, enhancing decision-making flexibility. Our experimental evaluations demonstrate improved performance metrics with the inclusion of the exploration factor ϵ.

2.3 Evaluations

2.3.1 ML Workload Profiling

We profiled five ML training workloads across various domains, including image classification, image segmentation, Reinforcement Learning (RL), Generative Adversarial Networks (GAN), and Natural Language Processing (NLP). For image classification, ML training workloads were profiled on architectures such as MobileNet [22], EfficientNetV2 [23], ResNet-50 [24], InceptionV3 [25], and NasnetMobile [26], using the German Traffic Sign Recognition Benchmark (GTSRB) dataset [27]. For image segmentation, we profiled the MobileNet architecture [22] on the Oxford-IIIT Pet dataset [28]. In the generative modeling domain, we profiled the Deep Convolutional Generative Adversarial Network (DCGAN) [29] on the MNIST

Table 1 Attribute description of our ML workload dataset. (Adapted from [21].)

Attribute	Attribute description
ID	Task ID
Model	Architecture of ML model used
Dataset	Dataset used for the task
Epoch	Number of iterations for training the ML model
BS	Batch Size
Run Env	Runtime Environment
CPU Util % ran	CPU Utilization percent range
Max CPU Util %	Maximum CPU Utilization percent
Min CPU Util %	Minimum CPU Utilization percent
GPU Util % ran	GPU Utilization percent range
MGUP	Maximum GPU Utilization percent
Min GPU Util %	Minimum GPU Utilization percent
Sys Mem Util % ran	System Memory Utilization percent range
Max Sys Mem Util %	Maximum System Memory Utilization percent
Min Sys Mem Util %	Minimum System Memory Utilization percent
MaPMiU(non swap)	Maximum Process Memory in Use (non swap)
PMiU(non swap) %	Process Memory in Use (non swap) percent
CPU Thds	CPU Threads
GPU Temp ran	GPU Temperature range
Max GPU Temp	Maximum GPU Temperature
Min GPU Temp	Minimum GPU Temperature
GTSAMPR	GPU Time spent accessing memory percent range
MaGTSAMP	Maximum GPU Time spent accessing memory percent
MiGTSAMP	Minimum GPU Time spent accessing memory percent
MGMAP	Maximum GPU Memory Allocated percent
GPUPR	GPU Power Usage percent range
MaGPUPR	Maximum GPU Power Usage percent
MiGPUPR	Minimum GPU Power Usage percent
SCT user	In user mode, execution time of normal processes
SCT nice	In user mode, execution time of priority processes
SCT sys	In kernel mode, execution time of processes
SCT idle	System idle time
SCT iowait	Input-output completion time not accounted in idle time counter
SCT irq	Time to service hardware interrupts
SCT softirq	Time to service software interrupts
SCT steal	Time consumed by operating systems (OS) running in virtualized environment
SCT guest	Time consumed to run a virtual CPU for guest OS under the control of the Linux kernel
SCT guest nice	Time consumed running a virtual CPU for guest OS while executing priority processes executing in user model

(continued)

Table 1 (continued)

Attribute	Attribute description
Cor in Sys	Number of cores in the system
CS ctx switches	Number of context switches since boot
CS interrupts	Number of interrupts since boot
CS soft interrupts	Number of software interrupts since boot
CS syscalls	Number of system calls since boot
SMU total	Total available physical memory (excluding swap)
SMU available	Memory that can be assigned to processes without any system swap
SMU percent	Percent of memory used
SMU used	Memory used
SMU free	Available memory
SMU active	Memory currently in use (or very recently used)
SMU inactive	Memory marked as unused
SMU buffers	Cache data like file system, storing metadata
SMU cached	Cached data
SMU shared	Memory that may be accessed by multiple processes
SMU slab	In-kernel data structures cache
DU total	Total disk space
DU used	Used disk space
DU free	Free disk space
DU percent	Disk usage percent
NIOB sent	Number of bytes sent
NIOB received	Number of bytes received
NIOP sent	Number of packets sent
NIOP received	Number of packets received
NIO errin	Total number of errors while receiving I/O signals
NIO errout	Total number of errors while sending I/O signals
NIO dropin	Total number of dropped incoming packets
NIO dropout	Number of dropped outgoing packets

dataset [30]. RL jobs were profiled using a deep Q-network model trained in OpenAI's gym environment, specifically for the cart-pole balancing task with restricted one-dimensional motion. For NLP, we fine-tuned BERT [31] models on the IMDb movie review dataset [32] for sentiment analysis.

Profiling was conducted on the Google Colab Pro[2] platform using Tesla T4 GPUs and monitored with 'Weights and Biases',[3] an MLOps tool. We collected sixty-two features, including model hyper-parameters, CPU and GPU utilization, memory usage, disk usage, network configurations, and kernel logs, among others. The ML workload dataset, comprising forty-nine ML training workloads and corresponding training scripts, is publicly available on our GitHub repository.[4] Table 1 illustrates the attributes of our dataset.

2.3.2 Feature Selection

Based on our analysis of a large cluster of K8S, containing 16 V100 GPUs, the primary limiting factors when allocating multiple ML training jobs on the same GPU within compute clusters are the maximum GPU utilization ($Max.GPU_{util}$) and maximum GPU memory allocation percentage ($Max.GPU_{memAlloc}$). To identify the most relevant features for our target variables, we employed Pearson's correlation coefficient for feature selection, alongside advanced feature extraction techniques such as random forests, ridge regression, and lasso regression. Among the sixty-two features analyzed, the most significant predictor was the maximum GPU time spent accessing memory percentage ($Max.GPU_{timeAccMemPer}$).

Feature selection scripts are publicly available on our GitHub repository.[5] A detailed ranking of features influencing $Max.GPU_{memAlloc}$, based on the selection methods, is also accessible in the repository. To meet scheduler-level objectives, the *JP* profiles $Max.GPU_{util}$, $Max.GPU_{memAlloc}$, and $Max.GPU_{timeAccMemPer}$ for each user-defined ML training workload, ensuring black-box access. These profiles are then forwarded to the *KB* for further processing and meta-data storage.

2.3.3 Generating Synthetic Workloads and Clustering

Synthetic Workload Generation: We expand our dataset fivefold by generating synthetic workloads based on the three selected features: $Max.GPU_{util}$, $Max.GPU_{memAlloc}$, and $Max.GPU_{timeAccMemPer}$. Using the synthetic workload generation methodology outlined in Sect. 2.1.1, synthetic data is generated from the real ground truth data (GT_{real}). To assess the quality of the synthetic samples, we blind the ground truth, and regenerate the original data using the synthetic samples as pseudo-ground truth, as illustrated in Fig. 2. The Root Mean Square Error (RMSE) between the regenerated ground truth and the original ground truth is used to evaluate the quality of the synthetic data. Table 2 presents the RMSE results for all synthetic data generation techniques.

---

[2] https://colab.research.google.com/.

[3] https://wandb.ai/site.

[4] https://github.com/CentaurusInfra/alnair.

[5] https://github.com/CentaurusInfra/alnair.

Table 2 Evaluation of synthetic data generation techniques. (Adapted from [20].)

ML Model/Task	Dataset	Number of synthetic data points generated	Interpolation technique	MGUP		MGMAP		MaGTSAMP	
				RMSE	MAE	RMSE	MAE	RMSE	MAE
MobileNet	GTSRB	16	Linear	2.25	1.99	5.20	3.73	2.18	1.95
NasnetMobile				4.64	4.16	8.48	6.28	3.79	3.25
EfficientNetV2				7.18	6.25	10.69	8.39	8.24	7.40
ResNet50				3.65	3.25	4.42	2.65	2.97	2.47
InceptionV3				1.98	1.82	4.43	2.65	3.17	2.44
Image Segmentation	Oxford-IIIT			3.98	3.65	16.27	11.71	3.04	2.66
RL	Cart-Pole			0.59	0.48	0.66	0.00	0.00	0.00
DCGAN	MNIST	17		4.77	4.40	1.47	1.24	1.59	1.40
MobileNet	GTSRB	16	Barycentric	1.68	1.06	1.89	1.20	1.39	0.88
NasnetMobile				2.73	1.73	7.95	5.03	2.99	1.89
EfficientNetV2				6.53	4.13	7.95	5.03	5.23	3.30
ResNet50				2.35	1.49	1.5e-11	9.5e-12	1.30	0.82
InceptionV3				0.63	0.40	7.4e-11	3.8e-11	1.88	1.84
Image Segmentation	Oxford-IIIT			3.20	10.28	15.91	10.06	2.73	1.73
RL	Cart-Pole			0.29	0.18	7.7e-13	3.8e-13	0.00	0.00
DCGAN	MNIST	17		5.9e-10	4.1e-10	9.5e-11	6.7e-11	2.5e-10	1.8e-10
MobileNet	GTSRB	16	Krogh	1.68	1.06	1.89	1.20	1.39	0.88
NasnetMobile				2.73	1.73	7.95	5.03	2.99	1.89
EfficientNetV2				6.53	4.13	7.95	5.03	5.23	3.30
ResNet50				2.35	1.49	1.7e-11	8.1e-12	1.30	0.82
InceptionV3				0.63	0.40	3.2e-11	1.7e-11	2.90	1.84
Image Segmentation	Oxford-IIIT			3.20	2.02	15.91	10.06	2.73	1.73
RL	Cart-Pole			0.29	0.18	1.6e-12	1.0e-12	0.00	0.00
DCGAN	MNIST	17		1.9e-10	1.3e-10	9.9e-11	7.0e-11	7.7e-11	5.4e-11

(continued)

Table 2 (continued)

ML Model/Task	Dataset	Number of synthetic data points generated	Interpolation technique	MGUP		MGMAP		MaGTSAMP	
				RMSE	MAE	RMSE	MAE	RMSE	MAE
MobileNet	GTSRB	16	Pchip	1.64	1.05	2.04	1.42	1.40	0.92
NasnetMobile				2.79	1.80	7.97	5.08	3.03	1.92
EfficientNetV2				6.39	4.31	8.11	5.25	5.04	3.69
ResNet50				2.41	1.54	0.48	0.30	1.34	0.88
InceptionV3				0.69	0.48	0.48	0.30	2.94	1.88
Image Segmentation	Oxford-IIIT			3.21	2.10	15.76	9.98	2.73	1.77
RL	Cart-Pole			0.31	0.20	0.00	0.00	0.00	0.00
DCGAN	MNIST	17		0.46	0.44	0.13	0.10	0.10	0.09
MobileNet	GTSRB	16	Cubicspline	1.64	1.05	1.90	1.20	1.39	0.87
NasnetMobile				2.74	1.73	7.95	5.03	3.00	1.89
EfficientNetV2				6.52	4.12	7.96	5.03	5.19	3.28
ResNet50				2.36	1.49	0.01	0.01	1.31	0.82
InceptionV3				0.63	0.40	0.01	0.01	2.91	1.84
Image Segmentation	Oxford-IIIT			3.20	2.02	15.90	10.05	2.73	1.72
RL	Cart-Pole			0.29	0.18	1.1e-5	5.1e-6	0.00	0.00
DCGAN	MNIST	17		4.4e-14	3.3e-14	8.1e-15	6.4e-15	3.7e-15	2.7e-15
MobileNet	GTSRB	16	Spline, order=2	1.55	1.03	4.62	3.66	2.53	2.41
NasnetMobile				4.61	4.12	8.66	5.89	4.09	3.17
EfficientNetV2				5.38	4.57	10.16	7.43	3.94	3.26
ResNet50				4.08	3.50	2.72	2.23	2.98	3.12
InceptionV3				2.82	2.65	2.72	2.23	18.63	2.64
Image Segmentation	Oxford-IIIT			4.35	4.09	15.44	11.48	3.30	3.08
RL	Cart-Pole			0.66	0.49	0.01	0.01	0.00	0.00
DCGAN	MNIST	17		2.93	2.90	2.40	2.38	2.52	2.50

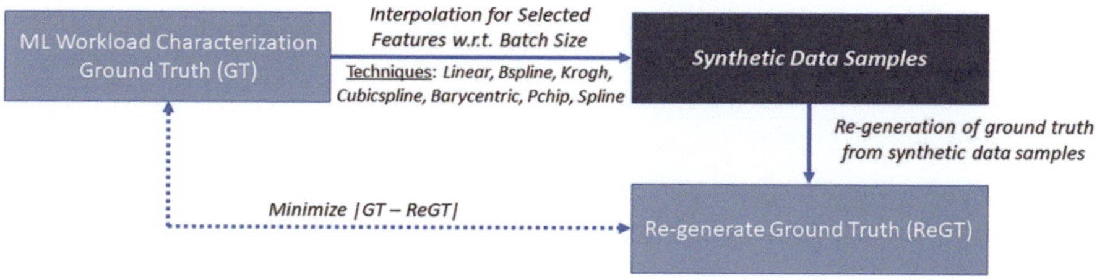

Fig. 2 Synthetic data generation strategy to augment ML workload dataset. (Adapted from [21].)

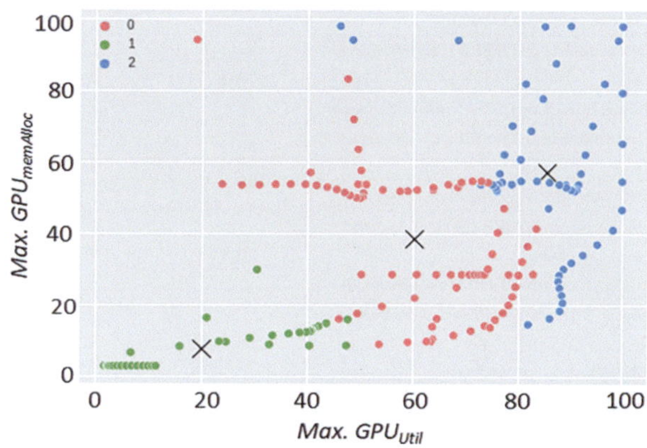

Fig. 3 Clustering of ML workloads based on $Max.GPU_{util}$, $Max.GPU_{memAlloc}$, and $Max.GPU_{timeAccMemPer}$. The black cross marks the cluster center. The farther a workload is from its assigned cluster center, the higher its associated uncertainty score. (Adapted from [33].)

Clustering: We cluster the real ML training workloads (GT_{real}) and the synthetic samples using the unsupervised k-means algorithm. Clustering decisions are guided by the elbow method and distortion scores of the k-means algorithm, utilizing $Max.GPU_{util}$, $Max.GPU_{memAlloc}$, and $Max.GPU_{timeAccMemPer}$. Jobs are grouped into clusters, and an uncertainty score is assigned based on the distance of each job from its cluster center. Figure 3 illustrates the clustering results. The *KB* maintains a privacy-preserving log of all user-defined ML training workloads. In an online setting, each job is assigned to an existing cluster in the *KB* with an accompanying uncertainty score based on the profiling results from the *JP*.

2.3.4 Task Interference Analysis

We conducted extensive experiments to analyze intra- and inter-cluster task interference for various ML workloads on 16 V100 GPUs on infrastructure provided by *Futurewei Technologies*.[6] A random selection of ML training workloads were profiled using the *JP*, clustered in the *KB* with assigned uncertainty scores, and then scheduled for analysis. Two ML jobs were scheduled concurrently on the same GPU via Argo workflow to assess inter- and intra-cluster task interference on the V100

[6] https://www.futurewei.com/.

Table 3 Inter cluster task interference for ML training workloads. (Adapted from [21].)

Task	Cluster ID	JCT (secs)	Ind_{SD}	Pac_{Sav}
MobileNet on GTSRB (*BS*=1024, *Epoch*=5) + DCGAN on MNIST (*BS*=230, *Epoch*=20)	0, 1	260.71, 292.14	7.5%, 12.3%	41.8%
NasnetMobile on GTSRB (*BS*=409, *Epoch*=5) + DCGAN on MNIST (*BS*=90, *Epoch*=20)	0, 1	268.26, 574.80	0.04%, 6.7%	28.6%
InceptionV3 on GTSRB (*BS*=20, *Epoch*=5) + RL (*BS*=128, *Epoch*=5000)	0, 1	317.16, 107.16	1.5%, 30.2%	19.6%
NasnetMobile on GTSRB (*BS*=154, *Epoch*=5) + Small BERT L2-H128-A2 (*BS*=256, *Epoch*=1)	0, 1	279.10, 270.72	1.1%, 2.2%	48.3%
NasnetMobile on GTSRB (*BS*=230, *Epoch*=5) + DCGAN on MNIST (*BS*=180, *Epoch*=20)	0, 1	276.87, 383.06	4.8%, 12.4%	36.6%
MobileNet on GTSRB (*BS*=869, *Epoch*=5) + DCGAN on MNIST (*BS*=154, *Epoch*=20)	0, 1	251.08, 392.68	5.8%, 3.1%	36.4%
Image Segmentation (*BS*=32, *Epoch*=20) + RL (*BS*=205, *Epoch*=5000)	0, 1	92.40, 104.88	11.4%, 14.6%	39.8%
Image Segmentation (*BS*=8, *Epoch*=20) + RL (*BS*=230, *Epoch*=5000)	0, 1	163.06, 114.38	2.1%, 32.8%	33.6%
MobileNet on GTSRB (*BS*=562, *Epoch*=5) + RL (*BS*=180, *Epoch*=5000)	0, 1	258.84, 93.50	7.1%, 8.2%	21.1%
Image Segmentation (*BS*=70, *Epoch*=20) + RL (*BS*=51, *Epoch*=5000)	0, 1	92.18, 109.56	10.6%, 40.4%	32.0%
MobileNet on GTSRB (*BS*=1024, *Epoch*=5) + ResNet50 on GTSRB (*BS*=1024, *Epoch*=5)	0, 2	253.99, 258.79	4.8%, 9.1%	43.3%
NasnetMobile on GTSRB (*BS*=409, *Epoch*=5) + InceptionV3 (*BS*=230, *Epoch*=5)	0, 2	269.87, 275.58	1.0%, 8.4%	47.1%
NasnetMobile on GTSRB (*BS*=154, *Epoch*=5) + Small BERT L4-H512-A8 (*BS*=256, *Epoch*=1)	0, 2	285.18, 318.65	3.3%, 5.2%	44.9%
NasnetMobile on GTSRB (*BS*=230, *Epoch*=5) + EfficientNetV2 on GTSRB (*BS*=256, *Epoch*=5)	0, 2	287.42, 3766.11	8.8%, 37.6%	29.9%
Image Segmentation (*BS*=32, *Epoch*=20) + InceptionV3 on GTSRB (*BS*=409, *Epoch*=5)	0, 2	118.16, 273.76	42.4%, 7.1%	19.1%
InceptionV3 on GTSRB (*BS*=20, *Epoch*=5) +	0, 2	348.13, 319.64	11.4%, 5.7%	43.3%

(continued)

Table 3 (continued)

Task	Cluster ID	JCT (secs)	Ind_{SD}	Pac_{Sav}
Small BERT L2-H768-A12 (*BS*=256, *Epoch*=1)				
MobileNet on GTSRB (*BS*=562, *Epoch*=5) + EfficientNetV2 on GTSRB (*BS*=562, *Epoch*=5)	0, 2	258.61, 318.69	6.9%, 15.7%	38.3%
Image Segmentation(*BS*=8, *Epoch*=20) + MobileNet on GTSRB (*BS*=869, *Epoch*=5)	0, 2	175.99, 258.75	10.2%, 9.1%	34.7%
EfficientNetV2 on GTSRB (*BS*=109, *Epoch*=5) + ResNet50 on GTSRB (*BS*=154, *Epoch*=5)	0, 2	472.88, 410.05	58.1%, 56.8%	15.6%
Image Segmentation (*BS*=70, *Epoch*=20) + ResNet-50 on GTSRB (*BS*=180, *Epoch*=5)	0, 2	170.98, 291.21	105.1%, 12.7%	14.7%
DCGAN on MNIST (*BS*=230, *Epoch*=20) + EfficientNetV2 on GTSRB (*BS*=109, *Epoch*=5)	1, 2	372.22, 406.37	43.1%, 35.8%	27.3%
DCGAN on MNIST (*BS*=230, *Epoch*=20) + EfficientNetV2 on GTSRB (*BS*=109, *Epoch*=5)	1, 2	372.22, 406.37	43.1%, 35.8%	27.3%
DCGAN on MNIST (*BS*=90, *Epoch*=20) + Small BERT L4-H512-A8 (*BS*=256, *Epoch*=1)	1, 2	552.29, 306.56	2.6%, 1.2%	34.3%
DCGAN on MNIST (*BS*=180, *Epoch*=20) + EfficientNetV2 on GTSRB (*BS*=256, *Epoch*=5)	1, 2	446.90, 382.49	31.1%, 39.9%	27.2%
Small BERT L2-H128-A2 (*BS*=256, *Epoch*=1) + ResNet-50 on GTSRB (*BS*=154, *Epoch*=5)	1, 2	296.76, 281.19	12.1%, 7.5%	43.5%
DCGAN on MNIST (*BS*=154, *Epoch*=20) + InceptionV3 on GTSRB (*BS*=230, *Epoch*=5)	1, 2	425.46, 270.28	11.6%, 6.3%	33.0%
RL (*BS*=180, *Epoch*=5000) + Small BERT L2-H768-A12 (*BS*=256, *Epoch*=1)	1, 2	91.94, 303.74	6.4%, 0.5%	21.8%
RL (*BS*=128, *Epoch*=5000) + EfficientNetV2 on GTSRB (*BS*=562, *Epoch*=5)	1, 2	129.54, 287.80	57.3%, 4.5%	19.4%
RL (*BS*=230, *Epoch*=5000) + InceptionV3 on GTSRB (*BS*=409, *Epoch*=5)	1, 2	130.56, 266.97	51.6%, 4.4%	21.8%
RL (*BS*=205, *Epoch*=5000) + Resnet-50 on GTSRB (*BS*=1024, *Epoch*=5)	1, 2	136.20, 280.88	48.8%, 7.2%	20.5%
RL (*BS*=51, *Epoch*=5000) + ResNet-50 on GTRSB (*BS*=180, *Epoch*=5000)	1, 2	100.15, 273.06	28.3%, 5.7%	18.7%

GPU. The results of the inter-cluster task interference are presented in Table 3, while Table 4 details the intra-cluster task interference results.

Interference is analyzed with respect to the defined metrics of $Ind_{SD}(j)$ and Pac_{Sav}. The relative ranking of inter- and intra-cluster task interference is logged within the *JRS* component and is dynamically updated whenever a set of workloads completes concurrent execution on the GPU. To achieve scheduler-level objectives, the *JRS* recommends competitor jobs from the job queue to the kube-scheduler for concurrent execution on hardware accelerators, managed through the Argo workflow. Figure 4 illustrates the relative ranking of the clusters with respect to the defined task interference metrics at a specific time instance.

2.3.5 Recommendation Service for Scheduling ML Training Jobs

The *JRS* component, which maintains the relative ranking of task interference, recommends competitor jobs from the user-defined job queue to the kube-scheduler in order to meet scheduler-level objectives. We randomly selected 24 ML training workloads, including autoencoder models, to simulate real-world conditions. Table 5 lists 15 unique workloads, each varied with respect to their hyper-parameters to emulate the dynamic nature of a user-defined job queue at any given time.

When a job enters the queue, it is profiled by the *JP* in the worker node. The *KB* then assigns these jobs to existing clusters and associates them with an uncertainty score. Based on the defined scheduler-level objectives, the *JRS* recommends suitable competitor jobs to the kube-scheduler, which places them on appropriate nodes. Concurrent execution of these jobs is managed through the Argo workflow framework.

- **Balancing between $Ind_{SD}(j)$ and Pac_{Sav}**: Compared to the kube-scheduler's random or first-come-first-serve approach to promoting multi-tenancy, decisions guided by our method show superior performance in terms of $Ind_{SD}(j)$ and Pac_{Sav}, as depicted in Fig. 5a. Table 6 lists the concurrent workloads used for performance testing of our presented design. The balancing of conflicting objectives in our design is further enhanced by introducing the exploration factor ϵ.
- **Emphasizing $Ind_{SD}(j)$**: In comparison to the kube-scheduler's random or first-come-first-serve approach, our method reduces $Ind_{SD}(j)$ by **3.79%** and **2.29%**, respectively, as shown in Fig. 5c. Table 7 details the concurrent workloads used for performance evaluation. Additionally, our presented design achieves a further reduction of $Ind_{SD}(j)$ by **5.08%** with the inclusion of ϵ.
- **Maximizing Pac_{Sav}**: Compared to the kube-scheduler's random or first-come-first-serve approach, our method demonstrates **7.74%** and **7.70%** improvements in Pac_{Sav}, respectively, for bin-packing via $Max.GPU_{util}$, as illustrated in Fig. 5b. Tables 2.8 and 2.9 outline the concurrent workloads used for performance testing. The presented design achieves an additional improvement of **10.82%** when bin-packing via $Max.GPU_{memAlloc}$.

Table 4 Intra cluster task interference for ML training workloads. (Adapted from [21].)

Task	Cluster ID	JCT (secs)	Ind_{SD}	Pac_{Sav}
MobileNet on GTSRB (*BS*=1024, *Epoch*=5) + MobileNet on GTSRB (*BS*=869, *Epoch*=20)	0, 0	253.81, 253.05	4.7%, 6.7%	47.1%
MobileNet on GTSRB (*BS*=562, *Epoch*=5) + Image Segmentation (*BS*=8, *Epoch*=20)	0, 0	254.39, 182.70	5.2%, 14.5%	36.6%
NasnetMobile on GTSRB (*BS*=409, *Epoch*=5) + InceptionV3 on GTSRB (*BS*=20, *Epoch*=5)	0, 0	270.65, 394.90	1.2%, 26.3%	31.8%
NasnetMobile on GTSRB (*BS*=230, *Epoch*=5) + Image Segmentation (*BS*=32, *Epoch*=20)	0, 0	278.34, 99.59	5.4%, 20.0%	19.7%
NasnetMobile on GTSRB (*BS*=154, *Epoch*=5) + Image Segmentation (*BS*=109, *Epoch*=20)	0, 0	322.51, 101.12	16.8%, 31.8%	8.5%
Image Segmentation (*BS*=70, *Epoch*=20) + NasnetMobile on GTSRB (*BS*=230, *Epoch*=5)	0, 0	100.05, 287.07	20.1%, 7.6%	18.2%
DCGAN on MNIST (*BS*=230, *Epoch*=20) + DCGAN on MNIST (*BS*=90, *Epoch*=20)	1, 1	296.67,588.42	14.1%, 9.3%	26.2%
RL (*BS*=230, *Epoch*=5000) + DCGAN on MNIST (*BS*=180, *Epoch*=20)	1, 1	112.62, 340.66	30.8%, 0.2%	19.9%
RL (*BS*=205, *Epoch*=5000) + RL (*BS*=51, *Epoch*=5000)	1, 1	92.61, 80.29	1.2%, 2.9%	45.3%
DCGAN on MNIST (*BS*=154, *Epoch*=20) + Small BERT L2-H128-A2 (*BS*=256, *Epoch*=1)	1, 1	372.88, 290.24	0.3%, 9.6%	40.8%
RL (*BS*=180, *Epoch*=5000) + RL (*BS*=128, *Epoch*=5000)	1, 1	84.26, 87.6	1.4%, 0.5%	48.1%
EfficientNetV2 on GTSRB (*BS*=562, *Epoch*=5) + ResNet-50 on GTSRB (*BS*=1024, *Epoch*=5)	2, 2	452.29, 424.10	64.3%, 61.9%	15.7%
EfficientNetV2 on GTSRB (*BS*=256, *Epoch*=5) + EfficientNetV2 on GTSRB (*BS*=109, *Epoch*=5)	2, 2	486.44, 526.50	77.9%, 76.1%	0.8%
InceptionV3 on GTSRB (*BS*=409, *Epoch*=5) + ResNet-50 on GTSRB (*BS*=180, *Epoch*=5)	2, 2	286.54, 360.76	12.1%, 39.7%	29.7%
ResNet-50 on GTSRB (*BS*=154, *Epoch*=5) + Small BERT L4-H512-A8 (*BS*=256, *Epoch*=1)	2, 2	304.51, 357.79	16.4%, 18.1%	36.6%
InceptionV3 on GTSRB (*BS*=230, *Epoch*=5) + Small BERT L2-H768-A12 (*BS*=256, *Epoch*=1)	2, 2	279.61, 328.48	9.9%, 8.7%	40.9%

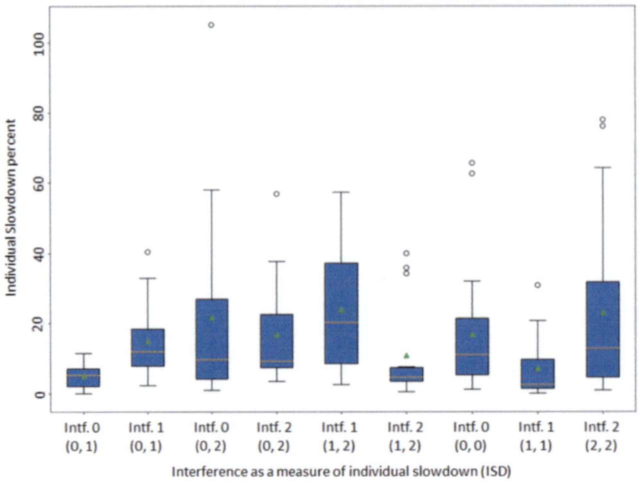

Cluster Interference	Median value	Relative Rank
Intf. 0 (0, 1)	5.3	3
Intf. 1 (0, 1)	11.9	7
Intf. 0 (0, 2)	9.6	5
Intf. 2 (0, 2)	9.1	4
Intf. 1 (1, 2)	20.3	9
Intf. 2 (1, 2)	4.5	2
Intf. 0 (0, 0)	11.1	6
Intf. 1 (1, 1)	2.7	1
Intf. 2 (2, 2)	12.9	8

(a)

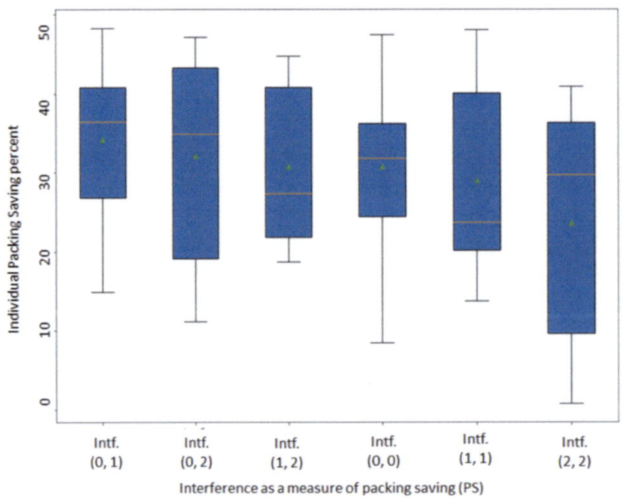

Cluster Interference	Median value	Relative Rank
Intf. (0, 1)	36.5	6
Intf. (0, 2)	34.9	5
Intf. (1, 2)	27.3	2
Intf. (0, 0)	31.8	4
Intf. (1, 1)	23.8	1
Intf. (2, 2)	29.7	3

(b)

Fig. 4 Relative ranking of **a** inter- and **b** intra-cluster task interference for the $Ind_{SD}(j)$ and Pac_{Sav} metrics. Figure on the left shows Intf.0(0,1), representing the interference of a job from cluster 0 when scheduled concurrently with a job from cluster 1 on the same GPU. Figure on the right shows Intf.(0,1), indicating the interference when jobs from clusters 0 and 1 are scheduled concurrently on the same GPU. (Adapted from [20].)

We also present the associated differential entropy to measure the certainty of the outcomes. A lower differential entropy indicates more favorable and reliable results (Table 8).

Table 5 Machine learning workloads for performance testing. (Adapted from [20].)

Category	Task ID	ML training job
Image classification	J1	MobileNet on GTSRB
	J2	EfficientNetV2 on GTSRB
	J3	ResNet50 on GTSRB
	J4	InceptionV3 on GTSRB
	J5	NasneMobile on GTSRB
Image segmentation	J6	MobileNet on Oxford-IIIT Pet Dataset
Generative adversarial networks	J7	DCGAN on MNIST
Autoencoders	J8	De-noising Autoencoder on CIFAR-10
	J9	De-noising Autoencoder on MNIST
	J10	De-noising Autoencoder on F-MNIST
Natural language processing	J11	Fine-tuning pre-trained Small BERT L2-H256-A4 model on IMDb movie review dataset
	J12	Fine-tuning pre-trained Small BERT L2-H128-A2 model on IMDb movie review dataset
	J13	Fine-tuning pre-trained Small BERT L2-H768-A12 model on IMDb movie review dataset
	J14	Fine-tuning pre-trained Small BERT L4-H256-A4 model on IMDb movie review dataset
Reinforcement learning	J15	Cart-pole

3 Inferencing Framework for Stable Diffusion Models

Developers are incrementally advancing the state-of-the-art SD models aimed at facilitating text-to-image conversion. For a variety of pre-trained SD models, each variant exhibits variations in resource footprints and processing time latencies across heterogeneous compute resources. In this Section, we present a model-less, privacy-preserving, low-latency inference framework to satisfy user-defined SLOs for SDaaS. Developers of SD models can register their models on our presented system through a declarative API. Users, on the other hand, can specify SLOs in terms of the style of the generated image, the requested latency, and the minimum requested CLIP score for inference through the user API. Our presented system manages model registration from the developers, and schedules volumes of user queries aimed to meet SLOs through an efficient deployment of the models onto hardware accelerators in the compute cluster (Table 9).

Fig. 5 Comparison of scheduler-level objective improvements for our three presented designs against the kube-scheduler using random and First Come First Serve (FCFS) scheduling decisions to promote multi-tenancy. The results highlight differential entropy, indicating the uncertainty in the scheduling outcomes. (Adapted from [20].)

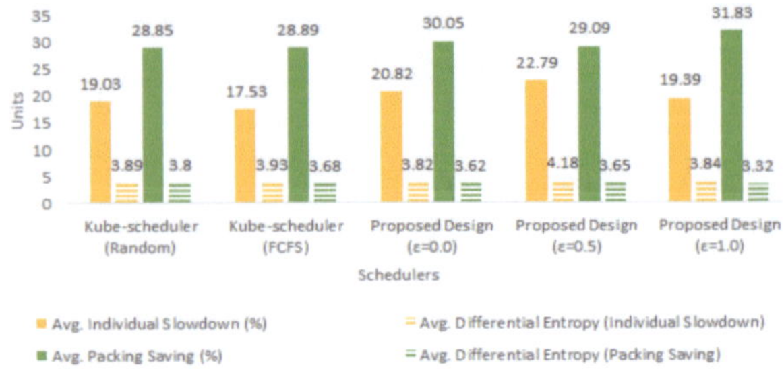

(a)

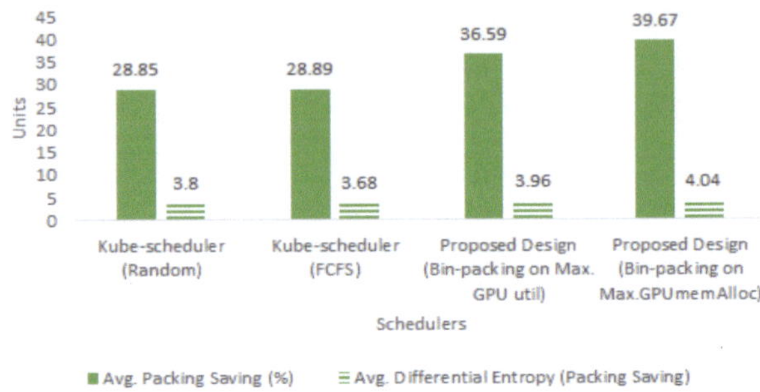

(b)

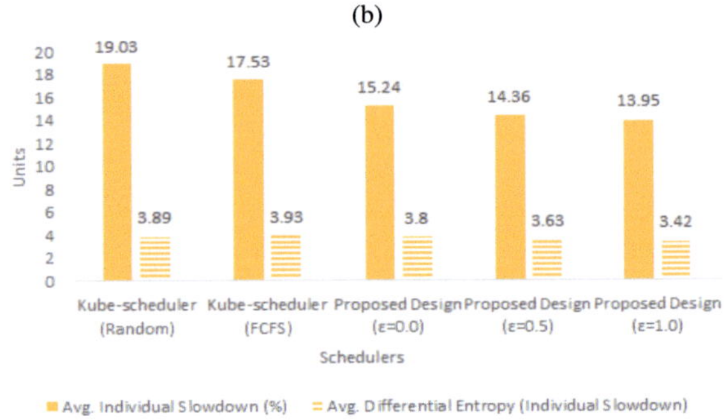

(c)

Table 6 Performance testing: balancing $Ind_{SD}(j)$ and Pac_{Sav}, $\epsilon = 0.0$. (Adapted from [20].)

Task	Cluster ID	Distance from cluster center	JCT (secs)	Ind_{SD}	Pac_{Sav}
J11 (*BS*=256, *Epoch*=1) + J4 (*BS*=409, *Epoch*=5)	1, 2	207.51, 122.85	288.91, 272.13	2.7%, 4.8%	46.5%
J7 (*BS*=205, *Epoch*=20) + J3 (*BS*=230, *Epoch*=5)	1, 2	133.11, 140.82	394.69, 342.09	40.3%, 31.7%	27.03%
J7 (*BS*=180, *Epoch*=20) + J6 (*BS*=716, *Epoch*=20)	1, 2	134.28, 169.33	361.85, 116.39	12.4%, 46.9%	9.7%
J7 (*BS*=109, *Epoch*=20) + J14 (*BS*=256, *Epoch*=1)	1, 2	135.99, 207.51	482.97, 289.89	3.4%, 7.7%	34.3%
J12(*BS*=256, *Epoch*=1) + J2 (*BS*=716, *Epoch*=5)	1, 2	150.17, 212.72	320.60, 316.84	23.2%, 14.9%	40.1%
J15 (*BS*=32, *Epoch*=5000) + J4 (*BS*=869, *Epoch*=5)	1, 2	198.65, 234.25	107.81, 265.76	45.8%, 3.4%	19.6%
J15 (*BS*=562, *Epoch*=5000) + J13 (*BS*=256, *Epoch*=1)	1, 2	208.97, 237.52	172.15, 293.24	11.8%, 0.96%	34.01%
J15 (*BS*=716, *Epoch*=5000) + J3 (*BS*=1024, *Epoch*=5)	1, 2	209.15, 245.53	256.16, 281.45	36.8%, 6.1%	37.7%
J1 (*BS*=205, *Epoch*=5) + J6 (*BS*=230, *Epoch*=20)	0, 0	114.8, 116.2	270.24, 99.97	5.6%, 26.3%	19.2%
J5 (*BS*=562, *Epoch*=5) + J8 (*BS*=256, *Epoch*=50)	0, 0	123.82, 124.15	273.01, 167.45	3.8%, 25.5%	31.1%
J7 (*BS*=1024, *Epoch*=20) + J8 (*BS*=1024, *Epoch*=50)	0, 0	125.03, 131.74	160.54, 150.94	35.9%, 50.5%	26.4%
J7 (*BS*=256, *Epoch*=50) + J10 (*BS*=256, *Epoch*=50)	0, 0	135.31, 134.81	122.71, 123.64	26.7%, 32.7%	34.9%

Table 7 Performance testing: minimizing individual slowdown, $\epsilon = 0.0$. (Adapted from [20].)

Task	Cluster ID	Dist. from cluster center	JCT (Secs)	Ind_{SD} (%)	Pac_{Sav} (%)
J11 (*BS*=256, *Epoch*=1) + J7 (*BS*=205, *Epoch*=20)	1, 1	133.07. 133.11	282.12, 292.51	0.2%, 3.9%	47.9%
J7 (*BS*=109, *Epoch*=20) + J7 (*BS*=180, *Epoch*=20)	1, 1	135.99, 134.28	533.61, 383.31	14.3%, 19.0%	32.3%
J12 (*BS*=256, *Epoch*=1) + J15 (*BS*=32, *Epoch*=5000)	1, 1	150.17, 198.35	261.29, 80.41	0.4%, 8.7%	21.7%
J15 (*BS*=562, *Epoch*=5000) + J15 (*BS*=716, *Epoch*=5000)	1, 1	208.97, 209.15	158.81, 188.69	3.1%, 0.7%	44.6%
J1 (*BS*=205, *Epoch*=5) + J4 (*BS*=409, *Epoch*=55)	0, 2	114.8, 122.85	287.27, 275.05	12.3%, 5.9%	44.2%
J6 (*BS*=230, *Epoch*=20) + J3 (*BS*=230, *Epoch*=5)	0, 2	116.2, 140.82	135.11, 313.33	70.7%, 20.6%	7.5%
J5 (*BS*=562, *Epoch*=5) + J6 (*BS*=716, *Epoch*=20)	0, 2	123.82, 169.33	304.29, 92.45	15.7%, 16.7%	11.0%
J8 (*BS*=256, *Epoch*=50) + J14 (*BS*=256, *Epoch*=1)	0, 2	124.15, 207.51	136.90, 283.88	2.6%, 5.5%	29.4%
J7 (*BS*=1024, *Epoch*=20) + J2 (*BS*=716, *Epoch*=5)	0, 2	125.03, 212.72	166.99, 354.50	41.4%, 28.6%	9.9%
J8 (*BS*=1024, *Epoch*=50) + J4 (*BS*=869, *Epoch*=5)	0, 2	131.74, 234.35	120.77, 305.94	20.4%, 19.1%	14.3%
J10 (*BS*=256, *Epoch*=50) + J13 (*BS*=256, *Epoch*=1)	0, 2	134.81, 237.52	98.08, 291.03	5.2%, 0.2%	24.1%
J9 (*BS*=256, *Epoch*=50) + J3 (*BS*=1024, *Epoch*=5)	0, 2	135.31, 245.53	128.12, 313.52	32.2%, 18.2%	13.3%

Table 8 Performance testing: maximizing packing saving. Bin packing via proportional resource distribution on $Max.GPU_{util}$, followed by *JRS* recommendations based on clustering. (Adapted from [20].)

Task	Cluster ID	Distance from cluster center	JCT (secs)	Ind_{SD}	Pac_{Sav}
J15(*BS*=716, *Epoch*=5000) + J15(*BS*=562, *Epoch*=5000) + J15(*BS*=32, *Epoch*=5000) + J12(*BS*=256, *Epoch*=1) + J11(*BS*=256, *Epoch*=1)	NA	NA	189, 167, 91, 281, 285	1.5% 9.1% 24.6% 8.1% 1.2%	70.1%
J7(*BS*=205, *Epoch*=20) + J7(*BS*=180, *Epoch*=20)	1, 1	133.11, 134.28	331.24, 329.24	17.7%, 2.5%	45.1%
J7(*BS*=109, *Epoch*=20) + J4(*BS*=409, *Epoch*=5)	1, 2	135.99, 122.85	520.04, 265.64	11.3%, 2.3%	28.4%
J3(*BS*=230, *Epoch*=5) + J6(*BS*=716, *Epoch*=20)	2, 2	140.82, 169.33	267.21, 82.14	2.9%, 3.7%	21.1%
J14(*BS*=256, *Epoch*=1) + J2(*BS*=716, *Epoch*=5)	2, 2	207.51, 212.72	270.29, 287.93	0.4%, 4.4%	47.1%
J4(*BS*=869, *Epoch*=5) + J13(*BS*=256, *Epoch*=1)	2, 2	234.35, 237.52	271.93, 313.96	5.8%, 8.1%	42.6%
J1(*BS*=205, *Epoch*=5) + J6(*BS*=230, *Epoch*=20)	0, 0	114.8, 116.2	270.24, 99.97	5.6%, 26.3%	19.2%
J5(*BS*=562, *Epoch*=5) + J8(*BS*=256, *Epoch*=50)	0, 0	123.82, 124.15	273.01, 167.45	3.8%, 25.5%	31.1%
J7(*BS*=1024, *Epoch*=50) + J8(*BS*=1024, *Epoch*=50)	0, 0	125.03, 131.74	160.54, 150.94	35.9%, 50.5%	26.4%
J10(*BS*=256, *Epoch*=50) + J9(*BS*=256, *Epoch*=50)	0, 0	134.81, 135.31	122.71, 123.64	26.7%, 32.7%	34.9%

Table 9 Performance testing: maximizing packing saving. Bin packing via proportional resource distribution on $Max.GPU_{memAlloc}$, followed by *JRS* recommendations based on clustering. (Adapted from [20].)

Task	Cluster ID	Distance from cluster center	JCT (secs)	Ind_{SD}	Pac_{Sav}
J15 (*BS*=562, *Epoch*=5000)			248,	62.0%	
+					
J15 (*BS*=32, *Epoch*=5000)			214,	193.1%	
+					
J15 (*BS*=716, *Epoch*=5000)			257,	37.4%	
+					
J9 (*BS*=256, *Epoch*=50)			175,	80.4%	
+					
J10 (*BS*=256, *Epoch*=50)			167,	79.5%	
+	NA	NA			69.2%
J7 (*BS*=109, *Epoch*=20)			556,	19.3%	
+					
J7 (*BS*=180, *Epoch*=20)			365,	13.7%	
+					
J7 (*BS*=205, *Epoch*=20)			334,	18.8%	
+					
J8 (*BS*=256, *Epoch*=50)			187,	40.6%	
J11 (*BS*=256, *Epoch*=1)					
+	1, 1	13.07, 150.17	262.92, 238.19	3.5%, 1.7%	46.1%
J12 (*BS*=256, *Epoch*=1)					
J4 (*BS*=409, *Epoch*=5)					
+	2, 2	122.85, 140.33	264.50, 309.80	1.8%, 19.3%	40.3%
J3 (*BS*=230, *Epoch*=5)					
J6 (*BS*=716, *Epoch*=20)					
+	2, 2	169.33, 207.51	81.51, 272.01	2.9%, 1.1%	21.8%
J14 (*BS*=256, *Epoch*=1)					
J2 (*BS*=716, *Epoch*=5)					
+	2, 2	212.72, 234.35	278.90, 274.01	1.2%, 6.6%	47.6%
J4 (*BS*=869, *Epoch*=5)					
J13 (*BS*=256, *Epoch*=1)					
+	2, 2	237.52, 245.53	295.13, 316.90	1.6%, 19.5%	42.9%
J3 (*BS*=1024, *Epoch*=5)					
J1 (*BS*=205, *Epoch*=5)					
+	0, 0	114.8, 116.2	264.85, 82.37	3.5%, 4.1%	20.1%
J6 (*BS*=230, *Epoch*=20)					
J5 (*BS*=562, *Epoch*=5)					
+	0, 0	123.82, 125.03	268.72, 121.72	2.2%, 3.1%	29.4%
J7 (*BS*=1024, *Epoch*=20)					

Table 10 Single step inference latency for stable diffusion models on CPU and GPU. (Adapted from [34].)

Stable diffusion models	Inference time (s)	
	CPU	GPU
DGSpitzer/Cyberpunk-Anime-Diffusion	22.47	0.317
dreamlike-art/dreamlike-anime-1.0	77.41	0.576
CompVis/stable-diffusion-v1-4	27.21	0.309
stabilityai/stable-diffusion-2-1-base	21.77	0.315
stabilityai/stable-diffusion-2-1	86.64	0.734
SG161222/Realistic_Vision_V1.4	35.34	0.304
hakurei/waifu-diffusion	24.94	0.328
Nilaier/Waifu-Diffusers	22.58	0.314
kohbanye/pixel-art-style	27.85	0.302

This Section is structured as follows: Sect. 3.1 provides the motivation for developing the inference-serving framework, highlighting the key findings that guide the development of our inference framework. Section 3.2 details the presented system architecture, followed by Sect. 3.3, which presents the evaluation results. Finally, Sect. 4 concludes the discussion on the presented inference framework for SDaaS.

3.1 Motivation

We evaluate the runtime latencies for a prompt across 9 pre-trained SD models[7] for text-to-image generation. We specifically investigate the performance discrepancy between CPU and GPU for a single inference step in the diffusion process. The aggregated findings for a randomly selected prompt 'an astronaut riding a cow', are presented in Table 10, utilizing a floating-point precision of 16 consistently across all SD models. We used an X86_64, employing 64-bit processing, powered by an Intel(R) Xeon(R) CPU @ 2.20GHz with an advertised frequency of 2.2000 GHz, and Tesla V100 as our CPU and GPU respectively. Our analysis reveals that for all the evaluated SD models, the average inference time on CPU is 92x higher, on average, compared to a GPU deployment. This highlights the need for scheduling SD inferencing workloads on GPU to achieve significantly lower latency.

From the data presented in Table 10, we observe the variability in inference times, when utilizing the same compute resource across a variety of SD models, for an

[7] https://huggingface.co/.

Table 11 GPU runtimes. (Adapted from [34].)

GPU	Inference time (s)
RTX4090	5.473
RTX3090	5.952
A100PCIE	3.684
RTX4080	7.417
RTXA6000	5.591
RTXA4000	9.601
TeslaV100	6.804
RTX3060	13.997
A40	6.214
RTX3080	7.211

identical query. This elucidates that every SD model exhibits significant variations in their resource requirements.

Next, we subject the identical inference workload for one SD model, across various hardware accelerators. In Table 11, we present the average inference time for 10 inference steps, for the SD model 'dreamlike-art/dreamlike-anime-1.0', while achieving a CLIP score of 0.37 on 'ViT-B/32'. This elucidates the pronounced variability in processing latencies across hardware accelerators in the compute cluster for SD models.

To evaluate the influence of floating point precision in SD models, we present in Table 12, the mean inference time over 10 inference steps, and CLIP score ('ViT-B/32') for the SD model 'dreamlike-art/dreamlike-anime-1.0', on the A100 PCIE GPU, for varying inference steps. With a higher floating point precision, the SD model achieves a marginally higher CLIP score in fewer inference steps, at the expense of higher latency per inference steps. This elucidates that there exists a marginal trade-off between the floating point precision of the SD model to the associated inference time latency and CLIP score.

To illustrate the diversity of styles in the output image, we present a selection of generated images from 3 SD models: namely, 'runwaymlstable-diffusion-v1-5', 'kohbanye/pixel-art-style', and 'DGSpitzer/Cyberpunk-Anime-Diffusion', across variable inference steps along with their corresponding CLIP Scores ('ViT-B/32'), as shown in Fig. 6. Generally, we anticipate that as the number of inference steps increases, the quality of the images will improve, leading to a proportional increase in the CLIP score. However, it is noteworthy that the variation in generated features within the latent information array, in response to the input prompt, leads to a non-monotonic relationship between the number of inference steps and the CLIP score.

To second this further, we extract the text captions from the Flickr8k [35] dataset and feed 10% text prompts into SD models to evaluate the runtime latency and CLIP score over 3 hardware accelerators. From Fig. 7, we observe a monotonic increase in

Table 12 Impact of inference time to CLIP score for floating point precision in stable diffusion models. (Adapted from [34].)

Inference steps	Half-precision (16FP)		Single-precision (32FP)	
	Latency (s)	CLIP score	Latency (s)	CLIP score
1	0.123	0.163	0.283	0.165
2	0.195	0.168	0.386	0.175
3	0.241	0.341	0.493	0.348
4	0.293	0.296	0.604	0.309
5	0.339	0.269	0.704	0.295
6	0.376	0.327	0.805	0.334
7	0.411	0.322	0.892	0.329
8	0.447	0.314	0.985	0.324

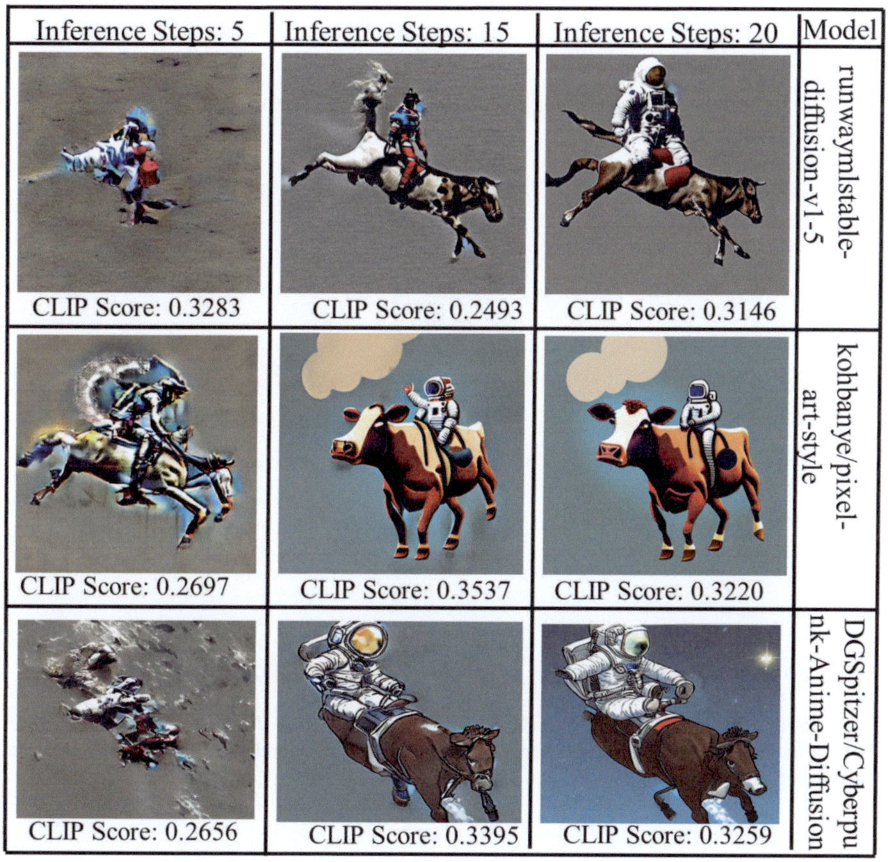

Fig. 6 Inference latency and CLIP Score ('ViT-B/32') for different styles of stable diffusion models, at seed 1024, for prompt 'an astronaut riding a cow'. (Adapted from [34].)

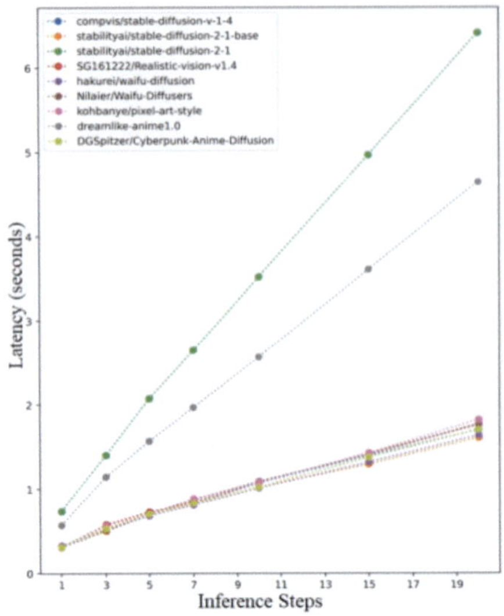

(a) Profiled on Tesla V100 GPU.

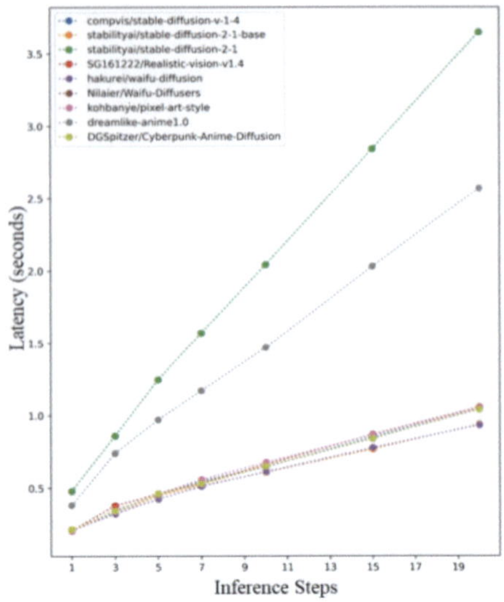

(b) Profiled on A100 PCIE GPU.

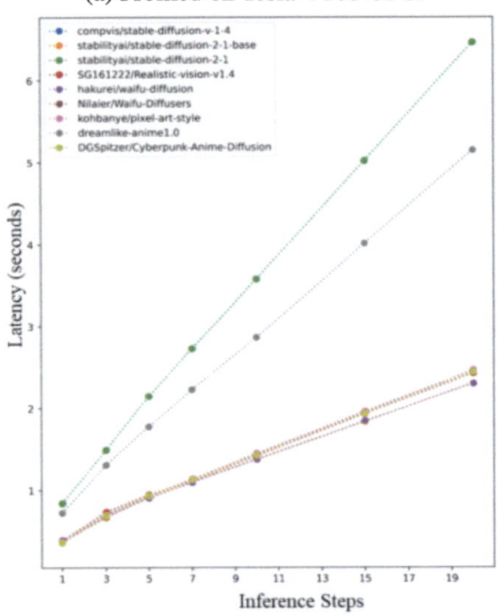

(c) Profiled on RTX4080 GPU.

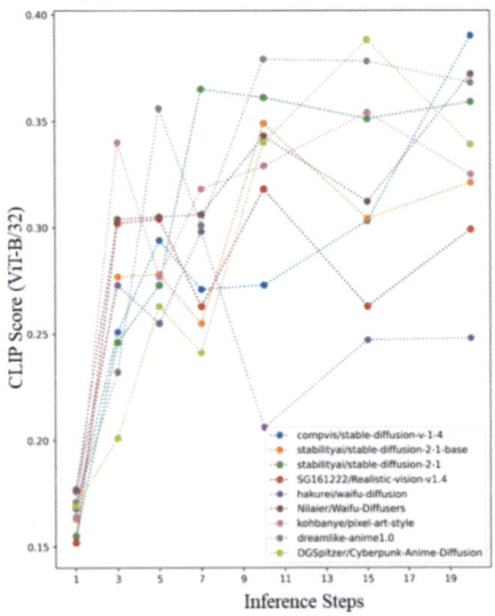

(d) CLIP Scores.

Fig. 7 Processing latency for SD models, over **a** Tesla V100, **b** A100, **c** RTX4080 hardware accelerators. **d** Represents the CLIP score ('ViT-B/32') of the SD models as a function of inference steps. (Adapted from [34].)

inference latency with respect to the inference steps, but a non-monotonic relationship between the number of inference steps and the CLIP score.

3.2 System Design

In this sub-section, we present the design of our model-less, privacy-preserving, low-latency inference framework to satisfy user-centric SLOs for SDaaS. The components of the presented system are outlined subsequently, and also illustrated in Fig. 8.

Design Principles. Developers of SD models can register their model on the presented system through a declarative API. Users, on the other hand, can specify SLOs in terms of the style of the generated image, the requested latency, and the minimum requested CLIP score for inference through the user API. Our presented system manages model registration from the developers, and schedules volumes of user queries to meet SLOs through an efficient deployment of the SD models to hardware accelerators in the compute cluster.

Model-less interface for inference. The front-end interface of the presented system involves model registration from the developers and the submission of user queries for inferencing on the registered models.

- *Model registration*: The developers can register their models using a declarative API. The API accommodates a model identifier designated by the developer and the trained weights of the SD model.
- *Query Submission*: Users can submit inference queries, mentioning the requested text prompt t_p to generate an output image, the style of the generated image s, and specify high-level performance criteria, like the requested CLIP score (SLO_{CLIP}). The user can also specify a requested SLO latency ($SLO_{latency}$) within which the request must be processed, alongside SLO_{CLIP}.

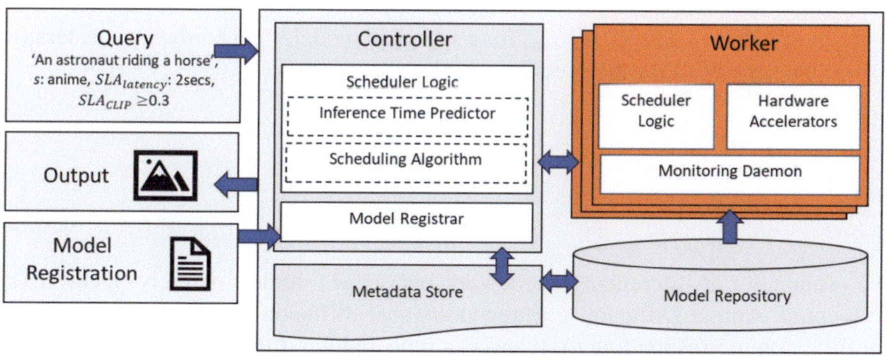

Fig. 8 Overview of the presented SDaaS system architecture. (Adapted from [34].)

Architecture. The controller, depending on the specified s loads the appropriate model from the model repository, and forwards the inference query to a worker machine with the scheduling logic. The workers execute the inference queries to hardware accelerators and subsequently respond with the inference results to the user. If the user-defined SLO could not be satisfied for a particular query, the system returns a message appropriately to the user.

- *Controller*: The centralized controller handles both model registration from developers, and SLO-based inference queries from the users. It comprises of two modules: (a) *scheduler logic block*, and (b) *model registrar*. The *scheduler logic block* consists of the inference time predictor and the scheduling algorithm. Assuming black-box access to the models to preserve developer privacy, all registered models are profiled on the worker nodes (hardware accelerators) over multiple text-to-image prompts for varying inference steps. Based on the profiling results, the inference time predictor heuristically predicts the number of inference steps, for a user-defined SLA_{CLIP} and $SLA_{latency}$, for a specified SD model, over a hardware accelerator, to satisfy the SLO for the specific query. The other module, *model registrar* is responsible for managing model registration from the developers to the *model repository* through an API.
- *Worker*: Worker nodes execute inference queries, following the scheduler logic, as directed by the controller, over the hardware accelerators. Hardware-specific execution daemons manage the deployment and execution of models.
- *Model Repository*: The *Model Repository* functions as a high-capacity, persistent storage, containing the trained SD models registered by the developers. The worker nodes can access this storage to execute SD models on hardware accelerators based on the user inference query.
- *Metadata Store*: Assuming black-box access to the models to preserve developer privacy, all registered models are profiled on the worker nodes over multiple text-to-image prompts for varying inference steps. The metadata store includes information concerning available models, and the profiled attributes of the number of inference steps to mean CLIP score and mean inference latency. The inference time predictor accesses this data to heuristically predict the number of inference steps, for the SLA_{CLIP} and $SLA_{latency}$ for a specified model, for a hardware accelerator, to satisfy the SLO for the specific query.

3.3 Evaluation

We evaluated our inferencing framework over 3 SD models, namely 'DGSpitzer/ Cyberpunk-Anime-Diffusion', 'runwayml/stable-diffusion-v1-5', and 'hakurei/ wa ifu-diffusion' corresponding to styles (s) anime, regular diffusion, and waifu respectively. To simulate the text prompts t_p in user queries, we randomly sampled 100 prompts from Flickr8k dataset [35], with a randomly assigned requested style s (35

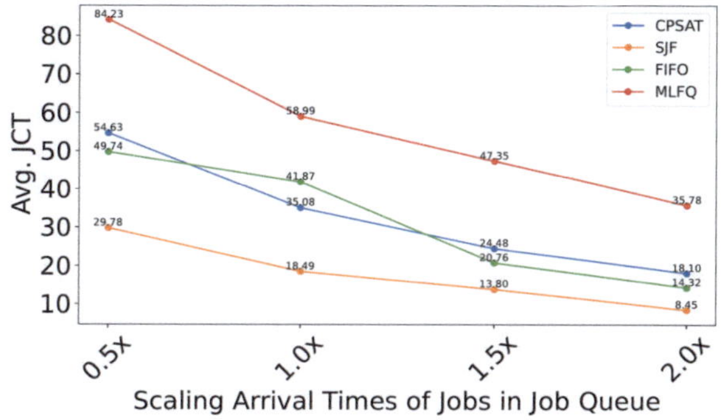

Fig. 9 Variation in job arrival rate to average job completion time. The presented inference time predictor in combination with SJF has the highest throughput. (Adapted from [34].)

t_p with s: regular diffusion, 33 t_p with s: anime, 32 t_p with s: waifu). The requested CLIP score (SLA_{CLIP}) is sampled from a normal distribution with μ: 0.3, σ: 0.02, while the requested latency ($SLA_{latency}$) is sampled from normal distribution with μ: 10, σ: 2, in the user query. Due to a lack of publicly available traces for generative AI workloads, we opted for increasing workload pattern [36], for the job arrival rate. We set a job arrival rate of 0.8 jobs/second, increasing linearly to 0.85 jobs/second over 100 s. All experiments were performed over NVIDIA T4, NVIDIA A10G, at a fixed floating point precision 16 for SD model. The CLIP score was estimated with a pre-trained 'ViT-B/32' model.

Assuming a black-box access, we extract the meta-data of the runtime latencies and the associated CLIP score of the 3 SD models in a privacy-preserving fashion. Utilizing this data, we design the inference predictor to heuristically predict the number of inference steps, for a user-defined SLO_{CLIP} and $SLO_{latency}$, for a specific SD model, for a hardware accelerator, to satisfy the SLO for the specific query. As a baseline, we use First In First Out (FIFO), no preemption Multi-Level Feedback Queue (MLFQ) used to schedule LLMs [37] (without KV-caching), constrained Integer Linear Programming[8] (CP-SAT) used to schedule traditional DNNs (in our case, we utilize the inference predictor to predict the number of inference steps for processing latency), and the presented Shortest Job First (SJF) algorithm in combination with the inference time predictor.

We first evaluate a scenario wherein the requested job queries are characterized by t_p, SLO_{CLIP} and s. We scale the job arrival rate to compute the average job completion time. Our results, as presented in Fig. 9 indicate the highest system throughput for the presented inference time predictor module in combination with SJF. Next, we evaluate a scenario wherein the jobs queries are characterized by t_p, SLO_{CLIP}, $SLO_{latency}$ and s. On scaling the job arrival rate and $SLO_{latency}$, our results indicate the presented inference time predictor in combination with SJF outperforms the baseline, as presented in Figs. 10, and 11 respectively.

[8] https://github.com/google/or-tools.

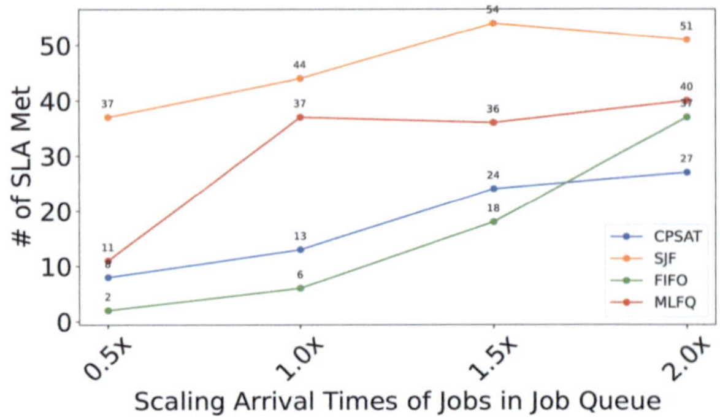

Fig. 10 Variation in job arrival rate to the number of SLOs met. The presented inference time predictor in combination with SJF satisfies the most number of SLOs. (Adapted from [34].)

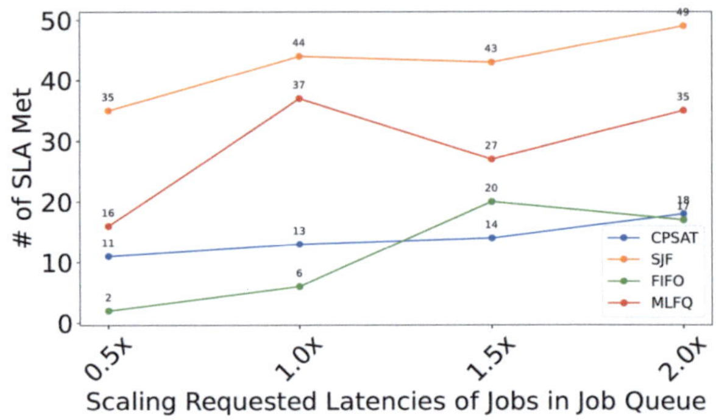

Fig. 11 Variation in $SLO_{latency}$ to the number of SLOs met. The presented inference time predictor in combination with SJF satisfies the most number of SLOs. (Adapted from [34].)

4 Conclusion

We addressed the challenges of multi-tenancy and privacy preservation in cloud environments for DNN and GenAI workloads. The presented fine-grain, online, privacy-preserving job scheduler for Kubernetes, incorporating JRS that intelligently selects competitor jobs based on task interference analysis to optimize scheduler-level objectives. We demonstrated significant improvements in performance metrics, including Ind_{SD} and Pac_{Sav}, compared to traditional scheduling methods.

Additionally, we introduced a model-less, privacy-preserving, low-latency inference framework specifically tailored for SDaaS. This novel approach addresses the unique inference demands of GenAI models, ensuring efficient resource utilization and reduced latency while safeguarding user privacy. We provide a scalable solution for enhancing cloud-based ML training and inference, setting the groundwork for further advancements in privacy-preserving multi-tenancy and resource management in ML-driven cloud infrastructures.

Acknowledgements The authors acknowledge the use of GenAI for editing this chapter.

References

1. Samira Pouyanfar, Saad Sadiq, Yilin Yan, Haiman Tian, Yudong Tao, Maria Presa Reyes, Mei-Ling Shyu, Shu-Ching Chen, and Sundaraja S Iyengar. A survey on deep learning: Algorithms, techniques, and applications. *ACM computing surveys (CSUR)*, 51(5):1–36, 2018.
2. GM Harshvardhan, Mahendra Kumar Gourisaria, Manjusha Pandey, and Siddharth Swarup Rautaray. A comprehensive survey and analysis of generative models in machine learning. *Computer Science Review*, 38:100285, 2020.
3. Fatemeh Khoda Parast, Chandni Sindhav, Seema Nikam, Hadiseh Izadi Yekta, Kenneth B Kent, and Saqib Hakak. Cloud computing security: A survey of service-based models. *Computers & Security*, 114:102580, 2022.
4. Daniel Mendoza, Francisco Romero, Qian Li, Neeraja J. Yadwadkar, and Christos Kozyrakis. Interference-aware scheduling for inference serving. New York, NY, USA, 2021. Association for Computing Machinery.
5. Fuxun Yu, Shawn Bray, Di Wang, Longfei Shangguan, Xulong Tang, Chenchen Liu, and Xiang Chen. Automated runtime-aware scheduling for multi-tenant dnn inference on gpu. In *2021 IEEE/ACM International Conference On Computer Aided Design (ICCAD)*, pages 1–9. IEEE, 2021.
6. Aditya Dhakal, Sameer G Kulkarni, and KK Ramakrishnan. Gslice: controlled spatial sharing of gpus for a scalable inference platform. In *Proceedings of the 11th ACM Symposium on Cloud Computing*, pages 492–506, 2020.
7. Cheng Tan, Zhichao Li, Jian Zhang, Yu Cao, Sikai Qi, Zherui Liu, Yibo Zhu, and Chuanxiong Guo. Serving dnn models with multi-instance gpus: A case of the reconfigurable machine scheduling problem. *arXiv preprint* arXiv:2109.11067, 2021.
8. Yujeong Choi and Minsoo Rhu. Prema: A predictive multi-task scheduling algorithm for preemptible neural processing units. In *2020 IEEE International Symposium on High Performance Computer Architecture (HPCA)*, pages 220–233. IEEE, 2020.
9. Xiaorui Wu, Hong Xu, and Yi Wang. Irina: Accelerating dnn inference with efficient online scheduling. In *4th Asia-Pacific Workshop on Networking*, pages 36–43, 2020.
10. Aditya Dhakal, Junguk Cho, Sameer G Kulkarni, KK Ramakrishnan, and Puneet Sharma. Spatial sharing of gpu for autotuning dnn models. *arXiv preprint* arXiv:2008.03602, 2020.
11. Wencong Xiao, Romil Bhardwaj, Ramachandran Ramjee, Muthian Sivathanu, Nipun Kwatra, Zhenhua Han, Pratyush Patel, Xuan Peng, Hanyu Zhao, Quanlu Zhang, et al. Gandiva: Introspective cluster scheduling for deep learning. In *13th USENIX Symposium on Operating Systems Design and Implementation (OSDI 18)*, pages 595–610, 2018.
12. Tan N Le, Xiao Sun, Mosharaf Chowdhury, and Zhenhua Liu. Allox: compute allocation in hybrid clusters. In *Proceedings of the Fifteenth European Conference on Computer Systems*, pages 1–16, 2020.
13. Jeff Zhang, Sameh Elnikety, Shuayb Zarar, Atul Gupta, and Siddharth Garg. {Model-Switching}: Dealing with fluctuating workloads in {Machine-Learning-as-a-Service} systems. In *12th USENIX Workshop on Hot Topics in Cloud Computing (HotCloud 20)*, 2020.
14. Francisco Romero, Qian Li, Neeraja J Yadwadkar, and Christos Kozyrakis. {INFaaS}: Automated model-less inference serving. In *2021 USENIX Annual Technical Conference (USENIX ATC 21)*, pages 397–411, 2021.
15. Jashwant Raj Gunasekaran, Cyan Subhra Mishra, Prashanth Thinakaran, Bikash Sharma, Mahmut Taylan Kandemir, and Chita R Das. Cocktail: A multidimensional optimization for model serving in cloud. In *19th USENIX Symposium on Networked Systems Design and Implementation (NSDI 22)*, pages 1041–1057, 2022.
16. Amirhossein Kazerouni, Ehsan Khodapanah Aghdam, Moein Heidari, Reza Azad, Mohsen Fayyaz, Ilker Hacihaliloglu, and Dorit Merhof. Diffusion models in medical imaging: A comprehensive survey. *Medical Image Analysis*, 88:102846, 2023.
17. OpenAI. Openai, 2024. Accessed: 2024.
18. Microsoft. Microsoft copilot, 2024. Accessed: 2024.

19. Carmen Carrión. Kubernetes scheduling: Taxonomy, ongoing issues and challenges. *ACM Computing Surveys*, 55(7):1–37, 2022.
20. Aritra Ray, Kyle Lafata, Zhaobo Zhang, Ying Xiong, and Krishnendu Chakrabarty. Job recommendation service for gpu sharing in kubernetes. In *2023 IEEE Cloud Summit*, pages 7–14. IEEE, 2023.
21. Aritra Ray, Zhaobo Zhang, Ying Xiong, and Krishnendu Chakrabarty. Prirect: Privacy-preserving job recommendation tool for gpu sharing. In *2022 IEEE Cloud Summit*, pages 97–104. IEEE, 2022.
22. Andrew G. Howard, Menglong Zhu, Bo Chen, Dmitry Kalenichenko, Weijun Wang, Tobias Weyand, Marco Andreetto, and Hartwig Adam. Mobilenets: Efficient convolutional neural networks for mobile vision applications, 2017.
23. Mingxing Tan and Quoc V. Le. Efficientnet: Rethinking model scaling for convolutional neural networks. 2019.
24. Kaiming He, Xiangyu Zhang, Shaoqing Ren, and Jian Sun. Deep residual learning for image recognition, 2015.
25. Christian Szegedy, Vincent Vanhoucke, Sergey Ioffe, Jonathon Shlens, and Zbigniew Wojna. Rethinking the inception architecture for computer vision, 2015.
26. Barret Zoph, Vijay Vasudevan, Jonathon Shlens, and Quoc V. Le. Learning transferable architectures for scalable image recognition, 2017.
27. J. Stallkamp, M. Schlipsing, J. Salmen, and C. Igel. Man vs. computer: Benchmarking machine learning algorithms for traffic sign recognition. *Neural Networks*, 32:323–332, 2012. Selected Papers from IJCNN 2011.
28. Omkar M. Parkhi, Andrea Vedaldi, Andrew Zisserman, and C. V. Jawahar. Cats and dogs. In *IEEE Conference on Computer Vision and Pattern Recognition*, 2012.
29. Alec Radford, Luke Metz, and Soumith Chintala. Unsupervised representation learning with deep convolutional generative adversarial networks, 2015.
30. Li Deng. The mnist database of handwritten digit images for machine learning research. *IEEE Signal Processing Magazine*, 29(6):141–142, 2012.
31. Jacob Devlin, Ming-Wei Chang, Kenton Lee, and Kristina Toutanova. Bert: Pre-training of deep bidirectional transformers for language understanding, 2018.
32. Andrew L. Maas, Raymond E. Daly, Peter T. Pham, Dan Huang, Andrew Y. Ng, and Christopher Potts. Learning word vectors for sentiment analysis. In *Proceedings of the 49th Annual Meeting of the Association for Computational Linguistics: Human Language Technologies*, pages 142–150, Portland, Oregon, USA, June 2011. Association for Computational Linguistics.
33. Aritra Ray, Kyle Lafata, Zhaobo Zhang, Ying Xiong, and Krishnendu Chakrabarty. Privacy-preserving job scheduler for gpu sharing. In *2023 IEEE/ACM 23rd International Symposium on Cluster, Cloud and Internet Computing Workshops (CCGridW)*, pages 337–339. IEEE, 2023.
34. Aritra Ray, Lukas Dannull, Farshad Firouzi, Kyle Lafata, and Krishnendu Chakrabarty. Inference serving system for stable diffusion as a service. In *2024 IEEE Cloud Summit*, pages 13–16. IEEE, 2024.
35. Micah Hodosh et. al. Framing image description as a ranking task: Data, models and evaluation metrics. *Journal of Artificial Intelligence Research 2013.*
36. In Kee Kim et. al. Forecasting cloud application workloads with cloudinsight for predictive resource management. *IEEE Transactions on Cloud Computing 2020.*
37. Bingyang a Wu et. al. Fast distributed inference serving for large language models. *arXiv preprint* arXiv:2305.05920, 2023.

Blockchain Revolution: Securing and Streamlining IoMT-Driven Healthcare Systems

Bahar Farahani, Farshad Firouzi, Avid Alizadeh, and Mahmoud Daneshmand

> *"Trust, but verify."*
>
> – Russian Proverb, popularized by Ronald Reagan

The digital revolution in healthcare is characterized by the convergence of technologies—from cloud computing to edge/fog computing, Artificial Intelligence (AI), big data, Intelligent Internet of Medical Things (IoMT), and Distributed Ledger Technology (DLT)—that is blurring the lines between physical and digital worlds. In this context, the IoMT tsunami, the ubiquitous adoption of intelligent connected devices, and the public embracement of DLT, of which blockchain is a popular example, are increasingly becoming an integral feature of many modern healthcare systems, particularly IoMT-based smart and connected healthcare. Although the IoMT and DLT/blockchain are two very different technologies and distinct from each other, the fusion of blockchain and IoMT technologies is an unprecedented paradigm shift that is expected to disrupt both current and future healthcare systems. Blockchain has exactly what is needed to fix the weaknesses and vulnerabilities of IoMT in healthcare. It solves the security fault line among intelligent IoMT, where most of the IoMT devices are connected to each other through a public trustless environment. Moreover, its distributed peer-to-peer nature can address the shortcomings of client/server models in Cloud-IoMT solutions. The integration of Blockchain-IoMT in healthcare

B. Farahani (✉)
Shahid Beheshti University, Tehran, Iran
e-mail: b_farahani@sbu.ac.ir

F. Firouzi
Arizona State University, Tempe, USA

A. Alizadeh
Queen Mary University of London, London, UK

M. Daneshmand
Stevens Institute of Technology, Hoboken, USA

F. Firouzi et al. (eds.), *Smart and Connected Health: AI, IoT, and Trustworthy Technologies*, https://doi.org/10.1007/978-3-032-06286-4_8

can enhance patient data security, ensure the integrity and confidentiality of medical records, and streamline data sharing across various stakeholders while maintaining patient privacy. The chapter presents foundational principles, reference architectures, recent technological advancements, and emerging application scenarios. It also discusses the promises and persistent challenges associated with this integration, with the aim of informing future research and development in Blockchain-enabled IoMT for healthcare.

1 Introduction

The healthcare and technology industries have been closely intertwined for a long time. However, the advent of smart and connected health technologies is significantly transforming healthcare processes [1, 2]. Innovative technologies like cloud computing and big data have already enhanced the Internet of Things (IoT) and the Internet of Medical Things (IoMT) by increasing availability, accessibility, personalization, precision, and reducing the cost of healthcare services. The next logical step in this evolution appears to be the integration of Distributed Ledger Technology (DLT) with IoMT and other smart health technologies [3, 4].

Traditionally, cloud solutions have provided secure data storage and interoperability for IoT applications. However, the Cloud IoT paradigm faces several shortcomings, particularly in security. DLT, including blockchain and Directed Acyclic Graph (DAG) technologies, is revolutionizing information dissemination [3, 5]. Its capability to build trust through distributed networks without relying on a third party is a significant advancement. This could transform various industries, especially IoT-based Smart and Connected Health, by addressing major shortcomings and vulnerabilities inherent in current client/server cloud IoT models. For example, cloud servers can be disabled by software bugs, cyber-attacks, or other mechanical issues. In contrast, IoT systems utilizing blockchain technology are not susceptible to a single point of failure, as identical data is maintained across multiple devices and computers [3, 6].

The combination of innovative technologies such as cloud computing and the IoT has proven invaluable, and blockchain is poised to further revolutionize the IoT by establishing a trustworthy information-sharing service that ensures data immutability and traceability [7, 8]. Blockchain can enhance IoT security by making data sources identifiable at any point, potentially solving the security vulnerabilities in cloud IoT systems. This is particularly crucial when IoT data must be securely distributed among many participants. For instance, ensuring food safety in the healthcare domain requires monitoring edible products, which involves the participation of distributors, manufacturers, farmers, and others. A data breach at any point could lead to fraud, complicate the search for infection sources, endanger consumer lives, and disrupt organizational, industry, and national economic interests. These scenarios illustrate how blockchain can augment the IoT, addressing issues such as IoT data reliability, scalability, and privacy [3, 9].

The massive growth of the IoT necessitates standardized protocols and mechanisms to reduce heterogeneity, which typically leads to the creation of vertical silos and hinders IoT adoption. Alongside improving integration and reducing heterogeneity, enhancing IoT data integrity is crucial. While people are likely to trust information provided by governmental agencies or financial institutions, it is essential to ensure that information from external entities and IoT devices is tamper-proof. This need is challenging to meet with centralized architectures, as unscrupulous institutions can alter data to suit their purposes, rendering it less reliable. This underscores the importance of verifying that data has not been altered [3, 10].

Recently, cloud computing technology has empowered the IoT to analyze data and translate information into real-time actions and up-to-the-minute knowledge. This unprecedented IoT expansion has created new methods of sharing and accessing information, including a foundational open data model. However, one of the primary vulnerabilities of these new information-sharing methods is a lack of confidence. While centralized architectures have significantly contributed to IoT growth, they often lack transparency, leaving participants uncertain about how their data will be used.

Currently, IoT security mainly concentrates on securing communication channels, leaving substantial gaps in the security of the data lifecycle, particularly in data sharing and access control. Traditional cloud-based models manage the connectivity, identification, and verification of IoT devices. However, this approach may not be the most effective solution because it ignores the physical location of the data and relies on the integration of a trusted third party. Rethinking the way IoT data is managed could enable users to depend on a decentralized, independent, and robust data management system that guarantees data ownership. Such a system would encompass several components [3, 11]: (1) access control, facilitated by a decentralized and auditable mechanism using DLT/blockchain to ensure data ownership and encrypted information sharing; (2) secure data storage, where data remains verifiable, immutable, and trusted; and (3) IoT compatibility, allowing data to be appended by a single writer and viewed by multiple readers [3].

Blockchain has the potential to greatly enhance IoT functionality from a security standpoint and drive further technological innovation. However, numerous research issues and challenges must be addressed to effectively combine blockchain and IoT. This field of research is just beginning, but successful integration is anticipated to provide several benefits, including decentralization, improved interoperability, enhanced privacy and security, reduced operating costs, better fault tolerance, greater resilience, faster transaction speeds, fair monetization, secure data sharing, and innovative IoT e-business models [3, 12].

The healthcare domain, with its interdisciplinary landscape, requires significant technological advancements in blockchain, cloud computing, and IoT to converge and synergize their efforts. This chapter focuses on presenting a comprehensive reference architecture for blockchain-IoT integration in healthcare, identifying new perspectives, and highlighting critical research issues and challenges necessary for the successful integration of these technologies [3].

2 Distributed Ledger Technologies (DLTs)

Distributed ledgers are a sophisticated technological framework designed for the distribution, exchange, and storage of data across a network of users, operating on either public or private networks. Essentially, a distributed ledger is a database that is replicated and synchronized across multiple nodes, each situated in different geographic locations. In this context, each participant or computer within the network is referred to as a node. Distributed ledgers can be thought of as shared databases that are maintained and updated across various distributed nodes, embodying the core principles of distributed systems. This technology is categorized into three primary forms based on its implementation: blockchain, which uses a linear chain of blocks to secure data; Directed Acyclic Graph (DAG), which allows for more complex data structures and potentially higher transaction throughput; and hybrid DLT [4, 5].

2.1 Types of DLT

2.1.1 Blockchain

The concept of blockchain was introduced in 2008 with the publication of a paper titled "Bitcoin: A Peer-to-Peer Electronic Cash System," likely authored pseudonymously by an individual or group known as Satoshi Nakamoto. At its core, blockchain is a continuously expanding list of records, or blocks, that are linked together using cryptographic techniques. In essence, a blockchain functions as a shared, decentralized, and immutable ledger that stores a time-stamped series of transaction records in a chain of blocks connected by hashes. Blockchain operates on a peer-to-peer (P2P) network topology and is managed by participants identified by private-public key pairs [3].

New records, or blocks, are added to the blockchain through a process called mining. Each block in a blockchain contains three critical elements: (i) the cryptographic hash of the previous block, (ii) a timestamp, and (iii) transaction data, typically represented by a Merkle tree. Once transaction data is recorded in a block, it cannot be altered without modifying all subsequent blocks, ensuring the integrity and security of the data (see Fig. 1) [4, 5].

In blockchain technology, the fundamental unit of record is a transaction. When a new transaction is created, it is broadcast to the entire blockchain network. Nodes, known as block miners, receive the transaction and authenticate it by verifying the attached digital signature. Once authenticated, the transaction is included in securely encrypted blocks through a process called mining. To generate a block, miners must solve a consensus problem using a distributed process. The miner that successfully solves this problem broadcasts the newly generated block to the entire network.

Upon receiving the new block, other miners who have not yet solved the problem add it to their local chains, acknowledging that the transaction has been authenticated and the block provides the solution to the consensus problem. Each new block

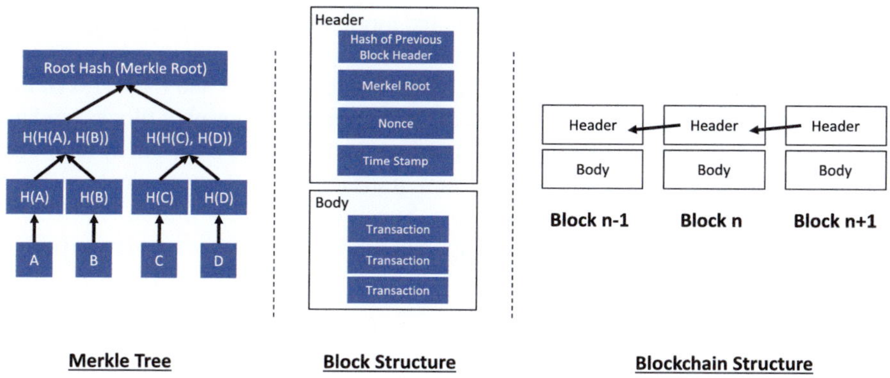

Fig. 1 Blockchain structure [3]

includes a cryptographic link to the previous block in the chain, ensuring the integrity and security of the blockchain. Block miners regularly synchronize their chains, and specific protocols ensure that the distributed ledger maintains only the longest chain in the event of discrepancies [4, 5].

Merkle trees are vital components of blockchain technology, enabling secure and efficient verification of the consistency and integrity of large datasets. They generate a digital fingerprint that uniquely identifies an entire set of transactions within a block, allowing users to verify whether a specific transaction is part of the block. Merkle trees are constructed from the bottom up, starting with the transaction IDs or pairs of hashed nodes, and continuing this process until a single hash, known as the Merkle Root or Root Hash, is obtained. Each leaf node contains a hash of transaction data, while non-leaf nodes contain hashes of their respective child nodes.

Since Merkle trees are binary, they require an even number of leaf nodes. If there is an odd number of transactions, the last hash is duplicated to ensure an even number. As illustrated in Fig. 1, four transactions (A, B, C, and D) are individually hashed to create Hash A, Hash B, Hash C, and Hash D, respectively. These pairs of leaf nodes are then combined and hashed together on parent nodes, resulting in Hash AB (from Hash A and Hash B) and Hash CD (from Hash C and Hash D). Finally, Hash AB and Hash CD are hashed together to produce the Merkle Root.

This hierarchical hashing process ensures data integrity and allows for rapid and reliable verification of data within the blockchain.

2.1.2 Directed Acyclic Graph (DAG)

Like blockchain, DAG technology is capable of storing transactions. In a DAG, each transaction is represented as a node that is linked to at least one, but potentially many other transactions. These links are specifically directed, pointing from an earlier transaction to a more recent one, following the topological order. Importantly, DAGs do not permit loops because they are acyclic [3].

A DAG functions as a graph where transactions act as nodes and the edges of the graph have directions. Unlike blockchain, DAGs are not composed of blocks, and no mining process is involved. Instead, transactions authenticate each other, although no transaction can validate itself. Each new transaction must authenticate at least one prior transaction when joining the DAG. Specifically, each new transaction must refer back to a parent transaction, sign the parent transaction's hashes, and incorporate those hashes into the new transaction [3–5].

2.1.3 Hybrid DLTs

Hybrid DLTs merge the strengths of both DAGs and blockchain technologies. A prime example of this is Bexam, a hybrid DLT that utilizes flexible chain structures and hierarchical nodes to deliver the security benefits of blockchain combined with the high transaction speed of a DAG, capable of processing approximately 40 million transactions per second.

Bexam's architecture is designed for seamless integration into enterprise-level infrastructures, offering significant scalability and flexibility. The platform's minimal computing resource and power requirements make it an efficient solution for large-scale deployments. Furthermore, Bexam employs token technology to facilitate and secure transactions within the network [3].

The hybrid approach of Bexam addresses several key challenges faced by traditional blockchain and DAG systems. By integrating the robust security features of blockchain with the high throughput capabilities of DAGs, Bexam provides a versatile and resilient framework suitable for a wide range of applications. This combination not only enhances transaction speed and efficiency but also ensures the integrity and security of the data, making it a promising solution for the evolving needs of modern healthcare systems [4, 5].

2.2 Permissioned Versus Permissionless

Blockchains can be categorized as either permissionless or permissioned. Permissioned blockchains can be further classified into federated or private types (see Table 1) [4, 5].

- **Permissionless (Public)**: also known as public blockchains, enable all transactions to be transparent to every node in the network. Any node can participate in the consensus process to authenticate transactions without requiring permission or being identified by the network. In a permissionless network like Bitcoin, nodes collaborate and operate on a large scale.

Table 1 Comparison of public, federated, and private blockchains [3]

	Public	Federated (consortium)	Private
Access	Anyone	Multiple selected organizaitons	Single organizaiton
Participants	Permissionless, anonymous	Permissioned, known identities	Permissioned, known identities
Decentralization	Decentralized	Partially decentralized	Centralized
Security & Consensus	Consensus mechanism, PoW or PoS	Pre-approved participants, voting/multi-party consensus, PBFT, PoET, PoA	Pre-approved participants, voting/multi-party consensus, Ripple
Immutability	Immutable	Partially immutable	Alterable
Transparency	Transparent	Partially transparent	Opaque
Transaction speed	Typically slow	Lighter and faster	Lighter and faster
Scalability and felixibility	Typically poor	Good	Supperior
Popular platforms	Bitcoin, Ethereum	Hyperledger	Multichain, GemOS

- **Permissioned**: These blockchain networks can be categorized as either Federated or Private.
  - **Federated (Consortium)**: These blockchains are managed by a specific group of users and do not allow participants outside the group to verify transactions. The public may be permitted to read transactions, but only members of the selected group can write transactions. Hyperledger Fabric is the most common and well-established federated blockchain.
  - **Private**: These blockchains are typically controlled by a single organization that has the authority to authenticate transactions. The public or approved parties may be allowed to read transactions. An example of a private blockchain is Multichain.

2.3 Distributed Ledger Platforms

Some of the most prominent DLT and blockchain platforms include [3]:

- **Bitcoin**: As the first and most valuable cryptocurrency, Bitcoin offers a secure, cost-effective, and efficient payment system, primarily used for financial transactions. Despite its limited integration with other IoT applications, Bitcoin's consensus algorithm and data structure provide foundational insights for many DLT-based IoT solutions.

- **Ethereum**: Ethereum stands as the second-largest cryptocurrency platform, distinguished by its smart contract system, which is programmed using the Solidity language. It also features the Ethereum Virtual Machine (EVM), a decentralized virtual machine capable of executing scripts. Ethereum transactions incur gas fees, which represent the computational effort required to process operations.
- **Hyperledger Fabric**: A leading member of the Hyperledger project, hosted by the Linux Foundation, Hyperledger Fabric is a modular blockchain framework designed for enterprise use. It supports smart contracts, known as chaincode, which can be developed in node.js or Go.
- **IOTA**: A platform focused on IoT, IOTA facilitates fee-less Machine-to-Machine (M2M) micropayments through its unique directed acyclic graph (DAG) structure called the tangle. Each IOTA node acts as a miner, and new transactions validate two previous ones, allowing for scalable network throughput without transaction fees.
- **MultiChain**: MultiChain is a private blockchain platform suitable for both intra- and inter-organizational applications. As an extended open-source fork of Bitcoin, MultiChain offers extensive configurability, rapid deployment, permissions management, native assets, and data streams.
- **Hydrachain**: An extension of the Ethereum platform, Hydrachain is designed to facilitate the creation of private ledgers.
- **IoT Chain**: A platform developed to meet the scalability and security requirements of IoT devices, IoT Chain combines a DAG with the Practical Byzantine Fault Tolerance (PBFT) consensus algorithm to achieve high transaction speeds.
- **HDAC**: A multichain platform that supports IoT contracts and M2M transactions, employing an energy-efficient version of the Proof-of-Work (PoW) consensus algorithm known as enhanced PoW (ePoW).
- **Hedera Hashgraph**: A distributed ledger technology based on a DAG structure for sequencing transactions, utilizing an asynchronous Byzantine Fault Tolerance (aBFT) consensus algorithm to ensure system security against attacks.
- **Atonomi**: A platform designed to create secure and trustworthy IoT systems by providing immutable identity management services, built on the Ethereum framework.
- **OriginTrail**: A decentralized platform based on Ethereum for sharing supply chain data. The network is structured into three layers: the data layer, the network layer, and the blockchain layer.
- **IoTeX**: blockchain-in-blockchain platform tailored for M2M transactions requiring privacy. Its architecture includes a public rootchain and multiple subchains, which can be either private or public blockchains. Communication between subchains occurs through cross-blockchain transactions mediated by the rootchain.
- **Streamr**: An open-source, off-chain network built on Ethereum for real-time data trading, featuring a global data marketplace where users can buy and sell data.
- **Grid+**: An Ethereum-based platform designed for energy markets, utilizing smart energy agents to handle real-time payment of electricity bills.

2.4 Validation Process

The process of blockchain verification, commonly known as mining, is generally carried out using a consensus algorithm. This algorithm sets the rules that nodes must follow to authenticate and validate blocks. For consensus to be achieved, participating nodes must receive confirmation from all other nodes by adhering to the specified transaction order. This collective agreement determines whether a block will be added to the blockchain. Despite the absence of a centralized authority, the consensus protocol ensures that all transactions are secure and accurately verified.

Consensus algorithms require all nodes in the network to reach a mutual agreement on the current state of the distributed ledger. This agreement guarantees the reliability of the blockchain and establishes trust among participants who may not know each other in a distributed network. In essence, consensus algorithms ensure that each new block added to the chain is the sole authenticated version accepted by all nodes.

Numerous consensus algorithms have been developed for blockchain applications, each with distinct strengths and weaknesses, making them suitable for different use cases [3–5, 7, 13–17].

- **Proof-of-Work (PoW)**: PoW is the most widely recognized consensus algorithm, employed to verify transactions and add newly authenticated blocks to the blockchain. In this system, miners compete to solve complex computational puzzles, known as proofs of work, to validate transactions, propose new blocks, and earn rewards. These puzzles are designed to be computationally intensive and time-consuming to solve, ensuring that the process requires significant effort and resources. However, once a solution is found, it is easy and quick for other nodes in the network to verify its correctness. Common PoW functions include Integer square root modulo a large prime, Weaken Fiat–Shamir signatures, Diffie–Hellman-based puzzles, Hash sequences, and Cuckoo Cycle. PoW is effective in addressing the Byzantine Generals problem and defending against Denial of Service (DoS) attacks because mounting a successful DoS attack would require an enormous amount of computing power. Despite its strengths, PoW blockchains are susceptible to a 51% attack, where an attacker gains control of more than half of the network's mining power. This majority control allows the attacker to manipulate the blockchain, potentially reversing transactions and double-spending coins. Additionally, PoW's requirement for massive computational power leads to high energy consumption, making it an environmentally and economically costly method for securing blockchain networks.
- **Proof-of-Stake (PoS)**: This algorithm was designed to address the weaknesses of Proof-of-Work (PoW). Specifically, PoW requires substantial computational power to solve complex puzzles, resulting in high energy consumption. PoS, in contrast, operates on the principle that nodes with a larger stake in the blockchain have a greater incentive to maintain the network's integrity and success. For example, if a node owns 30% of the total stakes in a blockchain, it has a 30% chance of being selected to validate transactions and create new blocks, theoretically allowing it to mine only 30% of the blocks. Unlike PoW, PoS does not require

significant energy consumption or highly powerful computing hardware, making PoS blockchains more energy-efficient and faster, as they do not rely on intensive mining processes. This efficiency also leads to quicker transaction times and lower operational costs. While PoS significantly reduces the risk of a 51% attack, as gaining such a stake would be prohibitively expensive, it does not entirely eliminate centralization. A limited number of nodes are responsible for creating new blocks, which can lead to a concentration of power among those with substantial stakes in the network. Despite this, PoS remains a popular alternative to PoW due to its enhanced energy efficiency and faster transaction processing capabilities.

- **Delegated Proof-of-Stake (DPoS)**: This well-known consensus algorithm, created by Daniel Larimer, involves three groups of entities: delegates, stakeholders, and witnesses. Stakeholders elect delegates and witnesses, who are then responsible for achieving consensus within the network. In the DPoS system, each stakeholder votes for witnesses by placing tokens on the names of their preferred candidates. The candidate with the highest number of votes is elected as a witness. Stakeholders continue to vote and increase the number of witnesses until at least 50% of the stakeholder group agrees that the blockchain is sufficiently decentralized. This voting process ensures that the elected witnesses, who are tasked with maintaining the blockchain, are accountable to the stakeholders, thereby enhancing the overall trust and security of the network.
- **Practical Byzantine Fault Tolerance (PBFT)**: PBFT is a popular consensus algorithm designed to function effectively in asynchronous systems, enabling blockchains to handle Byzantine faults. PBFT requires low latency and minimal overhead time. In this algorithm, all nodes in the blockchain are organized sequentially, with one node designated as the leader and the others as backups. When a node transmits a message, the remaining nodes exchange information to verify its integrity in case it has been compromised during transmission. Honest nodes reach a majority agreement regarding the blockchain's status, ensuring reliability and consistency. One of PBFT's key advantages is its ability to finalize blocks and transactions without needing additional confirmations, making it faster and more energy-efficient compared to Proof-of-Work (PoW). However, PBFT is best suited for smaller blockchains due to its communication method among nodes and its vulnerability to Sybil attacks, where an attacker creates multiple fake identities to gain influence in the network.
- **Proof-of-Burn (PoB)**: PoB is a consensus algorithm that does not require expensive hardware. Instead, it involves validators "burning" coins by irretrievably sending them to a specific address. By doing this, validators earn the right to mine on the network. Once they have burned their coins, validators are randomly selected to create new blocks. While validators incur a short-term loss by burning their coins, this process ensures their long-term commitment to the network. The act of burning coins demonstrates a validator's investment in the network's success and stability, thereby aligning their interests with the overall health and security of the blockchain.
- **Proof-of-Capacity (PoC)**: PoC is a consensus algorithm that requires validators to allocate hard drive space rather than burn coins or invest in expensive hardware.

Validators with more hard drive space have a higher probability of being selected to mine the next block and receive a reward. This approach leverages available storage capacity, making it a more energy-efficient and accessible alternative to traditional mining methods that rely on computational power.

- **Proof-of-Elapsed Time (PoET)**: PoET is regarded as one of the most equitable consensus algorithms and is commonly used in permissioned blockchain networks. It ensures that each validator has an equal opportunity to create a block. In PoET, every node waits for a randomly determined amount of time and then adds proof of this waiting period to the block. The generated blocks are broadcast across the network for review. The validator with the shortest waiting time, as indicated in the proof section of the block, is selected as the winner, and their block is added to the chain. The PoET algorithm includes mechanisms to prevent the same node from consistently winning or having the shortest waiting time, thereby maintaining fairness in the selection process. This ensures a balanced distribution of opportunities among validators and enhances the overall security and decentralization of the network.
- **Proof-of-Activity (PoA)**: PoA was proposed as an alternative to Bitcoin mining. This consensus algorithm combines elements of PoW and PoS to generate consensus, with a primary focus on rewarding active stakeholders. Initially, peers begin by mining, similar to PoW, by solving a computational puzzle against the block header, excluding the block transactions. Once the puzzle is solved, random validators are selected to vote on the validity of the mined block header. In PoA, the likelihood of a validator being chosen corresponds to their share of the network, akin to PoS. A block is verified and added to the blockchain if all selected validators agree on its validity. If some validators are offline, the next block and another group of verifiers are chosen to maintain the process. While PoA aims to reward active participation and improve consensus efficiency, it requires significant computing power and may be vulnerable to "nothing at stake" attacks if not implemented carefully. This vulnerability arises when validators have little incentive to act honestly, as they have nothing to lose by voting for multiple blockchain forks.
- **DAG-based Consensus**: Platforms such as IOTA, Byteball, Dagcoin, and Hashgraph utilize DAGs instead of traditional blockchain structures to distribute ledgers. An example of how consensus is achieved using a DAG distributed ledger can be seen in IOTA's tangle. In this system, a node within the tangle represents a transaction rather than a block. To issue a transaction in IOTA, a node must randomly approve two other unverified transactions through a lightweight Proof-of-Work. A subset of transactions within the tangle is then confirmed either directly or indirectly by subsequent transactions. As new transactions are added to the tangle, they validate previous transactions, creating a web of interlinked approvals. The confidence level of a transaction is determined by the fraction of new transactions that reference a specific transaction. This continuous process of validation ensures that older transactions become increasingly secure as more new transactions are added to the tangle.

2.5 Benefits of DLTs and Blockchain

The unique way in which blockchain is generated and maintained offers several advantages over traditional database systems. Key features such as consensus and decentralization are intrinsic to blockchain technology and provide a range of novel benefits, such as [3, 6–10]:

- **Decentralization**: Unlike centralized network infrastructures, blockchain-based systems allow two nodes to engage in transactions directly without relying on a central entity. This reduces service costs, alleviates performance bottlenecks, and mitigates the risks associated with single-point failures [18].
- **Immutability**: Blockchain technology eliminates the need for third-party interactions, thereby reducing the risk of unauthorized access or modification to stored data. Each new transaction must be approved by consensus, making the blockchain highly resistant to censorship and tampering. Additionally, previous records on the blockchain are immutable; altering any prior record would require an attacker to control the majority of the network's nodes, making it exceedingly difficult to change past data.
- **Pseudonymity**: While blockchain data is highly transparent, privacy is protected through the use of anonymous blockchain addresses. However, blockchain only safeguards privacy to a certain extent because addresses can be traced through inference, which can be useful in detecting illegal transactions or fraud. Thus, blockchain protects pseudonymity rather than complete privacy [7, 8, 19, 20].
- **Traceability**: All blockchain transactions are given a timestamp recorded at the time the transaction takes place. This feature makes it easy to confirm and trace the point of origin for historical data by reviewing the data with the appropriate timestamps.
- **Integrity and Security**: Each transaction is signed with a private key, creating a unique signature that can be verified by others using the corresponding public key. This ensures secure encryption, confirming the transaction's origin and maintaining its integrity. Blockchain's consensus mechanism enhances security by ensuring that transactions are not recorded until all participants agree, making it very difficult to hack.
- **Transparency**: In public blockchain systems such as Ethereum or Bitcoin, all users can interact with and access the network equally. Each new transaction is authenticated and added to the blockchain, making it visible to all users. This fosters greater transparency and builds trust among participants.
- **Permanency and Auditability**: Blockchain transactions are permanently stored, ensuring they are always available for audit or verification. Each participant has a copy of the blockchain and can access timestamped transactions, allowing for the verification of transactions involving specific addresses.
- **High Availability**: Blockchain utilizes a large network of nodes that operate on a peer-to-peer basis. Data is replicated and updated on each node, ensuring the system remains functional even if one node becomes inaccessible. This redundancy ensures high availability and resilience to failures.

- **Cost Savings**: Utilizing blockchain can significantly reduce transaction costs by enabling direct peer-to-peer or business-to-business transactions, eliminating the need for intermediaries such as banks or other third parties. Additionally, costs associated with governance, overhead, and auditing can be substantially lower, further contributing to overall savings.
- **Transaction Speed**: Blockchain accelerates transactions by eliminating lengthy authentication and clearance processes. All participating parties agree upon a single version of the data, which is readily available in the blockchain ledger, thereby speeding up the transaction process.
- **Smart Contracts**: Smart contracts facilitate agreements and manage digital asset transfers among participants under predefined conditions on the blockchain. They ensure transparent execution, permanence, and decentralization while providing secure and automated program execution.

3 Blockchain-IoT Integration

In this section, we will discuss the main challenges of IoT, such as security issues, and explore the revolutionary advantages and challenges of combining IoT with DLT/Blockchain technologies. Additionally, we will examine the corresponding architectures and integration techniques.

3.1 Challenges of IoT

IoT represents a natural progression in humanity's effort to use computers to connect globally. As the digital landscape continues to expand, there is a concerted effort to link physical objects worldwide, enabling them to exchange data and interact to make life safer and more efficient. Currently, the IoT ecosystem comprises physical devices, gateways, fog/edge nodes, the Cloud, and the Internet. The IoT enables physical devices to transmit data via a gateway device to fog nodes and Cloud servers for analysis and storage. These devices can also receive commands to perform actions from one another through the Cloud, often orchestrated by a central Cloud manager [3].

Standardized protocols and the IoT stack work together to develop architectural layers that provide services for physical devices within the IoT system. Presently, most IoT designs utilize a centralized architecture, which suffers from several significant weaknesses, including [3, 6–10]:

- **Heterogeneity**: IoT systems are inherently heterogeneous, featuring a wide range of communication mechanisms, data types (unstructured, structured, and semi-structured), and devices. This diversity presents significant challenges for privacy, interoperability, and security, complicating the integration and management of different components within the IoT ecosystem.

- **Lack of Standardization & Interoperability**: Numerous open solutions are facilitated by international governing groups, alliances, and independent bodies such as OMG, ITU-T, W3C, ETSI, IEEE, OneM2M, and OASIS IEC. These organizations cover a wide range of IoT objects, systems, services, and architectures. While some standards follow neutral, cross-domain approaches, others are specific to certain domains. The unregulated creation of these standards and the absence of universally accepted guidelines lead to fragmentation, hindering IoT adoption and integration across various domains.
- **Limited Scalability**: In traditional IoT systems, all data is transmitted from devices or objects to the Cloud for analysis, with responses returned from the Cloud to the objects as needed. As billions of additional devices and objects join IoT systems, scalability issues will intensify, making it increasingly challenging to manage the growing volume of data and interactions effectively.
- **Scalability in Data Management**: IoT devices generate vast amounts of data, creating significant challenges in terms of transmission, storage, and processing. To efficiently handle this high volume of data, scalable models and solutions are necessary to ensure effective management and utilization
- **Centralized Data Storage**: Most existing IoT architectures typically store only a single copy of data within the Cloud. If this copy is manipulated, it becomes impossible to identify what was altered. Additionally, there is no mechanism to intercept and prevent data tampering.
- **Single Point of Failure**: Each physical object in an IoT network represents a potential point of vulnerability that could compromise the security and functionality of the entire network. If one object fails, it can disrupt the network, leading to widespread issues and potential downtime.
- **Server Failure**: The entire infrastructure is at risk if a centralized server fails. A successful Denial of Service (DoS) attack can exploit this vulnerability, turning the server into a single point of failure and compromising the system's overall stability and functionality.
- **Security**: Security is widely recognized as the most challenging and critical weakness to address in IoT systems. Today's IoT systems are generally vulnerable to attacks targeting object identity, manipulation, services, or cryptanalysis. Beyond traditional web security issues, IoT faces additional threats due to the unique nature of its devices. IoT objects are often isolated hardware pieces that can be compromised in unforeseen ways by their creators. Additionally, these devices are frequently interconnected, complicating the governance of their interactions and making data manipulation attacks more difficult to defend against. Many IoT devices have limited computing power, which restricts the implementation of complex security measures. When interconnected via the Internet, IoT devices create a complex system that is challenging to secure, making them vulnerable to various web attacks such as traffic analysis, password attacks, eavesdropping, Sybil attacks, Distributed Denial of Service (DDoS) attacks, and message spoofing. Data alteration or spoofing can occur at any point in the IoT network, including within

physical devices, through communication networks, or during Cloud data processing and storage. Public concern about the security of data collected and stored in the Cloud is long-standing. IoT devices, such as smart actuators, can execute commands from the Cloud or other IoT devices. If such a command is intercepted and altered, the consequences could be severe, such as opening a home's door to a thief or causing fires or flooding due to inappropriate device actions. A generalized security model for IoT is difficult to implement. Effectively addressing IoT security requires innovative security paradigms that anticipate evolving threats and the development of best practices. These paradigms must integrate security-by-design principles with technologies that work across various stacks and processes to address IoT security comprehensively.

- **Privacy**: The massive amounts of data generated by IoT devices can reveal extensive sensitive information about users, such as their habits or locations. This data can be collected without explicit consent and exposed to third parties by IoT platforms, depriving users of control over their personal information. Additionally, data stored on centralized cloud servers can be analyzed to extract personally identifiable information (e.g., health data), and users often have little influence over who accesses their data and for what purposes. Although policies are in place to protect IoT data privacy, creating solutions that incorporate privacy safeguards by design remains a significant challenge.
- **Resource Constraints**: IoT systems, such as RFID tags, sensors, smart meters, and actuators, often depend on computing devices with limited resources, such as microcontrollers. These devices typically have restricted storage capacity and computing power, making it challenging to implement complex algorithms that protect data. As a result, IoT devices are particularly susceptible to cyberattacks due to their resource limitations.
- **Maintenance Difficulty**: Updating the software and firmware of IoT networks and endpoint devices is challenging due to the need to send updates to a vast number of objects dispersed across various physical locations. This complexity makes it difficult to ensure that all devices are consistently maintained and secure.
- **Immutable Records**: Most existing IoT networks do not maintain permanent records of interactions among physical devices, complicating efforts to trace and diagnose the root causes of issues. This lack of immutable records hinders effective troubleshooting and accountability.
- **Access Control Management (ACM)**: ACM refers to security protocols that regulate access to resources or services within a system. In IoT, access control and identity management are closely linked, with identity systems facilitating communication and managing resource and device authorization. Essentially, identities define access permissions between interacting parties and control access within the system. Traditional access control mechanisms, such as Role-based Access Control (RAC) and Access Control Lists (ACLs), are often ineffective for IoT technologies due to their reliance on centralized architectures and the continuous expansion of policies and roles. Additionally, the increasing number of factors and guidelines (such as location and time) complicates the development of access controls. Attribute-based Access Control (ABAC) models address some of these issues

but still face scalability problems due to their dependence on centralized identity providers. These models struggle with the growing complexity and scalability demands of IoT systems. Capability-based Access Control (CAC) offers adaptability but still requires third-party authentication, such as certificate authorities or identity providers. This reliance on third parties is incompatible with trustless IoT networks where user approval without third-party intervention is essential. Therefore, CAC models are only suitable for trusted environments. Given the challenges outlined above, key issues in developing IoT access control protocols include: (1) Designing an access control protocol that effectively manages access to diverse IoT components (services, users, objects, etc.) despite rapid growth in roles, policies, and users. (2) Building an access control mechanism suitable for trustless environments.

- **Data Quality and Consistency**: Ensuring that the data collected from various IoT devices is accurate, consistent, and reliable can be challenging, especially when dealing with large volumes of data from diverse sources.
- **Inefficiency**: As the IoT expands, centralized servers will struggle to handle the increased communication demands required for IoT automation, potentially restricting system scalability and growth.
- **Mobility versus Stable Topology**: IoT topologies can range from highly stable to highly mobile. For instance, smart home technologies usually exhibit a stable topology, while vehicle networks (VANETs) in transportation applications represent dynamic and mobile topologies. Stable networks, such as those in smart homes, have fixed devices, whereas mobile networks, like VANETs, require adaptable, time-variant topologies due to the constant movement of vehicles. This mobility poses significant challenges for network connectivity and entity management, making it difficult to maintain consistent and reliable communication across the network.
- **Legal and Regulatory Issues**: The deployment of IoT technologies often faces legal and regulatory hurdles, particularly concerning data privacy, security, and compliance with regional laws.

The challenges mentioned above necessitate a reevaluation of the current IoT structure. Blockchain technology, with its inherent security and distributed nature, presents a promising solution for addressing these issues. However, it is important to note that blockchain's history is relatively recent and less well-understood compared to the longstanding development of IoT.

3.2 IoT Security Threats

The most common attacks on IoT networks include [3, 10, 21–24]:

- **Physical Tampering**:Attackers physically tamper with or destroy IoT devices, disrupting network functionality and compromising data integrity. This type of attack can be particularly damaging in critical infrastructure and industrial IoT environments.

- **End Device Attacks**: These primarily focus on the exploitation of the device's software and stored data. Adversaries can capture and control a physical node. Information such as certificates and keys maintained on the captured nodes is then visible to the attacker. The information obtained can be used to masquerade as a valid node and perpetuate further attacks such as false data injection.
- **Communication Channel Attacks**: Attackers can eavesdrop on and tamper with transmission channels. Unencrypted radio signals can be easily intercepted due to their broadcast nature, allowing attackers to obtain sensitive information. Even encrypted signals can be analyzed to infer private details, such as the destination or source of the communication. Additionally, adversaries can disrupt communication by jamming wireless channels with noise.
- **Network Protocol Exploits**: Attackers can exploit vulnerabilities in network protocols through various methods such as man-in-the-middle, wormhole, Sybil, and replay attacks. In a Sybil attack, a device assumes multiple valid identities within the IoT system, undermining the integrity and efficiency of routing protocols and voting mechanisms. These attacks can severely compromise the overall security and functionality of the IoT network.
- **Sensory Data Manipulation**: Attackers can exploit ad hoc protocols to send messages hop-by-hop to their destination. During this process, adversaries can manipulate existing data or inject fake data, which is then forwarded through the network. When fake data is accepted, IoT applications may send incorrect instructions or provide erroneous services, undermining the reliability of IoT networks. For example, traffic jams can worsen if drivers receive incorrect driving assistance messages.
- **Denial of Service (DoS) Attacks**: DoS attacks aim to deplete resources and overwhelm IoT system services. One specific type, the sleep deprivation attack, disrupts the sleep routine of nodes or devices, keeping them awake until their battery power is exhausted. Given the limited communication and network resources of IoT devices, a DoS attack can severely impact network connectivity and reduce the network's overall lifespan.
- **Software Attacks**: Adversaries exploit software vulnerabilities using worms, viruses, or scripts to create backdoors, compromising the software and taking control of operations. Protecting IoT data and devices, such as actuator commands and sensor data, is essential for safeguarding IoT networks. Effective Intrusion Detection Systems (IDS) must be implemented to maintain the integrity, confidentiality, and verification of IoT communications and information. Proper identification of IoT devices ensures that data originates from legitimate sources, traditionally confirmed by a trusted Identity Provider. Encryption and authentication algorithms are crucial for safeguarding data integrity during transmission. Once data is transmitted to its storage location, its security relies on the robustness of the storage service provider's security measures.

To securely deploy IoT systems, it is essential to consider the following guidelines and mechanisms:

- **Physical and Hardware Security**: Physical security measures prevent unauthorized access or tampering with IoT devices. Secure boot mechanisms and hardware-based security features ensure that devices run only authorized and unaltered software.
- **Software and Firmware**: Ensuring that all IoT devices receive timely software and firmware updates is critical to patch vulnerabilities and protect against new threats.
- **Data Integrity, Confidentiality, and Privacy**: Appropriate encryption ensures confidentiality as IoT data moves through network hops. With the integration of diverse devices, networks, and services, the privacy of data maintained on devices becomes vulnerable to compromised IoT network nodes. Attackers can compromise data integrity by manipulating vulnerable IoT devices to alter stored data for malicious purposes.
- **End-to-End Encryption**: Robust encryption protocols for data transmission between devices and networks are crucial to protect against eavesdropping and tampering.
- **Accounting, Authentication, and Authorization**: Robust authentication protocols are essential for verifying the identity of communicating parties and devices, ensuring secure IoT communications. The diverse and heterogeneous nature of IoT architectures necessitates a variety of authentication protocols, complicating standardization efforts. Authorization protocols are crucial for granting system or information access exclusively to authorized parties. Proper implementation of these protocols fosters a reliable and secure communication environment. Additionally, comprehensive accounting, auditing, and reporting of resource utilization are vital for maintaining and protecting network management, ensuring transparency and accountability.
- **Access Control Management (ACM)**: Advanced ACM techniques such as Role-Based Access Control (RBAC) and Attribute-Based Access Control (ABAC) are essential to manage permissions and access rights effectively.
- **Service Availability**: IoT services can be disrupted by Denial of Service (DoS) attacks. Malicious tactics such as jamming, replay attacks, or sinkhole attacks can target various layers of IoT systems, aiming to degrade the Quality of Service (QoS) for users.
- **Resource Constraints and Energy Efficiency**: IoT devices often have limited resources, including minimal storage and power. Attacks on IoT systems can increase energy consumption by overwhelming the network with repetitive or false service requests, depleting available resources, and reducing the overall efficiency of the system.
- **Single Points of Failure**: The continuous expansion of IoT networks can expose numerous single points of failure, compromising IoT services. This issue highlights the need to create a robust environment for multiple devices and develop diverse protocols to ensure network operations are more fault-tolerant.

- **Network Segmentation**: Segmenting the IoT network can limit the impact of potential breaches by isolating different parts of the network, thus preventing attackers from moving laterally across the entire system.
- **Intrusion Detection Systems (IDS)**: Systems to monitor network traffic and detect any unusual or malicious activities that could indicate an attack are essential.

3.3 Benefits of Merging IoT and Blockchain

IoT is an innovative technology with immense potential for healthcare, although current IoT systems rely on devices with limited resources, making them vulnerable to cyberattacks. These networks are challenging to scale, difficult to maintain, and still face single points of failure. Additionally, IoT data is not permanent, and concerns about data security and privacy persist. Blockchain's decentralization, data creation and storage methods, and consensus mechanisms can address these weaknesses in centralized IoT systems (See Table 2).

Numerous use cases have demonstrated blockchain's applicability across all components of an IoT system. Blockchain can authenticate and encrypt data in communication networks, manage and store device identifications, and securely store data both in the Cloud and on distributed objects or devices. Moreover, blockchain can mitigate potential risks by using multiple nodes to exchange peer-to-peer data, making data nearly impossible to compromise. The consensus mechanism inherent in blockchain technology prevents compromised nodes from joining an IoT network and rejects data from compromised nodes, thus protecting data integrity.

In summary, blockchain offers the following benefits to IoT [3, 6–10]:

- **Decentralization**: Blockchain operates on a consensus mechanism where a majority of participants must verify and approve a transaction before it is added to the distributed ledger. This process ensures that no single entity has the authority to unilaterally verify transactions or establish rules for transaction acceptance. Trust is built through the collective agreement of the majority, providing a secure verification system. As a result, blockchain offers a robust platform for IoT devices, eliminating centralized traffic bottlenecks and mitigating single points of failure inherent in traditional centralized architectures.
- **Scalability**: A distributed peer-to-peer architecture overcomes performance bottlenecks and eliminates single points of failure inherent in centralized models. This approach prevents monopolization of data storage and processing by a few large entities. The decentralized nature of the system enhances scalability and fault tolerance, while also mitigating the formation of IoT silos.
- **Improved Interoperability**: Blockchain enables the conversion, processing, and storage of diverse IoT data within blocks. Its peer-to-peer network architecture supports integration across fragmented networks, allowing seamless data sharing among various business entities—such as hospitals, patients, pharmacies, and labs—while ensuring security and efficiency.

Table 2 Comparison between IoT and Blockchain [3]

Characteristic	IoT	Blockchain
Structure	Centralized (Cloud IoT model)	Decentralized
Reliability	Single Point of Failure (e.g., DoS attack to cloud servers)	More reliable as it eliminates single points of failure
Resources	Limited resources (IoT devices) with limited bandwidth and computational capabilities	Resource consuming and high bandwidth consumption
Scalability	Integrates billions of devices	Suffers from low throughput and low transaction speed
Automation	Typically needs human intervention	Autonomic interactions governed by smart contracts
Latency	Requires low latency	Mining is a time consuming task
Traceability	Low traceability	Tamper proof
Transparency	Low transparency	Trace any past transaction
Interoperability	Heterogeneous system with poor interoperability	Enhances interoperability because it can operate with diverse and fragmented networks with a universal internet access
Data sharing	Data exchanging is a challenging task	Data sharing can achieved through a P2P blockchain layer with uniform access across different IoT systems.
Privacy	Suffers from privacy vulnerability as confidential IoT data are uploaded to the third-party cloud servers	Supports P2P data sharing among connected devices without requiring communication through a centralized network
Security	Prone to many security vulnerabilities and malicious attacks	More secure as all transactions are encrypted and digitally-signed by cryptographic keys. Blockchain can be used for automatically-updating IoT device firmwares
Integrity	Infrastructure (e.g., Cloud) provider can alter the IoT data	Guarantees data integrity

- **Adaptability**: The diverse nature of IoT devices and algorithms often restricts interoperability. However, as a distributed database that operates independently of specific semantics, blockchain enhances the adaptability of IoT ecosystems. It has demonstrated compatibility across heterogeneous platforms, enabling IoT infrastructures based on blockchain to effectively adapt to various use cases and environments.
- **Coordination**: Blockchain facilitates rapid transaction processing and effective coordination among billions of devices.

- **Transparency**: Blockchain enables authorized network users to trace historical transactions, offering a reliable method for identifying and addressing data leaks.
- **Traceability**: Blockchain-based tracing systems offer stakeholders and participants the capability to verify data or product quality and origin. This technology facilitates the tracking of sensor data, ensuring accountability throughout the IoT ecosystem. Blockchain data remains identifiable at any time and from any location, with all historical transactions being verifiable. The immutable nature of blockchain ensures that stored transactions are nearly impossible to alter, allowing participants to authenticate and confirm the integrity of the data.
- **Autonomic Interactions**: Smart contracts enable blockchain to facilitate automatic interactions between IoT devices and subsystems without human intervention. This decentralized approach removes the need for traditional central roles, such as government authorities, allowing for seamless, self-executing operations.
- **Reduced Cost**: A notable benefit of blockchain is its potential to lower operating expenses. By facilitating peer-to-peer data transmission without requiring a central authority, blockchain reduces the need for costly infrastructure. Scaling a reliable, centralized system can be prohibitively expensive, whereas decentralization offers a more cost-effective solution for expanding IoT networks and mitigating single points of failure.
- **Streamlined Accounting**: Transparency provided by blockchain offers significant advantages for enterprise-level accounting departments. The ability to trace data sharing and fund transfers through a time-stamped, immutable chain enhances visibility and simplifies accounting processes. Even with the integration of a translation layer, the permanent nature of blockchain data facilitates easier accounting and auditing.
- **Accelerated Data Sharing**: Blockchain facilitates direct information exchange between connected devices and partners in the IoT ecosystem, bypassing the need for a centralized network such as the cloud. Although blockchain currently faces limitations in transaction speed, permissioned blockchains are capable of managing high data volumes, numerous devices, and rapid transaction rates required for enterprise applications. By employing trusted nodes, transaction verification times can be reduced, thereby addressing the performance needs essential for efficient IoT data exchange.
- **Contract Execution**: Smart contracts enable stakeholders to execute contract terms automatically when predefined conditions are satisfied. These contracts facilitate automatic authorization of payments and other actions once the contractual service requirements are met.
- **Greater Reliability**: Blockchain enhances the fault tolerance of IoT systems and eliminates single points of failure. By removing reliance on centralized servers, it prevents potential bottlenecks and protects users' personal data from third-party management. Each node in the blockchain maintains a complete copy of the ledger, which includes all network transactions, thereby increasing resilience against attacks. If one node is compromised, the integrity of the network is preserved by the remaining nodes. This comprehensive ledger replication also enhances the reliability and efficiency of IoT data sharing.

- **Byzantine Fault Tolerance**: IoT networks, comprised of constantly available smart devices that collect data and perform automated functions, require high availability for effective management. Centralized server architectures often struggle with this requirement. Blockchain offers resilience against Byzantine Faults through its distributed consensus and recordkeeping mechanisms, which enable it to recognize and handle failures effectively. This decentralized approach ensures that the system remains operational even in the presence of faults or malicious actors.
- **Trustless**: Participants and devices in an IoT network can establish trust without relying on a centralized authority to manage data storage or connectivity. This decentralization prevents third parties from accessing private data and facilitates quicker, automated payments for services.
- **Accelerated Transaction Speed**: Blockchain can significantly reduce transaction times, enabling IoT system services to evolve from days to near-instantaneous transactions.
- **Improved Security**: Blockchain enhances data security through the use of encryption and digital signatures (e.g., elliptic curve digital signature algorithm) to safeguard transactions. By enabling peer-to-peer information sharing among connected devices, blockchain mitigates vulnerabilities associated with centralized networks. Pseudonymity for addresses and distributed consensus for record permanence further bolster security. Data on a public blockchain is resistant to tampering due to its decentralized nature, and the computational or financial costs of new transactions help prevent DDoS and flooding attacks. Additionally, blockchain and smart contracts can enhance system security by automating device firmware updates to address vulnerabilities. In summary, blockchain strengthens both data and IoT stack security protocols as detailed below [7, 25–30].

  - **Highly Encrypted Transactions**: Transactions benefit from advanced encryption through cryptographic hashing, making them highly resistant to hacking and unauthorized access.
  - **No Single Point of Failure**: The decentralized architecture of blockchain mitigates risks associated with central points of failure, protecting the network from attacks such as DDoS and spoofing.
  - **Cryptographic Signatures**: Distributed cryptographic signatures across the network prevent man-in-the-middle attacks by ensuring that communications cannot be easily intercepted or altered.
  - **Node Status Tracking**: Continuous monitoring of node status aids in identifying and defending against potential security threats, enhancing the overall robustness of the IoT network.
  - **Data Integrity**: Blockchain provides a high level of data integrity by ensuring that data alterations can only be made by authorized parties. Tampering with blockchain transactions or creating fake blocks is nearly impossible due to the consensus mechanisms that validate and secure data.

  - **Confidentiality**: Confidentiality is maintained through public/private key encryption.
    Each transaction is secured with the sender's private key, safeguarding sensitive information from unauthorized access.
  - **Availability**: Blockchain integration enhances network availability by distributing data across multiple nodes, eliminating central points of failure, and ensuring uninterrupted access to data.
  - **Decentralized Trust**: Trust is established among participants without relying on a central authority. This decentralized approach reduces the risk of data manipulation and ensures more reliable and secure interactions.
  - **Immutable Records**: Once recorded on the blockchain, data becomes nearly impossible to alter or delete. This immutability enhances the reliability of data and ensures that historical records remain intact and verifiable.

- **Improved Privacy**: Maintaining user control over data access and ownership is crucial for privacy in IoT systems. Several methods enhance data privacy, including [7, 28, 31–38]:

  - Token-based Access Model: Data access requests are managed through tokens sent to the data owner. Data may be stored off-chain in distributed file systems (such as IPFS), distributed databases (like BigchainDB), or cloud storage. This method allows data owners to control every aspect of data sharing.
  - Associate Role-Based Model: Smart contracts can manage role-based data access privileges, granting users access based on predefined roles. Transactions are facilitated through monetary compensation or service exchanges.
  - Attribute-based Encryption (ABE): Data is encrypted according to an access policy defined by specific attributes. This model enables data owners, users, and consumers to include various attributes in decryption keys and access policies, providing fine-grained control over data access.
  - Tiered Architecture: This approach leverages a tiered blockchain structure, connecting multiple private blockchains to a public or federated blockchain. It forms a network where users across different blockchains can transfer data according to access control policies enforced by smart contracts. This model enhances data sharing flexibility and security, allowing for a scalable and controlled data exchange environment.

- **Improved Autonomy**: Decentralized peer-to-peer networks enhance the independence of IoT devices by eliminating the need for centralized servers in communication and service automation. Devices can independently authenticate received data and verify the sender's identity. Additionally, the secure nature of blockchain storage facilitates the deployment of software updates directly to IoT devices, further enhancing their operational autonomy.
- **Improved Functionality**: Blockchain enhances IoT functionality by leveraging the programmable logic of smart contracts, which enable automated interactions and transactions. These smart contracts also bolster security by managing confi-

dentiality, authentication, and access control, thus significantly improving overall IoT security.

- **Peer-to-Peer Data Sharing**: Cloud technology is crucial for processing, transmitting, and storing data, and has advanced IoT applications. However, it often leads to fragmented data silos that obstruct holistic IoT data analytics. In contrast, blockchain-based architecture supports secure, peer-to-peer data sharing, overcoming the limitations of cloud-based data fragmentation.
- **Decentralized IoT Data Marketplaces**: Traditional IoT service providers rely on centralized servers for data transmission. In contrast, public blockchain enables users to engage in data marketplaces, offering a platform to monetize their IoT data. Blockchain facilitates the creation of incentives, encouraging users to share their resources and contribute to a decentralized data economy.
- **Monetization**: IoT ecosystems generally consist of a three-tiered architecture: i) endpoint IoT devices that generate data, ii) gateways that transfer data to higher layers, and iii) cloud platforms that analyze and process the data to extract valuable insights. Monetizing IoT data is crucial for data sharing and ownership within these ecosystems. Traditionally, users must relinquish their data to third parties, such as cloud providers, which often leads to exposure of sensitive personal information, particularly in smart health applications. Blockchain technology offers a transformative approach to data monetization by eliminating reliance on central service providers. It facilitates peer-to-peer data sharing through secure access control and a robust storage layer. Data brokers and marketplaces can leverage blockchain to enable transactions directly between service providers, data customers, and data owners, bypassing the need for intermediaries. Smart contracts on the blockchain govern the level of data sharing, specifying what data can be shared and for how long. The distributed ledger enables data owners to monitor and trace the movement of their data across participants. Additionally, cryptocurrencies can be used to monetize data on the blockchain [7, 12].
- **Managing IoT Quality of Service (QoS)**: The rapid expansion of IoT presents challenges in maintaining Quality of Service (QoS) at both the edge of networks and in the Cloud. Traditional QoS measurement methods rely on centralized entities that use specific agents to collect and assess service performance. How-ever, these methods often struggle with the diverse and distributed nature of IoT services. Effective QoS management in IoT requires frequent and accurate data collection, access, and updates. Blockchain technology enhances this process by ensuring the integrity and trustworthiness of the data used to evaluate QoS, lever-aging its decentralized approach to maintain reliable and accurate service quality metrics [39].
- **Identity Management**: Blockchain-based systems can significantly enhance IoT security through advanced access and identity management. These systems securely handle digital rights, identities, credentials, and the provenance of goods. Blockchain's immutability ensures that once information is entered, it remains tamper-proof. However, a key challenge is ensuring the secure storage of identities, physical assets, and resources. This is often achievable for many IoT devices using private blockchains, which can store cryptographic hashes of device firmware.

Such systems create immutable records of device status and configuration, allowing for the verification of device authenticity and the integrity of settings and software. Devices must undergo authentication to connect with other services or devices, ensuring a secure and trustworthy IoT network.

- **Access Management**: As the IoT ecosystem rapidly evolves, managing billions of devices globally presents a significant challenge. Traditional access management technologies, which rely on centralized client/server paradigms, face limitations in handling the expansive and dynamic nature of IoT networks. These models, originally designed for conventional human-machine interactions within trusted domains, struggle with the mobility and multi-manager scenarios typical of IoT devices. Many IoT devices have constrained computing power, battery life, and resources, necessitating more innovative solutions. Researchers are exploring blockchain as a potential solution for access control in IoT environments. By leveraging blockchain's immutability and transparent auditing capabilities, access requests can be managed through smart contracts. This approach allows for a more secure and transparent access management system. However, challenges remain. Blockchain storage is limited and costly, making it impractical to store all access control policies on-chain. Private blockchain solutions could mitigate some storage issues, but they may hinder interoperability and broader adoption. To address these challenges, some researchers propose frameworks like FOCUS, which utilizes a three-dimensional social network model to manage access control across relationships between things, owners, and their interactions. Without blockchain, privacy preservation remains incomplete due to reliance on centralized identity providers, which can compromise the confidentiality of interactions. Blockchain-based identity management offers a shift towards user-controlled privacy, allowing individuals to decide who can access their personal information. Despite this, public blockchains may still expose some identity data. Therefore, incorporating privacy-preserving tools such as zero-knowledge proofs is crucial for selective data sharing and enhancing privacy in IoT systems.
- **Data Management**: The challenges of managing data in IoT systems encompass the storage, aggregation, and processing of information from smart objects. Effective management involves collecting online data, maintaining event logs, storing information, and auditing for offline analysis. IoT data management systems must handle complex and varied data with minimal semantics, as unprocessed data often lacks clear context. Additionally, the time-sensitive nature of many IoT applications necessitates rapid data processing, which is challenging due to the limited capabilities of some IoT devices. Traditional centralized models struggle to ensure trust and data integrity across diverse and heterogeneous datasets. Although blockchain technologies might face latency issues related to data storage, they offer a promising approach by ensuring global data integrity and reducing reliance on complex systems for managing IoT data [7].

3.4 DLT Stack for IoT

DLt/blockchain stack for IoT applications consists of multiple layers, each serving a distinct function to support the overall architecture. Figure 2 provides a visual representation of these layers, which are described in detail below [3, 8, 40]:

- **Device Layer**: The Device Layer includes all smart devices such as sensors, actuators, controllers, and edge/fog nodes. These devices are interconnected through various wireless and wired communication protocols, forming the basis of the Internet of Things.
- **Data Layer**: The Data Layer is responsible for collecting IoT data from the Device Layer. It processes and encrypts this data using digital signatures, asymmetric cryptographic algorithms, and hashing techniques to ensure security and integrity.
- **Network Layer**: The Network Layer provides a P2P network architecture that supports decentralized communication. This layer enables direct resource and information sharing among peers without requiring a central authority. Participants can serve various roles, such as database management, routing, mining, and wallet operations.
- **Consensus Layer**: This layer oversees the distributed consensus process essential for verifying the integrity of each block and ensuring that all peers maintain a consistent ledger. It addresses the challenge of differing perspectives on the system's state, such as forks, which can result from network faults, communication delays,

Fig. 2 DLT stack [3]

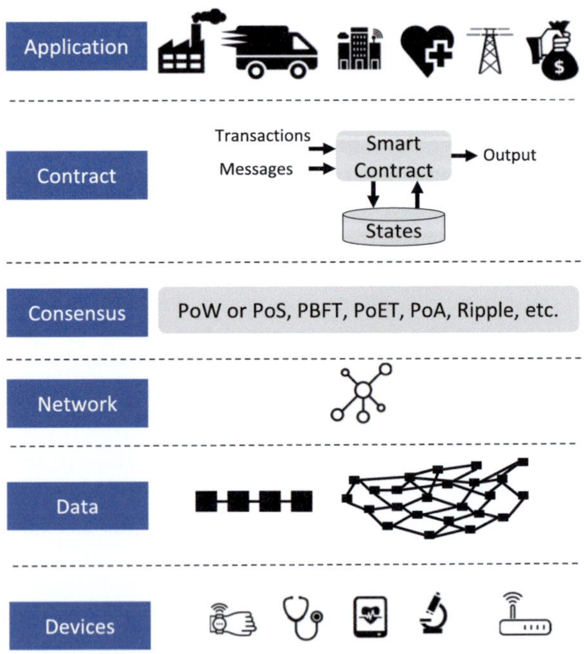

or malicious nodes. Effectively managing these discrepancies to avoid forks and uphold ledger consistency is a primary challenge of consensus algorithms.
- **Contract Layer**: The Contract Layer manages digital currencies and oversees the creation and execution of smart contracts. It facilitates automated transactions and the enforcement of predefined rules within the blockchain.
- **Application Layer**: This layer provides specialized services tailored to healthcare applications. By utilizing the capabilities of the underlying layers, it ensures the delivery of functional and secure solutions specifically designed to address healthcare needs. This includes applications such as patient monitoring, medical records management, and healthcare data analysis.

3.5 Blockchain and IoT Integration Schemes

The integration of blockchain and IoT can be achieved through several key approaches, each offering unique advantages for enhancing IoT systems. The main integration schemes are illustrated in Fig. 3 and include the following methods [3, 6, 7, 41, 42]:

- **Tight Integration**: This scheme requires that all communications among participants occur exclusively through the blockchain. In this model, every IoT device is considered a blockchain peer, and all interactions are recorded on the blockchain to ensure complete accountability. This approach allows for comprehensive tracking of all communications between specific IoT services and devices.
- **Loose Integration**: Recording every IoT interaction on the blockchain can lead to significant increases in storage and bandwidth requirements, which may be impractical for some applications. To address this, the loose integration scheme

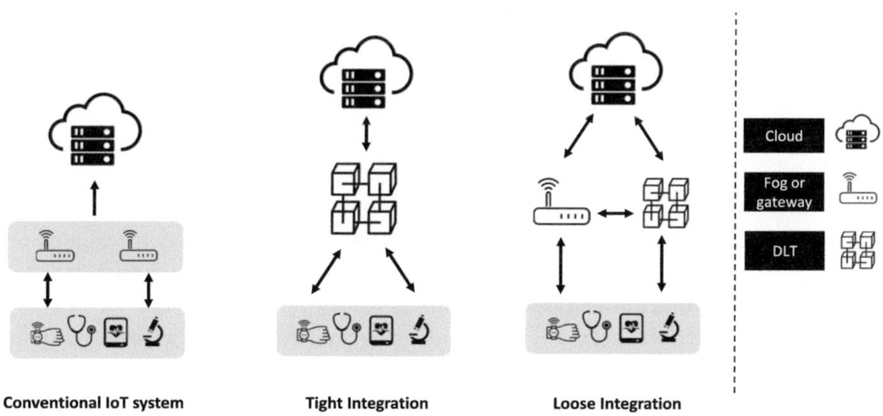

Fig. 3 Blockchain and IoT integration schemes [3]

uses resource-rich IoT devices, such as gateways or fog nodes, to handle consensus and other resource-intensive tasks. Resource-constrained devices are not required to maintain a full blockchain copy or process blockchain records. Instead, these devices can communicate off-chain or interact with the blockchain through the gateways. This approach necessitates additional discovery and routing protocols but offers low-latency communication and allows for selective recording of interactions on the blockchain. Essentially, not all data transfers need to be recorded on-chain. The blockchain serves as a control mechanism, with smart contracts enabling programmable logic for data transfers via peer-to-peer technology. Users can choose to use the blockchain for some interactions or rely on traditional cloud-based IoT protocols. This hybrid approach leverages the benefits of both decentralized recordkeeping and real-time IoT communication. It is particularly suitable for scenarios requiring high throughput, frequent interactions, reliable IoT data, and low latency. However, the level of decentralization in this scheme is not as strong as in tight integration, where devices communicate directly with the blockchain.

Note that transactions in a blockchain IoT system can be categorized as either on-chain or off-chain.

On-chain Transactions: These transactions are recorded on the distributed ledger and are accessible to all blockchain nodes. The validity of on-chain transactions is confirmed when they are mirrored on the public distributed ledger. An on-chain transaction must be authenticated by a specified number of blockchain participants, a process that can be time-consuming due to the multiple steps involved before it is considered successful. Additionally, the cost of such transactions may be unappealing to network participants.

Off-chain Transactions: In contrast, off-chain transactions occur outside the blockchain. The benefits of off-chain transactions include:

- **Lower Cost**: Transactions are generally free since they do not require validation by blockchain participants.
- **More Efficient**: Transactions are recorded instantaneously as they bypass the validation process.
- **Greater Privacy**: Transactions cannot be viewed by public blockchain participants.

Understanding the advantages and disadvantages of using blockchain for data storage involves considering the type of information stored. Blockchain operates as a distributed ledger, allowing all nodes to access the ledger, which contains all confirmed transactions in a block. Each transaction records details about who sent data to whom and under what conditions.

Since all transactions must be validated by every node in the network, smaller or less efficient devices are required to confirm and store transactions as quickly as larger, faster participants. This requirement can exponentially increase memory costs. Additionally, each block has a limited capacity, which restricts the amount of data that can be included in a single transaction. Transactions with large data volumes can slow down network performance, as subsequent transactions may not be processed within the same block.

Furthermore, as blockchains grow larger, centralization tends to increase because smaller devices have limited storage capacity. As a result, regular users might reconsider using blockchain technology if it causes significant storage issues.

Reducing costs and minimizing blockchain clutter can be achieved by using the off-chain technique of storing only hashes of data rather than the raw data itself on the blockchain. Blockchain's primary role is to establish trust among participants and provide proof of transactions, which does not require storing the actual data. However, for certain use cases where data security and privacy are paramount, it is essential to encrypt the data before hashing it. For scenarios where data distribution is necessary, the following off-chain solutions may be utilized:

- Distributed File systems: Facilitate file access in peer-to-peer networks. For example, IPFS (InterPlanetary File System) and Storj are two existing solutions in this domain.
- Distributed Databases: Address the need for high throughput, low latency, high capacity, and robust query support in data storage. Blockchain often does not meet these needs, but distributed databases like IPDB and BigchainDB provide solutions.

3.6 Blockchain and IoT Adoption Considerations

As discussed earlier, security architecture remains a major concern for current IoT systems. Blockchain enhances security by leveraging decentralization and consensus-based decision-making. Nonetheless, it is crucial to address the challenges involved in integrating a distributed ledger into the architecture [3, 6–10, 43]:

- **Scalability**: Large amounts of data from networked sensors and issues with latency or reduced transaction processing speed must be managed effectively in IoT systems. Establishing a clear data model in advance can prevent future problems and facilitate a smoother transition to production.
- **Network Privacy & Transaction Confidentiality**: The challenge of sharing a private IoT network's transaction history on a public blockchain lies in the potential for transaction pattern analysis to reveal user or device identities, even when protected by public keys. A thorough assessment of privacy needs is essential for determining whether a private or hybrid blockchain solution best fits the requirements.
- **Privacy**: The achievement of complete privacy is not feasible with blockchain due to its immutable nature. Data changes require confirmation by the majority of nodes in the network, ensuring the protection of transactions but preventing the removal of sensitive information.

- **Sensors**: The reliability of IoT sensors can be compromised by interference affecting the measurement criteria necessary for transaction execution. Safeguarding IoT device integrity is crucial to prevent compromise by third parties, thereby ensuring a secure environment for data collection and transaction completion.
- **Processing Power**: Many IoT devices lack the computational capacity needed to engage directly with blockchain networks.
- **Storage**: Although blockchain eliminates the need for a centralized server, the ledger must be maintained on each node. Over time, the ledger's size will increase, influenced by the volume of transactions and the number of nodes in the network.
- **Naming and Discovery**: Blockchain was not initially designed with IoT in mind, resulting in a lack of mechanisms for nodes to locate one another. For instance, Bitcoin embeds sender IP addresses within the client to establish network topology. This approach is inadequate for IoT environments, where many devices are mobile and frequently change positions, leading to a constantly shifting network topology.
- **Legal and Compliance Concerns**: The evolving nature of IoT presents challenges due to the lack of established regulations and legal precedents. This uncertainty can create difficulties for IoT service providers and device manufacturers, leading to hesitation in adopting blockchain technology.
- **Blockchain Security Vulnerabilities**: Despite its reputation for tamper-resistance and decentralization, blockchain technology is not immune to security vulnerabilities. Notable vulnerabilities include [10]:

  - **Double Spending**: Malicious actors attempt to deceive participants by creating conflicting transactions, such as spending the same cryptocurrency more than once. Adversaries may also pre-mine blocks to try and have conflicting transactions accepted by the blockchain.
  - **Consensus Protocol Attacks**: Adversaries may undermine consensus protocols by gaining control over a substantial portion of the network's computing power, thereby enabling the manipulation or reconstruction of the blockchain (e.g., through a 51% attack). An attacker who controls more than 50% of the network's hash power can coerce the blockchain into accepting invalid blocks by solving the consensus problem more rapidly than other participants. Moreover, it has been shown that possessing 33% of the hash power can significantly PoW blockchains.
  - **Eclipse Attacks**: Eclipse attacks target peer-to-peer networks by allowing adversaries to control all valid node connections, thereby isolating nodes from legitimate peers. The initial attempts at such attacks were observed in Bitcoin, where a randomized protocol was used to force a Bitcoin node to connect only with selected peers, disrupting peer-to-peer communication and blockchain operations. Ethereum has also been vulnerable to eclipse attacks, particularly through the Kademlia peer-to-peer protocol.
  - **Smart Contract Vulnerability**: The immutability and transparency of blockchain technology expose smart contracts to potential vulnerabilities. Both adversaries and the public can identify flaws or exploit vulnerabilities in smart contracts. Addressing issues or bugs within an implemented smart contract is

  challenging due to the immutable nature of the blockchain. A notable instance of such vulnerabilities is the 2016 Decentralized Autonomous Organization (DAO) attack, which led to a contentious split in the Ethereum blockchain.
  - **Programming Fraud**: Adversaries can exploit programming vulnerabilities or coding errors to compromise blockchain properties. For instance, the 2018 piracy attack, which involved exploiting code flaws, highlighted how programming weaknesses can be leveraged to undermine blockchain security.
  - **Distributed Denial of Service (DDoS) Attacks**: Adversaries launch DDoS attacks to overwhelm and deplete blockchain resources. For example, in 2016, attackers exploited underpriced Ethereum Virtual Machine (EVM) instructions to significantly slow down block processing speeds. This manipulation resulted in the creation of a large number of low-balance accounts, facilitating a DDoS attack.
  - **Private Key Leaks**: Adversaries can compromise an account by stealing its private key through physical access to the node or via traditional network attacks.

- **Storage Capacity & Scalability**: Blockchain's storage capability and scalability are contentious topics, and these challenges are amplified in IoT applications due to their inherent scalability and capacity limitations. While it may seem that blockchain is unsuitable for IoT applications, there are methods to mitigate or even eliminate these limitations. IoT devices generate gigabytes (GBs) of real-time data, posing a significant challenge to blockchain integration. Many current blockchain platforms can only process a limited number of transactions per second, potentially causing bottlenecks for IoT devices. Moreover, blockchain was not initially designed to handle the vast amount of data produced by IoT devices. However, combining these technologies can address these issues. Despite the large volume of data generated by IoT, only a small portion is used to gain insights or drive actions. Several methods have been proposed to filter, compress, or normalize IoT data. Given that the IoT ecosystem includes target services (such as cloud and blockchain), communication, and embedded devices, storing IoT data benefits many layers. Data compression lightens data processing, transmission, and storage. Additionally, transforming blockchain's consensus protocol to increase bandwidth and reduce transaction latency, as seen with Bitcoin-NG, can enhance IoT transactions.
- **Consensus Challenge**: The limited resources of IoT devices make them unsuitable for participating directly with consensus protocols (i.e., PoW). Many consensus mechanisms are in development, but they remain immature and untested. Resource requirements vary depending on the consensus mechanism used. Generally, solutions involve offloading tasks to gateways or more powerful devices. Off-chain options that transmit data outside the blockchain can reduce latency and improve functionality. While efforts aim to incorporate full blockchain nodes into IoT devices, mining remains a significant obstacle due to IoT resource limitations. Research continues to adapt consensus mechanisms to fully leverage the distributed potential of the IoT, such as the innovative Proof of Understanding

(PoU) used in Babelchain, which focuses on practical computation and key IoT communication issues.

- Secure Firmware Updates: Blockchain can be used to enhance smart manufacturing security. Centralized IoT systems experience a significant bottleneck by restricting factory upgrades. For example, addressing breaches in security requires that firmware be regularly upgraded. But, the majority of firmware updates must be downloaded via centralized servers and manually installed on IoT devices. This can be an inefficient and expensive process when working with distributed IoT systems.
- Enhancing Tracking and Tracing: Blockchain can be used by organizations to more securely, precisely, and effortlessly exchange data in complicated IoT-assisted supply chains. Blockchain has the capability of creating immutable records of products, materials, or parts, promoting total transparency and creating a primary information source for all chain participants. This can be highly valuable when supply chains include diverse IT systems, some participants are not trusted by all participants, or require constant inclusion of new users.
- Monetizing and Protecting Vital Intellectual Property: It is very important that manufacturing companies protect intellectual property. Blockchain can be used to definitively prove intellectual property ownership in the event of a dispute. For example, Bernstein Technologies has created blockchain-based web service that enables the registration of intellectual property and creating a certificate that confirms intellectual property ownership.
- Protecting and Simplifying Quality Checks: Companies can use blockchain and IoT to facilitate quality control and improve customer value. Without blockchain, providing complete transparency and accurate documentation regarding product or process quality can be an expensive endeavor, requiring support from centralized third parties. Blockchain is able to monitor and trace parts throughout the supply chain and provide permanent documentation of the production process and quality assurance. Blockchain is able to tag products and record each transaction, quality confirmation, or change in the blockchain. Utilizing this blockchain application requires that production incorporates automated quality inspections that create and record information directly to the chain. This allows multiple participants to access data, which removes the need to perform inbound quality checks to confirm supplier quality checks. This application may also decrease the need for centralized party or original manufacturer quality audits because participants could use blockchain's certificate management ability to view all documentation and ensure authenticity.
- Advancing Machines as a Service: Blockchain multiplies the possibilities of utilizing a pay-per-use model for IoT-based manufacturing machinery, referred to as Machines as a Service (MaaS). This paradigm allows providers of equipment to charge users based on the output generated rather than directly selling the machinery itself. For example, a compressor supplier could sell compressed air according to volume rather than selling the compressor itself. Using a MaaS paradigm would enable manufacturers to circumvent expensive preliminary investments and more easily upgrade machinery in order to utilize the most innovative technology. When

applied well, MaaS models empower manufacturers to expand production flexibility.

- Enabling Machine-Controlled Maintenance: Blockchain IoT model has the capability to facilitate innovative maintenance models and reduce time required for complete maintenance. Innovative manufacturing requires advancements in maintenance to keep pace with new, more complex machinery. Outsourcing maintenance requires users to provide installation information and service agreements to blockchain devices, providing a digital duplication (digital twin) of the device. Blockchain is then able to automatically execute and pay for machinery maintenance. If a piece of equipment needs maintenance, it can generate a service request as well as a smart contract for the maintenance. When completed, the payment is automatically processed and permanent documentation of the maintenance performed is added to the blockchain. These blockchain applications are currently in the development phase, but have the potential to increase equipment dependability, support equipment status monitoring, and develop auditable equipment health assessments. When maintenance is completed by in-house providers, blockchain can confirm that maintenance was completed in adherence to guarantee and warranty agreements. Permanent maintenance history documentation also supports sale of previously-owned machinery. Reduced product lifecycles and continuous changes in design will inspire manufacturers to upgrade equipment more often. Manufacturers selling used machinery can provide potential buyers with the blockchain records to confirm equipment has been appropriately maintained.

4 Blockchain for Smart and Connected Health

4.1 Challenges in Healthcare

Healthcare providers and researchers often face slow communications, workflow discrepancies, and data silos, yet they are frequently reluctant to share data. This reluctance stems from concerns about privacy regulations, potential liabilities, and financial repercussions. Additionally, incompatible health systems contribute to communication gaps, complicating the delivery of coordinated, patient-centered care. A significant challenge in healthcare is the inability to establish secure connections between disparate systems to form an end-to-end network while ensuring provider privacy. Standards like FHIR and HL7 offer basic interoperability, but their effectiveness is limited by the need for extensive data mapping and system adaptations whenever interfaces change [3, 5, 44].

4.2 Main Applications of DLT/Blockchain in Healthcare

Blockchain is already transforming the healthcare industry by enhancing data management and connectivity across diverse systems, thereby improving the accuracy of Electronic Health Records (EHR). Blockchain supports medical supply chains, facilitates drug prescriptions, safeguards system access, maintains risk data, supports data sharing, and creates audit trails for healthcare activities. Additionally, it enhances healthcare provider credentialing, medical billing, clinical trials, health record exchange, and industry contracts, supporting a patient-centered IoT-based model. Blockchain-based healthcare systems also improve data reliability and security by consolidating records and granting patients ownership of their health information [3].

Blockchain significantly reduces the need for centralized third parties to verify data ownership, ensure reliability, and manage transactions or data exchanges. It facilitates secure, pseudonymous transactions directly between participants. Blockchain's transparency, immutability, and decentralization address prevalent healthcare issues like inaccurate records and barriers to accessing personal health data. Effective healthcare systems rely on interoperability, enabling platforms and software applications to securely and efficiently share data across institutions. However, current healthcare systems are often siloed, with limited data access and slow communication, further hindering interoperability. Blockchain mitigates these issues by providing access to complete, tamper-proof, long-term medical records maintained in disparate systems, thereby improving communication and workflow integration [1, 3, 41, 42, 45–55].

The primary use cases and applications of DLT/blockchain in the healthcare domain include [3].

4.2.1 Healthcare Record Sharing

Sharing health records presents significant challenges due to the sensitive nature of patient information, yet blockchain technology has the potential to enhance this process. The security and accessibility of healthcare records are persistent industry concerns. Records are often dispersed across various systems, hindering both patients and providers from obtaining a comprehensive, real-time view of a patient's medical status. This dispersion makes linking records, even with common identifiers (e.g., Social Security Numbers), difficult due to interoperability issues. Consequently, the process of sharing health data remains inefficient and unproductive for the reasons outlined below [3]:

- **Slow Process Speed**: The process of creating and transferring health data copies is time-consuming.
- **Lack of Data Security**: Patient data copies are at risk of being lost or stolen during transfer.

- **Incomplete Records**: Medical histories are often fragmented across multiple systems. Without a universal storage system, patients must track where and when they received care to compile their current records.
- **Lack of Data Context**: The healthcare system is provider-centric, hindering patients from owning and controlling access to their healthcare records.

Blockchain technology addresses the communication gap among healthcare providers by eliminating the need for a centralized third party and enabling direct interactions among participants. However, blockchain alone cannot fully resolve the issue of data sharing; it must be integrated with existing disparate health systems and data standards. Recently, several architectures, such as MedRec and MedShare, have been proposed to enhance data sharing in the healthcare industry. These architectures aim to improve interoperability, security, and efficiency in managing and sharing health records [3]:

- MedRec Architecture: This blockchain-based architecture enhances the storage of health records by improving interoperability, data security, and data access. It utilizes permissioned peer-to-peer blockchain technology and smart contracts within the Ethereum framework to monitor and manage network status transitions.
- MedShare Architecture: This architecture comprises four layers: the user layer for data access via a graphical interface, the data query layer for processing system queries, the database infrastructure layer accessible only by specified institutions, and the data structuring and provenance layer that manages in-system processing using blockchain network structures, smart contracts, consensus mechanisms, and node authentication.

4.2.2 Log Management in Healthcare Systems

Log management is crucial for providing historical data to detect intrusions and analyze errors. Proper log management enhances user control over information access. However, traditional systems are vulnerable to data manipulation. Blockchain technology can ensure the immutability of log records. For example, [56] explored the application of blockchain in healthcare log management, using blockchain to manage information access logs, standardize data, and enable auditing through a permissioned blockchain. However, security audit logging can be complex due to potentially incomplete or irrelevant gathered information [3, 45].

4.2.3 Patient Monitoring

A Remote Patient Monitoring (RPM) system collects medical data from bio-sensors or IoT devices to monitor a patient's condition outside clinical settings. This approach is crucial in patient care, leveraging wearable healthcare devices and big data analytics to create new opportunities for remote healthcare services and alleviate the burden on clinical resources. However, data review poses security and privacy challenges. Blockchain technology offers a secure method for sharing, storing, and accessing collected data. For instance, smart IoT e-health devices implanted in patients can use blockchain to securely monitor and store personally identifiable information. Such technology must comply with HIPAA, GDPR, and other relevant data protection regulations [3].

Over recent years, several blockchain-based solutions have been proposed for RPM. Researchers have shown that Ethereum-based smart contracts can enable real-time monitoring and secure automated interventions. Others support using a Hyperledger-centered approach for blockchain data gathering and exchange among mobile healthcare stakeholders. Blockchain technology has also been used to develop SMEAD, a mobile monitoring device for diabetics, with mobile devices successfully transmitting data to blockchain applications on Hyperledger Fabric. Another innovative solution is the Patient Centric Agent (PCA), which leverages blockchain to ensure data privacy and security in continuous remote monitoring applications [3].

4.2.4 Tracing

Blockchain is also capable of tracking contagious patients suffering from illnesses including COVID-19, Severe Acute Respiratory Syndrome (SARS), or Middle East Respiratory Syndrome (MERS). An infected individual or suspected carrier wear IoT devices that continuously monitor their movements and health data. This data is then encrypted and securely recorded on the blockchain. The decentralized and immutable nature of blockchain ensures that the data remains tamper-proof and private. Health authorities can access and analyze this data to identify infection hotspots, trace contacts, and deploy targeted countermeasures, all while maintaining the anonymity and privacy of the patients involved [3].

4.2.5 Monetizing IoT eHealth Data and Services

Blockchain technology enables transparent interactions between IoT eHealth service providers and data owners without the need for intermediaries. Data owners can control the type and level of shared data and the duration of sharing through smart contracts. The distributed ledger of blockchain allows data owners to monitor data movement between entities, ensuring transparency. Additionally, cryptocurrencies facilitate the monetization of IoT eHealth data, providing an efficient and secure means of transaction and data exchange [3, 12].

4.2.6 Drug and Pharmaceutical Supply Chains

Blockchain technology can significantly enhance the management of healthcare supply chains, particularly for drugs and pharmaceutical products. Counterfeit or substandard drugs pose significant risks to patient safety. Blockchain technology can mitigate these risks by recording all prescription-related transactions on a decentralized ledger. This ledger connects key stakeholders, including physicians, manufacturers, patients, distributors, and pharmacists, ensuring transparency and alerting participants to any data tampering. Furthermore, some companies offer public access to immutable documentation for the IoT-based transportation of pharmaceutical products, thereby ensuring compliance with quality control standards.

4.2.7 Health Insurance Claims

Insurance companies assess claims to determine whether they should be fully or partially paid or denied. Partial payments often result from medical services being billed incorrectly or deemed unnecessary. Accurate coding is crucial, as minor errors, such as a misspelled name, can lead to claim denials. While most claims are processed through automated systems, complex cases are susceptible to fraud and errors. Currently, nearly a quarter of insurance claims are rejected due to simple errors like incorrect demographic details. IoT and smart contracts can significantly automate the decision-making process using a distributed network, enhancing transparency for both healthcare providers and insurers. This system allows for quicker detection and correction of fraud and errors. Additionally, smart contracts ensure that all network participants have access to up-to-date information and are informed of policy or guideline changes [3].

Blockchain can enhance insurance claim processing by ensuring data permanence, providing clear audit trails, improving data transparency, and facilitating decentralization. Patients would not need to manually collect paperwork for claims, as blockchain platforms could automatically link hospitals, healthcare providers, and insurers. This system enables insurers to authenticate claims and prevent fraud efficiently. Although promising, initial implementations have been limited. MIStore, a health insurance storage system on Ethereum, and Samsung SDS's Nexledger-based platform are examples of blockchain applications in health insurance claim processing, aiming to simplify claims and reduce fraud [3].

4.2.8 Fraud Detection

Blockchain technology can significantly aid in fraud detection by authenticating information to detect tampering and identifying malicious activities such as inserting false reviews in online systems. Researchers are also exploring its application in healthcare crowdfunding, a method for raising investment from many individuals or purchasing company stock to increase equity. Blockchain can simplify transactions and enhance the security and efficiency of fund transfers. It also provides an economical platform for registering shares or stocks of healthcare companies, enabling direct transactions between entrepreneurs and investors. Additionally, blockchain can facilitate shareholder voting systems for governance and enable regulators to monitor market status and prevent investment fraud.

4.2.9 Patient Digital Identity

Patient identity confirmation is critical for health data exchange. Unique datasets help locate patients within record databases. Enterprise Patient Master Index (EPMI) and Master Patient Index (MPI) manage patient identities within networks or institutions. Despite ongoing development, matching patient identities remains challenging, resulting in duplicate records and inaccurate data. The lack of standardized identifying information collection complicates this further, as requirements vary between organizations. Common identifiers like name, address, Social Security Number, and date of birth can be inconsistently formatted, and manual entry errors are frequent. Additionally, unique identification numbers often fail to work across multiple organizations. Blockchain technology can address these issues by providing a decentralized and immutable ledger for storing patient identities, ensuring consistency, and reducing the likelihood of errors or duplicate records [3].

4.2.10 Healthcare Interoperability

Interoperability in the healthcare industry refers to the ability of diverse software and IT systems, such as Electronic Health Records (EHRs), to share, communicate, and utilize exchanged information seamlessly. Enhanced interoperability within and among institutions elevates patient care by allowing providers to securely exchange records across various locations, regardless of their relationship status. A significant barrier to achieving interoperability is the reliance on centralized information storage, leading to fragmented health records, limited access, and reduced data quality for research. Millions of health records are stored in centralized databases at hospitals, making them susceptible to loss and inaccessibility. Blockchain technology addresses these challenges by ensuring the immutability and accessibility of health records, facilitating secure data sharing without centralized third parties, and improving overall data quality for research purposes [3].

4.2.11 Smart Manufacturing

Blockchain technology can be integrated into the smart manufacturing of healthcare products in several innovative ways [3, 57]:

- Secure Firmware Updates: Blockchain can significantly enhance the security and efficiency of firmware updates in smart manufacturing. Centralized IoT systems often face bottlenecks when deploying factory upgrades, particularly for security breaches requiring regular firmware updates. Typically, these updates are downloaded via centralized servers and manually installed on IoT devices, a process that can be both inefficient and costly. Blockchain enables decentralized, automated firmware updates, ensuring that all devices receive timely and secure updates, thereby reducing the risk of vulnerabilities in distributed IoT systems.
- Enhancing Tracking and Tracing: Blockchain enables organizations to securely and accurately exchange data within complex IoT-assisted supply chains. By creating immutable records of products, materials, or parts, blockchain fosters complete transparency and serves as a central information source for all participants in the supply chain. This capability is particularly valuable when dealing with diverse IT systems, participants who may not fully trust each other, or the constant inclusion of new users.
- Monetizing and Protecting Vital Intellectual Property: Manufacturing companies must safeguard their intellectual property. Blockchain can serve as a definitive method for proving intellectual property ownership in disputes. For example, Bernstein Technologies has developed a blockchain-based web service that enables the registration of intellectual property, generating a certificate that verifies ownership. This ensures that intellectual property is securely documented and traceable, protecting it from unauthorized use and providing clear proof of ownership.
- Protecting and Simplifying Quality Checks: Blockchain and IoT can streamline quality control processes, enhancing customer value. Traditionally, achieving full transparency and accurate documentation of product or process quality can be costly, often necessitating centralized third-party support. Blockchain can monitor and trace parts throughout the supply chain, providing permanent documentation of production processes and quality assurance. Each transaction, quality confirmation, or change can be recorded on the blockchain. This requires automated quality inspections that directly input data into the blockchain, allowing multiple participants to access this information and reducing the need for inbound quality checks and centralized audits. Participants can use blockchain's certificate management to view all documentation, ensuring authenticity without relying on original manufacturer quality audits.
- Advancing Machines as a Service: Blockchain significantly enhances the utilization of the pay-per-use model for IoT-based manufacturing machinery, known as Machines as a Service (MaaS). This model allows equipment providers to charge users based on output rather than selling the machinery. For instance, a compressor supplier could charge for compressed air by volume instead of selling the compressor. The MaaS paradigm helps manufacturers avoid expensive initial investments and facilitates easier upgrades to the latest technology, thereby increasing production flexibility.
- Enabling Machine-Controlled Maintenance: The blockchain IoT model facilitates innovative maintenance methods and reduces the time required for comprehensive maintenance. Advanced manufacturing necessitates keeping pace with new,

complex machinery. By outsourcing maintenance, users can provide installation details and service agreements to blockchain devices, creating a digital twin of the equipment. Blockchain can then automatically manage and pay for maintenance services. When maintenance is needed, equipment can generate a service request and a smart contract. Upon completion, payment is processed automatically, and a permanent record of the maintenance is stored on the blockchain. These applications, currently in development, have the potential to increase equipment reliability, support status monitoring, and create auditable health assessments. When maintenance is performed in-house, blockchain ensures adherence to warranty agreements. Permanent maintenance history documentation also facilitates the sale of used machinery by verifying proper upkeep. Reduced product lifecycles and continuous design changes will prompt manufacturers to upgrade equipment more frequently, and blockchain records can confirm appropriate maintenance for prospective buyers.

4.3 Main Challenges of Utilizing Blockchain in Healthcare

The integration of blockchain into the healthcare domain faces several significant challenges, as summarized below [3]:

- **Throughput**: High throughput is essential in healthcare systems to ensure timely access to lifesaving diagnoses and treatments. The increasing number of nodes and transactions within a blockchain network necessitates additional checks, potentially creating bottlenecks and delaying critical processes.
- **Latency**: Depending on the underlying DLT technology, authenticating a block typically might require several minutes, which can pose scalability issues as well as security risks. Healthcare systems need to be agile and constantly accessible; any delay in block authentication could adversely affect the timely analysis of medical exams and patient care.
- **Security**: Blockchain-based healthcare systems face various security challenges, as previously discussed. A notable example is the 51% attack, where if a malicious entity gains control of over half of the network's computational power, they can manipulate the blockchain and compromise its integrity. This can lead to significant security vulnerabilities, resulting in a loss of functionality and credibility. Therefore, it is crucial to implement robust security measures to ensure the reliability and trustworthiness of blockchain systems in healthcare. Effective measures are essential to protect against data breaches, unauthorized access, and other cyber threats.
- **Decentralization Challenges**: Although blockchain is inherently decentralized, certain implementations may centralize miners or validators e.g., for better scalability and resource efficiency, which decreases the network's reliability. This centralization creates single points of failure, making the system more vulnerable

to cyberattacks. Generally, designing and integrating blockchain into healthcare systems requires careful consideration of scalability. Effective architecture must balance decentralization, security, and resource efficiency to ensure the system can scale with growing demands while addressing other aspects like system reliability and security. Ignoring any of these aspects can undermine the system's overall effectiveness.

- **Privacy**: While blockchain is often touted as a guarantor of privacy, this assumption has been challenged. Blockchain's transparency can conflict with privacy requirements, especially under regulations like the GDPR. Effective privacy strategies that ensure compliance with privacy laws are essential to protect sensitive healthcare information. Implementing additional privacy-preserving techniques, such as zero-knowledge proofs and federated learning, can help enhance privacy in blockchain-based healthcare systems.
- **Regulatory Compliance**: Navigating the regulatory landscape for blockchain in healthcare is intricate, requiring adherence to various legal and compliance standards to ensure patient safety and data protection.
- **High Resource Demands**: Blockchain technology demands substantial resources, particularly during the mining process. Implementing blockchain in healthcare systems would result in high energy and computing costs due to the necessity of multiple devices, such as wearables, to monitor patients. This resource-intensive nature can pose a significant challenge in maintaining cost-effective and sustainable healthcare operations.
- **Device Limitations**: Many IoT healthcare devices, such as wearables, are constrained by their power, storage, memory, and processing capabilities. Consequently, these devices are unable to execute complex blockchain algorithms, posing a challenge for integrating blockchain technology into healthcare IoT systems.
- **Interoperability**: Integrating blockchain with existing, disparate healthcare systems and data standards is complex and often challenging, requiring significant efforts to achieve seamless interoperability.

5 Conclusion

In this chapter, we explored the intersection of IoT and DLT/blockchain technologies, with a specific focus on their applications in healthcare. While these technologies have evolved independently, their convergence in the IoT-Blockchain paradigm addresses critical challenges in healthcare, such as security, data integrity, auditability, transparency, reliability, secure peer-to-peer data sharing, and data monetization. We discussed how this integration can revolutionize healthcare by enhancing data management, patient care, and overall system efficiency. Finally, the chapter highlighted innovative applications and business models emerging from this convergence.

Acknowledgements The authors acknowledge the use of Generative AI (GenAI) in editing and assisting with the development of this book chapter.

References

1. F. Firouzi, B. Farahani, M. Ibrahim, and K. Chakrabarty, "Keynote paper: From eda to iot ehealth: Promise, challenges, and solutions," *IEEE Transactions on Computer-Aided Design of Integrated Circuits and Systems*, 2018.
2. B. Farahani, F. Firouzi, V. Chang, M. Badaroglu, N. Constant, and K. Mankodiya, "Towards fog-driven iot ehealth: Promises and challenges of iot in medicine and healthcare," *Future Generation Computer Systems*, vol. 78, pp. 659–676, 2018.
3. B. Farahani, F. Firouzi, and M. Luecking, "The convergence of iot and distributed ledger technologies (dlt): Opportunities, challenges, and solutions," *Journal of Network and Computer Applications*, vol. 177, p. 102936, 2021.
4. X. Liu, B. Farahani, and F. Firouzi, "Distributed ledger technology," in *Intelligent Internet of Things*. Springer, 2020, pp. 393–431.
5. F. Firouzi, K. Chakrabarty, and S. Nassif, *Intelligent Internet of Things: From Device to Fog and Cloud*. Springer, 2020.
6. Z. Xiong, Y. Zhang, N. C. Luong, D. Niyato, P. Wang, and N. Guizani, "The best of both worlds: A general architecture for data management in blockchain-enabled internet-of-things," *IEEE Network*, vol. 34, no. 1, pp. 166–173, 2020.
7. M. S. Ali, M. Vecchio, M. Pincheira, K. Dolui, F. Antonelli, and M. H. Rehmani, "Applications of blockchains in the internet of things: A comprehensive survey," *IEEE Communications Surveys & Tutorials*, vol. 21, no. 2, pp. 1676–1717, 2018.
8. H.-N. Dai, Z. Zheng, and Y. Zhang, "Blockchain for internet of things: A survey," *IEEE Internet of Things Journal*, vol. 6, no. 5, pp. 8076–8094, 2019.
9. W. Viriyasitavat, L. Da Xu, Z. Bi, and D. Hoonsopon, "Blockchain technology for applications in internet of things—mapping from system design perspective," *IEEE Internet of Things Journal*, vol. 6, no. 5, pp. 8155–8168, 2019.
10. X. Wang, X. Zha, W. Ni, R. P. Liu, Y. J. Guo, X. Niu, and K. Zheng, "Survey on blockchain for internet of things," *Computer Communications*, vol. 136, pp. 10–29, 2019.
11. H. Shafagh, L. Burkhalter, A. Hithnawi, and S. Duquennoy, "Towards blockchain-based auditable storage and sharing of iot data," in *Proceedings of the 2017 on Cloud Computing Security Workshop*, 2017, pp. 45–50.
12. F. Firouzi, B. Farahani, M. Barzegari, and M. Daneshmand, "Ai-driven data monetization: The other face of data in iot-based smart and connected health," *IEEE Internet of Things Journal*, 2020.
13. Proof-of-Work Explained, available at https://cointelegraph.com/explained/proof-of-work-explained
14. S. King and S. Nadal, "Ppcoin: Peer-to-peer crypto-currency with proof-of-stake," *self-published paper, August*, vol. 19, p. 1, 2012.
15. M. Castro, B. Liskov *et al.*, "Practical byzantine fault tolerance," in *OSDI*, vol. 99, no. 1999, 1999, pp. 173–186.
16. S. Nakamoto and A. Bitcoin, "A peer-to-peer electronic cash system," *Bitcoin.–URL:* https://bitcoin.org/bitcoin.pdf, 2008.
17. I. B. C. L. A. Mizrahi and M. Rosenfeld, "Proof of activity: Extending bitcoin's proof of work via proof of stake," 2014.
18. E. Bellini, Y. Iraqi, and E. Damiani, "Blockchain-based distributed trust and reputation management systems: a survey," *IEEE Access*, vol. 8, pp. 21 127–21 151, 2020.

19. G. Zyskind, O. Nathan *et al.*, "Decentralizing privacy: Using blockchain to protect personal data," in *2015 IEEE Security and Privacy Workshops*. IEEE, 2015, pp. 180–184.
20. S. S. Chawathe, "Clustering blockchain data," in *Clustering Methods for Big Data Analytics*. Springer, 2019, pp. 43–72.
21. X. Zha, K. Zheng, and D. Zhang, "Anti-pollution source location privacy preserving scheme in wireless sensor networks," in *2016 13th Annual IEEE International Conference on Sensing, Communication, and Networking (SECON)*. IEEE, 2016, pp. 1–8.
22. K. Zhang, X. Liang, R. Lu, and X. Shen, "Sybil attacks and their defenses in the internet of things," *IEEE Internet of Things Journal*, vol. 1, no. 5, pp. 372–383, 2014.
23. A. Mosenia and N. K. Jha, "A comprehensive study of security of internet-of-things," *IEEE Transactions on Emerging Topics in Computing*, vol. 5, no. 4, pp. 586–602, 2016.
24. E. Alsaadi and A. Tubaishat, "Internet of things: features, challenges, and vulnerabilities," *International Journal of Advanced Computer Science and Information Technology*, vol. 4, no. 1, pp. 1–13, 2015.
25. L. Axon and M. Goldsmith, "Pb-pki: A privacy-aware blockchain-based pki," 2016.
26. S. H. Hashemi, F. Faghri, P. Rausch, and R. H. Campbell, "World of empowered iot users," in *2016 IEEE First International Conference on Internet-of-Things Design and Implementation (IoTDI)*. IEEE, 2016, pp. 13–24.
27. J. H. Park and J. H. Park, "Blockchain security in cloud computing: Use cases, challenges, and solutions," *Symmetry*, vol. 9, no. 8, p. 164, 2017.
28. A. Ouaddah, A. Abou Elkalam, and A. Ait Ouahman, "Fairaccess: a new blockchain-based access control framework for the internet of things," *Security and Communication Networks*, vol. 9, no. 18, pp. 5943–5964, 2016.
29. O. Alphand, M. Amoretti, T. Claeys, S. Dall'Asta, A. Duda, G. Ferrari, F. Rousseau, B. Tourancheau, L. Veltri, and F. Zanichelli, "Iotchain: A blockchain security architecture for the internet of things," in *2018 IEEE wireless communications and networking conference (WCNC)*. IEEE, 2018, pp. 1–6.
30. A. Banafa, *Secure and Smart Internet of Things (IoT)*. River Publishers, 2018.
31. Y. Zhang and J. Wen, "The iot electric business model: Using blockchain technology for the internet of things," *Peer-to-Peer Networking and Applications*, vol. 10, no. 4, pp. 983–994, 2017.
32. A. Ouaddah, A. Abou Elkalam, and A. A. Ouahman, "Towards a novel privacy-preserving access control model based on blockchain technology in iot," in *Europe and MENA cooperation advances in information and communication technologies*. Springer, 2017, pp. 523–533.
33. G. Zyskind, O. Nathan, and A. Pentland, "Enigma: Decentralized computation platform with guaranteed privacy," *arXiv preprint* arXiv:1506.03471, 2015.
34. P. K. Sharma, M.-Y. Chen, and J. H. Park, "A software defined fog node based distributed blockchain cloud architecture for iot," *Ieee Access*, vol. 6, pp. 115–124, 2017.
35. T. Hardjono, N. Smith, and A. S. Pentland, "Anonymous identities for permissioned blockchains," 2014.
36. A. Kosba, A. Miller, E. Shi, Z. Wen, and C. Papamanthou, "Hawk: The blockchain model of cryptography and privacy-preserving smart contracts," in *2016 IEEE symposium on security and privacy (SP)*. IEEE, 2016, pp. 839–858.
37. S.-C. Cha, J.-F. Chen, C. Su, and K.-H. Yeh, "A blockchain connected gateway for ble-based devices in the internet of things," *IEEE Access*, vol. 6, pp. 24 639–24 649, 2018.
38. M. S. Ali, K. Dolui, and F. Antonelli, "Iot data privacy via blockchains and ipfs," in *Proceedings of the seventh international conference on the internet of things*, 2017, pp. 1–7.
39. W. Viriyasitavat, L. Da Xu, Z. Bi, D. Hoonsopon, and N. Charoenruk, "Managing qos of internet-of-things services using blockchain," *IEEE Transactions on Computational Social Systems*, vol. 6, no. 6, pp. 1357–1368, 2019.
40. Q. Zhu, S. W. Loke, R. Trujillo-Rasua, F. Jiang, and Y. Xiang, "Applications of distributed ledger technologies to the internet of things: A survey," *ACM Computing Surveys (CSUR)*, vol. 52, no. 6, pp. 1–34, 2019.

41. M. Naz, F. A. Al-zahrani, R. Khalid, N. Javaid, A. M. Qamar, M. K. Afzal, and M. Shafiq, "A Secure Data Sharing Platform Using Blockchain and Interplanetary File System," *Sustainability (Switzerland)*, vol. 11, no. 24, pp. 1–24, 2019.
42. N. B. Truong, K. Sun, G. M. Lee, and Y. Guo, "GDPR-Compliant Personal Data Management: A Blockchain-Based Solution," *IEEE Transactions on Information Forensics and Security*, vol. 15, no. OCTOBER, pp. 1746–1761, 2020.
43. Can blockchain accelerate Internet of Things (IoT) adoption?, available at https://www2.deloitte.com/ch/en/pages/innovation/articles/blockchain-accelerate-iot-adoption.html
44. B. Farahani, F. Firouzi, and K. Chakrabarty, "Healthcare iot," in *Intelligent Internet of Things*. Springer, 2020, pp. 515–545.
45. E. J. De Aguiar, B. S. Faiçal, B. Krishnamachari, and J. Ueyama, "A survey of blockchain-based strategies for healthcare," *ACM Computing Surveys (CSUR)*, vol. 53, no. 2, pp. 1–27, 2020.
46. B. Farahani, F. Firouzi, V. Chang, M. Badaroglu, N. Constant, and K. Mankodiya, "Towards fog-driven iot ehealth: promises and challenges of iot in medicine and healthcare," *Future Generation Computer Systems*, 2017.
47. F. Firouzi, A. M. Rahmani, K. Mankodiya, M. Badaroglu, G. V. Merrett, P. Wong, and B. Farahani, "Internet-of-things and big data for smarter healthcare: from device to architecture, applications and analytics," 2018.
48. B. Yan and G. Huang, "Supply chain information transmission based on rfid and internet of things," in *2009 ISECS International Colloquium on Computing, Communication, Control, and Management*, vol. 4. IEEE, 2009, pp. 166–169.
49. B. Houtan, A. S. Hafid, and D. Makrakis, "A survey on blockchain-based self-sovereign patient identity in healthcare," *IEEE Access*, vol. 8, pp. 90 478–90 494, 2020.
50. A. Kumar, D. Kumar Sharma, A. Nayyar, S. Singh, and B. Yoon, "Lightweight proof of game (lpog): A proof of work (pow)'s extended lightweight consensus algorithm for wearable kidneys," *Sensors*, vol. 20, no. 10, p. 2868, 2020.
51. A. M. Rahmani, T. N. Gia, B. Negash, A. Anzanpour, I. Azimi, M. Jiang, and P. Liljeberg, "Exploiting smart e-health gateways at the edge of healthcare internet-of-things: A fog computing approach," *Future Generation Computer Systems*, vol. 78, pp. 641–658, 2018.
52. M. A. Uddin, A. Stranieri, I. Gondal, and V. Balasubramanian, "Continuous patient monitoring with a patient centric agent: A block architecture," *IEEE Access*, vol. 6, pp. 32 700–32 726, 2018.
53. V. Patel, "A framework for secure and decentralized sharing of medical imaging data via blockchain consensus," *Health informatics journal*, vol. 25, no. 4, pp. 1398–1411, 2019.
54. Q. Xia, E. B. Sifah, K. O. Asamoah, J. Gao, X. Du, and M. Guizani, "Medshare: Trust-less medical data sharing among cloud service providers via blockchain," *IEEE Access*, vol. 5, pp. 14 757–14 767, 2017.
55. A. Dubovitskaya, Z. Xu, S. Ryu, M. Schumacher, and F. Wang, "Secure and trustable electronic medical records sharing using blockchain," in *AMIA annual symposium proceedings*, vol. 2017. American Medical Informatics Association, 2017, p. 650.
56. J. Anderson and S. Smith, "Securing standardizing and simplifying electronic health record audit logs through permissioned blockchain technology (technical report)," *Senior Honors Thesis, Dept. Comput. Sci., Dartmouth College, Hanover, NH, USA*, 2018.
57. S. Zhao, S. Li, and Y. Yao, "Blockchain enabled industrial internet of things technology," *IEEE Transactions on Computational Social Systems*, vol. 6, no. 6, pp. 1442–1453, 2019.

Adversarial Machine Learning: From Attack Surface to Defence Mechanism

Hossein Pourmehrani, Naghmeh Karimi, and Farshad Firouzi

"The only true wisdom is in knowing you know nothing."

– Socrates

1 Introduction

Machine learning has become a cornerstone of modern technology, transforming industries and driving innovation in fields such as healthcare, finance, autonomous systems, and security. Its capacity to learn from data and make predictions has enabled applications ranging from diagnosing diseases to detecting financial fraud. However, with this growing reliance on machine learning systems comes an increasing exposure to vulnerabilities. These systems, while powerful, are not inherently robust against malicious manipulation. Attacks on machine learning have highlighted the fragility of these systems, threatening their reliability and raising critical questions about their security [1].

In practice, machine learning systems are vulnerable to a variety of attacks targeting different components of their pipelines, reflecting the diverse and sophisticated nature of adversarial tactics. These include adversarial attacks, which manipulate

H. Pourmehrani (✉) · N. Karimi
University of Maryland Baltimore County, Baltimore, USA
e-mail: hosseip1@umbc.edu

N. Karimi
e-mail: nkarimi@umbc.edu

F. Firouzi
Arizona State University, Arizona, USA
e-mail: Farshad.Firouzi@asu.edu

F. Firouzi et al. (eds.), *Smart and Connected Health: AI, IoT, and Trustworthy Technologies*, https://doi.org/10.1007/978-3-032-06286-4_9

input data to deceive the model, such as evasion and poisoning [2], and data manipulation attacks that compromise the training process, such as label tampering [3]. Attacks on model inference, including membership and hyperparameter stealing [4], further highlight how adversaries can extract sensitive information from machine learning models. Model manipulation attacks, like bit flip and fault injection [5], exploit hardware vulnerabilities, while denial of service (DoS) attacks aim to disrupt system performance by overwhelming computational resources [6]. Privacy-focused threats, such as model inversion and training data reconstruction [7], alongside algorithmic bias exploitation and side-channel attacks leveraging power or timing analysis [8], further threaten system integrity. Moreover, emerging risks include supply chain attacks that target pre-trained models, environment exploitation through sensor spoofing [9], and federated learning-specific vulnerabilities, such as Byzantine attacks [10]. Reinforcement learning systems are not exempt, with attacks like policy poisoning and observation manipulation continuing to challenge their robustness [11]. Considering the importance of all these threats, there is an urgent need for robust defense mechanisms that can protect machine learning systems against such diverse and sophisticated adversarial tactics [12].

One of the most studied types of vulnerabilities is adversarial attacks, which exploit the weaknesses of machine learning models by crafting inputs specifically designed to manipulate the model's behavior, causing it to produce incorrect or unintended outputs. These attacks highlight a fundamental weakness in the robustness of machine learning systems, as even subtle perturbations to input data can lead to significant deviations in predictions. Such attacks can pose serious risks in critical applications such as autonomous vehicles, healthcare diagnostics, and cybersecurity, where reliability and accuracy are paramount. To improve the resilience of machine learning systems against such threats and ensure their safe and reliable deployment in real-world scenarios [13], a combination of techniques is implemented, including adversarial training, robust model architectures, and formal verification methods.

Data manipulation attacks also pose a significant threat to the integrity and reliability of machine learning models by directly altering input or training data. These attacks encompass various methods, such as label manipulation, where incorrect labels are assigned to data points in the training set to mislead the model during its learning process. Another common form is input tampering, which modifies input data to deceive the model during inference, often seen in adversarial examples that subtly change input features. Additionally, feature evasion involves altering specific features in the input data to bypass detection mechanisms, which is particularly concerning in security-sensitive applications such as intrusion detection systems. These attacks exploit the dependency of machine learning models on high-quality and representative data, highlighting the importance of robust data validation and pre-processing techniques to mitigate potential vulnerabilities [14].

Complementing these concerns, model inference attacks exploit the outputs or behaviors of machine learning models to extract sensitive information about the model or its training data. A notable example is membership inference attacks, where adversaries determine whether a specific data point was part of the model's training set, raising concerns for privacy-sensitive applications like healthcare and finance.

Similarly, model extraction attacks aim to replicate the functionality or architecture of a target model by systematically querying it, often resulting in intellectual property theft or unauthorized use. Additionally, hyperparameter stealing involves inferring the optimization settings of the model, such as learning rates or regularization parameters, which can compromise competitive advantages. These attacks highlight critical vulnerabilities in the deployment of machine learning models and necessitate the development of privacy-preserving techniques, such as differential privacy and query limitations, to mitigate risks [15].

Model manipulation attacks target the internal parameters or operations of machine learning models to degrade their performance, alter outputs, or disrupt functionality. Bit flip attacks are a prominent example of such an attack, where adversaries exploit vulnerabilities in hardware, such as memory faults, to flip specific bits in the model's weights, significantly impairing accuracy with minimal modifications. In general, fault injection attacks involve introducing deliberate errors into the computation process, often through physical means like electromagnetic interference or software-based techniques, to compromise model predictions. Another approach, weight erosion, gradually alters model parameters over time, leading to performance degradation while evading detection [16]. These attacks highlight the importance of secure hardware implementations, error detection mechanisms, and robust training techniques to safeguard machine learning systems against targeted manipulations [1].

Denial of service (DoS) attacks targeting machine learning systems aim to degrade performance or render the system unusable by overwhelming it with excessive demands or exploiting computational bottlenecks. In the context of ML, query flooding is a common technique where an attacker submits an excessive number of queries, exhausting computational resources and increasing latency for legitimate users. Similarly, resource exhaustion involves strategically targeting hardware or software limitations, such as memory or GPU utilization, to disrupt normal operation. These attacks are particularly concerning in real-time or critical applications, such as autonomous systems and online services, where downtime can have severe consequences. Countermeasures include rate limiting, query authentication, and resource monitoring to ensure resilience against such disruptions [17]. Another category of attacks that threaten machine learning models involves privacy attacks which exploit model behavior to infer sensitive information about the training data or input data. This attack may pose significant concerns in privacy-critical applications. One such attack is the model inversion attack that reconstructs input data, such as images or attributes, by leveraging the model's outputs, potentially exposing private or sensitive information. Membership inference attacks allow adversaries to determine whether a specific data sample was part of the model's training set, risking the confidentiality of datasets used in training. Furthermore, attribute inference attacks infer sensitive or hidden attributes of input data, even if the attributes were not explicitly part of the model's predictions. These attacks are especially problematic for models trained on confidential datasets, such as medical records or financial data. In practice, techniques like differential privacy, model obfuscation, and secure multi-party computation are crucial to mitigate these vulnerabilities and ensure privacy preservation in ML systems [18].

Adding to the spectrum of threats, side-channel attacks exploit indirect information, such as power consumption, timing, or electromagnetic emissions, to infer sensitive details about machine learning models or their inputs. Power analysis attacks monitor the energy usage of hardware running a model to extract information about its operations, such as weights or architecture. Similarly, timing attacks leverage variations in execution time, which can reveal patterns in input data or the computational graph. Electromagnetic (EM) analysis captures emissions from hardware devices during model inference to deduce internal parameters or input features. These attacks bypass traditional security measures by targeting physical or operational characteristics rather than the algorithm itself, making them highly effective in scenarios with physical access to the device. To mitigate these risks, techniques such as noise injection, constant-time operations, and hardware shielding are deployed [19].

Environment exploitation attacks target the physical or operational context in which a machine learning system operates, manipulating environmental factors to degrade model performance or alter outputs. One common method is sensor spoofing, where adversaries manipulate sensor inputs, such as tampering with GPS signals in autonomous vehicles or altering camera feeds in security systems, to deceive the model. Another form is data distribution shifts, where the adversary introduces changes to the input data statistical properties, rendering the model ineffective due to its reliance on prior distributions. These attacks are particularly concerning in real-world applications like robotics, surveillance, and critical infrastructure, where environmental manipulation can lead to catastrophic failures. Robust detection mechanisms, domain adaptation techniques, and adversarial training are essential countermeasures to mitigate vulnerabilities associated with environment exploitation [20].

In federated learning (FL), unique vulnerabilities arise due to its decentralized nature, making it susceptible to a variety of specific attacks. Byzantine attacks occur when malicious participants inject erroneous or harmful updates into the global model, degrading its overall performance. Similarly, gradient inversion attacks exploit shared gradient information to reconstruct private input data from local participants, violating privacy guarantees. Another concern is free-riding attacks, where participants benefit from the global model without contributing meaningful updates, reducing the effectiveness of collaboration. These attacks highlight the challenges in balancing model performance, security, and privacy in FL. In practice, robust aggregation techniques, such as Byzantine-resilient algorithms, along with encryption and differential privacy, are essential to mitigate these vulnerabilities and ensure the secure deployment of federated learning systems [21].

Note that reinforcement learning (RL) systems, which rely on interactions with an environment to optimize decision-making policies, are vulnerable to several unique types of attacks. Policy poisoning is a prominent threat where adversaries tamper with the reward signals or introduce adversarial perturbations into the environment, leading the agent to learn suboptimal or harmful policies. Observation manipulation involves altering the agent's perception of the environment, tricking it into making incorrect decisions. Similarly, exploration manipulation misguides the agent during the exploration phase, preventing it from discovering optimal strategies. These vulnerabilities are particularly concerning in high-stakes applications, such as autonomous vehicles and financial systems. Countermeasures like robust policy training, adversarial training, and environment validation are essential to enhance the resilience of RL systems against such attacks [22].

Among the attacks discussed earlier, adversarial attacks are particularly noteworthy due to their widespread implications. These attacks subtly perturb inputs in ways imperceptible to humans but devastating to machine learning models. The phenomenon exposes critical flaws in the decision boundaries of neural networks, making systems susceptible to misclassifications. As the content of this chapter elaborates, adversarial machine learning highlights how even state-of-the-art models falter under seemingly benign adversarial conditions. The subtle yet effective nature of adversarial perturbations underscores the need for robust defense mechanisms.

Adversarial examples challenge the stability of deep neural networks, demonstrating that slight, intentional modifications to input data can lead to significant errors. This not only jeopardizes tasks such as image classification and speech recognition but also raises concerns for applications in autonomous driving, where the misclassification of a stop sign could have dire consequences. As examined in this chapter, adversarial attacks extend beyond digital domains, posing threats to physical-world implementations. These attacks exploit inherent vulnerabilities in neural network architectures, highlighting the critical need for robust and secure defenses. Accordingly, the goal of this chapter is to provide a comprehensive exploration of adversarial machine learning attacks, their mechanisms, and their impact on system reliability. By detailing the adversarial landscape, with a specific focus on adversarial machine learning, this chapter aims to bridge theoretical understanding and practical defense strategies. The content highlights the importance of developing secure and robust machine learning models, equipping researchers and practitioners with the tools to anticipate, counter, and mitigate emerging threats in the evolving landscape of artificial intelligence.

2 Supervised Learning

Supervised learning is a subfield of artificial intelligence where a model learns from labeled data to make predictions or decisions. It is one of the most widely used machine learning paradigms due to its ability to map inputs to desired outputs effectively. For instance, to classify different kinds of animals, we can use a dataset containing a sufficient number of paired samples: images of animals and their corresponding labels indicating the animal type. Then using this labeled dataset, we can train a model, as illustrated in Fig. 1, to learn the relationship between the features in the images and their labels. Once the model is trained, during inference, if we input an image of a cat, the model can analyze the features and predict the correct class (e.g., "cat") with high confidence.

The dataset in supervised learning is typically structured as $D = \{(x_i, y_i)\}_{i=1}^{N}$, where $x_i \in \mathbb{R}^d$ denotes the input feature vector of the i-th sample, comprising d features. features. For instance, in image classification, x_i may represent the pixel

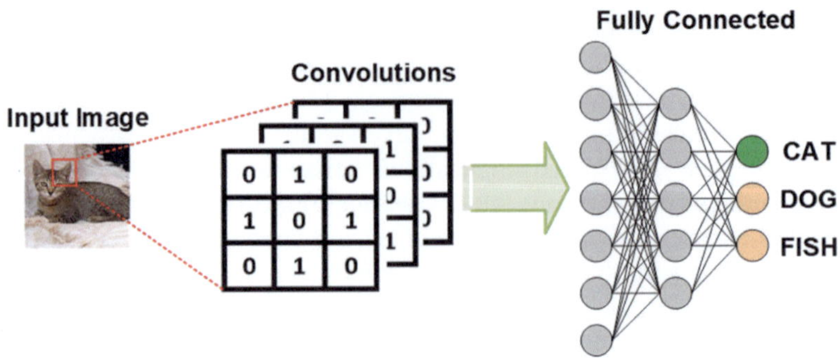

Fig. 1 The architecture of a convolutional neural network for image classification in supervised learning

values of an image. On the other hand, y_i is the corresponding label for x_i which can be a scalar (in regression tasks) or a categorical variable (in classification tasks). Here, N indicates the total number of samples in the dataset.

The model's goal in supervised learning is to find a mapping function $f(x; \theta)$, parameterized by θ, that predicts the label y for given input x, as shown in Fig. 1. This is expressed as:

$$\hat{y} = f(x; \theta) \tag{1}$$

In Eq. 1, $f()$ represents the model function (e.g., a linear regression model, neural network, or convolutional network), $\hat{y}$ denotes the predicted output, and θ refers to the model parameters, including weights and biases, which are to be trained.

A loss function $L(y, \hat{y})$ is defined to quantify the difference between the true labels y and the predicted labels $\hat{y}$. Minimizing this loss ensures that the model's prediction becomes increasingly accurate. The Mean Squared Error (MSE) is a commonly used loss function in machine learning models, defined by Eq. 2:

$$L(y, \hat{y}) = \frac{1}{N} \sum_{i=1}^{N} (y_i - \hat{y}_i)^2 \tag{2}$$

Here, y_i represents the true label, and $\hat{y}_i$ is the predicted value. The MSE penalizes larger errors more heavily, encouraging the model to make precise predictions. In classification tasks, however, the Cross-Entropy Loss is commonly used, as shown in Eq. 3

$$L(y, \hat{y}) = -\frac{1}{N} \sum_{i=1}^{N} \sum_{k=1}^{C} y_{i,k} \log(\hat{y}_{i,k}) \tag{3}$$

Where C is the number of classes, $y_{i,k}$ is a one-hot encoded label (1 for true class, 0 otherwise), and $\hat{y}_{i,k}$ is the predicted probability for class k. This loss measures the divergence between the true labels and the predicted probabilities, pushing the model to assign higher probabilities to the correct classes.

The objective of supervised learning is to minimize the loss function over the training data based on the following optimization function, where θ^* denotes the optimal model parameters and $L()$ is the loss function.

$$\theta^* = \arg\min_{\theta} L(y, \hat{y}) \tag{4}$$

Minimizing the loss function involves adjusting the model's parameters θ to reduce the error. This is achieved using optimization algorithms such as Gradient Descent. This parameter update rule is defined in Eq. 5, where η denotes the learning rate to control the step size for updates and $\nabla_\theta L(\theta)$ is the gradient of the loss function with respect to θ, indicating the direction of steepest increase.

$$\theta_{t+1} = \theta_t - \eta \nabla_\theta L(\theta) \tag{5}$$

Through iterative updates, the model learns to approximate the mapping function $f(x; \theta)$ that minimizes the loss. Once trained, the model is evaluated on a separate dataset called the test set, which contains samples unseen during training. The model's performance is assessed using metrics appropriate to the task, such as accuracy for classification. This evaluation ensures that the model generalizes well to new, unseen data.

In essence, supervised learning is a systematic approach that leverages labeled data, a well-defined loss function, and optimization techniques to build models capable of making accurate predictions. This concept is central to many practical applications, including image classification, speech recognition, and etc.

3 Adversarial Example

Adversarial examples are a phenomenon in machine learning where inputs are deliberately perturbed in ways that are imperceptible to humans but can cause machine learning models, particularly deep neural networks, to produce incorrect outputs. These inputs exploit weaknesses in the model's decision boundaries, exposing critical vulnerabilities in systems that rely on machine learning for tasks such as image recognition, natural language processing, and autonomous driving.

3.1 Definition and Formulation

Deep neural networks are highlighted for their ability to learn highly expressive and complex computations through supervised learning. This capability allows them to generalize well across a wide range of tasks, such as object detection and speech understanding, making them state-of-the-art tools in artificial intelligence [23]. Despite their high performance, neural networks have unexpected and counter-intuitive behaviors. One property is **Semantic Meaning of Individual Units**. Previous studies [23] assumed that the activations of specific units in the higher layers of neural networks carried distinct semantic information and the input patterns can maximize individual unit activations. However, authors in [23] show that this assumption is not completely correct, and semantic information is not localized to individual units. Rather, it is distributed across the entire activation space. This aligns with findings in word representations, where semantic relationships are encoded in the overall vector space rather than individual dimensions.

Another property is **Stability Against Small Perturbations**. In fact, neural networks are expected to be robust to minor changes in their input, especially in tasks like image recognition, where small perturbations do not alter the inherent class of an image. However, researchers [23] found that imperceptible, non-random perturbations can lead to significant errors in prediction. These perturbations, termed **adversarial examples**, can consistently mislead the model. This finding reveals a surprising instability in neural networks, contradicting the assumption that they generalize smoothly in the vicinity of training samples.

Deep neural networks are renowned for their ability to learn highly expressive and complex computations through supervised learning. This enables them to achieve state-of-the-art performance across a wide range of tasks, such as object detection and speech recognition, making them essential tools in artificial intelligence [23].

As previously discussed, DNNs are susceptible to input perturbations, which can lead them to misclassify input samples with high confidence. These perturbed inputs, referred to as **adversarial examples**, are formally defined in Eq. 6. In this equation, X is the original input, and δ represents the noise added to the input, resulting in X', the adversarial example:

$$X' = X + \delta \tag{6}$$

Adversarial examples have several key characteristics. First, the model's prediction for the adversarial example differs from that of the original image. If we define $f(x; \theta)$ as a model with parameters θ, then for X and X', $f(X; \theta) \neq f(X'; \theta)$. However, humans cannot perceive the difference between the original image and the adversarial example. The goal of the adversary is to estimate the noise δ to generate adversarial examples, ultimately leading to DNN misclassification while remaining indistinguishable from humans. The calculation of this perturbation is framed as an optimization problem, which is described in Eq. 7.

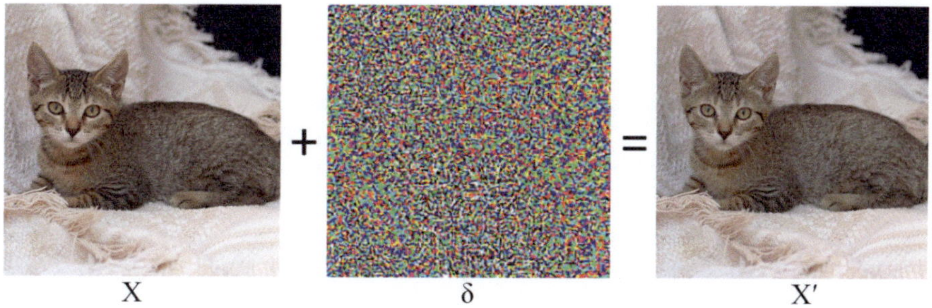

Fig. 2 The adversarial example is created by adding small perturbations to the original input image

$$\max_{\delta} L(f(X + \delta; \theta), Y) \tag{7}$$

In Eq. 7, $f()$ represents the model, X is the original input, δ is the perturbation to be determined, and L is the loss function. The goal is to increase the loss value to cause misclassification. There is also a constraint on δ, which will be discussed in the next section.

Figure 2 illustrates a simple perturbation method applied to the original image to create adversarial examples. The figure demonstrates that the adversarial example is visually almost identical to the original image, making the differences nearly impossible for humans to detect, yet leading to incorrect model predictions. Interestingly, the original input is an image of a cat, which the model classifies with a confidence of 97.99%. However, the adversarial example is misclassified as a tiger with 100% confidence.

The attacker's goal is to find δ to create an adversarial example by adding it to the input X. In practice, the attacker aims to determine the perturbation value that increases the loss, causing the model to misclassify a specific input. For this optimization problem, it is necessary to define a constraint.

To quantify the difference between the original input (X) and the adversarial example (X'), a suitable metric must be defined. The standard approach is to use the p-norm function, where p can take on values such as 0, 2, or ∞. Based on this definition, the difference between X and X' can be expressed by Eq. 8.

$$\|X - X'\|_p < \epsilon \tag{8}$$

In Eq. 8, p refers to the value used in the p-norm function, and ϵ represents the amount of noise that should be added to the image. This value is typically small and must be calculated appropriately, depending on the dataset being used. The value of ϵ often decreases as the complexity of the task increases [13].

In the context of adversarial attacks, various p-norm functions are commonly used. In the following sections, we will explain each of these norm functions and the corresponding attacks based on them.

3.2 L_0-Norm Attack

The L_0 norm is a mathematical concept often used in adversarial attacks to measure the sparsity of perturbations added to an input. The L_0 norm counts the number of non-zero elements in a vector. For a vector $x \in \mathbb{R}^n$, the L_0 norm is mathematically expressed in Eq. 9.

$$\|x\|_0 = (|x_1|^0 + |x_2|^0 + |x_3|^0 + ... + |x_n|^0) \tag{9}$$

Here x_1 to x_n are elements of the vector x and this equation counts the non-zero elements in the vector. For example, if $x = [1, 0, 0, 2]$ then $\|x\|_0 = 2$. This definition is standard and well-known in the mathematical literature, often discussed in the context of sparsity in vectors and models [24].

In an adversarial attack, the L_0 norm measures the sparsity of perturbations added to an input sample, such as an image. It helps quantify the number of features (e.g., pixels) altered to make a model misclassify the input [23]. In L_0-based attacks, the objective of an L_0 norm adversarial attack is to find the minimal number of features to change while ensuring that the perturbed input misleads the model. Formally, this can be written as Eq. 10 where δ denotes the perturbation, x is the original input, and ϵ indicates the constraint on the perturbation value.

$$\underset{\delta *}{\text{maximize}}\ f(x + \delta), \quad \text{subject to } \|\delta\|_0 \leq \epsilon \tag{10}$$

This optimization ensures that the attack is as sparse as possible, i.e., the perturbation impacts the fewest features [23].

In image classification, the L_0 norm represents the number of pixels altered to deceive the model. For example, starting with an image of a cat, a perturbed version could be generated by changing a few pixels, causing the model to misclassify it as a dog.

The L_0 norm is also used in NLP, where such attacks involve changing as few words as possible in a sentence to alter its classification. For instance, given the original sentence "The movie was fantastic" a perturbed version might be "The film was fantastic". This minor modification (replacing "movie" with "film") results in an L_0 norm of one and could potentially mislead sentiment analysis models.

L_0 norm is challenging because it is a combinatorial problem, making gradient-based optimization methods inapplicable. This is due to the non-differentiable nature of the L_0 norm. Researchers often resort to approximate or heuristic methods to identify the minimal set of features to perturb [24].

One notable example is the One Pixel Attack [25], which can fool deep neural networks by modifying just a single pixel in the image. This method leverages Eq.10 to determine two values: the dimensions that need to be perturbed and the corresponding modification strength for each dimension. In the case of a one-pixel attack, ϵ is set to one. Figure 3 illustrates a one-pixel attack scenario, which is a subset of the L_0 norm attack. These figures show examples of one-pixel attacks on

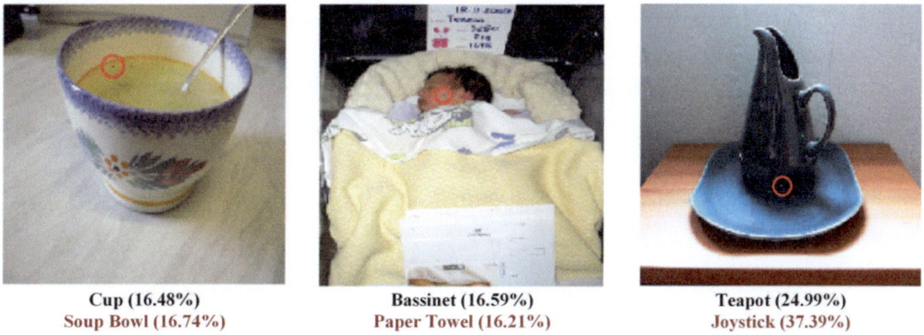

Fig. 3 The adversarial example is created by the One Pixel Attack scenario. Reprint form [25]

the ImageNet dataset, where the modified pixels are highlighted with red circles. The original class labels are in black, while the target class labels and their corresponding confidence scores are provided below. This experiment demonstrates the vulnerability of machine learning models to even small perturbations in the input.

3.3 L_2-Norm Attack

The L_2 norm also called the Euclidean norm, is a measure of the length or magnitude of a vector in space. For a vector $V = [v_1, v_2, ..., v_n]$, the L_2 norm is defined as Eq. 11

$$\|v\|_2 = \sqrt{\sum_{i=1}^{n} v_i^2} \tag{11}$$

In the context of the adversarial example, the L_2 norm measures the magnitude of the perturbation δ applied to the original input x. A smaller L_2 norm indicates that the perturbation is less perceptible to humans, which is a key characteristic of adversarial examples.

The L_2 norm is widely used in adversarial attacks for several reasons. First, it offers a natural interpretation since the L_2 norm corresponds to the Euclidean distance, a standard measure of similarity in many machine learning applications. This aligns well with human intuition regarding the closeness of two inputs in continuous space [24]. Second, it is optimization-friendly. The squared L_2 norm, $|\delta|_2^2$, is differentiable and smooth, making it ideal for gradient-based optimization techniques commonly used in adversarial attacks [13]. Finally, it has perceptual relevance, as small L_2 norm perturbations generally preserve the semantic content of the input. This ensures that adversarial examples are often imperceptible to humans [24, 26].

The L_2 norm attack has been widely used in the literature. For example, the Carlini & Wnager attack [24] is a powerful adversarial attack that minimizes the L_2 norm of the perturbation while ensuring that the perturbed input is misclassified. The problem is formulated in Eq. 12.

$$\underset{\delta}{\text{Minimize}} \quad \|\delta\|_2 + c \cdot g(x+\delta) \qquad \text{subject to } x+\delta \in [0,1]^n \tag{12}$$

Here, δ denotes the L_2 norm which measures the size of perturbation, g indicates an objective function that quantifies how much $x + \delta$ resembles the target class y, and c is a regularization constant balancing the perturbation size and the attack success.

The method leverages gradients to iteratively refine δ, ensuring efficient optimization. Additionally, the attacker can adjust parameters to specify the desired confidence level of the misclassification, enabling the generation of high-confidence adversarial examples. More details about this method as an attack scenario will be explained in Sect. 4.

Using the L_2 norm in adversarial attacks offers several advantages. Adversarial examples crafted using L_2 norm attacks are often visually or semantically indistinguishable from the original input, which is crucial in real-world attack scenarios [13, 24]. Additionally, these examples tend to transfer effectively across different models, particularly when the models share architectural similarities [24]. Lastly, the L_2 norm is continuous and differentiable, facilitating efficient optimization through gradient-based methods [13].

Note that the L_2 norm also has some limitations, including higher computational cost, reduced robustness against certain defense mechanisms, and limited applicability in specific domains. Minimizing the L_2 norm involves solving complex optimization problems, which can be computationally intensive [24, 26]. Some adversarial defenses explicitly target L_2 norm perturbations, making these attacks less effective against robustly trained models [27]. Finally, in domains such as text and audio, perceptual similarity does not always align well with L_2 norm minimization and alternative norms like L_∞ may be more suitable for such cases.

3.4 L_∞-Norm Attack

The last norm-based attack method is L_∞ norm attack which focuses on crafting perturbations that are imperceptible to humans by constraining the maximum change applied to any individual feature of the input. The L_∞ norm, also known as the maximum norm or Chebychev norm, measures the largest absolute difference between corresponding elements of two vectors. For a vector $X = [x_1, x_2, ..., x_n]$, the L_∞ norm is defined as Eq. 13.

$$\|X\|_\infty = \max_i |x_i| \tag{13}$$

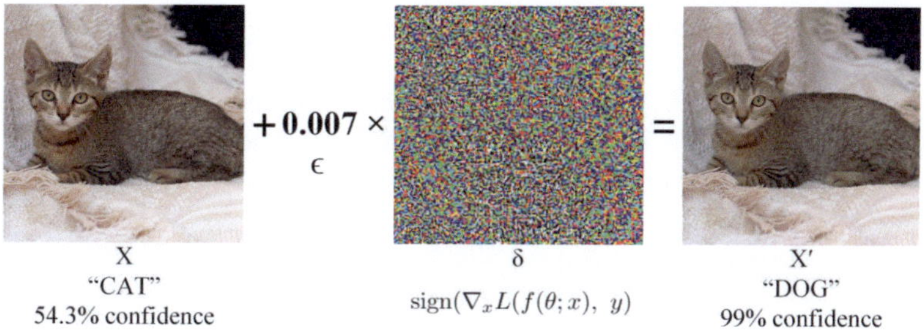

Fig. 4 A simple adversarial example generated by FGSM attack method

In the context of adversarial attacks, constraining the L_∞ norm ensures that no single feature of the input is altered beyond a specified threshold, maintaining the overall appearance of the input to human observers.

One of the oldest and most straightforward L_∞ norm attacks is **Fast Gradient Sign Method (FGSM)**, introduced in [13]. FGSM generates adversarial examples by adding a small perturbation to the input data in the direction that maximizes the model's loss. The value of the perturbation will be calculated based on Eq. 14 where $\nabla_x L$ denotes the gradient of the loss function L with respect to the input x, $Sign(.)$ indicates the element-wise sign function, and ϵ is a small scalar controlling the perturbation magnitude based on L_∞ norm.

$$\delta_{FGSM} = \underset{\|\delta\|_\infty < \epsilon}{\text{Maximize}} \ (\delta, \ \text{sign}(\nabla_x L(f(\theta; x), \ y)) = \epsilon.\text{sign}(\nabla_x L(f(\theta; x), \ y)) \quad (14)$$

This method leverages the model's gradients to determine the optimal direction for modifying each input feature, aiming to increase the loss and induce misclassification. Specifically, Eq. 14 seeks to find a small perturbation δ that, when added to the input x, generates an adversarial example x' that causes the model to misclassify with high confidence.

Figure 4 shows an example where a cat image is correctly classified as a cat with 54.3% confidence. By adding a small perturbation based on the FGSM method and using a small ϵ value, we can generate a visually similar image, as shown on the right side of Fig. 4. This modified image misleads the model into classifying it as a dog with high confidence, highlighting the model's vulnerability to this type of attack.

The L_∞ norm attack has both advantages and disadvantages. One key advantage is perceptual imperceptibility, which ensures that by constraining the maximum change to any single feature, L_∞ norm attacks produce perturbations that are often subtle and undetectable to humans. This property makes L_∞ norm attacks particularly effective for generating adversarial examples that appear identical to the original input [13]. However, this constraint does not consider the cumulative effect of perturbations across all features, which can limit the effectiveness of L_∞ norm attacks, especially in high-dimensional spaces or for models specifically designed to handle such perturbations [27].

Another advantage of L_∞ norm attacks is computational efficiency. In practice, L_∞ norm attacks are well suited for rapidly generating adversarial examples with minimal computational overhead, making them feasible for large-scale evaluations and adversarial training [28].

Lastly, L_∞ norm attacks play a significant role in enhancing robustness through adversarial training. Exposing models to adversarial examples during training can improve their robustness [27]. Defenses specifically focused on L_∞ norm attacks, such as adversarial training, reduce the effectiveness of these attacks as they become integrated into the defense mechanisms [13].

3.5 ϵ *Constraint*

In adversarial machine learning, **Epsilon** (ϵ) is a parameter that bounds the magnitude of perturbations applied to input data. It ensures that the crafted adversarial example remains visually or semantically similar to the original input, maintaining imperceptibility to humans.

The ϵ constraint limits the size of perturbations by specifying the allowable norm of the perturbation δ. The form of the constraint varies depending on the chosen norm (e.g., L_∞, L_2). For instance, in the L_∞ norm, $|\delta|_\infty < \epsilon$, where ϵ bounds the maximum change to any individual feature of the input x, ensuring that no single feature is altered by more than ϵ. In the L_2 norm, $|\delta|_2 < \epsilon$, where ϵ restricts the overall magnitude of the perturbation (Euclidean distance) across all features.

This constraint ensures that adversarial examples remain close to the original inputs in feature space, making them indistinguishable to human observers. For instance, small changes in pixel values of an image should not produce visible distortions [13]. Furthermore, by limiting the magnitude of perturbations, the ϵ constraint prevents adversarial attacks from introducing large, unrealistic modifications that could exploit specific patterns in the model [27]. Lastly, ϵ acts as a metric for evaluating a model's robustness under controlled levels of adversarial perturbation. It enables researchers to analyze the sensitivity of models to small, bounded perturbations [24].

The parameter ϵ represents a small value that defines the magnitude of permissible perturbations and varies based on the complexity of the dataset and the difficulty of the task. For simpler datasets, a relatively larger ϵ can be used without compromising the imperceptibility of adversarial examples. However, as the task becomes more challenging or the dataset increases in complexity, ϵ is typically reduced to maintain the subtlety of the perturbations.

In practice, prior research and experimentation inform the choice of ϵ. For instance, in the case of the MNIST dataset, which involves grayscale images of handwritten digits, ϵ is often set to 0.1. For more complex datasets like CIFAR-10 and CIFAR-100, which contain color images with greater diversity, ϵ is generally set to $\frac{8}{255}$. For the highly challenging ImageNet dataset, which features a vast number of classes and higher-resolution images, a smaller value of $\epsilon = \frac{4}{255}$ is commonly used to ensure the perturbations remain imperceptible [29].

The ϵ value strikes a balance between the effectiveness of adversarial attacks and the perceptual similarity of the adversarial examples to the original inputs. Researchers often adjust ϵ based on specific tasks, model robustness, and real-world constraints, highlighting its critical role in adversarial machine learning experiments.

3.6 Adversary Goals and Capacity

In general, an adversary may pursue one of the three goals discussed below, each reflecting a unique strategy to undermine the model's performance.

One goal can be **Confidence Reduction** which aims to reduce the confidence level of the model's predictions. For example, if a neural network predicts an image as a "cat" with 95% confidence, an adversarial attack might lower this confidence to 55%. Although the prediction remains "cat", the reduced confidence level makes the output less reliable. Such attacks are significant in applications where high-confidence predictions are critical, such as autonomous vehicles or medical diagnostics [29].

The adversary may also pursue **Untargeted Attacks** which are designed to mislead the model into predicting any incorrect class. The specific incorrect label is irrelevant as long as it differs from the original. For instance, an image originally classified as a "cat" might be misclassified as a "dog", "car", or any other label. These attacks are simpler to launch compared to targeted attacks and are commonly used to disrupt the reliability of a model's prediction.

An adversary may also intent **Targeted Attacks** which manipulates the model into predicting a specific incorrect class. For example, an attacker might modify an image of a "cat" so that it is classified as a "dog" irrespective of all other incorrect labels. These attacks are particularly dangerous in scenarios where specific misclassification can lead to significant consequences, such as biometric systems incorrectly identifying an unauthorized individual as an authorized user [29].

In sum, the aforementioned adversary goals showcase various ways in which attackers can exploit the vulnerabilities of machine learning models, either by reducing the reliability of predictions or by steering them toward specific incorrect outcomes.

The success of an adversarial attack also depends heavily on the adversary's capacity, which reflects their level of access and knowledge about the target model. This capacity can range from complete knowledge of the model to minimal information, leading to two primary classifications including **White-box** and **Black-box** as shown in Fig. 5.

In a white-box attack, the adversary has full access to the model, including its architecture, weights, biases, and training data. This complete knowledge allows the attacker to craft highly effective perturbations tailored specifically to the model. For example, if an attacker knows the exact layers and parameters of a neural network

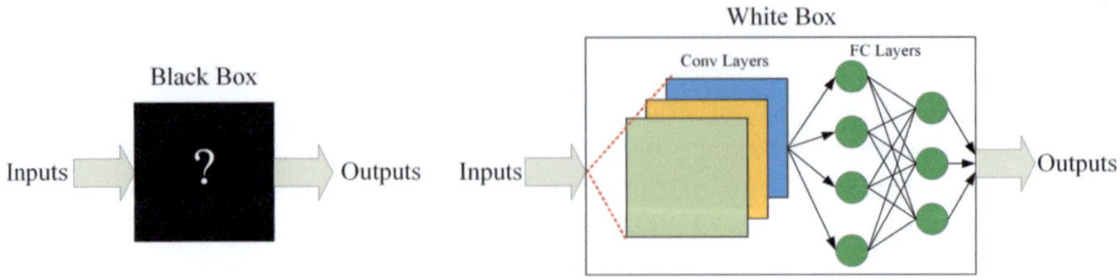

Fig. 5 The level of access in Black-box versus White-box attack

used in a self-driving car, they can design perturbations that misclassify stop signs as speed limit signs. While white-box attacks are powerful, their practical feasibility is limited to situations where attackers can obtain such extensive information.

Black-box attacks occur when the adversary has limited or no knowledge of the model's internals. Instead, they rely on querying the model (e.g., submitting inputs and observing outputs) or learning from pairs of input and corresponding output samples. For instance, an attacker might try to fool an online image recognition service by submitting modified images without knowing the underlying model architecture or parameters. These attacks are more challenging to execute but are often more realistic in practice, where models are typically deployed as closed systems.

To further refine these classifications, intermediate capacities can be considered, as defined in [30]. These include scenarios where the adversary has access to the training data only, the model architecture only, or an oracle that provides output responses to given inputs. Such nuanced distinctions allow for a more detailed understanding of the adversary's capabilities and constraints.

The interplay between adversary goals and capacity underscores the complexity of adversarial attacks. While the goals define the desired outcomes of an attack, the capacity dictates the feasibility and effectiveness of achieving those outcomes. By understanding these dimensions, researchers and practitioners can better anticipate and defend against potential adversarial threats.

4 Adversarial Attacks

Adversarial attacks in machine learning exploit model vulnerabilities by introducing small perturbations to inputs, causing misclassifications or other erroneous behaviors. Among the various classifications of adversarial attacks in the literature, one of the most effective categorizations divides them into **White-box**, **Black-box**, **Universal Adversarial Perturbations (UAPs)**, and **Auto Attacks**.

- **White-box** attacks assume the adversary has complete access to the model's architecture, parameters, and gradients. This allows precise crafting of perturbations using gradient-based methods.

- **Black-box** attacks operate under the assumption that the adversary has no direct access to the model's internals. Instead, adversaries query the model and analyze its outputs to infer its behavior.
- **Universal Adversarial Perturbations (UAPs)** focus on generating a single general perturbation that can be applied to multiple inputs. These attacks demonstrate that small, input-agnostic changes can consistently deceive models across diverse datasets.
- **Auto Attacks**, an advanced and automated category, combine several robust adversarial techniques to effectively evaluate and challenge model defenses.

These categories underscore the critical need to understand and address adversarial vulnerabilities in machine learning systems. In the following sections, we will introduce promising methods within each category to elucidate the concept of adversarial attacks further.

4.1 White-Box Attacks

Deep neural networks are known for their impressive ability to generalize and classify data. However, they are susceptible to a subtle vulnerability called **adversarial examples**, which are inputs modified by imperceptibly small perturbations to deceive the model. Adversarial examples were first proposed in [23], which used the box-constrained Limited-memory Broyden–Fletcher–Goldfarb–Shanno **(L-BFGS)** algorithm to demonstrate how such adversarial examples can be crafted effectively.

To create an adversarial example, we aim to minimally perturb an input x such that the model misclassifies the perturbed version $x + \delta$. Formally, this involves solving the constrained optimization problem mentioned in Eq. 15 which measures the magnitude of the perturbation. In Eq. 15, $\|\delta\|_2$ is the L_2-norm of the perturbation δ. Smaller δ implies less noticeable changes to the input. The function f denotes the classifier, mapping input x to a predicted label, and y is the target label we want the classifier to output after perturbation, intentionally different from the correct label of x. Here $[0, 1]^m$ ensures that the perturbed image remains valid (e.g., pixel intensities are within bounds).

$$\begin{aligned} \text{Minimize} \quad & \|\delta\|_2 \\ \text{subject to} \quad & f(x+\delta) = y, \\ & x + \delta \in [0, 1]^m. \end{aligned} \tag{15}$$

This optimization problem is non-trivial due to the non-linear and non-convex nature of neural networks. Directly solving the above problem is computationally expensive. Instead, we can reformulate Eq. 15 using a penalty-based approach, combining the original objective ($\|\delta\|_2$) with the loss function of the classifier as Eq. 16, where $L_f(x + \delta, y)$ measures the classification error of $x + \delta$ for the target label y. For example, this could be the cross-entropy loss commonly used in neural networks.

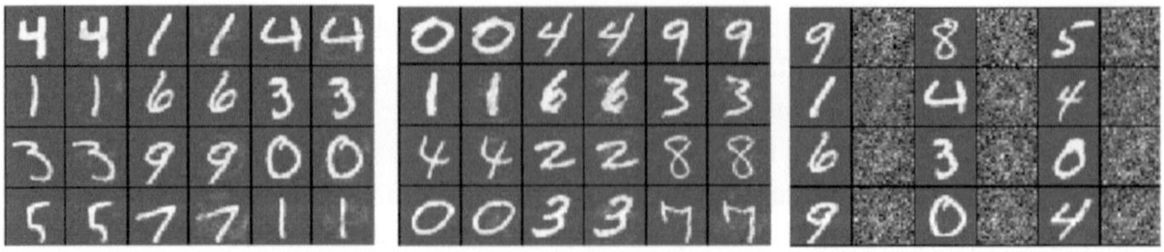

Even columns: adversarial examples for a linear classifier — Even columns: adversarial examples for a 200-200-10 sigmoid netwrok — Randomly distorted samples by Gaussian noise.

Fig. 6 Adversarial examples for a randomly chosen subset of MNIST compared with random distorted examples. Reprint from [23]

The constant value $c > 0$ is a regularization parameter that controls the trade-off between minimizing perturbation magnitude ($\|\delta\|_2$) and ensuring misclassification ($L_f(x+\delta, y)$). By gradually increasing c, this method ensures a balance between finding small perturbations and achieving the desired misclassification [23].

$$\begin{aligned} \text{Minimize} \quad & c\|\delta\|_2 + L_f(x+\delta, y) \\ \text{subject to} \quad & x+\delta \in [0,1]^m. \end{aligned} \tag{16}$$

The box-constrained L-BFGS algorithm is employed to solve Eq. 16. L-BFGS is a quasi-Newton method that approximates the second derivative (Hessian) to efficiently minimize the objective function, even in high dimensional spaces. This algorithm computes the gradient of the objective function with respect to the perturbation δ. This requires backpropagation through the neural network. Then, it iteratively adjusts the step size to minimize the objective function while ensuring the perturbation remains within bounds. After each iteration, the perturbation input $x + \delta$ is projected back into the valid range $[0, 1]^m$ to maintain image fidelity. The process terminates when the algorithm converges, producing a perturbation δ that effectively fools the classifier [23].

The L-BFGS-based method has demonstrated several intriguing results across datasets and models. The generated adversarial examples, based on this method, are visually indistinguishable from the original inputs. For example, as shown in Fig. 6, the odd columns relate to original images, and the even columns correspond to distorted counterparts. The adversarial examples generated for the specific model have an accuracy of 0% but the randomly distorted examples are hardly readable, yet classified correctly in half of the cases. Additionally, perturbations crafted by this method for one model often remain effective for other models, even with different architectures or hyperparameters. This suggests that adversarial examples exploit shared vulnerabilities in how models learn [23].

Although L-BFGS [23] can generate adversarial examples that are imperceptible to humans, it is computationally expensive. This limitation led to the development of a more efficient method known as the **Fast Gradient Sign Method (FGSM)** [31]. FGSM is a one-step approach for crafting adversarial examples by leveraging the model's cost function, its gradient, and a perturbation magnitude parameter, epsilon (ϵ). This method significantly reduces computational overhead while effectively generating adversarial perturbations [31].

The FGSM provides an efficient approach to crafting adversarial examples by exploiting the linear behavior of neural networks in high-dimensional spaces. To understand why adversarial examples exist, we must first grasp the concept of linearity in neural networks. Although neural networks are composed of layers with nonlinear activation functions, the operations in each layer (like matrix multiplications and additions) are inherently linear. This linearity plays a significant role, especially in high-dimensional spaces, where small changes in input can accumulate and cause large changes in the network's output [31].

Imagine the network as a system that makes predictions by weighting different features of the input. Each feature contributes to the output based on its weight. In high-dimensional spaces, even tiny changes to many features can collectively add up to a significant change in the output. This behavior is analogous to steering a ship, a slight shift in the rudder's angle (small input change) might lead to a dramatic course deviation (large output change) after traveling a long distance.

Consider the linear approximation of a neural network's prediction near a specific input x. The change in the network's prediction caused by small perturbation δ can be written as Eq. 17, where $\nabla_x L(\theta, x, y)$ is the gradient of the loss function L with respect to the input x, δ denotes a small perturbation applied to x, and θ represents the model parameters [31].

$$f(x+\delta) \approx f(x) + \nabla_x L(\theta, x, y)^{\top} \delta \tag{17}$$

Equation 17 highlights how the gradient $\nabla_x L$, which points in the direction of the steepest ascent in the loss, governs the sensitivity of the network's predictions to changes in the input. FGSM leverages this sensitivity to create perturbations that maximize the prediction error.

The FGSM generates adversarial examples by applying a calculated perturbation to the input. The perturbation is crafted to increase the model's prediction error as much as possible within a small, controlled limit. This method starts with an input x, its true label y, and a trained neural network parameterized by θ. It computes the loss $L(\theta, x, y)$, which measures how far the network's prediction for x is from y. Then, it calculates the gradient of this loss with respect to x ($\nabla_x L(\theta, x, y)$) [31].

In the next step, we use the sign of the gradient to create a perturbation δ as shown in Eq. 18, where ϵ is a small positive scalar controlling the size of the perturbation, and $sign(.)$ extracts the sign of each element in the gradient vector, ensuring that the perturbation is applied uniformly across all dimensions.

$$\delta = \epsilon \cdot \text{sign}(\nabla_x L(\theta, x, y)) \tag{18}$$

Finally, we add the perturbation δ to the original input to obtain the adversarial example x' ($x' = x + \delta$). When x' is passed through the neural network, the model is likely to misclassify it with high confidence, demonstrating its vulnerability.

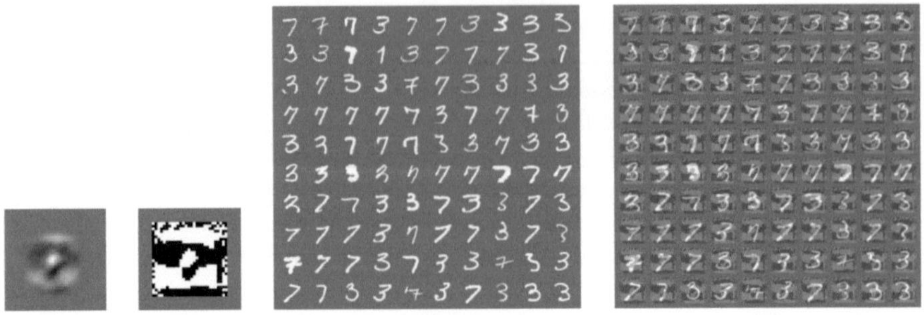

(a) (b) (c) (d)

Fig. 7 The fast gradient sign method applied to logistic regression. **a** The weights of a logistic regression model trained on MNIST; **b** The sign of weights; **c** The class of 3s and 7s in MNIST dataset; **d** FGSM applied to 3s and 7s classes. Reprint from [31]

FGSM leverages the linear nature of neural networks. By focusing on the gradient's direction, it ensures that the perturbation δ maximally increases the loss L. The use of the sign function ensures consistency across all dimensions, making the perturbation both computationally efficient and highly effective. The magnitude of the perturbation, ϵ, plays a critical role: small values of ϵ ensure the adversarial example remains nearly indistinguishable from the original input. However, due to the high dimensionality of the input space, even small perturbations can accumulate to cause significant changes in the network's predictions [31].

The FGSM method has been evaluated on various datasets and models, with Fig. 7 showcasing the results on the MNIST dataset. In Fig. 7, subfigure (b) displays the sign of the weights of a logistic regression model trained on MNIST, representing the optimal perturbation. Despite the model's low capacity and good fit, this perturbation is not easily recognizable to a human observer as being related to the relationship between 3s and 7s. Subfigure (c) illustrates the 3s and 7s classes, where the logistic regression model achieves a 1.6% error rate on the 3 versus 7 discrimination task. However, after applying FGSM with $\epsilon = 0.25$ to generate adversarial examples, the model's error rate increases dramatically to 99% on these examples. These results highlight how even imperceptible perturbations can lead to catastrophic failures in machine learning models.

Among many different adversarial attacks, **Carlini and Wagner (C&W)** [24] attack is one of the most powerful attacks, which is effective under three distance metrics including L_0, L_2, and L_∞ and provides a comprehensive evaluation of the robustness of the neural network.

The C&W attack [24] formulates the generation of adversarial examples as an optimization problem. For an input x with true label $C^*(x)$ and a target label $t \neq C^*(x)$, the objective is to find a perturbation δ such that the modified input $x' = x + \delta$ is classified as t, while minimizing the perceptual difference between x and x'. The optimization problem can be written as Eq. 19 where $\|\delta\|_p$ is the perturbation magnitude under the chosen L_p-norm, $f(x + \delta) = f(x')$ indicates an objective function ensuring x' is classified as t, and c is a regularization parameter balancing the trade-off between minimizing the perturbation and achieving the target misclassification. The specific L_p-norm used (L_0, L_2, and L_∞) determines the method and nature of the attack.

$$\begin{aligned} \underset{\delta}{\text{minimize}} \quad & \|\delta\|_p + c \cdot f(x+\delta) \\ \text{subject to} \quad & x+\delta \in [0,1]^n \end{aligned} \tag{19}$$

In the L_2 attack scenario, the L_2-norm measures the Euclidean distance between x and x', making it ideal for generating minimally visible perturbations. The optimization problem for the L_2 attack is shown in Eq. 20.

$$\begin{aligned} \underset{\delta}{\text{minimize}} \quad & \|\delta\|_2^2 + c \cdot f(x+\delta) \\ \text{subject to} \quad & x+\delta \in [0,1]^n \end{aligned} \tag{20}$$

The function $f(x')$ is defined in Eq. 21, where $Z(x')_i$ represents the logit for class i, and κ is a confidence parameter that controls how strongly x' is classified as t.

$$f(x') = \max\left([\max_{i \neq t} Z(x')_i] - Z(x')_t, -\kappa \right) \tag{21}$$

To handle the box constraint $x + \delta \in [0, 1]^n$, the attack uses a change of variable similar to Eq. 23, where w is an unconstrained variable optimized using gradient-based solvers like Adam. The attack also employs search over c to find the smallest perturbation δ that satisfies $f(x') \leq 0$.

$$\delta_i = \frac{1}{2}(\tanh(w_i) + 1) - x_i \tag{22}$$

Figure 8 demonstrates that the L_2 attack is particularly effective in generating imperceptible adversarial examples, as shown by its ability to achieve 100% success rate with minimal perturbation across datasets like MNIST. Additionally, the best way to choose c is to use the smallest value of c for which the resulting solution x' has $f(x') \leq 0$. This causes gradient descent to minimize both of the terms shown in Eq. 20 simultaneously instead of picking only one to optimize over first.

Another scenario is the L_0 attack which measures the number of pixels in x that are changed to create x'. This attack seeks to find the smallest number of modifications needed to mislead the classifier. The optimization problem for the L_0 attack is more challenging because the L_0 norm is non-differentiable. Instead of directly minimizing $\|\delta\|_0$, the attack uses an iterative algorithm. It starts with the full set of pixels that can be modified. At each iteration, it employs the L_2 attack to find an adversarial perturbation while restricting changes to the current set of allowed pixels. Finally, it identifies the pixel that contributes the least to the adversarial success and removes it

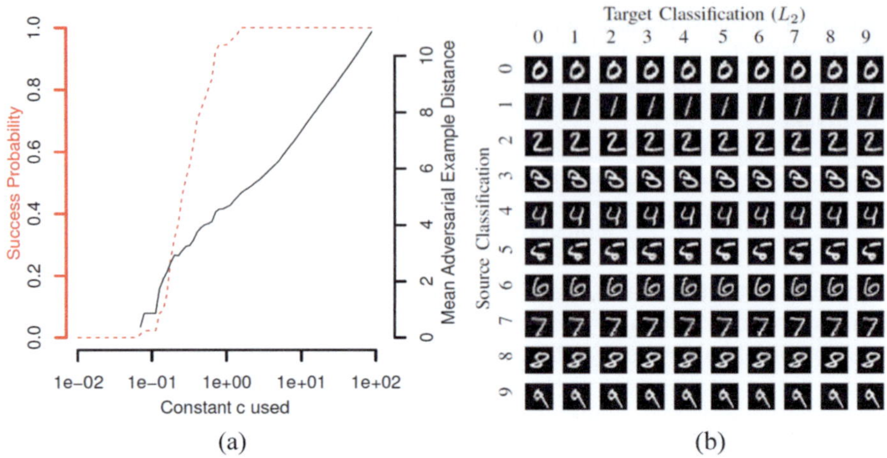

Fig. 8 The effect of the constant c on the L_2 attack and the adversarial examples generated using the L_2 attack in the C&W method. **a** Sensitivity to the constant c: The L_2 distance of the adversarial examples computed via gradient descent is plotted as a function of c. This demonstrates how the choice of c influences the magnitude of perturbations. **b** Application of the L_2 attack to the MNIST dataset: A targeted attack is performed for every source/target pair, showcasing the effectiveness of the L_2 attack in generating adversarial examples for different class mappings. Reprint from [24]

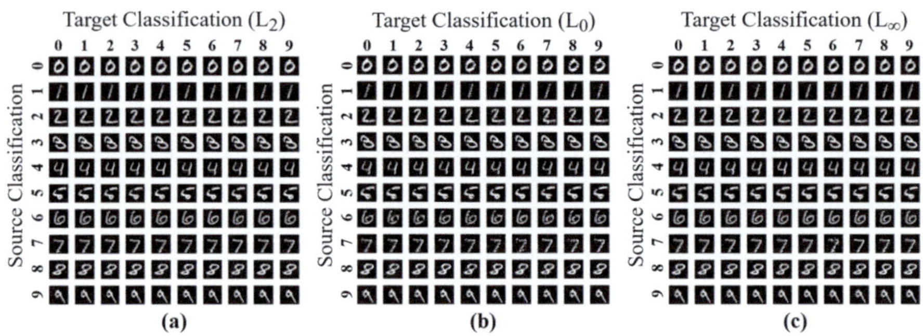

Fig. 9 The C&W method applied to MNIST dataset with different metrics. **a** L_2 attack applied to the MNIST dataset; **b** L_0 attack applied to the MNIST dataset; **c** L_∞ attack applied to the MNIST dataset performing a target attack for every source/target pair. Reprint from [24]

from the set of modified pixels. This process continues until no adversarial example can be found, ensuring that only the most critical pixels are modified. The final adversarial example achieves a high success rate while minimizing the number of altered pixels, making the changes more localized and less detectable. Figure 9b displays the adversarial examples generated under the L_0 attack scenario.

The final scenario in the C&W method [24] is the L_∞ attack, which measures the maximum change made to any single pixel. This attack focuses on limiting the largest individual modification, making it particularly relevant for scenarios where uniform perturbation bounds are critical. The optimization problem for L_∞ attack is given by Eq. 23.

$$\begin{aligned} &\text{Minimize } c \cdot f(x+\delta) + \max_i |\delta_i| \\ &\text{subject to} \quad x + \delta \in [0, 1]^n \end{aligned} \tag{23}$$

Note that directly optimizing this problem is difficult due to the non-smooth nature of the L_∞ norm. To address this, the C&W attack is used as an iterative approach. First, they replace the L_∞ norm term with $\sum \max(|\delta_i| - \tau, 0)$ as a penalty for any pixel changes exceeding a threshold τ. Then, they solve the modified optimization problem and gradually reduce τ in successive iterations. This approach ensures that the perturbation magnitude is uniformly bounded while effectively achieving the target misclassification. The L_∞ attack demonstrates remarkable success, even under stringent perturbation limits, such as flipping only the lowest bit of each pixel in ImageNet images.

The C&W attack [24] has consistently outperformed previous methods under all three distance metrics. On MNIST dataset, it achieved an average L_2 norm perturbation of 1.36, making changes imperceptible to humans. Under the L_0 attack, it modified as few as 8.5 pixels on average while maintaining 100% attack success. For L_∞, the attack achieved target misclassification with perturbations as small as 0.004 on ImageNet, corresponding to undetectable changes.

Many adversarial attacks have been proposed for the white-box attack scenario, including Projected Gradient Descent (PGD) [32], DeepFool [33], SmoothFool (SF) [34], Meta Gradient Adversarial Attack (MGAA) [35], and AdvGAN [36]. These methods aim to formulate and solve optimization problems to determine a minimal perturbation (δ) that, when added to the input x, generates a perturbed input x'. This perturbed input remains imperceptible to humans while successfully deceiving the model with high confidence.

4.2 *Black-Box Attacks*

In real-world applications, adversarial attacks often occur under black-box conditions where the attacker has limited or no knowledge of the model's architecture, parameters, or gradients. The **Black-box Limited Queries and Information Attack (LQIA)** [37] introduces novel methods for generating adversarial examples in three realistic black-box threat models, including the query-limited setting, partial-information setting, and label-only setting. These methods leverage gradient estimation techniques and optimization algorithms to efficiently generate adversarial examples even under strict constraints.

In a black-box setting, the attacker seeks to perturb an input x such that the modified input $x' = x + \delta$ is classified as a target class y, subject to a constraint

$\|\delta\|_\infty \leq \epsilon$. formally, the optimization problem is like Eq. 24, where $C(x)$ denotes the classifier's output for input x, and ϵ defines the maximum allowable perturbation.

$$\text{Find } x' \text{ such that } C(x') = y \text{ and } \|x' - x\|_\infty \leq \epsilon \tag{24}$$

To tackle this problem, LQIA [37] proposed three distinct threat models which include three settings: **Query-Limited**, **Partial-Information**, and **Label-Only**. In the Query-Limited setting, attackers are restricted to a finite number of queries L, necessitating efficient query usage; for instance, commercial APIs like Clarifai impose additional charges for exceeding query limits, discouraging extensive experimentation. In the Partial-Information setting, attackers receive only the top k predicted classes and their corresponding scores rather than the full probability distribution, mimicking systems like the Google Cloud Vision API that provide partial confidence scores. The most restrictive setting is the Label-Only setting, where attackers observe only the top k predicted labels without access to scores or probabilities. This setting is representative of applications such as image tagging systems that assign categorical labels.

In the query-limited setting, **Natural Evolutionary Strategies (NES)** is used to estimate the gradient without direct access to the model's gradient. NES uses a search distribution to approximate the gradient through query-based sampling. Gradient estimation using NES involves sampling random perturbations from a distribution and evaluating their impact on the objective function. The key idea is to approximate the gradient of the objective $F(x)$ (e.g., classification loss) at a given input x by considering how the function changes under small random perturbations [37].

The process begins with **perturbation sampling**, where a set of random perturbations $\delta_1, \delta_2, ..., \delta_n$ is drawn from a Gaussian distribution $\mathcal{N}(0, I)$, with I being the identity matrix. These perturbations represent small random changes applied to the input x, enabling an exploration of how the model's output changes in response to these perturbations.

To improve the quality of the gradient estimate, **antithetic sampling** is employed. This technique involves pairing each perturbation δ_i with its opposite $-\delta_i$, ensuring that both positive and negative directions are considered. By doing so, the algorithm cancels out symmetrical errors in the gradient computation, reducing the variance of the estimate and increasing its robustness [37]. Next, the gradient approximation step calculates the function value $F(x + \sigma\delta_i)$ for each perturbation, where σ is a small scaling factor controlling the magnitude of the perturbations. The gradient is approximated in Eq. 25, which leverages the sampled perturbations to estimate the gradient direction by observing how the function F changes in response to the perturbations.

$$\hat{\nabla} F(x) = \sigma \sum_{i=1}^{n} \delta_i \cdot F(x + \sigma\delta_i) \tag{25}$$

Finally, the **variance reduction** step incorporates both δ_i and $-\delta_i$, further refining the gradient estimate. The updated gradient formula is shown in Eq. 26 which ensures that symmetrical effects of the perturbations are fully exploited, leading to a more stable and accurate gradient estimate.

$$\hat{\nabla} F(x) = \frac{2\sigma}{n} \sum_{i=1}^{n} \delta_i \cdot (F(x + \sigma\delta_i) - F(x - \sigma\delta_i)) \tag{26}$$

Together, the aforementioned steps allow NES to efficiently approximate gradients in a query-limited black-box scenario, paving the way for iterative optimization methods like PGD [32] to generate adversarial examples.

PGD [32] is an iterative optimization method used to generate adversarial examples while ensuring the perturbation remains within a specified L_∞-bounded region. Once the gradient $\hat{\nabla} F(x)$ is estimated (e.g., using NES), PGD utilizes it to refine the adversarial example x' iteratively. The process alternates between taking a gradient-based step to maximize the adversarial objective and projecting the updated input back into the allowable perturbation space.

The first step in PGD is the gradient step, where the adversarial example $x^{(t)}$ is updated in the direction that maximizes the adversarial objective, such as increasing the likelihood of misclassification. This is achieved by taking a step proportional to the sign of the gradient estimate, as shown in Eq. 27, where η is the step size controlling the magnitude of the update. Using the sign function ensures that the update applies uniform changes to the input across all dimensions, maximizing efficiency while adhering to the L_∞-constraint [37].

$$x^{(t+1)} = x^{(t)} - \eta \cdot \text{sign}(\hat{\nabla} F(x^{(t)})) \tag{27}$$

Following the gradient step, a projection step ensures that the updated adversarial example remains within the allowable L_∞-bounded region, defined as $\|x' - x\|_\infty \leq \epsilon$. This projection clamps the values of the adversarial example to stay within the L_∞-ball around the original input as shown in Eq. 28.

$$x^{(t+1)} = \Pi_{[x-\epsilon, x+\epsilon]}(x^{(t+1)}) \tag{28}$$

The projection is implemented by clamping each pixel value of $x^{(t+1)}$ individually as displayed in Eq. 29, where x_i is the original value of the i-th pixel, and ϵ defines the maximum allowable perturbation. This ensures that each pixel's value remains close to its original value, adhering to the L_∞ norm constraint.

$$x_i(t+1) = \max\left(\min\left(x_i(t+1), x_i + \epsilon\right), x_i - \epsilon\right) \tag{29}$$

By alternating between the gradient and projection steps, PGD ensures that the adversarial example incrementally moves towards the desired objective (e.g., targeted misclassification) while staying within the prescribed perturbation limits. This iterative process continues until the adversarial example satisfies the attack goal or a predefined number of iterations is reached. PGD's ability to combine gradient-based optimization with strict constraint enforcement makes it a powerful method for generating robust adversarial examples in constrained settings [37].

Another threat model is **partial-information setting** which presents a more constrained attack scenario where the attacker only receives predictions for the top k classes instead of the full probability distribution across all classes. Despite this limitation, the attacker aims to generate an adversarial example x' such that the target class y is ranked within the top k predictions. The **Targeted Blending with Gradient Estimation** method addresses this challenge by alternating between projection and optimization steps [37].

The attack begins with an initial image x_0 sampled from the target class y. This initialization ensures that the adversarial class y has a high probability of appearing within the top k predictions for x_0. The initial image serves as a baseline for generating adversarial examples through iterative perturbations. The goal is to modify the original input x minimally while ensuring that $x' = x + \delta$ is adversarially classified into y [37].

The projection step then ensures that the perturbation remains within a decreasing L_∞-ball while keeping the target class y ranked within the top k predictions. The radius ϵ_t of the L_∞-ball is dynamically adjusted during the attack based on the rank of y. Formally, the radius ϵ_t is updated as Eq. 30.

$$\epsilon_t = \min\{\epsilon' : \text{rank}(y \mid x') < k\} \tag{30}$$

Such a constraint ensures that ϵ_t shrinks adaptively, allowing the attack to strike a balance between minimal perturbation and the adversarial objective. If the target class y falls outside the top k predictions during any iteration, the radius is expanded temporarily to restore its rank. This dynamic adjustment of ϵ_t is crucial for maintaining the adversarial class within the target prediction set while adhering to the L_∞ norm constraint [37].

Finally, the optimization step maximizes the probability of the target class y within the current L_∞-ball. Since the full gradient information is unavailable, the method employs NES to estimate the gradient. The estimated gradient $\hat{\nabla} F(x)$ is then used to iteratively refine x'. The optimization process is governed by Eq. 31 where η is the step size, dynamically adjusted using a line search method to ensure effective updates without overshooting.

$$x'^{(t+1)} = x'^{(t)} - \eta \cdot \hat{\nabla} F(x'^{(t)}) \tag{31}$$

The gradient estimate $\hat{\nabla} F(x)$ is computed using NES, which approximates the gradient by evaluating the changes in the model's output caused by small, randomly sampled perturbations. By iteratively applying this update, the adversarial example x' is refined to better satisfy the attack objective [37].

This approach ensures that adversarial example generation remains effective even under partial-information constraints. The dynamic adjustment of ϵ_t allows the attack to adapt to changes in the target class's rank during optimization, while NES enables efficient gradient estimation without full access to the model's probability distribution. Together, these methods demonstrate a robust capability to generate adversarial examples in scenarios where the attacker has access to only limited feedback from the target model [37].

The last threat model is **Label-only Setting** which presents one of the most restrictive scenarios for generating adversarial examples. In this setup, the attacker does not have access to probability distributions or confidence scores from the model. Instead, the attacker can only observe the predicted labels, typically limited to the top-ranked classes. This lack of detailed feedback poses a significant challenge for gradient-based optimization techniques, which traditionally rely on access to probabilities. To address this, a novel method based on proxy score estimation has been proposed, enabling effective adversarial optimization despite the absence of probabilities.

The proxy score $S(x)$ provides a substitute for softmax probabilities by focusing on the rank of the target label y in the predicted labels for input x. Intuitively, the rank $R(x)$ measures the position of y in the sorted list of predicted labels, with lower ranks indicating higher confidence for the target label. The proxy score is defined as the expected rank of y when small random perturbations δ are applied to x. Mathematically, it is expressed as Eq. 32 where δ represents random perturbations sampled uniformly from the interval $[-\mu, \mu]$, and μ is the sampling radius that controls the magnitude of the perturbations. This formulation evaluates the robustness of y's rank to random noise, serving as a proxy for the confidence of the target class [37].

$$S(x) = \mathbb{E}_{\delta \sim U[-\mu,\mu]}[R(x + \delta)] \tag{32}$$

Since $S(x)$ depends on the rank function $R(x)$, which is non-differentiable, its gradient cannot be directly computed. To estimate the gradient, Monte Carlo sampling has been employed. By drawing m random perturbations δ_i and evaluating the rank $R(x + \mu\delta_i)$, the proxy score can be approximated in Eq. 33.

$$\hat{S}(x) = \frac{1}{m} \sum_{i=1}^{m} R(x + \mu\delta_i) \tag{33}$$

This approximation enables the computation of a gradient estimate, which reflects how changes in x influence the rank of y. Despite its simplicity, this approach provides a powerful mechanism for optimizing adversarial examples in label-only settings. With the gradient estimate $\hat{S}(x)$ in hand, the adversarial input x' is optimized iteratively. The optimization process involves moving x' in the direction that minimizes the rank of y as shown in Eq. 31 previously. After each update, x' is projected back onto the L_∞-ball to ensure the perturbation remains bounded as Eq. 28.

Despite its computational cost, the proxy score estimation method achieves high success rates in generating adversarial examples. For example, experiments on ImageNet demonstrate a success rate of 90%, even under strict label-only constraints.

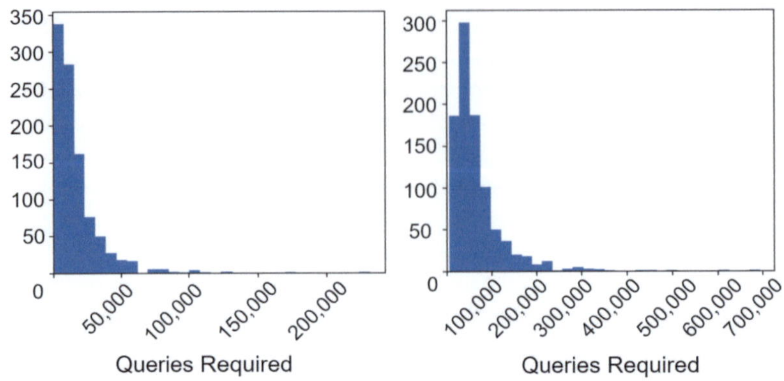

Fig. 10 The distribution of the number of queries for two threat models including query-limited (the right one) and partial-information (the left one) with $k = 1$. Reprint from [37]

This highlights the versatility of the approach and its applicability to real-world scenarios where access to probabilities is restricted. The method underscores the persistent vulnerability of neural networks, even in the most restrictive feedback settings, and sets a benchmark for adversarial attacks in label-only environments [37].

The effectiveness of the proposed methods was evaluated on an ImageNet classifier (InceptionV3) under these three black-box threat models. The results demonstrate the robustness and adaptability of the methods, even in highly constrained environments.

In the query-limited setting, where the attacker has access to probability distributions but is restricted by a query budget, the proposed method achieved an impressive 99.2% success rate. The median number of queries required to generate a successful adversarial example was 11,550. This demonstrates the efficiency of the gradient estimation technique, particularly in scenarios where minimizing queries is critical, such as when interacting with commercial APIs that charge per query [37].

The partial-information setting, where only the top k predicted classes and their probabilities are available, proved slightly more challenging. Despite this restriction, the attack maintained a high 93.6% success rate, requiring a median of 49,624 queries. This result highlights the strength of the targeted blending and gradient estimation method, which dynamically adjusts the perturbation radius and efficiently exploits the limited information provided by the model [37].

The label-only setting, which is the most restrictive scenario, achieved a 90% success rate, albeit with a significantly higher median query count of 2.7 million. This result underscores the computational cost of the label-only attack, as it relies on Monte Carlo sampling to estimate gradients based on proxy scores. Nevertheless, the high success rate demonstrates the resilience of the approach, even when faced with minimal feedback from the model [37].

To visualize the query efficiency and performance of the methods, Fig. 10 illustrates the distribution of queries required for successful attacks across query-limited

Fig. 11 Target adversarial examples. The top row includes the original images and the bottom row includes adversarial exmaples. Reprint from [37]

(right) and partial-information (left). Additionally, Fig. 11 presents examples of adversarial images generated under the $\epsilon = 0.05$ constraint, showcasing how small and imperceptible perturbations can effectively mislead the classifier.

In a real-world demonstration, these methods have been tested on the Google Cloud Vision API, which provides limited output information and variable-length prediction lists. Despite these constraints, the attack successfully manipulated the predictions. For example, an image originally labeled with skiing-related terms was adversarially modified to be classified as "dog". This result underscores the practicality and robustness of the proposed methods, revealing vulnerabilities in commercial machine-learning systems [37].

In the field of black-box settings, numerous methods have been proposed, including Feature-Guided Black-Box [38], Boundary Attack [39], HopSkipJump [40], Square Attack [41], Feature Importance-Aware Attack [42], and several others. The primary goal of these methods is to approximate adversarial examples using minimal information about the model, its parameters, or gradients. These techniques aim to deceive the model into misclassifying the adversarial example x' with the highest confidence level.

4.3 *Universal Adversarial Perturbation*

Universal Adversarial Perturbations (UAPs) [43] are a fascinating and alarming discovery in the field of deep learning. They represent a class of adversarial attacks capable of causing widespread misclassification in state-of-the-art deep neural networks. Unlike conventional adversarial perturbations, which are tailored to fool specific inputs, UAPs are image-agnostic, meaning a single perturbation vector can disrupt the classification of most natural images sampled from a data distribution. The existence of UAPs highlights the inherent vulnerabilities in the decision boundaries of neural networks, raising serious concerns about their deployment in real-world and potentially hostile environments.

A UAP [43] is defined as a perturbation vector $v \in \mathbb{R}^d$, where d is the dimensionality of the input space, satisfying two conditions. First, its magnitude must remain bounded by a pre-defined norm constraint, $\|v\|_p \leq \epsilon$, where p-norm (such as L_2 or L_∞) quantifies the size of the perturbation. This ensures that the perturbation remains quasi-imperceptible to human observers. Second, it must achieve a high fooling rate, i.e., it should cause a classifier $\hat{k}$ to output incorrect labels for a significant proportion of inputs from the data distribution μ. The fooling rate is expressed as Eq. 34, where δ is a small parameter controlling the acceptable error threshold. The surprising aspect of UAPs lies in their universal nature—they exploit the geometric redundancies of neural network decision boundaries in high-dimensional spaces, making them effective across diverse inputs.

$$P_{x\sim\mu}(\hat{k}(x+v) \neq \hat{k}(x)) \geq 1-\delta \tag{34}$$

To compute UAPs [43], a systematic algorithm is employed, leveraging an iterative process that gradually constructs the perturbation vector. The algorithm starts with an initial perturbation $v = 0$ and iteratively refines it by examining a set of training images, $X = \{x_1, x_2, \ldots, x_m\}$, sampled from the data distribution. For each image in X, if the current perturbed image $x_i + v$ is correctly classified by the model, the algorithm determines the minimal adjustment Δv_i required to push $x_i + v$ to the decision boundary of the classifier. This adjustment is computed using Eq. 35.

$$\Delta v_i = \arg\min_r \|r\|_2 \quad \text{subject to} \quad \hat{k}(x_i + v + r) \neq \hat{k}(x_i) \tag{35}$$

This optimization problem seeks the smallest possible perturbation r that changes the classifier's output for x_i. Once Δv_i is determined, it is added to the current perturbation v, and the result is projected onto a norm-constrained ball to ensure $\|v\|_p \leq \epsilon$. The projection operation is defined in Eq. 36. This ensures that the perturbation remains imperceptible while retaining its adversarial effectiveness. The algorithm continues to refine v by iterating through the dataset until the fooling rate surpasses a predefined threshold $1-\delta$, at which point the perturbation is finalized [43].

$$P_{p,\epsilon}(v) = \arg\min_{v'} \|v - v'\|_2 \quad \text{subject to} \quad \|v'\|_p \leq \epsilon \tag{36}$$

Empirical evaluations demonstrate the effectiveness and generalization capability of UAPs. For instance, perturbations computed on small training sets can generalize well to unseen validation data, achieving fooling rates exceeding 90% on complex datasets like ImageNet. This remarkable property arises because UAPs exploit corre-

Table 1 The information of fooling rate on the set X, and the validation set on various state-of-the-art models. Reprint from [43]

		CaffeNet (%)	VGG-F (%)	VGG-16 (%)	VGG-19 (%)	GoogLeNet (%)	ResNet-152 (%)
L_2	X	85.4	85.9	90.7	86.9	82.9	89.7
	Val.	85.6	87.0	90.3	84.5	82.0	88.5
L_∞	X	93.1	93.8	78.5	77.8	80.8	85.4
	Val.	93.3	93.7	78.3	77.8	78.9	84.0

lations in the decision boundaries of neural networks. Analysis using singular value decomposition (SVD) reveals that the normal vectors to decision boundaries near natural images often reside in a low-dimensional subspace. Perturbations confined to this subspace can efficiently fool a significant fraction of inputs [43].

The practical implications of UAPs extend beyond their ability to fool classifiers. They expose critical security vulnerabilities, particularly in applications where neural networks are deployed in sensitive environments. Visualization of perturbed inputs underscores their quasi-imperceptibility, raising concerns about their detectability. Furthermore, the universality of these perturbations extends across network architectures, meaning a UAP computed for one model, such as VGG-19, can successfully fool other models, including GoogLeNet or ResNet. Through both qualitative and quantitative analyses, we demonstrate how these perturbations exploit the vulnerabilities of neural networks [43].

Figure 12 provides a visual illustration of UAPs. The central panel shows a computed perturbation vector that, when added to natural images, induces misclassification. The left panel includes unperturbed natural images correctly classified by the neural network, while the right panel shows perturbed images with their altered labels. These examples highlight the quasi-imperceptibility of UAPs to the human eye, emphasizing their subtle yet highly disruptive nature.

To quantify their effectiveness, we compute the fooling rates of UAPs across several state-of-the-art neural network architectures. Table 1 summarizes these results under L_2 and L_∞-norm constraints. Remarkably, UAPs achieve fooling rates exceeding 90% for architectures like CaffeNet and VGG-F, even when constrained to small magnitudes. These results underscore the universality of the perturbations, as they generalize across diverse architectures and datasets [43].

Another intriguing property of UAPs is their ability to generalize from a small training set to unseen data points. Figure 13a illustrates the fooling rates achieved on the validation set as a function of the training set size X. Even with as few as 500 training images—far fewer than the total number of ImageNet classes—UAPs fool a significant proportion of validation images. This result demonstrates the perturbations' remarkable ability to extrapolate their adversarial effect to new, unseen samples [43].

To understand why UAPs [43] are so effective, we analyze the geometric correlations in neural network decision boundaries. Figure 13b visualizes the singular values of the matrix formed by the normal vectors to decision boundaries near data

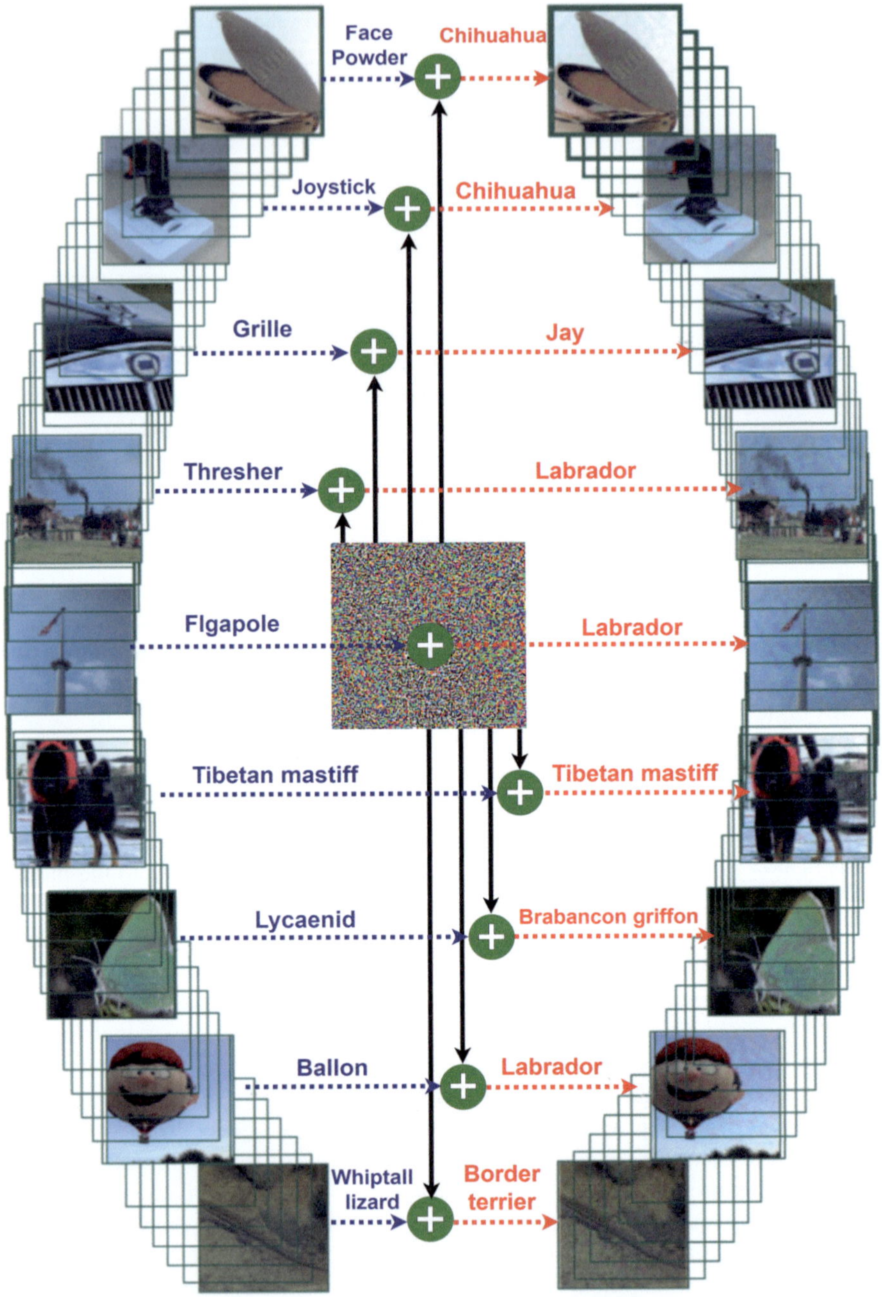

Fig. 12 A visual representation of the perturbation effect. This includes examples of unperturbed images, UAP, and the corresponding perturbed images with their misclassified labels. Reprint from [43]

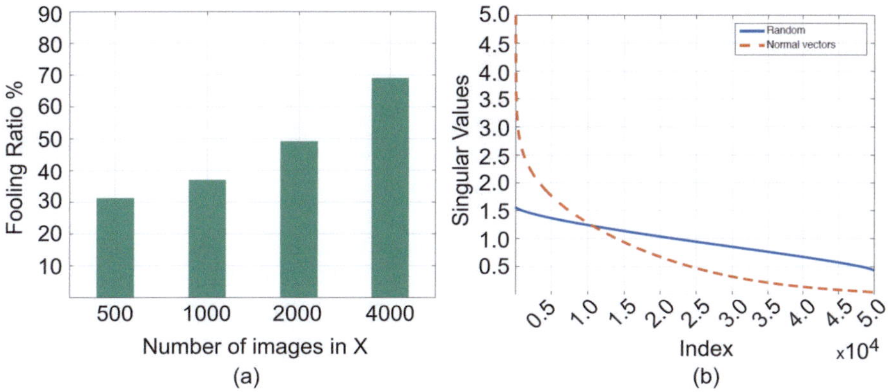

Fig. 13 **a** This graph depicts the variation of fooling rates with respect to the size of the training set X, highlighting the generalization ability of UAPs; **b** SVD analysis of decision boundary correlations, illustrating how normal vectors to the decision boundary form a low-dimensional subspace. Reprint from [43]

points. The rapid decay of these singular values indicates that the normals lie in a low-dimensional subspace. This finding suggests that UAPs exploit redundancies in the decision boundary geometry, enabling them to achieve high fooling rates with relatively small perturbations.

In summary, the results unequivocally demonstrate the effectiveness and generalization capabilities of UAPs. By exploiting geometric redundancies in decision boundaries, these perturbations pose significant challenges to neural network robustness. The figures collectively illustrate the subtle yet impactful nature of UAPs, their universality across architectures, and the underlying factors contributing to their success [43].

Adversarial examples, carefully crafted inputs designed to deceive a model, exploit weaknesses in DNN decision boundaries. While some attacks rely on imperceptible perturbations to fool models in digital domains, these methods often fail in physical-world scenarios due to environmental transformations like lighting, perspective shifts, and noise. Conversely, physical-world attacks tend to use larger perturbations that are more robust but often conspicuous, limiting their stealthiness. Balancing adversarial effectiveness with perceptual stealth remains a critical challenge in adversarial machine learning.

In the field of universal adversarial perturbation, the **Adversarial Camouflage (AdvCam)** [44] framework addresses this problem by combining adversarial attacks with neural style transfer to generate adversarial examples that are both effective and visually natural. Unlike traditional approaches that minimize perturbations, AdvCam transforms them into aesthetically plausible modifications, embedding adversarial patterns into natural styles like textures or patterns. This approach not only enhances the stealthiness of adversarial examples but also makes them adaptable to both digital and physical-world scenarios. By leveraging style transfer techniques, AdvCam camouflages perturbations in a way that is indistinguishable to human observers while maintaining their adversarial efficacy.

The problem formulation begins with an input image x with a true class label y, a DNN classifier F, and a target class $y' \neq y$. The goal is to generate an adversarial example x' such that $F(x') = y'$ with high confidence. At the same time, the generated perturbation should blend naturally into the image. The optimization problem is defined as Eq. 37, where $D(x, x')$ is a distance metric that quantifies the stealthiness of the adversarial perturbation, $L_{adv}(x')$ is the adversarial loss ensuring the attack's effectiveness, and λ balances these two objectives. Unlike traditional methods that use simple norms (e.g., L_2 or L_∞) for $D(x, x')$, AdvCam employs a style-based metric derived from neural style transfer [44].

$$\min\ D(x, x') + \lambda \cdot L_{\text{adv}}(x') \tag{37}$$

To achieve both camouflage and adversarial strength, AdvCam introduces a composite loss function as shown in Eq. 38, where L_s, L_c, and L_m represent style loss, content loss, and smoothness loss, respectively. These terms work in tandem to ensure that the adversarial perturbation is seamlessly integrated into the image while preserving its structural features [44].

$$L = (L_s + L_c + L_m) + \lambda \cdot L_{\text{adv}} \tag{38}$$

The style loss L_s is central to the AdvCam approach, ensuring that the adversarial perturbation adopts the texture and color patterns of a reference style image x_s. Using a feature extractor F_l, typically a pre-trained neural network, the style loss is defined in Eq. 39, where G represents the Gram matrix of feature activations, capturing correlations between different feature maps at each layer l. The set S_l includes layers selected to represent different aspects of style, such as color and texture [44].

$$L_s = \sum_{l \in S} \|G(F_l(x_s)) - G(F_l(x'))\|_2^2 \tag{39}$$

To preserve the structural integrity of the original image, AdvCam incorporates a content loss L_c, which measures the similarity between the original and adversarial images in a feature space as shown in Eq. 40, where C_l represents the set of deeper layers used to extract content features. This ensures that the adversarial image retains the recognizable structure of the original input [44].

$$L_c = \sum_{l \in C} \|F_l(x) - F_l(x')\|_2^2 \tag{40}$$

Smoothness loss L_m further enhances the camouflage by reducing pixel-wise variations, making the perturbation visually cohesive. It is defined in Eq. 41. This term is particularly important for physical-world attacks, as smooth perturbations

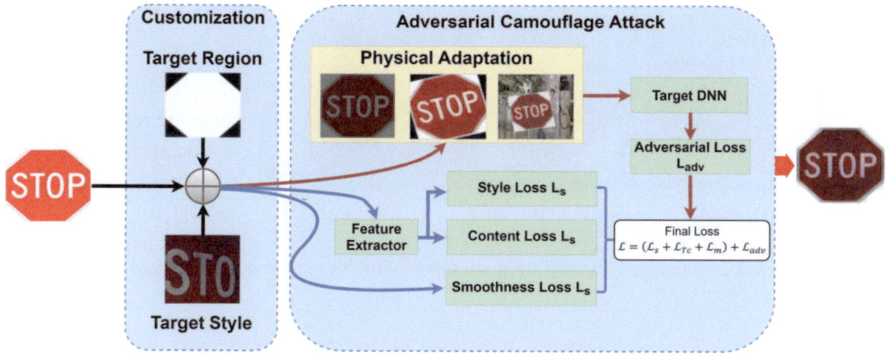

Fig. 14 The overview of the AdvCam method. Reprint from [44]

are less affected by environmental transformations and more likely to maintain their effectiveness under different conditions.

$$L_m = \sum_{i,j} \left((x'_{i,j} - x'_{i+1,j})^2 + (x'_{i,j} - x'_{i,j+1})^2 \right) \tag{41}$$

The adversarial loss L_{adv} drives the attack by ensuring that the generated image is classified into the target class y'. For targeted attacks, it is defined as Eq. 42, where $p_{y'}(x')$ and $p_y(x')$ are the classifier's confidence scores for the target and original classes, respectively.

$$L_{\text{adv}} = -\log(p_{y'}(x')) + \log(p_y(x')) \tag{42}$$

In physical-world settings, environmental factors such as lighting, perspective shifts, and noise can degrade the effectiveness of adversarial examples. To address this, AdvCam incorporates adaptations inspired by Expectation Over Transformation (EOT). The optimization simulates various transformations, such as rotations, scaling, and color shifts, during training. This ensures that the adversarial example remains effective across a range of physical conditions. Figure 14 provides an overview of all steps involved in generating the adversarial examples with AdvCam method [44].

The results of AdvCam demonstrate its unique ability to craft adversarial examples that are both highly effective and visually stealthy. Unlike traditional adversarial attacks, AdvCam successfully integrates perturbations into natural styles, making them indistinguishable from legitimate visual features while maintaining strong adversarial efficacy. The evaluation involves both digital and physical-world scenarios, ensuring a comprehensive understanding of AdvCam's capabilities. Here, we delve deeper into the results by analyzing success rates, stealthiness, human perception, and performance across various conditions [44].

Fig. 15 Examples generated by different methods with the successful physical-world attack. Reprint from [44]

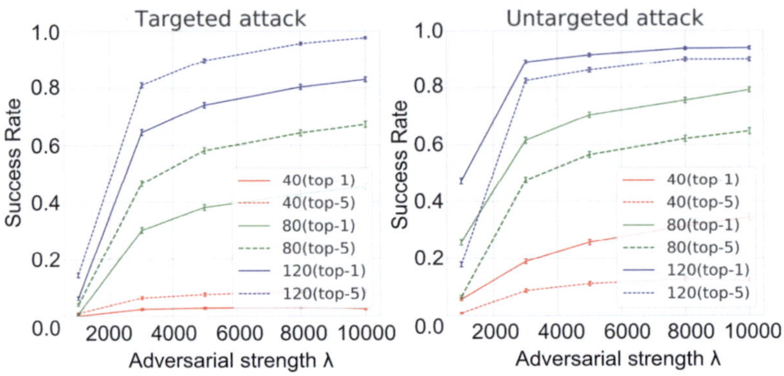

Fig. 16 The relationship between adversarial strength λ and attack success rates for targeted and untargeted attacks. Reprint from [44]

One of the significant aspects of AdvCam is its ability to camouflage perturbations into natural styles, as illustrated in Fig. 15 which provides a side-by-side comparison of adversarial examples crafted by AdvCam, PGD, and AdvPatch. PGD generates small, imperceptible perturbations but fails to create visually meaningful patterns. AdvPatch produces strong perturbations, but these are often conspicuous and artificial-looking. In contrast, AdvCam embeds perturbations into natural styles such as wood textures, graffiti, or fabric patterns, making them indistinguishable from real-world features. These camouflaged perturbations not only deceive neural networks but also appear legitimate to human observers, a critical advantage in both digital and physical settings [44].

One of the key findings of AdvCam is that it achieves high attack success rates across multiple neural network architectures. By adjusting the adversarial strength parameter λ and region size, AdvCam can control the success rate of both targeted and untargeted attacks. Figure 16 shows the relationship between λ and the success rate for both types of attacks. Larger values of λ increase the strength of the perturbation, leading to higher success rates. Similarly, larger attack regions result in more effective perturbations, as shown in the graph. For untargeted attacks, success rates exceed 90% even with moderate values of λ, while targeted attacks achieve comparable

Fig. 17 Examples of AdvCam applied to physical-world scenarios. Reprint from [44]

performance with slightly larger perturbations. The flexibility to control λ and region size allows AdvCam to balance stealthiness and adversarial effectiveness according to the application [44].

Figure 17 demonstrates AdvCam's ability to create camouflaged adversarial images by showing on-target and off-target examples alongside their original versions. In the on-target cases (first row), AdvCam embeds perturbations directly onto objects, blending them naturally into their surface, such as adding textures that mislead the classifier. In the off-target cases (second row), perturbations are applied to background elements. For instance, a price tag is altered to misclassify a revolver as toilet tissue, and floral patterns in the background cause a Blenheim spaniel to be identified as a bearskin. These examples highlight AdvCam's flexibility in crafting visually stealthy and highly effective adversarial attacks [44].

In summary, the results establish AdvCam as a robust and flexible framework for crafting stealthy adversarial examples. With high success rates, natural integration of perturbations, and superior performance in both digital and physical-world scenarios, AdvCam sets a new benchmark for evaluating the robustness of DNNs. Its ability to blend adversarial patterns into natural styles and deceive both machines and humans underscores the pressing need for more effective defense mechanisms against such advanced attacks.

Many other adversarial attacks have been proposed for the universal adversarial perturbation setting, including Universal Adversarial Network (UAN) [45], Stochastic Gradient Aggregation (SGA) [46], AdvEncoder [47], Basic Iterative Method [48], and Reflected Light Attack (RFLA) [49]. These methods aim to focus on generating a single, general perturbation that can be applied to multiple inputs. These attacks demonstrate that small, input-agnostic changes can consistently deceive models across diverse datasets.

4.4 Auto Attacks

Adversarial attacks highlight a critical vulnerability in DNNs: their susceptibility to small, carefully designed perturbations that lead to misclassification. Evaluating the robustness of DNNs against such attacks is essential, especially in applications like healthcare, autonomous driving, and cybersecurity, where incorrect predictions can have severe consequences. However, existing evaluations often overestimate robustness due to flawed testing protocols. These flaws include poorly tuned attack hyperparameters, reliance on single attack methods, and defenses that obscure vulnerabilities through techniques like gradient masking. To address these issues, **AutoAttack** [50] introduces a rigorous, parameter-free evaluation framework that combines multiple adversarial attack strategies into a single, reliable ensemble.

At the core of AutoAttack is a carefully designed ensemble of four adversarial attack methods, each contributing a unique perspective to robustness evaluation. These attacks include two variants of a method called Auto-PGD, the FAB attack [51], and the Square Attack [41]. Together, these methods ensure a thorough evaluation by covering different attack scenarios, including both white-box and black-box settings.

The first component of AutoAttack is **Auto-PGD (APGD)**, which builds on the widely used PGD attack [32]. PGD is an iterative white-box attack that modifies the input image step-by-step to maximize the likelihood of misclassification, constrained by a perturbation budget defined by a norm L_p (commonly L_∞ or L_2). However, PGD's effectiveness heavily depends on fixed hyperparameters, such as step size and the number of iterations, which may result in suboptimal results. APGD addresses this limitation by introducing an adaptive step size adjustment mechanism.

In Auto-PGD, the optimization is divided into two phases: exploration and exploitation. During the exploration phase, larger step sizes are used to broadly search for promising adversarial directions. In the exploitation phase, smaller step sizes refine the perturbation, ensuring that it remains within the perturbation budget while effectively fooling the network. The transition between these phases is governed by dynamic step size adjustments, which monitor the progress of the attack objective. If the improvement in adversarial success stalls, the step size is halved. This adaptive behavior ensures efficient convergence, allowing Auto-PGD to find adversarial examples more reliably than standard PGD, without requiring manual parameter tuning [50].

Auto-PGD includes two versions, each using a different loss function to guide the attack. The first version, APGD-CE, uses the cross-entropy (CE) loss, a commonly employed loss function in classification tasks. Cross-entropy measures the difference between the predicted class probabilities and the ground truth, encouraging the attack to maximize the likelihood of an incorrect class. However, CE loss has limitations in certain scenarios, particularly when gradient masking or obfuscation techniques are employed by the model. To address these issues, Auto-PGD introduces a second version called **APGD-DLR**, which uses the **Difference of Logits Ratio (DLR)** loss.

The DLR loss is a specially designed loss function that is both shift- and scale-invariant, making it robust to gradient masking. It is defined in Eq. 43, where z_y

represents the logit corresponding to the true class, $z_{max \neq y}$ is the highest logit among incorrect classes, and $z_{\pi 1}$, $z_{\pi 3}$ are the largest and third-largest logits, respectively. By focusing on the relative differences between logits, DLR ensures that the gradients remain informative, even when the network employs defensive mechanisms that obscure the CE loss.

$$\mathrm{DLR}(x, y) = \frac{-z_{\pi_1} - z_{\pi_3}}{z_y - z_{\max \neq y}} \tag{43}$$

The third component of AutoAttack is the **FAB** attack [51], which stands for Fast Adaptive Boundary. Unlike APGD, which maximizes the attack loss, FAB is designed to minimize the distance between the input image and the decision boundary of the classifier. This attack explicitly reduces the perturbation norm required to misclassify an image, making it highly effective in scenarios with strict L_p norm constraints. FAB iteratively adjusts the perturbation by projecting the input toward the nearest decision boundary, ensuring that the attack remains within the perturbation budget while achieving high success rates.

The fourth and final component of the ensemble is the **Square Attack** [41], a black-box attack that does not rely on gradient information. Square Attack uses a randomized search strategy to iteratively modify the input image. It perturbs small, square-shaped patches of the image, selecting the patches and the magnitude of the changes through a probabilistic mechanism. This approach is particularly valuable when gradient information is unavailable or unreliable, making Square Attack a crucial addition to the AutoAttack ensemble.

The combination of the aforementioned four methods creates a robust evaluation framework. APGD-CE and APGD-DLR provide complementary white-box attacks that address different loss function vulnerabilities, while FAB ensures tight control over perturbation norms. The inclusion of Square Attack adds diversity by covering black-box scenarios, ensuring that the ensemble evaluates defenses comprehensively under a wide range of conditions [50].

The goal of AutoAttack is not only to evaluate robustness but to standardize the process of adversarial testing. By being parameter-free, it eliminates the biases introduced by hyperparameter tuning, ensuring fair comparisons between models and defenses. Moreover, its ability to combine diverse attacks guarantees thorough evaluations, exposing vulnerabilities that would otherwise remain hidden [50].

The results of AutoAttack demonstrate its effectiveness in rigorously evaluating adversarial robustness across various datasets and models. By combining multiple complementary attack strategies into a single ensemble, AutoAttack systematically reveals vulnerabilities in defenses that are often overlooked by traditional methods. The evaluation shows significant reductions in reported robust accuracies compared to earlier evaluations, with reductions often exceeding 10-15% across widely studied datasets such as CIFAR-10, CIFAR-100, and ImageNet [50].

A key highlight of the results is the ability of Auto-PGD, with its adaptive step size mechanism, to outperform traditional fixed-step attacks by achieving better convergence and higher attack success rates. The dynamic adjustment of step sizes ensures that adversarial optimization proceeds efficiently, refining perturbations to maximize their impact without overshooting or getting stuck in suboptimal regions [50].

Compared to traditional loss functions like cross-entropy, DLR shows remarkable resilience against defenses that employ gradient masking or obfuscation. This makes APGD-DLR a critical component of the AutoAttack ensemble, capable of uncovering vulnerabilities in models designed to obscure their gradients.

In conclusion, AutoAttack represents a significant advancement in the evaluation of adversarial robustness. By combining adaptive optimization strategies, such as the dynamic step size adjustment of Auto-PGD, with robust loss functions like DLR and a diverse ensemble of attack methods, AutoAttack addresses the limitations of traditional adversarial testing. The method's parameter-free design ensures consistent and reliable evaluations across various models and datasets, eliminating biases introduced by manual hyperparameter tuning [50].

AutoAttack's rigorous and standardized framework not only challenges inflated robustness claims but also provides a critical tool for researchers and practitioners to assess and improve the resilience of machine learning models. As adversarial machine learning continues to evolve, methods like AutoAttack will play a pivotal role in driving the development of more robust and reliable AI systems [50].

5 Adversarial Defense

Adversarial attacks pose significant challenges to the robustness and reliability of machine learning models, particularly deep neural networks. To counteract these threats, various adversarial defense strategies have been developed, each aiming to enhance model resilience against malicious perturbations.

Adversarial Training involves augmenting the training dataset with adversarial examples—inputs intentionally crafted to deceive the model. By incorporating these perturbed samples during training, models learn to recognize and resist adversarial manipulations, thereby improving robustness. This method has been extensively studied and is considered one of the most effective defenses against adversarial attacks [52].

Training Process Modification encompasses adjustments to the learning paradigm, objective functions, or optimization algorithms to bolster model robustness. Techniques such as regularization methods, robust optimization, and alternative loss functions fall under this category. These modifications aim to make the training process inherently resistant to adversarial influences [53].

Supplementary Networks involve the integration of auxiliary models designed to detect or purify adversarial inputs before they reach the primary classifier. For instance, auxiliary networks can function as discriminators to identify and filter out adversarial perturbations, thereby preventing them from compromising the main model's performance [54].

Network Architecture Modification refers to altering the structure of neural networks to enhance their resilience. This can include designing architectures that are

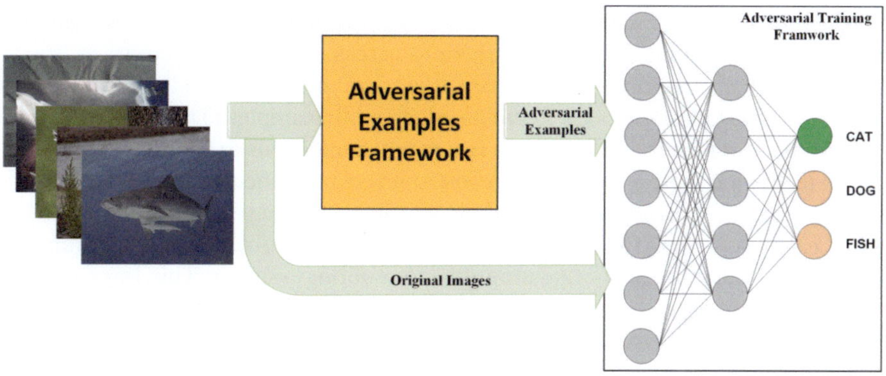

Fig. 18 Training the model with both original images and adversarial examples generated with different attack methods

inherently more robust to adversarial perturbations or incorporating layers specifically aimed at mitigating the effects of such attacks. By modifying the network's architecture, it becomes more difficult for adversaries to exploit vulnerabilities [53].

Adversarial Purification involves transforming potentially adversarial inputs back into a benign space before classification. Techniques such as denoising autoencoders or diffusion models are employed to 'purify' the input, removing adversarial noise while preserving the essential features necessary for accurate classification. This approach serves as a preprocessing step to safeguard the model from adversarial attacks [55].

Collectively, these strategies form a comprehensive defense framework against adversarial attacks, improving the security and reliability of machine learning systems across diverse applications. In this section, we will introduce some of the most promising methods in each category to provide a clear understanding of their effectiveness.

5.1 Adversarial Training

Adversarial training is a widely recognized strategy to improve the robustness of deep neural networks against adversarial attacks. This approach involves augmenting the training data with adversarial examples—perturbed inputs designed to mislead the model—and using them during the training process. By exposing the model to such examples, the network learns to correctly classify them, enhancing its ability to generalize and resist adversarial perturbations as shown in Fig. 18.

The main work by [23] highlighted the existence of adversarial examples that can cause significant misclassification errors despite being imperceptible to humans. This work demonstrated that these adversarial examples are not just artifacts of specific models but are somewhat universal, affecting various architectures trained on different subsets of data. This universality implies that adversarial training can benefit from generating diverse perturbations across models and datasets.

A key implementation detail of adversarial training, as explored by [23], involves iterative optimization methods to create adversarial examples. These examples are incorporated into the training process in a manner analogous to "hard-negative mining," a technique used in computer vision to prioritize difficult examples that challenge the model's current understanding. For instance, they successfully trained a neural network on the MNIST dataset by maintaining a pool of adversarial examples for each layer, continuously updating this pool during training. This layer-specific adversarial training was found to be particularly effective, with higher-layer examples contributing more to robustness improvements than lower-layer ones. Moreover, their experiments showed that adversarial training not only reduces test error but also helps improve generalization. This observation aligns with the hypothesis that adversarial training helps the model learn a more robust representation of the input space by addressing intrinsic blind spots in the data distribution.

In summary, adversarial training is a robust defense mechanism against adversarial attacks, benefiting from iterative optimization and systematic incorporation of challenging examples into the training dataset. By enhancing model resilience and generalization, this method provides a foundational defense strategy for securing neural networks against adversarial vulnerabilities.

As mentioned before, adversarial training is a cornerstone method for defending machine learning models against adversarial examples. It aims to improve robustness by incorporating adversarially perturbed inputs into the training process. These examples are generated to maximize the model's loss within a bounded perturbation size, ensuring the model learns to resist such malicious alterations. The process aligns with the framework of adversarial Empirical Risk Minimization (ERM) as shown in Eq. 44, where the optimization seeks to minimize the worst-case loss over these adversarial examples. In this equation, h indicates a model from a space H, and L is the loss function used to train the model (e.g., cross-entropy). While effective against white-box attacks, traditional adversarial training often struggles to scale to large datasets and falls short against black-box adversaries.

$$h^* = \arg\min_{h \in H} \mathbb{E}_{(x,y)\sim\mathcal{D}} \left[\max_{\|x'-x\|_\infty \leq \epsilon} L(h(x'), y) \right] \tag{44}$$

Ensemble Adversarial Training (EAT) [56] addresses these limitations by introducing a novel approach to adversarial defense. Instead of generating adversarial examples solely from the model being trained, EAT augments the training process with perturbations crafted using multiple static pre-trained models. This decoupling of adversarial example generation from the evolving parameters of the target model alleviates overfitting to specific attack methods and mitigates gradient masking, a common pitfall of traditional adversarial training.

The EAT process begins by randomly selecting a source model—either the model currently under training or one of the pre-trained models—for each batch. Adversarial examples are then generated using established methods such as Step-LL [56],

FGSM [31], or iterative attacks like PGD [32]. Half of the training examples in each batch are replaced with these adversarial examples, ensuring that the model is exposed to a diverse set of perturbations throughout the training process. This diversity significantly enhances the model's ability to generalize to unseen attacks, particularly those originating from black-box adversaries.

Empirical results demonstrate the efficacy of EAT in improving robustness. For instance, when tested on the ImageNet dataset, models trained with EAT show significantly lower error rates against adversarial examples transferred from holdout models. For example, the IRv2adv-ens model trained using EAT achieved a top-1 error rate of 27% under black-box attacks, compared to 34.5% for a traditionally trained IRv2adv model. These improvements are complemented by reductions in gradient masking, as evidenced by smoother decision boundaries in EAT-trained models. Unlike models trained with standard adversarial techniques, which often develop sharp curvatures in their loss landscapes, EAT-trained models exhibit a more linear and generalized structure, reducing susceptibility to adversarial perturbations [56].

In practice, the added robustness comes at a computational cost. EAT involves training on more complex adversarial examples and typically requires smaller batch sizes to accommodate the increased memory requirements of pre-trained models. This results in slower convergence compared to traditional adversarial training. For example, while a standard adversarially trained Inception v3 model converges after 187 epochs, its EAT-trained counterpart requires 280 epochs to achieve comparable performance. Despite this drawback, the benefits of increased robustness against both black-box and white-box adversaries justify the additional computational effort [56].

EAT also addresses certain vulnerabilities to white-box attacks, although iterative attacks like R+FGSM and Iter-LL remain effective to some extent. These attacks exploit localized sharp curvatures in the loss surface, which can still mislead gradient-based defenses. Nonetheless, EAT significantly reduces the dimensionality of adversarial subspaces, as shown by experiments measuring the number of orthogonal adversarial directions around input points. This reduction is particularly pronounced for high-magnitude perturbations, where traditional adversarial training often struggles to generalize [56].

Models trained with EAT outperformed simpler adversarial training strategies, achieving superior performance against black-box adversaries. This demonstrates the method's utility for real-world applications, where models must contend with a variety of adversarial attack scenarios. While the current implementation of EAT represents a significant step forward, ongoing research is necessary to optimize its scalability and address vulnerabilities to advanced iterative attacks. By continuing to refine ensemble methods and explore new strategies for adversarial defense, researchers can develop machine learning systems that are both robust and reliable in increasingly adversarial environments [56].

Another intriguing area is unsupervised machine learning, where adversarial defense techniques are also applied. One notable method in this field is **Unsupervised Adversarial Training (UAT)** [57], an innovative approach that leverages unlabeled data to train classifiers with enhanced robustness against adversarial attacks. This approach aims to minimize the adversarial risk, defined as Eq. 45, where

x' represents adversarially perturbed inputs within an L_∞-bounded neighborhood $\mathcal{N}_\epsilon(x) = \{x' : \|x' - x\|_\infty \leq \epsilon\}$, and ℓ is the loss function, such as cross-entropy.

$$L_{\text{adv}}(\theta) = \mathbb{E}_{P(X,Y)} \sup_{x' \in \mathcal{N}_\epsilon(x)} \ell(y, f_\theta(x')) \tag{45}$$

This method addresses a key limitation of adversarial training: its reliance on large quantities of labeled data, which are often expensive and time-consuming to obtain. UAT operates by focusing on two primary strategies: Unsupervised Adversarial Training with Online Targets **(UAT-OT)** and Unsupervised Adversarial Training with Fixed Targets **(UAT-FT)**, both of which are combined in the **UAT++** framework for enhanced performance [57].

UAT-OT is designed to minimize the model's sensitivity to adversarial perturbations without relying on labeled data. This is achieved by focusing on a smoothness loss that encourages the model's predictions to remain stable under small input perturbations. Formally, the smoothness loss is defined using the Kullback-Leibler (KL) divergence between the original and perturbed predictions as shown in Eq. 46, where $p_\theta(\cdot|x)$ represents the predicted probability distribution for the original input x, and $p_\theta(\cdot|x')$ represents the prediction for the perturbed input x'. The parameters θ are fixed during optimization to prevent the gradient from propagating through the reference model, ensuring stability in training. Perturbations x' are generated using PGD, a strong adversarial attack, to approximate the supremum over the adversarial neighborhood $\mathcal{N}_\epsilon(x)$. By minimizing this smoothness loss, UAT-OT effectively trains the model to generalize across a broader spectrum of adversarial examples without requiring labeled data [57].

$$L_{unsup}^{OT}(\theta) = \mathbb{E}_{x \sim P(X)} \sup_{x' \in N_\epsilon(x)} D_{KL}(p_\theta(\cdot|x), p_\theta(\cdot|x')) \tag{46}$$

In contrast, UAT-FT uses pseudo-labeling to leverage unlabeled data. This method begins by training a base classifier on a small labeled dataset. The trained model is then used to assign pseudo-labels to the unlabeled data. These pseudo-labels serve as targets for adversarial training, where adversarially perturbed inputs are generated from the unlabeled examples. The training objective for UAT-FT is expressed in Eq. 47, where $\hat{y}(x)$ represents the pseudo-label assigned to x by the base classifier, and $xnet$ is the cross-entropy loss. This strategy enables UAT-FT to utilize the large pool of unlabeled data for training while ensuring the robustness of the model to adversarial perturbations. Unlike UAT-OT, UAT-FT directly incorporates pseudo-labels into the adversarial training process, making it dependent on the accuracy of the initial base classifier. However, the method is robust to label noise, as demonstrated in the experiments [57].

$$L_{unsup}^{FT}(\theta) = \mathbb{E}_{x \sim P(X)} \sup_{x' \in \mathcal{N}_\epsilon(x)} \text{xent}(\hat{y}(x), p_\theta(\cdot \mid x')) \tag{47}$$

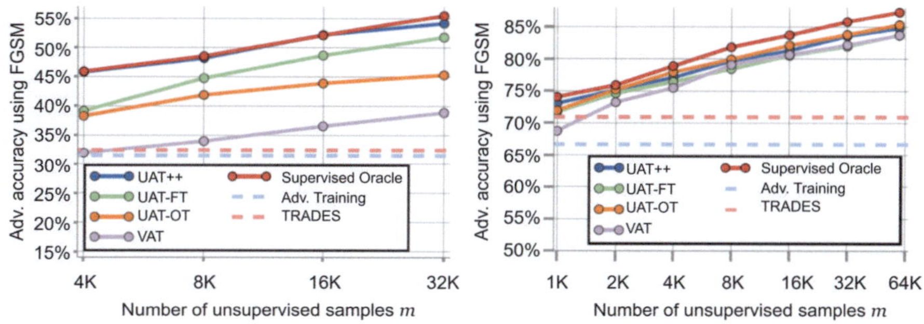

Fig. 19 Adversarial accuracy achieved by UAT variants (UAT-OT, UAT-FT, and UAT++) compared to baselines (TRADES and supervised oracle) on CIFAR-10 (left) and SVHN (right) datasets as the number of unsupervised samples increases. Reprint from [57]

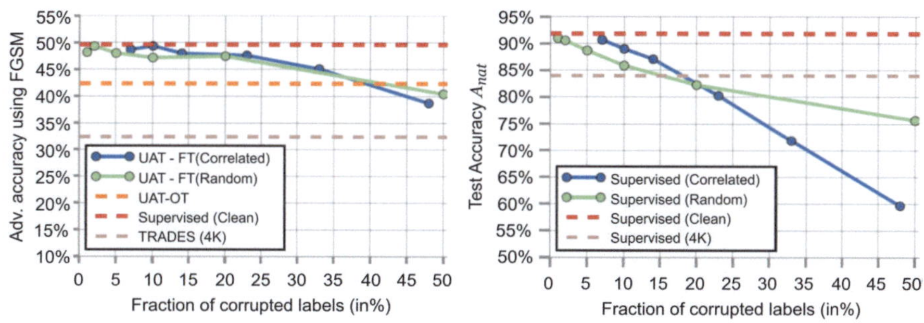

Fig. 20 Adversarial (left) and natural (right) accuracy on CIFAR-10 for UAT-FT under varying levels of label noise, compared to UAT-OT, supervised baselines, and TRADES. Reprint from [57]

UAT++ combines the strengths of UAT-OT and UAT-FT by simultaneously minimizing the smoothness loss and the pseudo-label-based adversarial loss. The overall objective for UAT++ is shown in Eq. 48, where $L_{\text{sup}}(\theta)$ represents the supervised loss on labeled examples, and λ is a hyperparameter controlling the weight of the unsupervised losses. By integrating both methods, UAT++ benefits from the label-independent smoothness properties of UAT-OT and the classification improvements of UAT-FT, achieving superior performance [57].

$$L(\theta) = L_{\text{sup}}(\theta) + \lambda \left(L_{\text{OT-unsup}}(\theta) + L_{\text{FT-unsup}}(\theta) \right) \tag{48}$$

The experimental results strongly support the efficacy of these methods. As shown in Fig. 19, UAT++ achieves robust accuracies of 54.1% and 84.4% on CIFAR-10 and SVHN, respectively, against FGSM attacks, nearly matching the performance of a supervised oracle that uses full labels. The robustness of UAT-FT to label noise is particularly notable; even when 50% of the pseudo-labels are corrupted, the model maintains an 8% improvement in adversarial accuracy over baseline adversarial training methods, as illustrated in Fig. 20.

The scalability of UAT is further highlighted through experiments on uncurated datasets, such as the 80 Million Tiny Images dataset. By leveraging up to 200K unlabeled samples, UAT++ achieves an adversarial accuracy of 63.65% under the MultiTargeted attack, outperforming state-of-the-art methods like TRADES [57]. This demonstrates the practicality of UAT in real-world settings where labeled data is scarce, opening new avenues for robust machine learning in domains such as medical diagnostics and autonomous systems.

Overall, UAT and its variants provide a comprehensive framework for adversarial training that effectively utilizes unlabeled data. By combining theoretical insights with strong empirical performance, these methods address key challenges in adversarial robustness, setting a new standard for scalable and efficient adversarial defense.

Adversarial training is one of the most extensively studied and widely adopted methods for defending against adversarial attacks. Over the years, many variations and enhancements to adversarial training have been proposed, each addressing specific challenges in improving model robustness. These methods aim to refine the standard adversarial training process, tackle computational inefficiencies, and explore new strategies to enhance defenses. Examples include **TRadeoff-inspired Adversarial DEfense via Surrogate-loss minimization (TRADES)** [58], which balances robustness and accuracy by optimizing a trade-off between natural error and adversarial robustness, and **Misclassification-Aware Adversarial Training (MART)** [59], which explicitly differentiates between correctly classified and misclassified examples, incorporating a misclassification-aware regularization term.

Another line of work focuses on making adversarial training computationally efficient. Methods like **Smooth Adversarial Training (SAT)** [60] employ curriculum learning principles, gradually increasing the difficulty of perturbations during training to stabilize the network. **Free Adversarial Training (Free-AT)** [61] eliminates the overhead of generating strong adversarial examples by reusing gradient information during weight updates. Similarly, **You Only Propagate Once (YOPO)** [62] limits backpropagation to the first layer of the network, significantly reducing computational costs without compromising robustness.

Some methods explore alternative ways to generate adversarial examples or use novel objectives to enhance training. **Latent Adversarial Training (LAT)** [63] focuses on fine-tuning adversarial-trained models at the latent layer to improve robustness at an intermediate level, addressing vulnerabilities to small-magnitude perturbations. **Helper-based Adversarial Training (HAT)** [64] incorporates additional wrongly labeled examples during training, achieving a balance between accuracy and robustness.

By integrating these innovative approaches, adversarial training has evolved into a versatile and powerful defense strategy. The development of these methods reflects ongoing efforts to refine adversarial defenses and address emerging challenges, demonstrating the dynamic nature of this critical field of research.

5.2 The Training Process Modification

This area encompasses adjustments to the learning paradigm, objective functions, or optimization algorithms to bolster model robustness and many methods have been proposed in this area.

Defensive Distillation (DD) [65] is a method designed to improve the robustness of DNNs against adversarial attacks by modifying the training process to reduce the sensitivity of the model to small, malicious perturbations. This method builds on the concept of knowledge transfer introduced in traditional distillation, where information is transferred from one model (often a larger model) to another (often a smaller model) to retain performance while reducing complexity. Defensive distillation repurposes this mechanism to enhance the security and resilience of DNNs, as opposed to optimizing their size or computational efficiency.

The process begins with training an initial model F using a dataset D. The output of the model for an input X is a probability distribution over the possible classes, calculated using the softmax function as shown in Eq. 49, where $z_i(X)$ is the logit (output of the last hidden layer) for class i, N is the total number of classes, and T is the temperature parameter. A high T smooths the probability distribution, making it less discrete and encoding more information about inter-class relationships [65].

$$F_i(X) = \frac{\exp(z_i(X)/T)}{\sum_{j=1}^{N} \exp(z_j(X)/T)} \tag{49}$$

Unlike a standard softmax, which outputs a single class with near certainty, the high-temperature softmax spreads the probability mass across all classes, providing richer information about the relative likelihoods of each class. For example, rather than outputting a classification of "cat" with a probability of 1 and all other classes as 0, a high-temperature softmax might output probabilities like 0.6 for "cat", 0.3 for "dog", and 0.1 for "rabbit", encoding the model's uncertainty and inter-class similarities.

The initial network is trained by minimizing the cross-entropy loss between the model's predicted probabilities $F(X)$ and the hard labels $Y(X)$ in the dataset as shown in Eq. 50. Here, $Y(X)$ is a one-hot encoded vector representing the true class of X, and $Y_i(X) = 1$ only for the correct class.

$$L = -\frac{1}{|D|} \sum_{X \in D} \sum_{i=1}^{N} Y_i(X) \log F_i(X) \tag{50}$$

After training the initial model, the outputs $F(X)$, which are soft probability distributions, are used to label the training data. These soft labels contain more nuanced information compared to the hard labels, as they reflect the model's belief about the likelihood of each class for a given input [65].

The second model F_d, which shares the same architecture as F, is trained using the soft-labeled dataset. The loss function in this step minimizes the cross-entropy between the soft labels $F(X)$ and the distilled model's predictions $F_d(X)$ as shown in Eq. 51. This equation ensures that the distilled model mimics the behavior of the original model while generalizing better, particularly for inputs near decision boundaries [65].

$$L_{\text{distilled}} = -\frac{1}{|D|} \sum_{X \in D} \sum_{i=1}^{N} F_i(X) \log F_{d,i}(X) \tag{51}$$

The robustness of the distilled model is linked to the sensitivity of its output to small input changes, quantified by the Jacobian matrix as shown in Eq. 52. Increasing the temperature T reduces the magnitude of the Jacobian components, making the model less sensitive to perturbations in the input. Specifically, the high temperature effectively normalizes logits, ensuring smoother decision boundaries and less vulnerability to adversarial attacks [65].

$$\frac{\partial F_i(X)}{\partial X_j} = \frac{\partial}{\partial X_j}\left(\frac{\exp(z_i(X)/T)}{\sum_{k=1}^{N} \exp(z_k(X)/T)}\right) \tag{52}$$

The key advantage of this approach is that the softer probability distributions carry additional information about the relationships between classes, which hard labels (one-hot encoding) fail to capture. This implicit regularization helps the distilled network F_d generalize more effectively, particularly in regions surrounding training samples, where adversarial perturbations are typically crafted. At test time, the softmax temperature is reset to $T = 1$, allowing the distilled network to make standard discrete predictions [65].

Empirical evaluations demonstrate the effectiveness of defensive distillation in significantly reducing the success rate of adversarial attacks while maintaining high classification accuracy. The experiments are conducted on two benchmark datasets: MNIST and CIFAR-10, using two 9-layer DNN architectures. These architectures differ primarily in their complexity, with the MNIST model tailored for grayscale handwritten digits and the CIFAR-10 model adapted for complex color images. Both models achieve baseline performances comparable to state-of-the-art DNNs without distillation [65].

Results in Fig. 21 clearly illustrate the relationship between the distillation temperature T and the success rate of adversarial attacks. For the MNIST network, defensive distillation reduces the attack success rate from 95.89% (non-distilled baseline) to 0.45% at a distillation temperature $T = 100$. Similarly, for the CIFAR-10 network, the success rate drops from 87.89% to 5.11%, highlighting a significant improvement in robustness. Figure 21 shows a consistent trend: increasing the distillation temperature smooths the decision boundaries of the model, making it more resistant to adversarial perturbations. The elbow point observed in Fig. 21 (around $T = 40$) indicates the temperature at which further increases yield diminishing returns, as the class probability distributions become sufficiently uniform.

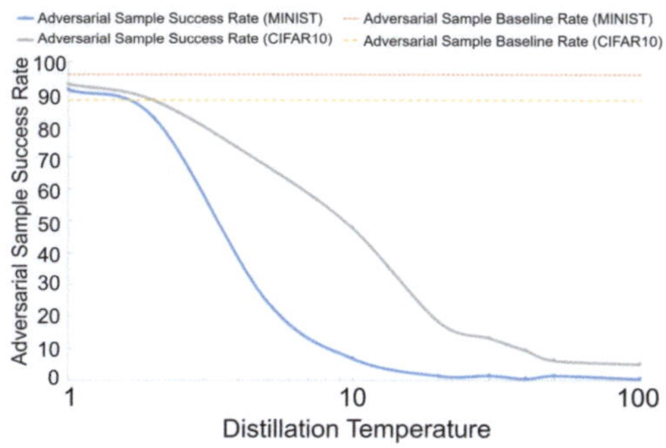

Distillation Temperature	MINST Adversarial Samples Success Rate (%)	CIFAR10 Adversarial Samples Success Rate (%)
1	91	92.78
2	82.23	87.67
5	24.67	67
10	6.78	47.56
20	1.34	18.23
30	1.44	13.23
40	0.45	9.34
50	1.45	6.23
100	0.45	5.11
No distillation	95.89	87.89

Fig. 21 Success rate of adversarial attacks for MNIST and CIFAR-10 models at varying distillation temperatures. Defensive distillation significantly reduces adversarial success, with an elbow point observed around $T = 40$. Reprint from [65]

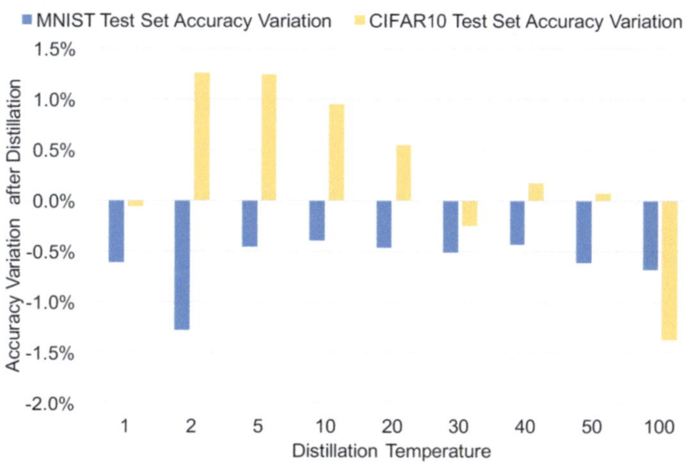

Fig. 22 Variation in classification accuracy for MNIST and CIFAR-10 models trained with defensive distillation at different temperatures. Accuracy remains stable, with variations below 1.37%. Reprint from [65]

To evaluate the impact of distillation on classification accuracy, Fig. 22 plots the variation in test accuracy for different distillation temperatures. The results show that distillation has negligible effects on accuracy, with variations of less than 1.37% for the CIFAR-10 network and 1.28% for the MNIST network across all tested temperatures. At $T = 20$, for example, the MNIST model retains an accuracy of 99.05%, while the CIFAR-10 model achieves 81.39%, both comparable to their non-distilled baselines. These findings confirm that defensive distillation enhances adversarial robustness without sacrificing the model's overall performance [65].

To better understand the mechanism behind this improvement, the gradient magnitudes of the network outputs with respect to the inputs have been analyzed. Figure 23 shows histograms of the gradient amplitudes for models trained with

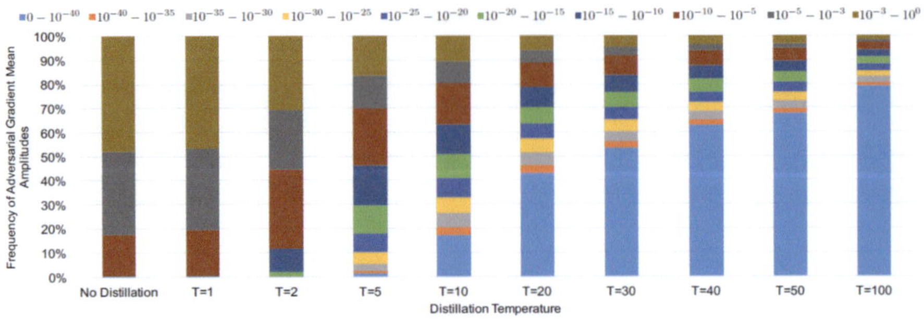

Fig. 23 Distribution of gradient magnitudes for models trained with and without defensive distillation. Distillation reduces gradient amplitudes, making the model less sensitive to adversarial perturbations. Reprint from [65]

and without distillation. For non-distilled models, the gradients are concentrated in higher magnitude ranges, which adversaries exploit to craft perturbations. In contrast, distilled models exhibit significantly reduced gradient amplitudes, with most values clustered near zero. This reduction, achieved through the high-temperature softmax during training, weakens the adversary's ability to induce misclassification with small perturbations. The figure highlights that defensive distillation reduces the gradient magnitude by up to 10^{30}, drastically increasing the effort required to craft adversarial examples [65].

Overall, these findings demonstrate that defensive distillation provides a robust and practical defense against adversarial attacks. By leveraging high-temperature softmax outputs during training, the method smooths the decision boundaries, reduces gradient sensitivity, and enhances the generalization of DNNs. The approach achieves substantial robustness improvements while maintaining classification accuracy and requiring no modifications to the underlying network architecture.

In addition to defensive distillation, several methods have been introduced to modify the training process to improve robustness against adversarial attacks. One key approach is **Robust Soft Label Adversarial Distillation (RSLAD)** [66], which enhances knowledge distillation by using robust soft labels from a teacher network to supervise the training of a student model. Unlike standard distillation, RSLAD incorporates both natural and adversarial examples to ensure that the student model generalizes well to adversarial perturbations while maintaining accuracy on clean data.

Another notable method is **Self-Adaptive Training (SAT)** [67], which combines the original training labels with the model's predictions to create dynamic, adaptive labels during training. This approach stabilizes the training process by gradually updating the training targets based on the model's performance, helping the network improve robustness over time. Similarly, **Robust Self-Training (RST)** [68] uses pseudo-labels generated by a pre-trained network to train another network, bridging the gap between clean and robust accuracy by incorporating unlabeled data effectively into the adversarial training process.

Regularization-based methods have also been explored to improve robustness by smoothing the loss surface. For example, **Curvature Regularization (CURE)** [69] penalizes large eigenvalues of the Hessian matrix to minimize sharp curvature in the loss landscape, reducing the sensitivity of the model to small input perturbations. Another technique, **Triplet Loss Adversarial (TLA)** [70], introduces a regularizer that clusters natural and adversarial examples of the same class while pushing examples from different classes further apart. This ensures that decision boundaries are well-separated, improving the model's resistance to adversarial attacks.

Data augmentation techniques have been widely used to modify the training process for better adversarial robustness. **Visual Causal Feature Learning** [71] generates augmented images with specific causal effects to strengthen the classifier's decision boundaries, while **Overfit Reduction (OR)** [72] combines semi-supervised data augmentation with early stopping to improve robustness without over-regularizing the model. **Bit Plane Feature Consistency (BPFC)** [73] introduces preprocessing by manipulating bit-plane representations of input images and enforcing consistency between the original and perturbed inputs, ensuring that higher-order features remain stable under adversarial perturbations.

Another promising direction involves manipulating the behavior of neurons within the model. **Sensitive Neuron Stabilizing (SNS)** [74] focuses on stabilizing the behavior of neurons that significantly contribute to predictions, ensuring that their activation patterns remain consistent across clean and adversarial examples. Similarly, **Controlling Neural Level Sets (CNLS)** [75] modifies the decision boundaries of networks by analyzing the sample position relative to network parameters, effectively minimizing the sensitivity of the classifier to adversarial perturbations.

The above methods, while distinct in their approaches—whether through label modification, regularization, data augmentation, or neuron behavior stabilization—collectively demonstrate the diversity of strategies for enhancing robustness by modifying the training process. By addressing different aspects of the adversarial vulnerability problem, these techniques complement adversarial training and provide a solid foundation for building robust and reliable deep learning systems.

5.3 Supplementary Networks

Supplementary Networks involve the integration of auxiliary models designed to detect or purify adversarial inputs before they reach the primary classifier. **MagNet** [76] is one of the methods in this area that addresses the problem of adversarial examples in neural networks by providing a defense mechanism that does not require modifying the target classifier or relying on adversarial examples for training. The primary insight behind MagNet is that adversarial examples often deviate from the manifold of legitimate inputs, which represents the space of natural, unperturbed data. By leveraging this observation, MagNet [76] uses two complementary components—a detector to identify adversarial inputs and a reformer to correct perturbed inputs—ensuring the model can effectively defend against a wide range of adversarial attacks.

The detector is responsible for identifying adversarial examples by determining how far an input lies from the data manifold. To achieve this, MagNet [76] employs autoencoders trained solely on clean, legitimate inputs. Autoencoders are neural networks that encode inputs into a lower-dimensional representation and then decode them back to their original space. During inference, when an input passes through the autoencoder, its reconstruction error serves as a measure of how "natural" the input is. Specifically, for an input x, the reconstruction error is calculated as shown in Eq. 53, where $AE(x)$ is the reconstructed output of the autoencoder and $\|.\|_p$ represents the L_p-norm.

$$E(x) = \|x - AE(x)\|_p \tag{53}$$

If the reconstruction error exceeds a predefined threshold t_{re}, the input is flagged as adversarial. The threshold t_{re} is chosen based on the reconstruction error distribution of clean inputs during validation. To ensure robustness to different perturbations, MagNet employs multiple detectors, where each autoencoder is trained with varying architectures or norms to provide diversity.

The adversarial examples that lie close to the data manifold may still evade detection. To address this, MagNet [76] introduces a reformer to project adversarial examples back onto the manifold of clean inputs, effectively "correcting" the perturbations. The reformer is another autoencoder, similar to the one used in the detector but optimized to reconstruct inputs in a way that aligns with the natural input space. For a given adversarial input x, the reformer outputs a reformed input x', defined as Eq. 54.

$$x' = AE_{\text{reformer}}(x) \tag{54}$$

The reformed input x' is then passed to the target classifier, where it is far less likely to be misclassified. This two-step process—detection and reforming—provides a robust defense pipeline that significantly reduces the success rate of adversarial attacks.

To evaluate MagNet's performance, extensive experiments were conducted on two benchmark datasets: MNIST and CIFAR-10. The target classifiers achieve high baseline accuracies of 99.4% on MNIST and 90.6% on CIFAR-10 when tested on clean data. Various attacks were considered in the experiments, including FGSM [31], Iterative Gradient Sign Method (IGSM) [77], DeepFool [33], and the powerful Carlini and Wagner (C&W) [24] attack, which is known for its ability to bypass many defenses.

The results demonstrate the significant improvement in robustness achieved by MagNet. On the MNIST dataset, the success rate of C&W L_2 attack drops from 100% (for the undefended model) to 0.45% when MagNet is deployed. Similarly, for the CIFAR-10 dataset, the success rate under the same attack decreases from 87.89% to 5.11%. This improvement highlights the effectiveness of the detector and reformer components in neutralizing adversarial perturbations [76].

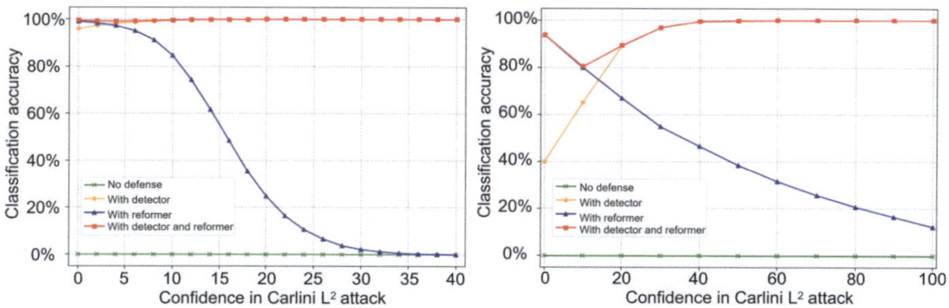

Fig. 24 Classification accuracy of MagNet on two different benchmarks including MNIST (left) and CIFAR-10 (right) under Carlini and Wagner's L_2 attack as attack confidence increases. MagNet maintains robust performance with both detection and reforming components. Reprint from [76]

To provide further insights, Fig. 24 illustrates the impact of MagNet's reformer under increasing attack strength (confidence levels) for C&W attack. As the attack confidence increases, the target classifier's accuracy drops sharply when no defense is applied. However, with MagNet in place, the accuracy remains consistently high, demonstrating the robustness provided by the reformer.

MagNet's approach [76] also extends to a scenario, where the attacker has partial knowledge of the defense system but lacks access to its specific parameters. To counteract such attacks, MagNet introduces a diversity strategy inspired by cryptographic principles. Specifically, multiple autoencoders with varying architectures and parameters are trained, and one is randomly selected during inference. This randomization prevents attackers from tailoring their perturbations to bypass a specific autoencoder, significantly increasing the difficulty of successfully crafting adversarial examples. Results show that MagNet maintains an accuracy of over 80% against this scenario, highlighting its robustness even when attackers attempt to adapt to the defense mechanism.

In summary, MagNet provides a robust and generalizable defense against adversarial attacks by detecting and reforming adversarial examples using autoencoders. The detector identifies inputs that deviate from the data manifold, while the reformer corrects perturbed inputs, ensuring accurate classification. However, it is not the only method to solve this issue. Many other methods have been proposed to address this problem. Another effective approach in this category is **Feature Squeezing** [78], which reduces the search space available to adversaries by compressing features in the input space. This method introduces a defense strategy that leverages color depth reduction and spatial smoothing as two main mechanisms of feature squeezing.

The Feature Squeezing process [78] operates by applying pre-processing steps to the input data before it is fed into a deep learning model. These steps are designed to "squeeze" redundant or adversarial features. Color depth reduction reduces the number of possible values for each pixel, limiting an adversary's ability to exploit small perturbations. For grayscale images, for example, pixel values originally represented with 8 bits can be compressed to 1 or 2 bits without significantly affecting

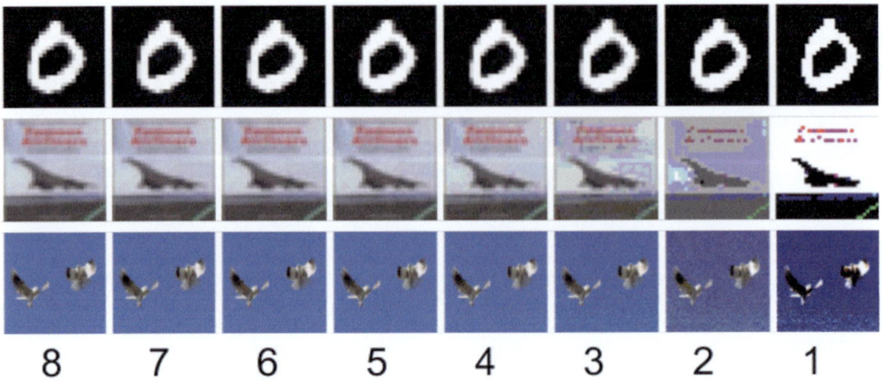

Fig. 25 Visualization of color depth reduction on MNIST and CIFAR-10 datasets. Reprint from [78]

recognizability to humans. In the experiments, Figure 25 illustrates this effect on the MNIST and CIFAR-10 datasets, where adversarial examples perturbed by state-of-the-art attacks appear much less altered after bit-depth reduction. This compression diminishes the adversary's opportunity to craft subtle perturbations while maintaining high model accuracy on clean inputs. Results confirm that squeezing to lower bit depths, such as 1-bit for MNIST and 4-bit for CIFAR-10, achieves robust performance against L_2 and L_∞ attacks.

Spatial smoothing focuses on reducing pixel variations using either local or non-local methods. Local smoothing applies filters such as the median filter to replace each pixel with the median value of its neighbors, which is particularly effective against sparse perturbations like salt-and-pepper noise generated by L_0 attacks. Non-local smoothing, in contrast, averages pixel values over larger image regions with similar patterns, effectively removing perturbations while preserving the edges of objects [78].

The method further presents joint detection by combining multiple squeezers. Predictions on the original input are compared against predictions on "squeezed" versions using the L_1 norm to quantify discrepancies. A significant difference indicates a likely adversarial input. Figure 26 visually demonstrates this joint detection framework, where the model predictions on squeezed inputs are cross-compared to flag inconsistencies. This strategy successfully achieves high detection rates across various attacks.

While Feature Squeezing [78] is computationally inexpensive compared to adversarial training, its simplicity makes it highly complementary to other defenses. For instance, combining adversarial training with feature squeezing boosts robustness, as illustrated in Fig. 27. Despite its effectiveness, the paper acknowledges that adaptive adversaries capable of accounting for feature squeezing might eventually succeed. However, such attacks often require significant computational resources, as demonstrated in follow-up evaluations with adaptive attacks.

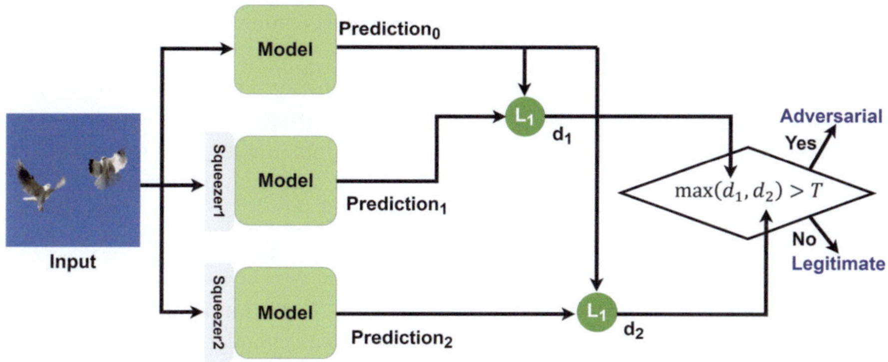

Fig. 26 The feature-squeezing framework detects adversarial examples by evaluating the model's predictions on both the original input and its pre-processed versions generated by feature squeezers. Reprint from [78]

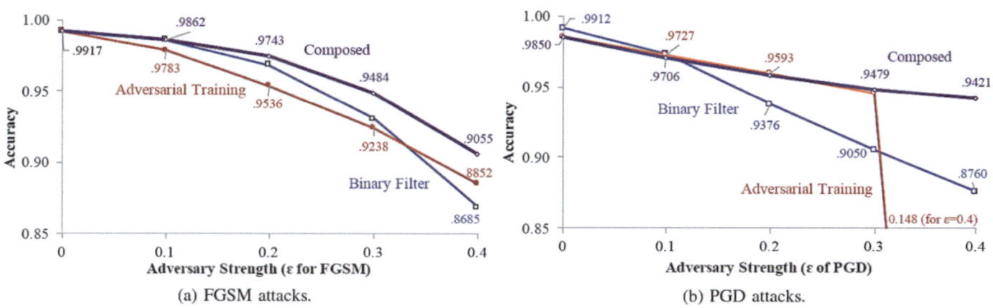

Fig. 27 Combining adversarial training with feature squeezing. The horizontal axis represents the adversary's strength (ϵ), increasing from left to right. Adversarial training is conducted with $\epsilon = 0.3$ for both FGSM and PGD attacks. Integrating the 1-bit filter with the adversarially trained model consistently achieves the best performance. Reprint from [78]

The experimental results affirm the robustness of Feature Squeezing methods, particularly in mitigating adversarial examples generated by state-of-the-art attacks. These findings establish feature squeezing as a lightweight yet powerful technique for adversarial defense, capable of significantly complicating adversarial efforts without sacrificing accuracy on legitimate inputs [78].

The use of supplementary networks in adversarial defense encompasses a variety of methods, each leveraging different strategies to detect or mitigate adversarial examples before they reach the classifier. These approaches primarily include adversarial detection, noise removal, and latent space modifications, providing flexibility and adaptability to various attack scenarios [29].

Among the notable methods, the **Adversary Detection Network (ADN)** [79] is designed to identify whether input data is adversarial or clean. ADN trains on adversarial images generated for a target classifier, distinguishing malicious perturbations through learned patterns. In contrast, the **Detector Graph (DG)** [80] employs a Latent Neighborhood Graph for each input image and utilizes Graph Neural Networks (GNNs) to distinguish between adversarial and clean examples based on pixel embeddings.

Denoising techniques offer another promising pathway. The **High-Level Representation Guided Denoiser (HGD)** [81] utilizes a denoising autoencoder trained to minimize the difference between adversarial and original images, suppressing adversarial perturbations while maintaining image fidelity. Similarly, **Defense-GAN** [82] leverages Wasserstein GANs (WGANs) to approximate clean image distributions and iteratively denoise adversarial examples by minimizing the L_2 distance between the input and the generated representation.

Some methods rely on latent space regularization. For example, the **Embedding Regularized Classifier (ER-Classifier)** [83] employs an encoder and a discriminator to project adversarial and clean images into a latent code space, enforcing regularity by aligning adversarial examples closer to clean data distributions.

Overall, these methods underscore the versatility of supplementary networks in adversarial defense, ranging from detection and denoising to latent space manipulation. Their ability to operate as modular components without altering the primary classifier makes them particularly useful in real-world scenarios. Such approaches continue to evolve, combining detection and purification mechanisms for increased robustness against both known and unknown adversarial attacks.

5.4 Network Architecture Modification

Neural network architectures play a critical role in determining a model's robustness against adversarial attacks. While most existing approaches focus on adversarial training, regularization, or preprocessing techniques, some methods focus on the architecture. **Robust Neural Architecture Search (RobNAS)** [84] is one of them which systematically explores the influence of neural network design on adversarial robustness. Through the adoption of a one-shot neural architecture search (NAS), a wide range of architectures are evaluated efficiently to identify key architectural patterns that enhance robustness. The authors propose a family of robust networks, termed **RobNets**, which exhibit superior resilience against both white-box and black-box adversarial attacks.

The method begins with a carefully constructed search space based on computation cells, which act as building blocks of the network. Each cell is represented as a directed acyclic graph (DAG), where nodes correspond to feature maps, and edges denote candidate operations. The operations considered are restricted to 3×3 separable convolutions, identity mappings, and zero operations, simplifying the search space without compromising the expressive capacity. A supernet is first trained, which subsumes all candidate architectures. Architectures are then sampled from the supernet and fine-tuned using adversarial training to evaluate their robustness efficiently.

To evaluate robustness, the authors employ PGD adversarial training, which follows a min-max optimization framework to produce adversarial examples during

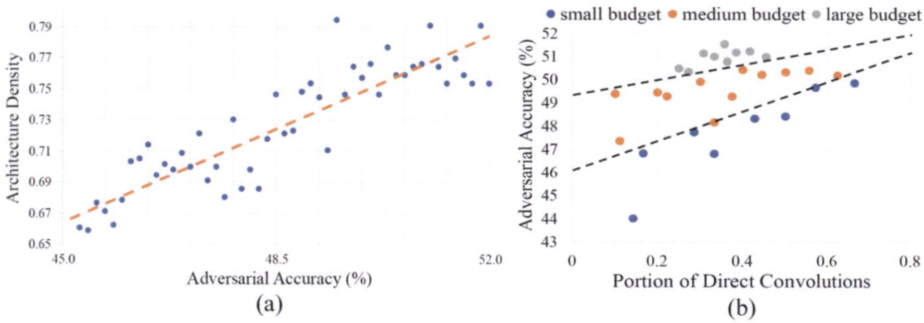

Fig. 28 **a** Correlation between architecture density and adversarial accuracy. Densely connected architectures exhibit higher robustness under adversarial attacks; **b** Effect of convolution operations under different computational budgets. Adding convolutions to direct edges improves robustness, especially for small and medium budgets. Reprint from [84]

training as shown in Eq. 55 where $S = \{x' |\ \|x - x'\|_\infty \leq \epsilon\}$, f is the neural network, and L is the classification loss. The adversarial accuracy of the architectures is then evaluated using white-box PGD attacks with up to 100 iterations.

$$\min_{\theta} \mathbb{E}_{(x,y)\sim D} \left[\max_{x' \in S} L(y, f(x'; \theta)) \right] \tag{55}$$

Extensive experiments reveal several key findings regarding the relationship between network design and adversarial robustness. First, densely connected patterns significantly improve robustness. Figure 28a illustrates a clear positive correlation between architecture density (defined as the proportion of connected edges) and adversarial accuracy. This observation explains why architectures like DenseNet, which feature dense connectivity, often outperform standard architectures like ResNet under adversarial attacks. Second, under limited computational budgets, convolution operations on direct edges (edges connecting consecutive nodes) are found to be more effective in improving robustness than operations on skip connections. Figure 28b highlights that for small and medium budgets, increasing the number of convolutions on direct edges results in substantial gains in adversarial accuracy [84].

RobNets method further explores feature flow between network layers to identify statistical indicators of robust architectures. By analyzing the Flow of Solution Procedure (FSP) matrix, which measures the correlations between input and output feature maps within each cell, they demonstrate that robust architectures exhibit smaller FSP distances between clean and adversarial examples in deeper layers. Specifically, for a given network, the FSP matrix loss is defined as Eq. 56, where $G(x)$ and $G(x')$ are the FSP matrices for clean and adversarial inputs, respectively [84].

$$L_{\mathrm{FSP}} = \frac{1}{N} \sum_{x} \|G(x) - G(x')\|_2^2 \tag{56}$$

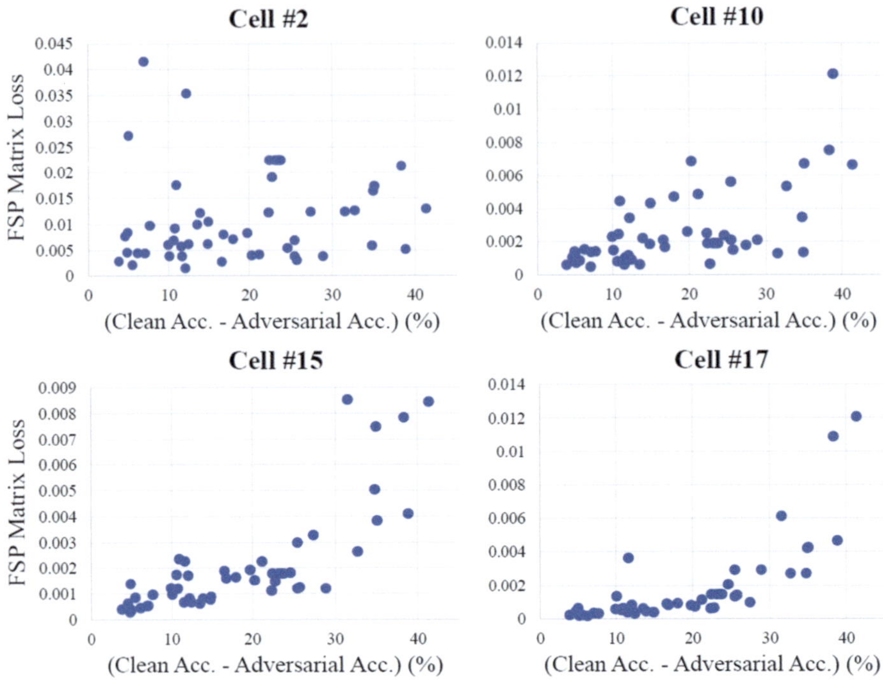

Fig. 29 Analysis of the FSP matrix distance. Robust networks show lower FSP matrix loss in deeper cells, correlating with improved adversarial robustness. Reprint from [84]

Figure 29 demonstrates a strong correlation between lower FSP loss and higher adversarial robustness, providing an efficient metric for identifying robust architectures without extensive fine-tuning.

The resulting RobNet family [84] consists of architectures optimized for robustness under varying computational budgets. Experimental evaluations on CIFAR-10, SVHN, Tiny-ImageNet, and ImageNet confirm the effectiveness of RobNets. RobNets achieve a significant improvement in adversarial accuracy, outperforming widely-used architectures such as ResNet-18, ResNet-50, WideResNet, and DenseNet, even with fewer parameters. For instance, RobNet-large achieves 49.24% robust accuracy under PGD-100 attacks on CIFAR-10, compared to 47.46% for DenseNet-121 and 45.10% for ResNet-18.

In summary, this method provides a systematic study of the relationship between network architecture and adversarial robustness. By leveraging one-shot NAS and analyzing architectural properties, the authors identify key patterns, such as dense connectivity and effective feature flow, that contribute to robustness. The resulting RobNet family not only achieves state-of-the-art adversarial performance under both white-box and black-box attacks but also demonstrates generalizability across datasets, offering a promising direction for designing robust neural networks.

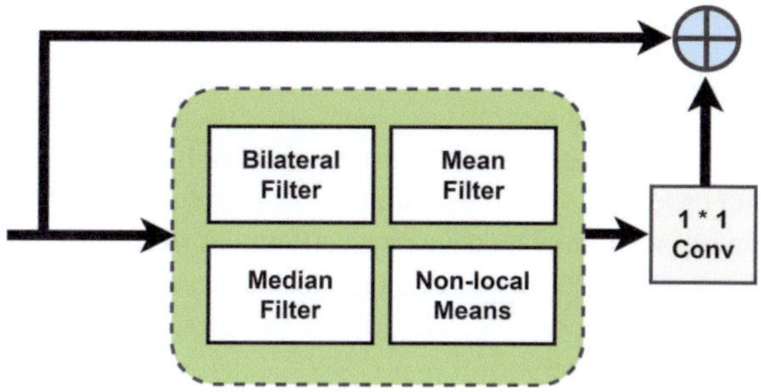

Fig. 30 An overview of feature denoising block. Reprint from [29]

In addition to RobNets, which leverage neural architecture search for adversarial robustness, several methods focus on modifying the network architecture to improve resilience to adversarial attacks. These approaches aim to embed robustness directly into the design of the network, reducing its susceptibility to adversarial perturbations without relying solely on adversarial training. Among these methods, **HYDRA** [85] stands out as an adversarial pruning framework that removes connections detrimental to robustness. Unlike conventional pruning, HYDRA incorporates robust training objectives to ensure that the resulting compressed models maintain competitive standards and adversarial accuracy.

Another notable approach is **Dynamic Network Rewiring (DNR)** [86], which utilizes a unified constrained optimization framework to prune and rewire neural networks. This method employs a hybrid loss function that balances model compression and robust adversarial training. By dynamically rewiring the network's connectivity, DNR achieves high robust accuracy while preserving performance on clean data.

Feature Denoising (FD) [87] is another method designed to counter adversarial perturbations by introducing feature-denoising blocks into intermediate layers of CNNs. These blocks apply denoising operations to reduce adversarial noise, as illustrated in Fig. 30. This strategy makes the network more robust without significant overhead, as the denoising blocks seamlessly integrate into existing architectures.

To address adversarial perturbations at the input stage, **Input Randomization** [88] introduces layers that perform random resizing and padding of input images. By randomly scaling and zero-padding images before feeding them into the network, Input Randomization disrupts the adversary's ability to craft effective perturbations. Similarly, the **Sparse Transformation Layer (STL)** [89] transforms input images into a low-dimensional quasi-natural representation. This transformation suppresses adversarial perturbations and improves the network's resilience to attacks.

The above methods, while diverse in their approaches, share a common goal of creating inherently robust architectures by strategically modifying layers, connections, and activations. By integrating these changes at the architectural level, they provide complementary solutions to adversarial training and supplementary network methods, further enhancing the robustness of deep neural networks.

5.5 *Perform Network Validation*

Deep learning systems deployed in critical applications, such as autonomous driving and malware detection, must be rigorously tested to ensure their reliability under corner cases and adversarial inputs. Many methods have been developed for this purpose, one of which is **DeepXplore** [90] which introduces a novel white-box testing framework designed to validate DNNs systematically by uncovering erroneous behaviors and ensuring broader coverage of the network's internal logic. The key motivation lies in the observation that DNNs, despite their impressive performance, can exhibit unexpected and unsafe behaviors under corner-case inputs or slight perturbations. Traditional testing methods, such as manual input generation or black-box adversarial attacks, fail to provide systematic guarantees and often only cover a small portion of the model's decision space. DeepXplore overcomes these limitations by combining differential testing and neuron coverage maximization into a unified approach.

The core challenge in validating DNNs is twofold: identifying inputs that (1) activate diverse parts of the network and (2) cause incorrect behavior. To solve this, DeepXplore [90] introduces the concept of neuron coverage, inspired by traditional software testing metrics like statement or branch coverage. Neuron coverage measures the percentage of neurons activated beyond a predefined threshold t, which quantifies how much of the network's decision space has been exercised. Given a set of test inputs T, the neuron coverage is formally defined as Eq. 57, where N is the total number of neurons in the network, and $out(n, x)$ is the activation output of neuron n for input x. Maximizing neuron coverage ensures that the generated inputs explore different regions of the model, helping to uncover hidden flaws that might not manifest under traditional testing.

$$\mathrm{NCov}(T) = \frac{|\{n \mid \forall x \in T, \mathrm{out}(n, x) > t\}|}{|N|} \tag{57}$$

To identify erroneous behaviors, DeepXplore employs differential testing by leveraging multiple DNNs trained for the same task as cross-referencing oracles. The intuition is that while a single model might behave unpredictably for corner cases, different models trained on the same task are unlikely to agree on the same erroneous predictions. For an input x, if two DNNs F_i and F_j produce conflicting outputs, the input is flagged as a "difference-inducing input" and considered as potentially adversarial or erroneous. Formally, this is defined as Eq. 58, where $F_i(x)$ represents the prediction of model i. This cross-referencing method eliminates the need for a manual oracle or labeled ground-truth data during testing.

$$\mathrm{Diff}(x) = \exists i, j \text{ such that } F_i(x) \neq F_j(x), \;\; i \neq j \tag{58}$$

To efficiently generate inputs that maximize both neuron coverage and the likelihood of difference-inducing behaviors, DeepXplore [90] formulates the process as a joint optimization problem. The goal is to perturb inputs in such a way that they activate new neurons while inducing inconsistent behaviors across models. This is

Fig. 31 An example of erroneous behavior uncovered by DeepXplore on Nvidia's DAVE-2 self-driving car model. A slight reduction in brightness causes the car to steer incorrectly, leading to a crash. Reprint from [90]

achieved using gradient-based search, where gradients of the model outputs with respect to the input are used to iteratively modify the inputs. The optimization objective is expressed as Eq. 59, where $F_i(x)[c]$ is the confidence of model F_i for class c, and λ_1 and λ_2 are hyperparameters that balance the trade-off between inducing differences and maximizing neuron coverage. The gradient ascent process ensures that inputs are iteratively perturbed while adhering to domain-specific constraints, such as maintaining valid pixel values for images.

$$\text{obj}_{\text{joint}} = \sum_{i \neq j} \left[F_i(x)[c] - \lambda_1 F_j(x)[c] \right] + \lambda_2 \cdot \text{Neuron Coverage}(x) \tag{59}$$

The DeepXplore method offers several advantages. First, its white-box nature allows the gradients of the network to be directly leveraged, ensuring efficiency in input generation. Second, by maximizing neuron coverage, the framework systematically explores different parts of the network's decision space, uncovering hidden vulnerabilities. Finally, the use of differential testing provides an automated way to identify erroneous behaviors without relying on labeled datasets or predefined rules.

Once the method is clearly defined, DeepXplore [90] is evaluated across multiple datasets and tasks to demonstrate its effectiveness. For example, experiments on the Udacity self-driving car dataset reveal critical flaws in Nvidia's DAVE-2 model, where minor perturbations in lighting cause the model to produce dangerous mispredictions. In Fig. 31, a slight reduction in brightness causes the car to incorrectly steer toward a guardrail, illustrating how small changes can lead to significant failures. These real-world examples highlight the sensitivity of deep learning models to subtle input variations and the necessity of systematic testing frameworks like DeepXplore.

The evaluation further includes experiments on standard image classification datasets such as MNIST and ImageNet. The results show that DeepXplore [90] generates thousands of difference-inducing inputs across the tested models. Compared to adversarial testing and random input selection, DeepXplore achieves up to 34.4% higher neuron coverage, demonstrating its ability to explore more regions of the network's decision space. This increased coverage translates to improved detection of errors and adversarial behaviors that would otherwise remain hidden.

In conclusion, DeepXplore [90] provides a powerful and systematic framework for validating deep learning models. By combining neuron coverage maximization with differential testing, it efficiently uncovers erroneous behaviors and adversarial vulnerabilities while ensuring a broad exploration of the network's internal logic. The ability to generalize across tasks, improve robustness through retraining, and operate efficiently makes DeepXplore a valuable tool for ensuring the safety and reliability of DNNs in critical applications.

Several techniques have been proposed to validate neural networks against adversarial attacks, ensuring their reliability and robustness. One prominent method in this area is **DeepGauge** [91], which focuses on creating comprehensive multi-granularity testing criteria. DeepGauge evaluates both neuron-level and layer-level coverage during testing. This multi-faceted testing considers not only the primary function of a network but also its behavior under extreme corner-case inputs. By identifying neuron activation patterns, DeepGauge can reveal weaknesses that adversarial inputs might exploit, making it a robust method for assessing network reliability.

Another noteworthy approach is **Surprise Adequacy for Deep Learning Systems (SADL)** [92]. This method introduces the concept of "Surprise Adequacy", which measures the behavior difference between a given input and the data already learned by the network. SADL quantifies how "surprising" an input is to the model, ensuring that the input falls within the expected range. This is particularly effective for ensuring the comprehensive coverage of inputs during validation. SADL uses this metric as an adequacy criterion to assess the robustness of a neural network.

Formal verification methods have also emerged as critical tools in network validation. One such method involves **SMT (Satisfiability Modulo Theory)-based solvers** [93], such as the Reluplex solver. SMT solvers mathematically validate the safety properties of neural networks by encoding network behaviors into logical constraints. This method can rigorously verify the absence of adversarial behaviors under predefined conditions, offering provable guarantees about the model's robustness.

These methods highlight the diversity of validation techniques, each addressing different aspects of adversarial defense. While methods like DeepGauge and SADL focus on behavior-driven metrics, formal verification, and abstract interpretation emphasize rigorous mathematical guarantees, ensuring comprehensive protection against adversarial inputs. Together, these strategies play a crucial role in enhancing the reliability and robustness of deep learning systems.

5.6 Adversarial Purification

Deep learning models have demonstrated remarkable accuracy on clean datasets; however, their vulnerability to adversarial perturbations remains a significant challenge. Adversarial training, one of the most effective defense strategies, typically requires large amounts of labeled data to enhance robustness. Unfortunately, real-world datasets are often limited in size and diversity.

To address this limitation, a promising approach known as Adversarial Purification has emerged. Among the methods in this area, one innovative solution leverages data generated by generative models to augment the training process [94]. This approach enables the creation of robust classifiers without relying on external, real-world datasets. Instead, it generates synthetic data that approximates the original data distribution. Crucially, the synthetic data is pseudo-labeled using a pre-trained non-robust classifier, allowing seamless integration into the adversarial training pipeline.

The method begins with the training of a generative model on the original training dataset. This generative model learns to approximate the data distribution $P_{real}(x)$, capturing the visual features and diversity of the original inputs. Various generative models, including Denoising Diffusion Probabilistic Models (DDPM) [95], BigGAN [96], StyleGAN2 [97], and Very Deep Variational Autoencoders (VDVAE) [94], are explored to produce high-quality synthetic samples. Among these, DDPM is highlighted as particularly effective due to its ability to generate diverse, high-fidelity images that closely align with the real data distribution. The synthetic data generated by these models is critical to addressing the limitations of small datasets by introducing additional training samples.

To integrate the generated data into the training process, a non-robust classifier $f_N R$, pre-trained on the original dataset, is used to assign pseudo-labels to the synthetic data. These pseudo-labels approximate the ground-truth labels of the generated samples and ensure that the classifier can incorporate the new data without requiring manual labeling. The generative model outputs a dataset $\hat{D}$ of synthetic samples, where each sample x' is associated with a pseudo-label $f_{NR}(x')$. The pseudo-labeled generated data is then combined with the original dataset to create a comprehensive training set for adversarial training [94].

The adversarial training procedure is conducted using both real and generated data, aiming to optimize a loss function that balances the contributions of the two data sources. Formally, the objective is defined as Eq. 60, where L_{ce} is the cross-entropy loss, θ represents the robust classifier's parameters, α balances the contribution of real data and generated data, S defines the adversarial perturbation set, and f_{NR} is the pre-trained non-robust classifier. The generated dataset $\hat{D}$ is pre-sampled offline, and pseudo-labels are assigned to these generated samples by the non-robust classifier [94].

$$\begin{aligned} \arg\min_{\theta}\ & \alpha \cdot \mathbb{E}_{(x,y)\in D_{\text{train}}}\left[\max_{\delta\in S} L_{\text{ce}}(f(x+\delta;\theta), y)\right] \\ & + (1-\alpha)\cdot \mathbb{E}_{x\sim\hat{D}}\left[\max_{\delta\in S} L_{\text{ce}}(f(x+\delta;\theta), f_{\text{NR}}(x))\right] \end{aligned} \tag{60}$$

The first term corresponds to adversarial training on real data, while the second term performs adversarial training on generated data, with pseudo-labels assigned by

f_{NR}. The hyperparameter α controls the relative contribution of real and generated data during training.

During the training process, adversarial examples $x + \theta$ are generated using PGD, one of the most powerful adversarial attack methods. The inner maximization problem searches for the worst-case perturbation θ that maximizes the loss while adhering to the predefined perturbation set S. The outer minimization ensures that the model parameters θ are updated to minimize the adversarial loss for both real and generated data (Fig. 32).

A critical insight of this approach is that even low-quality generated data can contribute to robustness. For example, class-conditional Gaussian noise models, which produce simple and low-fidelity synthetic samples, still provide meaningful benefits when incorporated into adversarial training. This observation challenges the common assumption that generative models must produce high-fidelity outputs to be effective. This method further demonstrates that as the quality and diversity of the generated data improve, so does the model's robust accuracy, emphasizing the importance of using advanced generative models like DDPM [94].

During the training process, adversarial examples $x + \theta$ are generated using PGD, one of the most powerful adversarial attack methods. The inner maximization problem searches for the worst-case perturbation θ that maximizes the loss while adhering to the predefined perturbation set S. The outer minimization ensures that the model parameters θ are updated to minimize the adversarial loss for both real and generated data.

The method introduces an optimal mixing ratio α, which balances the use of real and generated data. Through extensive experimentation, it is shown that an α value of approximately 0.8 (80% real data, 20% generated data) provides the best results, achieving a strong trade-off between maintaining clean accuracy and improving robustness. The generated data enhances the coverage of the input space, helping adversarial training identify smoother and more robust decision boundaries [94].

The proposed approach also includes a theoretical analysis that establishes sufficient conditions for generated data to improve robustness. The analysis highlights that as long as the generative model approximates the true data distribution reasonably well and the pseudo-labeling classifier achieves sufficient accuracy, the generated data will complement the real data in training. This explains why even imperfectly labeled synthetic samples can provide significant robustness improvements [94].

One of the most interesting findings of this method [94] is that even low-quality synthetic data, such as samples from a class-conditional Gaussian model, can improve robustness. This challenges the assumption that high visual fidelity is a strict requirement for adversarial robustness. For example, as shown in Fig. 33a, augmenting the CIFAR-10 dataset with Gaussian-generated data improves robust accuracy by up to 0.93% compared to training with real data alone. However, higher-quality generative models, such as DDPM, yield significantly better results due to their ability to generate diverse and complementary samples.

The experiments demonstrate that combining real and generated data during adversarial training leads to substantial improvements in robust accuracy. On the CIFAR-10 dataset, against L_∞ norm perturbations of size $\epsilon = 8/255$, the proposed

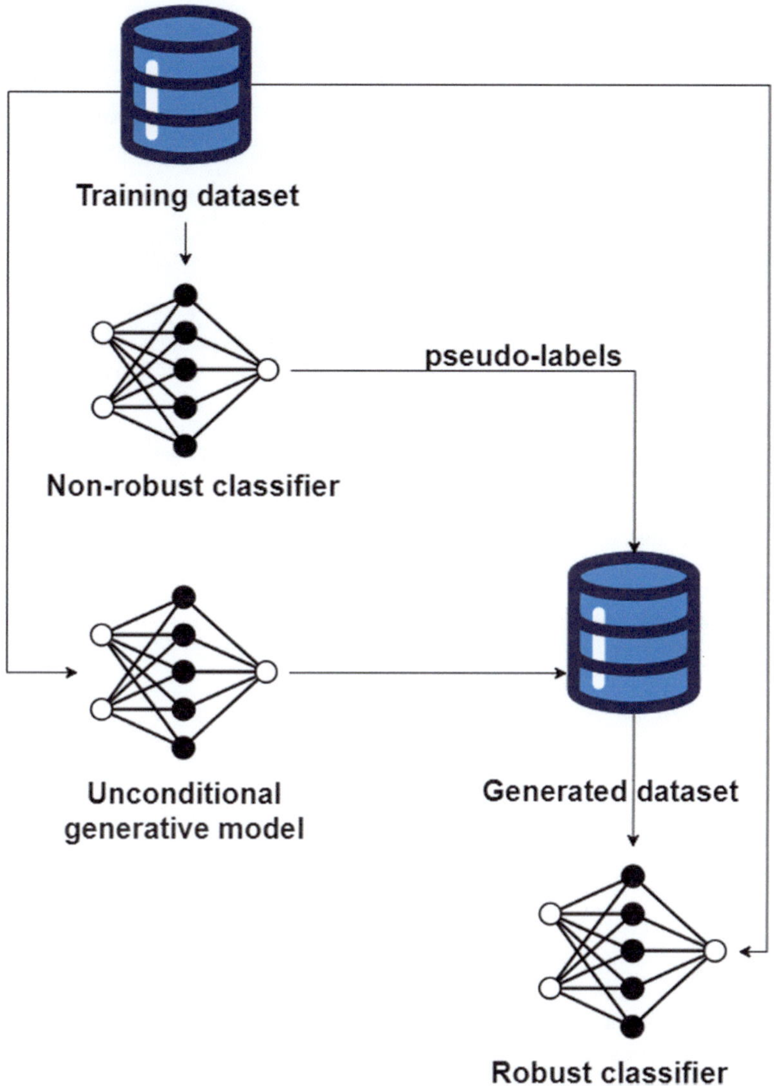

Fig. 32 Overview of the proposed method. A generative model produces synthetic data, which is pseudo-labeled using a pre-trained non-robust classifier. The generated data is then combined with real data for adversarial training. Reprint from [94]

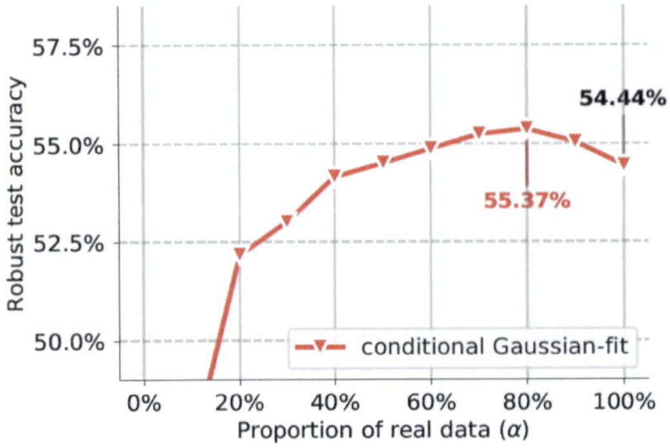

(a)

(b)

Fig. 33 Robust test accuracy on CIFAR-10 when mixing real data with Gaussian-generated samples. Even low-quality generated data improves robustness, with optimal gains observed at an 80:20 mix. Reprint from [94]

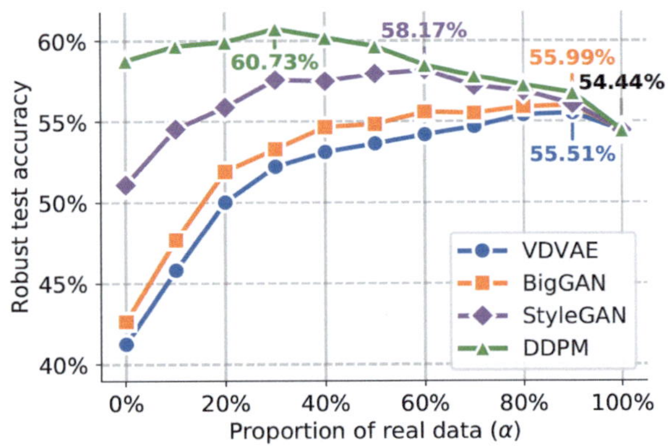

Fig. 34 Impact of the mixing factor α on robust accuracy for CIFAR-10 using DDPM-generated data. The best performance is achieved with 80% real data and 20% generated data. Reprint from [94]

method achieves a robust accuracy of 66.10%, representing an 8.96% improvement over prior state-of-the-art approaches. Similarly, robust accuracy against L_2 norm perturbations reaches 78.31%, outperforming existing methods. Figure 34 shows the impact of the mixing parameter α, revealing that an optimal mix of 80% real data and 20% generated data yields the highest robust accuracy [94].

In conclusion, this method introduces an innovative approach to improving adversarial robustness by leveraging generated data. By using generative models to produce synthetic samples and assigning pseudo-labels with a non-robust classifier, the method effectively augments the training set and enhances robust generalization. The use of high-quality generative models, such as DDPM, further amplifies the gains, enabling state-of-the-art robust accuracy across multiple datasets and perturbation types. This approach provides a scalable and efficient solution to overcoming the limitations of data scarcity in adversarial training.

Adversarial purification is a promising strategy in adversarial defense that aims to remove adversarial perturbations from input data using various mechanisms, primarily centered around generative and denoising models. While we previously detailed the method leveraging generated data to improve robustness, other notable techniques have been introduced under the umbrella of adversarial purification, each with unique approaches and methodologies.

One method, **DiffPure** [98], employs Denoising Diffusion Probabilistic Models (DDPM) to purify adversarial examples. The process involves reversing the diffusion of noise added to adversarial inputs. By leveraging a trained DDPM, DiffPure iteratively denoises adversarial inputs back to their clean form. This reverse generative process significantly mitigates the effect of adversarial perturbations while maintaining high fidelity in the purified images. This method highlights the efficacy of diffusion-based approaches in adversarial purification by offering robust protection against a wide range of attacks.

Another approach, **DensePure** [99], builds upon iterative denoising techniques. In DensePure, an input is repeatedly denoised using varying random seeds to produce multiple purified versions. These samples are then passed through the classifier, and the final prediction is determined via majority voting. This method enhances robustness by exploring diverse purifications of the same input, ensuring that residual adversarial perturbations are minimized effectively.

Finally, **Guided Diffusion Model for Adversarial Purification (GDMAP)** [100] introduces a slightly different strategy. Instead of starting with an adversarial input, this method begins with pure Gaussian noise and iteratively denoises it under the guidance of an adversarial input. By steering the generative process toward the target image distribution, GDMAP achieves purification while maintaining computational efficiency.

Adversarial purification methods share the common goal of restoring clean data from adversarially perturbed inputs while maintaining computational feasibility. These approaches capitalize on advanced generative modeling techniques, such as diffusion models and denoising autoencoders, to mitigate adversarial effects effectively. Together, they form a robust line of defense against adversarial attacks, showing great promise for real-world applications where reliability and robustness are paramount.

6 Applications of Adversarial Machine Learning

As previously discussed, Adversarial Machine Learning (AML) is a specialized field within artificial intelligence that explores how adversaries can exploit machine learning models by introducing deceptive inputs, known as adversarial examples. These carefully crafted inputs are designed to cause models to make errors, posing serious challenges across a wide range of applications, including healthcare, autonomous vehicles, military systems, and many other domains.

In the realm of computer vision, adversarial attacks can subtly alter images in ways that are imperceptible to humans but lead models to misclassify them. For instance, a slight modification to a stop sign image might cause an autonomous vehicle's recognition system to perceive it as a yield sign, potentially leading to hazardous situations. This vulnerability underscores the critical need for robust defenses in safety-critical applications [101].

Natural Language Processing (NLP) systems are also susceptible to adversarial attacks. By making minor changes to input text, such as substituting synonyms or altering punctuation, adversaries can cause sentiment analysis tools to produce incorrect evaluations or deceive spam filters into misclassifying malicious content as benign. These manipulations can compromise the reliability of NLP applications in real-world scenarios [102].

In cybersecurity, adversarial techniques are employed to test and enhance the resilience of machine learning-based Intrusion Detection Systems (IDS). By simulating sophisticated attacks, researchers can identify vulnerabilities and strengthen IDS defenses, ensuring robust performance against real-world threats [103].

The healthcare sector faces significant risks from adversarial attacks, particularly in medical imaging. Subtle perturbations to medical images, such as CT scans or MRIs, can lead diagnostic models to incorrect conclusions, potentially resulting in misdiagnoses and inappropriate treatments. This highlights the necessity for developing robust models that can withstand adversarial manipulations in critical healthcare applications [104].

In this section, we explore several intriguing domains where adversarial attacks have a significant impact. We delve into the challenges posed by adversarial attacks in each field, highlighting how these manipulations can undermine decision-making processes and compromise the integrity of systems. Furthermore, we discuss the potential benefits of adversarial attacks, demonstrating how they can be leveraged to enhance system robustness and foster innovation. Ultimately, adversarial machine learning presents both challenges and opportunities across diverse applications. A thorough understanding of these attacks and effective mitigation strategies is essential for developing resilient AI systems capable of maintaining performance and reliability in the face of adversarial threats.

6.1 Adversarial Robustness in Security-Critical Applications

Adversarial robustness is a critical concern in security-sensitive applications, where the integrity and reliability of machine learning models are paramount. Adversarial attacks involve subtle perturbations to input data, often imperceptible to humans, that can lead models to make incorrect predictions. In security-critical domains, such vulnerabilities can have severe consequences, making it essential to develop models that are resilient to such manipulations.

Identification applications, such as biometric authentication and object recognition, play a critical role in security, access control, and surveillance systems. These

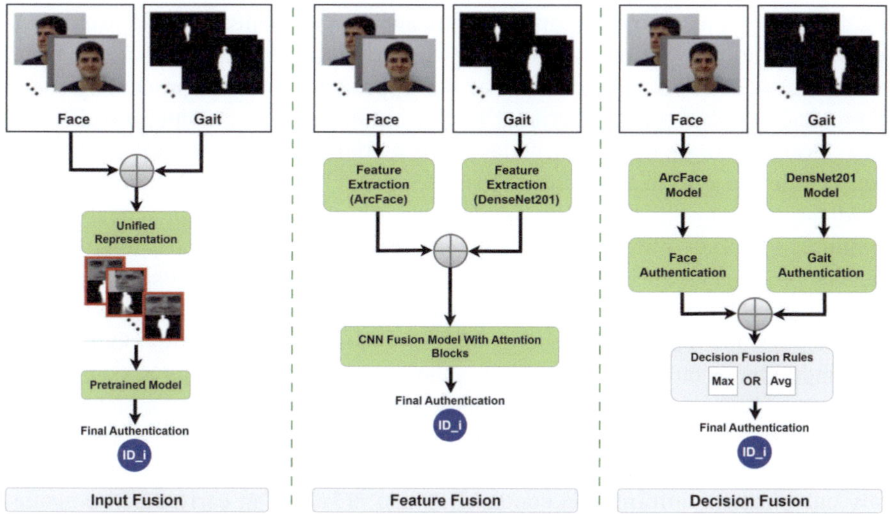

Fig. 35 Representation of multimodal fusion across input, feature, and decision levels using facial recognition and gait energy images (GEIs). The input fusion integrates raw biometric data, feature fusion combines extracted features, and decision fusion consolidates model outputs to enhance system robustness. Reprint from [106]

systems rely on machine learning models to accurately identify individuals or objects based on features extracted from data, such as facial images, fingerprints, or iris scans. Adversarial examples, however, pose significant challenges to the reliability and integrity of these applications, introducing vulnerabilities that can lead to severe consequences.

Biometric authentication systems, which utilize unique physiological or behavioral traits such as fingerprints, facial features, or voice patterns, are integral to modern security infrastructures. However, these systems are susceptible to adversarial attacks, where malicious inputs are crafted to deceive the model into incorrect classifications. For instance, adversaries can create subtle perturbations in fingerprint images that are imperceptible to humans but cause the system to misidentify the individual, potentially granting unauthorized access [105].

To mitigate such vulnerabilities, researchers have proposed various defense mechanisms. One effective approach involves integrating denoising mechanisms into the authentication process to filter out adversarial noise. Additionally, implementing robust training methods that expose the model to adversarial examples during the learning phase can enhance resilience. A notable study introduces a defense strategy that combines denoising with adversarial training, resulting in improved robustness against adversarial fingerprint attacks [105].

Figure 35 illustrates the architecture of a multimodal biometric system that integrates facial recognition, fingerprint scanning, and voice recognition to enhance security. The integration of multiple biometric modalities, such as facial recognition and gait energy images (GEIs), offers significant advancements in the robustness and accuracy of biometric systems, particularly in addressing vulnerabilities to adversarial attacks. A detailed representation of this approach shows the processing stages where fusion occurs at three critical levels: input, feature, and decision. Each of these levels leverages unique attributes of the modalities to enhance the overall performance of the system and mitigate the risks posed by adversarial examples [106].

At the input level, raw biometric data from both face and gait modalities are combined to create a unified representation. This integration occurs at the earliest stage of processing, allowing the system to exploit complementary spatial features inherent to each modality. For example, while facial recognition data provides rich detail about static features, gait energy images capture dynamic and temporal aspects of an individual's movement. Combining these datasets ensures a more comprehensive representation of the individual. This fused input is then processed by deep learning models, such as ArcFace and DenseNet201, which extract complex features that contribute to the robustness of the system. The input fusion process not only enhances the information content but also provides an early defense against adversarial manipulations that may target specific modalities [106].

The next stage, feature fusion, focuses on combining the feature vectors extracted from the individual modalities. Deep learning architectures independently process facial and gait data to extract their respective features, which are then concatenated to form a unified feature set. This stage often includes mechanisms like Squeeze-and-Excitation (SE) blocks, which apply attention mechanisms to prioritize the most critical features while suppressing irrelevant or redundant information. The resulting feature representation leverages the strengths of both modalities, creating a robust vector that enhances the biometric system's resistance to adversarial attacks. By integrating features at this level, the system becomes more adaptive and resilient to subtle perturbations that might otherwise compromise a single modality [106].

The final stage, decision fusion, consolidates the outputs of individual models trained on the separate modalities. In this stage, the system combines decisions from the facial recognition model (e.g., ArcFace) and the gait recognition model (e.g., DenseNet201) using aggregation rules such as maximum confidence scores or weighted averages. This process ensures that the strengths of each modality are maximally utilized while compensating for weaknesses in individual classifiers. For example, even if one modality is successfully attacked or produces erroneous outputs, the final decision reflects the combined strength of both, significantly reducing the likelihood of misclassification [106].

This multimodal fusion strategy, as visualized in the referenced figure, underscores the importance of integrating diverse biometric data to bolster system robustness. By distributing the authentication process across multiple levels of integration, the system becomes highly resistant to adversarial attacks targeting any single modality. Furthermore, the visual representation effectively illustrates how each level of fusion contributes uniquely to the system's defense mechanisms, providing valuable insights into the design and implementation of advanced biometric systems in security-critical applications [106].

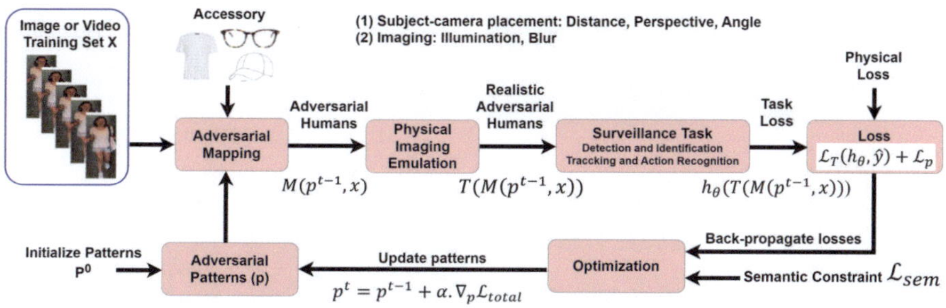

Fig. 36 Proposed framework for analyzing modern physical adversarial attacks in surveillance systems, illustrating the processes for both frame-based detection and identification and video-based tracking and action recognition tasks. Reprint from [108]

In surveillance and law enforcement, automated systems rely heavily on deep learning models for tasks such as object detection, identification, and tracking. These systems are vulnerable to physical adversarial attacks, where adversaries alter their appearance using specially designed accessories like adversarial t-shirts or glasses to evade detection. Such manipulations can lead to significant security breaches, allowing individuals to avoid surveillance or mislead law enforcement agencies [107].

To counter these threats, enhancing the security of surveillance systems is imperative. Implementing robust real-time video object detection systems that can identify and mitigate adversarial perturbations is a promising solution. For example, integrating adversarial training and detection mechanisms into the surveillance pipeline can improve the system's ability to recognize and resist adversarial inputs. A recent study presents a comprehensive survey on physical adversarial attacks in surveillance, discussing various defense strategies to build resilience within these systems.

Figure 36 provides a comprehensive framework for understanding physical adversarial attacks targeting surveillance systems. This framework is designed to analyze the stages of attack generation and their impact on tasks such as detection, identification, tracking, and action recognition. It highlights the optimization process of generating adversarial patterns that mislead these systems. For example, adversarial accessories like t-shirts or glasses are crafted to introduce physical perturbations, effectively deceiving surveillance algorithms. By applying this framework, researchers can systematically evaluate the vulnerabilities of different surveillance tasks and propose targeted defenses, ensuring the adaptability and resilience of these systems against real-world adversarial scenarios.

Figure 37 illustrates a variety of human-designed physical adversarial attacks aimed at evading detection and identification in surveillance systems. Examples include 3-D face masks that obscure identity, adversarial glasses designed to trick facial recognition algorithms, and camouflage makeup that alters facial symmetry to bypass detection. Adversarial fashion patterns and artifacts like LED glasses further demonstrate the diversity of techniques employed to mislead machine learning models. This figure underscores the ingenuity behind such attacks, showing how

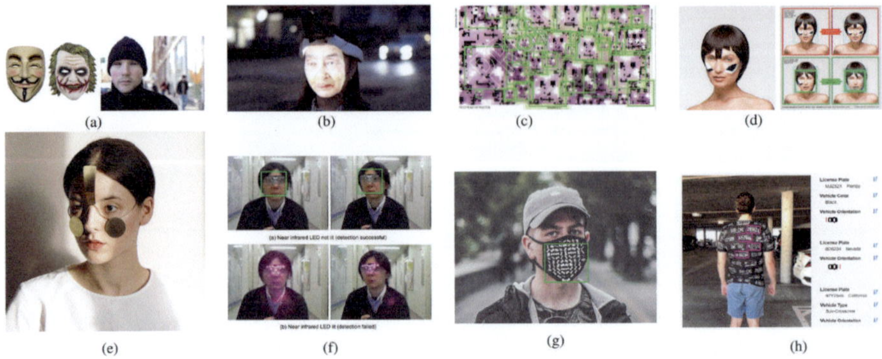

Fig. 37 Examples of human-designed physical adversarial attacks showcasing various accessories and techniques, including **a** 3-D face masks, **b** Face projector, **c** Face patterns scarf, **d** Camouflage makeup, **e** Incognito facial artifact, **f** Adversarial LED glasses, **g** Face mask, and **h** adversarial fashion, used to deceive surveillance systems. Reprint from [108]

attackers leverage physical and digital elements to evade state-of-the-art systems, highlighting the need for robust countermeasures.

In retail and consumer systems, Internet of Things (IoT) devices are increasingly employed for identification and personalized services. Machine learning models used in these systems are susceptible to adversarial attacks, where malicious inputs can lead to misclassification of devices, resulting in unauthorized access or service disruptions. For instance, adversaries can craft network traffic patterns that deceive IoT device identification systems, posing significant security risks [109].

Addressing these challenges requires the development of robust IoT identification systems capable of withstanding adversarial attacks. Implementing adversarial training, where models are trained with adversarial examples, can enhance resilience. Additionally, designing detection mechanisms that monitor for abnormal patterns indicative of adversarial behavior is crucial. A study on adversarial attacks against IoT identification systems highlights the need for tailored defense strategies in this domain [109].

In conclusion, adversarial attacks pose significant challenges across various identification applications, including biometric authentication, surveillance, and IoT-based retail systems. Developing and implementing robust defense mechanisms, such as adversarial training and detection systems, are essential steps toward enhancing the security and reliability of these critical applications.

6.2 Medical Diagnostics and Patient Monitoring

Adversarial attacks in medical diagnostics and patient monitoring present a significant threat to the reliability and safety of machine learning systems used in health-

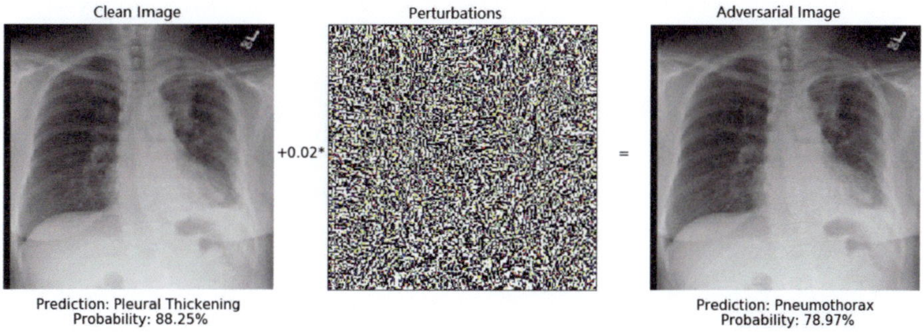

Fig. 38 An example of adversarial attack applied on medical images with FGSM method. Reprint from [110]

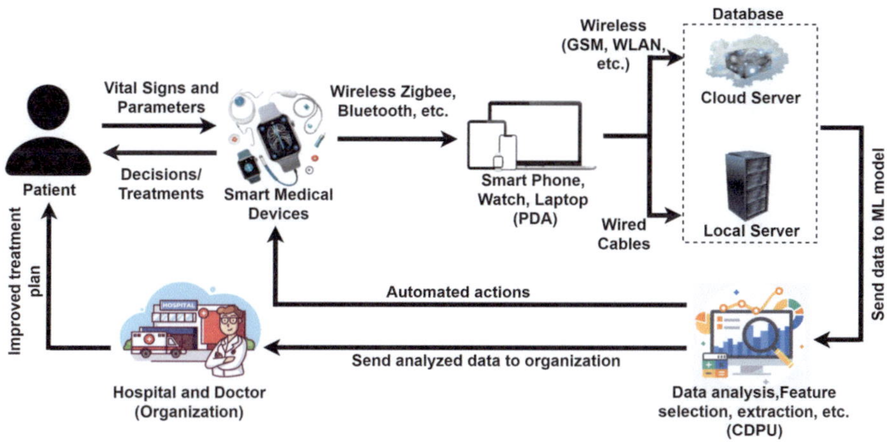

Fig. 39 An example of smart monitoring of healthcare system. Reprint from [111]

care. These attacks involve imperceptible perturbations to input data, such as medical images or physiological signals, which can lead to incorrect predictions by machine learning models. The consequences of such manipulations are dire, potentially resulting in misdiagnoses, delayed treatments, or inappropriate medical interventions, ultimately jeopardizing patient safety.

In medical imaging, adversarial attacks have been shown to drastically alter the outputs of diagnostic models while leaving the input image nearly identical to the human eye. For example, Fig. 38 [110] demonstrates this phenomenon. The figure juxtaposes an original CT scan image with its adversarially perturbed counterpart. While both images appear visually identical, the deep learning model misclassifies the condition in the manipulated image, categorizing a Pleural Thickening as Pneumothorax. This example highlights how adversarial perturbations can deceive even sophisticated diagnostic models, emphasizing the need for robust defenses [110].

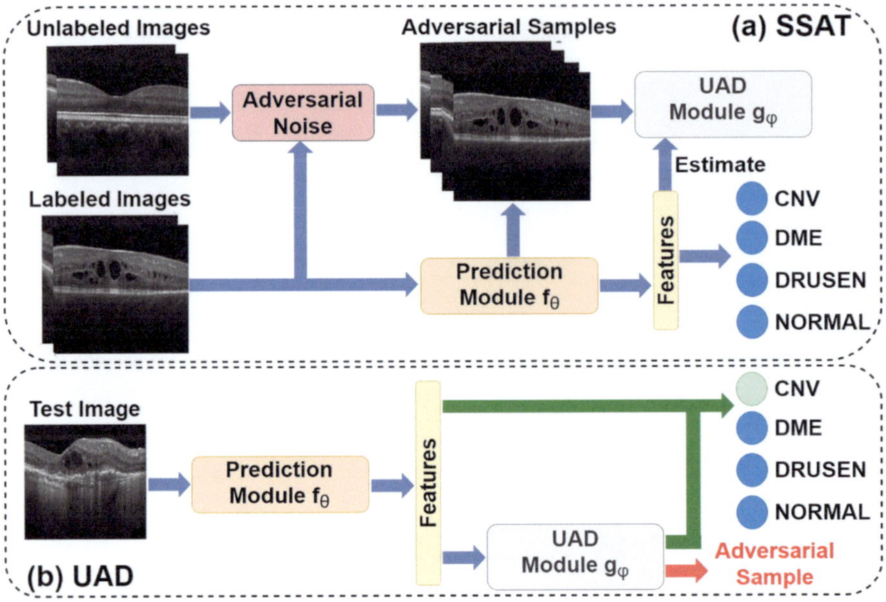

Fig. 40 A robust system for medical image classification equiped with SSAT and UAD modules. Reprint from [112]

Patient monitoring systems as shown in Fig. 39, including wearable devices and real-time vital sign trackers, are similarly vulnerable to adversarial attacks. An attacker could, for instance, manipulate electrocardiogram (ECG) data to mimic arrhythmias, potentially leading to unnecessary treatments or alarms [111]. In patient monitoring systems, adversarial attacks can target wearable devices or sensors that track vital signs, leading to false readings and potentially triggering inappropriate medical responses. Implementing robust signal processing techniques and secure data transmission protocols are essential steps in mitigating such risks. Moreover, continuous monitoring systems should incorporate anomaly detection algorithms capable of identifying and flagging irregular patterns indicative of adversarial interference.

To address these challenges, several defense mechanisms have been proposed. Adversarial training, a process in which models are exposed to adversarial examples during training, has been demonstrated to improve robustness. Figure 40 provides an overview of an adversarial defense framework. During the training phase Fig. 40a, the system employs SSAT to learn robust feature representations, which are used by both the prediction and UAD modules. In the inference phase Fig. 40b, when an unseen test image is provided, the system extracts its features as input to the UAD module. If the image is identified as an adversarial sample, it is rejected; otherwise, it proceeds to the loss layer for class label prediction [112].

Ensemble approaches have been also proposed to mitigate the effects of adversarial attacks. By combining predictions from multiple models, the overall system becomes more resilient, as it is unlikely that all models would fail on the same adversarial input [113]. In sum, adversarial attacks represent a substantial challenge in medical diagnostics and patient monitoring, with potential risks to patient safety and system reliability. By adopting strategies such as adversarial training, anomaly detection, and ensemble methods, healthcare systems can enhance their robustness and mitigate the impact of adversarial manipulations.

6.3 Autonomous Vehicles and Robotics

Adversarial attacks on autonomous vehicles and robotics represent a significant threat to the safety and reliability of these systems, primarily due to their dependence on deep learning (DL) models for perception and decision-making. Autonomous driving systems (ADS) utilize sensors such as cameras, LiDAR, and radars to gather data about their surroundings. These inputs are processed by DL models, which are unfortunately vulnerable to adversarial perturbations. These perturbations, often imperceptible to humans, are carefully crafted to manipulate DL models' predictions, potentially leading to misclassifications and erroneous decisions. For instance, adversarial examples can cause traffic signs to be misinterpreted, resulting in unsafe driving behaviors such as accelerating when a stop sign is misclassified as a speed limit sign [114–116].

In practice, physical adversarial attacks [114] pose unique challenges as they exploit real-world scenarios to deceive perception systems. For example, adversarial patches added to objects in the environment, such as stickers on road signs or manipulated textures on vehicles, can trick neural networks into misidentifying critical objects. Figure 41 comprehensively illustrates different methods for physical adversarial attacks targeting traffic sign recognition systems. Figure 41a showcases adversarial stickers applied to traffic signs. These stickers are optimized to introduce perturbations in sensitive regions of the signs, leading to significant misclassifications by deep learning models. This method is straightforward to implement but highly effective. Figure 41b demonstrates the use of adversarial scratches, a subtle approach where marks or abrasions on the surface of the sign alter the reflection and visual patterns detected by the system. This method enhances stealth by resembling natural wear and tear. Figure 41c highlights translucent patches designed to degrade the system's ability to classify traffic signs. These patches achieve adversarial effects while being less noticeable to human observers, improving attack stealth and reducing detection likelihood. Figure 41d shows the use of color films to simulate environmental distortions like haze or poor lighting conditions. These films disrupt the perception model's understanding of the sign without leaving permanent alterations. Lastly, Fig. 41e depicts signal-triggered adversarial patches that activate under specific conditions, such as responding to sound signals. These dynamic perturbations showcase advanced attack methodologies that interact with both the environment and system inputs. The diversity of these methods emphasizes the adaptability of

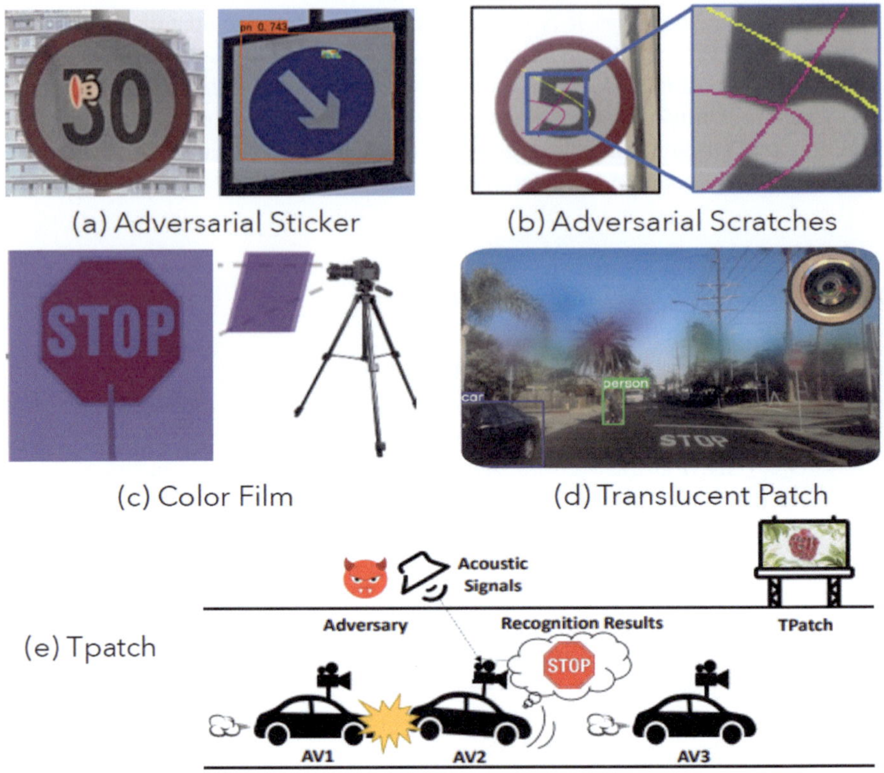

Fig. 41 Examples of adversarial attacks on traffic sign recognition using various physical perturbation methods. These include **a** adversarial stickers strategically placed to deceive perception systems, **b** adversarial scratches to alter sign recognition subtly, **c** translucent patches disrupting detection while maintaining low visibility, **d** color films mimicking environmental distortions, and **e** signal-triggered patches activating under specific conditions. Reprint from [114]

adversarial attacks in physical contexts, highlighting the pressing need for robust, multi-modal defensive strategies in autonomous driving systems.

One notable study [116] investigates the relationship between the accuracy decline of deep learning-based 3D object detection models under adversarial attacks and the corresponding impact on driving safety. The study focuses on two state-of-the-art models, Stereo R-CNN and DSGN, evaluating their robustness under perturbation and patch attacks. It introduces an end-to-end driving safety evaluation framework, which incorporates motion-planning and trajectory analysis, to assess driving safety metrics like successful planning rate, collision rate, and safe driving rate. Notably, the results reveal a decoupling effect where adversarial attacks causing significant accuracy degradation in object detection do not always translate into safety risks, and vice versa. The method also identifies DSGN as more robust than Stereo R-CNN, attributing this to architectural differences, such as the inclusion of a Spatial Pyramid Pooling (SPP) module in DSGN.

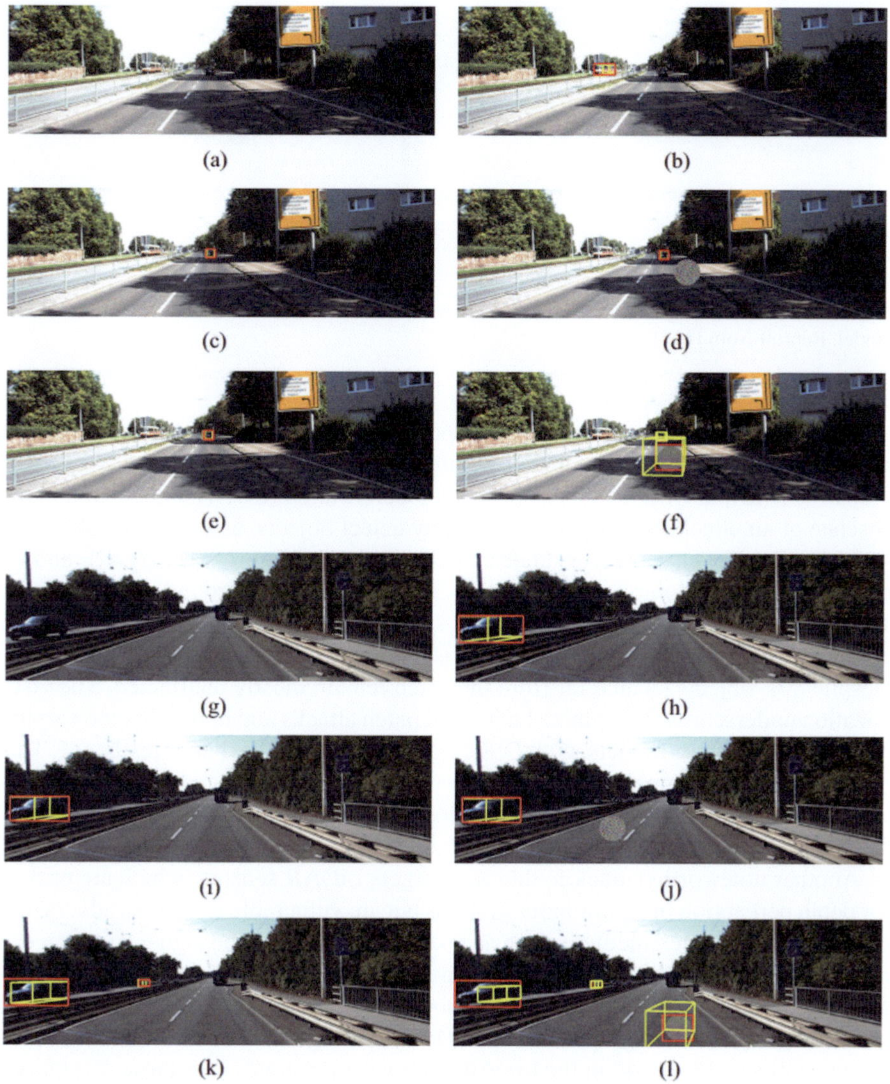

Fig. 42 Visualization of the impact of patch attacks on object detection results. Ghost objects are generated near the patch, influencing detection but sparing objects far from the patch's area. Reprint from [116]

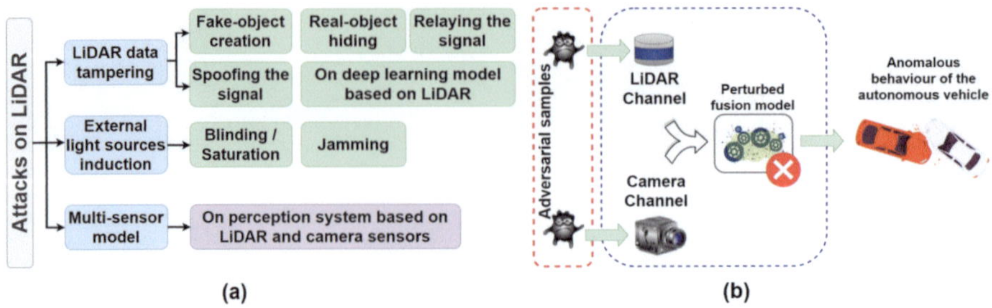

Fig. 43 **a** Various attack scenarios on LiDAR sensor; **b** An overview of attack on the multi-sensor model. Reprint from [117]

Figure 42 vividly demonstrates how patch attacks manipulate object detection systems by introducing ghost objects within the patch's vicinity. The figure contrasts scenarios with and without attacks for two models, Stereo R-CNN and DSGN. In the absence of an attack, both models accurately detect objects, as shown in subfigures (c), (e), (i), and (k). However, under patch attacks, subfigures (d), (f), (j), and (l) reveal the introduction of false bounding boxes, particularly around the patch area. This indicates that the attacks can exploit specific vulnerabilities in object detection algorithms, causing them to detect non-existent objects or misinterpret existing ones. Despite this, objects located far from the patch remain mostly unaffected. The visualization underscores the localized nature of patch attacks and highlights the varying robustness of the two models, with DSGN generally exhibiting better resilience. This figure serves as a compelling visual aid in understanding how patch-based adversarial attacks manifest in real-world scenarios and their implications for autonomous driving systems [116].

Another noteworthy attack in this area targets LiDAR sensors, which are critical for depth perception in autonomous systems but are vulnerable to various adversarial manipulations. Attack methods such as fake-object creation and signal spoofing can inject false data points into the LiDAR point cloud, causing the vehicle to misinterpret its environment. For example, a malicious actor could project laser beams mimicking legitimate signals into the LiDAR, leading the vehicle to perceive non-existent objects. Figure 43 in the LiDAR review demonstrates how these disruptions create phantom objects or hide real ones, emphasizing the potential for catastrophic consequences if these vulnerabilities are exploited. Fusion attacks that target both image and LiDAR data further amplify these risks by exploiting inter-modal correlations to bypass single-sensor defenses. Figure 5 from the same study showcases the effectiveness of such multi-source adversarial attacks in confusing perception systems [117].

In response, autonomous vehicles are increasingly employing countermeasures to mitigate the risks posed by adversarial attacks, leveraging methods such as data modification, model enhancements, and auxiliary defensive models. Data modification techniques, including feature denoising and input transformation, focus on cleaning adversarial noise from sensor data before it is processed by deep learning models. For example, adversarial training introduces adversarial examples into the training dataset, enabling the model to learn to identify and resist these manipulations. Input transformation techniques, such as JPEG compression or bit-depth reduction, alter data in ways that hinder the success of adversarial attacks by making perturbations less effective without degrading the model's ability to handle clean data. Such methods have proven effective in defending against white-box and black-box attacks [118].

Additionally, defensive strategies include adding auxiliary models like adversarial detectors and utilizing ensemble methods to bolster the primary model's resilience. Adversarial detectors act as binary classifiers that determine whether incoming data contains adversarial perturbations, rejecting malicious inputs before they can influence decision-making. Ensemble methods combine multiple models to achieve robustness, as an attacker must compromise all ensemble components to succeed. Beyond these techniques, modifying existing models directly, such as through gradient hiding or defensive distillation, has shown promise in enhancing their inherent resistance to adversarial examples. For instance, defensive distillation trains models to improve stability by reducing sensitivity to small input changes, thereby mitigating adversarial perturbations. However, challenges such as computational overhead and generalization to unseen attacks remain key areas for future research [118].

6.4 Positive Use Cases of Adversarial Examples

Adversarial examples, traditionally regarded as security threats, have evolved into powerful tools for enhancing the performance and robustness of machine learning models when applied constructively. By introducing small, imperceptible perturbations to input data, adversarial examples can reveal weaknesses in model decision boundaries and help refine training processes. These perturbations can adjust decision boundaries for imbalanced datasets, enhance model robustness against noisy or challenging inputs, and improve generalization in image recognition tasks. For example, adversarial training with adversarial examples has been applied to correct biased decision boundaries in imbalanced data classification, significantly boosting minority class accuracy while maintaining overall performance [119]. Similarly, in image recognition, the AdvProp method treats adversarial examples as additional training data, improving accuracy across datasets, including challenging benchmarks like ImageNet-C [120]. These use cases exemplify a shift from defensive adversarial methods to proactive applications that exploit adversarial perturbations for good.

An impactful application of adversarial examples lies in addressing class imbalance—a prevalent issue in real-world datasets where certain classes dominate in frequency while others are underrepresented. A notable study proposed **Guiding Adversarial Examples (GAEs)** [119] as a solution to this challenge. GAEs are specifically crafted to correct biased decision boundaries between majority and minority classes. This method generates adversarial examples near the decision boundary of minority classes, enabling the model to learn a more balanced and equitable boundary.

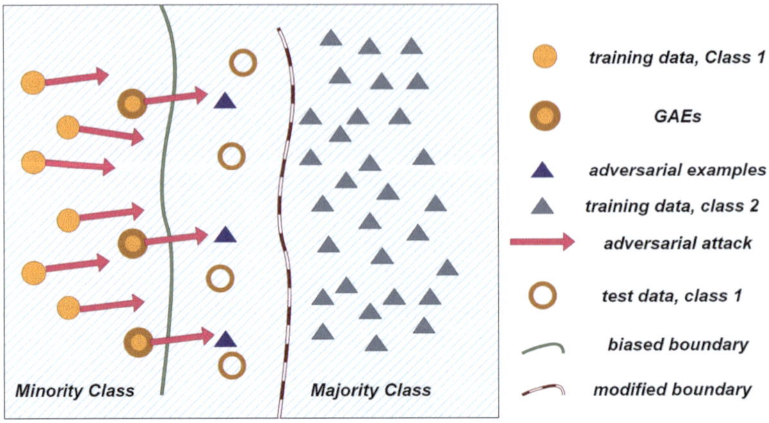

Fig. 44 An overview of Guiding Adversarial Examples (GAE). Reprint from [119]

Figure 44 provides a conceptual visualization of GAEs and their impact on decision boundaries in imbalanced learning. The diagram highlights the biased decision boundary that often forms in datasets with significant class imbalance, where the majority class dominates the feature space. GAEs are strategically crafted adversarial examples generated near this biased boundary, specifically targeting instances in minority classes. By using these examples during training, the boundary shifts to better include minority class instances, resulting in a more balanced model. This adjustment ensures that the minority class receives more representation without significantly compromising the performance on the majority class. Through this approach, models achieve higher accuracy for minority classes with only a minimal sacrifice in the performance of majority classes, presenting a novel solution for long-tailed data distributions [119].

This method was tested across datasets like CIFAR-10 and SVHN, demonstrating consistent improvements in class-specific accuracy. For instance, Fig. 45a, b compares the accuracy of each class achieved by the proposed GAE-based method and the baseline Cross-Entropy (CE) approach across imbalanced datasets like CIFAR-10 and SVHN. The bar chart distinctly shows that the GAE-based approach provides a substantial accuracy boost for the minority classes while maintaining competitive accuracy for majority classes. This improvement addresses one of the core challenges in long-tailed datasets, where traditional methods fail to learn adequate representations for underrepresented classes. The results demonstrate that GAEs are particularly effective in guiding the learning process, especially for the tail-end classes that typically suffer from poor performance due to limited samples. This figure reinforces the practical efficacy of the proposed approach, validating its benefits across different dataset configurations [119].

Additionally, the T-SNE visualizations in Fig. 45c compares the feature space of models trained using the standard Cross-Entropy approach versus those trained with GAEs. The scatter plots provide a vivid representation of the clustering effect for each

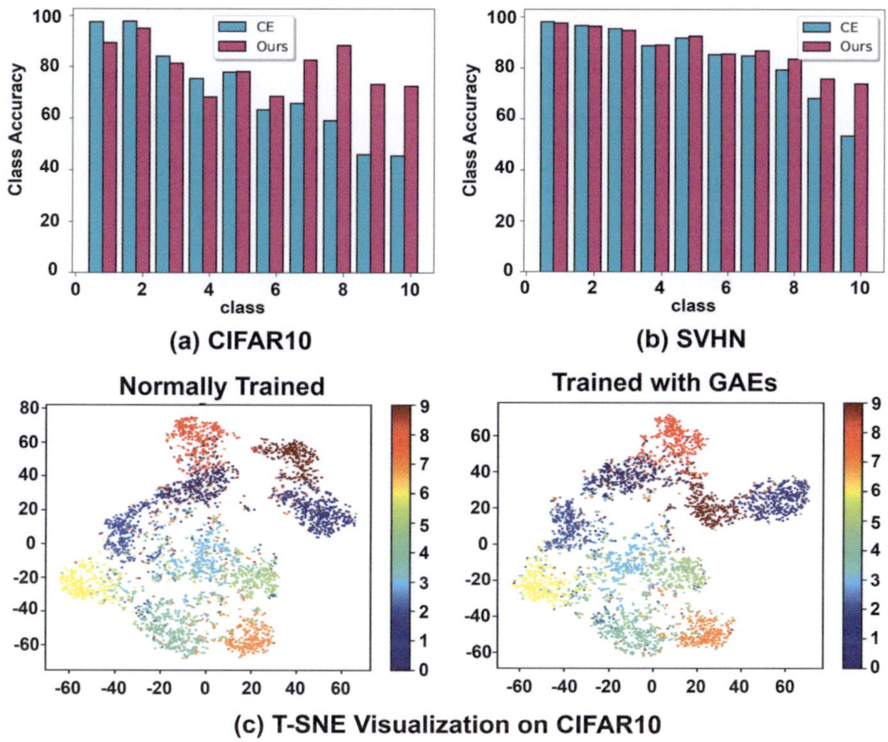

Fig. 45 Some experiments for GAE. **a** Comparing the accuracy on CIFAR-10 dataset; **b** Comparing the accuracy on SVHN dataset; **c** The t-sne visualization of standard vs GAE methods on CIFAR-10 dataset. Reprint from [119]

class. In the standard training approach, data points from minority classes are often poorly separated and overlap significantly with majority class points, reflecting the biased decision boundary. In contrast, the model trained with GAEs exhibits distinct and well-separated clusters for both majority and minority classes. This improved separation is a direct result of the decision boundary adjustments driven by GAEs. The visualization not only supports the quantitative results but also offers an intuitive understanding of how GAEs enhance feature space representation, leading to better classification performance [119].

In sum, the study demonstrates the transformative potential of adversarial examples as tools for addressing class imbalance in machine learning. By introducing GAEs to adjust biased decision boundaries, the proposed method significantly improves minority class accuracy while maintaining overall model performance. Extensive experiments across various datasets validate the efficacy of this approach, showcasing its ability to balance decision-making in long-tailed data distributions. The research highlights a novel and practical use case for adversarial examples, paving the way for future work in leveraging adversarial techniques to enhance fairness and robustness in machine learning models [119].

Another groundbreaking study [120] explored how adversarial examples can enhance model accuracy in supervised learning tasks, even on challenging datasets. This study [120] introduced AdvProp, a novel adversarial training method that disentangles the distributions of adversarial examples and clean inputs by employing separate batch normalization layers. This approach effectively treats adversarial examples as additional training samples, addressing the performance degradation typically observed in standard adversarial training. The study tested AdvProp on EfficientNet models and found substantial improvements across datasets, including ImageNet, ImageNet-C (corrupted images), ImageNet-A (challenging natural images), and Stylized-ImageNet (texture-agnostic images).

Figure 46 illustrates the effectiveness of the AdvProp method in enhancing model performance across datasets. AdvProp leverages adversarial examples as auxiliary training data, which allows models to better generalize, especially under challenging conditions like corrupted or stylized inputs. The figure compares the top-1 accuracy improvements across datasets such as ImageNet-C (mean corruption error reduced by 6.5%) and Stylized-ImageNet (+4.8%). This visual highlights that AdvProp is not only beneficial for improving accuracy on standard datasets but also excels in tasks that require robustness to distortions and adversarial perturbations [120].

A critical aspect of AdvProp is its ability to disentangle the distributions of adversarial and clean inputs. AdvProp demonstrated scalability with model size, delivering greater accuracy gains with larger networks. Their result highlights the improvements AdvProp brings to EfficientNet models across different scales. The result compares the top-1 accuracy of models trained with and without AdvProp, showing significant gains, particularly for larger models like EfficientNet-B7, which achieved an accuracy of 85.2% on ImageNet. This improvement underscores AdvProp's scalability and its ability to capitalize on the increased capacity of larger architectures [120].

The study introduces AdvProp, a novel adversarial training method that effectively utilizes adversarial examples as additional training data while disentangling their distributions from clean examples through auxiliary batch normalization. This approach significantly enhances accuracy and robustness across various datasets, including challenging benchmarks like ImageNet-C and Stylized-ImageNet. By scaling performance with model size and resolving distribution mismatches, AdvProp sets a new standard for integrating adversarial examples into supervised learning frameworks, demonstrating its potential to improve both generalization and robustness in image recognition models [120].

In sum, the positive use cases of adversarial examples highlight their dual role in machine learning. Beyond being obstacles to overcome, they serve as innovative tools for addressing critical challenges such as class imbalance, generalization, and robustness. Integrating adversarial examples into the training process not only enhances model security but also improves performance on real-world data. These advancements redefine adversarial examples from mere threats to valuable opportunities, paving the way for new research and practical applications.

Fig. 46 AdvProp improves model accuracy across multiple datasets, including ImageNet and challenging variants like ImageNet-C, ImageNet-A, and Stylized-ImageNet, by leveraging adversarial examples as additional training data while disentangling their distributions from clean images. Reprint from [120]

7 Conclusion

Adversarial machine learning has fundamentally reshaped the understanding of vulnerabilities in artificial intelligence systems, particularly those reliant on DNNs. These systems, despite their impressive capabilities, exhibit a troubling susceptibility to adversarial examples—inputs subtly altered to deceive models while remaining imperceptible to humans. Such inputs expose critical weaknesses in decision boundaries, highlighting the fragility of machine learning models in real-world applications. This vulnerability is particularly concerning in safety-critical domains like autonomous vehicles, biometric authentication, and medical diagnostics, where even minor perturbations can lead to catastrophic failures.

The exploration of adversarial examples showcases the diverse strategies employed by attackers. White-box attacks, leveraging complete knowledge of the model's architecture and parameters, demonstrate the precision with which perturbations can be crafted to induce misclassifications. Conversely, black-box attacks, operating with minimal knowledge, rely on techniques like gradient estimation to achieve similar objectives. Universal Adversarial Perturbations (UAPs) further underscore the risks, with their remarkable ability to deceive models across varied inputs and architectures. These perturbations reveal a systemic weakness in neural networks, stemming from the geometric redundancies in their decision boundaries.

The implications of these attacks extend beyond theoretical concerns. In real-world scenarios, adversarial examples have been shown to mislead autonomous vehicles, causing misinterpretations of traffic signs, or to compromise facial recognition systems, granting unauthorized access. Many results illustrating the success of adversarial attacks, such as targeted perturbations that achieve near 100% misclassification rates, emphasize the tangible risks posed by these techniques. Such examples reinforce the urgent need for robust defenses to safeguard AI systems from malicious exploitation.

Defense mechanisms, while promising, must contend with the evolving sophistication of adversarial strategies. Adversarial training, one of the most studied approaches, involves augmenting training data with adversarial examples to improve model robustness. This technique has shown significant success in enhancing the resilience of models in tasks like image recognition. However, its computational demands and potential overfitting to specific attack types highlight the need for complementary strategies. Techniques like defensive distillation, which smoothens model outputs to reduce sensitivity to perturbations, and input transformation methods, such as feature denoising, further enrich the defenses.

An intriguing dimension of adversarial machine learning lies in the adaptability of attacks and defenses. Black-box scenarios, where attackers operate without direct access to model parameters, showcase the resilience of adversarial techniques. Methods like query-limited and label-only attacks demonstrate how even minimal feedback can be exploited to craft effective adversarial examples. These scenarios emphasize the importance of dynamic defenses that can respond to evolving threat landscapes. Gradient masking and ensemble methods have emerged as critical tools in this context, providing layers of protection against sophisticated black-box attacks.

Beyond defense, adversarial techniques have also been repurposed for constructive applications. Adversarial examples, when used as part of training, have been shown to enhance model generalization and robustness. For instance, GAEs have been effectively used to address class imbalance in datasets, improving the representation of minority classes in model decision-making. Similarly, methods like AdvProp, which disentangle adversarial and clean data distributions during training, have yielded significant performance gains across challenging datasets.

The study of adversarial machine learning also raises ethical considerations. While the development of adversarial techniques has primarily focused on exposing and mitigating vulnerabilities, their misuse for malicious purposes cannot be ignored. This dual-edged nature necessitates a balanced approach, where the potential for harm is mitigated through responsible research and transparent collaboration. The field's emphasis on building robust and fair AI systems aligns with the broader goal of ensuring trust in artificial intelligence technologies.

Future directions in adversarial research highlight the need for scalable, real-time defenses. Current methods, while effective, often struggle to balance robustness with efficiency, limiting their applicability in resource-constrained settings. Advancements in model interpretability and explainability could offer new avenues for detecting and mitigating adversarial attacks. By understanding the decision-making processes of models, researchers can design systems that are inherently more resilient to adversarial manipulations.

In essence, adversarial machine learning underscores the importance of viewing challenges as opportunities for innovation. The vulnerabilities exposed by adversarial examples serve as a catalyst for developing smarter, safer, and more adaptive AI systems. As the field continues to evolve, fostering collaboration across domains and stakeholders will be key to navigating the complexities of adversarial threats and ensuring the widespread trust and reliability of artificial intelligence.

The authors acknowledge the use of GenAI for editing this chapter.

References

1. H. Pourmehrani, J. Bahrami, P. Nooralinejad, H. Pirsiavash and N. Karimi, "FAT-RABBIT: Fault-Aware Training towards Robustness Against Bit-flip Based Attacks in Deep Neural Networks," IEEE International Test Conference (ITC), San Diego, CA, USA, 2024, pp. 106-110.
2. Y. Wang et al., "Adversarial Attacks and Defenses in Machine Learning-Empowered Communication Systems and Networks: A Contemporary Survey," in IEEE Communications Surveys & Tutorials, vol. 25, no. 4, pp. 2245-2298.
3. M. Jagielski, A. Oprea, B. Biggio, C. Liu, C. Nita-Rotaru and B. Li, "Manipulating Machine Learning: Poisoning Attacks and Countermeasures for Regression Learning," IEEE Symposium on Security and Privacy (SP), San Francisco, CA, USA, 2018, pp. 19-35.
4. M. Xue, C. Yuan, H. Wu, Y. Zhang and W. Liu, "Machine Learning Security: Threats, Countermeasures, and Evaluations," in IEEE Access, vol. 8, pp. 74720-74742, 2020.

5. S. Kashyap, A. Sharma, S. Gautam, R. Sharma, S. Chauhan and Simran, "Adversarial Attacks and Defenses in Deep Learning," International Conference on Emerging Innovations and Advanced Computing (INNOCOMP), Sonipat, India, 2024, pp. 318-323.
6. M. Goldblum et al., "Dataset Security for Machine Learning: Data Poisoning, Backdoor Attacks, and Defenses," in IEEE Transactions on Pattern Analysis and Machine Intelligence, vol. 45, no. 2, pp. 1563-1580, 1 Feb. 2023.
7. S. Chahar, S. Gupta, I. Dhingra and K. S. Kaswan, "Adversarial Threats in Machine Learning: A Critical Analysis," International Conference on Computational Intelligence and Computing Applications (ICCICA), Samalkha, India, 2024, pp. 253-258.
8. M. Aljanabi et al., "Data Poisoning: Issues, Challenges, and Needs," IET Smart Cities Symposium (SCS), Hybrid Conference, Bahrain, 2023, pp. 359-363.
9. A. Guesmi, K. N. Khasawneh, N. Abu-Ghazaleh and I. Alouani, "ROOM: Adversarial Machine Learning Attacks Under Real-Time Constraints," International Joint Conference on Neural Networks (IJCNN), Padua, Italy, 2022, pp. 1-10.
10. M. Abrishami et al., "Classification and Analysis of Adversarial Machine Learning Attacks in IoT: a Label Flipping Attack Case Study," Conference of Open Innovations Association (FRUCT), Tampere, Finland, 2022, pp. 3-14.
11. C. Koball, Y. Wang, B. P. Rimal and V. Vaidyan, "Machine Learning Security: Threat Model, Attacks, and Challenges," in Computer, vol. 57, no. 10, pp. 26-35, 2024.
12. A. Thangaraju and C. Merkel, "Exploring Adversarial Attacks and Defenses in Deep Learning," IEEE International Conference on Electronics, Computing and Communication Technologies (CONECCT), Bangalore, India, 2022, pp. 1-6.
13. I. J. Goodfellow, J. Shlens, and C. Szegedy, "Explaining and Harnessing Adversarial Examples," in Proceedings of the 3rd International Conference on Learning Representations (ICLR), San Diego, CA, USA, 2015, pp. 1–11.
14. B. Biggio et al., "Poisoning Attacks Against Support Vector Machines," in Proceedings of the 29th International Conference on Machine Learning (ICML), Edinburgh, UK, 2012, pp. 1467–1474.
15. M. Fredrikson et al., "Model Inversion Attacks that Exploit Confidence Information and Basic Countermeasures," in Proceedings of the 22nd ACM SIGSAC Conference on Computer and Communications Security (CCS), Denver, CO, USA, 2015, pp. 1322–1333.
16. H. Li et al., "Practical Bit-Flip Attack on Deep Neural Network," in Proceedings of the IEEE International Conference on Computer Design (ICCD), Abu Dhabi, UAE, 2020, pp. 1–9.
17. A. Shumailov et al., "Sponge Examples: Energy-Latency Attacks on Neural Networks," in Proceedings of the 2021 IEEE Symposium on Security and Privacy (SP), San Francisco, CA, USA, 2021, pp. 503–518.
18. R. Shokri et al., "Membership Inference Attacks Against Machine Learning Models," in Proceedings of the IEEE Symposium on Security and Privacy (SP), San Jose, CA, USA, 2017, pp. 3–18.
19. Y. Batina et al., "CSI NN: Reverse Engineering of Neural Network Architectures Through Electromagnetic Side-Channel Analysis," in Proceedings of the USENIX Security Symposium, 2019, pp. 515–532.
20. H. Zhang et al., "Adversarial Sensor Attack on LiDAR-Based Perception in Autonomous Driving," in Proceedings of the IEEE/CVF Conference on Computer Vision and Pattern Recognition (CVPR), Virtual, 2021, pp. 2983–2991.
21. P. Kairouz et al., "Advances and Open Problems in Federated Learning," Foundations and Trends® in Machine Learning, vol. 14, no. 1, pp. 1–210, 2021.
22. H. Zhang et al., "Adversarial Attacks on Deep Reinforcement Learning Policies," in Proceedings of the 32nd AAAI Conference on Artificial Intelligence (AAAI), New Orleans, LA, USA, 2018, pp. 4574–4581.
23. C. Szegedy,W. Zaremba, I. Sutskever, J. Bruna, D. Erhan, I. Goodfellow, and R. Fergus, "Intriguing Properties of Neural Networks," 2013, arXiv:1312.6199.
24. N. Carlini and D. Wagner, "Towards Evaluating the Robustness of Neural Networks," in Proc. IEEE Symp. Secur. Privacy (SP), May 2017, pp. 39–57.

25. J. Su, D. V. Vargas and K. Sakurai, "One Pixel Attack for Fooling Deep Neural Networks," in IEEE Transactions on Evolutionary Computation, vol. 23, no. 5, pp. 828-841, 2019.
26. J. Rony, L. G. Hafemann, L. S. Oliveira, I. Ben Ayed, R. Sabourin and E. Granger, "Decoupling Direction and Norm for Efficient Gradient-Based L2 Adversarial Attacks and Defenses," IEEE/CVF Conference on Computer Vision and Pattern Recognition (CVPR), Long Beach, CA, USA, 2019, pp. 4317-4325.
27. Madry, Aleksander et al. "Towards Deep Learning Models Resistant to Adversarial Attacks," ArXiv abs/1706.06083 (2017).
28. A. Kurakin, I. Goodfellow, and S. Bengio, "Adversarial Examples in the Physical World," 2016, arXiv 1607.02533.
29. J. C. Costa, T. Roxo, H. Proença and P. R. M. Inácio, "How Deep Learning Sees the World: A Survey on Adversarial Attacks & Defenses," in IEEE Access, vol. 12, pp. 61113-61136, 2024.
30. N. Papernot, P. McDaniel, I. Goodfellow, S. Jha, Z. B. Celik, and A. Swami, "Practical Black-Box Attacks against Machine Learning," ASIA CCS, 2017, Association for Computing Machinery, USA, 506–519.
31. I. J. Goodfellow, J. Shlens, and C. Szegedy, "Explaining and Harnessing Adversarial Examples," 2014, arXiv:1412.6572.
32. A. Madry, A. Makelov, L. Schmidt, D. Tsipras, and A. Vladu, "Towards Deep Learning Models Resistant to Adversarial Attacks," in Proc. Int. Conf. Learn. Represent., 2018, pp. 1–23.
33. S.-M. Moosavi-Dezfooli, A. Fawzi, and P. Frossard, "DeepFool: A Simple and Accurate Method to Fool Deep Neural Networks," in Proc. IEEE Conf. Comput. Vis. Pattern Recognit. (CVPR), Jun. 2016, pp. 2574–2582.
34. A. Dabouei, S. Soleymani, F. Taherkhani, J. M. Dawson, and N. M. Nasrabadi, "Smoothfool: An Efficient Framework for Computing Smooth dversarial perturbations," in Proc. IEEE WACV, Oct. 2020, pp. 2654–2663.
35. Z. Yuan, J. Zhang, Y. Jia, C. Tan, T. Xue, and S. Shan, "Meta Gradient Adversarial Attack," in Proc. IEEE/CVF Int. Conf. Comput. Vis. (ICCV), 2021, pp. 7748–7757.
36. C. Xiao, B. Li, J.-Y. Zhu, W. He, M. Liu, and D. Song, "Generating Adversarial Examples with Adversarial Networks," in Proc. 27th Int. Joint Conf. Artif. Intell., 2018, pp. 3905–3911.
37. A. Ilyas, L. Engstrom, A. Athalye, and J. Lin, "Black-box Adversarial Attacks with Limited Queries and Information," in Proc. ICML, 2018, pp. 2137–2146.
38. M. Wicker, X. Huang, and M. Kwiatkowska, "Feature-guided Blackbox Safety Testing of Deep Neural Networks," in Proc. TACAS, 2018, pp. 408–426.
39. W. Brendel, J. Rauber, and M. Bethge, "Decision-based Adversarial Attacks: Reliable Attacks Against Black-box Machine Learning models," in Proc. Int. Conf. Learn. Represent., 2018, pp. 1–12.
40. J. Chen, M. I. Jordan, and M. J. Wainwright, "Hopskipjumpattack: A Query Efficient Decision-based Attack," in Proc. IEEE Symp. Secur. Privacy (SP), Sep. 2020, pp. 1277 1294.
41. M. Andriushchenko, F. Croce, N. Flammarion, and M. Hein, "Square Attack: A Query-efficient Black-box Adversarial Attack via Random Search," in Proc. Eur. Conf. Comput. Vis. Glasgow, U.K.: Springer, 2020, pp. 484–501.
42. Z. Wang, H. Guo, Z. Zhang, W. Liu, Z. Qin, and K. Ren, "Feature Importance-Aware Transferable Adversarial Attacks," in Proc. IEEE/CVF Int. Conf. Comput. Vis. (ICCV), Oct. 2021, pp. 7619–7628.
43. S.-M. Moosavi-Dezfooli, A. Fawzi, O. Fawzi, and P. Frossard, "Universal Adversarial Perturbations," in Proc. IEEE Conf. Comput. Vis. Pattern Recognit. (CVPR), Jul. 2017, pp. 86–94.
44. R. Duan, X. Ma, Y.Wang, J. Bailey, A. K. Qin, and Y. Yang, "Adversarial Camouflage: Hiding Physical-world Attacks with Natural Styles," in Proc. IEEE/CVF Conf. Comput. Vis. Pattern Recognit. (CVPR), Jun. 2020, pp. 997–1005.
45. J. Hayes and G. Danezis, "Learning Universal Adversarial Perturbations with Generative Models," in Proc. IEEE Secur. PrivacyWorkshops (SPW), May 2018, pp. 43–49.

46. X. Liu, Y. Zhong, Y. Zhang, L. Qin, and W. Deng, "Enhancing Generalization of Universal Adversarial Perturbation through Gradient Aggregation," in Proc. IEEE/CVF Int. Conf. Comput. Vis. (ICCV), Oct. 2023, pp. 4435–4444.
47. Z. Zhou, S. Hu, R. Zhao, Q. Wang, L. Y. Zhang, J. Hou, and H. Jin, "Downstream-agnostic Adversarial Examples," in Proc. IEEE/CVF Int. Conf. Comput. Vis., Oct. 2023, pp. 4345–4355.
48. A. Kurakin, I. J. Goodfellow, and S. Bengio, "Adversarial EExamples in the Physical World," in Proc. Artif. Intell. Saf. Secur. Boca Raton, FL, USA: CRC Press, 2018, pp. 99–112.
49. D. Wang, W. Yao, T. Jiang, C. Li, and X. Chen, "RFLA: A Stealthy Reflected Light Adversarial Attack in the Physical World," in Proc. IEEE/CVF Int. Conf. Comput. Vis. (ICCV), Oct. 2023, pp. 4455–4465.
50. F. Croce and M. Hein, "Reliable Evaluation of Adversarial Robustness with an Ensemble of Diverse Parameter-free Attacks," in Proc. Int. Conf. Mach. Learn., 2020, pp. 2206–2216.
51. F. Croce and M. Hein, "Minimally Distorted Adversarial Examples with a Fast Adaptive Boundary Attack," In ICML, 2020.
52. W. Zhao, S. Alwidian, and Q. H. Mahmoud, "Adversarial Training Methods for Deep Learning: A Systematic Review," Algorithms, vol. 15, no. 8, pp. 283, 2022.
53. B. Wu, S. Wei, M. Zhu, M. Zheng, Z. Zhu, M. Zhang, H. Chen, D. Yuan, L. Liu, and Q. Liu, "Defenses in Adversarial Machine Learning: A Survey," ArXiv preprint arXiv:2312.08890, 2023.
54. D. Hwang, E. Lee, and W. Rhee, "AID-Purifier: A Light Auxiliary Network for Boosting Adversarial Defense," ICML 2021 Workshop on A Blessing in Disguise: The Prospects and Perils of Adversarial Machine Learning, 2021.
55. D. Kalaria, A. Hazra, and P. P. Chakrabarti, "Towards Adversarial Purification using Denoising AutoEncoders," ArXiv preprint arXiv:2208.13838, 2022.
56. F. Tramer, A. Kurakin, N. Papernot, I. Goodfellow, D. Boneh, and P. McDaniel, "Ensemble Adversarial Training: Attacks and Defenses," in Proc. Int. Conf. Learn. Represent., 2018, pp. 1–11.
57. J.-B. Alayrac, J. Uesato, P.-S. Huang, A. Fawzi, R. Stanforth, and P.Kohli, "Are Labels Required for Improving Adversarial Robustness?" in Proc. Adv. Neural Inf. Process. Syst., vol. 32, 2019, pp. 1–10.
58. H. Zhang, Y. Yu, J. Jiao, E. Xing, L. El Ghaoui, and M. Jordan, "Theoretically Principled Trade-off Between Robustness and Accuracy," in Proc. Int. Conf. Mach. Learn., 2019, pp. 7472–7482.
59. Y. Wang, D. Zou, J. Yi, J. Bailey, X. Ma, and Q. Gu, "Improving Adversarial Robustness Requires Revisiting Misclassified Examples," in Proc. Int. Conf. Learn. Represent., 2020, pp. 1–14.
60. C. Sitawarin, S. Chakraborty, and D. Wagner, "SAT: Improving Adversarial Training via Curriculum-based Loss Smoothing," 2020, arXiv:2003.09347.
61. A. Shafahi, M. Najibi, M. A. Ghiasi, Z. Xu, J. Dickerson, C. Studer, L. S. Davis, G. Taylor, and T. Goldstein, "Adversarial Training for Tree!" in Proc. Adv. Neural Inf. Process. Syst., vol. 32, 2019, pp. 1–12.
62. D. Zhang, T. Zhang, Y. Lu, Z. Zhu, and B. Dong, "You only propagate once: Accelerating Adversarial Training via Maximal Principle," in Proc. Adv. Neural Inf. Process. Syst., vol. 32, 2019, pp. 1–12.
63. N. Kumari, M. Singh, A. Sinha, H. Machiraju, B. Krishnamurthy, and V. N. Balasubramanian, "Harnessing the Vulnerability of Latent Layers in Adversarially Trained models," in Proc. 28th Int. Joint Conf. Artif. Intell., Aug. 2019, pp. 2779–2785.
64. R. Rade and S.-M. Moosavi-Dezfooli, "Helper-based Adversarial Training: Reducing Excessive Margin to Achieve a Better Accuracy vs. Robustness Trade-off," in Proc. ICML, 2021, pp. 1–10.
65. N. Papernot, P. McDaniel, X. Wu, S. Jha, and A. Swami, "Distillation as a Defense to Adversarial Perturbations Against Deep Neural Networks," in Proc. IEEE Symp. Secur. Privacy (SP), May 2016, pp. 582–597.

66. B. Zi, S. Zhao, X. Ma, and Y.-G. Jiang, "Revisiting Adversarial Robustness Distillation: Robust Soft Labels Make Student Better," in Proc. ICCV, Oct. 2021, pp. 16443–16452.
67. L. Huang, C. Zhang, and H. Zhang, "Self-adaptive Training: Beyond Empirical Risk Minimization," in Proc. Adv. Neural Inf. Process. Syst., vol. 33, 2020, pp. 19365–19376.
68. Y. Carmon, A. Raghunathan, L. Schmidt, J. C. Duchi, and P. S. Liang, "Unlabeled Data Improves Adversarial Robustness," in Proc. Adv. Neural Inf. Process. Syst., vol. 32, 2019, pp. 1–12.
69. S.-M. Moosavi-Dezfooli, A. Fawzi, J. Uesato, and P. Frossard, "Robustness via Curvature Regularization, and Vice Versa," in Proc. IEEE/CVF Conf. (CVPR), Jun. 2019, pp. 9078–9086.
70. C. Mao, Z. Zhong, J. Yang, C. Vondrick, and B. Ray, "Metric Learning for Adversarial Robustness," in Proc. Adv. Neural Inf. Process. Syst., vol. 32, 2019, pp. 1–12.
71. K. Chalupka, P. Perona, and F. Eberhardt, "Visual Causal Feature Learning," in Proc. UAI, 2015, pp. 1–15.
72. L. Rice, E.Wong, and Z. Kolter, "Overfitting in Adversarially Robust Deep Learning," in Proc. 37th Int. Conf. Mach. Learn., 2020, pp. 8093–8104.
73. S. Addepalli, B. S. Vivek, A. Baburaj, G. Sriramanan, and R. V. Babu, "Towards Achieving Adversarial Robustness by Enforcing Feature Consistency Across Bit Planes," in Proc. IEEE/CVF Conf. Comput. Vis. Pattern Recognit. (CVPR), Jun. 2020, pp. 1017–1026.
74. C. Zhang, A. Liu, X. Liu, Y. Xu, H. Yu, Y. Ma, and T. Li, "Interpreting and Improving Adversarial Robustness of Deep Neural Networks with Neuron Sensitivity," IEEE Trans. Image Process., vol. 30, pp. 1291–1304, 2021.
75. M. Atzmon, N. Haim, L. Yariv, O. Israelov, H. Maron, and Y. Lipman, "Controlling Neural Level Sets," in Proc. Adv. Neural Inf. Process. Syst., vol. 32, 2019, pp. 1–10.
76. D. Meng and H. Chen, "MagNet: A Two-pronged Defense Against Adversarial Examples," in Proc. ACM SIGSAC Conf. Comput. Commun. Secur., Oct. 2017.
77. Y. Dong, F. Liao, T. Pang, H. Su, J. Zhu, X. Hu, and J. Li, "Boosting Adversarial Attacks with Momentum," in Proc. CVPR, Jun. 2018, pp. 9185–9193.
78. W. Xu, D. Evans, and Y. Qi, "Feature Squeezing: Detecting Adversarial Examples in Deep Neural Networks," in Proc. Netw. Distrib. Syst. Secur. Symp., 2018, pp. 1–15.
79. J. H. Metzen, T. Genewein, V. Fischer, and B. Bischoff, "On Detecting Adversarial Perturbations," in Proc. Int. Conf. Learn. Represent., 2016, pp. 1–12.
80. A. Abusnaina, Y. Wu, S. Arora, Y. Wang, F. Wang, H. Yang, and D. Mohaisen, "Adversarial Example Detection Using Latent Neighborhood Graph," in Proc. IEEE/CVF (ICCV), Oct. 2021, pp. 7687–7696.
81. F. Liao, M. Liang, Y. Dong, T. Pang, X. Hu, and J. Zhu, "Defense Against Adversarial Attacks Using High-level Representation Guided Denoiser," in Proc. IEEE/CVF Conf. Comput. Vis. Pattern Recognit., Jun. 2018, pp. 1778–1787.
82. P. Samangouei, M. Kabkab, and R. Chellappa, "Defense-GAN: Protecting Classifiers Against Adversarial Attacks Using Generative Models," in Proc. Int. Conf. Learn. Represent., 2018, pp. 1–17.
83. Y. Li, M. R. Min, T. Lee, W. Yu, E. Kruus, W. Wang, and C.-J. Hsich, "Towards Robustness of Deep Neural Networks via Regularization," in Proc. IEEE/CVF (ICCV), Oct. 2021, pp. 7496–7505.
84. M. Guo, Y. Yang, R. Xu, Z. Liu, and D. Lin, "When NAS Meets Robustness: In Search of Robust Architectures Against Adversarial Attacks," in Proc. IEEE/CVF Conf. Comput. Vis. Pattern Recognit. (CVPR), Jun. 2020, pp. 628–637.
85. V. Sehwag, S.Wang, P. Mittal, and S. Jana, "Hydra: Pruning Adversarially Robust Neural Networks," in Proc. Adv. Neural Inf. Process. Syst., vol. 33, 2020, pp. 19655–19666.
86. S. Kundu, M. Nazemi, P. A. Beerel, and M. Pedram, "DNR: A Tunable Robust Pruning Framework Through Dynamic Network Rewiring of DNNs," in Proc. 26th Asia South Pacific Design Autom. Conf. (ASP-DAC), Jan. 2021, pp. 344–350.
87. C. Xie, Y. Wu, L. V. D. Maaten, A. L. Yuille, and K. He, "Feature Denoising for Improving Adversarial Robustness," in Proc. IEEE/CVF Conf. Comput. Vis. Pattern Recognit. (CVPR), Jun. 2019, pp. 501–509.

88. C. Xie, J.Wang, Z. Zhang, Z. Ren, and A. Yuille, "Mitigating Adversarial Effects Through Randomization," in Proc. Int. Conf. Learn. Represent., 2018, pp. 1–16.
89. B. Sun, N.-H. Tsai, F. Liu, R. Yu, and H. Su, "Adversarial Defense by Stratified Convolutional Sparse Coding," in Proc. IEEE/CVF Conf. Comput. Vis. Pattern Recognit. (CVPR), Jun. 2019, pp. 11439–11448.
90. K. Pei,Y. Cao, J.Yang, and S. S. Jana, "Deepxplore: Automated Whitebox Testing of Deep Learning Systems," in Proc. 26th Symp. Operating Syst. Princ., 2017, pp. 1–18.
91. L. Ma, F. Juefei-Xu, F. Zhang, J. Sun, M. Xue, B. Li, C. Chen, T. Su, L. Li, Y. Liu, J. Zhao, and Y. Wang, "DeepGauge: Multi-granularity Testing Criteria for Deep Learning Systems," in Proc. 33rd IEEE/ACM Int. Conf. ASE, Sep. 2018, pp. 120–131.
92. J. Kim, R. Feldt, and S. Yoo, "Guiding Deep Learning System Testing Using Surprise Adequacy," in Proc. IEEE/ACM 41st Int. Conf. Softw. Eng. (ICSE), May 2019, pp. 1039–1049.
93. X. Huang, M. Kwiatkowska, S. Wang, and M. Wu, "Safety Verification of Deep Neural Networks," in Proc. CAV, 2017, pp. 3–29.
94. S. Gowal, S.-A. Rebuffi, O. Wiles, F. Stimberg, D. A. Calian, and T. A. Mann, "Improving Robustness Using Generated Data," in Proc. Adv. Neural Inf. Process. Syst., vol. 34, 2021, pp. 4218–4233.
95. J. Ho, A. Jain, and P. Abbeel, "Denoising Diffusion Probabilistic Models," in Proc. Adv. Neural Inf. Process. Syst., vol. 33, 2020, pp. 6840–6851.
96. A. Brock, J. Donahue, and K. Simonyan, "Large Scale GAN Training for High Fidelity Natural Image Synthesis," In International Conference on Learning Representations, 2019.
97. T. Karras, M. Aittala, J. Hellsten, S. Laine, J. Lehtinen, and T. Aila, "Training Generative Adversarial Networks with Limited Data," 2020, arXiv:2006.06676v1.
98. W. Nie, B. Guo, Y. Huang, C. Xiao, A. Vahdat, and A. Anandkumar, "Diffusion Models for Adversarial Purification," in Proc. Int. Conf. Mach. Learn. (ICML), 2022, pp. 16805–16827.
99. Z. Chen, K. Jin, J. Wang, W. Nie, M. Liu, A. Anandkumar, B. Li, and D. Song, "DensePure: Understanding Diffusion Models Towards Adversarial Robustness," in Proc. NIPS, 2022, pp. 1–20.
100. Q. Wu, H. Ye, and Y. Gu, "Guided Diffusion Model for Adversarial Purification from Random Noise," 2022, arXiv:2206.10875.
101. H.-Y. Lin and B. Biggio, "Adversarial Machine Learning: Attacks From Laboratories to the Real World," IEEE Computer, vol. 54, no. 5, pp. 56–60, 2021.
102. I. Alsmadi, N. Aljaafari, M. Nazzal, S. Alhamed, A. H. Sawalmeh, C. P. Vizcarra, A. Khreishah, M. Anan, A. Algosaibi, M. A. Al-Naeem, A. Aldalbahi, and A. Al-Humam, "Adversarial Machine Learning in Text Processing: A Literature Survey," IEEE Access, vol. 10, pp. 17043-17058, 2022.
103. L. Pajola and M. Conti, "Lights Toward Adversarial Machine Learning: The Achilles Heel of Artificial Intelligence," IEEE Intelligent Systems, vol. 39, no. 5, pp. 20–24, 2024.
104. J. Liu, M. Nogueira, J. Fernandes, and B. Kantarci, "Adversarial Machine Learning: A Multi-layer Review of the State-of-the-Art and Challenges for Wireless and Mobile Systems," IEEE Communications Surveys & Tutorials, vol. 24, no. 1, pp. 123–138, 2022.
105. H. Yoo, P. M. Hong, T. Kim, J. W. Yoon, and Y. K. Lee, "Defending Against Adversarial Fingerprint Attacks Based on Deep Image Prior," IEEE Access, vol. 11, pp. 78713-78725, 2023.
106. S. M. Alghamdi, S. K. Jarraya, and F. Kateb, "Enhancing Security in Multimodal Biometric Fusion: Analyzing Adversarial Attacks," IEEE Access, vol. 12, pp. 106133-106145, 2024.
107. B. Biggio, G. Fumera, P. Russu, L. Didaci, and F. Roli, "Adversarial Biometric Recognition: A Review on Biometric System Security from the Adversarial Machine-Learning Perspective," IEEE Signal Processing Magazine, vol. 32, no. 5, pp. 31–41, 2015.
108. K. Nguyen, T. Fernando, C. Fookes, and S. Sridharan, "Physical Adversarial Attacks for Surveillance: A Survey," IEEE Transactions on Neural Networks and Learning Systems, vol. 35, no. 12, pp. 17036-17049, 2024.

109. N. Bhartiya, N. Jangid, and S. Jannu, "Biometric Authentication Systems: Security Concerns and Solutions," 3rd International Conference for Convergence in Technology (I2CT), Pune, India, 2018, pp. 1-6.
110. X. Li and D. Zhu, "Robust Detection of Adversarial Attacks on Medical Images," IEEE 17th International Symposium on Biomedical Imaging (ISBI), USA, 2020, pp. 1154-1158.
111. A. I. Newaz, N. I. Haque, A. K. Sikder, M. A. Rahman, and A. S. Uluagac, "Adversarial Attacks to Machine Learning-Based Smart Healthcare Systems," IEEE Global Communications Conference (GLOBECOM), Taiwan, 2020, pp. 1-6.
112. X. Li, D. Pan, and D. Zhu, "Defending Against Adversarial Attacks on Medical Imaging AI Systems: Classification or Detection?" IEEE 18th International Symposium on Biomedical Imaging (ISBI), France, 2021, pp. 1677-1681.
113. S. M., A. Kumar Sagar, and V. Khemchandani, "Adversarial Attack and Defense Mechanisms in Medical Imaging: A Comprehensive Review," International Conference on Computing, Power, and Communication Technologies (IC2PCT), India, 2024, pp. 1657-1661.
114. L. Chi, M. Msahli, Q. Zhang, H. Qiu, T. Zhang, G. Memmi, and M. Qiu, "Adversarial Attacks on Autonomous Driving Systems in the Physical World: A Survey," IEEE Transactions on Intelligent Vehicles, pp. 1–50, 2024.
115. M. Girdhar, J. Hong, and J. Moore, "Cybersecurity of Autonomous Vehicles: A Systematic Literature Review of Adversarial Attacks and Defense Models," IEEE Open Journal of Vehicular Technology, vol. 4, pp. 417-431, 2023.
116. J. Zhang, Y. Lou, J. Wang, K. Wu, K. Lu, and X. Jia, "Evaluating Adversarial Attacks on Driving Safety in Vision-Based Autonomous Vehicles," IEEE Internet of Things Journal, vol. 9, no. 5, pp. 3443–3456, 2022.
117. S. A. Salguero-Luna, K. A. Ramírez-Gutiérrez, and A. Martínez-Cruz, "A State-of-the-Art Review on Attacks and Defense Mechanisms for LiDAR on Autonomous Vehicles," IEEE Transactions on Intelligent Transportation Systems, pp. 1–50, 2024.
118. A. Kloukiniotis, A. Papandreou, A. Lalos, P. Kapsalas, D.-V. Nguyen, and K. Moustakas, "Countering Adversarial Attacks on Autonomous Vehicles Using Denoising Techniques: A Review," IEEE Open Journal of Intelligent Transportation Systems, vol. 3, pp. 61-78, 2022.
119. J. Zhang, L. Zhang, G. Li, and C. Wu, "Adversarial Examples for Good: Adversarial Examples Guided Imbalanced Learning," IEEE International Conference on Image Processing (ICIP), France, 2022, pp. 1369–1373.
120. C. Xie, M. Tan, B. Gong, J. Wang, A. Yuille, and Q. V. Le, "Adversarial Examples Improve Image Recognition," IEEE/CVF Conference on Computer Vision and Pattern Recognition (CVPR), USA, 2020, pp. 1-9.

MIX
Papier aus verantwortungsvollen Quellen
Paper from responsible sources
FSC® C105338

If you have any concerns about our products,
you can contact us on
ProductSafety@springernature.com

In case Publisher is established outside the EU,
the EU authorized representative is:
Springer Nature Customer Service Center GmbH
Europaplatz 3, 69115 Heidelberg, Germany

Printed by Libri Plureos GmbH
in Hamburg, Germany